Communications
in Computer and Information Science   1962

Rationale

The CCIS series is devoted to the publication of proceedings of computer science conferences. Its aim is to efficiently disseminate original research results in informatics in printed and electronic form. While the focus is on publication of peer-reviewed full papers presenting mature work, inclusion of reviewed short papers reporting on work in progress is welcome, too. Besides globally relevant meetings with internationally representative program committees guaranteeing a strict peer-reviewing and paper selection process, conferences run by societies or of high regional or national relevance are also considered for publication.

Topics

The topical scope of CCIS spans the entire spectrum of informatics ranging from foundational topics in the theory of computing to information and communications science and technology and a broad variety of interdisciplinary application fields.

Information for Volume Editors and Authors

Publication in CCIS is free of charge. No royalties are paid, however, we offer registered conference participants temporary free access to the online version of the conference proceedings on SpringerLink (http://link.springer.com) by means of an http referrer from the conference website and/or a number of complimentary printed copies, as specified in the official acceptance email of the event.

CCIS proceedings can be published in time for distribution at conferences or as postproceedings, and delivered in the form of printed books and/or electronically as USBs and/or e-content licenses for accessing proceedings at SpringerLink. Furthermore, CCIS proceedings are included in the CCIS electronic book series hosted in the SpringerLink digital library at http://link.springer.com/bookseries/7899. Conferences publishing in CCIS are allowed to use Online Conference Service (OCS) for managing the whole proceedings lifecycle (from submission and reviewing to preparing for publication) free of charge.

Publication process

The language of publication is exclusively English. Authors publishing in CCIS have to sign the Springer CCIS copyright transfer form, however, they are free to use their material published in CCIS for substantially changed, more elaborate subsequent publications elsewhere. For the preparation of the camera-ready papers/files, authors have to strictly adhere to the Springer CCIS Authors' Instructions and are strongly encouraged to use the CCIS LaTeX style files or templates.

Abstracting/Indexing

CCIS is abstracted/indexed in DBLP, Google Scholar, EI-Compendex, Mathematical Reviews, SCImago, Scopus. CCIS volumes are also submitted for the inclusion in ISI Proceedings.

How to start

To start the evaluation of your proposal for inclusion in the CCIS series, please send an e-mail to ccis@springer.com.

Biao Luo · Long Cheng · Zheng-Guang Wu ·
Hongyi Li · Chaojie Li

Editors

Neural
Information Processing

30th International Conference, ICONIP 2023
Changsha, China, November 20–23, 2023
Proceedings, Part VIII

 Springer

Editors
Biao Luo (iD)
School of Automation
Central South University
Changsha, China

Long Cheng (iD)
Institute of Automation
Chinese Academy of Sciences
Beijing, China

Zheng-Guang Wu (iD)
Institute of Cyber-Systems and Control
Zhejiang University
Hangzhou, China

Hongyi Li (iD)
School of Automation
Guangdong University of Technology
Guangzhou, China

Chaojie Li (iD)
School of Electrical Engineering
and Telecommunications
UNSW Sydney
Sydney, NSW, Australia

ISSN 1865-0929   ISSN 1865-0937 (electronic)
Communications in Computer and Information Science
ISBN 978-981-99-8131-1   ISBN 978-981-99-8132-8 (eBook)
https://doi.org/10.1007/978-981-99-8132-8

This Springer imprint is published by the registered company Springer Nature Singapore Pte Ltd.
The registered company address is: 152 Beach Road, #21-01/04 Gateway East, Singapore 189721, Singapore

Paper in this product is recyclable.

Preface

Welcome to the 30th International Conference on Neural Information Processing (ICONIP2023) of the Asia-Pacific Neural Network Society (APNNS), held in Changsha, China, November 20–23, 2023.

The mission of the Asia-Pacific Neural Network Society is to promote active interactions among researchers, scientists, and industry professionals who are working in neural networks and related fields in the Asia-Pacific region. APNNS has Governing Board Members from 13 countries/regions – Australia, China, Hong Kong, India, Japan, Malaysia, New Zealand, Singapore, South Korea, Qatar, Taiwan, Thailand, and Turkey. The society's flagship annual conference is the International Conference of Neural Information Processing (ICONIP). The ICONIP conference aims to provide a leading international forum for researchers, scientists, and industry professionals who are working in neuroscience, neural networks, deep learning, and related fields to share their new ideas, progress, and achievements.

ICONIP2023 received 1274 papers, of which 394 papers were accepted for publication in Communications in Computer and Information Science (CCIS), representing an acceptance rate of 30.93% and reflecting the increasingly high quality of research in neural networks and related areas. The conference focused on four main areas, i.e., "Theory and Algorithms", "Cognitive Neurosciences", "Human-Centered Computing", and "Applications". All the submissions were rigorously reviewed by the conference Program Committee (PC), comprising 258 PC members, and they ensured that every paper had at least two high-quality single-blind reviews. In fact, 5270 reviews were provided by 2145 reviewers. On average, each paper received 4.14 reviews.

We would like to take this opportunity to thank all the authors for submitting their papers to our conference, and our great appreciation goes to the Program Committee members and the reviewers who devoted their time and effort to our rigorous peer-review process; their insightful reviews and timely feedback ensured the high quality of the papers accepted for publication. We hope you enjoyed the research program at the conference.

October 2023

Biao Luo
Long Cheng
Zheng-Guang Wu
Hongyi Li
Chaojie Li

Organization

Honorary Chair

Weihua Gui	Central South University, China

Advisory Chairs

Jonathan Chan	King Mongkut's University of Technology Thonburi, Thailand
Zeng-Guang Hou	Chinese Academy of Sciences, China
Nikola Kasabov	Auckland University of Technology, New Zealand
Derong Liu	Southern University of Science and Technology, China
Seiichi Ozawa	Kobe University, Japan
Kevin Wong	Murdoch University, Australia

General Chairs

Tingwen Huang	Texas A&M University at Qatar, Qatar
Chunhua Yang	Central South University, China

Program Chairs

Biao Luo	Central South University, China
Long Cheng	Chinese Academy of Sciences, China
Zheng-Guang Wu	Zhejiang University, China
Hongyi Li	Guangdong University of Technology, China
Chaojie Li	University of New South Wales, Australia

Technical Chairs

Xing He	Southwest University, China
Keke Huang	Central South University, China
Huaqing Li	Southwest University, China
Qi Zhou	Guangdong University of Technology, China

Local Arrangement Chairs

Wenfeng Hu	Central South University, China
Bei Sun	Central South University, China

Finance Chairs

Fanbiao Li	Central South University, China
Hayaru Shouno	University of Electro-Communications, Japan
Xiaojun Zhou	Central South University, China

Special Session Chairs

Hongjing Liang	University of Electronic Science and Technology, China
Paul S. Pang	Federation University, Australia
Qiankun Song	Chongqing Jiaotong University, China
Lin Xiao	Hunan Normal University, China

Tutorial Chairs

Min Liu	Hunan University, China
M. Tanveer	Indian Institute of Technology Indore, India
Guanghui Wen	Southeast University, China

Publicity Chairs

Sabri Arik	Istanbul University-Cerrahpaşa, Turkey
Sung-Bae Cho	Yonsei University, South Korea
Maryam Doborjeh	Auckland University of Technology, New Zealand
El-Sayed M. El-Alfy	King Fahd University of Petroleum and Minerals, Saudi Arabia
Ashish Ghosh	Indian Statistical Institute, India
Chuandong Li	Southwest University, China
Weng Kin Lai	Tunku Abdul Rahman University of Management & Technology, Malaysia
Chu Kiong Loo	University of Malaya, Malaysia
Qinmin Yang	Zhejiang University, China
Zhigang Zeng	Huazhong University of Science and Technology, China

Publication Chairs

Zhiwen Chen	Central South University, China
Andrew Chi-Sing Leung	City University of Hong Kong, China
Xin Wang	Southwest University, China
Xiaofeng Yuan	Central South University, China

Secretaries

Yun Feng	Hunan University, China
Bingchuan Wang	Central South University, China

Webmasters

Tianmeng Hu	Central South University, China
Xianzhe Liu	Xiangtan University, China

Program Committee

Rohit Agarwal	UiT The Arctic University of Norway, Norway
Hasin Ahmed	Gauhati University, India
Harith Al-Sahaf	Victoria University of Wellington, New Zealand
Brad Alexander	University of Adelaide, Australia
Mashaan Alshammari	Independent Researcher, Saudi Arabia
Sabri Arik	Istanbul University, Turkey
Ravneet Singh Arora	Block Inc., USA
Zeyar Aung	Khalifa University of Science and Technology, UAE
Monowar Bhuyan	Umeå University, Sweden
Jingguo Bi	Beijing University of Posts and Telecommunications, China
Xu Bin	Northwestern Polytechnical University, China
Marcin Blachnik	Silesian University of Technology, Poland
Paul Black	Federation University, Australia
Anoop C. S.	Govt. Engineering College, India
Ning Cai	Beijing University of Posts and Telecommunications, China
Siripinyo Chantamunee	Walailak University, Thailand
Hangjun Che	City University of Hong Kong, China

Wei-Wei Che	Qingdao University, China
Huabin Chen	Nanchang University, China
Jinpeng Chen	Beijing University of Posts & Telecommunications, China
Ke-Jia Chen	Nanjing University of Posts and Telecommunications, China
Lv Chen	Shandong Normal University, China
Qiuyuan Chen	Tencent Technology, China
Wei-Neng Chen	South China University of Technology, China
Yufei Chen	Tongji University, China
Long Cheng	Institute of Automation, China
Yongli Cheng	Fuzhou University, China
Sung-Bae Cho	Yonsei University, South Korea
Ruikai Cui	Australian National University, Australia
Jianhua Dai	Hunan Normal University, China
Tao Dai	Tsinghua University, China
Yuxin Ding	Harbin Institute of Technology, China
Bo Dong	Xi'an Jiaotong University, China
Shanling Dong	Zhejiang University, China
Sidong Feng	Monash University, Australia
Yuming Feng	Chongqing Three Gorges University, China
Yun Feng	Hunan University, China
Junjie Fu	Southeast University, China
Yanggeng Fu	Fuzhou University, China
Ninnart Fuengfusin	Kyushu Institute of Technology, Japan
Thippa Reddy Gadekallu	VIT University, India
Ruobin Gao	Nanyang Technological University, Singapore
Tom Gedeon	Curtin University, Australia
Kam Meng Goh	Tunku Abdul Rahman University of Management and Technology, Malaysia
Zbigniew Gomolka	University of Rzeszow, Poland
Shengrong Gong	Changshu Institute of Technology, China
Xiaodong Gu	Fudan University, China
Zhihao Gu	Shanghai Jiao Tong University, China
Changlu Guo	Budapest University of Technology and Economics, Hungary
Weixin Han	Northwestern Polytechnical University, China
Xing He	Southwest University, China
Akira Hirose	University of Tokyo, Japan
Yin Hongwei	Huzhou Normal University, China
Md Zakir Hossain	Curtin University, Australia
Zengguang Hou	Chinese Academy of Sciences, China

Lu Hu   Jiangsu University, China
Zeke Zexi Hu   University of Sydney, Australia
He Huang   Soochow University, China
Junjian Huang   Chongqing University of Education, China
Kaizhu Huang   Duke Kunshan University, China
David Iclanzan   Sapientia University, Romania
Radu Tudor Ionescu   University of Bucharest, Romania
Asim Iqbal   Cornell University, USA
Syed Islam   Edith Cowan University, Australia
Kazunori Iwata   Hiroshima City University, Japan
Junkai Ji   Shenzhen University, China
Yi Ji   Soochow University, China
Canghong Jin   Zhejiang University, China
Xiaoyang Kang   Fudan University, China
Mutsumi Kimura   Ryukoku University, Japan
Masahiro Kohjima   NTT, Japan
Damian Kordos   Rzeszow University of Technology, Poland
Marek Kraft   Poznań University of Technology, Poland
Lov Kumar   NIT Kurukshetra, India
Weng Kin Lai   Tunku Abdul Rahman University of
                           Management & Technology, Malaysia
Xinyi Le   Shanghai Jiao Tong University, China
Bin Li   University of Science and Technology of China,
                           China
Hongfei Li   Xinjiang University, China
Houcheng Li   Chinese Academy of Sciences, China
Huaqing Li   Southwest University, China
Jianfeng Li   Southwest University, China
Jun Li   Nanjing Normal University, China
Kan Li   Beijing Institute of Technology, China
Peifeng Li   Soochow University, China
Wenye Li   Chinese University of Hong Kong, China
Xiangyu Li   Beijing Jiaotong University, China
Yantao Li   Chongqing University, China
Yaoman Li   Chinese University of Hong Kong, China
Yinlin Li   Chinese Academy of Sciences, China
Yuan Li   Academy of Military Science, China
Yun Li   Nanjing University of Posts and
                           Telecommunications, China
Zhidong Li   University of Technology Sydney, Australia
Zhixin Li   Guangxi Normal University, China
Zhongyi Li   Beihang University, China

Toshiaki Omori	Kobe University, Japan
Babatunde Onasanya	University of Ibadan, Nigeria
Manisha Padala	Indian Institute of Science, India
Sarbani Palit	Indian Statistical Institute, India
Paul Pang	Federation University, Australia
Rasmita Panigrahi	Giet University, India
Kitsuchart Pasupa	King Mongkut's Institute of Technology Ladkrabang, Thailand
Dipanjyoti Paul	Ohio State University, USA
Hu Peng	Jiujiang University, China
Kebin Peng	University of Texas at San Antonio, USA
Dawid Połap	Silesian University of Technology, Poland
Zhong Qian	Soochow University, China
Sitian Qin	Harbin Institute of Technology at Weihai, China
Toshimichi Saito	Hosei University, Japan
Fumiaki Saitoh	Chiba Institute of Technology, Japan
Naoyuki Sato	Future University Hakodate, Japan
Chandni Saxena	Chinese University of Hong Kong, China
Jiaxing Shang	Chongqing University, China
Lin Shang	Nanjing University, China
Jie Shao	University of Science and Technology of China, China
Yin Sheng	Huazhong University of Science and Technology, China
Liu Sheng-Lan	Dalian University of Technology, China
Hayaru Shouno	University of Electro-Communications, Japan
Gautam Srivastava	Brandon University, Canada
Jianbo Su	Shanghai Jiao Tong University, China
Jianhua Su	Institute of Automation, China
Xiangdong Su	Inner Mongolia University, China
Daiki Suehiro	Kyushu University, Japan
Basem Suleiman	University of New South Wales, Australia
Ning Sun	Shandong Normal University, China
Shiliang Sun	East China Normal University, China
Chunyu Tan	Anhui University, China
Gouhei Tanaka	University of Tokyo, Japan
Maolin Tang	Queensland University of Technology, Australia
Shu Tian	University of Science and Technology Beijing, China
Shikui Tu	Shanghai Jiao Tong University, China
Nancy Victor	Vellore Institute of Technology, India
Petra Vidnerová	Institute of Computer Science, Czech Republic

Shanchuan Wan	University of Tokyo, Japan
Tao Wan	Beihang University, China
Ying Wan	Southeast University, China
Bangjun Wang	Soochow University, China
Hao Wang	Shanghai University, China
Huamin Wang	Southwest University, China
Hui Wang	Nanchang Institute of Technology, China
Huiwei Wang	Southwest University, China
Jianzong Wang	Ping An Technology, China
Lei Wang	National University of Defense Technology, China
Lin Wang	University of Jinan, China
Shi Lin Wang	Shanghai Jiao Tong University, China
Wei Wang	Shenzhen MSU-BIT University, China
Weiqun Wang	Chinese Academy of Sciences, China
Xiaoyu Wang	Tokyo Institute of Technology, Japan
Xin Wang	Southwest University, China
Xin Wang	Southwest University, China
Yan Wang	Chinese Academy of Sciences, China
Yan Wang	Sichuan University, China
Yonghua Wang	Guangdong University of Technology, China
Yongyu Wang	JD Logistics, China
Zhenhua Wang	Northwest A&F University, China
Zi-Peng Wang	Beijing University of Technology, China
Hongxi Wei	Inner Mongolia University, China
Guanghui Wen	Southeast University, China
Guoguang Wen	Beijing Jiaotong University, China
Ka-Chun Wong	City University of Hong Kong, China
Anna Wróblewska	Warsaw University of Technology, Poland
Fengge Wu	Institute of Software, Chinese Academy of Sciences, China
Ji Wu	Tsinghua University, China
Wei Wu	Inner Mongolia University, China
Yue Wu	Shanghai Jiao Tong University, China
Likun Xia	Capital Normal University, China
Lin Xiao	Hunan Normal University, China
Qiang Xiao	Huazhong University of Science and Technology, China
Hao Xiong	Macquarie University, Australia
Dongpo Xu	Northeast Normal University, China
Hua Xu	Tsinghua University, China
Jianhua Xu	Nanjing Normal University, China

Xinyue Xu	Hong Kong University of Science and Technology, China
Yong Xu	Beijing Institute of Technology, China
Ngo Xuan Bach	Posts and Telecommunications Institute of Technology, Vietnam
Hao Xue	University of New South Wales, Australia
Yang Xujun	Chongqing Jiaotong University, China
Haitian Yang	Chinese Academy of Sciences, China
Jie Yang	Shanghai Jiao Tong University, China
Minghao Yang	Chinese Academy of Sciences, China
Peipei Yang	Chinese Academy of Science, China
Zhiyuan Yang	City University of Hong Kong, China
Wangshu Yao	Soochow University, China
Ming Yin	Guangdong University of Technology, China
Qiang Yu	Tianjin University, China
Wenxin Yu	Southwest University of Science and Technology, China
Yun-Hao Yuan	Yangzhou University, China
Xiaodong Yue	Shanghai University, China
Paweł Zawistowski	Warsaw University of Technology, Poland
Hui Zeng	Southwest University of Science and Technology, China
Wang Zengyunwang	Hunan First Normal University, China
Daren Zha	Institute of Information Engineering, China
Zhi-Hui Zhan	South China University of Technology, China
Baojie Zhang	Chongqing Three Gorges University, China
Canlong Zhang	Guangxi Normal University, China
Guixuan Zhang	Chinese Academy of Science, China
Jianming Zhang	Changsha University of Science and Technology, China
Li Zhang	Soochow University, China
Wei Zhang	Southwest University, China
Wenbing Zhang	Yangzhou University, China
Xiang Zhang	National University of Defense Technology, China
Xiaofang Zhang	Soochow University, China
Xiaowang Zhang	Tianjin University, China
Xinglong Zhang	National University of Defense Technology, China
Dongdong Zhao	Wuhan University of Technology, China
Xiang Zhao	National University of Defense Technology, China
Xu Zhao	Shanghai Jiao Tong University, China

Contents – Part VIII

Theory and Algorithms

Multi-model Smart Contract Vulnerability Detection Based on BiGRU

Shuxiao Song, Xiao Yu$^{(\boxtimes)}$, Yuexuan Ma, Jiale Li, and Jie Yu

School of Computer Science and Technology, Shandong University of Technology,
Zibo 255049, Shandong, China
yuxiao8907118@163.com

Abstract. Smart contracts have been under constant attack from outside, with frequent security problems causing great economic losses to the virtual currency market, and their security research has attracted much attention in the academic community. Traditional smart contract detection methods rely heavily on expert rules, resulting in low detection precision and efficiency. This paper explores the effectiveness of deep learning methods on smart contract detection and propose a multi-model smart contract detection method, which is based on a multi-model vulnerability detection method combining Bi-directional Gated Recurrent Unit (BiGRU) and Synthetic Minority Over-sampling Technique (SMOTE) for smart contract vulnerability detection. Through a comparative study on the vulnerability detection of 10312 smart contract codes, the method can achieve an identification accuracy of 90.17% and a recall rate of 97.7%. Compared with other deep network models, the method used in this paper has superior performance in terms of recall and accuracy.

Keywords: Block Chain · Smart Contracts · Deep Learning · Vulnerability

1 Introduction

Bugs in software can harm it significantly and even kill it in some cases. The same is true for smart contracts on the blockchain; they are more susceptible to hacking since they frequently include economic qualities, and once there are exploitable weaknesses, they often lead to significant financial losses.

On significant blockchain platforms like Ethereum, EOS, and VNT Chain, tens of thousands of smart contracts are already in use, and the number is constantly increasing. However, as the quantity of smart contracts increases, security breaches in smart contracts have also been reported. In 2016, hackers exploited a reentrancy vulnerability in the DAO to steal 3.6 million Ether, resulting in a loss of over $60 million. In 2017, a multi-signature wallet vulnerability in Parity was attacked, resulting in the freezing of $300 million in Ethereum. In 2022, an access control vulnerability in the $CF token contract resulted in a $1.9 million loss. Smart contract security has drawn much attention and has become a prominent area of research.

Once a smart contract is deployed on the blockchain, it cannot be modified, so it is especially important to audit it before it goes on the chain [1]. To find security vulnerabilities in smart contracts before they are uploaded, Professionals must complete the audit of smart contracts. That requires a lot of human resources.

To save resources, this paper takes the Solidity smart contract source code as the research object and propose the S-BiGRU smart contract detection method, a multi-model smart contract vulnerability detection method based on BiGRU. Since the key code locations of the smart contract source code are strongly correlated with the contract functionality, the capture of contextual information is very important for smart contract vulnerability detection work. The method uses the SMOTE algorithm to balance the positive and negative class imbalance problems existing in the dataset. It uses the BiGRU model to effectively capture the feature and contextual information in the smart contract source code to identify vulnerabilities. In the following, this method is referred to as the S-BiGRU method. The multi-model detection mentioned therein refers to training its independent binary classification model for each detectable vulnerability and integrating the output of multiple models in the vulnerability detection phase to derive the final smart contract vulnerability detection results. This approach leverages the variability between different vulnerability types, and by training specialized models for each vulnerability type, the features and patterns of a particular vulnerability can be captured and identified more accurately. The final output integration can improve the overall vulnerability detection performance and accuracy.

This paper contributes as follows:

- Detection of Solidity code for smart contracts using the method combining BiGRU and SMOTE, achieving vulnerability identification accuracy of 90.17% on 10312 samples;
- The SMOTE algorithm is proposed to deal with the problem of severely skewed positive and negative classes in the smart contract dataset and achieve better results.

2 Related Work

The DAO incident in 2016 brought a huge impact on Ethereum, which not only exposed the smart contract security issue but also drew the attention of the majority of blockchain-related people.

Compared to traditional software flaws, smart contracts are hidden and modifiable. As a program running on a decentralized blockchain, the lifecycle of a smart contract and the nature of the development language provide credibility and opportunities for hackers to exploit them. Unlike traditional programs, smart contracts cannot be patched to fix vulnerabilities due to their tamper-evident nature. Therefore, detecting vulnerability before a smart contract is put on the chain is especially important.

In recent years, automated smart contract detection tools have been gradually introduced. The main methods used are symbolic execution [2–4], fuzzy testing [5, 6], formal verification [7, 8], deep learning [9, 10], etc. The following is a brief introduction and comparison of the above methods.

Symbolic execution is a static analysis method that explores different execution paths of a smart contract by symbolizing the variables in the smart contract and constructing constraints. The advantage of symbolic execution is its ability to find accurate vulnerabilities and anomalous behavior. Still, its disadvantage is that it is prone to path explosion problems, leading to very time-consuming analysis for large contracts. **Fuzzy testing** is a dynamic approach simulating contract execution by generating many random or variant inputs to find potential vulnerabilities. The advantage of fuzzy testing is that it can quickly generate test cases and find unknown vulnerabilities. Still, its disadvantage is that it is difficult to cover complex execution paths and does not provide root cause analysis of contract vulnerabilities. **Formal verification** converts the concepts, judgments, and reasoning in a contract into a formal model through formal language to eliminate ambiguities in the contract. Then it verifies the correctness and security of the functions in the smart contract with rigorous logic and proofs. The advantage of formal verification is that it can provide rigorous mathematical logic proofs to ensure the contract satisfies the statute and security properties. Still, its disadvantage is that the statute and proofs can be very cumbersome for complex contracts. **Deep learning** uses multi-layer neural networks to extract and learn features in smart contracts for risk and vulnerability prediction. The advantage of deep learning is that it can handle large amounts of data and complex features. Still, its disadvantage is that it requires large amounts of training data and computational resources, and it may be difficult to explain the reasons for the prediction results.

3   S-BiGRU Smart Contracts Detection Method

In this section, a basic overview of the S-BiGRU smart contract detection model is first presented in 3.1, then the training and prediction parts of the model are described in detail in 3.2.

3.1   S-BiGRU Multi-model Smart Contracts Vulnerability Detection Framework

To detect vulnerabilities in Ethereum smart contracts, this paper proposes a BiGRU-based smart contract vulnerability detection scheme, named the S-BiGRU multi-model smart contract detection method, and its overall architecture is shown in **Fig. 1**.

The figure is divided into two parts, and the top half shows the S-BiGRU model training dominated by blue arrows: first, the dataset is preprocessed, such as removing blank lines, non-ASCII characters, and other irrelevant annotations. Second, the preprocessed Solidity source code is processed and parsed into a series of Tokens embedded into feature vectors using Word2Vec, denoted as word vectors. Then, minority-classified samples are generated using the SMOTE algorithm to solve the data imbalance problem. Finally, these word vectors are fed into the model for training to obtain a model capable of detecting smart contract vulnerabilities for subsequent detection work.

Red arrows dominate the bottom half of the figure and show the vulnerability detection part of the smart contract. The data to be detected is pre-processed to obtain the corresponding word vectors, fed into the different vulnerability detection models trained to detect the results and output the integrated results.

3.2   Model Training

Feature Extraction. Smart contracts are generally multi-line code programs written in solidity, an Ethereum high-level language. However, some lines of code in a contract

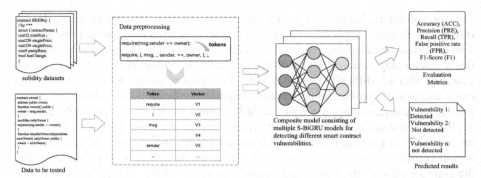

Fig. 1. S-BiGRU multi-model smart contract vulnerability detection framework.

may be irrelevant to the contract itself, such as empty lines of code or comments with explanatory meaning, so code cleaning is needed to extract features more precisely.

Since this paper uses recurrent networks as training models, their inputs must be represented as vectors, so the smart contract Solidity source code is to be described as semantically meaningful vectors, as shown in Data processing in Fig. 1. First, the cleaned Solidity code is taken as input and decomposed into identifiers, variables, keywords, and operators. After that, this paper divides it into a sequence of tokens via lexical analysis, preserving the order of their occurrence. The whole code fragment is marked in this way to join each line as a long list. To avoid the matrix sparsity problem, a word vector is obtained by pre-training using a word embedding model. Eventually, the model and the buffered list are removed, and only the word vectors are kept as input to the neural network.

To better capture the feature information in Solidity code, this paper uses Word2Vec. This paper uses CBOW in Word2Vec to train word vectors in the smart contract Solidity source code by mapping a token to an integer and then transforming it into a fixed-dimension vector. Since each long input list contains different numbers of tokens, the transformed vectors are of different lengths. At this time, the following adjustments will be made: (1) when the length is less than the fixed dimension, pad zero at the end of the vector; (2) discard the end of the vector when the length exceeds the fixed dimension.

Eliminating Data Imbalance Problems. From Table 1, it can be seen that there is a strong imbalance in the experimental samples. This imbalance cannot be eliminated by using the multi-label training method. Therefore, this paper simplifies it into multiple binary classification problems and trains multiple binary classification models for identifying smart contract-related vulnerabilities separately so that the imbalance problem of each model can be treated independently.

This study uses the SMOTE [11] method to deal with the sample imbalance problem of smart contracts. The basic idea of the SMOTE algorithm is to analyze minority-classified samples and artificially synthesize new samples to add to the dataset based on the minority-classified samples to balance the dataset. The algorithm first finds the Euclidean distance from each minority-classified sample x to all others to get its k-nearest neighbors. Then, it sets a sampling ratio according to the sample imbalance ratio to determine the sampling multiplicity N. For each minority-classified sample x, several samples are randomly selected from their k-nearest neighbors. Assuming that the selected nearest neighbors are o, and for each randomly selected nearest neighbor o, Construct a new sample x_{new} with the original sample x according to the following formula.

$$x_{new} = x + rand(0, 1) * |x - o| \qquad (1)$$

In this way, SMOTE can effectively increase the number of minority-classified samples, thus improving the classifier's performance when dealing with unbalanced data sets.

Model Training. In this paper, the word vectors generated by word2vec are trained using the s-bigru model. Since there is more than one vulnerability to be detected and they are independent of each other, each vulnerability will be trained independently as a binary classification training, considering that the Multi-Label classification transformation method has problems such as multiple and complex class labels, overfitting, and unbalanced classification.

Figure 2 depicts the training process of the S-BiGRU method, which is the part of the composite model unfolded in Fig. 1. The word vector is generated by Word2Vec from the Solidity dataset as input. The SMOTE algorithm generates the augmented sample of minority-classified samples. After that, the samples are divided into training and validation sets by 8:2, and the training set is input to the BiGRU layer for training. Then the output is expanded to the FC layer and Softmax, and the model detection results are finally output.

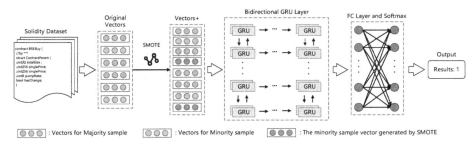

Fig. 2. S-BiGRU method training process.

The model is based on BiGRU, a type of sequence model that consists of a GRU network that processes sequences in the forward direction and a GRU network that processes sequences in the reverse direction, as shown in Fig. 3, which can capture the contextual information of sequence features. GRU [12] was proposed by Cho et al. in 2014 as a type of recurrent neural network, which is similar to the LSTM structure but easier to compute and can improve training efficiency to a great extent.

In the figure, $\overrightarrow{h}_t$ is the hidden layer state obtained from the operation in the forward layer at the time of t, and $\overleftarrow{h}_t$ is the hidden layer state obtained from the operation in the

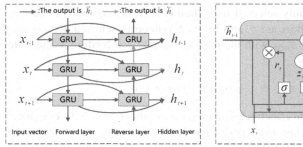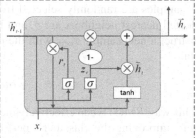

(a) BiGRU structure diagram   (b) GRU unit structure diagram

Fig. 3. BiGRU network structure diagram.

reverse layer; the forward and reverse hidden layer states of the input vector at the time of t can be expressed as follows:

$$\begin{cases} \overrightarrow{h}_t = \overrightarrow{g}\left(\overrightarrow{h}_{t-1}, x_t\right) \\ \overleftarrow{h}_t = \overleftarrow{g}\left(\overleftarrow{h}_{t-1}, x_t\right) \end{cases} \tag{2}$$

Figure 3. **(b)** shows the GRU cell structure, which contains the update gate z_t and the reset gate r_t.

$$\begin{cases} r_t = \sigma\left(W_r \cdot \left[\overrightarrow{h}_{t-1}, x_t\right]\right) \\ z_t = \sigma\left(W_z \cdot \left[\overrightarrow{h}_{t-1}, x_t\right]\right) \\ \widetilde{h}_t = \tanh\left(W_h \cdot \left[r_t \times \overrightarrow{h}_{t-1}, x_t\right]\right) \\ h_t = (I - z_t) \times h_{t-1} + z_t \times \widetilde{h}_t \end{cases} \tag{3}$$

where x_t is the current input software source code vector. h_t and h_{t-1} denote the state memory variables of the hidden layer. $\widetilde{h}_t$ denotes the candidate set state and the output state of y_t. W_r, W_z, W_h denote the trainable weight parameter matrices. I denotes the unit matrix and σ denotes the sigmoid activation function. The expressions for *sigmoid* and *tanh* are the following.

$$\mathrm{sigmoid}(x) = \frac{1}{1+e^{-x}} \tag{4}$$

$$\tanh(x) = \frac{e^x - e^{-x}}{e^x + e^{-x}} \tag{5}$$

The output state of the model will be determined by the output states of both forward and reverse GRUs. The BiGRU forward and reverse hidden layer states will be superimposed to obtain the total hidden layer state of BiGRU h_t.

$$h_t = \overrightarrow{h}_t + \overleftarrow{h}_t \tag{6}$$

Using the above model, the prediction models for different smart contract vulnerabilities are trained separately to obtain the prediction models for this vulnerability. After training, the prediction models for different smart contract vulnerabilities are obtained.

4 Experiment and Analysis

4.1 Evaluation Indicators

To evaluate the performance of the model, this paper uses evaluation metrics commonly used in machine learning: Accuracy (ACC), False Positive Rate (FPR), Recall (TPR), and F1-score(F1). Also, to facilitate observation, this paper uses receiver operating Characteristic (ROC) curves and Area Under Curve (AUC) values.

4.2 Smart Contract Dataset

As of April 2023, 430,490 smart contracts have been validated on etherscan. Currently, etherscan [13] only shows the source code of the last 500 verified contracts. To get the source code of other smart contracts, this paper must get their contract addresses for easy searching.

Due to the small number of contracts displayed by Etherscan, this paper uses the SoliAudit [14] method, retaining the 10,312 contracts that can be validly labeled as the experimental dataset. This dataset was created using Oyente [15] and Remix [16] 13 categories of smart contract vulnerabilities are labeled. Table 1 shows the relevant vulnerabilities and the number of vulnerabilities included in the dataset.

Table 1. Details of the SoliAudit-based dataset.

Tools	Vulnerabilities	Descriptions	Positive samples
Oyente	Underflow	Integer underflow	6421
	Overflow	Integer overflow	9246
	CallDepth	Use send or call cmd, but do not check the cmd result	229
	TOD	State will depend on the tx order	1518
	TimeDep	State will depend on the timestamp	675
	Reentrancy	Contract contains reentrancy function	295
Remix	AssertFail	Contract contains the condition of assert fail	4407
	Tx.origin	Contract use tx.origin	83
	CheckEffects	Contract checks if the state has been updated before the transaction or not	3908
	InlineAssembly	Contract uses assembly code	769
	BlockTimestamp	Contract uses block.timestamp	3250
	LowlevelCalls	Contract uses send or call, not transfer	2972
	SelfDestruct	Contract uses selfdestruct	761

4.3 Model Parameter Selection

To verify the validity of the models, the training and test sets were split out in a ratio of 8:2. For each vulnerability model, a 10-f old cross-validation method is used to select and optimize the parameters. The gradient descent algorithm used to learn all models by optimizing the cross-entropy loss is Adam.

As to parameter, experiments are conducted in this paper for different models. The appropriate learning rate lr is selected in [0.0001, 0.0005, 0.001, 0.002, 0.005], the dropout rate dr is selected in [0.2, 0.4, 0.6, 0.8], the number of decimal places in the feature vector representation is set to 100, the batch size is set to 64, and the vector dimension vm is chosen among [200,300,400,500]. After pre-experiments, the final parameters were set as follows: $lr = 0.002$, $dr = 0.2$, and $vm = 300$.

4.4 Experimental Design and Results

In this section, since it is not guaranteed that all vulnerability features can be accurately identified by deep learning, firstly, smart contract vulnerabilities are screened; secondly, the performance of the method proposed in this paper is compared with other deep learning models for smart contract vulnerability detection; and finally, to test the necessity of balancing the data, the training effect of the unbalanced dataset is compared with the processed one. All the data below are obtained by averaging the results of 10 experiments to make the experimental results more reliable.

Firstly, the effectiveness of this method for detecting smart contract vulnerabilities is discussed. Secondly, on the basis of the data set processed by SMOTE algorithm, the model validity of BiGRU is compared with other sequential models. For ease of description, the dataset that did not use SMOTE is called SCD, and the dataset that did use SMOTE is called SSCD. Then, in order to verify the validity of SMOTE, compare the effect of sequential model on SCD and SSCD respectively.

Effectiveness of Smart Contract Vulnerability Detection. Since some smart contract vulnerabilities do not show obvious features in the source code, the deep learning model does not detect them well. This paper experimentally screens some of the vulnerabilities that can be predicted: Assertfail, BlockTimestamp, CheckEfects, Inlineassembly, and Reentrancy to vulnerabilities and analyze their detection results.

Table 2. Vulnerability detection results

	ACC(%)	FPR(%)	TPR(%)	PRE(%)	F1
AssertFail	89.71	14.03	93.52	86.77	90.02
BlockTimestamp	77.98	35.04	91.65	71.36	80.24
CheckEfects	78.73	35.19	92.47	72.68	81.39
InlineAssembly	88.24	23.13	99.68	81.06	89.41
Reentrancy	92.56	14.80	99.90	87.14	93.08

Table 2 shows the results of using this paper to detect the above five vulnerabilities, and it can be observed from the table that the models differ in their detection effects for different vulnerabilities. The accuracy of the model in identifying AssertFail, InlineAssembly, and Reentrancy is above 88%, and the F1 score is around 90. Especially for Reentrancy, the accuracy can reach 92.56%, and there is a low false alarm rate. While identifying BlockTimestamp and CheckEfects, the model performs poorly, with an accuracy of less than 80% for both vulnerabilities and a high false alarm rate with low accuracy. Since the positive and negative sample distributions of these two vulnerabilities and AssertFail are similar, the influence caused by the imbalance in the data processed by the SMOTE algorithm can be excluded. This may be because the source code features of the vulnerability are less obvious and more difficult to extract, making it difficult to distinguish them accurately. Therefore, in the following study, the three vulnerability types that perform better in the sequence model, AssertFail, InlineAssembly, and Reentrancy, are mainly elaborated.

Model Performance Evaluation. To evaluate the performance of S-BiGRU model proposed in this paper, the prediction performance of the present model is further evaluated for the vulnerabilities that performed well in the three previous experiments. This paper conducts a series of comparison experiments to compare the present model with the sequence models RNN, LSTM, GRU, and BiLSTM. The experimental Results Are Shown in Table 3, it can be seen from the table that BiGRU shows a higher performance compared to the other four deep models in that the accuracy can reach more than 90.17% and has a recall rate of 97.7%. In order to compare the differences between the models, the visualization results of the models are given in Fig. 4. It is observed from Fig. 4(a) that the BiGRU Model is due to the other models in all scores. Figure 4(B) Shows the ROC curve visualization of the model, the model used in this paper has the largest area under the curve, which also implies the largest auc value of over 90%, which again illustrates the excellent performance of the model.

Table 3. Model comparison

SSCD	ACC(%)	FPR(%)	TPR(%)	PRE(%)	F1
Simple_RNN	84.53	26.18	95.21	78.93	86.06
LSTM	87.29	22.55	97.06	81.79	88.61
GRU	86.99	23.32	97.32	80.82	88.18
BiLSTM	88.93	18.69	96.57	83.83	89.64
BiGRU	90.17	17.32	97.7	84.99	90.84

SMOTE Algorithm Validation. Table 4 shows the effects of the sequence model on the data set SCD and the data set SSCD respectively.

The model handles SSCD much better than the results shown on SCD. Comparing the sequential models used in this paper alone, using the SMOTE algorithm to balance the dataset resulted in a significant improvement in model effectiveness, with 29.98%

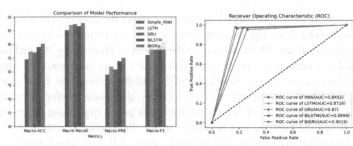

(a) Sequential model evaluation index comparison   (b) ROC curves and AUC values of sequential models

Fig. 4. Model performance visualization results (The models in the figure all combine SMOTE algorithm, therefore, the BiGRU marked in the figure is the S-BiGRU model).

and 29.76% improvements in accuracy and precision, respectively. In the combined view of several training models, SSCD improves by 23.17%, 130.29%, 5.20%, 21.62%, and 15.35% over SCD inaccuracy, false positive rate, recall, precision, and F1 score, respectively.

Table 4. Experimental results of the original sequence model.

SCD	ACC(%)	FPR(%)	TPR(%)	PRE(%)	F1
Simple_RNN	71.31	47.87	90.5	67.23	76.5
LSTM	70.77	45.08	86.63	67.6	75.24
GRU	69.69	55.19	94.58	66.89	76.79
BiLSTM	74.4	43.05	91.85	70.2	78.87
BiGRU	69.37	57.66	96.4	65.5	76.95
SSCD	ACC(%)	FPR(%)	TPR(%)	PRE(%)	F1
Simple_RNN	84.53	26.18	95.21	78.93	86.06
LSTM	87.29	22.55	97.06	81.79	88.61
GRU	86.99	23.32	97.32	80.82	88.18
BiLSTM	88.93	18.69	96.57	83.83	89.64
BiGRU	**90.17**	**17.32**	**97.7**	**84.99**	**90.84**

Data imbalance affects the training effect of the model, especially when the data ratio is severely skewed. The neural network is less effective in recognizing minority-classified samples because the model will tend to learn the features of a larger number of classes. Among the three vulnerabilities AssertFail, InlineAssembly, and Reentrancy focused on detection in this paper, the latter two of them have such a severely skewed ratio of positive and negative classes, so the SMOTE algorithm balancing the data set is a very effective method for this situation.

5   Conclusion

This paper presents a multi-model smart contract vulnerability detection method based on BiGRU for detecting vulnerabilities in the smart contract Solidity source code. The method utilizes the SMOTE algorithm combined with the BiGRU model, which is able to capture information in smart contracts while balancing the dataset. Experimental results show that the method excels in smart contract vulnerability detection, with both ACC and F1 values reaching over 90%, while having a relatively low FPR. It indicates that the method can effectively detect vulnerabilities in smart contracts with high accuracy and robustness. This suggests that this method can effectively detect vulnerabilities in smart contracts with high accuracy and robustness. Therefore, the BiGRU-based deep learning method used in this paper provides an effective solution for smart contract vulnerability detection and important technical support for smart contract development and security assurance.

Acknowledgments. This study was supported by the National Key Research and Development Program of China (2020YFB1005704).

References

1. Qian, P., Liu, Z., He, Q., et al.: Smart contract vulnerability detection technique: a survey. J. Softw. **33**(8), 3059–3085 (2022)
2. Chen, J., Xia, X., Lo, D., et al.: Defectchecker: automated smart contract defect detection by analyzing EVM bytecode. IEEE Trans. Softw. Eng. **48**(7), 2189–2207 (2021)
3. Zheng, P., Zheng, Z., Luo, X.: Park: accelerating smart contract vulnerability detection via parallel-fork symbolic execution. In: Proceedings of the 31st ACM SIGSOFT International Symposium on Software Testing and Analysis, pp. 740–751. Association for Computing Machinery (ACM), New York, NY, USA (2022)
4. Zhao, W., Zhang, W., Wang, J., et al.: Smart contract vulnerability detection scheme based on symbol execution. J. Comput. Appl. **40**(4), 947–953 (2020)
5. Choi, J., Kim, D., Kim, S., et al.: Smartian: enhancing smart contract fuzzing with static and dynamic data-flow analyses. In: 2021 36th IEEE/ACM International Conference on Automated Software Engineering (ASE), pp. 227–239. IEEE, Melbourne, Australia (2021)
6. Jiang, B., Liu, Y., Chan W.: Contractfuzzer: fuzzing smart contracts for vulnerability detection. In: Proceedings of the 33rd ACM/IEEE International Conference on Automated Software Engineering, pp. 259–269. Association for Computing Machinery (ACM), New York, NY, USA (2018)
7. Zhao, Y., Zhu, X., Li, G., Bao, Y.: Time constraint patterns of smart contracts and their formal verification. J. Softw. **33**(8), 2875–2895 (2022)
8. Li, Z., Lu, S., Zhang, R., et al.: SmartFast: an accurate and robust formal analysis tool for Ethereum smart contracts. Empir. Softw. Eng. **27**(7), 197 (2022)
9. Qian, P., Liu, Z., He, Q., et al.: Towards automated reentrancy detection for smart contracts based on sequential models. IEEE Access **8**, 19685–19695 (2020)
10. Zhang, G., Liu, Y., Wang, H., Yu, N.: Contract vulnerability detection scheme based on BiLSTM and attention mechanism. Netinfo Secur. **22**(09), 46–54 (2022)
11. Chawla, N., Bowyer, K., Hall, L., et al.: SMOTE: synthetic minority over-sampling technique. J. Artif. Intell. Res. **16**, 321–357 (2002)

12. Chung, J., Gulcehre, C., Cho, K., et al.: Empirical evaluation of gated recurrent neural networks on sequence modeling, arXiv preprint arXiv:1412.3555 (2014)
13. Etherscan Homepage,https://etherscan.io, Accessed 1 May 2023
14. Liao, J., Tsai, T., He, C., et al.: Soliaudit: smart contract vulnerability assessment based on machine learning and fuzz testing. In: Sixth International Conference on Internet of Things: Systems, Management and Security (IOTSMS), pp. 458–465 (2019)
15. Oyente-project, https://github.com/enzymefinance/oyente, Accessed 10 May 2023
16. Remix-project, https://github.com/ethereum/remix-project, Accessed 10 May 2023

Time-Warp-Invariant Processing with Multi-spike Learning

Xiaohan Zhou[1,2], Yuzhe Liu[2], Wei Sun[2], and Qiang Yu[2(✉)]

[1] Tianjin International Engineering Institute, Tianjin University, Tianjin, China
[2] Tianjin Key Laboratory of Cognitive Computing and Application, College of Intelligence and Computing, Tianjin University, Tianjin, China
{xzhou,lyz2022,issun2020,yuqiang}@tju.edu.cn

Abstract. Sensory signals are encoded and processed by neurons in the brain in a form of action potentials, also called spikes that carry clue information across both spatial and temporal dimensions. Learning of such a clue information could be challenging, especially considering the case of long-delayed reward. This temporal credit assignment problem has been solved by a new concept of aggregate-label learning that motivates the development of a family of multi-spike learning algorithms whose remarkable learning performance has been demonstrated. However, most of the current spike-based learning methods are developed without consideration of input temporal fluctuations that constitute a common source of variability in sensory signals such as speech. Therefore, robust spike-based learning under fluctuations of both compression and dilation remains intriguing for exploration. In this paper, we first show the time-warp invariant characteristic of a conductance-based neuron model, based on which we then develop a new multi-spike learning algorithm for time-warp-invariant processing. Experimental results for speech recognition highlight the outstanding robustness of our algorithm against temporal distortions as compared with other relevant spike-based methods. Therefore, our study successfully confirms the effectiveness of multi-spike learning for time-warp robustness, extending a new scope for spike-based processing and learning.

Keywords: Time-Warp-Invariant Processing · Spiking Neural Networks · Speech Recognition · Multi-Spike Learning

1 Introduction

Artificial neural networks (ANNs) have shown remarkable performance in computer vision and speech recognition [10], but they are data-dependent and computationally intensive compared to the brain. To address this, biologically plausible spiking neural networks (SNNs) have been proposed as an efficient brain-like processing [13,16].

B. Luo et al. (Eds.): ICONIP 2023, CCIS 1962, pp. 15–25, 2024.
https://doi.org/10.1007/978-981-99-8132-8_2

SNNs differ from ANNs by using discrete spikes for computation, resembling biological systems [9]. However, effectively utilizing spiking neurons for learning and decision-making while maintaining biological realism remains challenging.

Many spiking neuron models have been proposed to simulate the ability of biological neurons, including the Hodgkin-Huxley (HH) model [7], spike response (SRM) model [3], and leaky integrate-and-fire (LIF) model [2]. Among these, LIF and SRM models are widely used due to their simplicity and efficiency.

Various learning rules have been proposed for the neuron models mentioned above. The Tempotron [5] effectively trains neurons to respond with binary states: fire or not fire, but this binary response lacks robustness and constrains the ability to utilize the temporal structure of the output [20]. To address this, supervised learning rules like ReSuMe [15] and PSD [21] have been proposed. ReSuMe combines STDP and anti-STDP to train the neuron to respond to the specific input pattern. PSD can produce multiple spikes at precise time [21]. Associating firing times with different classes, neurons trained by these rules can successfully perform the challenging multi-classification. However, how the biological systems overcome temporal fluctuations with precise instruction signals is still not clear, and these rules solely rely on temporal coding rather than rate coding, which encodes information through spike frequencies [1].

Recent advancements have led to the development of membrane potential-driven learning rules for multi-category classification using rate and temporal coding schemes. The Multi-spike Tempotron (MST) rule [4] trains neurons to produce specific spike patterns by combining firing thresholds and output spikes. Other multi-spike learning rules inspired by MST have been proposed, such as TDP1 and TDP2 [20]. TDP1 and TDP2 [20] reduce the calculation and improve simplicity by assuming the linear state at the moment of threshold crossing. In [14], synaptic weights are updated based on local maximum points to achieve desired spike number. EML [12] simplifies the LIF model and learning process, and EMLC [12] simplifies computation using the membrane potential state instead of a complex STS function. However, these rules neglect input temporal fluctuations, which are essential mechanisms in the brain.

In this paper, we propose a multi-spike learning rule that effectively processes temporal fluctuation signals. Firstly, we demonstrate the time-warp invariant characteristic of a conductance-based neuron model over the membrane potential when spike patterns are compressed or stretched. Based on this, we develop the new multi-spike learning rule. Our method exhibits excellent word recognition performance, maintaining stable recognition accuracy even with highly variable test speech samples. In summary, our proposed method contributes to advancements in neuromorphic computing.

2   Method

In this section, the neuron model, time-warp invariance and a new multi-spike learning rule are introduced.

2.1   Neuron Model

We use the conductance-based leaky integrate-and-fire (LIF) model due to its simplicity and computational efficiency [6]. The differential formula of neuron membrane potential $V(t)$ is as follows

$$\frac{dV(t)}{dt} = -V(t)(g_{leak} + G(t)) + I(t) \tag{1}$$

where $G(t)$ and $I(t)$ are the total synaptic conductance and the total synaptic current respectively. g_{leak} is the leak conductance. These variables are defined as

$$G(t) = \sum_{i=1}^{N} \sum_{t_i^j < t} g_i(t - t_i^j) \tag{2}$$

$$I(t) = \sum_{i=1}^{N} V_i^{\text{rev}} \sum_{t_i^j < t} g_i(t - t_i^j) \tag{3}$$

V_i^{rev} represents the reversal potential of the i-th synapse. N signifies the number of presynaptic afferents, while t_i^j corresponds to the arrival time of the j-th pre-synaptic spike of the i-th afferent. $g_i(t) = g_i^{max} exp(-\frac{t}{\tau_s})$ describes the exponentially decaying synaptic conductance. Here, g_i^{max} represents the peak conductance of the i-th synapse, which must be maintained as $g_i^{max} > 0$ during the computation of the membrane potential. τ_s is the synaptic time constant. If there is a change in the sign of g_i^{max}, the reversal potential needs to be adjusted accordingly.

Due to the complexity of obtaining an analytic expression for Eq.(1), we compute $V(t)$ using forward Euler with a step size Δt, which is given by

$$V(t) = V(t - \Delta t) + \Delta t \frac{dV(t - \Delta t)}{dt} \tag{4}$$

At each time step, the membrane potential is calculated and compared to a threshold ϑ to determine if a spike is fired. If the membrane potential at the current moment surpasses the threshold, it is reset to 0 at the next time step. The choice of time step Δt is crucial as it affects the accuracy and speed of the simulation, necessitating the selection of an appropriate value.

2.2   Time-Warp Invariance

In this section, we demonstrate the time-warp invariance of the conductance-based LIF neuron model over membrane potential trace.

A given spike pattern can be compressed or stretched (see Fig. 1(a)-(c)). Here, we set β as the scaling factor, where $\beta < 1$ is compression and $\beta > 1$ is stretch. $I(t)$ and $G(t)$ can be changed as

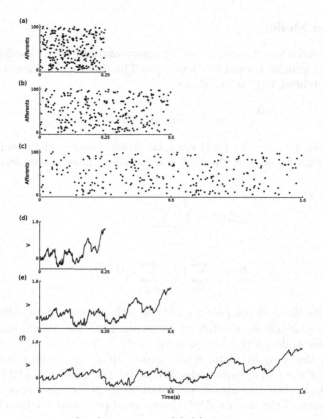

Fig. 1. Spike patterns and voltage traces. (a)-(c) denote spikes patterns with $\beta = 0.5, 1, 1.5$, respectively. (d)-(f) denote voltage traces corresponded to (a)-(c).

$$I(\beta t, \beta) = \sum_{i=1}^{N} V_i^{\text{rev}} \sum_{t_i^j < t} g_i(\beta t - \beta t_i^j) \tag{5}$$

and

$$G(\beta t, \beta) = \sum_{i=1}^{N} \sum_{t_i^j < t} g_i(\beta t - \beta t_i^j) \tag{6}$$

The total synaptic current integrates the incoming spikes of all synapses and delivers a total charge, remaining unaffected when changing the rate of the incoming spikes [6]. Thus, the total charge can be obtained by integrating $I(t, 1)$ as

$$\int I(t, 1)dt = \int \sum_{i=1}^{N} g_i^{max} V_i^{\text{rev}} \sum_{t_i^j < t} exp(-\frac{t - t_i^j}{\tau_{\text{s}}})dt \tag{7}$$

By ignoring the influence of τ_s, which is short relative to the time scale of spike pattern, we get

$$\int I(t,1)dt \approx \sum_{i=1}^{N} g_i^{max} V_i^{rev} \sum_{t_i^j < t} \tau_s \tag{8}$$

When the rate of incoming spikes is changed, we can get the charge by integrating $I(\beta t, \beta)$ as

$$\int I(\beta t, \beta)d(\beta t) \approx \beta \sum_{i=1}^{N} g_i^{max} V_i^{rev} \sum_{t_i^j < t} \frac{\tau_s}{\beta} \tag{9}$$

Following as $I(t)$, $G(t,1)$ and $G(\beta t, \beta)$ are integrated as

$$\int G(t,1)dt \approx \sum_{i=1}^{N} g_i^{max} \sum_{t_i^j < t} \tau_s \tag{10}$$

$$\int G(\beta t, \beta)d(\beta t) \approx \beta \sum_{i=1}^{N} g_i^{max} \sum_{t_i^j < t} \frac{\tau_s}{\beta} \tag{11}$$

So we can get

$$\int I(\beta t, \beta)dt \approx \beta^{-1} \int I(t,1)dt \tag{12}$$

$$\int G(\beta t, \beta)dt \approx \beta^{-1} \int G(t,1)dt \tag{13}$$

According to Eq.(1), $V(t,1)$ and $V(\beta t, \beta)$ are given as

$$V(t,1) = e^{-\int (g_{leak}+G(t,1))dt}\left(c + \int I(t,1)e^{\int (g_{leak}+G(t,1))dt}dt\right) \tag{14}$$

and

$$V(\beta t, \beta) = e^{-\beta \int (g_{leak}+G(\beta t,\beta))dt}\left(c + \beta \int I(\beta t, \beta)e^{\beta \int (g_{leak}+G(\beta t,\beta))dt}dt\right) \tag{15}$$

When $g_{leak} \ll G(t)$, the membrane potential traces exhibit approximate time-warp invariance, which can be expressed as

$$V(\beta t, \beta) = V(t,1) \tag{16}$$

As is shown in Fig. 1, when warping the spike pattern (see Fig. 1(a)-(c)), it can be observed that the membrane potentials at βt and t exhibit a high degree of similarity (see Fig. 1(d)-(f)). This result indicates that the membrane potential traces exhibit time-warp invariance, regardless of the absence or presence of spike firing.

2.3 Time-Warp-Invariant Multi-spike Learning Rule

Learning is an essential issue in SNNs research. Recently, MST [4] has been proposed to fire the desired number of spikes for the target pattern. Inspired by this, several methods have been developed, such as TDP [20], which simplifies the search process by a linear assumption for threshold crossing. All of these methods are based on the spike-threshold-surface (STS) function, which describes the relationship between the threshold and the number of output spikes. However, finding critical thresholds using the STS function is relatively inefficient. Therefore, we aim to develop an efficient learning method that relies solely on the membrane potential state, rather than the STS.

For a given spike pattern, the membrane potential is directly modified to fire desired output spikes by updating the synaptic peak conductance. The derivative of our method can be expressed as

$$\frac{\partial V(t)}{\partial g_i^{max}} = \frac{\partial V(t - \Delta t)}{\partial g_i^{max}} + \Delta t \frac{\partial}{\partial g_i^{max}} \left(\frac{dV(t - \Delta t)}{dt} \right) \tag{17}$$

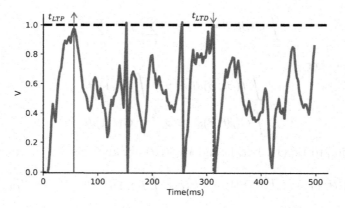

Fig. 2. Learning rule including long-term potentiation and long-term depression. The solid line is t_{LTP} and the dotted line is t_{LTD}.

Here, two time points, t_{LTP} and t_{LTD}, are defined to update synaptic peak conductance. If the desired number of spikes n_d is greater than the actual number n_o, time point t_{LTP} is selected for the long-term potentiation (LTP). Conversely, If $n_d < n_o$, t_{LTD} is selected for a long-term depression (LTD). The time point t_{LTP} is the position of the maximum subthreshold voltage value (see Fig. 2 solid), while t_{LTD} is the position of the minimum voltage value of firing spikes (see Fig. 2 dotted). The adjusting formula of synaptic peak conductance is given as

$$\Delta g_i^{max} = \begin{cases} -\eta \frac{\partial V(t_{LTD})}{\partial g_i^{max}} & if \ n_o > n_d \\ \eta \frac{\partial V(t_{LTP})}{\partial g_i^{max}} & if \ n_o < n_d \end{cases} \tag{18}$$

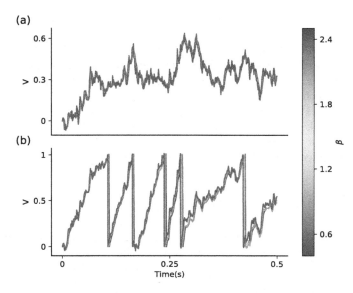

Fig. 3. Membrane potential traces of the same spike patterns warped by different β. Different voltage traces are normalized to the same length and superimposed. (a) Without spike. (b) With desired spike numbers.

where $\eta > 0$ is the learning rate controlling the updating size. With our learning rule, named TWI-ML, the membrane potential is not only invariant with no spike (see Fig. 3(a)), but also invariant after updating to firing target output spikes (see Fig. 3(b)) for a given pattern that is warped in time with different β.

3   Application to Speech Recognition

SNNs are suitable for processing temporal-rich information due to their rich spatio-temporal characteristics. In this section, we apply our method to word recognition to evaluate its capability for solving realistic recognition tasks. Furthermore, considering the capability of time-warp invariance, we investigate the robustness of our method and several other multi-spike rules on the real-world task.

The conversion of raw speech into spike patterns is achieved through a coding frontend, with the utilization of the threshold-coding scheme [6]. This coding process enables the transformation of speech signals into spike representations. Subsequently, these spike patterns are forwarded to the subsequent layer for the purpose of classification. Here, we train each neuron to fire 6 spikes on Ti46 and 5 spikes on TIDIGITS for a given class, while remaining silent for other classes. During testing, if a neuron spikes more than others for the given class, we believe that classification is successfully completed.

3.1 Dataset

- **Ti46** [8]. The dataset includes 46 words spoken by 16 adult speakers (8 males and 8 females). Here, its subset has been used, which contains 10 words (i.e., 'zero' to 'nine') and is partitioned into train and test sets, consisting of 1592 and 2542 speech utterances.
- **TIDIGITS** [11]. The corpus contains isolated 11 words (i.e., 'zero' to 'nine and 'oh'), spoken by 225 speakers from 22 different dialectical regions. In this experiment, during training and testing, we use consist of 2464 and 2486 speech utterances, respectively.

3.2 Result

Table 1 presents the classification result of different SNN-based models. Our method achieves the highest accuracy of 98.29% on the Ti46 dataset compared with other classifiers. Additionally, our method achieves an accuracy of 95.38% on the TIDIGITS. Although the accuracy of SOM and tempotron-like classifier [19] is higher, the number of samples used for training is more. And compared with MPD-AL [23] applying additional decoding algorithms, our framework is simpler. Moreover, the performance of TWI-ML is significantly better than other multi-spike rules. This reconfirms the advantage of time-warp-invariant multi-spike learning method on word recognition tasks.

Table 1. Classification accuracy of two datasets.

Model	Dataset	Accuracy(%)
Liquid State Machine [24]	The subset of Ti46	92.30
SWAT [18]	The subset of Ti46	95.25
MST [4]	The subset of Ti46	97.42
TDP [20]	The subset of Ti46	97.64
EML [12]	The subset of Ti46	96.24
TWI-ML (Ours)	**The subset of Ti46**	**98.29**
Spike signature and SVM [17]	Aurora	91.00
MPD-AL with $N_d = 3$ [23]	TIDIGITS	95.35
ResuMe-DW [22]	TIDIGITS	92.45
MST [4]	TIDIGITS	93.08
TDP [20]	TIDIGITS	93.41
EML [12]	TIDIGITS	90.09
TWI-ML (Ours)	**TIDIGITS**	**95.38**

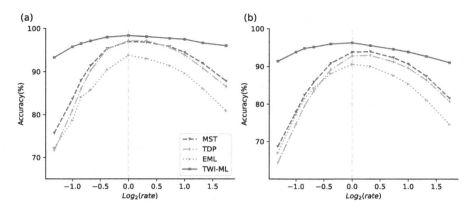

Fig. 4. The classification accuracy. (a) and (b) show the accuracy with different speed rates of Ti46 and TIDIGITS datasets of different methods.

Furthermore, we design a task to show the advanced performance of our method. In this task, neurons are trained by normal training samples, and tested by different time-warped test datasets, which means the rate of test speech samples is changed. According to Fig. 4, our proposed model maintains a high classification accuracy and stability even when the test samples are warped. Conversely, the performance of other multi-spike learning rules deteriorates as the warping level increases. It is proved that our algorithm is superior in processing temporal fluctuated information.

4   Conclusion

In this work, a time-warp-invariant multi-spike learning method is proposed, which has a strong robustness to time-warp information. We investigate the performance of our method on the real-world tasks. Experimental result shows that our method has better performance than other multi-spike rules. Significantly, our method demonstrates exceptional capability in recognizing time-warped speech, even when trained with a limited amount of data. In contrast, other multi-spike learning rules and two ANN models exhibit a noticeable decline in recognition performance. In summary, our method offers an effective solution for processing time-warped signals without depending on a great number of label data. This characteristic enhances the robustness and efficiency of neuromorphic computing.

Acknowledgement. This work was supported by the National Natural Science Foundation of China under Grant 62176179.

References

1. Brette, R.: Philosophy of the spike: rate-based vs spike-based theories of the brain. Front. Syst. Neurosci. **9**, 151 (2015)

2. Burkitt, A.N.: A review of the integrate-and-fire neuron model: I. homogeneous synaptic input. Biol. Cybern. **95**(1), 1–19 (2006). https://doi.org/10.1007/s00422-006-0068-6

3. Gerstner, W., Kistler, W.M.: Spiking neuron models: single neurons, populations, plasticity. Cambridge University Press (2002)

4. Gütig, R.: Spiking neurons can discover predictive features by aggregate-label learning. Science **351**(6277), aab4113 (2016)

5. Gütig, R., Sompolinsky, H.: The tempotron: a neuron that learns spike timing-based decisions. Nat. Neurosci. **9**(3), 420–428 (2006)

6. Gütig, R., Sompolinsky, H.: Time-warp-invariant neuronal processing. PLoS Biol. **7**(7), e1000141 (2009)

7. Hodgkin, A.L., Huxley, A.F.: A quantitative description of membrane current and its application to conduction and excitation in nerve. J. Physiol. **117**(4), 500–544 (1952)

8. Instruments, T.: TI 46-word speaker-dependent isolated word corpus (cd-rom). NIST, Gaithersburg (1991)

9. Kandel, E.R., Schwartz, J.H., Jessell, T.M., Siegelbaum, S., Hudspeth, A.J., Mack, S.: Principles of neural science, vol. 4. McGraw-hill New York (2000)

10. LeCun, Y., Bengio, Y., Hinton, G.: Deep learning. Nature **521**(7553), 436–444 (2015)

11. Leonard, R.G., Doddington, G.: TIDIGITS speech corpus. Texas Instruments, Inc. (1993)

12. Li, S., Yu, Q.: New efficient multi-spike learning for fast processing and robust learning. In: Proceedings of the AAAI Conference on Artificial Intelligence, vol. 34, pp. 4650–4657 (2020)

13. Maass, W.: Networks of spiking neurons: the third generation of neural network models. Neural Netw. **10**(9), 1659–1671 (1997)

14. Miao, Y., Tang, H., Pan, G.: A supervised multi-spike learning algorithm for spiking neural networks. In: 2018 International Joint Conference on Neural Networks (IJCNN), pp. 1–7. IEEE (2018)

15. Ponulak, F.: ReSuMe-new supervised learning method for spiking neural networks. Poznan University of Technology, Institute of Control and Information Engineering (2005)

16. Roy, K., Jaiswal, A., Panda, P.: Towards spike-based machine intelligence with neuromorphic computing. Nature **575**(7784), 607–617 (2019)

17. Tavanaei, A., Maida, A.S.: A spiking network that learns to extract spike signatures from speech signals. Neurocomputing **240**, 191–199 (2017)

18. Wade, J.J., McDaid, L.J., Santos, J.A., Sayers, H.M.: Swat: a spiking neural network training algorithm for classification problems. IEEE Trans. Neural Networks **21**(11), 1817–1830 (2010)

19. Wu, J., Chua, Y., Li, H.: A biologically plausible speech recognition framework based on spiking neural networks. In: 2018 International Joint Conference on Neural Networks (IJCNN), pp. 1–8. IEEE (2018)

20. Yu, Q., Li, H., Tan, K.C.: Spike timing or rate? Neurons learn to make decisions for both through threshold-driven plasticity. IEEE Trans. Cybern. **49**(6), 2178–2189 (2018)

21. Yu, Q., Tang, H., Tan, K.C., Li, H.: Precise-spike-driven synaptic plasticity: learning hetero-association of spatiotemporal spike patterns. PLoS ONE **8**(11), e78318 (2013)

22. Zhang, M., et al.: Supervised learning in spiking neural networks with synaptic delay-weight plasticity. Neurocomputing **409**, 103–118 (2020)

23. Zhang, M., et al.: MPD-AL: an efficient membrane potential driven aggregate-label learning algorithm for spiking neurons. In: Proceedings of the AAAI Conference on Artificial Intelligence, vol. 33, pp. 1327–1334 (2019)
24. Zhang, Y., Li, P., Jin, Y., Choe, Y.: A digital liquid state machine with biologically inspired learning and its application to speech recognition. IEEE Trans. Neural Netw. Learn. Syst. **26**(11), 2635–2649 (2015)

ECOST: Enhanced CoST Framework for Fast and Accurate Time Series Forecasting

Yao Wang[1], Chuang Gao[2]($\boxtimes$) (iD), and Haifeng Yu[1]

[1] ShanghaiTech University, Shanghai, China
{wangyao2,yuhf}@shanghaitech.edu.cn
[2] Shanghai Advanced Research Institute, Chinese Academy of Sciences, Shanghai, China
gaoc@sari.ac.cn

Abstract. We introduce an enhanced forecasting framework for time series using contrastive learning. This method builds on the existing CoST (Contrastive Learning of Disentangled Seasonal-Trend Representations) framework, which, despite its promising performance, is still encumbered by the challenges posed by the intricate nature of time series data. In our proposed framework, we bypass the need for a backbone encoder and directly perform time series decomposition to extract the trend and detrended subseries. These components subsequently undergo independent trend and seasonal feature extraction. This approach ensures a more robust, efficient, and direct representation of inherent time series characteristics. We incorporate Reversible Instance Normalization (RevIN) to improve forecasting accuracy and account for potential distribution bias. Additionally, a new concept, the 'trend queue', is proposed for storing past trend features, improving the learning of trend nuances. Our ECoST model has shown significant improvements, with an increase in prediction accuracy by 8.5% and a 74% enhancement in training time efficiency compared to the CoST model. These results were validated through experiments conducted on several real-world time series datasets. This underscores the effectiveness of our approach in providing a more robust and efficient time series forecasting methodology, thereby setting a new benchmark in the field of contrastive learning for time series data.

Keywords: Time Series Forecasting · Contrast Learning · Feature Extraction

1 Introduction

Time series data are chronologically ordered observations that possess significant time dependence. Time series forecasting involves the analysis of time series data using statistical and modeling techniques to predict future trends

B. Luo et al. (Eds.): ICONIP 2023, CCIS 1962, pp. 26–36, 2024.
https://doi.org/10.1007/978-981-99-8132-8_3

and inform decision-making. It has a broad spectrum of applications in various fields such as stock market analysis [1], electricity load forecasting [2], and traffic road forecasting [3]. However, the complexity and diversity of time series data pose challenges to the design and training of predictive models. On one hand, time series data are typically high-dimensional, comprise long series, and exhibit non-linearity and non-smoothness. This makes it difficult for traditional linear models to capture their intrinsic dynamic laws, especially for multivariate long series time series forecasting problems [4]. On the other hand, time series data often contain multiple components such as trends, seasonality, periodicity, and noise. These components may influence each other or exist independently, possessing different time-domain and frequency-domain characteristics. This makes it challenging for a single model to simultaneously extract information from these components [5]. Consequently, it is both a valuable and challenging research problem to effectively extract useful features from time series data for accurate and reliable forecasting.

Several researchers have found that forecasting can be effectively performed using time series decomposition and feature extraction [6]. Time series decomposition involves decomposing time series data into different components such as trend, seasonality, and periodicity, simplifying the data structure and complexity. Feature extraction entails transforming time series data into representative and distinguishing feature vectors, thereby reducing data dimensionality and noise while enhancing the model's generalization ability. Effective utilization of time series data features can significantly improve prediction performance.

In 2021, Wu Haixu et al. [7] proposed Autoformer, which combined the Transformer architecture with a straightforward seasonal trend decomposition architecture, demonstrating exceptional performance on multiple publicly available datasets. In 2022, Tian Zhou et al. [8] proposed FEDformer, a model that integrated Fourier analysis with Autoformer based on the Transformer. This helped capture the global features of time series more effectively. In the same year, Gerald Woo et al. [9] proposed CoST, a model that combined contrastive learning with time series forecasting by learning separate trend and seasonal features for time series forecasting. These methods have shown that using the components and features of a robust time series can effectively improve forecasting performance. However, these methods also have some limitations. Firstly, some methods require a longer running time or larger computational resources, which may limit their feasibility in practical applications. Secondly, some methods employ a simplistic or fixed approach to extract or process time series components and features, which can result in inflexible or underrefined models. Finally, some methods do not fully utilize contrastive learning techniques to enhance the model's understanding of differences and similarities among time series features, which could affect the model's generalizability and robustness.

To overcome these limitations and challenges, we propose a time series forecasting framework based on contrastive learning, improving upon the CoST framework [9] for more effective and efficient extraction of trend and seasonal features from time series. The CoST framework is a time series forecasting framework that employs contrastive learning methods to learn separate trend and seasonal feature representations. It consists of a backbone encoder and two branch encoders for extracting trend and seasonal features respectively. It then uses contrastive loss functions in the time and frequency domains to train the model, such that trend and seasonal features are more discriminative between different time series. Finally, it uses a simple regressor to make predictions based on the learned feature representations. While we believe that the CoST framework is a promising paradigm for time series forecasting, we identify some areas for improvement. Specifically, we propose four enhancements to the CoST framework as follows:

1. We eliminate the use of a backbone encoder and instead directly decompose the time series into trended and detrended subseries using a straightforward linear time series decomposition technique. We then perform feature extraction on both subseries separately. This approach reduces the model's computational complexity and parameter count while more efficiently extracting trend and seasonal components.

2. We integrate Reversible Instance Normalization (RevIN) [10] into our model to enhance its robustness against shifts in the distribution of time series data. RevIN is a normalization technique that maintains a constant relative distance between inputs and outputs. This effectively minimizes the discrepancies between different time series, thereby improving the model's generalization ability.

3. To augment the model's learning of trend properties, we introduce the concept of a trend queue. This queue stores past trend characteristics, enabling comparisons not only within the current batch but also between trend characteristics of current and past batches during the calculation of trend contrastive loss.

4. Our model, named the Enhanced CoST Framework for Fast and Accurate Time Series Forecasting (ECOST), has been rigorously tested on five real-world benchmark datasets. ECOST not only surpasses state-of-the-art models by an impressive 8.5% in multivariate prediction but also significantly reduces the training time by 74%, highlighting its potential as a highly efficient tool for time series prediction.

2   Proposed Method

We propose a contrastive learning-based model for time series forecasting, aimed at learning decoupled seasonal and trend representations from time series data to enhance forecasting performance. After the input layer, our model incorporates a temporal decomposition module, which separates the time series into two components: seasonality and trend. For each of these components, we conduct feature extraction, employing a Convolutional Neural Network (CNN) to learn the trend features and a Band Fourier Layer (BFL) to discern seasonal features. We further integrate a RevIN layer to normalize the outputs derived from the feature extraction layer, thereby resolving the distribution bias. We optimize our model using a contrastive loss function, ensuring high similarity between seasonal and trend representations within the same time series and low similarity between those from different time series. To further bolster trend feature learning, we incorporate a trend queue into the contrastive loss function. Our model demonstrates its adaptability by its compatibility with various regressors, making it suited for a range of time series forecasting tasks. Figures 1 and 2 depict our model's overall architecture. We first introduce the time series decomposition module, followed by the feature extraction module and contrastive loss function. The prediction module is discussed last.

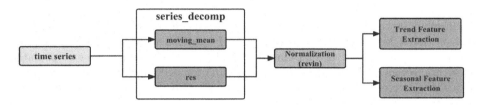

Fig. 1. Time series decomposition and feature extraction

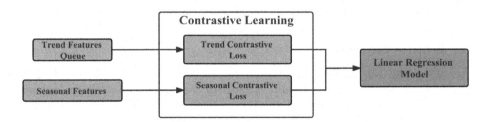

Fig. 2. Contrast learning and regression prediction

2.1 Timing Decomposition

Time series decomposition, a standard method in time series analysis, allows for the breakdown of time series into different components, such as trend, seasonal, periodic, and stochastic terms [11]. These components reflect various characteristics of the time series, including long-term changes, cyclical fluctuations, and short-term fluctuations. Time series decomposition enables a deeper understanding of the structure and patterns of time series and provides improved predictions for future changes.

Our model employs a simple yet effective method for time series decomposition called the Moving Average method (MA). The moving average is a smoothing technique that reduces short-term fluctuations in a time series, resulting in a more stable trend term. Specifically, for a given time series $x = (x_1, x_2, ..., x_T)$, we use a window of size k sliding over the time series and compute the average value within each window to establish the trend term t_i:

$$t_i = \frac{1}{k} \sum_{j=i-k+1}^{i} x_j, \quad i = k, k+1, ..., T \tag{1}$$

We then subtract the trend term from the original time series to get the seasonal term s_i, eliminating the long-term variation:

$$s_i = x_i - t_i, \quad i = k, k+1, ..., T \tag{2}$$

To implement the moving average method, we use a convolutional layer with a convolutional kernel size of k, a step size of 1, and a weight of $\frac{1}{k}$. We refer to this convolutional layer as a moving average layer (MAL), which we encapsulate as a time series decomposition module (SDM) for decomposing the input time series into two components: seasonal and trend. Our time series decomposition module can be expressed as follows:

$$(s, t) = \text{SDM}(x) = (x - \text{MAL}(x), \text{MAL}(x)) \tag{3}$$

Here, s and t represent the seasonal and trend representations, respectively, which share the same dimension as the input time series. Figure 2 shows a schematic diagram of the time series decomposition module.

2.2 Feature Extraction

After obtaining the seasonal and trend representations, we apply the RevIN layer, which calculates the mean and variance of the inputs and performs reverse normalization. Given the input x, the RevIN layer is computed as follows:

1. Compute the mean μ and standard deviation σ of the input:

$$\mu = \text{mean}(x)$$
$$sigma = \sqrt{\text{var}(x) + \epsilon} \tag{4}$$

2. Perform reverse normalization:

$$x' = (x - \mu)/\sigma \tag{5}$$

3. If learnable weights and biases exist, perform an affine transformation:

$$x' = x' \cdot w + b \tag{6}$$

Thus, we define the RevIN-processed t and s as follows:

$$t^* = \text{RevIN}(t), s^* = \text{RevIN}(s) \tag{7}$$

Following that, we perform feature extraction on the s^* and t^* processed by RevIN to acquire higher-level and more abstract feature representations. As seasonal and trend representations exhibit different characteristics, we utilize different feature extraction modules to process them.

For trend representation t, we use CNN for feature extraction.CNN is a commonly used deep learning model for processing data with spatial or temporal structure, such as images, speech, text, etc. CNN extracts local features in the data by using multiple convolutional layers and reduces the dimensionality and complexity of the data by pooling layers or stepwise convolution. CNN has good parameter sharing and translation invariance properties, which can effectively handle long-term dependencies and nonlinear relationships in time series.

We use multiple one-dimensional convolutional layers to construct the trend feature extractor (TFE) and perform softmax weighted summation on the output of each convolutional layer:

$$v_T = \text{TFE}(t^*) = \text{SoftmaxWeightedSum}(\text{CNN}_1(t^*), \text{CNN}_2(t^*), ..., \text{CNN}_K(t^*)) \tag{8}$$

For the seasonal representation s^*, we use a Band Fourier Layer (BFL) for feature extraction. We utilize multiple Band Fourier Layers to construct the seasonal feature extractor (SFE):

$$v_S = \text{SFE}(s^*) = \text{Concat}(\text{BFL}_1(s^*), \text{BFL}_2(s^*), ..., \text{BFL}_L(s^*)) \tag{9}$$

In this way, we extract the corresponding trend and seasonal features from the original time series, which can further be utilized in subsequent time series forecasting tasks.

2.3 Contrast Learning

First, we introduce a trend queue to enhance the model's ability to learn trend characteristics. This queue stores past trend characteristics, enabling us to compare not only within the current batch but also the trend characteristics of current and past batches when calculating trend comparison loss. We define the comparison loss as follows:

$$L = -\mathbb{E}(q_t, k_t) \sim_{\mathcal{D}} \left[\log \frac{\exp(q_t^T k_t / T)}{\exp(q_t^T k_t / T) + \sum_{k_{\text{neg}} \in \text{TrendQueue}} \exp(q_t^T k_{\text{neg}} / T)} \right] \tag{10}$$

where q_t and k_t are the query and key of the current batch, k_{neg} is a negative sample in the trend queue, and T is a temperature parameter. This loss function will drive the model to learn the trend properties better because it requires the model to make the positive sample in the current batch score higher than the negative sample in the trend queue.

Second, we follow the treatment of seasonal characteristics in the original paper, i.e., we transform the time-series signal to the frequency domain via the Fourier transform and learn the amplitude and phase of the signal in the contrast. We define the seasonal contrast loss as

$$L_{\text{seasonal}} = \alpha \left(L_{\text{instance}}(\text{Amp}(q_s), \text{Amp}(k_s)) + L_{\text{instance}}(\text{Phase}(q_s), \text{Phase}(k_s)) \right) / 2 \tag{11}$$

where q_s and k_s are the query and key of the current batch, Amp(·) and Phase(·) denote the amplitude and phase after Fourier transform, respectively, α is a weight parameter, and L_{instance} denotes the instance_contrastive_loss function.

Overall, our approach improves the performance of the model by introducing a trend queue that better captures the complex dynamic properties of the time series. At the same time, we also retain the treatment of seasonal properties in the original paper.

2.4 Prediction

The prediction module is the downstream part of our model for time series forecasting using our decoupled representations v_T^* and v_S^*. We use a simple linear regression layer as the prediction module whose input is a decoupled representation of shape (B, T, d), where $d = d_T + d_S$ is the dimensionality of the decoupled representation. We use a weight vector w of shape $(d, 1)$ and a bias term b to compute the predicted values $\hat{x}_t$ for each time step:

$$\hat{x}_t = w^\top [v_{T_t^*}; v_{S_t^*}] + b$$

where $[v_{T_t^*}; v_{S_t^*}]$ is the decoupled representation of the tth time step. We use the mean squared error (MSE) as the loss function of the prediction module and fine-tune it at the end of training using the decoupled representations v_T^* and v_S^*. We use mean absolute error (MAE) and mean squared error (MSE) as evaluation metrics and report the results of each method on the test set.

3 Experiments

This section presents a detailed empirical analysis of CoST. We compare it against various time series representation learning approaches and end-to-end supervised forecasting methods.

3.1 Datasets

(1) *ETT (Electricity Transformer Temperature)*: This dataset consists of temperature and load data from power transformers in two different regions of China, suitable for exploring long series time series prediction problems [12]. The dataset spans two years, with recordings every minute. It is divided into 1-hour level subsets (ETTh1 and ETTh2) and 15-minute level subsets (ETTm1 and ETTm2)[1].

(2) *Weather*: This dataset includes 4 years of data from nearly 1,600 locations in the U.S., recorded hourly with 11 features per time point[2]. We use the "wet bulb" feature as the target value for univariate forecasting.

---

[1] https://github.com/zhouhaoyi/ETDataset
[2] https://www.ncei.noaa.gov/data/local-climatological-data/

We compare our Enhanced CoST (ECoST) against the latest state-of-the-art methods for time series modeling and forecasting tasks from two categories: (1) representation learning techniques, including CoST [9], TS2Vec [13,14], and TNC [15]; (2) end-to-end forecasting models, including Informer [12], and TCN [16].

3.2   Evaluation Setup

We use MSE and MAE as evaluation metrics, and apply a 60/20/20 training/validation/test split. We report the test set results when the model achieves peak performance on the validation set.

3.3   Implemention Details

ECoST adopts the same parameter settings as CoST, including a batch size of 256, a learning rate of 1E-3, a momentum of 0.9, and a weight decay of 1E-4 across all datasets. We implement an SGD optimizer and cosine annealing. For datasets with fewer than 100,000 samples, we conduct 200 training iterations. For larger datasets, we carry out 600 iterations.

4   Results and Analyses

For the task of multivariate prediction, ECoST demonstrates superior performance on all five benchmark datasets, as shown in Table 1. The proposed ECoST model achieves an overall relative MSE reduction of 8.5% compared to CoST. We sourced the results for CoST and other benchmark models from CoST [9], except for the ETTm2 dataset results, which we derived from SimTS [17].

In the above experimental results, our ECoST model surpasses the benchmark model in terms of both MAE and MSE on each dataset. This suggests improvements in both accuracy and stability of prediction. Next, by considering the running time, we further evaluate the efficiency of our improved model.

In our study, as shown in Table 2, we conducted an in-depth comparison and analysis of three methods: ECoST, CoST, and TS2vec. In terms of training time, although TS2vec performed best with a time of 1 min and 25 s, ECoST's training time of 2 min and 13 s represented an efficiency improvement of approximately 74% compared to CoST's 8 min and 32 s. This makes ECoST a more efficient choice for practical applications.

In terms of encoder inference time, ECoST performed best with a time of 1.83 s, which is about 64% and 3% more efficient than CoST (5.07 s) and TS2vec (1.89 s) respectively. This further demonstrates ECoST's advantage in handling complex prediction tasks.

In terms of LR inference time, while CoST performed well within shorter time frames (24 and 48), ECoST stood out in longer time frames (168, 336, and 720), with efficiency improvements of approximately 63%, 50%, and 40% compared

Table 1. Multivariate long-term series forecasting results on five datasets with prediction length $O \in \{24, 48, 168, 336, 720\}$ (For ETTm2 dataset, we use prediction length $O \in \{24, 48, 96, 288, 672\}$). A lower MSE indicates better performance, and the best results are highlighted in bold.

| Methods | | Representation Learning | | | | | | | | End-to-end Forecasting | | | |
| | | ECoST | | CoST | | TS2Vec | | TNC | | Informer | | TCN | |
Metrics		MSE	MAE	MSE	MAE	MSE	MAE	MSE	MAE	MSE	MAE	MSE	MAE
ETTh1	24	**0.256**	**0.349**	0.386	0.429	0.590	0.531	0.708	0.592	0.577	0.549	0.583	0.547
	48	**0.328**	**0.404**	0.437	0.464	0.624	0.555	0.749	0.619	0.685	0.625	0.670	0.606
	168	**0.571**	**0.557**	0.643	0.582	0.762	0.639	0.884	0.699	0.931	0.752	0.811	0.680
	336	**0.767**	0.684	0.812	**0.679**	0.931	0.728	1.020	0.768	1.128	0.873	1.132	0.815
	720	**0.912**	0.779	0.970	**0.771**	1.063	0.799	1.157	0.830	1.215	0.896	1.165	0.813
ETTh2	24	**0.331**	**0.426**	0.447	0.502	0.423	0.489	0.612	0.595	0.720	0.665	0.935	0.754
	48	**0.572**	**0.580**	0.699	0.637	0.619	0.605	0.840	0.716	1.457	1.001	1.300	0.911
	168	**1.484**	**0.954**	1.549	0.982	1.845	1.074	2.359	1.213	3.489	1.515	4.017	1.579
	336	1.760	1.044	**1.749**	**1.042**	2.194	1.197	2.782	1.349	2.723	1.340	3.460	1.456
	720	**1.962**	1.103	1.971	**1.092**	2.636	1.307	2.753	1.394	3.467	1.473	3.106	1.381
ETTm1	24	**0.207**	**0.307**	0.246	0.329	0.453	0.444	0.522	0.472	0.323	0.369	0.363	0.397
	48	**0.304**	**0.378**	0.331	0.386	0.592	0.521	0.695	0.567	0.494	0.503	0.542	0.508
	168	0.399	0.447	**0.378**	**0.419**	0.635	0.554	0.731	0.595	0.678	0.614	0.666	0.578
	336	**0.465**	0.494	0.472	**0.486**	0.693	0.597	0.818	0.649	1.056	0.786	0.991	0.735
	720	**0.594**	**0.571**	0.620	0.574	0.782	0.653	0.932	0.712	1.192	0.926	1.032	0.756
ETTm2	24	**0.108**	**0.221**	0.122	0.244	0.180	0.293	0.185	0.297	0.173	0.301	0.180	0.324
	48	**0.158**	**0.279**	0.183	0.305	0.244	0.350	0.264	0.360	0.303	0.409	0.204	0.327
	96	**0.252**	**0.364**	0.294	0.394	0.360	0.427	0.389	0.458	0.365	0.453	3.041	1.330
	288	**0.679**	**0.629**	0.723	0.652	0.723	0.639	0.920	0.788	1.047	0.804	3.162	1.337
	672	**1.548**	**0.962**	1.899	1.073	1.753	1.007	2.164	1.135	3.126	1.302	3.624	1.484
Weather	24	**0.217**	**0.309**	0.298	0.360	0.307	0.363	0.320	0.373	0.335	0.381	0.321	0.367
	48	**0.298**	**0.374**	0.359	0.411	0.374	0.418	0.380	0.421	0.395	0.459	0.386	0.423
	168	**0.443**	**0.483**	0.464	0.491	0.491	0.506	0.479	0.495	0.608	0.567	0.491	0.501
	336	**0.488**	**0.515**	0.497	0.517	0.525	0.530	0.505	0.514	0.702	0.620	0.502	0.507
	720	**0.530**	0.544	0.533	**0.542**	0.556	0.552	0.519	0.525	0.831	0.731	0.498	0.508
Avg.		**0.625**	**0.550**	0.683	0.575	0.814	0.631	0.947	0.685	1.121	0.757	1.327	0.785

to CoST. Although TS2vec performed well in some aspects, it did not surpass ECoST and CoST in terms of LR inference time over longer time frames.

Most importantly, despite not being the best in terms of training time, ECoST performed best in prediction accuracy. This indicates that while training time and inference time are important performance indicators, prediction accuracy remains the key determinant of a model's quality. Therefore, our research results support ECoST as an efficient and accurate prediction method.

In summary, our research results show that ECoST has significant advantages in terms of time efficiency and prediction accuracy. Therefore, we recommend considering ECoST as the primary choice when facing complex and large-scale prediction tasks.

Table 2. Comparison of Time Efficiency

Method	Training Time	Encoder Inference Time	LR Inference Time
ECoST	0:02:13	1.83	24: 0.0121
			48: 0.0171
			168: 0.0837
			336: 0.1364
			720: 0.2136
CoST	0:08:32	5.07	24: 0.005
			48: 0.0147
			168: 0.0311
			336: 0.0689
			720: 0.1282
TS2vec	0:01:25	1.89	24: 0.008
			48: 0.0073
			168: 0.0533
			336: 0.0693
			720: 0.143

5 Conclusion

In this paper, we propose a long-term time series forecasting model based on contrastive learning that achieves state-of-the-art performance while significantly enhancing operational efficiency. Our work demonstrates that a backbone encoder is not necessary, and that a simple time series decomposition method can be employed to improve efficiency while maintaining high performance. Finally, comprehensive experiments show that our model achieves the best performance on five benchmark datasets.

Acknowledgement. This work is funded by the Youth Program of Chinese Academy of Sciences (No. E224821231).

References

1. Rojo-Álvarez, J.L., Martínez-Ramón, M., de Prado-Cumplido, M., et al.: Support vector method for robust ARMA system identification. IEEE Trans. Signal Process. **52**(1), 155–164 (2004)
2. Hong, T., Fan, S.: Probabilistic electric load forecasting: a tutorial review. Int. J. Forecast. **32**(3), 914–938 (2016)
3. Shao, H., Soong, B.H.: Traffic flow prediction with long short-term memory networks (LSTMs). In: 2016 IEEE Region 10 Conference (TENCON), pp. 2986–2989. IEEE (2016)

4. Long, L., Liu, Q., Peng, H., et al.: Multivariate time series forecasting method based on nonlinear spiking neural P systems and non-subsampled shearlet transform. Neural Netw. **152**, 300–310 (2022)

5. Brownlee, J.: How to decompose time series data into trend and seasonality. Machinelearningmastery.com (2017)

6. Shen, L., Wei, Y., Wang, Y.: Respecting time series properties makes deep time series forecasting perfect. arXiv preprint arXiv:2207.10941 (2022)

7. Wu, H., Xu, J., Wang, J., et al.: Autoformer: decomposition transformers with auto-correlation for long-term series forecasting. Adv. Neural. Inf. Process. Syst. **34**, 22419–22430 (2021)

8. Zhou, T., Ma, Z., Wen, Q., et al.: FEDformer: frequency enhanced decomposed transformer for long-term series forecasting. In: International Conference on Machine Learning, pp. 27268–27286. PMLR (2022)

9. Woo, G., Liu, C., Sahoo, D., et al.: CoST: contrastive learning of disentangled seasonal-trend representations for time series forecasting. arXiv preprint arXiv:2202.01575 (2022)

10. Kim, T., Kim, J., Tae, Y., et al.: Reversible instance normalization for accurate time-series forecasting against distribution shift. In: International Conference on Learning Representations (2021)

11. Hyndman, R.J., Athanasopoulos, G.: Forecasting: principles and practice. OTexts (2018)

12. Zhou, H., Zhang, S., Peng, J., et al.: Informer: beyond efficient transformer for long sequence time-series forecasting. In: Proceedings of the AAAI Conference on Artificial Intelligence, vol. 35, no. 12, pp. 11106–11115 (2021)

13. Yue, Z.: TS2Vec: towards universal representation of time series. In: Proceedings of the AAAI Conference on Artificial Intelligence, vol. 36, no. 8, pp. 8980–8987 (2022)

14. https://github.com/yuezhihan/ts2vec . Accessed 2 June 2023

15. Tonekaboni, S., Eytan, D., Goldenberg, A.: Unsupervised representation learning for time series with temporal neighborhood coding. In: International Conference on Learning Representations (2021)

16. Lea, C., Vidal, R., Reiter, A., Hager, G.D.: Temporal convolutional networks: a unified approach to action segmentation. In: Hua, G., Jégou, H. (eds.) ECCV 2016. LNCS, vol. 9915, pp. 47–54. Springer, Cham (2016). https://doi.org/10.1007/978-3-319-49409-8_7

17. Zheng, X., Chen, X., Schürch, M., et al.: SimTS: rethinking contrastive representation learning for time series forecasting. arXiv preprint arXiv:2303.18205 (2023)

Adaptive CNN-Based Image Compression Model for Improved Remote Desktop Experience

Hejun Wang, Kai Deng$^{(\boxtimes)}$, Yubing Duan, Mingyong Yin, Yulong Wang, and Fanzhi Meng

China Academy of Engineering Physics, Institute of Computer Application, Mianyang, China
{dengkai,yinmy,wangyulong}@caep.cn

Abstract. This paper addresses the optimization of desktop image presentation in remote desktop scenarios. Remote desktop tools, essential for work efficiency, often employ image compression to manage bandwidth. While JPEG is a prevalent choice due to its efficiency in eliminating redundancy, it can introduce artifacts as compression increases. Recently, deep learning-based compression techniques have emerged, rivaling traditional methods like JPEG. This research introduces a convolutional neural network-based model for image compression and reconstruction, emphasizing human visual perception. By integrating adaptive spatial and channel attention mechanisms, it ensures better preservation of text and texture. This method outperforms JPEG and other deep learning algorithms in image quality and compression ratio.

Keywords: Image Compression · Convolutional Neural Network · Remote Desktop

1 Introduction

Remote desktops, essential in various sectors such as cloud computing [19,20], face challenges with the growth of display resolutions. For instance, a 4K resolution desktop image, uncompressed, demands about 23.73 MB per frame. At 60 Hz, this equates to a bandwidth of roughly 1.39 GB/s. Such high bandwidth consumption can degrade user experience. VNC, a popular remote desktop tool, uses the RFB protocol with its Tight encoding mode employing JPEG for image compression. This approach conserves bandwidth. However, JPEG's fixed parameters in its discrete cosine transform may not always suit the image's features. Moreover, its block processing can introduce block effects and artifacts.

Some researchers have proposed different solutions for desktop image processing methods in remote desktop application scenarios. Lin et al. proposed a composite image compression algorithm specifically for remote desktop image transmission [9]. This algorithm uses a method similar to the block partitioning

B. Luo et al. (Eds.): ICONIP 2023, CCIS 1962, pp. 37–52, 2024.
https://doi.org/10.1007/978-981-99-8132-8_4

of desktop images in the JPEG compression algorithm. On this basis, the algorithm divides the text and graphics in the blocks and uses different compression methods for them than for images. For text and graphics, the color values can be indexed and compressed based on the index for text and graphics areas, while for areas with complex color information such as images, the JPEG compression algorithm is used for compression. However, this method still does not solve the shadow and blur problems caused by the JPEG compression algorithm for desktop images under low-quality parameters. Wang et al. proposed a composite image compression method and joint coding (UC) based on several effective text/graphics and natural image compression algorithms [16,17]. This method mainly uses lossy intra-frame hybrid coding tools or their variants to compress natural images, while for text/graphics, it mainly uses dictionary entropy coding, run-length coding (RLE) in the RFB protocol, Hextile in the RFB protocol, and PNG and other encoding tools as candidates for text/graphics compression. By appropriately combining these lossless tools with intra-frame hybrid coding, high compression performance is achieved for the text/graphics part of the composite image. This method uses an optimizer based on Rate-Distortion cost to separate natural images and text/graphics, and finally selects lossy and lossless coding tools for encoding respectively. However, this desktop image coding scheme still has some drawbacks. Firstly, the image to be encoded needs to be divided into small pixel blocks. Secondly, iterative selection is required for different block coding, which may affect the real-time performance of image coding. Lastly, the support for high-resolution images is not good enough, which cannot adapt well to the current era of high-resolution desktop images.

Several researchers have optimized desktop image coding methods for specific remote desktop usage scenarios. Sazawa et al. proposed a desktop image coding scheme for engineering applications such as 2D-CAD and CAE: Remote Virtual Environment Computing (RVEC) [13]. RVEC combines movie compression and compression algorithms with small static image loss, which greatly reduces the bandwidth of image transmission without compromising the acceptable engineering standards. In addition, RVEC uses a new lossless compression algorithm, the graphic compression algorithm, to compress static images applied to graphics. The graphic compression algorithm uses vector features in the displayed image to obtain better compression ratios without affecting the compression speed. Shimada et al. later proposed a high-compression method for graphic images in 3D-CAD, CAE software, and other engineering applications based on Sazawa's research [14]. This method uses the characteristic that the pixel values in artificial images do not change locally and extracts constant gradients through frequency transformation, fully utilizing the characteristics of graphic images. It provides reasonable actual compression time, compression size, and image quality for engineering applications in cloud environments. However, although the image coding scheme proposed by Sazawa, Shimada, and others performs well in the field of engineering applications, it is not universal and therefore difficult to apply to general remote desktop usage scenarios.

Deep learning's advancements offer promising applications in image processing. Its ability to discern image features can reduce data redundancy, optimizing bandwidth for remote desktop transmission. Given that users focus on specific desktop areas, deep learning can enhance the reconstruction of text and graphics, improving visual experience.

Deep learning-based image compression methods have been proposed, showing performance approaching or surpassing traditional methods [11]. This potential paves the way for its application in remote desktop image transmission. This paper focuses on constructing an algorithm that is more suitable for remote desktops, replicating the method proposed by Ballé et al. [3], with Mean Square Error (MSE) as the optimization target. Despite optimizing Peak Signal-to-Noise Ratio (PSNR) and Structural Similarity (SSIM) of images [21], this method presents some limitations [5]. Considering the importance of users' subjective perception of desktop image quality in performance evaluation, this paper introduces Multi-Scale Structural Similarity (MS-SSIM) as an additional optimization target. Moreover, an adaptive attention mechanism module is introduced, combining spatial and channel attention, which serves as an image encoder enhancement module, applying attention weights to the latent representations of images [4,6,12,18]. Experimental results suggest that the model, with MS-SSIM and MSE as joint optimization targets, improves the MS-SSIM score of the reconstructed image, while better preserving text and graphic information in desktop images. The incorporated attention mechanism module also assists in improving the PSNR performance of the algorithm, while mitigating the loss of MS-SSIM performance.

The remainder of this paper is organized as follows: Sect. 2 reviews related work, while Sect. 3 analyzes the differences between various image quality assessment methods. Section 4 presents the proposed method in this work. Section 5 presents the experimental results and analysis, and Sect. 6 presents the conclusion.

2   Related Works

Most image compression methods based on deep learning are lossy image compression methods [22]. Due to the robust modeling capabilities of deep learning, these algorithms have gradually approached or even surpassed the performance of traditional image compression methods. Currently, the primary deep learning-based image compression techniques involve the integration of convolutional neural networks (CNNs), recurrent neural networks (RNNs), and generative adversarial networks (GANs). CNNs, in particular, have seen rapid development in image processing, excelling in tasks such as object detection, image classification, and semantic segmentation. The sparse connectivity and parameter sharing properties of CNN convolution operations have demonstrated advantages in image compression. In 2016, Ballé et al. [1] introduced a parametric nonlinear model for Gaussianizing data derived from natural images. Their primary contribution was the development of a normalization layer - the Generalized Divisive Normalization (GDN) layer - optimized for image compression and reconstruction tasks. This layer effectively reduces random noise introduced by traditional

Batch Normalization (BN) layers. Later, Ballé et al. [2] proposed an image compression technique based on nonlinear transformation and uniform quantization. This approach leverages a CNN to extract latent feature representations from images and implements local gain via the previously proposed GDN layer to reduce additive noise. This marked the first integration of CNNs with image compression techniques, establishing a foundation for the development of end-to-end CNN-based image compression methods. Subsequently, Ballé et al. [3] introduced an end-to-end image compression model that integrates hyper-prior encoders with decoders. This model effectively captures spatial dependencies within latent feature representations and eliminates redundancy via Gaussian distribution modeling, resulting in lower bit-rate image compression. Jiang et al. [7] introduced an end-to-end CNN-based image compression encoder-decoder. This approach utilizes a neural network in conjunction with traditional image processing techniques to process images. Initially, a CNN-based image encoder is employed to extract a compact representation of the image, effectively shrinking the original image along its height and width dimensions. Subsequently, traditional image compression methods (e.g., JPEG or BPG) are utilized to encode this compact representation of the image. At the decoding stage, traditional image compression encoding is first applied to restore the compact representation of the image. Then, a CNN-based image decoder is used to scale up this compact representation and restore it back into its original form as an image. Zhao et al. [24] argued that integrating traditional forms of image compression may not be optimal for deep learning-based approaches to this task. As such, they proposed a method that employs a virtual encoder to connect the encoding and decoding stages during training. This virtual encoder - also based on CNNs - can directly map compact representations onto image code streams transmitted to the decoding stage via nonlinear mapping. This approach effectively links neural network-based encoding and decoding stages and can be integrated with traditional encoders to produce high-quality reconstructed images. It can also be extended to other end-to-end CNN-based image compression architectures. To enhance the quality of reconstructed images, Liu et al. [7] introduced a decoding-stage enhancement module capable of learning residual information between reconstructed and original images via neural networks. By leveraging this residual information to improve decoding performance, the module further enhances the quality of images reconstructed by decoders.

3   Algorithm for Remote Desktop Image and Compression Quality Evaluation

Image compression algorithms are typically evaluated by applying image quality standards to measure the extent of quality degradation after compression. These evaluation techniques can be categorized into subjective and objective assessments. Subjective assessment involves viewer-based scoring, wherein several evaluators grade the reconstructed image against the original one, with the mean score serving as the final evaluation. In contrast, objective assessment calculates the differences between images using mathematical models. Although subjective assessments are time-consuming and affected by numerous factors, objective assessments offer automated scoring that is not influenced by viewer bias.

Common objective image quality metrics include Peak Signal to Noise Ratio (PSNR), widely used in the field of image and video processing. PSNR, calculated conveniently through Mean Square Error (MSE), quantifies image distortion. The computation formula for MSE is presented in Eq. 1, and substituting it into Eq. 2 gives the PSNR value for an image, where n represents the number of pixels in the image.

$$\text{MSE} = \frac{1}{n} \sum_{i=1}^{n} (x_i - \hat{x}_i)^2 \tag{1}$$

$$\text{PSNR} = 10 \times \log_{10}(\frac{(2^n - 1)^2}{MSE}) \tag{2}$$

In optimizing the quality of reconstructed desktop images, the Mean Squared Error (MSE) method presents certain limitations. MSE, offering equal weight to all pixels, may not aptly assess reconstructed desktop images. Even with identical MSEs, perceptual differences between images can be significant, indicating that the Peak Signal-to-Noise Ratio (PSNR), derived from MSE, may not accurately represent human perception. When viewing desktop images, the human eye often concentrates on specific work areas and is more sensitive to noise in these sections than in other areas. Furthermore, the human eye has greater sensitivity to luminance than color, a facet overlooked when merely calculating MSE between the reconstructed image and the original. In contrast, Structural Similarity Index (SSIM) estimates luminance similarity, contrast, and structure similarity between two images by calculating their means, variance, and covariance. Formulas 3, 4, and 5 respectively represent the calculations for luminance similarity, contrast, and structure similarity.

$$l(x, y) = \frac{2\mu_x\mu_y + c_1}{\mu_x^2 + \mu_y^2 + c_1} \tag{3}$$

$$c(x, y) = \frac{2\sigma_x\sigma_y + c_2}{\sigma_x^2 + \sigma_y^2 + c_2} \tag{4}$$

$$s(x, y) = \frac{\sigma_{xy} + c_3}{\sigma_x\sigma_y + c_3} \tag{5}$$

Here, $C_1 = (K_1L)^2$, $C_2 = (K_2L)^2$, $C_3 = \frac{C_2}{2}$ with $K_1 = 0.01$, $K_2 = 0.03$, and $L = 2^B - 1$, where B denotes the pixel depth of the image. The structural similarity between the distorted and original images can be calculated by substituting formulas 3, 4, 5 into formula 6, with $\alpha = \beta = \gamma = 1$. Optimal performance of structural similarity requires specific configuration. In contrast, Multi-Scale Structural Similarity (MS-SSIM) calculates the degree of structural similarity at different scales between the distorted and original images through multiple low-pass filtering and down-sampling processes, maintaining good performance across different image resolutions. Formula 7 describes the calculation of structural similarity at multiple scales, where $\alpha_j = \beta_j = \gamma_j$, and $\sum_{j=1}^{M} \gamma_j = 1$.

$$\text{SSIM}(x, y) = [l(x,y)]^\alpha [c(x,y)]^\beta [s(x,y)]^\gamma \tag{6}$$

This study aims to optimize the performance of end-to-end image compression models using Mean Squared Error (MSE) combined with Multi-Scale Structural Similarity (MS-SSIM) as the loss function. By disregarding the quality of monotonic or non-working areas that are less relevant to human perception, we ensure a more complete preservation of image information in the working area of the reconstructed desktop image, thereby enhancing the comprehensive performance of the proposed method in the field of desktop image compression.

$$\text{SSIM}(x, y) = [l_M(x,y)]^\alpha M \prod_{j=1}^{M} [c_j(x,y)]^{\beta_j} [s_j(x,y)]^{\gamma_j} \tag{7}$$

4   Convolutional Neural Network-Based Image Compression Codec

4.1   Overview

As depicted in Fig. 1, the proposed model includes an image encoder, image decoder, and a hyper-prior encoder-decoder. The input digital image x is transformed into latent representation y via the encoding network and then quantized to $\hat{y}$ by the quantizer Q. In the decoding stage, $\hat{y}$ is used to reconstruct image $\hat{x}$. The hyper-prior encoder-decoder extracts edge information from the latent representation, reducing redundancy and facilitating the entropy coding process, thereby achieving a shorter coding length and higher compression ratio.

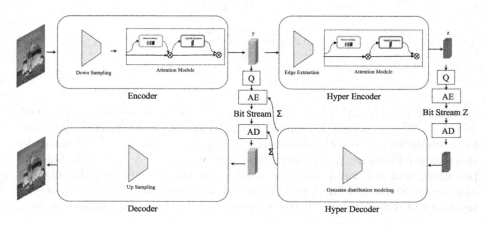

Fig. 1. The architecture of proposed End-to-End image compression model.

The encoder, primarily comprising a downsampling module and a spatial-channel attention module, extracts latent feature representation. Under the attention module, the final latent feature representation y is obtained with applied channel and spatial attention.

The downsampling module consists of four convolutional layers and a Generalized Divisive Normalization (GDN) layer between each pair. It extracts the latent feature representation by downsampling the input image x. If a higher compression rate is required at some quality loss, 128 convolution kernels per convolutional layer are sufficient. For higher quality reconstructed images, 192 convolution kernels per layer are needed.

Between each pair of convolutional layers, a normalization layer, the GDN, is applied. While Batch Normalization (BN) is typically used in deep learning-based image processing tasks, in this study, the advantages of BN become disadvantages. Hence, BN is unsuitable for this application. Conversely, the GDN layer, more appropriate for image reconstruction, eliminates additive noise brought by BN. During training, the GDN layer normalizes the feature map through unsupervised learning, performing Gaussianization on the symbolic data. The GDN layer is expressed as in Eq. (2-8).

$$y_i = \frac{x_i}{(\beta_i + \sum_i \gamma_i \times x_i^2)^{\frac{1}{2}}} \tag{8}$$

In the quantization stage, this paper employs Eqs. 9 and 10 to quantize the latent representation y into $\hat{y}$. This approach adds uniform additive noise to the latent feature representation instead of directly truncating the fractional part, ensuring quantization of the feature representation while preserving gradients.

$$\hat{y} = y + \Delta y \tag{9}$$

$$\Delta y \sim \mu(-0.5, 0.5) \tag{10}$$

The decoder utilizes quantized latent features as input, reconstructing the image through a number of transposed convolutional layers mirroring the encoder. Most layers match the encoder's convolutional kernel size, quantity, and stride: 128 5×5 kernels with a stride of 2 under low-quality encoding, and 192 5×5 kernels with the same stride under high-quality conditions. An exception is the final layer, utilizing three 5×5 kernels to reconstruct the image's three color channels. Between every two transposed convolutional layers in the decoder, an "Inversed Generalized Divisive Normalization" (IGDN) layer is inserted to reverse the GDN layer's output, assisting image reconstruction. The calculation process for the IGDN layer aligns with Eq. 11.

$$y_i = x_i \times (\beta_i + \sum_i \gamma_i \times x_i^2)^{\frac{1}{2}} \tag{11}$$

The hyperprior encoder h_a extracts edge information from latent features y, yielding edge representation z. This is quantized to $\hat{z}$, aiding encoding of y. The hyperprior decoder h_s uses $\hat{z}$ to estimate the standard deviation distribution $\hat{\sigma}$

of $\hat{y}$ and encode it. $\hat{z}$, also needed at the decoding stage for $\hat{\sigma}$ estimation and entropy decoding, must be compressed and transmitted from the encoder to the decoder. Edge information is extracted from y via a convolutional layer and a nonlinear activation function in h_a, while h_s uses a transposed convolution and a nonlinear activation function to estimate $\hat{\sigma}$. The standard deviation $\hat{\sigma}$ models $\hat{y}$ using a Gaussian distribution (Eq. 12). Substituting Eq. 12 and $\hat{\sigma}$ into Eq. 13 yields the probability distribution of any symbol in $\hat{y}$. During model training iterations, symbols with minimal impact on image quality approach zero, increasing their probability and shortening the encoding length during entropy coding, thus removing redundancies.

$$N(x|\mu, \sigma) = \frac{1}{\sqrt{2\pi}\sigma} \exp\left(-\frac{(x-\mu)^2}{2\sigma^2}\right) \tag{12}$$

$$p_{\hat{y}_i}(\hat{y}_i|\hat{\sigma}_i) = \int_{\hat{y}_i - \frac{1}{2}}^{\hat{y}_i + \frac{1}{2}} N(\hat{y}_i|0, \hat{\sigma}_i)\, dy \tag{13}$$

4.2   Attention Mechanism

This paper introduces the Convolutional Block Attention Module (CBAM) [23], a mechanism that combines spatial and channel attention to adaptively learn the importance of different channels and spatial positions within feature maps. By weighting feature maps in multiple dimensions, it achieves improved compression and reconstruction. Its detailed structure is presented in Fig. 2.

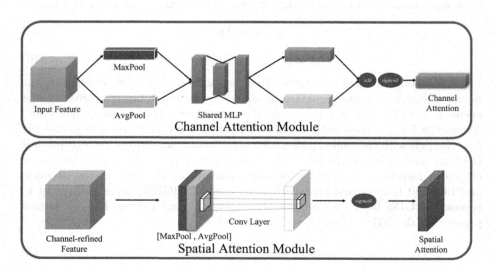

Fig. 2. Attention mechanism.

For channel attention, CBAM initially extracts weights from different channels within feature maps through max pooling and average pooling. Subsequently, it applies two fully connected layers to perform nonlinear transformations on these pooled results. To obtain the final channel weights, it first sums the output from the fully connected layers and then applies another nonlinear transformation. The input to this operation is $F_{c_{in}} \in \mathbb{R}^{C \times H \times W}$, with the computation process as shown in Eqs. 14 and 15.

$$M_c(F) = \sigma(\text{FC}(\text{AvgPool}(F)) + \text{FC}(\text{MaxPool}(F))) \tag{14}$$

$$\text{FC}(F) = W_1(\sigma(W_0(F))) \tag{15}$$

In terms of spatial attention, the input feature map undergoes max pooling and average pooling along the channel axis, thereby converting the input dimensions from $F_{s_{in}} \in \mathbb{R}^{C \times H \times W}$ to $2 \times H \times W$. A convolution operation is then performed on this dimensionally-transformed feature map to extract weights from different spaces, as shown in Eq. 16. Finally, the effects of channel attention and spatial attention are collectively applied to the feature map, as described by Eq. 17.

$$M_s(F) = \sigma(\text{Conv}(\theta, [\text{AvgPool}(F); \text{MaxPool}(F)])) \tag{16}$$

$$F' = M_s(M_c(F) \otimes F) \otimes (M_c(F) \otimes F) \tag{17}$$

4.3 Loss Function

The end-to-end image compression model balances distortion and bitrate according to Eq. 18, where λ—analogous to the quality parameter QP in JPEG—controls reconstructed image quality and storage space requirements. The distortion function, $d(\cdot)$, quantifies the distortion between the input and reconstructed image (Eq. 19), considering both MS-SSIM (L_{SSIM}, Eq. 20) and L2 error (L_{MSE}, Eq. 21) with parameter α balancing attention to these aspects. The rate estimation function, $H(\cdot)$, includes the bitrates required for encoding the feature representation $\hat{y}$ and the auxiliary information $\hat{z}$ (Eqs. 22 and 23). The probability distribution function for $\hat{z}$ and its cumulative distribution function are given in Eqs. 24 to 28, with parameters a, h, and b following a normal distribution $N(0, 0.01)$, and K is set to 4 in this study.

$$L = \lambda D + R = \lambda d(\hat{x}, x) + H(\hat{y}) + H(\hat{z}) \tag{18}$$

$$d(\hat{x}, x) = \alpha \times L_{SSIM}(\hat{x}, x) + (1 - \alpha) \times L_{MSE}(\hat{x}, x) \tag{19}$$

$$L_{SSIM} = 1 - SSIM(\hat{x}, x) \tag{20}$$

$$L_{MSE} = \frac{1}{N} \sum_{i=1}^{N} ||\hat{x}_i - x_i||^2 \tag{21}$$

$$H(\hat{y}) = -\sum_i \log_2 P_{\hat{y}}(\hat{y}|\hat{z}) \tag{22}$$

$$H(\hat{z}) = -\sum_i \log_2 P_{\hat{z}}(\hat{z}) \tag{23}$$

$$p_{\hat{z}_i}(\hat{z}_i) = \text{BitEstimator}(\hat{z}_i + \frac{1}{2}) - \text{BitEstimator}(\hat{z}_i - \frac{1}{2}) \tag{24}$$

$$\text{BitEstimator}(x) = f_K \circ f_{K-1} \circ \ldots \circ f_1 = f_K(f_{K-1}(\ldots f_1(x))) \tag{25}$$

$$f_K(x) = \sigma(x \odot \zeta(h) \oplus b) \tag{26}$$

$$f_k(x) = g_k(x \odot \zeta(h) \oplus b) \tag{27}$$

$$g_k(x) = x \oplus \tanh(x) \odot \tanh(a) \tag{28}$$

5   Experiments and Analysis

5.1   Data Set and Parameters

This paper first selects 20745 images from the Flickr dataset[1] as the training set to train the model. These images have high resolution, and in order to speed up the training process, we randomly cropped and flipped all images, resulting in a training set with image sizes of 256 pixels × 256 pixels. To evaluate performance, we selected the Kodak image dataset as the validation set to demonstrate the effectiveness of the model proposed in this chapter. In addition to using natural image datasets, we also collected an additional dataset consisting mainly of desktop images. Compared to natural images, desktop images have more prominent text and texture features and are more suitable for the model proposed in this study.

The experimental results presented in this section demonstrate the effectiveness of the proposed end-to-end deep image codec. The proposed end-to-end image compression model is implemented on the PyTorch platform. All experiments were conducted on a server with GPU with 24GB of video memory. Different λ values (64, 128, 256, 512, 1024, 2048) were used as the distortion-rate balance control parameter in the loss function to train the model. For gradient updates, the Adam algorithm [6] was chosen as the optimizer to update gradients. The entire model learning process required a total of 2,500,000 iterations. In the first 2,000,000 iterations, the learning rate parameter was set to 1×10^{-4}, and in the remaining iterations, the learning rate parameter was set to 1×10^{-5}. For high-bitrate image compression models (with 192 convolutional kernels in

---

[1] https://www.flickr.com/.

the encoder's convolutional layer), a model with a distortion-rate balance control parameter of 2048 was first trained as a pre-training model for other models. For models with λ values of 512 and 1024, the weights of the model with a distortion-rate balance control parameter of 2048 at iteration 500000 were first loaded as pre-training parameters for the model and then completed on the training set. For low-bitrate compression models, a model with a distortion-rate balance control parameter of 256 was first trained as a pre-training model for models with distortion-rate balance control parameters of 64 and 128. Similarly, the weights at step 500000 were selected as pre-training weight parameters. The value of α for controlling MS-SSIM and MSE balance is chosen to be 0.1.

5.2   Experimental Results

This paper first compares the proposed model with traditional image compression methods, as well as some deep learning-based image compression methods [3,8,10,15], using the Kodak dataset as the validation set. Figure 3 shows the rate-distortion curves of different methods on the Kodak dataset[2]. Here, Peak Signal-to-Noise Ratio (PSNR) and Multi-Scale Structural Similarity (MS-SSIM) are used as image evaluation standards to assess the distortion level of images. As shown in Fig. 3a, the algorithm proposed in this paper surpasses most of the image compression methods used for comparison in terms of the PSNR evaluation index, and is significantly superior to the JPEG compression algorithm, achieving a 7.5% improvement in PSNR performance at low bit rates compared to JPEG. As for MS-SSIM, as shown in Fig. 3b, the algorithm proposed in this paper has an advantage, in the range of $0.4 < $ bpp $ < 0.7$, it performs equally well with the model proposed by Lee et al., and outperforms the model proposed by Lee et al. in terms of PSNR evaluation method.

Table 1 presents the performance data of the model proposed in this paper compared with other compression methods. In addition, this paper also adds different modules to the baseline to evaluate the performance of models with different methods added. The baseline is a reproduction of the code and results of Ballé [3]. Here, Peak Signal-to-Noise Ratio (PSNR) and Multi-Scale Structural Similarity (MS-SSIM) are used as image evaluation standards to assess the distortion level of reconstructed images. Figure 4 presents the rate-distortion curves of methods using different strategies. As can be seen from Fig. 4a, the method with the attention mechanism added improves the PSNR performance and reduces the loss on MSE compared to the method optimized only for the mixed function of MS-SSIM and MSE. Figure 4b shows that the method proposed in this paper improves the performance in terms of MS-SSIM compared to the baseline and the method that only applies the attention mechanism.

---

[2] https://r0k.us/graphics/kodak/.

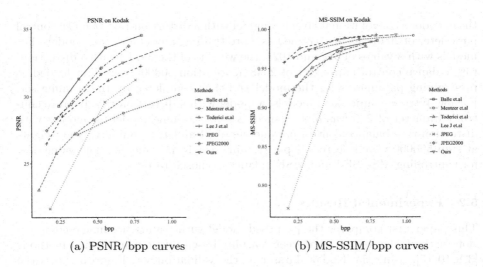

(a) PSNR/bpp curves   (b) MS-SSIM/bpp curves

Fig. 3. Rate-Distortion curves of different methods.

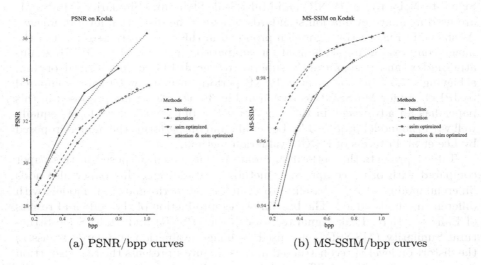

(a) PSNR/bpp curves   (b) MS-SSIM/bpp curves

Fig. 4. Comparison curves of different methods against the baseline.

Figure 5 illustrates a comparative analysis of the reconstructed images of common desktop visuals using the baseline and the method proposed in this study. As discernible from Fig. 5, the proposed method lays enhanced emphasis on the preservation of intrinsic image attributes such as structure and contrast. Consequently, in certain scenarios, the proposed method exhibits superior preservation of detailed elements including the edge textures of text, icons, and graphics in desktop images, leading to a more comprehensive and clear representation.

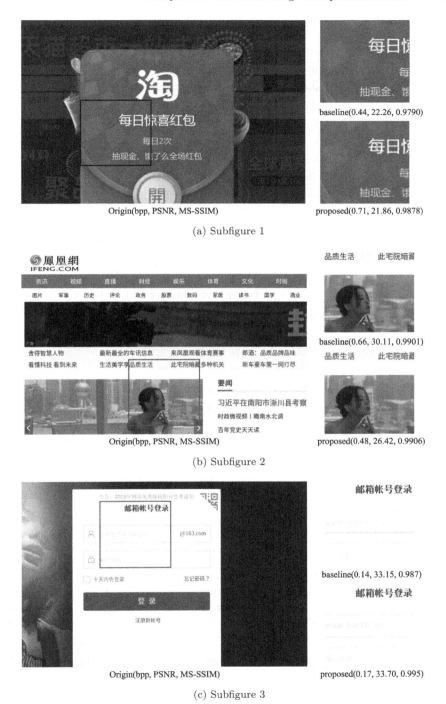

(a) Subfigure 1

(b) Subfigure 2

(c) Subfigure 3

Fig. 5. Comparison of reconstructed images between baseline and proposed methods.

Table 1. Comparison of rate-distortion data for different methods.

methods	bpp	PSNR	MS-SSIM
Ballé et.al	0.24	29.30	0.940
Ballé et.al	0.38	31.31	0.963
Ballé et.al	0.56	33.61	0.976
Mentzer et.al	0.37	27.17	0.975
Mentzer et.al	0.68	28.77	0.987
Mentzer et.al	1.05	30.14	0.992
Toderici et.al	0.11	23.01	0.841
Toderici et.al	0.22	25.76	0.923
Toderici et.al	0.35	27.21	0.951
Toderici et.a	0.72	30.14	0.978
Lee J et.al	0.15	26.48	0.958
Lee J et.al	0.26	28.56	0.976
Lee J et.al	0.57	31.21	0.989
Lee J et.al	0.79	32.23	0.993
JPEG	0.18	21.61	0.771
JPEG	0.55	29.59	0.972
JPEG	0.76	31.22	0.984
JPEG2000	0.16	27.48	0.916
JPEG2000	0.53	32.11	0.972
JPEG2000	0.71	33.72	0.981
Proposed	0.22	28.39	0.964
Proposed	0.32	29.45	0.974
Proposed	0.53	31.67	0.987
Proposed	0.67	32.50	0.990

6   Conclusion

This paper proposes an end-to-end deep learning image compression method for remote desktops, optimized with a mixed MS-SSIM and MSE loss function and an adaptive channel and spatial attention mechanism. Empirical evidence shows that deep learning-based image compression encoders and decoders outperform traditional encoding schemes in terms of reconstructed image quality and compression rate performance metrics. Furthermore, when applied to compress desktop images containing more text and texture information, the reconstructed images from the proposed deep learning-based image compression model outperform JPEG compression algorithms and most deep learning-based image compression models proposed in recent years in terms of multi-scale structural similarity, an objective evaluation standard. Importantly, it retains more image and texture details in areas of high user attention.

References

1. Ballé, J., Laparra, V., Simoncelli, E.P.: Density modeling of images using a generalized normalization transformation. arXiv preprint arXiv:1511.06281 (2015)
2. Ballé, J., Laparra, V., Simoncelli, E.P.: End-to-end optimized image compression. arXiv preprint arXiv:1611.01704 (2016)
3. Ballé, J., Minnen, D., Singh, S., Hwang, S.J., Johnston, N.: Variational image compression with a scale hyperprior. arXiv preprint arXiv:1802.01436 (2018)
4. Bi, Q., Qin, K., Zhang, H., Li, Z., Xu, K.: RADC-Net: a residual attention based convolution network for aerial scene classification. Neurocomputing **377**, 345–359 (2020)
5. Hore, A., Ziou, D.: Image quality metrics: PSNR vs SSIM. In: 2010 20th International Conference on Pattern Recognition, pp. 2366–2369. IEEE (2010)
6. Hu, J., Shen, L., Sun, G.: Squeeze-and-excitation networks. In: Proceedings of the IEEE Conference on Computer Vision and Pattern Recognition, pp. 7132–7141 (2018)
7. Jiang, F., Tao, W., Liu, S., Ren, J., Guo, X., Zhao, D.: An end-to-end compression framework based on convolutional neural networks. IEEE Trans. Circuits Syst. Video Technol. **28**(10), 3007–3018 (2017)
8. Lee, J., Cho, S., Beack, S.K.: Context-adaptive entropy model for end-to-end optimized image compression. arXiv preprint arXiv:1809.10452 (2018)
9. Lin, T., Hao, P.: Compound image compression for real-time computer screen image transmission. IEEE Trans. Image Process. **14**(8), 993–1005 (2005)
10. Mentzer, F., Agustsson, E., Tschannen, M., Timofte, R., Van Gool, L.: Conditional probability models for deep image compression. In: Proceedings of the IEEE Conference on Computer Vision and Pattern Recognition, pp. 4394–4402 (2018)
11. Mishra, D., Singh, S.K., Singh, R.K.: Deep architectures for image compression: a critical review. Signal Process. **191**, 108346 (2022)
12. Mnih, V., Heess, N., Graves, A., et al.: Recurrent models of visual attention. In: Advances in Neural Information Processing Systems 27 (2014)
13. Sazawa, S., Hashima, M., Sato, Y., Horio, K., Matsui, K.: RVEC: efficient remote desktop for the engineering cloud. In: 2012 26th International Conference on Advanced Information Networking and Applications Workshops, pp. 1081–1088. IEEE (2012)
14. Shimada, D., Hashima, M., Sato, Y.: Image compression for remote desktop for engineering cloud. In: 2014 IEEE International Conference on Cloud Engineering, pp. 478–483. IEEE (2014)
15. Toderici, G., et al.: Full resolution image compression with recurrent neural networks. In: Proceedings of the IEEE Conference on Computer Vision and Pattern Recognition, pp. 5306–5314 (2017)
16. Wang, S., Lin, T.: United coding for compound image compression. In: 2010 3rd International Congress on Image and Signal Processing, vol. 2, pp. 566–570. IEEE (2010)
17. Wang, S., Lin, T.: United coding method for compound image compression. Multimedia Tools Appl. **71**, 1263–1282 (2014)

18. Wang, X., Girshick, R., Gupta, A., He, K.: Non-local neural networks. In: Proceedings of the IEEE Conference on Computer Vision and Pattern Recognition, pp. 7794–7803 (2018)
19. Wang, Y., Chen, X., Wang, Q., Yang, R., Xin, B.: Unsupervised anomaly detection for container cloud via BILSTM-based variational auto-encoder. In: ICASSP 2022-2022 IEEE International Conference on Acoustics, Speech and Signal Processing (ICASSP), pp. 3024–3028. IEEE (2022)
20. Wang, Y., Wang, Q., Qin, X., Chen, X., Xin, B., Yang, R.: DockerWatch: a two-phase hybrid detection of malware using various static features in container cloud. Soft. Comput. **27**(2), 1015–1031 (2023)
21. Wang, Z., Bovik, A.C., Sheikh, H.R., Simoncelli, E.P.: Image quality assessment: from error visibility to structural similarity. IEEE Trans. Image Process. **13**(4), 600–612 (2004)
22. Weinberger, M.J., Seroussi, G., Sapiro, G.: The LOCO-I lossless image compression algorithm: Principles and standardization into JPEG-LS. IEEE Trans. Image Process. **9**(8), 1309–1324 (2000)
23. Woo, S., Park, J., Lee, J.-Y., Kweon, I.S.: CBAM: convolutional block attention module. In: Ferrari, V., Hebert, M., Sminchisescu, C., Weiss, Y. (eds.) ECCV 2018. LNCS, vol. 11211, pp. 3–19. Springer, Cham (2018). https://doi.org/10.1007/978-3-030-01234-2_1
24. Zhao, L., Bai, H., Wang, A., Zhao, Y.: Learning a virtual codec based on deep convolutional neural network to compress image. J. Vis. Commun. Image Represent. **63**, 102589 (2019)

LCformer: Linear Convolutional Decomposed Transformer for Long-Term Series Forecasting

Jiaji Qin[✉], Chao Gao, and Dingkun Wang

Harbin Engineering University, Harbin 150001, China
henry@hrbeu.edu.cn

Abstract. Transformer-based methods have shown excellent results in long-term series forecasting, but they still suffer from high time and space costs; difficulties in analysing sequence correlation due to entanglement of the original sequence; bottleneck in information utilisation due to the dot-product pattern of the attention mechanism. To address these problems, we propose a sequence decomposition architecture to identify the different features of sub-series decomposed from the original time series. We then utilize causal convolution to solve the information bottleneck problem caused by the attention mechanism's dot-product pattern. To further improve the efficiency of the model in handling long-term series forecasting, we propose the Linear Convolution Transformer (LCformer) based on a linear self-attention mechanism with O(n) complexity, which exhibits superior prediction performance and lower consumption on long-term series prediction problems. Experimental results on two different types of benchmark datasets show that the LCformer exhibits better prediction performance compared to those of the state-of-the-art Transformer-based methods, and exhibits near linear complexity for long series prediction.

Keywords: Linear Complexity · Long-Term Series Forecasting · Sequence Decomposition · Local Convolution

1 Introduction

Time series prediction plays an important role in modern life, helping people make informed decisions and manage resources. Long-term prediction problems are of particular concern. Recently, Transformer [1] has been introduced as a new architecture for time series prediction to capture the long-term dependencies of sequences. Unlike RNN and its variants, Transformer employs self-attention mechanisms and has shown excellent results [2]. However, due to the quadratic complexity of sequence length, Transformer is difficult to apply to long-term sequence prediction. Current improvements of Transformer mainly focus on improving the self-attention mechanism. Although the complexity of the model can be reduced, these methods still rely on dot-product attention, and the sparse dot-product connectivity sacrifices the model's utilization of information, thereby affecting performance in long-term prediction problems. At the same time, this dot-product attention pattern also makes it difficult for the model to identify the global features of long-term time series.

In addressing challenges such as long-term energy utilization planning and anomalous weather forecasting, there is an imperative need to extend the duration of reliable predictions. The Autoformer [3] has demonstrated commendable performance in long-term time series forecasting, leveraging its decomposition structure and auto-correlation mechanism. However, a conspicuous limitation is its neglect of the pronounced continuous relationships inherent in the original series. Furthermore, despite its capabilities, the Autoformer still experiences considerable time overhead when tasked with long-term series predictions.

To solve these issues, we propose Linear Convolution Transformer (LCformer) for long-term time series prediction. Based on the Autoformer architecture [3], LCformer introduces a multi-class decomposition transformer structure that allows the model to alternate between processing and refining intermediate results during the prediction process. This solves the problem of difficult analysis of sequence correlation due to entanglement patterns, effectively capturing the global features of time series. Meanwhile, the conventional self-attention mechanism, which analyzes series based on the sparse dot-product attention, also overlooks the continuous interplay between series data points. To address this, the LCformer integrates a local convolution module. By employing the causal convolution approach, the LCformer emphasizes the local context within the series, thereby accentuating the continuous relationship between a current observation and its nearby data points. This enhancement markedly elevates the model's accuracy in long-term series forecasting. While the Autoformer model utilizes the autocorrelation mechanism based on the Fast Fourier Transform, achieving $O(n(logn))$ time complexity, it still grapples with significant time overhead in long-term series forecasting. To ameliorate this, LCformer incorporates the linear autocorrelation mechanism that optimises the model's temporal efficiency for long-term series forecasting, endowing it with near-linear complexity. Consequently, this significantly diminishes the model's time overhead for such forecasting challenges. The contributions are as follows:

1 To better capture the local characteristics of long-term time series, we propose local convolution modules which use causal convolution to focus on the sequence's local context. This breaks the information bottleneck caused by traditional dot-product attention mechanisms and obtains more information for accurate prediction.
2 To address the problem of difficult analysis of sequence correlation due to entanglement patterns, we propose a multi-class decomposition transformer structure based on the Autoformer architecture. This structure decomposes the original time series into trend, seasonal, and periodic sub-sequences and processes them separately.
3 We propose a linear self-attention mechanism with $O(n)$ complexity, which can replace the sparse attention mechanisms or autocorrelation mechanisms with $O(n(logn))$ complexity and significantly reduce time costs in long sequence prediction tasks.
4 We conducted experiments on two benchmark datasets with different types and features. The experimental results show that the proposed LCformer performs at least 4.82% better than advanced models such as Autoformer in multivariate prediction and exhibits near-linear time complexity $O(n)$ in long sequence prediction.

2   Relative Works

Traditional time series forecasting methods, such as autoregressive models and state space models [4], require redesigning to fit each type of sequence and involve a high degree of human involvement. As a result, they have gradually been abandoned in modern time series forecasting tasks. Neural network models have been proposed as a widely used alternative, with recurrent neural networks (RNN) being widely used, but suffering from serious gradient vanishing and explosion problems [5]. Its variant, LSTM, also has limitations when facing long sequence prediction problems; for example, Zhao et al. [6] found that LSTM has difficulty in capturing the long-term dependence of sequences. DeepAR [7] combines RNN and autoregressive methods to model the probability distribution of sequences. LSTNet [8] introduces convolutional neural networks (CNN) connected by recursive jumps in order to capture both short-term and long-term temporal patterns. In addition, Shun-Yao [9] models temporal relations based on temporal convolutional networks (TCN) for those with recursive connectivity, causal convolution or temporal attention.

The Transformer based on self-attention mechanism has shown excellent performance in processing sequence data [10], such as NLP [11–14], computer vision [15–17] and audio recognition and processing [18–20]. However, limited by the quadratic complexity of Transformer in time and space, it is difficult to directly apply it to long-term time series prediction problems. LogSparse Transformer [21] introduces CNN and LogSparse attention mechanism into Transformer, reducing the model complexity to $O(n(logn)^2)$. Reformer [22] further reduces the model complexity to $O(n(logn))$ by introducing the local sensitive hash (LSH) attention mechanism. Informer [2] extends Transformer based on the ProbeSparse attention mechanism and also achieves $O(n(logn))$ complexity. Autoformer [3] proposes a decomposition architecture and uses autocorrelation mechanism instead of attention mechanism, achieving a significant improvement in prediction performance. Chen [23] improved the Transofmer structure based on the Square Subsequent Masking approach, enabling the model to show excellent performance in long-term prediction of traffic flows.

3   Approach

3.1   LCformer

To address the issues of local insensitivity and storage bottleneck in traditional Transformer models, we propose the LCformer model by combining the local convolution model and the linear self-attention mechanism in Linformer. The model has a sequence decomposition module that can decompose sequence information from input time series to show different temporal features of the sequence. By introducing the linear self-attention mechanism, the memory bottleneck problem of Transformer is optimized, and theoretically, the time complexity and space complexity of the model can reach $O(n)$. The structure of the LCformer model proposed in this paper is shown in Fig. 1, which includes an input layer, a sequence decomposition layer, a local convolution layer, Encoder layers, Decoder layers, and an output layer.

The LCformer first takes the time series as input and preserves the order feature of the sequence through position embedding. Then, the input sequence is decomposed into trend, seasonal, and periodic parts using the sequence decomposition module. In the Encoder, the local convolution module is used to transform the input into Q, K, V for better perception of local contextual information. The trend part is gradually eliminated through the sequence decomposition module to better focus on the seasonality and cycle parts. In the Decoder, the local convolution and linear attention mechanism are used to process the seasonal and periodic parts. For the trend part, a cumulative structure is used to extract the trend information from the predicted hidden variables.

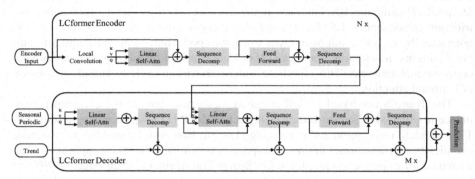

Fig. 1. LCformer structure

Based on this architecture, the model can progressively decompose the hidden variables during the prediction process, and obtain the prediction results through local convolution, linear self-attention mechanism, and accumulation structure.

3.2 Local Convolution Module

Due to various events such as natural disasters and public holidays, time series often have multiple patterns of change. Therefore, whether an observation point belongs to an anomaly or change point largely depends on its local context information. However, as shown in Fig. 2-(b), traditional Transformer models perform Q, K, V calculations separately for each observation point without fully utilizing the local context information around it. This can lead to confusion between different patterns of change, as shown in Fig. 2-(a), where two observation points (red dots) with different features on the time series (one has a slow change while the other has a rapid change) may be identified as the same change pattern due to their equal absolute values. Similarly, in Fig. 2-(c), two regions with similar local context features may be identified as different change patterns due to differences in their absolute values, leading to potential issues in prediction performance.

To fully utilize the local context features of the sequence for the Q, K calculations in Transformer, a causal convolution with stride 1 and kernel size ke is used. Through the local convolution module, Q, K can better understand the local context information of the current observation point, and calculate the pattern similarity between each observation

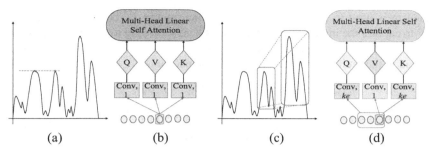

Fig. 2. Working mechanism of Local Convlution Module

point using local shape information, thus avoiding confusion caused by absolute values alone. When *ke* is set to 1, the local convolution module will degenerate to the traditional Transformer, making it a generalization process of Transformer.

3.3 Linear Self Attention Mechanism

Linformer [24] uses a linear attention mechanism to decompose the scaled dot-product attention pattern of the original self-attention mechanism into several smaller patterns using linear projection. These smaller attentions are combined to form a low-rank factorization of the original self-attention mechanism. The Linformer method can theoretically achieve time and space complexity of $O(n)$ when dealing with long sequence problems, surpassing the $O(n(logn))$ of Auto-Corretion. Below is a brief introduction to the linear attention mechanism (Fig. 3).

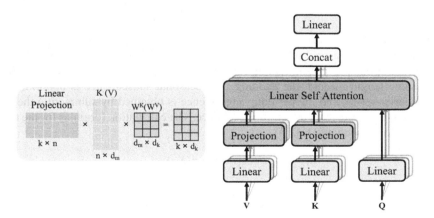

Fig. 3. Architecture of Linear Self Attention

The core idea of the linear self-attention mechanism is to add two linear projection matrices $E_i, F_i \in R^{n \times k}$ when computing V and K. Firstly project the original $(n \times d)$-dimensional K and V into $(k \times d)$-dimension. Then, the scaled dot-product attention mechanism is used to compute the n-dimensional matrix $\overline{P}$. The Transformer model uses $\overline{P}$ to capture the input context of a given token, according to the combination of all tokens in the sequence. Finally, Eq. (1) is used to calculate the context embeddings for each head in the multi-head linear self-attention mechanism.

$$\overline{head}_i = Attention\left(QW_i^Q, E_iKW_i^K, F_iVW_i^V\right)$$
$$= softmax\left(\frac{QW_i^Q\left(E_iKW_i^K\right)^T}{\sqrt{d_k}}\right) \cdot F_iVW_i^V \tag{1}$$

where $Q, K, V \in R^{n \times d_m}$ represents the input embedding matrix, n represents the length of the input sequence, and h is the number of heads in the multi-head attention mechanism. $W_i^Q, W_i^K \in R^{d_m \times d_k}, W_i^V \in R^{d_m \times d_v}, W^O \in R^{hd_v \times d_m}$ are the learning matrixs and d_k, d_v denote the hidden dimension of the cast shadow subspace. The time and space complexity of these operations is $O(nk)$. When the projection dimension k is appropriate, the linear attention mechanism could significantly reduce the model's consumption both in time and memory.

3.4 Sequence Decomposition Structure

To learn the different features of the original time series in the prediction process, the model utilizes the decomposition method based on Fourier transform and sliding average ideas to decompose the sequence into trend, seasonal, and periodic components. Assuming that a time series is composed of multiple components, it can be represented by Eq. (2):

$$X = X_t + X_s + X_c \tag{2}$$

where X represents the original time series data, X_t, X_s, X_c respectively represent the trend subsequence, seasonal subsequence, and periodic subsequence obtained by decomposition. The trend component is processed using the AvgPool operation and maintained at the same length through the Padding. The seasonal subsequence is approximated using Fourier series, and the periodic subsequence is obtained by subtracting the first two subsequence data from the original sequence. The calculation formulas for the three subsequence are shown in Eq. (3).

$$X_t = AvgPool(Padding(X))$$
$$X_s = \sum_{i=1}^{N}\left(a_i\cos\frac{2\pi lt}{p} + b_i\sin\frac{2\pi lt}{p}\right) \tag{3}$$
$$X_c = X - X_t - X_s$$

where X is the original input series, i is the count value, p is the annual data and $p = 365$, a_i and b_i are coefficients.

3.5 Encoder

As shown in Fig. 1, the Encoder module consists of local convolutional module, sequence decomposition module, linear attention mechanism, and feedforward neural network. The input to the encoder includes trend subsequence X_{ent}, seasonal subsequence X_{ens}, and periodic subsequence X_{enc}.

$$X_{ens}, X_{enc}, X_{ent} = Decomp(X_{en}) \tag{4}$$

where X_{en} represents the original time series of the input and $X_{ens}, X_{enc}, X_{ent}$ are the subsequences obtained by decomposing X_{en}. The Encoder focuses on the identification of seasonal and periodic information, and its output contains the seasonality and periodicity of the sequence, which will be used to assist the Decoder in making predictions. Assuming a total of N layers of Encoder, the specific formula is given in Eq. (5).

$$
\begin{aligned}
S_{en}^{l,1}, C_{en,-}^{l,1} &= Decomp\left(X_{en}^{l-1}\right) \\
S_{en}^{l,2}, C_{en,-}^{l,2} &= Decomp\left(FeedForward\left(S_{en}^{l,1}, C_{en}^{l,1}\right) + S_{en}^{l,1}, C_{en}^{l,1}\right)
\end{aligned}
\tag{5}
$$

where $S_{en}^{l,2}, C_{en}^{l,2}, l \in \{1, \cdots, N\}$ denote the seasonal part, the periodic part, and "_" is the excluded trend part obtained from the decomposition. $S_{en}^{l,i}, C_{en}^{l,i}, i \in \{1, 2\}$ then represent the seasonal part and the periodic part respectively after the ith sequence decomposition module in the Encoder at layer l. Let the output of the Encoder at layer l be denoted as $X_{en}^l = Encoder\left(X_{en}^{l-1}\right)$. Then $X_{en}^l = S_{en}^{l,2} + C_{en}^{l,2}, l \in \{1, \cdots, M\}$.

3.6 Decoder

The Decoder module consists of two parts: a stacked self-attention mechanism containing the seasonal and periodic components, and a cumulative structure containing the trend component. Each decoder layer contains a local convolutional module, linear self-attention mechanism, and sequence decomposition module, which can refine predictions and combine past seasonal and periodic information. This allows the model to extract potential changes in the hidden variables and eliminate long-term errors accumulated based on cycle dependencies in the linear self-attention mechanism. The input to the Decoder module includes trend subsequence X_{det}, seasonal subsequence X_{des}, and periodic subsequence X_{dec}.

$$
\begin{aligned}
X_{des} &= Concat(X_{ens}, X_0) \\
X_{dec} &= Concat(X_{ent}, X_0) \\
X_{det} &= Concat(X_{enh}, X_{Mean})
\end{aligned}
\tag{6}
$$

where X_0 represents a placeholder filled with zeros, and X_{Mean} represents the mean of X_{en}. Assuming there are M layers in the Decoder module, the specific calculation formula is shown in Eq. (7).

$$S_{de}^{l,1}, C_{de}^{l,1}, T_{de}^{l,1} = Decomp\left(Linear\ SelfAttention\left(X_{de}^{l-1}\right) + X_{de}^{l-1}\right)$$

$$S_{de}^{l,2}, C_{de}^{l,2}, T_{de}^{l,2} = Decomp\left(Linear\ SelfAttention\left(S_{de}^{l,1}, C_{de}^{l,1}, X_{en}^{N}\right) + S_{de}^{l,1}, C_{de}^{l,1}\right) \triangleright$$

$$S_{de}^{l,3}, C_{de}^{l,3}, T_{de}^{l,3} = Decomp\left(FeedForward\left(S_{de}^{l,2}, C_{de}^{l,2}\right) + S_{de}^{l,2}, C_{de}^{l,2}\right)$$

$$T_{de}^{l} = T_{de}^{l-1} + W_{l,1} * T_{de}^{l,1} + W_{l,2} * T_{de}^{l,2} + W_{l,3} * T_{de}^{l,3} \tag{7}$$

where $S_{de}^{l,3}, C_{de}^{l,3}, T_{de}^{l,3}, l \in \{1, \cdots, M\}$ denotes the seasonal, periodic and trend parts obtained from the decomposition at layer l. $S_{de}^{l,i}, C_{de}^{l,i}, T_{de}^{l,i}, i \in \{1, 2, 3\}$ denote the seasonal, periodic and trend components respectively after the i th sequence decomposition module in the Decoder at layer l. $W_{l,i}, i \in \{1, 2, 3\}$ denotes the projection of the trend $T_{de}^{l,i}$ from the i th decomposition.

Let the output of the Encoder at layer l denote as $X_{de}^{l} = Decoder\left(X_{de}^{l-1} + X_{en}^{N}\right)$. Then $X_{de}^{l} = S_{de}^{l,3} + C_{de}^{l,3}, l \in \{1, \cdots, M\}$. The final prediction of the LCformer model is shown in Eq. (8).

$$Pre = W_S * X_{de}^{M} + T_{de}^{M} \tag{8}$$

where Pre is the final prediction of the model and $X_{de}^{M} + T_{de}^{M}$ is the sum of the two processing components of Decoder at layer M, where W_S projects the processed seasonal and periodic components onto the target dimension.

4 Experiments

4.1 Data Set and Evaluation Metrics

To evaluate the proposed LCformer model in this paper, we conducted experiments on two classic real-world datasets, namely ETT [2] and Exchange [8]. (1) The Electricity Transformer Temperature (ETT) dataset contains data collected every 15 min from two independent counties' power transformers in China between July 2016 and 2018. Each data point consists of oil temperature and 6 power load features. (2) The Exchange dataset contains daily exchange rate information from eight different countries between 1990 and 2016. We divided the two datasets into training, validation, and testing sets in a 6:2:2 ratio.

The difference between these two datasets is twofold, firstly ETT has more covariates to assist the model for analysis and secondly Exchange does not exhibit significant periodic characteristics compared to general time series, therefore the experimental results obtained in these two datasets reflect the comprehensive performance of LCformer.

To overcome the limitations associated with a single evaluation index, this study uses two performance evaluation metrics, namely Mean Absolute Error (MAE) and Mean Square Error (MSE), to evaluate the stability and effectiveness of various models. The formulas for calculating these indices are as follows:

$$MSE = \frac{1}{m} \sum_{i=1}^{m} (y_i - \widehat{y}_i)^2 \tag{9}$$

$$MAE = \frac{1}{n} \sum_{i=1}^{n} |\widehat{y_i} - y_i| \tag{10}$$

In the formula, the ranges for MAE and MSE are from 0 to infinity, where a smaller MAE value indicates better model stability, while a smaller MSE value indicates better predictive performance of the model.

4.2 Implementation Settings

Our proposed method is optimized with the Adam optimizer and its initial learning rate of $10e^{-4}$. The batch size is set to 32, the number of layers in the Encoder module is set to 2, the number of layers in the Decoder module is set to 1, and the number of heads in the multi-head attention mechanism is set to 8. Secondly, two important parameters in the LCformer need to be determined: the convolution kernel size ke of the local convolution module and the projection dimension k in the linear self-attention mechanism.

***ke* of the Local Convolution Module.** To set an appropriate value for ke, we conducted experiments on the ETT and Exchange datasets using different sizes of ke. The results of the model under different sizes of ke are shown in Table 1.

Table 1. Model performance with different sizes of ke

DataSet	Metric	ke = 1	ke = 2	ke = 3	ke = 6	ke = 9
ETT	MSE	0.229	0.226	**0.224**	**0.224**	0.225
	MAE	0.316	**0.309**	**0.309**	0.311	0.312
Exchange	MSE	0.208	0.201	0.199	0.198	**0.196**
	MAE	0.331	0.319	0.315	0.315	**0.313**

On the ETT dataset, the LCformer model performed well when ke was equal to 3 and 6. We also observed that the change in ke had little impact on the model's performance on this dataset. We believe that this is because the rich covariates in the ETT dataset have already provided sufficient information for the model to make accurate predictions. Therefore, using a larger value of ke may not help the model learn more local context. However, on the Exchange dataset with less auxiliary information, the model with a larger ke showed better prediction performance, with a relative performance improvement of up to 5.4%. This is due to the fact that a larger ke allows the model to focus on more local context. Therefore, we set ke to 3 on the ETT dataset and 9 on the Exchange dataset.

***k* in the Linear Self-Attention Mechanism.** To determine an appropriate projection dimension k, we conducted experiments using different values of k in the LCformer model and compared them with full self-attention in the Transformer model. The results are shown in Fig. 4.

Clearly, as the projection dimension k increases, the performance of the model improves. When k equals to 128 and 256, the performance of the model is already

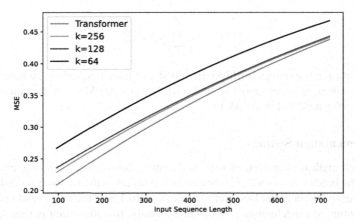

Fig. 4. Performance of the model under different k

close to that of the full self-attention mechanism in the Transformer model. However, when k is set to 64, the performance of the model is relatively poor. Therefore, in this paper, we set the projection dimension k to 128.

4.3 Performance Comparison

To verify the performance of LCformer, we selected several advanced Transformer-based models and traditional RNN-based models as baselines. In order to compare the performance of each model under different prediction output lengths, we fixed the length of the input sequence to 96 and the length of the training and evaluation prediction outputs to 96, 192, 336, and 720, respectively. The experimental results are shown in Table 2.

It can be observed that the proposed LCformer performs the best on both the ETTm2 and Exchange datasets. Compared with Autoformer, which also uses a decomposition structure, LCformer reduces the MSE by 12.2% on the ETTm2 dataset when the prediction length is 96. Overall, the relative improvements of LCformer in terms of MSE and MAE are 4.82% and 2.92%, respectively, compared to the performance-competitive Autoformer.

Table 2. Time series forecasting results of all methods on two datasets

DataSet		ETTm2				Exchange			
Prediction Length		96	192	336	720	96	192	336	720
LCformer	MSE	**0.224**	**0.274**	**0.328**	**0.420**	**0.196**	**0.204**	**0.218**	**0.244**
	MAE	**0.309**	**0.335**	**0.369**	**0.416**	**0.313**	**0.317**	**0.328**	**0.353**

(continued)

Table 2. (*continued*)

DataSet		ETTm2				Exchange			
Prediction Length		96	192	336	720	96	192	336	720
Autformer [3]	MSE	0.255	0.281	0.339	0.422	0.201	0.222	0.231	0.254
	MAE	0.339	0.340	0.372	0.419	0.317	0.334	0.338	0.361
Informer [2]	MSE	0.365	0.533	1.363	3.379	0.274	0.296	0.300	0.373
	MAE	0.453	0.563	0.887	1.338	0.368	0.386	0.394	0.439
LogTrans [21]	MSE	0.768	0.989	1.334	3.048	0.258	0.266	0.280	0.283
	MAE	0.642	0.757	0.872	1.328	0.357	0.368	0.380	0.376
Reformer [22]	MSE	0.658	1.078	1.549	2.631	0.312	0.348	0.350	0.340
	MAE	0.619	0.827	0.972	1.242	0.402	0.433	0.433	0.420
LSTM [6]	MSE	2.041	2.249	2.568	2.720	1.453	1.846	2.136	2.984
	MAE	1.073	1.112	1.238	1.287	1.049	1.179	1.231	1.427

In addition, LCformer still shows excellent performance on the Exchange dataset without clear periodicity features, indicating the model's generalisability. As the prediction sequence length increases, the performance change of LCformer is more stable compared to models such as Informer and LogTrans. Our results indicate that the proposed LCformer has significantly outperformed other state-of-the-art methods for long-term series forecasting.

4.4 Case Study

We performed visualization on the experimental results of the Exchange dataset. For ease of observation, we only show the LSTM and Autoformer models. The experimental results are shown in Fig. 5.

Figure 5 shows that the proposed LCformer performs better than other prediction methods. This is partly because the local convolutional module allows the model to focus on local context and obtain more historical information to assist prediction, especially for data like the Exchange dataset that lacks covariates. The use of a sequence decomposition module to decompose the original time series into subsequences with different features also enables the model to better focus on the different features of the original sequence, which optimizes the performance of the model.

5 Model Analysis

5.1 Ablation Studies

We conducted ablation experiments on the ETTm2 dataset to demonstrate the impact of the local convolutional module and linear self-attention mechanism on LCformer. We mainly tested two variants of LCformer: (1) LCformer_1, which only removes the

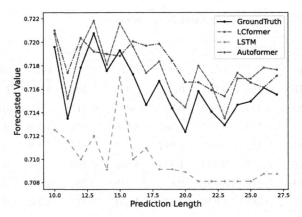

Fig. 5. Visualization results for case study on Exchange

local convolutional module in LCformer; (2) LCformer_2, which only replaces the linear self-attention mechanism with the autocorrelation mechanism in Autoformer. The experimental results are shown in Table 3.

Table 3. Ablation results of the proposed LCformer.

DataSet		ETTm2			
Prediction Length		96	192	336	720
LCformer	MSE	0.224	0.274	0.328	0.417
	MAE	0.309	0.335	0.369	0.415
LCformer_1	MSE	0.259	0.289	0.355	0.454
	MAE	0.345	0.351	0.391	0.451
LCformer_2	MSE	0.218	0.272	0.325	0.415
	MAE	0.296	0.332	0.371	0.413
Autoformer [3]	MSE	0.255	0.281	0.339	0.422
	MAE	0.339	0.340	0.372	0.419
Informer [2]	MSE	0.227	0.300	0.382	1.637
	MAE	0.305	0.360	0.410	0.794
Transformer	MSE	0.268	0.304	0.365	0.475
	MAE	0.346	0.355	0.400	0.466

As shown in the table, after removing the local convolutional module, LCformer_1 has higher MSE and MAE than LCformer, with an average increase of 9.6% and 7.8%, respectively. However, it is still more competitive than the Transformer model because the model retains the sequence decomposition structure to identify different features of the time series. On the other hand, replacing the linear self-attention mechanism in LCformer with the autocorrelation mechanism resulted in a slight improvement in performance, with LCformer_2 reducing the MSE and MAE of LCformer by an average of 1.2% and 1.3%, respectively. This is because the autocorrelation mechanism learns more information about the sequence at a higher time complexity level. However, this small improvement comes at the cost of increased time and memory consumption, which does not have practical value compared to the linear time complexity of LCformer in long-term series prediction tasks.

5.2 Time Complexity Analysis

To study the time performance of LCformer under different input sequence lengths, we compared it with Autoformer, which has the best time performance among the baselines. The experimental results are shown in Fig. 6.

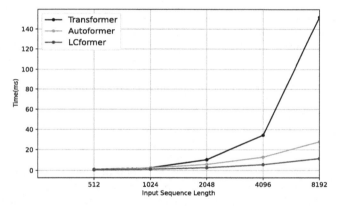

Fig. 6. Running time efficiency analysis

Figure 6 shows that Autoformer roughly exhibits O($n(logn)$) time complexity as the input sequence length keeps increasing. However, the LCformer proposed in this paper exhibits an approximately linear time complexity. The experimental results demonstrate that LCformer has an approximate O(n) time complexity when dealing with long sequence problems.

6   Conclusion

In this paper, we proposed a linear convolution model based on decomposition structure (LCformer) for long-term series forecasting. The model identifies the correlation between sub-sequences by decomposing the original sequence and solves the information bottleneck problem caused by the dot-product attention pattern of the traditional attention mechanism through the use of causal convolution. In addition, the use of linear self-attention mechanism enables the model to achieve linear complexity in long-term series forecasting. Experimental results on two different types of datasets show that

LCformer achieves better predictive performance and significant complexity reduction compared to other state-of-the-art methods for long-term series forecasting such as Autoformer and Informer. The proposed method can be applied to long-term series forecasting, such as extreme weather warnings, to reduce human labour and improve forecast accuracy.

References

1. Vaswani, A., et al.: Attention is all you need. In: Proceedings of the 31st International Conference on Neural Information Processing Systems, pp. 6000–6010. Curran Associates Inc. (2017)
2. Zhou, H., Zhang, S., Peng, J., et al.: Informer: beyond efficient transformer for long sequence time-series forecasting. In: Proceedings of the 35th AAAI Conference on Artificial Intelligence, vol. 35, no. 12, pp. 11106–11115. Association for the Advancement of Artificial Intelligence (AAAI) (2021)
3. Wu, H., Xu, J., Wang, J., et al.: Autoformer: Decomposition transformers with auto-correlation for long-term series forecasting. Adv. Neural. Inf. Process. Syst. **34**(1), 22419–22430 (2021)
4. Durbin, J., Koopman, S.J.: Time Series Analysis by State Space Methods: Second Edition. Oxford University Press (2012). https://doi.org/10.1093/acprof:oso/9780199641178.001.0001
5. Pascanu, R., Mikolov, T., Bengio, Y.: On the difficulty of training recurrent neural networks. In: 30th International Conference on Machine Learning, pp. 1310–1318. Association for Computing and Machinery (ACM) (2013)
6. Zhao, J., Huang, F., Lv, J., et al.: Do RNN and LSTM have long memory? In: Proceedings of the 37th International Conference on Machine Learning, PMLR, pp. 11365–11375. Association for Computing and Machinery (ACM) (2020)
7. Salinas, D., Flunkert, V., Gasthaus, J., et al.: DeepAR: probabilistic forecasting with autoregressive recurrent networks. Int. J. Forecast. **36**(3), 1181–1191 (2020)
8. Lai, G., Chang, W.C., Yang, Y., et al.: Modeling long-and short-term temporal patterns with deep neural networks. In: 41st International ACM SIGIR Conference on Research & Development in Information Retrieval, pp. 95–104. Association for Computing Machinery (ACM) (2018)
9. Shih, S.Y., Sun, F.K., Lee, H.: Temporal pattern attention for multivariate time series forecasting. Mach. Learn. **108**, 1421–1441 (2019)
10. Brown, T., Mann, B., Ryder, N., et al.: Language models are few-shot learners. Adv. Neural. Inf. Process. Syst. **33**, 1877–1901 (2020)
11. Wang, J., Jin, L., Ding, K.: LiLT: a simple yet effective language-independent layout transformer for structured document understanding. In: Proceedings of the 60th Annual Meeting of the Association for Computational Linguistics, vol. 1, pp. 7747–7757. Association for Computational Linguistics (ACL) (2022)
12. Peer, D., Stabinger, S., Engl, S., et al.: Greedy-layer pruning: speeding up transformer models for natural language processing. Pattern Recogn. Lett. **157**, 76–82 (2022)
13. Kjell, O.N.E., Sikström, S., Kjell, K., et al.: Natural language analyzed with AI-based transformers predict traditional subjective well-being measures approaching the theoretical upper limits in accuracy. Sci. Rep. **12**(1), 3918 (2022)
14. Von der Mosel, J., Trautsch, A., Herbold, S.: On the validity of pre-trained transformers for natural language processing in the software engineering domain. IEEE Trans. Software Eng. **49**(4), 1487–1507 (2023)

15. Dong, X., Bao, J., Chen, D., et al.: CSWin transformer: a general vision transformer backbone with cross-shaped windows. In: Proceedings of the IEEE/CVF Conference on Computer Vision and Pattern Recognition, pp. 12124–12134. Institute of Electrical and Electronics Engineers (IEEE) (2022)

16. Lee, Y., Kim, J., Willette, J., et al.: MPViT: multi-path vision transformer for dense prediction. In: Proceedings of the IEEE/CVF Conference on Computer Vision and Pattern Recognition, pp. 7287–7296. Institute of Electrical and Electronics Engineers (IEEE) (2022)

17. Zhengzhong, T., et al.: Maxvit: Multi-axis vision transformer. In: Avidan, S., Brostow, G., Cissé, M., Farinella, G.M., Hassner, T. (eds.) Computer Vision – ECCV 2022: 17th European Conference, Tel Aviv, Israel, October 23–27, 2022, Proceedings, Part XXIV, pp. 459–479. Springer Nature Switzerland, Cham (2022). https://doi.org/10.1007/978-3-031-20053-3_27

18. Li, B., Zhao, Y., Zhelun, S., et al.: DanceFormer: music conditioned 3D dance generation with parametric motion transformer. In: Proceedings of the 36th AAAI Conference on Artificial Intelligence, vol. 36, no. 2, pp. 1272–1279. Association for the Advancement of Artificial Intelligence (AAAI) (2022)

19. Di, S., Jiang, Z., Liu, S., et al.: Video background music generation with controllable music transformer. In: Proceedings of the 29th ACM International Conference on Multimedia, pp. 2037–2045. Association for Computing Machinery (ACM) (2021)

20. Hernandez-Olivan, C., Beltrán, J.R.: Music composition with deep learning: a review. In: Biswas, A., Wennekes, E., Wieczorkowska, A., Laskar, R.H. (eds.) Advances in Speech and Music Technology: Computational Aspects and Applications, pp. 25–50. Springer International Publishing, Cham (2023). https://doi.org/10.1007/978-3-031-18444-4_2

21. Li, S., Jin, X., Xuan, Y., et al.: Enhancing the locality and breaking the memory bottleneck of transformer on time series forecasting. In: Proceedings of the 33rd International Conference on Neural Information Processing Systems, pp. 5243–5253. Curran Associates Inc. (2019)

22. Yang, X., Liu, Y., Wang, X.: ReFormer: the relational transformer for image captioning. In: Proceedings of the 30th ACM International Conference on Multimedia, pp. 5398–5406. Association for Computing Machinery (ACM) (2022)

23. Chen, C., Liu, Y., Chen, L., et al.: Bidirectional spatial-temporal adaptive transformer for urban traffic flow forecasting. IEEE Trans. Neural Networks Learn. Syst. **34**, 6913–6925 (2022)

24. Dao, T., Fu, D., Ermon, S., et al.: FlashAttention: fast and memory-efficient exact attention with IO-awareness. Adv. Neural. Inf. Process. Syst. **35**, 16344–16359 (2022)

Fashion Trend Forecasting Based on Multivariate Attention Fusion

Jia Chen[1,2], Yi Zhao[1], Saishang Zhong[1,2(✉)], and Xinrong Hu[1,2]

[1] Wuhan Textile University, Wuhan 430200, China
[2] Engineering Research Center of Hubei Province for Clothing Information,
Wuhan 430200, Hubei, China
`saishang@cug.edu.cn`

Abstract. The objective of garment fashion trend prediction is to capture the trends of different garment attributes such as round necks and camouflage, enabling forecasts of their future popularity. Existing fashion trend prediction models have not sufficiently integrated comments on current social media networks and user preferences. Thus affecting the accuracy of garment popularity prediction. To address this issue, this paper proposes a fashion popularity prediction model based on multivariate attention fusion(MAFT). It combines diverse information posted by users on fashion platforms like Chictopia, uses GLU modules and dilated convolutions to preprocess multivariate features, enhances context feature extraction on sequence data, and suppresses irrelevant information. Subsequently, a multivariate attention fusion block is designed to capture the mapping relationship between dynamic and static variables in the input. After feature fusion, trend prediction for the garment is achieved through a Transformer layer. Experimental results demonstrate that this method accurately predicts future trends on the SFS, FIT, and Geo datasets, with improvements of 8.79% and 11.77% in MAE and MAPE evaluation metrics, respectively, compared to the best existing fashion trends prediction models.

Keywords: Fashion trend · Multivariate time series forecasting · Transformer · Attention mechanism · Deep learning

1 Introduction

Fashion signifies changeability. Garment fashion trend prediction is of paramount importance in the fashion industry as it aims to comprehend the patterns of change in diverse garment attributes, both in the past and in the future. These predictions not only help fashion companies to develop new products but also help consumers to find the hottest models of the season. In recent years, the rapid growth of data from fashion platforms on the Internet has accelerated changes in garment trends, making trend forecasting even more challenging. This data includes static variables that do not change over time, such as age, gender and fashion brands associated with users. The objective of this research, as shown in Fig. 1, is to extract valuable information from static and dynamic multivariate

B. Luo et al. (Eds.): ICONIP 2023, CCIS 1962, pp. 68–81, 2024.
https://doi.org/10.1007/978-981-99-8132-8_6

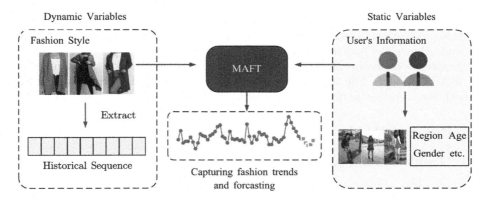

Fig. 1. Accurate fashion trend prediction by using the user information of fashion platforms and the historical sequences of fashion elements.

fashion data posted by users on fashion platforms and predict the future trends of different garment elements (such as sweaters, hoodies, etc.). This task faces two main challenges:

Firstly, due to the complex composition of fashion style data, effectively utilizing this information is challenging. Simply inputting them into nonlinear processing models may not always yield the best computational results. Therefore, researching a preprocessing method for static and dynamic variables is essential. This study treats static user information and dynamic style trends as multivariate sequences. Processing multivariate data has always been a major challenge. Each variable depends not only on its past values but also on other variables, such as trends, seasonality, and other time patterns. This study proposes a preprocessing method for handling dynamic and static multivariate features to process sequential inputs and capture inherent information in the sequences effectively.

Second, building a predictive model that integrates multivariate features and has excellent predictive capabilities is a great challenge. In recent years, Transformers have shown excellent performance in predicting multivariate time series problems. They are suitable for understanding repeated patterns with long-term dependencies. Inspired by Transformer architectures based on self-attention mechanisms [1,2], this study introduces a new approach called Multivariate Fusion Attention(MFA) and proposes an attention mechanism for fusing dynamic trends and static information. Additionally, a model named Multivariate Attention Fusion Transformer (MAFT) is designed to predict garment fashion trends, enabling information exchange between different types of variables.

In summary, the work presented in this paper can be summarized as follows:

- We propose a sequence preprocessing scheme for multivariate garment data, aiming to better capture the temporal dynamics of fashion trend sequences while enhancing the flexibility of the model in capturing static features.

- We introduce the MAFT model, which utilizes a Multivariable Fusion Attention mechanism to focus on the interaction between different types of variables and achieve the fusion of multivariate features.
- Extensive experiments are conducted on apparel trend prediction using the MAFT model. The experimental results show that the proposed model exhibits superior performance in predicting future apparel fashion trends.

2   Related Work

2.1   Fashion Trend Analysis

Fashion plays a significant role in people's lives, and an increasing number of researchers are focusing on various issues in the fashion domain, including garment style discovery [3–5], virtual try-on [6–8]. Meanwhile, the analysis and prediction of garment fashion trends have been a crucial research topic in the fashion industry because they drive innovation. Al-Halah et al. [9] developed models based on influential patterns to capture the popularity of specific styles in any city's future. Hsiao et al. [10] introduced a multimodal statistical model for detecting global events and their impact on garment choices. However, factors influencing fashion trend changes typically involve multiple variables, and these methods lack the integration of multivariate factors, thereby being unable to harness this knowledge to capture complex trends.

2.2   Fashion Trend Forecasting Model

The essence of garment fashion trend prediction lies in the domain of multivariable time series forecasting problems. Most of the existing time series forecasting models heavily rely on statistical models. For instance, matrix factorization methods [11] simulate sequence data by learning information within matrices derived from the modelling process. In recent years, deep neural networks, particularly recurrent neural networks (RNNs) [12], have been proposed to handle time series data and achieve sequential prediction, showcasing the exceptional performance of deep learning models in solving time series prediction problems.

Recently, several Transformer-based models have been applied to time series prediction problems. Li et al. [13] introduced the LogSparse Transformer for time series tasks, which exhibited exceptional performance on four common prediction datasets. Wu et al. [14] employed the Transformer to predict the prevalence of influenza and demonstrated superior performance over ARIMA and LSTM. These methods validate the feasibility of Transformers in multivariable prediction problems. However, in the domain of garment fashion trend prediction, the feasibility of Transformer models has yet to be verified.

3   Methodology

This section describes the proposed Multivariable Attention Fusion Transformer model (MAFT) for modelling the fashion element trend sequences, as shown

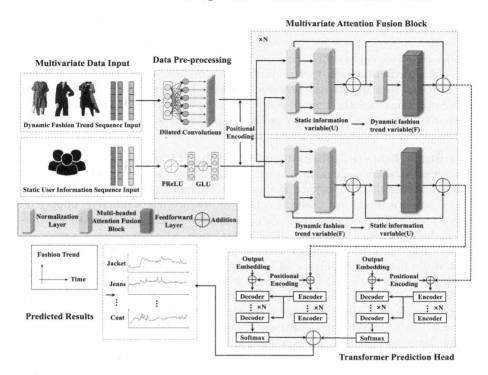

Fig. 2. The framework of MAFT.

in Fig. 2. MAFT combines multivariate sequences through the pairwise fusion process of the Multivariate Attention Fusion Transformer.

3.1 Multivariate Sequence Pre-processing

Before predicting fashion element trends, it is necessary to preprocess the input trend data for the model. In this study, different processing methods are used to process the trend sequence data and rich static feature information before forecasting.

Pre-processing Module of Fashion Element Trend Sequence. n order to enhance the model's understanding of the temporal dynamics of fashion trend collections, this study first created a vector representation of the input time series data using a fully connected layer. This layer maps the data linearly to the specified dimensions. Variations and anomalies in the time series are highly dependent on the contextual data points. However, the self-concern mechanism in Transformers is not sensitive to the local context when executing queries. This can lead to model anomalies and underlying optimization issues.

To address this problem, this study employs Dilated Convolutions [15], which is a method for transforming time series data. The Dilated Convolutions have a broader receptive field for historical data in time series, making them suitable for time series prediction tasks that require long-term memory. By using the Dilated Convolutions, the model can preserve sequential information when modelling time data. Thus making it more sensitive to the local context information and ultimately improving the model's predictive performance.

User Static Feature Information Processing Module. In addition to the time series data for each garment element, the input data also includes various static variables that do not exhibit periodic changes over time, such as user gender, age, location, and brand. Each static feature has a different impact on fashion trends. Furthermore, the covariates composed of numerous static features also include variables unrelated to fashion trends, which may have a negative impact on the model's predictive ability. Therefore, this study preprocesses the input static features, as illustrated in Fig. 3.

First, various static features undergo nonlinear processing. For different types of static features, after entering the first hidden layer for fully connected processing, this study uses the PReLU (parameterized rectified linear unit) activation function [16] to nonlinearly transform them. Then, the output enters the next hidden layer to compute new hidden states. Finally, the results are normalized. This is done to prevent significant differences in weight magnitudes when connecting with the processed fashion element time series data. The process representation of the PReLU activation function is shown in Eq. 1:

$$f(x) = \begin{cases} x, x > 0 \\ wx, x \leq 0 \end{cases} \tag{1}$$

Here w is the parameter that needs to be learned and is responsible for controlling the slope of the negative semi-axis.

In order to suppress the impact of irrelevant variables on model predictions, this study draws inspiration from the Gated Linear Unit (GLU) [17] and employs a gating mechanism to selectively ignore past information. During the training

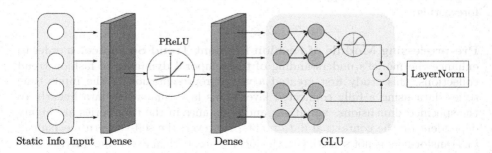

Static Info Input   Dense   Dense   GLU

Fig. 3. The User static feature information processing module

process, the sigmoid function's S-shaped output values for irrelevant static features approach 0. The process of handling static features is shown in Eqs. 2, 3, 4:

$$hid_1 = \text{PReLU}\left(w_1 + b_1\right) \tag{2}$$

$$hid_2 = w_2 hid_1 + b_2 \tag{3}$$

$$H_{GLU} = \sigma\left(w_3 hid_2 + b_3\right) \odot \left(w_4 hid_2 + b_4\right) \tag{4}$$

where σ is the S-shaped activation function, w and b are weights and deviations, $\odot$ is the elemental Hadamard product, and hid_1 and hid_2 are intermediate states. Equation 4 is a component gating layer based on the gated linear unit (GLU).

3.2 MAFT Model

This section will provide a detailed description of the components of the MAFT model and how the multivariable attention fusion block is designed.

Positional Embedding. To enable the sequences to carry temporal information, we augment positional embedding (PE) to $\hat{Z}_{\{U,F\}}$:

$$L^{[0]}_{\{U,F\}} = \hat{Z}_{\{U,F\}} + \text{PE}\left(T_{\{U,F\}}, d\right) \tag{5}$$

where $\text{PE}\left(T_{\{U,F\}}, d\right) \in \mathbb{R}^{T_{\{U,F\}} \times d}$ computes the embeddings for each position index, and $L^{[0]}_{\{U,F\}}$ are the resulting low-level position-aware features for different variable type.

Multivariate Attention Fusion Block. This section introduces the proposed multivariable attention fusion block. In order to facilitate the interaction and fusion between the two types of input variables, namely the static variables representing user information (U) and the dynamic variables representing garment fashion trend (F), a multivariable fusion attention mechanism called MFA is proposed. Each type of feature variable has two sequences represented $Z_F \in \mathbb{R}^{T_F \times d_F}$ and $Z_U \in \mathbb{R}^{T_U \times d_U}$ ($T_{(.)}$ and $d_{(.)}$ represent the sequence length and feature dimension, respectively). It has been shown in previous work [18] that identifying potential adaptations between these two types of variables can better facilitate attention between different variable types. Therefore, taking the attention calculation from U to F as an example, the query Q, key K, and value V are defined as follows: $Q_F = Z_F W_{Q_F}, K_U = Z_U W_{K_U}, V_U = Z_U W_{V_U}$. where $W_{Q_F} \in \mathbb{R}^{d_F \times d_k}, W_{K_U} \in \mathbb{R}^{d_U \times d_k}$ and W are weights.

The potential adaptability from U to F, denoted as $MFA_{U \to F}$, is computed through the multivariable fusion attention mechanism, and the calculation process is defined by Eq. 6:

$$\begin{aligned} Att_F &= \text{MFA}_{U \to F}\left(Z_F, Z_U\right) \\ &= \text{softmax}\left(\frac{Q_F K_U^\top}{\sqrt{d_k}}\right) V_U \end{aligned} \tag{6}$$

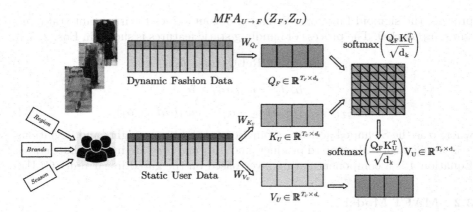

Fig. 4. Multivariate Fusion Attention $MFA_{U \to F}(Z_F, Z_U)$ between sequences Z_F, Z_U.

Att_F has the same length as Q_F, but is meanwhile represented in the feature space of V_U. Specifically, the scaled (by $\sqrt{d_k}$) softmax in Eq. 6 computes a score matrix $\mathrm{softmax}(\cdot) \in \mathbb{R}^{T_F \times T_U}$. We call this Eq. 6 multivariate fusion attention(MFA), which is illustrated in Fig. 4.

Multivariate Attention Fusion Transformer. Based on the multivariable attention fusion block, this paper designs the multivariable attention fusion Transformer model, which allows the two types of variable features to learn the feature information that is passed between them. In the following text, an example of passing the user's static information (U) to the fashion trend (F) is denoted as "$U \to F$". The dimensionality d of each attention fusion block is fixed at N. Each multivariable attention fusion Transformer consists of N layers of attention fusion blocks (as shown in the multivariable attention fusion block section of Fig. 2). Formally, for the $i = 1, \ldots, N$ layer:

$$\hat{L}_{U \to F}^{[i]} = \mathrm{MFA}_{U \to F}^{[i], \mathrm{mul}} \left(\mathrm{LN} \left(L_{U \to F}^{[i-1]} \right), \mathrm{LN} \left(L_U^{[0]} \right) \right) \\ + \mathrm{LN} \left(L_{U \to F}^{[i-1]} \right) \tag{7}$$

$$L_{U \to F}^{[i]} = f_{\theta_{U \to L}^{[i]}} \left(\mathrm{LN} \left(\hat{L}_{U \to F}^{[i]} \right) \right) + \mathrm{LN} \left(\hat{L}_{U \to F}^{[i]} \right) \tag{8}$$

Where f_θ is the position-wise feed-forward sublayer parameterized by θ, and $\mathrm{UFM}_{U \to F}^{[i], \mathrm{mul}}$ represents the position-wise feed-forward sublayer parameterized by $\mathrm{UFM}_{U \to F}$ at layer i. LN denotes the normalization layer.

In this process, each variable continuously updates its sequence through the low-level external information provided by the multivariate attention fusion module. Therefore, considering two types of variables $(F$ and $U)$. This paper designs two Multivariate Attention Fusion Transformers, each responsible for computing feature fusion information between dynamic fashion trends and static information.

Transformer Prediction Head. As the final step, this paper concatenates the output of the multivariable attention fusion Transformer that transforms dynamic and static variables, in order to generate the final prediction results $L_{\{U,F\}} \in \mathbb{R}^{T_{\{U,F\}} \times 2d}$.

$$L_{\text{multi}} = \left[L_{U \to F}^{[N]}, L_{F \to U}^{[N]} \right] \tag{9}$$

Then, each of them is fed into a sequence model to gather temporal information for prediction. This paper chooses the Transformer based on the self-attention mechanism [19]. Using the historical trend sequence of each garment element as the output embedding part, the last element of the sequence model is extracted by the Transformer and predicted using the fully connected layer.

4 Experiments

In this section, we verify the superiority of our model through extensive and rigorous experiments.

4.1 Datasets and Setup

The experiments were conducted using three datasets: Street Fashion Style (SFS) dataset [20], Fashion Image Text (FIT) dataset collected from Instagram [21], and fashion images dataset (Geo) collected from both Instagram and Flickr [22]. These datasets contain not only style, category and other descriptive attributes of garments images, but also a large number of static user attributes collected from fashion platforms in online media. Taking the SFS dataset as an example, it is a new street-style photo dataset collected from Chictopia, consisting of a total of 293,105 user posts. In each post, users typically share photos of their outfits along with relevant tags. These tags often include current seasons, suitable occasions, fashion styles, detailed garment information (such as category, colour, and brand), geographical and temporal information, and more.

For the experimental dataset settings, this study defines the temporal trend of a fashion element $f \in \mathcal{F}$ with respect to a given user group $u \in \mathcal{U}$. The trend is represented as a time series $\boldsymbol{y}_u^f = (y_1, \cdots, y_t, \cdots)$, where $\mathcal{F}$ represents the set of all fashion elements and $\mathcal{U}$ represents the set of all user information. The value of the time series at each time step is defined as $y_t = N_t^{u,f} / N_t^u$, where $N_t^{u,f}$ is the count of fashion element f observed for user u at time point t. Furthermore, $y_t = N_t^u$ represents the count of all fashion items (such as garments, shoes, etc.) observed for user u at time point t.

The three datasets used in this study span from 2014 to 2020. Two prediction experiment setups were conducted on the datasets:

1. Using data from 2015-2019 as input to predict future trend for the following year and comparing the predicted values with the actual trend trends in the dataset for the year 2020.

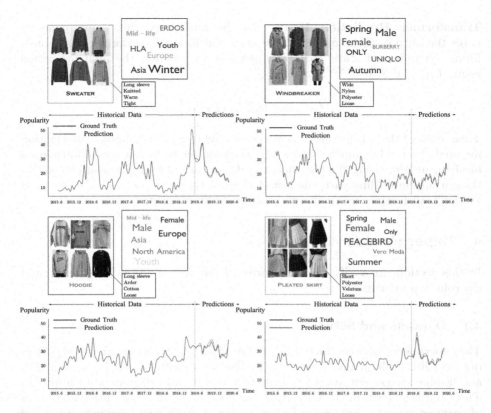

Fig. 5. The MAFT model predicts the trends of multiple garment styles(the **Blue** curve indicates historical data, the **Orange** curve is the ground truth from datasets, and the **Green** curve is the predicted value). The fashion styles from the left side of the first row to the right side of the second row are sweaters, trench coats, down jackets, hoodies, pleated skirts, and wide-legged pants.

2. Using data from the last six years as input to predict the trend for the next three years. Google Trends, which provides trend values for trend words, was used as a comparison reference during the prediction process.

In this study, a sliding window strategy was applied to the datasets to generate training and testing samples. The Mean Absolute Error (MAE) and Mean Absolute Percentage Error (MAPE) were used as evaluation metrics.

All experiments in this study were conducted in a Python 3.8 environment, using a 32GB NVIDIA Tesla V100 GPU.

4.2   Analysis of Fashion Trends

To further illustrate the effectiveness of the MAFT model in fashion prediction, this section presents additional visual results. In this section, the comparison is

Table 1. MAFT model results of MAE and MAPE on three data sets compared with other methods

Dataset	SFS		FIT		GEO	
Method	MAE	MAPE	MAE	MAPE	MAE	MAPE
Mean	0.0366	32.67	0.137	63.55	0.0293	25.82
Last	0.0312	28.75	0.143	53.46	0.0222	21.11
Var	0.0224	23.27	0.128	48.21	0.0151	17.97
Linear	0.0482	31.17	0.136	46.16	0.0367	24.38
Geostyle	0.0215	22.89	0.145	52.26	0.0152	16.13
Transformer	0.0208	18.78	0.107	39.77	0.0142	15.02
KERN	0.0182	20.37	0.094	33.45	0.0135	14.26
MAFT	**0.0166**	**16.57**	**0.082**	**29.23**	**0.0128**	**13.97**

made between the predictions made by the MAFT model for four representative garment elements and the actual values in the dataset, as shown in Fig. 5. Figure 5 also displays user information associated with each garment element. It can be observed that MAFT not only predicts the outcomes but also integrates with a wealth of static user information. Generally, MAFT can predict the trends of garment elements well, even for predictions with quite complex patterns.

4.3   Discussion of Results

In this experiment, to predict the fashion trend of garment elements in the coming year using the experimental setup 1. We use MAFT to compare the test with seven prediction methods in the past, which are Mean and Last(Using the average value or the last data point of the input historical data as the prediction value), the VAR (Vector Autoregressive) [23] model in the traditional method, the Linear model mentioned in Geo [22], and GeoStyle [22], KERN [21] and the original Transformer model [19].

Through experiments, we obtained the MAE and MAPE values for the different comparison models on the three datasets, as shown in Table 1. Traditional forecasting methods such as Mean and Last face difficulties in predicting long time series problems due to the extended span and large data volume of the input fashion trend sequence, resulting in low accuracy. The knowledge-enhanced model used by KERN improves the prediction of more realistic fashion trends significantly, but it still has shortcomings in long-sequence prediction. Transformer models, on the other hand, can only predict the future values of fashion trend data and cannot achieve interaction and fusion of multivariate data. MAFT, as proposed in this paper, excels in feature fusion of dynamic and static multivariate data, leverages the excellent performance of Transformer in long sequence prediction, and ultimately achieves an improvement in accuracy by 8.79% and 11.77% compared to the current best model, KERN.

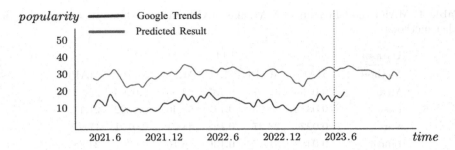

Fig. 6. Predictions for future dress trends on SFS

Additionally, using experimental setup 2, the predicted fashion trend values for dresses are compared with the keyword trend provided by Google Trends, as shown in Fig. 6. It can be observed that the fluctuations of the two roughly coincide, representing the reliability of the prediction results of the method proposed in this paper. Furthermore, in Fig. 6, the method presented in this paper can also predict the trend curve for the next six months.

4.4 Ablation Studies

To further investigate the impact and effectiveness of each component in MAFT, this section conducts ablation experiments.

Firstly, the preprocessing schemes for static and dynamic variables are removed in this study. The input part is directly encoded with positional encoding before entering the subsequent steps (this model is denoted as MAFT-P). The results are compared using MAE and MAPE as evaluation metrics. According to Table 2, the preprocessing methods for static and dynamic variables outperform the more straightforward approach of using positional encoding only. Furthermore, as shown in Fig. 7(a), the loss curve during prediction without preprocessing work progresses slightly slower, indicating that the preprocessing of multiple variables effectively improves the predictive performance of the model.

Furthermore, to verify the effect of the multivariable attention fusion on predicting future trends, the performance of the model is compared when using separate transformer models for static user feature variables and dynamic garment trend feature variables. The model with only dynamic features is denoted

Table 2. Percentage reduction of MAE and MAPE results after ablation of MAFT Module

Ablation Module	Model	MAE	MAPE
Multivariate Pre-processing Module	MAFT-P	0.202	20.32
Multivariate Attention Fusion Module	MAFT-D	0.191	19.34
	MAFT-S	0.221	23.07

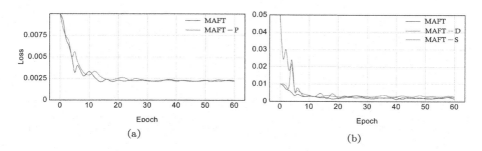

Fig. 7. Comparison of loss function curves before and after removal of each module. (a) Preprocessing module; (b) Multivariate Attention Fusion Module.

as MAFT-D, and the model with only static features is denoted as MAFT-S. In Fig. 7(b), the loss curve of MAFT-D fluctuates significantly in the first 20 epochs and then quickly converges to a stable value. However, the training loss of MAFT-S increases, indicating that the training performance is no longer as excellent as in the initial stages. From Table 2, it can be seen that the MAE of the model with only dynamic variables decreases from 0.01228 to 0.0137, and the MAPE decreases from 13.67 to 14.66, while the model with only static variables performs poorly. This shows that the multivariate attention fusion model significantly improves the prediction results. The implementation of multivariate feature interaction and fusion proposed in this method is confirmed.

5 Conclusions

In this paper, we explore the problem of predicting fashion trends based on multivariate attention fusion in the context of garments. In order to fuse multivariate fashion data, we propose a multivariate fusion attention mechanism after preprocessing the multivariate data separately, which effectively fuses different types of data. We propose an effective model called Multivariate Attention Fusion Transformer (MAFT) that accurately predicts future trends. Despite the considerable efforts and satisfactory results, there are still areas that can be further improved in the future. We should also consider exploring more user information, such as the number of followers or occupation of users on social platforms.

Acknowledgements. Chen's research was sponsored by the National Natural Science Foundation of China(Grant No.62202345).

References

1. Zerveas, G., Jayaraman, S., Patel, D., Bhamidipaty, A., Eickhoff, C.: A transformer-based framework for multivariate time series representation learning. In: Proceedings of the 27th ACM SIGKDD Conference on Knowledge Discovery & Data Mining, pp. 2114–2124 (2021)

2. Ye, Y., Lu, J.: A fusion transformer for multivariable time series forecasting: the mooney viscosity prediction case. Entropy **24**(4), 528 (2022)
3. Kiapour, M.H., Yamaguchi, K., Berg, A.C., Berg, T.L.: Hipster wars: discovering elements of fashion styles. In: Fleet, D., Pajdla, T., Schiele, B., Tuytelaars, T. (eds.) ECCV 2014. LNCS, vol. 8689, pp. 472–488. Springer, Cham (2014). https://doi.org/10.1007/978-3-319-10590-1_31
4. Jia, M., et al.: Fashionpedia: ontology, segmentation, and an attribute localization dataset. In: Vedaldi, A., Bischof, H., Brox, T., Frahm, J.-M. (eds.) ECCV 2020. LNCS, vol. 12346, pp. 316–332. Springer, Cham (2020). https://doi.org/10.1007/978-3-030-58452-8_19
5. Zou, X., Pang, K., Zhang, W., Wong, W.: How good is aesthetic ability of a fashion model? In: Proceedings of the IEEE/CVF Conference on Computer Vision and Pattern Recognition, pp. 21200–21209 (2022)
6. Wang, B., Zheng, H., Liang, X., Chen, Y., Lin, L., Yang, M.: Toward characteristic-preserving image-based virtual try-on network. In: Proceedings of the European Conference on Computer Vision (ECCV), pp. 589–604 (2018)
7. Dong, H., et al.: Towards multi-pose guided virtual try-on network. In: Proceedings of the IEEE/CVF International Conference on Computer Vision, pp. 9026–9035 (2019)
8. Hsieh, C.-W., Chen, C.-Y., Chou, C.-L., Shuai, H.-H., Liu, J., Cheng, W.-H.: Fashionon: semantic-guided image-based virtual try-on with detailed human and clothing information. In: Proceedings of the 27th ACM International Conference on Multimedia, pp. 275–283 (2019)
9. Al-Halah, Z., Grauman, K.: From Paris to Berlin: discovering fashion style influences around the world. In: Proceedings of the IEEE/CVF Conference on Computer Vision and Pattern Recognition, pp. 10136–10145 (2020)
10. Hsiao, W.-L., Grauman, K.: From culture to clothing: discovering the world events behind a century of fashion images. In: Proceedings of the IEEE/CVF International Conference on Computer Vision, pp. 1066–1075 (2021)
11. Yu, H.-F., Rao, N., Dhillon, I.S.: Temporal regularized matrix factorization for high-dimensional time series prediction. In: Advances in Neural Information Processing Systems, vol. 29 (2016)
12. Tokgöz, A., Ünal, G.: A RNN based time series approach for forecasting Turkish electricity load. In: 2018 26th Signal Processing and Communications Applications Conference (SIU), pp. 1–4. IEEE (2018)
13. Li, S., et al.: Enhancing the locality and breaking the memory bottleneck of transformer on time series forecasting. In: Advances in Neural Information Processing Systems, vol. 32 (2019)
14. Wu, N., Green, B., Ben, X., O'Banion, S.: Deep transformer models for time series forecasting: the influenza prevalence case. arXiv preprint arXiv:2001.08317 (2020)
15. van den Oord, A., et al.: WaveNet: a generative model for raw audio. arXiv preprint arXiv:1609.03499 (2016)
16. He, K., Zhang, X., Ren, S., Sun, J.: Delving deep into rectifiers: Surpassing human-level performance on ImageNet classification. In: Proceedings of the IEEE International Conference on Computer Vision, pp. 1026–1034 (2015)
17. Salinas, D., Valentin, F., Jan, G., Tim, J.: Deepar: probabilistic forecasting with autoregressive recurrent networks. Int. J. Forecast. **36**(3), 1181–1191 (2020)
18. Hubert Tsai, Y.-H., Bai, S., Pu Liang, P., Zico Kolter, J., Morency, L.-P., Salakhutdinov, R.: Multimodal transformer for unaligned multimodal language sequences. In: Proceedings of the conference. Association for Computational Linguistics. Meeting, vol. 2019, pp. 6558. NIH Public Access (2019)

19. Vaswani, A., et al.: Attention is all you need. In: Advances in Neural Information Processing Systems, vol. 30 (2017)
20. Gu, X., Wong, Y., Peng, P., Shou, L., Chen, G., Kankanhalli, M.S.: Understanding fashion trends from street photos via neighbor-constrained embedding learning. In: Proceedings of the 25th ACM international conference on Multimedia, pp. 190–198 (2017)
21. Ma, Y., Ding, Y., Yang, X., Liao, L., Wong, W.K., Chua, T.-S.: Knowledge enhanced neural fashion trend forecasting. In: Proceedings of the 2020 International Conference on Multimedia Retrieval, pp. 82–90 (2020)
22. Mall, U., Matzen, K., Hariharan, B., Snavely, N., Bala, K.: Geostyle: discovering fashion trends and events. In: Proceedings of the IEEE/CVF International Conference on Computer Vision, pp. 411–420 (2019)
23. Christopher, A.: Sims. Macroeconomics and reality. Econometrica **48**, 1–48 (1980)

On Searching for Minimal Integer Representation of Undirected Graphs

Victor Parque$^{(\boxtimes)}$ (iD) and Tomoyuki Miyashita (iD)

Department of Modern Mechanical Engineering, Waseda University, 3-4-1 Okubo, Shinjuku, Tokyo 169-8555, Japan
parque@aoni.waseda.jp

Abstract. Minimal and efficient graph representations are key to store, communicate, and sample the search space of graphs and networks while meeting user-defined criteria. In this paper, we investigate the feasibility of gradient-free optimization heuristics based on Differential Evolution to search for minimal integer representations of undirected graphs. The class of Differential Evolution algorithms are population-based gradient-free optimization heuristics having found a relevant attention in the nonconvex and nonlinear optimization communities. Our computational experiments using eight classes of Differential Evolution schemes and graph instances with varying degrees of sparsity have shown the merit of attaining minimal numbers for graph encoding/representation rendered by exploration-oriented strategies within few function evaluations. Our results have the potential to elucidate new number-based encoding and sample-based algorithms for graph representation, network design and optimization.

Keywords: undirected graphs · graph representation · enumerative representation · differential evolution · optimization

1 Introduction

Graphs allow to compute with dependencies in several fields. Devising succinct graph representations is essential to realize efficient mechanisms for storing, transmission and sampling graphical structures ubiquitously. Although matrices and lists are useful data structures allowing the simple and straightforward representation of graphs, they are unable to meet the information-theoretic tight bounds. In a seminal paper, Turan outlined the concept of succinctness when dealing with graph representations, and proposed the succinct representations of both labeled and unlabeled planar graphs using size up to $12n$ bits (n is the number of nodes in the graph) [30]. And Farzan and Munro proposed the representation of arbitrary graphs with the size being a multiplicative factor $1 + \epsilon$ above the information-theoretic minimum, for small positive ϵ [10]. Finding efficient representation of arbitrary graphs implies finding regularities in graphs that can be polynomially encoded/decoded. As such, the community has used regular features in the graph topology such as planarity [1, 2, 14, 16, 30], triangularity [1, 2, 15], separability [3, 4, 15] and symmetry [17]. When using regularity in graph representation, it is often useful find label encodings that allow tests for node adjacency [9, 11]. Generally speaking, when regularity in the graph topology is known, the search space of tailored graph representations is reduced; thus it becomes feasible to investigate tight bounds on the representation mechanism.

The community has also explored the feasibility of compressing real-world networks [5–8], in which compression efficiency is often reported in bits per edge. It is also possible to realize the number-based representation of graphs by using combinatorial ideas [18, 20, 21, 25], in which a graph coding scheme computes an integer number representing the graph, and a decoding scheme generates the graph topology from the integer number. Representing graphs with numbers is advantageous when a-priori knowledge of graph regularity is unknown. Also, the number representing the graph meets the information-theoretic tight bound over all possible graph representations in the same class. On the other hand, when the problem is to find an optimal graph topology, the number based representation enables the one-dimensional (integer) search space and the implicit parallelization of search.

One of the key questions in number-based graph representation is whether exploring the landscape of enumerative encodings would render representations with fewer digits, and thus smallest possible integers. Although approaches for succinct graph encodings have relied on apriori knowledge of graph invariants such as planarity, separability, symmetry, and sparsity, the study of representing arbitrary graphs with the smallest possible integers has received little attention in the literature. In this paper, we study the feasibility of using gradient-free population-based optimization heuristics inspired by Differential Evolution for smallest/minimal integer representation of graphs, and present an analysis of the convergence performance. In particular, our contributions are as follows: (1) We formulate the problem of searching for minimal integer representation of graphs, portray examples of the search space, and evaluate the performance of eight gradient-free optimization algorithms with varying forms of exploration-exploitation based on Differential Evolution schemes. (2) Our computational experiments using graph instances with varying degrees of sparsity has shown the merit of exploration strategies to attain better convergence with few function evaluations.

The rest of this paper is organized as follows. Section 2 provides the preliminaries on enumerative and minimal graph representation, Sect. 3 describes the computational experiments, and Sect. 4 concludes our paper.

2   Graphs with Minimal Integers

2.1   Enumerative Representation of Graphs

Let $\mathscr{G}_{n,m}$ be the class of undirected graphs with n nodes and m edges. When graphs in the class $\mathscr{G}_{n,m}$ are represented by integer numbers g, the following condition holds: $g \in [0, |\mathscr{G}_{n,m}| - 1]$, in which,

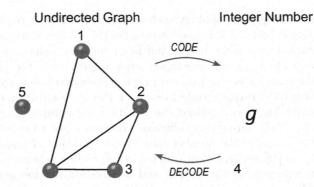

Fig. 1. Basic concept of representing graphs as numbers. Given an graph, the *CODE* function generates the number g representing the graph and, viceversa, the *DECODE* function renders the graph from the integer number g.

$$|\mathscr{G}_{n,m}| = \binom{\binom{n}{2}}{m}. \tag{1}$$

The above implies that graphs in the class $\mathscr{G}_{n,m}$ are combinatorial structures that involves the combinations of $\binom{n}{2}$ edges taken m at the time. Let the graph instance $G = (V, E) \in \mathscr{G}_{n,m}$, with $n = |V|$ and $m = |E|$; a numerical representation (or numerical encoding) of G is the tuple $(g, CODE, DECODE)$ as shown by Fig. 1:

$$CODE : G \to g, \tag{2}$$

$$DECODE(CODE(g)) \to G. \tag{3}$$

For a defined labeling of nodes with the tuple $(1, 2, 3, ..., n)$, it is possible to define the *CODE* function as the following algebraic relation [18,20]:

$$g(x) = \sum_{c=1}^{m} (-1)^{m-c} \left[\binom{i_c}{c} - 1 \right], \tag{4}$$

where i_c is the numerical label (integer number) of the c-th edge $(u, v) \in E$, computed as follows [20]:

$$i_c = \binom{u-1}{2} + v, \text{ for } u > v, \tag{5}$$

where $u \in [1, n]$ and $v \in [1, n]$ are labels of nodes from the set V.

The reader may note that the number representation of a graph is contingent upon the values of n and m, and the above-mentioned definitions consider a defined labeling order of the set of nodes V. It is possible to define the *DECODE* function with a decoding algorithm based on the revolving door order [18,20].

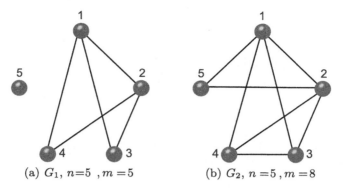

(a) G_1, $n=5$, $m=5$ (b) G_2, $n=5$, $m=8$

Fig. 2. Example of undirected graph topologies.

2.2 Minimal Integer Representation

Graphs can be represented with minimal integers by solving the following:

$$\text{Minimize}_{x} \quad L(x)$$
$$\text{subject to} \quad x \in \mathbf{F}(n) \tag{6}$$

where the objective function is

$$L(x) = \log(g(x)), \tag{7}$$

where x denotes the representation of the ordering of labels in the graph, and $\mathbf{F}(n)$ is the search space related to the factorial numbering system of n. The problem described above is NP-hard, and without further knowledge of the graph topology or regularities in the structure, computing the gradient of L is unfeasible; thus, the use of a gradient-free and stochastic optimization heuristic is a desirable choice to sample the search space of $F(x)$. The logarithmic function in Eq. (7) facilitates computing the number of digits in number-based representations of graphs, thus it provides an intuitive idea on the space needed to encode undirected graph. Furthermore, the use of the factorial (or *factoradic*) space $\mathbf{F}(n)$ in Eq. (6) is due to the permutation of node labels in terms of $n!$.

To exemplify the landscape of the above-mentioned objective function, we consider the following examples of graphs:

- G_1: an undirected graph with $n = 5$ nodes and $m = 5$ edges, and
- G_2: an undirected graph with $n = 5$ nodes and $m = 8$ edges.

Figure 2 shows the topologies of graphs G_1 and G_2. For both graphs, $n = 5$, yet the reader may note that a node with label 5 in G_1 is not connected to any other node. Then, one can compute the objective function L over the factorial space $\mathbf{F}(n)$, for each permutation of the node labels. For $n = 5$, there exists $5! = 120$ possible orders of the labels, thus 120 possible values of the metric L.

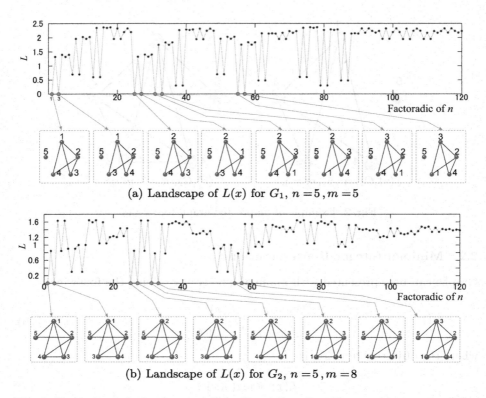

(a) Landscape of $L(x)$ for G_1, $n = 5$, $m = 5$

(b) Landscape of $L(x)$ for G_2, $n = 5$, $m = 8$

Fig. 3. Example of the landscapes of the fitness functions for representing of graphs by smallest numbers. Green marks imply the lowest smallest integers.

In order to show the landscape of the objective function, Fig. 3 shows the rendering of the metric L for both graphs G_1 and G_2 when evaluating all possible 120 permutations of labels. Here, the x-axis shows the factorial number system corresponding to the ordering of the label (for $n = 5$, the x-axis shows numbers in the range 1–120), and the y-axis shows the metric L. By observing Fig. 3, one can note that the landscape is nonsmooth, and it is possible to find regions with smallest values of L. Such regions (labeled with green color) are of interest to find node and edge labels that render smallest/minimal integer representation. Examples of such node labels are provided in Fig. 3-(a) and 3-(b), for each graph G_1 and G_2, respectively. Sampling through the factorial number system allows to explore possible configurations of graphs whose representation may lead to the use of fewer number of digits.

In order to show the overall landscape for distinct and arbitrary number of edge configurations, Fig. 4 shows the landscapes of the metric L for arbitrary graphs when considering varying number of edges. Here, we show the case of undirected graphs with $n = 5$, $m \in [2,8]$, and the case of undirected graphs with $n = 6$ and number of edges $m \in [2,14]$. Compared to Fig. 3, the land-

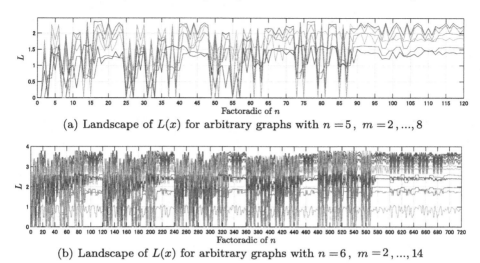

(a) Landscape of $L(x)$ for arbitrary graphs with $n = 5$, $m = 2, ..., 8$

(b) Landscape of $L(x)$ for arbitrary graphs with $n = 6$, $m = 2, ..., 14$

Fig. 4. Examples of landscapes of the function L for arbitrary graphs. (a) The case of $n = 5, m \in [2,8]$. (b) The case of $n = 6$ and $m \in [2,14]$.

scape of in Fig. 4 shows the larger number of instances in which it is possible to attain smallest integer representation. This is due to the graph configurations arising from larger number of edges and nodes. Also, sparse undirected graphs with fewer edges compared to number of nodes will allow the larger number of combinatorial permutations with smallest integer representation. Furthermore, although the minimization of the objective function in Eq. (7) implies searching for smallest numbers, it is possible to explore other objective functions to search for graphs with tailored properties (e.g., by using factorization and the fundamental theorem of arithmetic).

3   Computational Experiments

To evaluate the performance of the proposed approach, we conducted computational experiments using diverse graph topologies, configuration of nodes and edges, with varying degrees of sparsity. Figure 5 shows the set of graph instances. For ease of reference, plots correspond to arbitrary ordering/labeling of nodes (in the range of 10 to 20), considering the modeling of labeled trees as well. The key motivation for using the settings in Fig. 5 is due to our interest in finding compact representations for multiagent communication networks, where the number of agents is expected to be distributed in indoor environments (e.g., robots, drones, sensors distributed in the map). As such, exploring the compact representation of graphs is desirable to communicate and share the topology of the graph among members of a robot swarm. The reader may also note that some graphs in Fig. 5 are not connected to the entire network. This observation (and situation) is especially relevant when agents (nodes) are independent of the

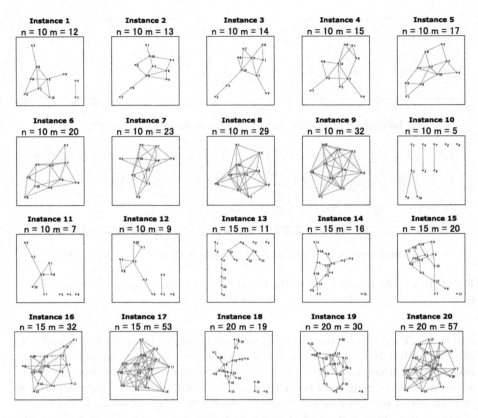

Fig. 5. Graph instances used in this study.

group (swarm); thus, communication and information gathering is not essential at all times.

To evaluate the feasibility in attaining quick convergence towards graph representations with smaller integers, we use a representative class of optimization algorithms derived from Differential Evolution, which is a population-based gradient-free optimization algorithm extending the difference of vectors:

1. DESPS: Differential Evolution with Successful Parent Selection [13]
2. OBDE: Differential Evolution with Opposition-based Learning [26]
3. JADE: Adaptive Differential Evolution External Archive [31]
4. DCMAEA: Differential Covariance Matrix Adaptation [12]
5. DERAND: DE/rand/1/bin Strategy [28],
6. DEBEST: DE/best/1/bin Strategy [28],
7. RBDE: Rank-based Differential Evolution [29],
8. DESIM: Differential Evolution with Similarity Based Mutation [27].

The key motivation of using the above set is due to our interest in evaluating distinct forms of selection pressure, parameter adaptation, exploration and exploitation mechanisms during search. As for dimensionality of the problem, since $x \in \mathbf{F}(n)$, the dimensionality is related to n due to node labelings in the factorial search space. For simplicity and without loss of generality, we used the factoradic representation to encode solutions in the factorial search space. As for algorithm parameters, we used the probability of crossover $CR = 0.5$, scaling factor $F = 0.7$, population sizes $POP = 10$, the bias term in RBDE $\rho = 3$, and the termination criterion is $\Omega = 1000$ function evaluations. In DESIM, the similarity is based on the Euclidean distance, the learning parameters F and CR are adapted using JADE [31], and initialization uses Opposition Based Learning [26].

Also, due to the stochastic nature of the above mentioned metaheuristics, 10 independent runs were evaluated for each algorithm and each configuration. Other parameters followed the suggested values of the above-mentioned references. The fine-tuning of optimization parameters is out of the scope of this paper. The key motivations for using the above parameters are as follows: Crossover probability with $CR = 0.5$ implements the equal importance and consideration to historical search directions up to the current number of iterations t. Small population size $NP = 10$ and number of evaluations up to 1000 allow evaluating the (frontier) performance of the gradient-free algorithms under tight computational budgets.

Computational experiments were conducted in Matlab 2018b. Since computing the number g that represents a graph involves the summation of binomial coefficients, we rather use the logarithm of g to efficiently compute the reduction of large binomial coefficients. In this paper, the computation of logarithmic function of the number g is realized through the parallel reductions in a GPU (NVIDIA GeForce GTX TITAN X). The reason of using a GPU is due to the efficiency in the reduction scheme; however, since we are constrained by the GPU hardware, we can only explore the possibility of computing the logarithms within the bound of `realmax` in Matlab (1.8×10^{308}). As such, exploring the compact number-based representations for large graphs is out of the scope of this paper (due to hardware limitations) and is left for future work in our agenda (potentially through a parallel distributed computing environment).

3.1   Results and Discussion

Figure 6 shows the mean convergence performance through distinct graph instances and independent runs. Also, the x-axis of each plot in Fig. 6 shows the number of function evaluations, whereas the y-axis shows the value of the objective function L. By observing Fig. 6, we note the following facts: (1) convergence and stagnation in flat regions often occur after 200 function evaluations with graphs with ten nodes and 20 edges, whereas convergence occurs after 400 functions in other graph instances; (2) the minimal objective values corresponding to each graph instance differ from one to another. It is possible to observe the pattern that a large (small) number of edges implies higher (lower) objective values. This is due to the fact that graphs encode dependencies; as such, more information will be necessary to encode graphs with a larger number of edges, and the integer number representing the graph will be larger due to Eq. 4 depending on m on the summation component.

Among all studied graph instances, the objective value of graph instance 10–11 is lower compared to the objective values of other graph instances. Yet, for all cases, it is possible to attain lower objective values for all cases. Overall cases, the algorithms relying on exploration, such as DERAND, and the algorithm relying on heavy exploration during the early stages of the search, such as DESIM show better convergence performance. This observation is due to the noisy landscape of the problem, such as those depicted in Figs. 3 and 4. Despite the large factorial search space, the convergence figures show that it is possible to find potential areas of low objective values using few function evaluations.

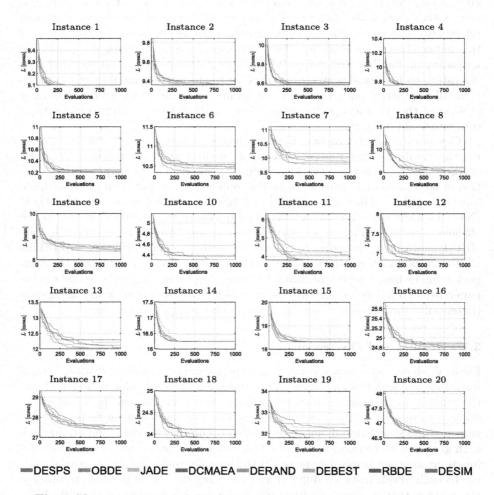

Fig. 6. Mean convergence of the objective function L across independent runs.

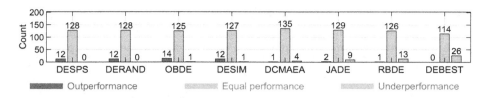

Fig. 7. Number of cases of outperformance, equal performance and underperformance derived from pair-wise Wilcoxon tests at 5% significance level.

The observations above show it is possible to find convergence regions quickly and that exploration strategies are relevant for better performance. To show the comparison of the performance of the algorithms across graph instances, we performed statistical test at 5% significance level (Wilcoxon). Here, our goal is to compute instances in which an algorithm is significantly better(+), equal(=), or worse(-) compared to another algorithm. As such, Fig. 7 shows the overall performance comparison among algorithms based on the statistical significance. Here, the bars in the plot show the number of times an algorithm is significantly better (outperformance), equal (equal performance), or worse (underperformance) compared to other algorithms across all graph instances overall independent runs. For clarity of exposition, the counts (numbers depicted over each bar) are computed by accumulating the number of events in which [+], [=], and [-] occur from the Wilcoxon test. By observing Fig. 7, we note that algorithms with an explicit exploration strategy outperform other algorithms. For instance, DESPS, DERAND, OBDE and DESIM outperform other algorithms in 12–14 problem instances. Also, algorithms with exploitation strategies such as DEBEST and RBDE perform equally in most problem scenarios, yet underperform in 13–26 instances.

The above results show that using gradient-free heuristics can enable the search for compact integer representations of graphs using few evaluations, which may find use in modeling and sampling graphs and networks with utmost efficiency. The future work will aim at studying the tailored optimization algorithms extending adaptation and explorative strategies, and the modular schemes for large graphs [23]. It also remains to be answered whether different labeling strategies than the revolving door [20] are able to achieve smaller values of the objective function (smaller numbers representing graphs). Furthermore, it remains to be answered how to achieve minimal number representation for weighted graphs, directed graphs [21], loopy graphs [22], and other relevant graph architectures. On the application side, we aim at using the insights from this study to extrapolate towards large-scale minimal networks in the plane [19,24], and the further works in network design and optimization.

4 Conclusion

We have studied the feasibility of gradient-free population-based optimization heuristics based on Differential Evolution variants for minimal integer representations of graphs. Our computational experiments using graph instances with varying degrees of sparsity and eight optimization algorithms with distinct forms of adaptation and exploration-exploitation have shown the merit of exploration-oriented strategies to attain better convergence with fewer function evaluations. Investigating the role of adaptation, exploration and modular strategies in large graphs, and evaluating other forms of graph encoding strategies is left for future work. Our results have the potential to elucidate new number-based approaches for graph representation, network design and optimization.

Acknowledgment. This work was supported by JSPS KAKENHI Grant Number 20K11998.

References

1. Aleardi, L.C., Devillers, O., Schaeffer, G.: Optimal succinct representations of planar maps. In: Proceedings of the Twenty-Second Annual Symposium on Computational Geometry, pp. 309–318 (2006)
2. Barbay, J., Castelli Aleardi, L., He, M., Munro, J.I.: Succinct representation of labeled graphs. Algorithmica **62**, 224–257 (2012)
3. Blandford, D.K., Blelloch, G.E., Kash, I.A.: Compact representations of separable graphs. In: Proceedings of the Fourteenth Annual ACM-SIAM Symposium on Discrete Algorithms, pp. 679–688. SODA 2003, Society for Industrial and Applied Mathematics, USA (2003)
4. Blelloch, G.E., Farzan, A.: Succinct representations of separable graphs. In: Amir, A., Parida, L. (eds.) CPM 2010. LNCS, vol. 6129, pp. 138–150. Springer, Heidelberg (2010). https://doi.org/10.1007/978-3-642-13509-5_13
5. Boldi, P., Rosa, M., Santini, M., Vigna, S.: Layered label propagation: a multiresolution coordinate-free ordering for compressing social networks. In: Proceedings of the 20th International Conference on World Wide Web, pp. 587–596 (2011)
6. Brisaboa, N.R., Ladra, S., Navarro, G.: Compact representation of web graphs with extended functionality. Inf. Syst. **39**, 152–174 (2014)
7. Buehrer, G., Chellapilla, K.: A scalable pattern mining approach to web graph compression with communities. In: Proceedings of the 2008 International Conference on Web Search and Data Mining, pp. 95–106 (2008)
8. Claude, F., Navarro, G.: Fast and compact web graph representations. ACM Trans. Web (TWEB) **4**(4), 1–31 (2010)
9. Erdös, P., Evans, A.B.: Representations of graphs and orthogonal Latin square graphs. J. Graph Theor. **13**(5), 593–595 (1989)
10. Farzan, A., Munro, J.I.: Succinct representations of arbitrary graphs. In: Halperin, D., Mehlhorn, K. (eds.) ESA 2008. LNCS, vol. 5193, pp. 393–404. Springer, Heidelberg (2008). https://doi.org/10.1007/978-3-540-87744-8_33
11. Gallian, J.A.: A dynamic survey of graph labeling. Electron. J. Comb. **1**, DS6 (2018)
12. Ghosh, S., Das, S., Roy, S., Minhazul Islam, S., Suganthan, P.: A differential covariance matrix adaptation evolutionary algorithm for real parameter optimization. Inf. Sci. **182**(1), 199–219 (2012), nature-Inspired Collective Intelligence in Theory and Practice

13. Guo, S.M., Yang, C.C., Hsu, P.H., Tsai, J.S.H.: Improving differential evolution with a successful-parent-selecting framework. IEEE Trans. Evol. Comput. **19**(5), 717–730 (2015). https://doi.org/10.1109/TEVC.2014.2375933
14. He, X., Kao, M.Y., Lu, H.I.: Linear-time succinct encodings of planar graphs via canonical orderings. SIAM J. Discret. Math. **12**(3), 317–325 (1999)
15. He, X., Kao, M.Y., Lu, H.I.: A fast general methodology for information-theoretically optimal encodings of graphs. SIAM J. Comput. **30**(3), 838–846 (2000)
16. Jacobson, G.: Space-efficient static trees and graphs. In: 30th Annual Symposium on Foundations of Computer Science, pp. 549–554. IEEE Computer Society (1989)
17. Katebi, H., Sakallah, K.A., Markov, I.L.: Graph symmetry detection and canonical labeling: differences and synergies. In: Voronkov, A. (ed.) Turing-100 - The Alan Turing Centenary, Manchester, UK, June 22–25, 2012. EPiC Series in Computing, vol. 10, pp. 181–195. EasyChair (2012). https://doi.org/10.29007/gzc1
18. Kreher, D.L., Stinson, D.R.: Combinatorial algorithms: generation, enumeration, and search. ACM SIGACT News **30**(1), 33–35 (1999)
19. Parque, V.: On hybrid heuristics for Steiner trees on the plane with obstacles. In: Zarges, C., Verel, S. (eds.) Evolutionary Computation in Combinatorial Optimization – 21st European Conference, EvoCOP 2021, Held as Part of EvoStar 2021, Virtual Event, April 7–9, 2021, Proceedings. Lecture Notes in Computer Science, vol. 12692, pp. 120–135. Springer, Cham (2021). https://doi.org/10.1007/978-3-030-72904-2_8
20. Parque, V., Kobayashi, M., Higashi, M.: Bijections for the numeric representation of labeled graphs. In: 2014 IEEE International Conference on Systems, Man, and Cybernetics, SMC 2014, San Diego, CA, USA, October 5–8, 2014, pp. 447–452. IEEE (2014)
21. Parque, V., Miyashita, T.: On succinct representation of directed graphs. In: 2017 IEEE International Conference on Big Data and Smart Computing, BigComp 2017, Jeju Island, South Korea, February 13–16, 2017, pp. 199–205. IEEE (2017)
22. Parque, V., Miyashita, T.: On the numerical representation of labeled graphs with self-loops. In: 29th IEEE International Conference on Tools with Artificial Intelligence, ICTAI 2017, Boston, MA, USA, November 6–8, 2017, pp. 342–349. IEEE Computer Society (2017). https://doi.org/10.1109/ICTAI.2017.00061
23. Parque, V., Miyashita, T.: Numerical representation of modular graphs. In: 2018 IEEE 42nd Annual Computer Software and Applications Conference, COMPSAC 2018, Tokyo, Japan, 23–27 July 2018, vol. 1, pp. 819–820. IEEE Computer Society (2018). https://doi.org/10.1109/COMPSAC.2018.00136
24. Parque, V., Miyashita, T.: Obstacle-avoiding Euclidean Steiner trees by n-star bundles. In: IEEE 30th International Conference on Tools with Artificial Intelligence, ICTAI 2018, 5–7 November 2018, Volos, Greece, pp. 315–319. IEEE (2018). https://doi.org/10.1109/ICTAI.2018.00057
25. Parque, V., Miyashita, T.: On graph representation with smallest numerical encoding. In: 2018 IEEE 42nd Annual Computer Software and Applications Conference, COMPSAC 2018, Tokyo, Japan, 23–27 July 2018, vol. 1, pp. 817–818. IEEE Computer Society (2018). https://doi.org/10.1109/COMPSAC.2018.00135
26. Rahnamayan, S., Tizhoosh, H.R., Salama, M.M.: Opposition-based differential evolution. IEEE Trans. Evol. Comput. **12**(1), 64–79 (2008)
27. Segredo, E., Lalla-Ruiz, E., Hart, E.: A novel similarity-based mutant vector generation strategy for differential evolution. In: Proceedings of the Genetic and Evolutionary Computation Conference, pp. 881–888 (2018)
28. Storn, R., Price, K.: Differential evolution-a simple and efficient heuristic for global optimization over continuous spaces. J. Global Optim. **11**(4), 341 (1997)

29. Sutton, A.M., Lunacek, M., Whitley, L.D.: Differential evolution and non-separability: using selective pressure to focus search. In: Proceedings of the 9th Annual Conference on Genetic and Evolutionary Computation, pp. 1428–1435 (2007)
30. Turán, G.: On the succinct representation of graphs. Discret. Appl. Math. **8**(3), 289–294 (1984)
31. Zhang, J., Sanderson, A.C.: Jade: adaptive differential evolution with optional external archive. IEEE Trans. Evol. Comput. **13**(5), 945–958 (2009)

Multi-scale Multi-step Dependency Graph Neural Network for Multivariate Time-Series Forecasting

Wenchang Zhang[1], Kaiqiang Zhang[2], Linling Jiang[1], and Fan Zhang[1,3]($\boxtimes$) (iD)

[1] School of Computer Science and Technology, Shandong Technology and Business University, Yantai 264005, China
[2] Shandong Qilu Big Data Research Institute, Jinan 250101, China
[3] Shandong Future Intelligent Financial Engineering Laboratory, Yantai 264005, China
zhangfan@sdtbu.edu.cn

Abstract. This study addressed the limitations of existing graph neural network methods in time-series prediction, specifically the inability to establish strong dependencies between variables and the weak correlation in time-series across different time scales. To overcome these challenges, we proposed a graph neural network-based multi-scale multi-step dependency (GMSSD) model. To capture temporal dependencies in time-series data, we first designed a temporal convolution module that learns multi-scale representations between sequences. We extracted features at multiple scales using dilated convolutions and a gated linear unit (GLU) while controlling the information flow, thereby capturing temporal dependencies in time-series data. Furthermore, we employed a gated recurrent unit (GRU) and fully connected layers to derive the graph structure and capture the complex relationships between variables in the sequence data. In particular, existing graph neural network methods have a strong dependence on graph structures and are unable to adapt to complex and dynamic graph structures. They also have limitations in capturing long-range dependency relationships within the graph. Therefore, a graph convolution module is designed to explore the current node information and its neighbor information. It has the capability to integrate information contributions from different time steps, effectively capturing the spatial dependencies among nodes. The experimental results show that the proposed model outperformed existing methods in both single-step and multi-step prediction tasks. This study provided a novel approach for time-series forecasting and achieved significant improvements.

Keywords: Graph Neural Network · Multi-Step Dependency · Multi-Scale · Graph Structure Learning · Multivariate Time-Series Forecasting

1 Introduction

Multivariate time-series forecasting is an extremely important task that has been widely applied in various domains such as finance [9,23], weather [1,12], and traffic [7,19]. For example, in traffic prediction, future traffic flow can be forecasted by analyzing factors such as historical data, the surrounding environment, and upcoming holidays. With the continuous increase in traffic-related datasets and in their heterogeneity, multivariate time-series forecasting plays an increasingly significant role in improving traffic prediction accuracy.

In early statistics, autoregressive models were commonly used for prediction because of their favorable mathematical properties and efficiency [25]. However, the use of such methods requires several assumptions, such as the requirement for the predicted sequence to be stationary. Moreover, in fields such as solar and wind energy, these methods are primarily suitable for univariate forecasting problems and may not yield significant results. Owing to the complex correlations between multiple variables and the dynamic spatial and temporal dependencies between them, accurately capturing these relationships is challenging.

Recently, deep learning techniques have gained widespread application in time-series forecasting. Convolutional neural networks (CNNs), recurrent neural networks (RNNs) and Transformer-based methods [16,22] are commonly used deep learning approaches to capture dependencies in time-series data. CNNs can learn important features from sequential data for predictions. M-TCN [15], through the connection of multiple TCN layers and the use of innovative multichannel and asymmetric residual block networks, constructs sequence-to-sequence frameworks to handle non-periodic data. However, CNNs neglect global sequence features when dealing with long sequences, limiting their ability to capture overall sequence characteristics. Although, certain RNN-based variants [2,10] can capture dependencies between multiple variables in long sequences, they only consider the relationships between current time step information and disregard the influence of historical inputs. In addition, they require multiple iterations, which may lead to errors, explosions, or vanishing gradients. Therefore, solely relying on RNNs to extract historical information from long multivariate sequences is insufficient.

To overcome these limitations, researchers have introduced Transformer models into time-series forecasting, achieving significant progress. Transformer models utilize self-attention mechanisms to capture the complex dependencies between sequences. TPA-LSTM [14] integrates convolutional neural networks, recurrent neural networks, and attention mechanisms to address the issue of correlations among multivariate sequences. Some studies [8,11,24] achieved promising results using Transformer models to predict multiple time-series by taking the observed values of multiple time-series as inputs. However, despite the advantages of Transformers in time-series forecasting, there are also some drawbacks that need to be considered. Transformers still face challenges in modeling long-term dependencies, which can result in inefficient information propagation or gradient vanishing issues, particularly when the time span of the time-series is large.

Recently, because of the rapid development of graph neural networks (GNNs), they have been widely applied in various fields and have achieved remarkable results in time-series forecasting. StemGNNs [3] combine graph Fourier transform (GFT) and discrete Fourier transform (DFT) to obtain spectrograms with clear patterns. By leveraging convolutional and sequential learning modules, StemGNNs achieve effective predictions. Another approach, MRA-BGCN [4], has been proposed for predicting data with continuously growing trends. It uses a novel GNN structure consisting of a time extractor and graph convolutional filters to capture both temporal and spatial correlations. ESG [20] employs a recursive approach to learn the interrelationships between graph nodes and utilizes graph convolution to capture spatial dependencies. GNNs can handle nonlinear relationships and complex dependencies, thereby overcoming the limitations of traditional models. GNNs can accurately model nonlinear features in data by capturing the complex relationships between nodes, thereby improving the accuracy and stability of predictions. However, there are still some limitations that need to be addressed:

1. Information features vary across different observation scales. The correlations between variables in the short term may differ from those in the long term. For example, in the field of weather forecasting, in the short term, the weather in the next few days may be consistently rainy due to factors such as typhoons. However, in the long term, the presence of a typhoon does not necessarily change the local seasonal climate.
2. The relationships between sequences change over time, making the structure of the graph complex and dynamic. Existing research mostly assumes a fixed and static graph structure, which clearly cannot handle such complex scenarios.
3. The dynamic nature of the graph structure affects the exploration of spatial dependencies between graph nodes and their neighbors in the graph convolutional module. This dependence on the graph structure poses a challenge for graph convolution methods. Moreover, graph convolutional layers typically only consider the information from neighboring nodes and fail to capture long-range dependencies in the graph.

To address the aforementioned challenges, we propose a novel approach called graph neural network-based multi-scale multi-step dependency model that combines GNNs, CNNs, RNNs, and attention mechanisms. In the GMSSD model, we leverage various techniques to effectively capture the dependencies in multi-variable time-series. Firstly, we employ dilated convolutions with different kernel sizes to extract features at multiple scales, capturing the correlations within the multi-variable time-series. Additionally, we integrate GNNs by utilizing the GRU to learn the dependencies between nodes, continuously updating the node features, and constructing a dynamic graph structure. This dynamic graph structure allows for a more effective representation of the temporal dependencies compared to fixed graph structures. Furthermore, we introduce a multi-layered architecture to reduce the impact of noise. Additionally, we incorporate attention mechanisms to better utilize the contributions from previous time steps. By combining the strengths of GNNs, CNNs, RNNs, and attention mechanisms, our proposed model is particularly well-suited for handling long sequential data.

The contributions of this study are as follows:

1. We investigated how the proposed model leverages important historical information for multivariable time-series prediction. By employing attention mechanisms, the model automatically learns weight allocation, allowing more attention to be focused on crucial historical time steps. This reduces reliance on irrelevant time steps. Furthermore, by integrating attention mechanisms with graph convolution, we designed the graph convolution module and multi-step dependency propagation layer, enabling the fusion of node information and neighbor information based on different temporal contexts.
2. We designed a time convolution module that divides the time-series data into multiple scales. Dilated convolutions are utilized to extract features at different scales. Additionally, we introduced the GLU to control the amount of information propagation, facilitating the learning of temporal dependencies within the time-series data.
3. A graph structure learner was developed to improve the representation of the graph's structural information. By utilizing the GRU and fully connected layers, we learned the adjacency matrix between nodes more effectively. Unlike existing methods that often employ fixed graph structures, our approach accurately captures the complex and dynamic nature of time-series data.
4. We extensively evaluated our proposed model on multiple well-known datasets and compared it with state-of-the-art methods. The experimental results demonstrated that our model, GMSSD, achieved significant performance improvements across various datasets, demonstrating its superior predictive and modeling capabilities. Additionally, ablation experiments were conducted to validate the effectiveness of the model structure and assess the contributions of each component.

2   Preliminary

2.1   Definition of Graph Neural Network

The relationships between variables can be described using a graph $G = (V, E, A)$, where V represents the set of nodes in the entire graph, E represents the set of edges connecting the nodes, and A represents the adjacency matrix of the graph.

2.2   Definition of the Problem

$x_t = \left\{ x^{1,t}, x^{2,t}, x^{3,t}, \cdots, x^{N,t} \right\} \in R^{N \times C}$, where x_t represents the value of a variable at time step t, where C represents the feature dimension of the variable. The prediction problem can be divided into two steps: single and multi-step prediction. In single-step prediction, the prediction $Y(t + \theta)$ is obtained from historical sequence with a backtracking window τ and prediction horizon of θ.

In multi-step prediction, predictions for multiple horizons are made, resulting in $Y(t+1:t+\theta)$, as shown in Eq. (1).

$$[X_{t-\tau+1}, \cdots, X_t] \xrightarrow{\Psi(\cdot)} Y_{t+\theta}$$
$$[X_{t-\tau+1}, \cdots, X_t] \xrightarrow{\Psi(\cdot)} [Y_{t+1}, \cdots, Y_{t+\theta}] \tag{1}$$

where $[X_{t-\tau+1}, \cdots, X_t]$ represents a historical sequence of features, and $\Psi(\cdot)$ is a mapping function that enables the prediction of future-step features based on the historical sequence.

3 Methodology

3.1 Overview

This study presents the GMSSD model, which is composed of multiple components: temporal convolutional modules, graph structure learners, graph convolutional modules, skip-connection layers, and an output module. To extract the features of long sequential redundant information, the time convolution module employs dilated convolutions with filters of different sizes and utilizes gating units to control the flow of information. The graph structure learner generates a graph adjacency matrix based on the connections between nodes, which is then combined with the hidden state from the previous time step and fed into the graph convolution layer, emphasizing both the intrinsic features of the node and the contributions from different time steps. The graph and time convolution layers capture spatial and temporal dependencies, respectively. To mitigate overfitting, residual connections were introduced after the time convolution module, where a residual is added to the output of the graph convolution layer at each iteration. Finally, skip connection layer transfers information from each iteration to the output module. The overall framework, shown in Fig. 1.

3.2 Temporal Convolution Module

The temporal convolution module initially consists of two dilated convolutions. To better capture long-term sequential features and obtain temporal correlations between sequences, the dilated convolutions employ filters with varying kernel sizes. During the embedding training of the i-th time step, it is crucial to consider the input sequence at the current time step. To address this objective, we introduced a dilated causal convolution and utilized a GLU to regulate the amount of information in the output. Furthermore, as the depth increased, we progressively expanded the receptive field for feature extraction to achieve more comprehensive information capture, model long-term dependencies, and enhance the sensitivity to large-scale features. These enhancements lead to an improved performance and generalization ability of the model, as described in Eq. (2):

$$w^{(i)} = concat\left(w^{(i)} \otimes f_{1 \times k_1}(t), \cdots, w^{(i)} \otimes f_{1 \times k_4}(t)\right) \tag{2}$$

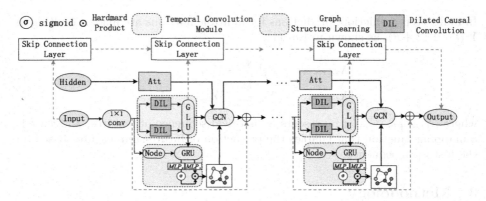

Fig. 1. GMSSD model.

The $[k_1, k_2, k_3, k_4]$ represent convolutional kernels of various sizes. This module comprises two dilated convolutional layers with their outputs fed into a GLU to regulate the amount of information transmission. The outputs are then multiplied element-wise to obtain information feature $w^{(i)}$ at the i-th time step.

Because the different filters may have outputs of varying lengths, the outputs are truncated to the length of the shortest filter to ensure compatibility. Subsequently, they are concatenated along the time dimension. The concatenated output is then passed through a GLU to control the amount of information in the output. The GLU consists of an input gate and sigmoid activation function. Given the two outputs $P^{(i)}$ and $Q^{(i)}$ of the dilated convolutional layer at the i-th time step, the GLU can be represented by Eq. (3).

$$GLU = P^{(i)} \odot \sigma\left(Q^{(i)}\right) \tag{3}$$

where σ is the sigmoid function, $\odot$ represents the Hadamard product. $P^{(i)}$ and $Q^{(i)}$ are fed into the GLU, resulting in an output, which is then passed into both the graph convolutional module and the graph structure learning module.

3.3   Graph Structure Learning

Several practical applications in real-life scenarios involve complex networks of relationships that evolve over time. We designed a graph structure learner to capture the dynamic relationships between nodes and enhance the representation of attributes by leveraging information from neighboring nodes. This learner enables learning of the adjacency matrix of the graph based on the correlations between nodes. In addition, the individual features of the nodes are crucial for learning the node correlations. Therefore, we designed a node extraction function to extract static features from the input information, as described in Eq. (4) below:

$$\delta_i = F_a\left(x^*\right) \tag{4}$$

δ_i represents the extracted static-node features. F_a is a node extraction function consisting of multiple one-dimensional convolutions, Batch Normalization, and an activation function. $x^{(i)}$ denotes the input data for the i-th time step. To leverage the correlations between nodes and form the adjacency matrix of the graph, we employed a GRU. The input X sequence is divided along the time dimension into $[X_1, X_2, ..., X_t]$ and fed to the GRU. The initial hidden layer H_0 is denoted by δ_i. The update process of GRU is depicted in Eq. (5):

$$
\begin{aligned}
u_t &= sigmoid\left([X_t, H_{t-1}] + b_u\right) \\
r_t &= sigmoid\left([X_t, H_{t-1}] + b_r\right) \\
C_t &= tanh\left([X_t, (r_t \odot H_{t-1})] + b_c\right) \\
H_t &= u_t \odot H_{t-1} + (1 - u_t) \odot C_t \\
a_t &= \beta\left(H_t\right)
\end{aligned} \tag{5}
$$

where u, r, and C represent the update gate, reset gate, and candidate state, respectively. The function β represents the Rectified Linear Unit (ReLU), and H_t corresponds to the hidden layer. We hypothesize that in a graph, the information from the previous node can influence the information of the succeeding node, which does not have a reciprocal effect on the previous node. For instance, considering the relationship between the stock prices of a parent company and its subsidiaries, changes in the parent company's stock prices can impact those of its subsidiaries, but changes in subsidiaries' stock prices can barely affect those of the parent company. Therefore, we employed a ReLU activation function to facilitate one-way information propagation.

We represent node features using two nodes a_i and a_j, and connect them in series to derive the graph structure using fully connected layers. To control the amount of information output, we utilize a mask. As shown in Eq. (6):

$$
\begin{aligned}
A_{ij} &= MLP_a\left(a_i, a_j\right) \\
Mask_{ij} &= MLP_m\left(a_i, a_j\right) \\
A &= A_{ij} \odot \sigma\left(Mask_{ij}\right)
\end{aligned} \tag{6}
$$

The function σ represents the sigmoid function, A represents the final derived graph structure, and $\odot$ is Hadamard product.

3.4 Graph Convolution Module

We propose a graph convolution operation to address the spatial dependence issue in graph data and fuse node information with neighbor information. Our graph convolution module consists of multi-step dependency propagation layers, where each layer includes information propagation and information selection.

The graph convolution layer heavily relies on the structure of the graph and the adjacency matrix. The adjacency matrix defines the connection relationships between nodes in the graph and determines how information is propagated between nodes in the graph convolution layer. Consequently, any perturbation or change to the graph structure directly affects the behavior of the model. For instance, adding or removing nodes or edges results in a change in the adjacency matrix, which in turn alters the way the model processes the input data. Such changes may cause the model to struggle in adapting to the new graph structure, ultimately reducing its robustness and generalization ability. Additionally, despite designing a dynamic graph structure where the graph evolves over time, the graph convolution model may require frequent updates to the adjacency matrix in order to adapt to the graph's evolution, leading to increased complexity and computational cost.

To overcome these issues, we employ a layer-by-layer transfer strategy that proportionally propagates the relationship between the original node inputs and their spatial neighbors, reducing the dependence on the graph structure. This approach makes the graph convolution layer less sensitive to noise present in the input adjacency matrix. Traditional graph convolution layers typically consider only information from neighboring nodes, lacking direct access to global information. This limitation in utilizing only local information may prevent the model from capturing long-range dependencies or global patterns within the graph. To address this problem, we introduce an attention mechanism that allows the model to focus on useful node information and assign different weights to different neighboring nodes during the graph convolution process. By doing so, the model effectively captures the importance of node information at different time steps for a specific node representation, achieving spatial dependencies between propagating nodes that maintain both local and long-range dependencies. Figure 2 shows the structure of the graph convolution and the multi-step dependency propagation layer.

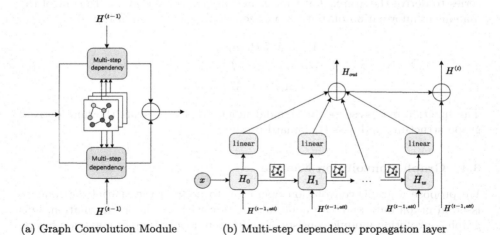

(a) Graph Convolution Module   (b) Multi-step dependency propagation layer

Fig. 2. Graph convolution module and multi-step dependency propagation layer.

During the information propagation step, we recursively propagate node information, graph structure, and information from the previous time step. The Information propagation can be described by Eq. (7):

$$H^{(t-1,att)} = F_{att}\left(H^{(t-1)}\right)$$
$$H_w^{(t)} = \alpha x^{(t)} + (1 - \alpha)\, AH_{w-1}^{(t)} + H^{(t-1,att)}$$

$$(7)$$

where $x^{(t)}$ denotes the input to the graph convolution, A represents a series of graph adjacency matrices generated by a graph structure learning module, $H_w^{(t)}$ denotes the output of the w-th step at time step t, $H^{(t-1)}$ corresponds to the node information at time step t-1, $H^{(t-1,att)}$ is the contribution of the node information at time step t-1, α is used to control the output between different messages. We utilize F_{att} to control the proportion of contributions from previous time steps and aggregate the contributions from each previous time step into Eq. (8),

$$F_{att} = \frac{exp\left(H^{(t-1)} + q^{(l)}\right)}{\sum_{l=1}^{t} exp\left(H^{(t-1)} + q^{(l)}\right)}$$

$$(8)$$

We define a learnable parameter matrix, denoted as $q^{(l)}$, to represent the node information from the previous t time steps and its corresponding dependencies at the current time step. To enhance the representational capacity of each layer and eliminate irrelevant information, an information selection step is introduced. The information selection can be described by Eq. (9).

$$H_{out}^{(t)} = \sum_{w=0}^{w} H_w^{(t)} W_w^{(t)}$$
$$H^{(t)} = concat\left(H_{out}^{(t)}, H^{(t-1,att)}\right)$$

$$(9)$$

where w represents the propagation depth, and the parameter matrix W_w acts as an information selector that chooses relevant information and emphasizes critical details. $H_{out}^{(t)}$ represents the output of the final Multi-step Dependency Propagation layer, and $H^{(t)}$ denotes the node information at time step t. Here, we concatenated $H_{out}^{(t)}$ and $H^{(t-1,att)}$ along the time dimension to obtain $H^{(t)}$, which is then fed into the next time step's graph convolutional Module for further learning.

3.5   Skip Connection Layer and Output Module

The overall framework involves multiple iterations of sequence inputs into the model, and each iteration passes through a skip connection layer. In the experiment, we configure the skip connection layers to consist of four layers. The skip connection layer comprises a 2D convolutional layer that allows forwarding information to the final output layer. In the final output layer, the outputs from each skip connection layer and from the last iteration of the overall framework get concatenated along the time dimension. The final output layer consists of two 1×1 standard convolutional layers and two ReLU functions, which convert the channel dimension to the desired output dimension. For single-step prediction, we aim to predict the future at a specific time step, requiring a one-dimensional output. In multi-step prediction, where the goal is to predict future V time steps, the necessary output dimension becomes V.

4   Experiments

We rigorously validated the accuracy of the proposed model using both single and multi-step predictions.

4.1   Experimental Setup

To evaluate the performance of the proposed GMSSD model and demonstrate its generalizability, we conducted experimental validation using six popular datasets.

Datasets and Evaluation Metrics. In this study, we utilized multiple popular datasets to evaluate the performance of both single and multi-step prediction tasks. For the single-step prediction task, the following four datasets were selected for evaluation:

- **Electricity**: This dataset [6] recorded the hourly power consumption of 321 customers from 2012 to 2014. We used this dataset to analyze and predict single-step variations in power consumption.
- **Solar Energy**: This dataset [6] contained solar energy generation collected every 10 min. We utilized this dataset for single-step prediction to investigate the trends in solar energy generation.
- **Exchange Rate**: This dataset [6] contained exchange rate data from 1990 to 2016 with daily time interval. We employed this dataset for single-step predictions of fluctuations in currency exchange rates.
- **Wind**: This dataset [17] included hourly estimates of the wind power potential in a specific region with a daily time interval. We used this dataset for single-step predictions of variations in wind power.

Additionally, for the multi-step prediction task, we selected two publicly available datasets from New York City.

- **NYC-Taxi**: This dataset [21] contained information on taxi trips in New York City, with a time interval of 30 min. We utilized this dataset for multi-step prediction of trends in taxi demand.
- **NYC-Bike**: This dataset [21] included records of bike trips in New York City with a time interval of 30 min. We used this dataset for multi-step predictions of patterns of bike usage.

For both the single-step and multi-step prediction tasks, we employed the following evaluation metrics to assess the performance of the model:

Relative squared error (RSE): RSE measures the discrepancy between predicted values and actual observations, with lower values indicating better model performance.

Empirical correlation coefficient (CORR): CORR quantifies the correlation between predicted values and actual observations, with values closer to 1 indicating better model performance.

For the multi-step prediction task, we introduced the following additional evaluation metrics:

Mean absolute error (MAE): MAE describes the difference between predicted and true values, with lower values indicating better model performance.

Their definitions are as follows:

$$RMSE = \sqrt{\sum_{t=0}^{\beta} \left(\hat{y}^{(t)} - y^{(t)} \right)^2} \tag{10}$$

$$CORR = \frac{1}{\theta} \sum_{n=1}^{\theta} \frac{\sum_{t=0}^{\beta} \left(\hat{y}_n^{(t)} - \bar{\hat{y}}_n \right) \left(y_n^{(t)} - \bar{y}_n \right)}{\sqrt{\sum_{t=0}^{\beta} \left(\hat{y}_n^{(t)} - \bar{\hat{y}}_n \right)^2 \left(y_n^{(t)} - \bar{y}_n \right)^2}} \tag{11}$$

$$MAE = \sum_{t=0}^{\beta} \left| y^{(t)} - \hat{y}^{(t)} \right| \tag{12}$$

Comparison Models. In this study, we compared the proposed model with several existing multivariate prediction models commonly used in the field. The following models were selected for comparison:

- **LSTNet**: LSTNet [6] utilizes a combination of CNN to extract short-term local dependency patterns between variables and RNN to capture long-term patterns in time-series trends.
- **TPA-LSTM**: TPA-LSTM incorporates a temporal pattern attention (TPA) mechanism to operate on the hidden layer outputs of an LSTM model.
- **MTGNN**: MTGNN [18] introduces graph learning layers, graph convolutional modules, and time learners to capture the relationships between variables.
- **StemGNN**: StemGNN employs Fourier Transform to capture temporal dependencies in the data.
- **STID**: STID [13] combines spatial information with two types of time information to propose a simple and efficient model.
- **MTGODE**: MTGODE [5] proposes a continuous graph propagation mechanism and a graph structure learning module to capture long-term spatial correlations among time-series.
- **ESG**: ESG improves the MTGNN by incorporating an RNN to enhance the graph learning layers.

Hyperparameters. After conducting multiple experiments, we determined the optimal parameters as follows. For the previous K time steps, we selected a value from a range of [32, 64]. The time dimension was set to 181 and the feature dimension was determined based on the number of features in the dataset. During the training process, we used the Adam optimizer with a learning rate range of [0.001, 0.0001], and the batch size was selected from a range of [8, 16]. In addition, we implemented early stopping with a checkpoint frequency of 30 epochs, If the validation loss converges within 30 epochs, the optimizer will stop early. Otherwise, it will stop after 300 epochs.

4.2   Main Results

In the single-step prediction experiments, we compared the performance of different methods listed in Table 1. The bold lines indicates the best performance. The GMSSD model outperformed the other baseline methods on most evaluation metrics. Whereas the ESG model utilizes an RNN to construct a self-learning adjacency matrix to capture the correlations between sequences and achieve satisfactory results, the GMSSD model demonstrated significant improvements in utilizing historical node information. In particularly, in terms of the relative root RSE at prediction intervals of 6 and 12 for the solar energy dataset, the GMSSD model exhibited improvements of 2.8% and 3.9%, respectively, compared to state-of-the-art methods. Furthermore, for the power dataset at a prediction interval of 6, the GMSSD model exhibited a 3.0% improvement in RSE compared to state-of-the-art methods.

Table 1. Comparative experiment of single-step prediction.

Dataset	Metrics	Electricity				Solar-Energy				Exchange Rate				Wind			
		3	6	12	24	3	6	12	24	3	6	12	24	3	6	12	24
LSTNet	RSE	0.0864	0.0931	0.1007	0.1007	0.1843	0.2559	0.3254	0.4643	0.0226	0.0280	0.0356	0.0449	0.6079	0.6262	0.6279	0.6257
	CORR	0.9283	0.9135	0.9077	0.9119	0.9843	0.9690	0.9467	0.8870	0.9735	0.9658	0.9511	0.9354	0.7436	0.7275	0.7249	0.7284
TPA-LSTM	RSE	0.0823	0.0916	0.0964	0.1006	0.1803	0.2347	0.3234	0.4389	0.0174	**0.0241**	0.0341	0.0444	0.6093	0.6292	0.6290	0.6335
	CORR	0.9439	0.9337	0.9250	0.9133	0.985	0.9742	0.9487	0.9081	0.9790	0.9709	0.9564	0.9381	0.7433	0.7240	0.7235	0.7202
MTGNN	RSE	0.0745	0.0878	0.0916	**0.0953**	0.1778	0.2348	0.3109	0.4270	0.0194	0.0259	0.0349	0.0456	0.6204	0.6346	0.6363	0.6426
	CORR	0.9474	0.9316	0.9278	0.9234	0.9852	0.9726	0.9509	0.9031	0.9786	0.9708	0.9551	0.9372	0.7414	0.7164	0.7134	
StemGNN	RSE	0.0799	0.0909	0.0989	0.1019	0.1839	0.2612	0.3564	0.4768	0.0506	0.0674	0.0676	0.0685	0.6197	0.6358	0.6243	0.6379
	CORR	0.9490	**0.9397**	**0.9342**	0.9209	0.9841	0.9679	0.9395	0.8740	0.8871	0.8703	0.8499	0.8738	0.7282	0.7202	0.7228	0.7130
STID	RSE	0.0854	0.1237	0.1016	0.1040	0.2224	0.3196	0.4492	0.4257	0.0205	0.0256	0.0346	0.0433	0.6314	0.6535	0.6645	0.6832
	CORR	0.9331	0.9116	0.9014	0.9028	0.9763	0.9472	0.8895	0.9029	0.9765	0.9699	0.9553	0.9386	0.7203	0.7008	0.6904	0.6742
MTGODE	RSE	0.0810	0.1028	0.0933	0.0970	0.1816	0.2416	0.3347	0.4320	0.0204	0.0289	0.0351	0.0530	0.6283	0.6350	0.6481	0.6647
	CORR	0.9294	0.9109	0.9060	0.9080	0.9847	0.9713	0.9440	0.9018	0.9758	0.9687	0.9548	0.9316	0.7004	0.6974	0.6920	
ESG	RSE	0.0718	0.0844	0.0898	0.0962	0.1708	0.2278	0.3073	0.4101	0.0181	0.0246	0.0345	0.0468	0.6118	0.6250	0.6272	0.6298
	CORR	**0.9494**	0.9372	0.9321	**0.9279**	0.9865	0.9743	0.9519	0.9100	**0.9792**	**0.9717**	**0.9564**	**0.9392**	0.7417	0.7281	0.7258	0.7225
GMSSD	RSE	**0.0713**	**0.0818**	**0.0893**	0.0957	**0.1653**	**0.2214**	**0.2953**	**0.4010**	**0.0172**	0.0242	**0.0334**	**0.0431**	0.6084	0.6217	0.6245	0.6286
	CORR	0.9470	0.9366	0.9302	0.9226	**0.9874**	**0.9760**	**0.9565**	**0.9136**	0.9766	0.9705	0.9551	0.9355	**0.7435**	0.7295	**0.7258**	0.7235

In the multi-step prediction experiments, we compared several different spatiotemporal prediction baseline methods, as shown in Table 2, where the bold lines indicate the best performance. The GMSSD model achieved near-optimal performance within the prediction horizons of 3, 6 and 12, and the average over all levels, validating the effectiveness of the our proposed method. The

STID model encodes spatial information and two types of temporal information, resulting in faster predictions for multivariate time-series data. The MTGNN, CCRNN, and GTS models address the issue of predefined graph structures for spatiotemporal sequence prediction, exhibiting significant performance improvements compared to previous methods. The MTGODE model uses dynamic graph neural ordinary differential equations to overcome the limitations of model complexity and static graph structures. The ESG model uses evolving graph structure, which captures the correlations between sequences better than static graph structures. However, these models have fallen short in effectively capturing historical information and have not adequately emphasized the importance of historical information. As a result, their overall performance is inferior to ours.

Table 2. Comparison test of multi-step prediction.

Dataset	Metrics	Horizon 3			Horizon 6			Horizon 12			All		
		RMSE	MAE	CORR	RMSE	MAE	CORR	RMSE	MAE	CORR	RMSE	MAE	CORR
NYC-Bike	MTGNN	2.5962	1.5668	0.7626	2.7588	1.6525	0.7447	3.3068	1.7892	0.6931	2.7791	1.6595	0.7353
	CCRNN	2.6538	1.6565	0.7534	2.7561	1.7061	0.7411	2.9436	1.8040	0.7029	2.7674	1.7133	0.7333
	GTS	2.7628	1.7159	0.7248	2.9287	1.7769	0.7007	3.1649	1.8905	0.6622	2.9258	1.7798	0.6985
	STID	3.9492	2.2161	0.4939	4.5093	2.5615	0.3252	4.8765	2.8146	0.2692	4.3866	2.4901	0.3796
	ESG	2.5529	1.5483	**0.7638**	2.6484	1.6026	0.7511	2.8778	1.7173	0.7152	2.6727	1.6129	0.7449
	GMSSD	**2.5330**	**1.5454**	0.7636	**2.6327**	**1.5939**	**0.7512**	**2.8553**	**1.6961**	**0.7168**	2.6900	**1.6085**	**0.745**
NYC-Taxi	MTGNN	10.3394	5.6775	0.8374	10.7534	5.8168	0.8312	12.5164	6.5285	0.7972	10.9472	5.9192	0.8249
	CCRNN	9.3033	5.4586	0.8529	9.7794	5.6362	0.8438	10.9585	6.1416	0.8186	9.8744	5.6636	0.8416
	GTS	10.7796	6.2337	0.7974	13.0215	7.3251	0.7299	14.9906	8.5328	0.6524	12.7511	7.2095	0.7348
	STID	15.8537	8.1636	0.705	22.006	11.2516	0.4946	27.0769	15.0924	0.1306	21.6494	11.1606	0.4901
	ESG	8.5745	4.8750	0.8656	9.0125	5.0500	0.8592	9.7857	5.4019	0.8450	**8.9759**	5.0344	0.8592
	GMSSD	**8.3593**	**4.7823**	**0.8689**	**8.8783**	**4.9880**	**0.8599**	**9.7014**	**5.3146**	**0.8475**	8.8885	**4.9653**	**0.8607**

4.3 Ablation Study

We conducted an ablation study on the NYC-taxi dataset to validate the effectiveness of key components. The variants of the GMSSD model were created and named as follows:

- **GMSSD w/o GRU**: This variant excludes the GRU in the graph-learning module and extracts static features to capture correlations among sequences.
- **GMSSD w/ GRU**: In this variant, the GRU is used in the graph-learning module to learn the correlations among sequences.
- **GMSSD w/o attention**: The attention mechanism is removed from the graph convolution module.
- **GMSSD w/o scale**: In this variant, Only a series of graph structures are generated from the raw input data, which are later used for graph convolution, without considering the variations of correlations at different time scales.
- **GMSSD w/o dilated Convolution**: The dilated convolution is replaced with one-dimensional convolution to extract multi-scale features.
- **GMSSD w/o residual Connection**: The residual connection is removed in this variant.

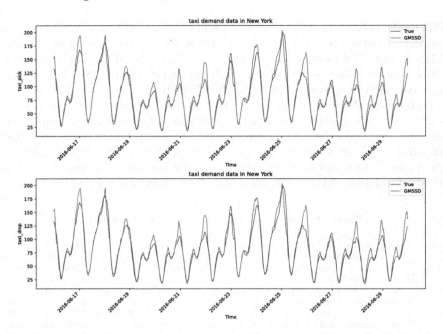

Fig. 3. Presents the visualization of the predicted taxi demand data in New York.

To conduct the ablation experiments, we repeated each experiment three times; the results of different models are listed in Table 3. We denote GMSSD w/ GRU as GMSSD. Table 3 show that the GMSSD model achieved the best performance. The introduction of an attention mechanism improves the effectiveness of the model in capturing historical information. When the GRU was removed from the graph learning module, GMSSD outperformed GMSSD w/o GRU in the multistep prediction dataset. Despite the slight performance drop after removing the GRU, good performance was still achieved, indicating that the graph-learning module continues to play a positive role.

When the multi-scale component was removed from the temporal convolution module, GMSSD significantly outperformed GMSSD w/o GRU, suggesting the crucial role of multi-scale information features in capturing correlations. After removing the residual connection and dilated convolution, there was a slight decrease in the performance, indicating their relatively minor roles in the model.

These experimental results validated the significance of each module in the proposed model. They are an inseparable collections, and each module plays a critical role in the overall performance.

4.4   Visual Experiment

In this study, we presented a GMSSD model and validated its performance using the NYC-taxi dataset. Figure 3 show the actual demand for taxi pick-ups and

Table 3. Presents the ablation experiments conducted on the NYC-taxi dataset.

Method	MAE	RMSE	CORR
GMSSD w/o GRU	5.1483 ± 0.0096	9.1260 ± 0.0270	0.8555 ± 0.0019
GMSSD w/o Attention	5.1306 ± 0.0419	9.1534 ± 0.0787	0.8543 ± 0.0015
GMSSD w/o Scale	5.5072 ± 0.1047	9.8494 ± 0.2853	0.8345 ± 0.0051
GMSSD w/o Residual Connection	5.1179 ± 0.1001	9.1416 ± 0.2286	0.8567 ± 0.0038
GMSSD w/o Dilated Convolution	5.0763 ± 0.0455	9.0722 ± 0.0459	0.8584 ± 0.0024
GMSSD	**4.9860 ± 0.1310**	**8.9532 ± 0.1283**	**0.8585 ± 0.0021**

drop-offs in New York City from June 16th to June 30th and the prediction results generated by the proposed model. The graph clearly demonstrates the effectiveness of the GMSSD model in providing accurate predictions.

Furthermore, we evaluated the GMSSD model using a currency exchange rate dataset. Figure 4 shows the real values of the daily exchange rates in Australia as well as the predicted results of the GMSSD model for a horizon of 3. The graph shows a close fit between the predictions of the proposed model and the actual values, thus confirming the effectiveness of the model.

Fig. 4. Displays the visualization of the predicted exchange rates.

Overall, the GMSSD model exhibited excellent prediction performance in both single-step and multi-step forecasting tasks.

5   Conclusions

This study investigates the problem of multivariate time-series forecasting and proposes a novel multivariate time-series prediction model called GMSSD. We design time convolution module, graph structure learnering, and graph convolution module to effectively extract features at different scales, learn the correlations between nodes, and capture the historical information and spatiotemporal characteristics in multivariate time-series. Notably, we incorporate a multi-step dependency layer in the graph convolution module, enabling the fusion of node information and neighbor information based on their contributions at different time steps. Finally, we combine these modules into a unified model. Compared to existing models, our proposed model demonstrates outstanding performance on several popular datasets.

However, there is still room for improvement in our model. Future work will focus on exploring more sophisticated designs for the graph structure learner and enhancing the methods for capturing historical information in the graph convolution module. Through further research and experimentation, we aim to further enhance the performance of our model and explore its applications in a broader range of domains.

Acknowledgment. This work is supported by the National Natural Science Foundation of China (62272281), the Special Funds for Taishan Scholars Project (tsqn202306274), and the Youth Innovation Technology Project of Higher School in Shandong Province (2019KJN042).

References

1. Alley, R.B., Emanuel, K.A., Zhang, F.: Advances in weather prediction. Science **363**(6425), 342–344 (2019)
2. Bandara, K., Shi, P., Bergmeir, C., Hewamalage, H., Tran, Q., Seaman, B.: Sales demand forecast in E-commerce using a long short-term memory neural network methodology. In: Gedeon, T., Wong, K.W., Lee, M. (eds.) ICONIP 2019. LNCS, vol. 11955, pp. 462–474. Springer, Cham (2019). https://doi.org/10.1007/978-3-030-36718-3_39
3. Cao, D., et al.: Spectral temporal graph neural network for multivariate time-series forecasting. Adv. Neural. Inf. Process. Syst. **33**, 17766–17778 (2020)
4. Chen, W., Chen, L., Xie, Y., Cao, W., Gao, Y., Feng, X.: Multi-range attentive bicomponent graph convolutional network for traffic forecasting. In: Proceedings of the AAAI Conference on Artificial Intelligence, vol. 34, pp. 3529–3536 (2020)
5. Jin, M., Zheng, Y., Li, Y.F., Chen, S., Yang, B., Pan, S.: Multivariate time series forecasting with dynamic graph neural odes. IEEE Trans. Knowl. Data Eng. **35**, 9168–9180 (2022)
6. Lai, G., Chang, W.C., Yang, Y., Liu, H.: Modeling long-and short-term temporal patterns with deep neural networks. In: The 41st international ACM SIGIR Conference on Research & Development in Information Retrieval, pp. 95–104 (2018)
7. Li, Y., Yu, R., Shahabi, C., Liu, Y.: Diffusion convolutional recurrent neural network: Data-driven traffic forecasting. In: ICLR (2018)
8. Lin, Y., Koprinska, I., Rana, M.: SpringNet: transformer and spring DTW for time series forecasting. In: Yang, H., Pasupa, K., Leung, A.C.-S., Kwok, J.T., Chan, J.H., King, I. (eds.) ICONIP 2020. LNCS, vol. 12534, pp. 616–628. Springer, Cham (2020). https://doi.org/10.1007/978-3-030-63836-8_51
9. Liu, X., Guo, J., Wang, H., Zhang, F.: Prediction of stock market index based on ISSA-BP neural network. Expert Syst. Appl. **204**, 117604 (2022)
10. Liu, Y., Gong, C., Yang, L., Chen, Y.: DSTP-RNN: a dual-stage two-phase attention-based recurrent neural network for long-term and multivariate time series prediction. Expert Syst. Appl. **143**, 113082 (2020)

11. Rußwurm, M., Körner, M.: Self-attention for raw optical satellite time series classification. ISPRS J. Photogramm. Remote. Sens. **169**, 421–435 (2020)
12. Sanhudo, L., Rodrigues, J., Vasconcelos Filho, E.: Multivariate time series clustering and forecasting for building energy analysis: application to weather data quality control. J. Build. Eng. **35**, 101996 (2021)
13. Shao, Z., Zhang, Z., Wang, F., Wei, W., Xu, Y.: Spatial-temporal identity: a simple yet effective baseline for multivariate time series forecasting. In: Proceedings of the 31st ACM International Conference on Information & Knowledge Management, pp. 4454–4458 (2022)
14. Shih, S.Y., Sun, F.K., Lee, H.Y.: Temporal pattern attention for multivariate time series forecasting. Mach. Learn. **108**, 1421–1441 (2019)
15. Wan, R., Mei, S., Wang, J., Liu, M., Yang, F.: Multivariate temporal convolutional network: a deep neural networks approach for multivariate time series forecasting. Electronics **8**(8), 876 (2019)
16. Wu, H., Xu, J., Wang, J., Long, M.: Autoformer: decomposition transformers with auto-correlation for long-term series forecasting. Adv. Neural. Inf. Process. Syst. **34**, 22419–22430 (2021)
17. Wu, S., Xiao, X., Ding, Q., Zhao, P., Wei, Y., Huang, J.: Adversarial sparse transformer for time series forecasting. Adv. Neural. Inf. Process. Syst. **33**, 17105–17115 (2020)
18. Wu, Z., Pan, S., Long, G., Jiang, J., Chang, X., Zhang, C.: Connecting the dots: multivariate time series forecasting with graph neural networks. In: Proceedings of the 26th ACM SIGKDD International Conference on Knowledge Discovery & Data Mining, pp. 753–763 (2020)
19. Yan, Y., Zhang, S., Tang, J., Wang, X.: Understanding characteristics in multivariate traffic flow time series from complex network structure. Phys. A **477**, 149–160 (2017)
20. Ye, J., Liu, Z., Du, B., Sun, L., Li, W., Fu, Y., Xiong, H.: Learning the evolutionary and multi-scale graph structure for multivariate time series forecasting. In: Proceedings of the 28th ACM SIGKDD Conference on Knowledge Discovery and Data Mining. pp. 2296–2306 (2022)
21. Ye, J., Sun, L., Du, B., Fu, Y., Xiong, H.: Coupled layer-wise graph convolution for transportation demand prediction. In: Proceedings of the AAAI Conference on Artificial Intelligence, vol. 35, pp. 4617–4625 (2021)
22. Zhang, F., Chen, G., Wang, H., Li, J., Zhang, C.: Multi-scale video super-resolution transformer with polynomial approximation. IEEE Trans. Circuits Syst. Video Technol. **33**, 4496–4506 (2023)
23. Zhang, F., Guo, T., Wang, H.: DFNet: decomposition fusion model for long sequence time-series forecasting. Knowl.-Based Syst. **277**, 110794 (2023)
24. Zhou, H., et al.: Informer: beyond efficient transformer for long sequence time-series forecasting. In: Proceedings of the AAAI Conference on Artificial Intelligence, vol. 35, pp. 11106–11115 (2021)
25. Zhu, L., Wang, Y., Fan, Q.: MODWT-ARMA model for time series prediction. Appl. Math. Model. **38**(5–6), 1859–1865 (2014)

Q-Learning Based Adaptive Scheduling Method for Hospital Outpatient Clinics

Wenlong Ni[1](✉), Lingyue Lai[1], Xuan Zhao[1], and Jue Wang[2]

[1] School of Computer Information Engineering, JiangXi Normal University,
Nanchang, China
{wni,laily,xuanzhao}@jxnu.edu.cn
[2] Department of Rehabilitation, The Second Affiliated Hospital of NanChang
University, Nanchang, China
cny9707@dingtalk.com

Abstract. Proper selection of the number of Service Providers (SPs) such as doctors, registration windows, and examination equipments in outpatient clinics can improve the efficiency of services and promote the sharing and effective use of medical resources. In this paper an adaptive scheduling model for hospital outpatient clinics on the number of SPs while minimizing total cost is proposed. Firstly, the M/G/K model of the outpatient queuing process is constructed based on queuing theory, where M denotes that the Poisson process of patient arrivals, the service time follows the general distribution defined as G, and K is the number of SPs. Secondly, the objective function of minimizing cost such as waiting, setup, SP usage is established. The optimal number of SP is solved with the Q-learning (QL) algorithm in reinforcement learning (RL). Finally, the simulation verifies that as the cost of service gradually increases, the system will favor fewer SPs to perform the service to reduce the total cost. This scheduling model can not only adjust the scheduling scheme according to the different service costs to maximize the economic efficiency of the hospital, but also can be used to manage the hospital staffing.

Keywords: General Distribution · Queuing Theory · Adaptive Scheduling · Q-Learning

1 Introduction

Hospitals are generally under pressure to reduce costs and improve service quality, and a proper and rational queuing system can bring better benefits to hospitals [1]. Queuing systems are widely used in theoretical research and practical system design, such as service systems, transportation systems, and medical systems, and queuing theory is used to establish a rational allocation system [2–4].

This Work Was Supported in Part by JiangXi Education Department Under Grant No. GJJ191688.

The random nature of people's demand and patient flow makes the structure of queuing systems more complex and variable. Many studies on hospital queuing are based on analytical software such as Arena and Spss [5–7], which are used to model and simulate queuing systems to propose solutions to reduce queuing and waiting times. And the cost was introduced in the queuing system for the same type of patients to determine the optimal number of service stations [8,9]. Feldman J et al. [10] developed a dynamic model that changes according to the appointment status and used a heuristic algorithm to solve for the optimal strategy. Hulshof et al. [11] takes into account the randomisation factor, the approximate dynamic programming model is used to study the allocation problem of multiple resources and the scheduling problem of multiple types of patients, and the superiority of the approximate dynamic programming method in dealing with the patient scheduling problem is demonstrated.

With the continuous improvement of computer technology, RL has attracted great attention in the problem of task schedualing [12,13], it is an adaptable environment-based machine learning method that uses environmental feedback as input, and is a branch of artificial intelligence (AI) [14], whose main idea is to continuously interact with the environment by trial and error and use evaluative feedback signals to achieve decision optimization. RL consists of four important objects: Agent, State, Action and Reward Function. Agent will choose some actions to change the state, feedback gets positive or negative reward, it will select the action and move to the next state according to the principle of maximum reward value, and repeat this process until the end of learning. QL is an efficient algorithm for task scheduling problem in RL [15,16]. Queuing theory, as an optimization theory specializing in the study of congestion generation with random factors, allows the optimal selection of quantitative such as the number of SPs in hospital outpatient queuing systems, and few scholars have used the idea of reinforcement learning to deal with the scheduling problem in hospital outpatient queuing systems.

To address the randomness and structural complexity of the queuing system, this paper combines the characteristics of trial and error of continuous interaction between the agent and the environment, and uses RL to dynamically schedule queuing. The contributions of this paper are as follows: (1) For the hospital queuing problem, an M/G/K simulation model is proposed in this paper. In order to better generate a set of random numbers that conform to the hospital service rule with the lognormal distribution. And the theoretical results are used to verify the accuracy of this model; (2) Dynamic scheduling of queuing system is implemented based on QL algorithm. Aim to minimise the system cost. Dynamic operations of the number of SPs, such as decreasing or increasing; (3) The experiment analyses the effect of different costs on the choice of actions, and the system can dynamically adjust the SP according to the change of cost. It is shown that the adaptive nature of the model proposed in this paper.

The rest of the paper is organized as follows. The second part presents the system model, which includes the sysytem model and the QL scheduling model.

The third part shows the experimental simulations. Finally, the paper is summarized in the fourth section.

2   System Model

The system model consists of two parts, one is the M/G/K queuing simulation model, and the other is the QL dynamic scheduling model. As shown in Fig 1.

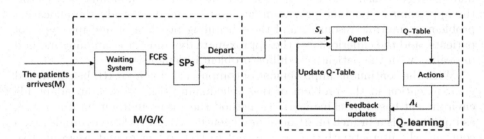

Fig. 1. System Framework

To facilitate the analysis of the study, the system model is described as follows:

- **Assumption 1** *Limited available medical resources in the system and limited maximum number of patients per day.*
- **Assumption 2** *All SPs have the same capacity and priority and serve at most one patient at a time. Patients are served by any available SP, and patients do not leave in the middle of the service, but only leave after receiving the service.*
- **Assumption 3** *Patients randomly arrive and enter the system to wait for service according to the first come first served (FCFS) rule. Two scenarios can occur: one is that there are SPs available at the time of arrival, and patients do not need to wait for immediate service; the other is that all SPs are occupied at the time of arrival, and patients enter the queue to wait for the resulting waiting time.*
- **Assumption 4** *The system contains three types of costs, the waiting cost required for patients to wait in the system, the service cost consumed by SPs to provide the service and the cost of setup one SP, and the goal of this paper is to minimize the total cost.*

2.1   M/G/K Queuing Model

M in M/G/K indicates that patient arrivals obey a Poisson distribution with an arrival rate of λ, G indicates that the services received by patients obey a general distribution with a service rate of μ, and K for the number of SPs.

Currently there is no formulaic derivation process for the M/G/K queuing model in operation researchs, but many scholars such as T. Kimura [17] have done a lot of profound research on the approximation formulas for system performance indicators such as average waiting time in the M/G/K queuing model, and the research results have been widely use multi-service desk queuing scenarios. The expected value of the average waiting time is:

$$E(W_q) = \frac{V + E^2}{2E(K - \rho)}[1 + \sum_{i=0}^{K-1} \frac{(K-1)!(K-\rho)}{i!\rho^{K-i}}]^{-1}, \tag{1}$$

where, E is the expected value of the general distribution G, V is the variance, $\rho = \lambda E$, and the condition for Eq. (1) to hold is $K > \rho$.

In this paper, the M/G/K model will be simulated using time series, $[arr_1, arr_2, ..., arr_N]$ denotes the arrival time of each patient to the system, $[ser_1, ser_2, ..., ser_N]$ denotes the service time received by each patient. Whether or not the patient experiences a queue, which will directly affect the start time of the service for subsequent patients. The need for patients to queue will be analyzed for two scenarios:

- **Empty Queues**

   At this point there are some SPs availabily, the patient receives immediate service after arrival, and the waiting time for the i-th patient is:

$$w_i = 0, \tag{2}$$

   The time of arrival is the same as the start time of the patient receiving the service, then the time the patient leaves the system is:

$$d_i = arr_i + ser_i. \tag{3}$$

- **Not Empty Queues**

   The service times of patients are generated randomly, then patient selection of SPs is also a random process. When the i-th patient arrives, the number of people $q \geq K$ in the system. These randomness will disrupt the order of patient departures, some patients who arrive later but have a shorter service time will leave the system earlier, and will re-order the patient departure times in chronological order, denoted by D. Then, the service time of the i-th patient starts from D_{i-1}, then the departure time of the i-th patient is:

$$d_i = D_{i-1} + ser_i, \tag{4}$$

   The waiting time for the i-th patient is:

$$w_i = D_{i-1} - arr_i. \tag{5}$$

In summary, the average patient waiting time for the system was obtained by summing up the waiting time for each patient:

$$W_q = \frac{\sum_{i=1}^{N} w_i}{N}, \tag{6}$$

The average waiting captain is:

$$L_q = \begin{cases} L_s - K, & L_s > K, \\ 0, & L_s \le K, \end{cases} \tag{7}$$

where $L_s = \frac{\sum_{i=1}^{N} L_i}{N}$, is the average number of people staying in the system, L_i is the number of patients staying in the ith moment system.

2.2   Work Flow

The selection of the number of SPs at the arrival of each patient is important and will affect subsequent selections and cost, and the conditions for changes in the number of SPs are given in Fig 2.

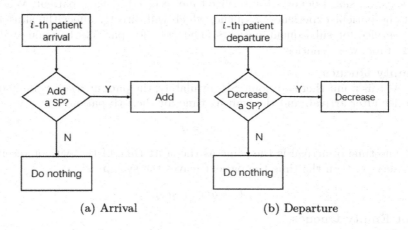

(a) Arrival   (b) Departure

Fig. 2. Arrivals/Departures Workflow

From Fig 2, the arrival and departure of the i-th patient will choose different actions. And how to select the optimal number of SPs to make the system obtain the minimum long-term cost is an important problem to be solved in this paper. Q-learning as an efficient scheduling algorithm in RL, can obtain the optimal long-term reward. This algorithm will be used to solve for the optimal action selection in M/G/K.

2.3   Q-Learning Dynamic Scheduling Model

The QL algorithm is a pattern-free RL algorithm that provides the agent with the ability to learn and select the best action based on historical experience in a Markovian environment. The three main parts of their algorithm are represented as follows:

- **System States**: The system state can be expressed as a parameter or a set of parameters used to describe the system. For the definition of the state in which the intelligences in this paper are located, the state is described according to the simulation environment established by M/G/K as:

$$s = (n_1, n_2, e), \quad 0 \le n_1 \le N, 0 \le n_2 \le C, \tag{8}$$

where n_1 is the number of patients in the system, the maximum number of patients that can be accommodated is N, n_2 is the number of SPs, and C is the maximum number of SPs that the hospital can provide to the outpatient clinic. e is the set of events generated in each state, A and D represent the actions that the patient can select under different events, and the set of events is represented as $e \in \{A, D\}$, with these descriptions, the state space is:
$$S = \{s | s = \langle s_1, s_2, ..., s_i, ..., s_N, e \rangle = \langle \hat{s}, e \rangle\}.$$

- **Actions**: Using the relationship between n1 and n2 as a decision point to determine the set of events available to the agent in the current state. When $n_1 \ge n_2$, it indicates that all SPs are in the service of the patient, such as reducing the number of SPs will make the patients in the queue dissatisfied, so the number of SPs cannot be reduced, and the action event at this time is A; when $n_1 < n_2$, the number of SPs can be reduced when the SPs are idle, and the action event at this time is D. The action space is expressed as $A_s = \{-1, 0, 1\}$, where $+1$ means add one SP, -1 is decreasing one SP, 0 means no change, and the set of actions $A(e)$ for different events is:

$$A(e) = \begin{cases} \{0, 1\}, & e = A, \\ \{-1, 0\}, & e = D. \end{cases} \tag{9}$$

- **Reward Function**: The reward function is the key to choose the optimal solution, and after multiple choices is able to accumulate the reward and finally the optimization result of scheduling. The scheduling goal in this paper is to maximize the hospital's efficiency. Each adjustment of the number of SPs by agent will change the departure time of the patients who are queuing in the queue, and different departure times will generate different waiting queues, and the system will pay the corresponding waiting cost for each waiting patient and the service cost consumed by each SP during this time. In this paper, we construct the reward function $r(s, a)$ based on the local cost in each state.

$$\begin{aligned} r(s, a) &= w(s, a) - c(s, a) \\ &= w(s, a) - t(L_q C_1 + K C_2), \end{aligned} \tag{10}$$

where $w(s, a)$ denotes the immediate cost contingent on state s having performed action a, t is the transfer time from the current state to the next state, related to the patient's arrival time to the system, $c(s, a)$ is the cost in the next this state, the patient's waiting cost and the SP's service cost, respectively, C_1 is the unit waiting cost and C_2 is the unit service cost, setting $w(s, a)$ to:

$$w(s, a) = \begin{cases} 0, & a = 0, -1, \\ , & a = +1, \end{cases} \tag{11}$$

Here C_0 is expressed as a fixed constant for the cost required to setup a SP.

- **ϵ-greedy Strategy**: In this paper, we use the ϵ-greedy policy for action selection. The rule of ϵ-greedy for action selection is shown in Eq.(12), when the agent observes state s, the action corresponding to the largest value of $Q(s, a)$ is selected in the Q-table with probability 1-ϵ, and the action is selected randomly with probability ϵ. This method guarantees that all actions in the Q-table will be updated if the number of iterations is large enough to praeserve a certain level of exploration while keeping the Agent moving in a better direction.

$$A(s_i|a_i) = \begin{cases} randomA(s_i), & rand > 1 - \epsilon, \\ max, & others, \end{cases} \tag{12}$$

where a_{max} is the action with the largest Q value under state s_i, A(s_i) represents the set of all selectable actions in the s_i state, and rand is a randomly sampled decimal.

QL solves the optimal scheduling scheme in the queuing model M/G/K, which uses each state-action in the system to obtain Q-table, Q-table is used to record how good or bad an agent takes a certain action in a certain state, which can indirectly decide what decision the agent takes, $Q(s, a)$ is the state action value function, which is the state s the expected value of the reward that can be obtained after taking action a. Firstly, all the values in Q-table are initialized to 0. As each patient arrives, the agent gets the state at this point, selects the action and moves to the next state according to the state and the reward function, and the Q-table gets feedback and is updated. Finally, the action with the highest value of Q(s,a) is declared as the optimal action for that state. The algorithmic flow for solving hospital queueing scheduling with the QL algorithm is given as follows:

Step 1 : Determine the learning rate α, discount factor γ, ϵ and the number of iterations, initialize the agent, status, number of service desks K, and Q-table, and initialize the value in Q-table to 0.

Step 2 The current state s_i of the ith patient is used to determine the set of actions that the Agent can take at the current moment to satisfy the constraints, and then ϵ-greedy is used to select the action a_i from the action set.

Step 3 The agent obtains the set A_{i+1} of actions for which the constraint is satisfied by the next patient through s_i and a_i and the immediate reward$r(s_i, a_i)$. The value of $Q(s_i, a_i)$ is updated according to the Q-table and the resulting A_{i+1}, and the update function is:

$$Q(s_i, a_i) \leftarrow Q(s_i, a_i) + \alpha[r(s_i, a_i) + \gamma \max_{a'} Q' - Q(s_i, a')], \tag{13}$$

where $\max\limits_{a'} Q' = \max\limits_{a'} Q'(s_{i+1}, a')$ is the action with the largest value selected for the next patient state s_{i+1}.

Step 4 After the Q value of the current patient has been updated, the agent will continue to update the status of the next patient, going back to step 2, until the status of the last patient has been updated.

Step 5 When the Agent has reached the maximum number of training rounds, it exits training and obtains the optimal Q-table as well as the optimal scheduling scheme.

QL can adapt to the dynamic queuing process of a changing hospital and find the optimal strategy to minimum the total cost by the rewards obtained by representing the pseudo code as follows:

Algorithm 1. Q-table update process

Input: Q-table,s_i
$\quad$ //α,γ are predefined learning parameters
$\quad$ **repeat**
$\qquad$ a_i=Chooes Action(Q-table,s_i) //Use ϵ-greedy
$\qquad$ Take action a_i
$\qquad$ r=Get Reward(s_i,a_i)
$\qquad$ $Q(s_i, a_i) \leftarrow Q(s_i, a_i) + \alpha[r + \gamma \max_{a'} Q' - Q(s_i, a_i)]$
$\qquad$ $s_i \rightarrow s_{i+1}$ //The next one arrives at the patient's state
$\quad$ **until** All patients leave
Output: Q-table //Updated Q-table

---

3 Experimental Analysis

G in M/G/K is represented as a general distribution, Latruwe et al. [18] through simulation found that the service time distribution tends to be more consistent with the lognormal distribution, this paper will use the lognormal distribution. The model parameters used in the experiment are shown in Table 1.

Table 1. Parameter values

Parameter	Value	Memo
N	10	Total number of Patients
λ	1	Arrival Rate
α	0.1	Learning Rate
γ	0.9	Discount Factor
ϵ	0.9	Greedy selection Rate

3.1 Waiting Time

For the correctness and validity of the M/G/K simulation model, the approximate solution theory Eq.(1) was used to compare with the simulation model constructed in this paper. According to the condition that Eq.(1) holds: $K¿\rho$, when K=1, V=0.9, E=$1/\mu$ <1. The comparison result is expressed as Table 2.

Table 2. Comparison of results

E	1/2	1/3	1/4	1/5
Theoretical value	0.800	0.818	0.842	0.862
Simulation value	0.804	0.816	0.836	0.865
Absolute error	0.004	0.002	0.006	0.003

As shown in the table, the simulation results of the M/G/K model constructed in this paper have a small error with the theoretical results, with a maximum of only 0.006. This result shows that the M/G/K model can achieve an approximate solution to the theoretical values. The theoretical values hold only within the constraints, while the simulation can efficiently simulate many different cases.

3.2 Adaptive Scheduling

The interrelated and dynamically changing states in the queuing system eventually form a three-dimensional Q-table, which is not conducive to maintain and causes data redundancy. To reduce the dimensionality of the data for analysis, the Q-table is divided into different event sets, when $n1 \geq n2$, the Q-table is represented as $Q(n_1, n_2, A)$; and when $n_1 < n_2$, it is represented as $Q(n_1, n_2, D)$. As shown in the Table 3, $\mu=1/3$, $\varnothing$ means an empty state.

Table 3. Q-table

(a) $Q(n_1, n_2, A)$

n_1 \ n_2	0	1	2	3	4	5
0	-4.85	$\varnothing$	$\varnothing$	$\varnothing$	$\varnothing$	$\varnothing$
1	$\varnothing$	-5.53	$\varnothing$	$\varnothing$	$\varnothing$	$\varnothing$
2	$\varnothing$	-5.59	-5.80	$\varnothing$	$\varnothing$	$\varnothing$
3	$\varnothing$	-6.41	-5.52	-5.71	$\varnothing$	$\varnothing$
4	$\varnothing$	-7.01	-6.10	-5.58	-5.68	$\varnothing$
5	$\varnothing$	-7.61	-6.62	-5.97	-5.65	-5.78
6	$\varnothing$	-8.21	-7.16	-6.36	-6.01	-5.86
7	$\varnothing$	-8.81	-7.74	-6.81	-6.32	-6.11
8	$\varnothing$	-9.38	-8.29	-7.29	-6.71	-6.39

(b) $Q(n_1, n_2, D)$

n_1 \ n_2	0	1	2	3	4	5
0	$\varnothing$	$\varnothing$	$\varnothing$	$\varnothing$	$\varnothing$	$\varnothing$
1	$\varnothing$	-6.91	$\varnothing$	$\varnothing$	$\varnothing$	$\varnothing$
2	$\varnothing$	-6.61	-7.77	$\varnothing$	$\varnothing$	$\varnothing$
3	$\varnothing$	-8.20	-6.77	-7.19	$\varnothing$	$\varnothing$
4	$\varnothing$	-8.80	-7.72	-6.43	-6.89	$\varnothing$
5	$\varnothing$	-9.35	-8.28	-7.28	-6.44	-6.69
6	$\varnothing$	-9.79	-8.85	-7.80	-6.96	-6.63
7	$\varnothing$	-10.16	-9.41	-8.33	-7.37	-6.88
8	$\varnothing$	-10.48	-9.87	-8.83	-7.82	-7.20

The model can choose the optimal SP number according to the size relationship between different costs, and simplify the cost value per unit time. Among them, C_0 is the cost of setup one SP, C_1 is the unit time waiting cost, and C_2 is the service cost. The cost in Table 3 is set to $C_0=1$, $C_1=1$, $C_2=1$. The agent selects the maximum action according to the maximum action in the Q table, and its maximum Q value corresponds to the minimum cost value, and different action choices will affect the subsequent reward value, so different sets of events will receive different values.

The initial state of the system is $Q(0, 0, A)$, and when a patient arrives, the agent will setup a SP for the patient, and the subsequent number of SPs is at least one regardless of whether there is a patient or not. So there is no situation where the number of SP is 0 when there is a patient in the system. From the Table 3, the overall cost in the corresponding state of $Q(n_1, n_2, A)$ are slightly smaller than those in $Q(n_1, n_2, D)$. When a patient arrives, n_1 is greater than n_2 and the number of SPs can be kept constant or increased by 1. When a patient leaves, the number of SPs can only be kept constant or decreased by 1. The waiting cost in the system can be reduced in $Q(n_1, n_2, A)$ by increasing the SPs, thus reducing the total system cost.

When n_1 is the same and $n_2 < n_1$, the larger the n_2 the lower the waiting cost of the queue, and the overall cost gradually decreases with the increase of n_2. When n_2 is the same, the overall service efficiency of the system is the same, and the larger n_1 is, the longer the waiting time of patients in the system, and the waiting cost increases, leading to an increase in the overall cost of the system.

Event D only reduces the number of SPs when there are no patients in the queue, and the action selection process of event A better reflects the agent's change in action after weighing costs. So, $Q(n_1, n_2, A)$ will be used to analyze the adaptability of the model after the cost change, and several sets of experiments with different cost sizes were used for comparative analysis, as shown in Table 4.

Table 4. Actions at different C_0

(a) $C_0 = 0.1, C_2 = 1$

n_1 \ n_2	0	1	2	3	4	5
0	1	∅	∅	∅	∅	∅
1	∅	1	∅	∅	∅	∅
2	∅	1	1	∅	∅	∅
3	∅	1	1	1	∅	∅
4	∅	1	1	1	1	∅
5	∅	1	1	1	1	1
6	∅	1	1	1	1	1

(b) $C_0 = 1, C_2 = 1$

n_1 \ n_2	0	1	2	3	4	5
0	1	∅	∅	∅	∅	∅
1	∅	0	∅	∅	∅	∅
2	∅	0	0	∅	∅	∅
3	∅	0	0	0	∅	∅
4	∅	1	0	0	0	∅
5	∅	1	0	0	0	0
6	∅	1	0	0	0	0

Table 4 indicates the effect of C_0 on action selection for the same C_1 and C_2. Regardless of the cost change, in the initial state, the agent will choose to add an SP. From (a), it follows that when C_0 is small, the system will keep increasing the number of SPs, which will minimize the total cost by reducing the waiting cost in the system. (b) indicates that when C_0 increases to 1, the system will hardly increase the number of SPs again and keep it constant, only when n_1 is much larger than n_2.

Table 5. Actions at different C_0

(a) $C_0 = 1, C_2 = 0.5$

n_1 \ n_2	0	1	2	3	4	5
0	1	∅	∅	∅	∅	∅
1	∅	0	∅	∅	∅	∅
2	∅	0	0	∅	∅	∅
3	∅	1	0	0	∅	∅
4	∅	1	0	0	0	∅
5	∅	1	1	1	0	0
6	∅	1	1	1	1	1

(b) $C_0 = 1, C_2 = 1$

n_1 \ n_2	0	1	2	3	4	5
0	1	∅	∅	∅	∅	∅
1	∅	0	∅	∅	∅	∅
2	∅	0	0	∅	∅	∅
3	∅	0	0	0	∅	∅
4	∅	1	0	0	0	∅
5	∅	1	0	0	0	0
6	∅	1	0	0	0	0

Table 5 represents the variation of action selection for different unit service costs C_2 when the same C_0 and C_1. Setting C_2 to 0.3 and 1, respectively, the action selection results are shown in (a) and (b). (a) indicates that when the cost is low, the system chooses to increase the number of SPs, and as n_2 increases to be close to n_1, the system considers the setup cost and does not increase the number of SPs. (b) shows that when both C_0 and C_2 are high, the system weighs the waiting cost and service cost. Then the number of SPs is basically unchanged; while when n_1 is much larger than n_2, the number of SPs needs to be increased to reduce the waiting cost. It can be seen that the model adaptively selects the optimal number of SPs as the cost changes.

4   Conclusion

In this paper, we study the optimal scheduling problem of hospital outpatient clinics, where patients arrive at the hospital is a stochastic process, and this stochastic phenomenon often results in overcrowding as well as inefficient service. In order to determine when the number of SPs can be adjusted so that the hospital can provide efficient services with the highest economic efficiency, this paper proposes an adaptive scheduling model based on QL, which treats the state space as a dynamically changing captain and the current number of SPs, M/G/K simulation is used to simulate the hospital queuing process and obtain the system performance indexes in each state, and QL is used to select the optimal number of SPs for each state by using the state and reward function. From the experimental analysis, it can be concluded that this model can help hospitals to give the most economical scheduling solution after weighing the service cost and waiting time cost in different states.

References

1. Marynissen, J., Demeulemeester, E.: Literature review on integrated hospital scheduling problems. KU Leuven Fac. Econ. Bus. KBI-1627 (2016)
2. Angst, C.M., Devaraj, S., Queenan, C.C., Greenwood, B.: Performance effects related to the sequence of integration of healthcare technologies. Prod. Oper. Manag. **20**(3), 319–333 (2011)
3. Yang, O.: The inquiry in hospital beds arrangement based on queuing theory. In: 2011 6th International Conference on Computer Science & Education (ICCSE), pp. 406-410. IEEE (2011)
4. Vass, H., Szabo, Z.K.: Application of queuing model to patient flow in emergency department. Case study. Proc. Econ. Finan. **32**, 479–487 (2015)
5. Folake, A.O., Agu, M.N., Okebanama, U.F.: Application of queue model in health care sector. Int. Res. J. Adv. Eng. Sci. **5**(3), 48–50 (2020)
6. Aziati, A.N., Hamdan, N.S.B.: Application of queuing theory model and simulation to patient flow at the outpatient department. In: Proceedings of the International Conference on Industrial Engineering and Operations Management Bandung, Indonesia, pp. 3016–3028 (2018)
7. Haghighinejad, H.A., et al.: Using queuing theory and simulation modelling to reduce waiting times in an Iranian emergency department. Int. J. Commun. Based Nurs. Midwifery **4**(1), 11 (2016)
8. Prasad, S.V., Donthi, R., Challa, M.K.: The Sensitivity Analysis of Service and Waiting Costs of A Multi Server Queuing Model. In: IOP Conference Series: Materials Science and Engineering, vol. 993, no. 1, p. 012107. IOP Publishing (2020)
9. Kembe, M.M., Onah, E.S., Iorkegh, S.: A study of waiting and service costs of a multi-server queuing model in a specialist hospital. Int. J. Sci. Technol. Res. **1**(8), 19–23 (2012)
10. Feldman, J., Liu, N., Topaloglu, H., Ziya, S.: Appointment scheduling under patient preference and no-show behavior. Oper. Res. **62**(4), 794–811 (2014)
11. Hulshof, P.J., Mes, M.R., Boucherie, R.J., Hans, E.W.: Patient admission planning using approximate dynamic programming. Flex. Serv. Manuf. J. **28**, 30–61 (2016)
12. Tong, Z., Xiao, Z., Li, K., Li, K.: Proactive scheduling in distributed computing–A reinforcement learning approach. J. Parallel Distrib. Comput. **74**(7), 2662–2672 (2014)
13. Xiao, Z., Tong, Z., Li, K., Li, K.: Learning non-cooperative game for load balancing under self-interested distributed environment. Appl. Soft Comput. **52**, 376–386 (2017)
14. Barto, A.G., Sutton, R.S.: Reinforcement learning: an introduction (adaptive computation and machine learning). MIT press (1998)
15. Peng, Z., Cui, D., Zuo, J., Li, Q., Xu, B., Lin, W.: Random task scheduling scheme based on reinforcement learning in cloud computing. Clust. Comput. **18**, 1595–1607 (2015)
16. Wei, Z., Zhang, Y., Xu, X., Shi, L., Feng, L.: A task scheduling algorithm based on Q-learning and shared value function for WSNs. Comput. Netw. **126**, 141–149 (2017)
17. Kimura, T.: A transform-free approximation for the finite capacity M/G/s queue. Oper. Res. **44**(6), 984–988 (1996)
18. Latruwe, T., Van der Wee, M., Vanleenhove, P., Devriese, J., Verbrugge, S., Colle, D.: A long-term forecasting and simulation model for strategic planning of hospital bed capacity. Oper. Res. Health Care **36**, 100375 (2023)

Temporal Task Graph Based Dynamic Agent Allocation for Applying Behavior Trees in Multi-agent Games

Xianglong Li, Yuan Li$^{(\boxtimes)}$, Jieyuan Zhang$^{(\boxtimes)}$, Xinhai Xu, and Donghong Liu

Academy of Military Sciences, Beijing 100091, China
nudt_liyuan@163.com, 13007145224@163.com

Abstract. Behavior Trees have gained widespread applications across diverse domains, facilitating the decomposition of complex tasks into manageable subtasks. However, an inherent challenge in maximizing the performance of BTs lies in dynamically allocating agents to subtasks as time progresses. This allocation predicament is compounded by the intricate nature of game states and the temporal variations in subtask activation. In this paper, we propose a novel approach that combines temporal task graphs with reinforcement learning to dynamically allocate agents among subtasks in BT. We employ a temporal task graph to model the dynamic activation of subtasks, where encoded vectors are multiplied by the agent's encoded observation. This enables each agent to be assigned to a specific subtask while considering comprehensive information about all subtasks. Moreover, we aggregate the Q-values of selected subtasks for all agents, leveraging this information to compute a total loss for updating the entire network. To evaluate the efficacy of our approach, we conducted extensive experiments on the challenging benchmark provided by Google Research Football. The results clearly demonstrate a significant performance improvement in BTs when leveraging our proposed framework.

Keywords: Behavior Trees · Dynamic Agent Allocation · Temporal Task Graph

1 Introduction

Behavior Trees (BTs) have been used for many different applications, including real-time games [6], manipulators [13], mobile robots [14] and so on. By decomposing complex processes into simple subtasks, BTs can effectively help designers understand the problem space and clarify the function of the system [17]. However, it needs much human knowledge and takes lots of effort to design a good one for a specific scenario. A large number of works have been done to combine the design of BTs with reinforcement learning [7]. They mainly concentrate on using learning methods to solve certain subtasks [10] or optimizing the order of nodes in BTs [4]. Optimizing the allocation of resources among subtasks in BTs has not been paid much attention.

B. Luo et al. (Eds.): ICONIP 2023, CCIS 1962, pp. 124–135, 2024.
https://doi.org/10.1007/978-981-99-8132-8_10

Agent is an important resource in games, for example, in the StarCraft II, Lancers should be properly allocated to the defense task and the attacking task. Normally, the design of BTs makes a static allocation of agents to different subtasks, which seriously restricts the whole performance [10]. Finding a good allocating scheme is not easy in games. Firstly, the state of a game is dynamically changed with the time going on. The allocation scheme should be adjusted according to the change of states. Secondly, the strategy of opponents always changes which also poses challenges for the allocation. Lastly, the allocation should consider the coordination among different subtasks. Therefore, a dynamic not static allocation strategy should be invented for applying BTs to games.

It is clear to see that traditional assignment methods are hard to be used for the considered problem due to the dynamic environment in multi-agent games. Some recent works design different reinforcement learning methods for allocation problems in different domains such as combinatorial optimization [5], resource allocation [19] and some complex real-life allocation problems [1], etc. The work in [11] proposes a heterogeneous DDPG-based method to alleviate the spatiotemporal mismatch between the supply and demand distributions of vehicles and passengers. The work in [2] proposes an RL-based intelligent resource allocation algorithm to minimize the energy consumption of each user subject to the latency requirement. However, such problems are different from ours since there are no confrontations between two opposing parties, which poses great challenges for the learning method. It can be explicitly formulated and the state space is simpler compared to games. Moreover, there have been some studies to solve dynamic allocation problem. The work in [3] proposes a dynamic approach to task allocation for multi-robot systems, considering both uncertainty and temporal constraints. The work in [15] proposes a new graph neural network architecture called CapAM, to solve multi-robot task allocation problems. However, these studies made the assumption that tasks remain active continuously, which may not hold true in multi-agent games. For instance, in a soccer game, the penalty defense task is activated repeatedly only when the ball enters the penalty area.

In this paper, we propose a temporal task graph based reinforcement learning framework for dynamic agent allocation in games. Given a set of subtasks, we use the graph to model their cooperation relations. Nodes represent subtasks and edges are set up between simultaneously activated subtasks. Since subtasks are activated dynamically, the structure of the graph varies with the time going on. For such time-varying graphs, we introduce the temporal graph neural network (TGN) [16] to extract features of subtasks. TGN uses a memory module to capture long-term dependencies and an embedding module to generate representations incorporating both current and historical information, enabling effective modeling of temporal dynamics. It employs message passing to incorporate the influence of neighboring nodes at different time points, facilitating learning performance on dynamic graphs. With TGN, we design a novel mechanism for allocating agents to subtasks. We multiply the encoded observation of an agent with extracted features of TGN, and computing Q-values corresponding to all subtasks. The agent will be assigned to the subtask with the maximal Q-value. To enhance overall allocation performance, we use a mix network to combine Q-values for all agents together. A total loss is computed based on the output of the mix network, which is used to update the whole neural networks.

Finally, we conduct extensive experiments on full 11 vs 11 matches in Google Research Football and compare our method with some recent works. The results show that our method could make proper allocation of agents in real-time and gain much better performance than other similar works. It is worth noting that the proposed dynamic allocation method gets around a significant 98.71% improvement in win rate compared to a static BT in certain scenarios.

2   Background

2.1   Behavior Trees

A BT consists of nodes that perform specific actions or conditions. These nodes are categorized into three main types: control nodes, task nodes, and condition nodes. The control nodes control the flow of execution in the BT, which is divided into three categories, i.e., *Sequence, Fallback* and *Parallel*. The Sequence node executes its child nodes in order until one succeeds or all fail. The Fallback node executes its child nodes in order until one fails or all succeed. The Parallel node executes all of its child nodes simultaneously. It returns Success if all children return Success. The task nodes represent specific actions that an agent can perform. The condition node corresponds to a proposition, which will return Success or Failure depending on if the proposition holds or not.

The execution of a BT begins at the root node, which is the topmost node in the tree. A tick signal is generated and then propagated through the tree branches according to each node type [8]. A node is executed if, and only if, it receives the tick. The child node immediately returns Running to the parent, if its execution is underway, Success if it has achieved its goal, or Failure otherwise. The tick signal allows a BT to respond and adapt to changes in the environment or the agent's state in real-time. By regularly updating the tree's nodes, the agent can make informed decisions and perform appropriate actions based on the current state. Suppose there are M subtasks in a BT. At each tick time t, the BT gets a state S_t from the environment. A tick signal is generated and activates the execution process of BT. The set of agents associated with task m is denoted by $\mathcal{N}_m^t$, which will be dynamically changed with time going on.

2.2   Dynamic Graph Embedding

Dynamic graphs, where nodes and edges change over time, mainly consist of two models: discrete-time dynamic graph (DTDG) and continuous-time dynamic graph (CTDG). DTDG represents sequences of static graph snapshots taken at intervals in time. It is modeled as $\mathcal{G} = \mathcal{G}^{T_1}, \mathcal{G}^{T_2}, ...$, and the embedding approach learns a sequence of mappings $\mathcal{F} = f^{T_1}, f^{T_2}, ...$, where each mapping $f^{T_i} \in \mathcal{F}$

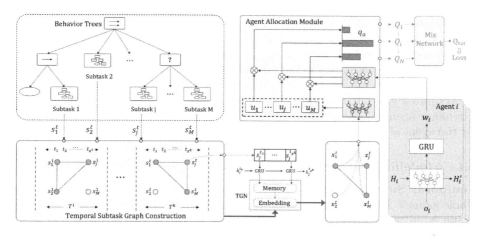

Fig. 1. Temporal subtask graph based reinforcement learning framework for dynamic agent allocation in BT. Gray circles represent inactive subtasks, red circles indicate active subtasks. (Color figure online)

encodes each node v_i in $\mathcal{G}^{T_i}$ into a high-level representation space. CTDG is modeled as a sequence of time-stamped events $\mathcal{G} = x(t_1), x(t_2), ...$, where $x(t_1)$ is a list of events including edge addition or deletion, node addition or deletion, and node or edge feature transformations. The dynamic graph embedding approach learns the node embedding for each time t. TGN [16] is a generic framework for deep learning on dynamic graphs represented as sequences of timed events. It contains two modules: memory and embedding operation. The embedding module generates the temporal embedding of a node at any time t, while the memory has the capability to memorize long-term dependencies for each node in the graph. However, message aggregation for the memory in [16] keeps only the most recent message or averages all messages for a given node. In this paper, we propose using a local GRU module to aggregate a node's local history message. Moreover, we model the subtasks of BTs through continuous-time dynamic graphs. The vertex represents a subtask in BT, and the edge connects two subtasks. We use an RNN model to aggregate messages.

3   Method

In this section, we introduce a temporal subtask graph based reinforcement learning framework (TSGraph-RL) for dynamic agent allocation in multi-agent games. The entire framework is illustrated in Fig. 1. The upper-left section illustrates the set of subtasks organized by BT. It shows the number of activated subtasks in different time slots in the game. The temporal subtask graph construction module is used to model the dynamic relationships among subtasks in BT. Features for such relationships are extracted through the TGN, which is shown in the bottom-right section. The upper-right part shows the agent allocation module, where each agent is associated with an agent network that encodes observations as w_i. The agent allocation module is responsible for assigning agents to appropriate subtasks. The output of each allocation network are Q-values for all possible subtasks, based on which a proper subtask is selected for each agent. Then Q_{tot} is calculated across Q-values for all agents. In the following, we introduce the construction of the temporal subtask graph to capture collaborations among subtasks. Then, we describe a dynamic agent allocation mechanism. Finally, we present the training procedure for the agent allocation learning.

3.1   Modelling Dynamic Subtask with Temporal Graph Network

Temporal Subtask Graph Construction. Assume the total time for the whole game is $\bar{T}$ and the number of subtasks in BT is M. The temporal subtask graph is an undirected dynamic graph represented as $G^k = \{V, E^k, T^k\}$. $V = \{v_1, v_2, ..., v_M\}$ represents the set of subtasks in BT, which does not change with the time going on. $E^k = e_{ij}(i \leq M, j \leq M, i \neq j)$ denotes the set of edges connecting simultaneously active subtasks, which run as the parallel mode in BT. T^k includes several time slots, i.e., $T^k = \{t_1, t_2, ...t_{z^k}\}$, during which the structure of the graph does not change and z^k is the maximum number of time slots.

The construction of the TSGraph is introduced in Algorithm 1, which is named as TSGC. The input is the current running status of BT, the current time slot t and TSGraph G^k at the last time slot. The output is the TSGraph for the current time slot. Steps 1–5 involve constructing the set of edges based on the operational state of the BT. If parallel subtasks are detected between any two subtasks, edges are established and added to E. If the new edge set E matches the edge set of the previous TSGraph, indicating no change in the TSGraph structure, we simply add the current time slot to T^k and return the last TSGraph G^k. However, if the new edge set E differs from the edge set of the last TSGraph, indicating a change in the TSGraph structure, we construct a new TSGraph as outlined in Steps 9–12.

Subtask Node Embedding. Based on the temporal subtask graph G^k of each time slot, Temporal Graph Network (TGN) is used to extract features indicated in the graph. The TGN utilizes a memory module to capture long-term dependencies and an embedding module to generate node representations that include both current and historical information. Specifically, for each TSGraph $G^k = \{V, E^k, T^k\}$, we apply the GRU module in the memory module to acquire the local historical information of each subtask, which can be described by Eq. (1). Note that $T^k = \{t_1, t_2, ..., t_{z^k}\}$ and t_{z^k} is the current time slot. $s^{|T^k|}$ represents the set of node states in G^k over the time period T^k, i.e., $s^{|T^k|} = \{s^{t_1}, s^{t_2}, ..., s^{t_{z^k}}\}$.

$$s^{t_{z^k}} = GRU(h^{t_{z^k}-1}, s^{|T^k|}) \tag{1}$$

Together with $s^{|T^k|}$, temporal subtask graph $G^1, G^2, ..., G^k$ are all fed into TGN to compute final node feature vectors $X = x_1, x_2, ..., x_M$. Formally, the output of

---

Algorithm 1: TSGC

Input: Current BT, current time slot t, last TSGraph $G^k = \{V, E^k, T^k\}$
Output: Current TSGraph

1 Set $E = \phi$;
2 **for** $v \in V$ **do**
3   **for** $v' \in V$ **do**
4   **if** v and v' are parallel subtasks in current BT and $v \neq v'$ **then**
5   Add edge $e_{v,v'}$ to E;

6 **if** $E == E^k$ **then**
7   Add t to T^k;

8 **else**
9       $k = k + 1$;
10      $T^k = \{t\}$;
11      $E^k = E$;
12      $G^k = \{V, E^k, T^k\}$;

13 **return** G^k;

---

the subtask embedding module is defined in Eq. (2), where A^k is the adjacency matrix of G^k, W represents the parameters of TGN, and σ denotes the non-linear ReLU function. Subsequently, all subtask features are encoded as $u = u_1, u_2, ..., u_M$ and used as input to the agent allocation module.

$$X^{t_{z^k}} = TGN(G^1, G^2, ..., G^k, s^{t_{z^k}}) = \sigma(A^k S^k W) \tag{2}$$

3.2 Agent-Allocation Mechanism

The agent allocation module is used to select specific subtasks for each agent from a set of available subtasks. The output is the action space for the agent, i.e., $a = \{a_1, a_2, ..., a_M\}$. As part of the input, the agent's observation o_i is encoded into w_i, which is then fed into the agent allocation module. w_i is processed by another MLP module, and its output is multiplied by the encoded subtask state variables $v_j, j = 1, 2, ..., M$. The resulting output consists of Q-values associated with all possible subtask selections, denoted as $q_1, q_2, ..., q_M$, where M represents the number of subtasks. The subtask selection follows the well-known ϵ-greedy principle [18], ensuring a balance between exploration and exploitation. In the execution mode, the index corresponds to the maximum Q-value is the subtask that should be assigned the agent i. The maximum Q-value of all agents is denoted as $Q_1, Q_2, ..., Q_N$, where N is the number of agents. Then they are passed to the mixing network to compute the Q_{tot}. For agent allocation training, the loss at each decision time t is computed according to the Eq. (3), where r^t is the global reward, o denotes joint observation of agents and a denotes the joint action of agents. $v = \{v_j, j = 1, 2, ..., M\}$ represents states of all subtasks.

$$Loss = r^t + \gamma \max_{a'} Q_{tot}(o^{t+1}, v^{t+1}, a') - Q_{tot}(o^t, v^t, a) \tag{3}$$

3.3   The Overall Training

In this section, we present the training procedure of TSGraph-RL for dynamic agent allocation in multi-games, which is outlined in Algorithm 2. Each episode corresponds to a game round, where K denotes the maximum number of training episodes. The maximum time slots of the game is $\bar{T}$. For each time slot t, getting observations of all agents are encoded to o, and states of all subtasks s. With Algorithm 1, TSGraph is constructed based on running status of subtasks in BT, see Step 7. Then, TGN is used to extract features reflecting collaborative information among subtasks in Step 8. Steps 9–12 show the process of allocating agents to subtasks, and related samples are generated. Steps 14–15 show the updating of the whole neural network, and the keyword "UPDATE" is used to control the frequency of the updating.

Algorithm 2: TSGraph-RL

Input: A BT with the set of subtasks $\mathcal{M}$, set of agents $\mathcal{N}$
Output: Parameters of whole neural networks

1 **for** *episode* $= 1, 2, \ldots, L$ **do**
2 $\quad$ Reset the Environment and BT parameters;
3 $\quad$ Set $k = 0$ and initialize TSGraph $G^k = (V, E^k, T^k)$, i.e., $V = \{v_1, v_2, ..., v_M\}$, $E^k = \phi$, $T^k = \phi$;
4 $\quad$ **while** $t \leq \bar{T}$ **do**
5 $\quad\quad$ Collecting observations o_i^t and encoded vectors w_i^t for all agents $i \in \mathcal{N}$;
6 $\quad\quad$ Collecting all subtasks information $s_1^t, s_2^t, ..., s_M^t$;
7 $\quad\quad$ Constructing TSGraph by Algorithm 1, i.e., $G^k = TSGC(G^k, t)$;
8 $\quad\quad$ Getting features of subtasks with TGN, i.e., $u = MLP(TGN(G^0, ..., G^k, s^{t_{z^k}}))$;
9 $\quad\quad$ Computing Q-values $q_1, q_2, ..., q_M$ based on u and $w = MLP(o)$;
10 $\quad\quad$ Getting subtask a_i^t assigned to each agent i;
11 $\quad\quad$ Running the BT with allocation results, getting the game reward r^t;
12 $\quad\quad$ Storing the sample $(o_i^t, a_i^t, o_i^{t-1}, r^t)$ to buffer D;
13 $\quad$ **if** *UPDATE* **then**
14 $\quad\quad$ Randomly select samples from D;
15 $\quad\quad$ Computing *Loss* following (3), and perform gradient descent over the whole network.

---

4   Experiments

In this section, we evaluate the performance of our proposed method TSGraph-RL on a challenging task, i.e., Google Research Football (GRF) [9]. We specifically focus on the full game scenario, where 11 players compete against an opposing team consisting of 11 players, simulating a realistic football match. The GRF

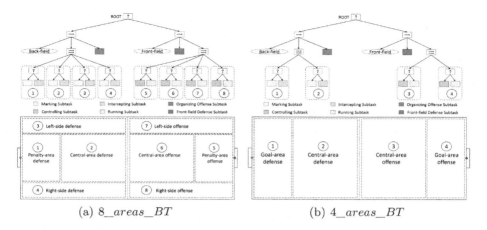

(a) 8_areas_BT   (b) 4_areas_BT

Fig. 2. Two scenarios with different number of subtasks used in the experiment.

environment presents unique and complex challenges, requiring a delicate balance between short-term control tasks such as passing, dribbling, and shooting, and long-term strategic planning. The observations of each player contain the positions and movement directions of teammates, opponents, and the ball. In our experiment, we divide the football ground into two zones, i.e., Back-field and Front-field, with different subtasks assigned to each area.

We have constructed two distinct Behavior Trees (BTs) consisting of different numbers of subtasks: namely, 8_areas_BT and 4_areas_BT, as illustrated in Fig. 2. In the 8_areas_BT scenario (Fig. 2(a)), the Back-field consists of one organizing offense subtask and eight defensive subtasks. Specifically, the eight defensive subtasks are responsible for executing marking and interception subtasks in the left-side, right-side, central-area, and penalty-area defensive areas. Upon gaining possession of the ball in the Back-field, the organizing offense subtask is activated. In the Front-field, we have one front-field defense subtask and eight offensive subtasks, each associated with performing controlling and passing subtasks in the offensive areas of the left-side, right-side, central-area, and penalty-area, respectively. The setting of subtasks in scenario 4_areas_BT is similar to that in scenario 8_areas_BT, with differences in the partition of areas, as shown in Fig. 2(b). We apply the proposed framework TSGraph-RL for the Backfield and Front-field, respectively.

All experiments were conducted on a computer with an i7-11700F CPU, RTX3060 GPU, and 32 GB of RAM. The discount factor was set to $\gamma = 0.99$. Optimization was performed using Adam with a learning rate of 5×10^{-4}. For exploration, we employed the $\epsilon-$greedy strategy with ϵ linearly annealed from 1.0 to 0.05 over 50,000 time steps and then kept constant for the remainder of the training.

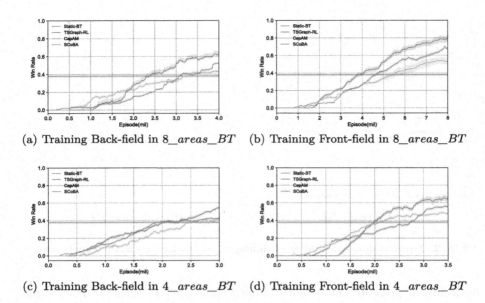

(a) Training Back-field in 8_*areas*_*BT*   (b) Training Front-field in 8_*areas*_*BT*

(c) Training Back-field in 4_*areas*_*BT*   (d) Training Front-field in 4_*areas*_*BT*

Fig. 3. Training performance comparison on two scenarios.

4.1   Performance Analysis

We evaluate the performance of TSGraph-RL by comparing with three base-lines in two scenarios. **Static-BT**: each player is controlled by the rules within the GRF platform; **SCoBA**: stochastic conflict-based allocation [3] aims to effi-ciently allocate tasks to robots in real-time while considering uncertain task durations and temporal constraints. **CapAM**: Capsule Attention Network [15] combines capsule networks and attention mechanisms to learn scalable policies for efficiently allocating tasks to multiple robots in complex environments. For each scenario, we adopt all methods to train each part of the BT, i.e., Front-field and Back-field, respectively.

Comparison of the Training Performance: The experimental results are shown in Fig. 3, where the x-axis represents the training set in millions, and the y-axis represents the win rate. For all curves, we can see that TSGraph-RL achieves the best performance over all scenarios. Figures 3(a) and 3(b) show the performance of all methods in the Back-field and Front-field of the 8_*areas*_*BT* scenario, respectively. As we can see, the win rate of TSGraph-RL, ScoBA and CapAM gradually increase with the training goes on. After a certain amount of training, The performance of TSGraph-RL is much better than that of Static-BT, the win rate increases by about 52.23% in Back-field, while 98.71% in Front-field. Moreover, ScoBA and CapAM are also better than static-BT, but not more than TSGraph-RL, which illustrates the effectiveness of TSGraph-RL. ScoBA outperforms CapAM, the result is suitable for dynamic tasks. We also note that the performance in the Front-field is slightly superior to that in the Back-field,

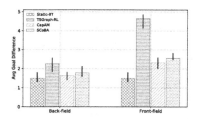   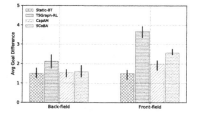

(a) Performance in 8_*areas_BT*   (b) Performance in 4_*areas_BT*

Fig. 4. Testing performance of average goal difference on two scenarios.

as it directly contributes to increasing the number of victories, while in the Back-field, the win rate is enhanced by improving the success rate of defense.

Furthermore, we conduct the same experiment on the 4_*areas_BT* scenario, which reduces number of subtasks compared to 8_*areas_BT* scenarios. The results are shown in Fig. 3(c) and Fig. 3(d). As we can see, similar conclusions are also held. Another notice is that TSGraph-RL performs worse in 4_*areas_BT* scenario than the scenario 8_*areas_BT*. This can be attributed to the fact that having more subtasks allows the agents to focus more on their specific responsibilities.

Comparison of the Evaluating Performance: We further investigate the impact of all methods on the overall performance of the 11vs11 full game. We run the game around 100 matches with the trained model. We assess performance using the average goal difference, which represents the average margin of goals scored or conceded by a team compared to their opponents. The results for all methods under two scenarios are shown in Fig. 4. The results clearly demonstrate that TSGraph-RL outperforms the other baselines in terms of the average goal difference in both scenarios. Specifically, in the Front-field of the 8_*areas_BT* scenario, TSGraph-RL achieves a remarkable increase of 4.61 in the average goal difference. These findings provide substantial evidence that TSGraph-RL significantly enhances the performance of the BT in the game.

4.2 Ablation Studies

In this section, we conducted ablation experiments to investigate the influence of the temporal subtask graph on TSGraph-RL in the 8_*areas_BT* scenario. We introduced three variants of TSGraph-RL. For the first one, we remove the temporal subtask graph and the TGN from the method, i.e., using state of subtasks s to replace x as the input the agent allocation module. It is named as Raw-RL. For the second one, we use Dynamic Graph Convolutional Network [12] to replace TCN in the method and we call it DGCN-RL. For the third one, we remove the GRU from the input of the memory of TGN, and use states of subtasks directly. It is named as TGN-noGRU-RL. The agent allocation module remains the same across all methods. The results of the ablation

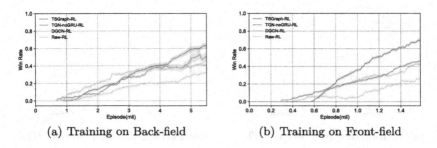

(a) Training on Back-field   (b) Training on Front-field

Fig. 5. Ablation studies in 8_areas_BT.

experiments are presented in Fig. 5. As we can see, TSGraph-RL outperforms Raw-RL, DGCN-RL, and TGN-RL, which shows that it is necessary to consider the cooperative relationship between subtasks for dynamic agent allocation in BT. Notably, TSGraph-RL outperforms TGN-RL, providing favorable evidence that local GRU modules help capture local state changes in subtasks.

5   Conclusion

In this paper, we have presented TSGraph-RL, a novel reinforcement learning framework based on temporal subtask graphs, designed to tackle the challenge of dynamic agent allocation in BT. TSGraph-RL utilizes a temporal subtask graph to effectively model the relationships among dynamically activated subtasks. And, the TGN network is introduced to extract key features from these subtasks. Moreover, we have proposed a novel agent allocation mechanism that has empowered each agent to make decisions regarding subtask selection. To validate our approach, we have conducted experiments on the GRF game, and the empirical results have demonstrated a significant performance enhancement in BTs when employing our framework. The proposed framework has served as a crucial complement to BTs, offering considerable time savings for the design of BTs. As future work, we aim to explore the optimization of various resource types for BTs and extend the applicability of our approach to large-scale games.

Acknowledgments. This work was financially supported by the National Natural Science Foundation of China Youth Science Foundation under Grants No. 61902425, No. 62102444.

References

1. Bitsakos, C., Konstantinou, I., Koziris, N.: DERP: a deep reinforcement learning cloud system for elastic resource provisioning. In: 2018 IEEE International Conference on Cloud Computing Technology and Science (CloudCom), pp. 21–29. IEEE (2018)

2. Chen, X., Liu, G.: Energy-efficient task offloading and resource allocation via deep reinforcement learning for augmented reality in mobile edge networks. IEEE Internet Things J. **8**(13), 10843–10856 (2021)
3. Choudhury, S., Gupta, J.K., Kochenderfer, M.J., Sadigh, D., Bohg, J.: Dynamic multi-robot task allocation under uncertainty and temporal constraints. Auton. Robot. **46**, 231–247 (2020)
4. Dey, R., Child, C.: QL-BT: enhancing behaviour tree design and implementation with Q-learning. In: 2013 IEEE Conference on Computational Intelligence in Games (CIG), pp. 1–8. IEEE (2013)
5. Groß, J., Gros, T.P.: Using deep reinforcement learning to optimize assignment problems (2021)
6. Hoff, J.W., Christensen, H.J.: Evolving behaviour trees:-automatic generation of AI opponents for real-time strategy games. Master's thesis, NTNU (2016)
7. Hu, H., Jia, X., Liu, K., Sun, B.: Self-adaptive traffic control model with behavior trees and reinforcement learning for AGV in industry 4.0. IEEE Trans. Ind. Inf. **17**(12), 7968–7979 (2021)
8. Iovino, M., Scukins, E., Styrud, J., Ögren, P., Smith, C.: A survey of behavior trees in robotics and AI. Robot. Auton. Syst. **154**, 104096 (2022)
9. Kurach, K., et al.: Google research football: a novel reinforcement learning environment. In: Proceedings of the AAAI Conference on Artificial Intelligence, vol. 34, pp. 4501–4510 (2020)
10. Li, L., Wang, L., Li, Y., Sheng, J.: Mixed deep reinforcement learning-behavior tree for intelligent agents design. In: ICAART, vol. 1, pp. 113–124 (2021)
11. Liu, C., Chen, C.X., Chen, C.: META: a city-wide taxi repositioning framework based on multi-agent reinforcement learning. IEEE Trans. Intell. Transp. Syst. (2021)
12. Manessi, F., Rozza, A., Manzo, M.: Dynamic graph convolutional networks. Pattern Recogn. **97**, 107000 (2020)
13. Marzinotto, A.: Flexible robot to object interactions through rigid and deformable cages. Ph.D. thesis, KTH Royal Institute of Technology (2017)
14. Neupane, A., Goodrich, M.A.: Designing emergent swarm behaviors using behavior trees and grammatical evolution. In: Proceedings of the 18th International Conference on Autonomous Agents and MultiAgent Systems, pp. 2138–2140 (2019)
15. Paul, S., Ghassemi, P., Chowdhury, S.: Learning scalable policies over graphs for multi-robot task allocation using capsule attention networks. In: 2022 International Conference on Robotics and Automation (ICRA), pp. 8815–8822 (2022)
16. Rossi, E., Chamberlain, B., Frasca, F., Eynard, D., Monti, F., Bronstein, M.: Temporal graph networks for deep learning on dynamic graphs. arXiv preprint arXiv:2006.10637 (2020)
17. Rovida, F., Grossmann, B., Krüger, V.: Extended behavior trees for quick definition of flexible robotic tasks. In: 2017 IEEE/RSJ International Conference on Intelligent Robots and Systems (IROS), pp. 6793–6800. IEEE (2017)
18. Tokic, M., Palm, G.: Value-difference based exploration: adaptive control between epsilon-greedy and softmax. In: Bach, J., Edelkamp, S. (eds.) KI 2011. LNCS (LNAI), vol. 7006, pp. 335–346. Springer, Heidelberg (2011). https://doi.org/10.1007/978-3-642-24455-1_33
19. Wang, J., Zhao, L., Liu, J., Kato, N.: Smart resource allocation for mobile edge computing: a deep reinforcement learning approach. IEEE Trans. Emerg. Top. Comput. **9**(3), 1529–1541 (2019)

A Distributed kWTA for Decentralized Auctions

Gary Sum[1], John Sum[2(✉)], Andrew Chi-Sing Leung[3], and Janet C. C. Chang[2]

[1] Swiss Capital Group, Hong Kong, China
gary.sum@swisscapitalasia.com
[2] ITM National Chung Hsing University, Taichung, Taiwan
pfsum@nchu.edu.tw
[3] EE City University of Hong Kong, Kowloon, Hong Kong
eeleungc@cityu.edu.hk

Abstract. A distributed k-Winner-Take-All (kWTA) is presented in this paper. Its state-space model is given by

$$\frac{d}{dt}x_i(t) = -\left[h(x_i(t) + u_i) - \frac{k}{n} + \beta \sum_{j \in \mathcal{N}_i} (x_i(t) - x_j(t)) \right]$$

$$z_i(x_i(t)) = h(x_i(t) + u_i) = \begin{cases} 1 \text{ if } x_i(t) + u_i > 0 \\ 0 \text{ if } x_i(t) + u_i \leq 0 \end{cases}$$

for $i = 1, \cdots, n$. Here, u_is and z_is are the inputs and the outputs; β is a positive constant and $\mathcal{N}_i$ is the neighbor set of the ith node. If $\beta \to 1$, both $x_i(t)$ and $z_i(t)$ converge in finite-time; $z_i = 1$ if and only if u_i is one of the k largest inputs; and $x_i \to -u_{\pi_{n-k}}$ (resp. $-u_{\pi_{n-k+1}}$) if $x_i(0) \ll -1$ (resp. $x_i(0) = 0$). Accordingly, our kWTA is the best algorithm to be applied as a decentralized mechanism for sealed-bid first (resp. second) price auction if $\beta \to \infty$, $k = 1$ and u_is are the bid prices. Compared with the conventional mechanism, our novel mechanism is able to protect the privacy of the bidders. Auctioneer can be ripped out. Bidders can only know the value of the payment and losing bidders cannot know who is the winner.

Keywords: Differential Inclusion · Distributed kWTA · Finite-Time Convergence · Sealed-Bid Auction

1 Introduction

For decades, many Winner-Take-All (WTA) and k-Winner-Take-All (kWTA) models have been developed [1–17]. A kWTA network is a system with n inputs, $u_1, \ldots, u_n$, and n outputs, $z_1, \cdots, z_n$. Given $u_1, \cdots, u_n$, the purpose of a kWTA is to find which k inputs are the k largest. One representation is to set $z_i = 1$ if u_i is one of the k largest inputs, i.e.

$$z_i = \begin{cases} 1 \text{ if } u_i \text{ is one of the } k \text{ largest inputs,} \\ 0 \text{ otherwise.} \end{cases}$$

© The Author(s), under exclusive license to Springer Nature Singapore Pte Ltd. 2024
B. Luo et al. (Eds.): ICONIP 2023, CCIS 1962, pp. 136–147, 2024.
https://doi.org/10.1007/978-981-99-8132-8_11

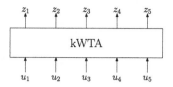

Fig. 1. An kWTA network with five inputs $u_1, \cdots, u_5$ and five outputs $z_1, \cdots, u_5$.

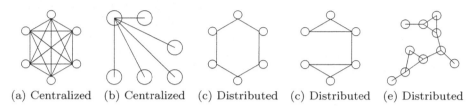

(a) Centralized   (b) Centralized   (c) Distributed   (c) Distributed   (e) Distributed

Fig. 2. Centralized structure (a, b) versus distributed structure (c, d, e).

Figure 1 shows a schematic diagram for a kWTA network with five inputs and five outputs. kWTA has been succeeded in solving various engineering problems, its application in auction has yet to be investigated. In this paper, a distributed kWTA[1] is introduced and its awesome application as a decentralized mechanism for sealed-bid first (resp. second) price auction with privacy protection is revealed.

To start with, let us highlight the differences between the centralized kWTA and the distributed kWTA. Our kWTA is a distributed model for solving the kWTA problem (resp. sealed-bid auction) in a distributed network of agents (resp. bidders). Figure 2 shows the topological differences between centralized and distributed kWTA models. For the centralized models Fig. 2a, b, at least one node is connected to all others. For the distributed models, Fig. 2c, d, e, no any node is able to connect to all other nodes. Under such topological constraint, algorithm design for solving the kWTA problem is already a challenge. The design of a decentralized mechanism for sealed-bid auction with privacy protection is more challenging.

1.1   Centralized kWTA Network

Inspired by the Hopfield network [18], early kWTA models [1,6–8] were developed. They are fully connected recurrent networks with state-space models generally given by $\frac{dx_i}{dt} = f(u_i, \mathbf{x}, \mathbf{z})$ and $z_i = h(x_i)$ for $i = 1, \cdots, n$, where $f(\cdot)$ is a function of u_i, $\mathbf{x}$ and $\mathbf{z}$. $z_i = h(x_i)$ is the ith output and $h(\cdot)$ is a scalar function.

Later, by formulating the kWTA problem as a constraint optimization problem, various kWTA were proposed [9,11–13,19]. These models cannot be associ-

---

[1] In this paper, kWTA network, kWTA model and kWTA algorithm are used interchangeably.

ated with any neuronal network. Like the model introduced [12], its state-space model is given by

$$\frac{dy}{dt} = \sum_{i=1}^{n} h(u_i - y(t)) - k \text{ and } z_i = h(u_i - y(t)) = \begin{cases} 1 \text{ if } u_i - y(t) > 0 \\ 0 \text{ if } u_i - y(t) \leq 0. \end{cases} \quad (1)$$

For the models as introduced in [13,19], their state-space models even more complicated as they consist of multiple state equations. $dx/dt = F(\mathbf{u}, \mathbf{x}, \mathbf{y}, \mathbf{z})$, $dy/dt = G(\mathbf{u}, \mathbf{x}, \mathbf{y}, \mathbf{z})$ and $\mathbf{z} = H(\mathbf{u}, \mathbf{x}, \mathbf{y})$, where $F(\cdot)$, $G(\cdot)$ and $H(\cdot)$ are nonlinear vector functions. A state vector could be the Lagrangian multiplier. As the dynamics of each of these state vectors $\mathbf{x}$ (and/or $\mathbf{y}$) depends on state information of all the nodes, they are centralized kWTA [20].

1.2   Distributed kWTA Network

In contrast to *centralized k WTA*, *distributed k WTA* is a distributed algorithm to be implemented in a distributed network of agents. One reason is due to an increasing attention on the control problems on a distributed network of agents (like mobile sensors and unmanned vehicles) [21–23] and the kWTA problem is a key problem to be solved for the target tracking/elimination problems [20,24,25]. In general form, their dynamics could simply be stated as a nonlinear state-space model. $dx_i/dt = f(x_j \mid j \in i \cup \mathcal{N}_i)$ and $z_i = h(x_i)$ for $i = 1, \cdots, n$ and $f(\cdot)$ is nonlinear scalar function of u_i and x_j for $j \in i \cup \mathcal{N}_i$. $h(\cdot)$ is a nonlinear function of x_i only. One should be noted that their nonlinear models [20,24,25] consist of at least two state equations. As the update of an agent's state variables depends on the information of its neighbor agents, these models are coined the *distributed k WTA models* [20].

1.3   Decentralized Auction

In conventional design for sealed-bid auctions [26], bidders submit their bid prices to an auctioneer. The auctioneer then determines which bidder is the winner (i.e. *winner determination*) and how much the winner has to pay (i.e. *payment determination*) for the item. Under such circumstance, the auctioneer will get all the information of the bid prices. One problem of particular concern is that the losers might not want their bid prices to be disclosed. Thus, there is an interest in developing decentralized auction mechanisms to rip out the auctioneer and to have privacy protection. This leads to recent researches on the development of decentralized auctions on blockchain platforms [27,28].

Instead of using a blockchain platform, we consider a network of n bidders whose cell phones are wireless connected and each cell phone has been installed a software agent for auction. As each cell phone can only communicate with the cell phones within its communication range, each agent can only exchange information among the agents within its communication range. In the end, the algorithm to be implemented in an agent must be in distributed algorithm.

If (i) each bidder submits the bid via the software agent, (ii) the bid prices are denoted as u_is and (iii) $k = 1$, the distributed kWTA models [20,24,25] can be applied in *winner determination* for a sealed-bid first (resp. second) price auction. However, they cannot be applied in *payment determination*. Therefore, it is interesting to develop a distributed kWTA and applied in decentralized auctions.

1.4   Organization of the Paper

In this regards, this paper introduces a distributed kWTA model which is particularly suitable to be applied in the sealed-bid first (resp. second) price auctions. Our distributed kWTA consists of only one state equation in contrast to multiple state equations in the models introduced in [20,24,25]. *Applying our distributed kWTA as a decentralized implementation of an auction, not just the winner can be determined by our proposed kWTA, but also the payment can be determined.* The contributions of this paper are two folds. First, a distributed kWTA with Heaviside outputs and single state equation is developed. Second, its applications in the first price auction and the second price auction are shown.

In the next section, the distributed kWTA is introduced and its properties are sketched. The applications of the distributed kWTA for auctions are presented in Sect. 3. Illustrative examples are given in the section to highlight the limitations of the applications. Finally, the conclusions of the paper are presented in Sect. 4.

2   Our Distributed kWTA

Our proposed distributed kWTA has n nodes its dynamics is given by the following state-space model.

$$\frac{dx_i(t)}{dt} = -\left[z_i(x_i(t)) - \frac{k}{n} + \beta \sum_{j \in \mathcal{N}_i} (x_i(t) - x_j(t)) \right] \tag{2}$$

$$z_i(x_i(t)) = h(x_i(t) + u_i) = \begin{cases} 1 \text{ if } x_i(t) + u_i > 0, \\ 0 \text{ if } x_i(t) + u_i \le 0, \end{cases} \tag{3}$$

for $i = 1, \cdots, n$, where $\mathcal{N}_i$ is the set of the neighbors of the ith node and $\beta \gg 1$.

In vector form, (2) could be equivalently written as follows:

$$\frac{d}{dt}\mathbf{x} = -\left[\mathbf{z}(\mathbf{x}) - \frac{k}{n}\mathbf{1} + \beta \mathbf{L}\mathbf{x} \right] \text{ and } \mathbf{z}(\mathbf{x}(t)) = \mathbf{h}(\mathbf{x}(t) + \mathbf{u}), \tag{4}$$

$\beta \gg 1$, $\mathbf{1} = (1, \cdots, 1)^T$. $\mathbf{L}$ is a graph Laplacian matrix, i.e. $(\mathbf{L})_{ij} = |\mathcal{N}_i|$ if $i = j$ and $(\mathbf{L})_{ij} = -1$ if $j \in \mathcal{N}_i$. $\mathbf{h}(\mathbf{x}(t) + \mathbf{u}) = (h(x_1(t) + u_1), \cdots, h(x_n(t) + u_n))^T$. It should also be noted that $\mathbf{z}(\mathbf{x})$ (equivalently, $\mathbf{h}(\mathbf{x}+\mathbf{u})$) is a discontinuous function of $\mathbf{x}$. Thus, state Eq. (2) (resp. (4)) is a differential equation with discontinuous right hand side [29,30]. Analyzing the properties of (2) can be investigated via the following differential inclusion:

$$\frac{d\mathbf{x}(t)}{dt} \in \mathcal{F}(\mathbf{x}(t)), \tag{5}$$

where $\mathcal{F}(\mathbf{x})$ is a set-valued function defined as follows:

$$\mathcal{F}(\mathbf{x}) \triangleq \bigcap_{\delta > 0} \bigcap_{\mu(\mathcal{N})=0} \overline{co} \left\{ \mathbf{f}(B(\mathbf{x}, \delta) \setminus \mathcal{N}) \right\} \text{ and } \mathbf{f}(\mathbf{x}) = -\mathbf{h}(\mathbf{x}+\mathbf{u}) + \frac{k}{n}\mathbf{1} - \beta \mathbf{L}\mathbf{x}, \tag{6}$$

where $B(\mathbf{x}, \delta) = \{\mathbf{x}' \in R^n \mid \|\mathbf{x}' - \mathbf{x}\| \leq \delta\}$. $\mathbf{f}(B(\mathbf{x}, \delta) \setminus \mathcal{N})$ is the image of $\mathbf{f}(\cdot)$ on the domain $B(\mathbf{x}, \delta) \setminus \mathcal{N}$ and $\mathcal{N} = \bigcup_{i=1}^{n} \{ \mathbf{x} \in R^n \text{ and } x_i + u_i = 0 \}$. Clearly, the Lebesgue measure $\mu(\mathcal{N})$ is zero. $\overline{co}$ denotes convex closure.

2.1   Properties

By the theory in discontinuous dynamic systems and differential inclusions [29–31], the properties for (5) can be analyzed.

Theorem 1 (Existence and Uniqueness of Filippov Solution). *There exists a unique absolutely continuous Filippov solution $\mathbf{x}(t)$ for the kWTA (4) for any initial condition $\mathbf{x}(0)$.*

Poof: As $\mathcal{F}(\mathbf{x})$ as defined in (6) is upper semicontinuous and by [30, P.70, Theorem 1], there exists a unique absolutely continuous Filippov solution $\mathbf{x}(t)$ for the kWTA (4) for any initial condition $\mathbf{x}(0)$. **Q.E.D.**

Theorem 2 (Energy Function). *The energy function $V(\mathbf{x})$ of the distributed kWTA (4) is given by*

$$V(\mathbf{x}) = \sum_{i=1}^{n} \max\{0, x_i + u_i\} - \frac{k}{n}\sum_{i=1}^{n} x_i + \frac{\beta}{2}\sum_{i=1}^{n}\sum_{j=1}^{n} L_{ij}x_i x_j. \tag{7}$$

Hence, for all $0 \leq t \leq t'$, $V(\mathbf{x}(t')) \leq V(\mathbf{x}(t))$.

Poof: Note that $\mathcal{F}(\mathbf{x}) = -\partial V(\mathbf{x})$. $\partial V(\mathbf{x})$ is the subgradient of $V(\mathbf{x})$ at $\mathbf{x}$. The differential inclusion (5) is in fact a gradient differential inclusion its energy function is $V(\mathbf{x})$. Thus, $V(\mathbf{x}(t')) \leq V(\mathbf{x}(t))$ for all $t' \leq t$. **Q.E.D.**

Note that $V(\mathbf{x})$ is a nonsmooth convex function, the set of minimum points Ω is also a convex set. For $\beta \to \infty$, it can be shown that the network gives correct output, i.e. $z_i = 1$ if and only if u_i is one of the k largest inputs. To be precise, the definition of correct output is stated below.

Definition 1 (Correct Output). A kWTA network has correct output if and only if $z_i = 1$ and u_i is one of the k largest inputs.

For the following analysis, we need the definitions for the set Ω, $\mathcal{R}_*$ and the vector $\mathbf{b}_*$.

Definition 2 $(\Omega, \mathcal{R}_\star, \mathbf{b}_\star)$.

$$\Omega = \left\{ \mathbf{x} \mid \mathbf{h}(\mathbf{x} + \mathbf{u}) - \frac{k}{n}\mathbf{1} + \beta\mathbf{Lx} = \mathbf{0} \right\} \tag{8}$$

$$\mathbf{b}_\star = \{\mathbf{h}(\mathbf{x} + \mathbf{u}) \mid \mathbf{x} \in \Omega\} \tag{9}$$

$$\mathcal{R}_\star = \left\{ \mathbf{x} \mid \mathbf{b}_\star - \frac{k}{n}\mathbf{1} + \beta\mathbf{Lx} = \mathbf{0} \right\}. \tag{10}$$

Theorem 3. *If $\beta \to \infty$, the distributed kWTA (4) gives correct output.*

Poof: For the equilibrium set Ω, we can get that

$$\mathbf{h}(\mathbf{x} + \mathbf{u}) - \frac{k}{n}\mathbf{1} + \beta\mathbf{Lx} = \mathbf{0} \quad \forall \mathbf{x} \in \Omega. \tag{11}$$

By Theorem 2, (5) is a gradient differential inclusion and $\mathbf{x}(t)$ must converge to Ω. That is to say, $\mathbf{h}(\mathbf{x} + \mathbf{u}) \to \mathbf{b}_\star$. The left hand side of (11) can be rewritten as follows:

$$\mathbf{h}(\mathbf{x} + \mathbf{u}) - \frac{k}{n}\mathbf{1} + \beta\mathbf{Lx} = \mathbf{b}_\star - \frac{k}{n}\mathbf{1} + \beta\mathbf{Lx} \tag{12}$$

$$= \sum_{i=1}^{n} \mathbf{v}_i \mathbf{v}_i^T \left(\mathbf{b}_\star - \frac{k}{n}\mathbf{1} \right) + \beta \sum_{i=2}^{n} \lambda_i \mathbf{v}_i \mathbf{v}_i^T \mathbf{x}, \tag{13}$$

where λ_i and $\mathbf{v}_i$ are i^{th} eigenvalues and the ith eigenvector of the Laplacian matric $\mathbf{L}$. Moreover, $\mathbf{v}_1^T \left(\mathbf{b}_\star - \frac{k}{n}\mathbf{1} \right) = 0$ and $\mathbf{v}_i^T \left(\mathbf{b}_\star - \frac{k}{n}\mathbf{1} + \beta\lambda_i\mathbf{x} \right) = 0$ for $i = 2, \cdots, n$. In other words, $\sum_{i=1}^{n}(\mathbf{b}_\star)_i = k$ and $\mathbf{v}_i^T\mathbf{x} = (\beta\lambda_i)^{-1}\mathbf{v}_i^T \left(\mathbf{b}_\star - \frac{k}{n} \right) \to 0$ for $\beta \to \infty$ and $i = 2, \cdots, n$. $\mathbf{x}$ is almost orthogonal to the eigenvectors $\mathbf{v}_2, \cdots, \mathbf{v}_n$ and hence $\mathbf{x}$ is almost in parallel with $\mathbf{v}_1$. That is to say, $x_i \to \bar{x}$ for $i = 1, \cdots, n$ and $\bar{x}$ falls in the range of $(-u_{\pi_{n-k+1}})^+$ and $(-u_{\pi_{n-k}})$. As a result, the outputs of the kWTA is correct because $z_{\pi_{n-k+1}} = \cdots = z_{\pi_n} = 1$ and $z_{\pi_1} = \cdots = z_{\pi_{n-k}} = 0$. **Q.E.D.**

Theorem 4 (Finite-Time Convergence). *If $\beta \to \infty$, $dV(\mathbf{x}(t))/dt \leq -1/n$ for $\mathbf{x}(t) \in \mathcal{R}_\star$. Thus, $\mathbf{x}(t)$ reaches $\mathcal{R}_\star$ in finite-time. The output vector $\mathbf{z}(\mathbf{x}(t))$ converges to $\mathbf{b}_\star$ in finite-time and hence our kWTA gives correct output in finite-time. Moreover, $\mathbf{x}(t)$ converges in finite-time.*

Poof: From Theorem 2, it is clear that the kWTA is a gradient differential inclusion, i.e. $d\mathbf{x}/dt \in -\partial V(\mathbf{x})$. For $\mathbf{x}(t) \in \mathcal{R}_\star$, we get that

$$\frac{dV(\mathbf{x}(t))}{dt} = -\left\| \mathbf{h}(\mathbf{x} + \mathbf{u}) - \frac{k}{n}\mathbf{1} + \beta\mathbf{Lx} \right\|^2 = -\|\mathbf{b}_\star - \mathbf{h}(\mathbf{x} + \mathbf{u}) + \beta\mathbf{L}(\mathbf{x}_\star - \mathbf{x})\|^2$$

$$< -\left\{ (\mathbf{b}_\star - \mathbf{h}(\mathbf{x} + \mathbf{u}))^T \mathbf{v}_1 \right\}^2 < -1/n.$$

The last inequality is due to the fact that $\mathbf{b}_\star$ and $\mathbf{h}(\mathbf{x}+\mathbf{u})$ must have at least one element in difference. As $V(\mathbf{x})$ is lower bounded and $\dot{\mathbf{x}}(t)$ is a gradient differential inclusion, $\mathbf{x}(t)$ must reach $\mathcal{R}_\star$ in finite-time. The output vector $\mathbf{z}(\mathbf{x}(t))$ converges to $\mathbf{b}_\star$ in finite-time and hence our kWTA gives correct output in finite-time.

Once $\mathbf{x}(t)$ has entered $\mathcal{R}_\star$, the state equation can be written as follows:

$$\frac{d\mathbf{x}}{dt} = -\left[\mathbf{b}_\star - \frac{k}{n}\mathbf{1} + \beta\mathbf{L}\mathbf{x}(t)\right], \tag{14}$$

where $\mathbf{b}_\star$ is the output vector of correct output. Let $y_i(t) = \mathbf{v}_i^T\mathbf{x}(t)$, we get that

$$\frac{dy_1}{dt} = 0 \text{ and } \frac{dy_i}{dt} = -\beta\lambda_i\left[y_i(t) + (\beta\lambda_i)^{-1}\mathbf{v}_i^T\left(\mathbf{b}_\star - \frac{k}{n}\mathbf{1}\right)\right]$$

for $i = 2,\cdots,n$. Let $\tilde{y}_i(t)$ be $\left[y_i(t) + (\beta\lambda_i)^{-1}\mathbf{v}_i^T\left(\mathbf{b}_\star - \frac{k}{n}\mathbf{1}\right)\right]$. For $\beta \to \infty$, $\tilde{y}_i(t)$ converges almost instantaneously and hence $\mathbf{x}(t)$ converges almost instantaneously once $\mathbf{x}(t)$ has entered $\mathcal{R}_\star$. $\mathbf{x}(t)$ converges in finite-time and its time of convergent is the time it arrives $\mathcal{R}_\star$. The proof is completed. **Q.E.D.**

Theorem 5 (Upper bound on $\mathcal{T}_\star$). *If (i) $\mathbf{x}(0) = \mathbf{0}$ and (ii) $u_i > 0$ for $i = 1,\cdots,n$, the proposed kWTA gives correct output before the time $\mathcal{T}_\star$, where $\mathcal{T}_\star < n(2n + k)\max_i\{u_i\}$.*

Poof: Trivial! **Q.E.D.**

For $\mathbf{x} = \alpha\mathbf{v}_1 - \beta^{-1}\sum_{i=2}^n \lambda^{-1}\mathbf{v}_i\mathbf{v}_i^T((k/n)\mathbf{1} - \mathbf{h})$, where α is a constant and $\mathbf{v}_i$ is the eigenvector of the corresponding eigenvalue λ_i. For $\beta \to \infty$, it can be shown that $x_1(t) = \cdots = x_n(t)$ for $t \to \infty$ and thus the dynamics of (4) resembles a distributed implementation of the Wang kWTA [12].

Theorem 6. *Let $0 \le u_i \le M$. If $x_i(0) > 0$ for $i = 1,\cdots n$, $\lim_{t\to\infty} x_i(t) \approx -\left(u_{\pi_{n-k}}\right)$ for $i = 1,\cdots,n$. If $x_i(0) < -M$ for $i = 1,\cdots n$, $\lim_{t\to\infty} x_i(t) \approx -\left(u_{\pi_{n-k+1}}\right)$ for $i = 1,\cdots,n$.*

Poof: Trivial! **Q.E.D.**

2.2   Illustration

To illustrate this property, two simulated experiments have been conducted for $n = 10$, $\beta = 100$ and u_is are random integers ranging from 0 to 20, i.e. $M = 20$. The network is set to be a circular network. That is to say, the Laplacian matrix is given by

$$\mathbf{L} = \begin{bmatrix} 2 & -1 & 0 & \cdots & 0 & -1 \\ -1 & 2 & -1 & \cdots & 0 & 0 \\ 0 & -1 & 2 & \cdots & 0 & 0 \\ \vdots & \vdots & \vdots & \ddots & \vdots & \vdots \\ -1 & 0 & 0 & \cdots & -1 & 2 \end{bmatrix}. \tag{15}$$

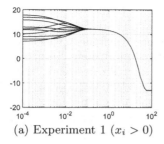
(a) Experiment 1 ($x_i > 0$)

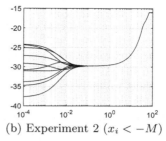
(b) Experiment 2 ($x_i < -M$)

Fig. 3. Illustrative examples for Theorem 6. The time scale is in second.

For the first experiment, $x_i(0)$s are initialized as random integers ranging from 0 to 20 and the inputs u_is are given by {7, 13, 18, 7, 16, 18, 2, 11, 12, 8}. For the second experiment, $x_i(0)$s are initialized as random integers ranging from -40 to -20 and the inputs u_is are given by {16, 19, 20, 3, 8, 2, 5, 8, 12, 3}. Figure 3 shows the changes of the values in the vector $\mathbf{x}(t)$. Clearly, vector $\mathbf{x}(t)$ converges in both experiments. Besides, the values of $x_i(t)$s converge to a similar value after $t = 0.1$.

For the first experiment, $x_i(t)$ converges to somewhere close to -12 and 12 is the value of u_{π_7}. If we take a close look at the values of $x_i(\infty)$s, one will notice that their values are slightly different as depicted below.

x_1	x_2	x_3	x_4	x_5	x_6	x_7	x_8	x_9	x_{10}
-12.995	-13.000	-13.009	-13.010	-13.015	-13.012	-13.003	-12.996	-12.993	-12.992

For the second experiment, $x_i(t)$ converges to somewhere close to -16 and 16 is the value of u_{π_8}. Again, if we take a close look at the values of $x_i(\infty)$s, one will notice that their values are slightly different as depicted below.

x_1	x_2	x_3	x_4	x_5	x_6	x_7	x_8	x_9	x_{10}
-16.000	-16.004	-16.000	-15.990	-15.982	-15.978	-15.976	-15.978	-15.982	-15.990

While not shown in this paper, one would imagine that the deviation of $x_i(\infty)$s decreases as β increases.

3   Decentralized Auction Mechanisms

Auction is considered as a fair mechanism for selling a commodity. In the conventional sealed-bid auction mechanism setting, all the bidders submit their *sealed bids* to the auctioneer on an item. The auctioneer will allocate the item to the bidder with the highest bid price and subsequently determine the payment the winner has to pay for the item.

3.1 Second Price Auction

Sealed-bid second-price auction [32] is an auction mechanism which could encourage the bidders to reveal their true valuations on an item. In accordance with the Theorem 6, the states $x_i(t)$ for $i = 1, \cdots, n$ will converge to the negative of the second highest value, i.e. $-u_{\pi_{n-1}}$, if $x_i(0) \geq 0$ for $i = 1, \cdots, n$. In this regard, the proposed kWTA with $k = 1$ could be applied as a decentralized sealed-bid second price auction mechanism. If we restrict that the bid prices must be integers ranging from 0 and M; and assume that all the bid prices are different. Our proposed kWTA could be applied to solve the winner determination problem in sealed-bid second price auction and at the same time the winning bidder is able to know how much he/she has to pay to the auctioneer.

S1. For $i = 1, \cdots, n$, the ith bidder sets the bid price as u_i.
S2. For $i = 1, \cdots, n$, the ith bidder initializes a random positive integer for $x_i(0)$.
S3. Updating $x_1(t), \cdots, x_n(t)$ for $t = 0, \Delta t, \cdots, 10^3$.
S3.1. For $i = 1, \cdots, n$, the ith bidder communicates with his/her neighbors the value of $x_i(t)$.
S3.2. For $i = 1, \cdots, n$, the ith bidder updates $x_i(t)$ based on (4), i.e.

$$x_i(t + \Delta t) = x_i(t) - f_i(\mathbf{x}(t))\Delta t, \tag{16}$$

$$(\mathbf{x}(t)) = h(x_i(t) + u_i) - \frac{1}{n} + \beta \sum_{j \in \mathcal{N}_j} (x_i(t) - x_j(t)). \tag{17}$$

S4. The bidder with $h(x_i + u_i) = 1$ is the winner. The payment is $-x_i(t)$.

As $\lim_{t \to \infty} x_i(t) = -u_{\pi_{n-1}}$, the winner can know how much he/she has to pay for the item. An advantage of this auction mechanism is that all the bidders can only know (i) the information of the second highest bid price based on the value of $-x_i(\infty)$ and (ii) the information if he/she is a winner based on the value of $z_i(\infty)$. The information of the bid prices of the other bidders is unknown to every bidder. Even the item seller, i.e. the auctioneer, cannot have the information about the bidding prices.

For illustration, it is assumed that there are five bidders and each bidder holds a mobile device. The devices have been connected via Bluetooth to form a ring network. Each bidder submits his/her bid via an APP running in the device. Their bid prices are depicted as below.

Bidders	1	2	3	4	5
Bids	68	89	99	3	30
$-x_i(\infty)$	88.998	89.000	89.004	89.000	88.998
$-[x_i(\infty)]$	89	89	89	89	89
$z_i(\infty)$	0	0	1	0	0

The change of $\mathbf{x}(t)$ against time is shown in Fig. 4. Clearly, the values of $x_i(t)$s converge almost to same value after $t \geq 0.03$ s. Around 300 s, the values of $-x_i(\infty)$ for $i = 1, \cdots, n$ converge to a value close to 89. It is the second highest bid price.

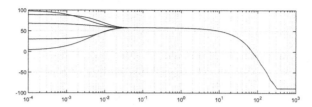

Fig. 4. Change of $\mathbf{x}(t)$ against time t. Here, $n = 5$, $\beta = 100$ and $\Delta t = 10^{-4}$. The bid prices are 68, 89, 99, 3 and 30. For $t > 300$, $-x_i(t)$s converge and $-x_i(t) \approx 89$.

3.2 First Price Auction

In the setting of a sealed-bid first price auction, all bidders submit their bids for an item. An auctioneer will determine the winner to be the one with the highest bid. The winner will have to pay for the item with the highest price.

The principle of applying the distributed kWTA, as stated in (2) and (4), for First price auction is essentially the same as for the case of sealed-bid second price auction. Assuming that all the bids are positive integers ranging from 0 to M, the value of k is set to one and the initial conditions of $x_i(0)$s are set to be ranging from $-2M$ to $-M$. In this regard, the $z_i(\infty)$ of the winner will be one and the $x_i(\infty)$s will converge to the value close to the highest bid. Rounding the value $-x_i(\infty)$ to its nearest integer, the winner will know how much to pay for the bidding item.

4   Conclusions

In this paper, various kWTA models have been introduced and the limitation of applying exiting kWTA models in solving decentralized auction problems in a distributed network of agents have been elucidated in Sect. 1. In sequel, a simple distributed kWTA has been presented. Some of its key properties, including (i) finite-time convergence and (ii) $-x_i(t) \to u_{\pi_n}$ (resp. $-x_i(t) \to u_{\pi_{n-1}}$) for $i = 1, \cdots, n$, are analyzed in Sect. 2. In accordance with these nice properties, the proposed kWTA is particular useful for implementation as a decentralized mechanism for sealed-bid first (resp. second) price auctions. Subsequently, its applications in sealed-bid auctions with privacy protection have been presented in Sect. 3. As a matter of fact, our distributed kWTA algorithm can be applied in other auction mechanisms, such as uniform price auction and discriminatory auction. Those results will be presented elsewhere.

Acknowledgements. The work presented in this paper is supported in part by research grants from the Taiwan Ministry of Science and Technology (MOST) and National Science and Technology Council (NSTC) (110-2221-E-005-053, 111-2221-E-005-084 and 112-2221-E-005-076); and a *Sustainable Technology Fund* (STF) from Swiss Capital Group, Hong Kong.

References

1. Lazzaro, J., Ryckebusch, S., Mahowald, M., Mead, C.: Winner-take-all networks of $\mathcal{O}(N)$ complexity. In: Advances in Neural Information Processing Systems. Morgan Kaufmann, pp. 703–711 (1989)
2. Seiler, G., Nossek, J.A.: Winner-take-all cellular neural networks. IEEE Trans. Circuits Syst. II Analog Digit. Sig. Process. **40**(3), 184–190 (1993)
3. Andrew, L.L.: Improving the robustness of winner-take-all cellular neural networks. IEEE Trans. Circuits Syst. II Analog Digit. Sig. Process. **43**(4), 329–334 (1996)
4. Sum, J.P., Leung, C.-S., Tam, P.K., Young, G.H., Kan, W.-K., Chan, L.-W.: Analysis for a class of winner-take-all model. IEEE Trans. Neural Networks **10**(1), 64–71 (1999)
5. Marinov, C.A., Costea, R.L.: Time-oriented synthesis for a WTA continuous-time neural network affected by capacitive cross-coupling. IEEE Trans. Circuits Syst. I Regul. Pap. **57**(6), 1358–1370 (2010)
6. Majani, E., Erlanson, R., Abu-Mostafa, Y.S.: On the K-winners-take-all network. In: Touretzky, D.S. (ed.) Advances in Neural Information Processing Systems, vol. 1, pp. 634–642. Morgan-Kaufmann (1988). http://papers.nips.cc/paper/157-on-the-k-winners-take-all-network.pdf
7. Calvert, B.D., Marinov, C.A.: Another k-winners-take-all analog neural network. IEEE Trans. Neural Networks **11**(4), 829–838 (2000)
8. Marinov, C.A., Calvert, B.D.: Performance analysis for a k-winners-take-all analog neural network: basic theory. IEEE Trans. Neural Networks **14**(4), 766–780 (2003)
9. Hu, X., Wang, J.: An improved dual neural network for solving a classs of quadratic programming problems with its k-winners-take-all application. IEEE Trans. Neural Networks **19**(12), 2022–2031 (2008)
10. Liu, Q., Wang, J.: Two k-winners-take-all networks with discontinuous activation functions. Neural Netw. **21**(2–3), 406–413 (2008)
11. Liu, Q., Dang, C., Cao, J.: A novel recurrent neural network with one neuron and finite-time convergence for k-winners-take-all operation. IEEE Trans. Neural Networks **21**(7), 1140–1148 (2010)
12. Wang, J.: Analysis and design of k-winners-take-all model with a single state variable and the heaviside step activation function. IEEE Trans. Neural Networks **21**(9), 1496–1506 (2010)
13. Zhang, Y., Li, S., Geng, G.: Initialization-based k-winners-take-all neural network model using modified gradient descent. IEEE Trans. Neural Networks Learn. Syst. (2021). Early access
14. Qi, Y., Jin, L., Luo, X., Shi, Y., Liu, M.: Robust k-WTA network generation, analysis, and applications to multiagent coordination. IEEE Trans. Cybern. **52**(8), 8515–8527 (2022)
15. Qi, Y., Jin, L., Luo, X., Zhou, M.: Recurrent neural dynamics models for perturbed nonstationary quadratic programs: a control-theoretical perspective. IEEE Trans. Neural Networks Learn. Syst. **33**(3), 1216–1227 (2022)

16. Liu, M., Zhang, X., Shang, M., Jin, L.: Gradient-based differential kWTA network with application to competitive coordination of multiple robots. IEEE/CAA J. Automatica Sinica **9**(8), 1452–1463 (2022)
17. Liang, S., Peng, B., Stanimirović, P.S., Jin, L.: Design, analysis, and application of projected k-winner-take-all network. Inf. Sci. **621**, 74–87 (2023)
18. Hopfield, J.J.: Neural networks and physical systems with emergent collective computational abilities. Proc. Natl. Acad. Sci. **79**(8), 2554–2558 (1982)
19. Qi, Y., Jin, L., Luo, X., Shi, Y., Liu, M.: Robust k-WTA network generation, analysis, and applications to multiagent coordination. IEEE Trans. Cybern. **52**(8), 8515–8527 (2021)
20. Zhang, Y., Li, S., Xu, B., Yang, Y.: Analysis and design of a distributed k-winners-take-all model. Automatica **115**, 108868 (2020)
21. Fax, J.A., Murray, R.M.: Information flow and cooperative control of vehicle formations. IEEE Trans. Autom. Control **49**(9), 1465–1476 (2004)
22. Ren, W., Beard, R.W., Atkins, E.M.: A survey of consensus problems in multi-agent coordination. In: Proceedings of the 2005 American Control Conference, pp. 1859–1864. IEEE (2005)
23. Cortés, J.: Finite-time convergent gradient flows with applications to network consensus. Automatica **42**(11), 1993–2000 (2006)
24. Li, S., Zhou, M., Luo, X., You, Z.-H.: Distributed winner-take-all in dynamic networks. IEEE Trans. Autom. Control **62**(2), 577–589 (2017)
25. Zhang, Y., Li, S., Weng, J.: Distributed k-winners-take-all network: an optimization perspective. IEEE Trans. Cybern. (2022). Early access
26. Krishna, V.: Auction Theory. Academic Press (2009)
27. Shi, Z., de Laat, C., Grosso, P., Zhao, Z.: When blockchain meets auction models: a survey, some applications, and challenges. arXiv e-prints, arXiv-2110 (2021)
28. Shi, Z., de Laat, C., Grosso, P., Zhao, Z.: Integration of blockchain and auction models: a survey, some applications, and challenges. IEEE Commun. Surv. Tutorials **25**(1), 497–537 (2023)
29. Filippov, A.F.: Differential equations with discontinuous righthand side. Mat. Sb. (N.S.) **93**(1), 99–128 (1960). (in Russian)
30. Filippov, A.F.: Differential Equations with Discontinuous Righthand Sides. Kluwer Academic Publishers (1988)
31. Cortés, J.: Discontinuous dynamical systems: a tutorial on solutions, nonsmooth analysis and stability. IEEE Control Syst. Mag. **28**(3), 36–73 (2008)
32. Vickrey, W.: Counterspeculation, auctions, and competitive sealed tenders. J. Financ. **16**(1), 8–37 (1961)

Leveraging Hierarchical Similarities
for Contrastive Clustering

Yuanshu Li[1], Yubin Xiao[1], Xuan Wu[1], Lei Song[1], Yanchun Liang[2],
and You Zhou[1(✉)]

[1] Key Laboratory of Symbolic Computation and Knowledge Engineering of Ministry
of Education, College of Computer Science and Technology, Jilin University,
Changchun, China
zyou@jlu.edu.cn
[2] School of Computer Science, Zhuhai College of Science and Technology, Zhuhai,
China

Abstract. Recently, contrastive clustering has demonstrated high performance in the field of deep clustering due to its powerful feature extraction capabilities. However, existing contrastive clustering methods suffer from inter-class conflicts and often produce suboptimal clustering outcomes due to the disregard of latent class information. To address this issue, we propose a novel method called Contrastive learning using Hierarchical data similarities for Deep Clustering (CHDC), consisting of three modules, namely the inter-class separation enhancer, the intra-class compactness enhancer, and the clustering module. Specifically, to induct the latent class information by utilizing the sample pairs with data similarities, the inter-class separation enhancer and the intra-class compactness enhancer handle negative and positive sample pairs, respectively, with distinct hierarchical similarities. Additionally, the clustering module aims to ensure the alignment of cluster assignments between samples and their neighboring samples. By designing these three modules that work collaboratively, inter-class conflicts are alleviated, allowing CHDC to learn more discriminative features. Lastly, we design a novel update method for positive sample pairs to reduce the likelihood of introducing erroneous information. To evaluate the performance of CHDC, we conduct extensive experiments on five widely adopted image classification datasets. The experimental results demonstrate the superiority of CHDC compared to state-of-the-art methods. Moreover, ablation studies demonstrate the effectiveness of the proposed modules.

Keywords: Deep Clustering · Contrastive Learning · Unsupervised
Learning

1 Introduction

Clustering is a kind of important unsupervised learning algorithm and can be used as a standalone tool or as a pre-processing part of other algorithms. Therefore, clustering has a large number of applications in various fields, such as facial recognition [36], gene sequencing [21], and so on.

© The Author(s), under exclusive license to Springer Nature Singapore Pte Ltd. 2024
B. Luo et al. (Eds.): ICONIP 2023, CCIS 1962, pp. 148–168, 2024.
https://doi.org/10.1007/978-981-99-8132-8_12

Traditional clustering algorithms (e.g., K-means [27] and Gaussian Mixture Models [33]) are effective only when the features are representative [32]. However, when confronted with high-dimensional raw data, traditional clustering algorithms may not perform well [32]. To address this problem, many clustering methods focus on utilizing dimensionality reduction techniques to map the raw data into a lower-dimensional feature space, facilitating more effective clustering. Specifically, common dimensionality reduction techniques include Principal Component Analysis [1], MultiDimensional Scaling [23], Spectral Clustering [30], Kernel methods [24], Deep Neural Networks (DNNs), and so on. Among these techniques, DNNs have gained increasing attention due to their exceptional ability to perform nonlinear mappings and their flexibility in capturing complex data patterns [40,43].

Therefore, a series of methods have been proposed for clustering using DNNs, aka deep clustering. Early deep clustering methods [44] mostly utilized autoencoder [2] to learn and cluster the embedded features. In recent years, numerous Contrastive Learning-based methods have been proposed, such as Contrastive Clustering (CC) [25] and Semantic Clustering by Adopting Nearest neighbors (SCAN) [38]. That is because Contrastive Learning(CL) methods have powerful feature extraction capabilities, and have achieved great success in many fields [29,35]. Specifically, CL generates two augmented versions of the input sample, and then learns the most representative feature representation of the sample by minimizing the distance between two augmented visions of the input sample and maximizing the distance between visions of other samples. However, these methods follow the basic framework of CL, with lack of incorporating potential class information into clustering or considering the correlation between features of different images belonging to the same class. Consequently, these methods may lead to inter-class conflict during the training processes by pushing away samples that should actually belong to the same class [26]. Therefore, the performance for clustering might be limited [35].

To better address inter-class conflicts in CL, samples belonging to the same semantic class should share similar feature representations and cluster assignments, while samples from different classes should have distinct representations and assignments [19]. Therefore, we use pseudo-labels based on hierarchical clustering to prevent the samples of the same class from being pushed away, and encourage samples from different classes to be separate. Specifically, we design three modules, namely the inter-class separation enhancer, the intra-class compactness enhancer, and the clustering module. In inter-class separation enhancer, we construct negative sample pairs using the obtained pseudo-labels to increase the distance between samples from different classes. In the intra-class compactness enhancer, we construct positive sample pairs to decrease the distance between samples from the same class. By combining the intra-class compactness enhancer and inter-class separation enhancer, we effectively mitigate inter-class conflicts and facilitate the learning of more discriminative feature representations. The clustering module aims to learn more compact cluster assignments by encouraging samples and their neighbors to have similar cluster assignments. Finally, we also design a novel update method of positive sample pairs to reduce the probability of introducing error information. We name the overall method Contrastive learning using Hierarchical data similarities for Deep Clustering (CHDC). By combining these three modules, CHDC effectively promotes the learning of discriminative features, and improves cluster assignments (see Sect. 4.3).

To comprehensively assess the effectiveness of CHDC, we conduct extensive experiments on five image datasets. The experimental results compared to baseline methods, including K-means [27], SC [30], AC [10], NMF [4], AE [2], DAE [39], DCGAN [28], VAE [20], JULE [45], DEC [44], DAC [7], ADC [14], DDC [6], DCCM [41], IIC [17], PICA [16], CC [25], LNSCC [26], ConCURL [31], and GCC [46], demonstrate that our method achieves state-of-the-art results. In addition, we conduct a series of ablation studies to analyze the impact of different components in our method, further confirming the rationality and effectiveness of our method. Lastly, we perform qualitative analysis to showcase the superiority of our model visually.

The key contributions of this work are as follows:

- By combining hierarchical clustering with CL, we introduce hierarchical data similarities and latent information of semantic categories into CHDC. Subsequently, we define pseudo-labels using hierarchical clustering, which enhances versatility and reduces the likelihood of introducing erroneous information. Moreover, we utilize K-means to constrain sample pairs to further reduce the probability of introducing error information.
- To leverage the incorporated latent class information, CHDC is designed to be composed of three modules. The inter-class separation enhancer facilitates the separation of dissimilar samples, while the intra-class compactness enhancer promotes the cohesiveness of clustering within each class. The clustering module fosters well-structured clustering tasks.
- We conduct extensive experiments on image clustering, evaluating our proposed method on various datasets. The experimental results show improvement compared to existing approaches. Additionally, we perform detailed ablation studies to analyze the effectiveness of our method.

2   Related Work

In this section, we first introduce the development of deep clustering. Then, we introduce hierarchical clustering.

2.1   Deep Clustering

Recently, DNNs have garnered significant interest for their remarkable capability to create intricate mappings in data [42]. The rapid advancement of Deep Learning has led to the emergence of numerous deep clustering methods. For instance, Xie *et al.* [44] introduced Deep Embedded Clustering (DEC), where an autoencoder is employed as the underlying network structure, and clustering performance is enhanced through iterative minimization of Kullback-Leibler Divergence. Building upon DEC, a number of variants have been proposed. For example, Improved Deep Embedded Clustering [11] and Deep Convolution Embedded Clustering [12] leverage the principles of DEC to achieve superior performance. However, conventional autoencoders encode input data into discrete latent variables, which cannot effectively capture the continuous variations in the data, limiting their performance in feature extraction and clustering. To solve this problem, Jiang *et al.* [18] introduced variational autoencoder (VAE) [20] into deep clustering, which better capture the continuous variations by introducing randomness in the latent space, achieved better clustering outcomes. Finally, apart from AE and its variants, there have been several works [15,37] that explored various neural network frameworks for deep clustering and made great progress in clustering tasks.

In recent years, CL is widely used in deep clustering because of its outstanding ability of feature extraction. In clustering, CL aims to reduce the distances between samples of the same class and increase the distances between samples of different classes. For instance, CC [25] enhances clustering performance by jointly leveraging instance-level and clustering-level CL. However, traditional CL-based methods solely focus on reducing the distance between two augmented views of the same sample as positive sample pairs, while treating augmented views of all other samples as negative sample pairs and pushing them apart. Thus, though the clustering methods utilized CL demonstrate the capability to learn effective feature representations. However, this kind of methods may separate samples belonging to the same semantic class, resulting in inter-class conflicts, which may lead to the limited performance.

To address these limitations, we propose an end-to-end method. Specifically, we leverage the information shared among different samples, and introduce pseudo-labels for positive and negative pairs using hierarchical clustering, integrating the learning of cluster assignments and feature representations jointly. Therefore, this method can mitigate inter-class conflicts and enhance the clustering performance (see Sect. 3.2 for more details).

2.2   Hierarchical Clustering

In hierarchical clustering methods, the clustering process starts by considering each input sample point as an individual cluster, and then iteratively merges pairs of similar clusters based on a similarity metric. This process is repeated until all sample points belong to the same cluster. For instance, Density-Based Spatial Clustering of Applications with Noise (DBSCAN) [3] defines a radius and categorizes points by drawing neighborhood circles based on this radius. However, DBSCAN assigns all points within a specified radius to the same category, necessitating subjective judgment and trial-and-error to determine an appropriate radius distance. In contrast, the algorithm First Integer Neighbor indices Clustering Hierarchy (FINCH) [34] adopts a different strategy. Specifically, it only selects the nearest neighbor of each data point to be in the same category, avoiding the impact of setting hyperparameters on the clustering effect and ensuring the purity of the clusters. FINCH returns a partition hierarchy structure. However, for different datasets, which partition level is closest to the true number of clusters is uncertain, and requires extra judgment for making a selection.

To address this limitation, we propose a method that utilizes the partitions in FINCH to define pseudo-labels for clustering. This method leverages the purity and inclusiveness of the first and last partitions without requiring additional selection steps, thus increasing the algorithm's generality. Besides, we also design a novel update method of positive sample pairs to reduce the probability of introducing error information.

3 Contrastive Learning Using Hierarchical Data Similarities for Deep Clustering

To alleviate inter-class conflict produced by traditional CL-based methods, we propose a novel method Contrastive learning using Hierarchical data similarities for Deep Clustering. In Sect. 3.1, we provide an initial explanation of leveraging the latent information of the data to assign pseudo-labels. Subsequently, we generate positive and negative sample pairs based on these assigned pseudo-labels. Furthermore, in Sect. 3.2, we introduce a novel approach incorporating the obtained sample pair information into a contrastive loss function. By designing this loss function, semantically similar data points are grouped more closely together, while semantically dissimilar points are pushed away from the remaining batch samples in the whole clustering processes.

3.1 Construction of the Positive and Negative Sample Pairs

The achievements of DeepCluster [5] and Hierarchical Contrastive Selective Coding (HCSC) [13] demonstrate that combining pseudo-labels with latent label information has the potential to improve clustering performance. Specifically, DeepCluster typically employs K-means to generate pseudo-labels. Different with [5], HCSC constructs and dynamically updates a set of hierarchical prototypes to represent the hierarchical semantic structures, where the prototypes of each higher hierarchy are derived by iteratively applying K-means clustering to the prototypes of the hierarchy below. HCSC generates pseudo-labels based on the hierarchical semantic structures. Nevertheless, both methods are subject to the limitations inherent in the K-means, including the significant impact of initial centroid selection on final clustering results and the challenge of determining suitable initial centroid positions.

In contrast to earlier researches [5,13] that utilize K-means, our approach involves using hierarchical clustering to generate pseudo-labels. We only focus on whether two data points belong to a positive sample pair or a negative sample pair, without considering their specific prototypes produced in K-means. By

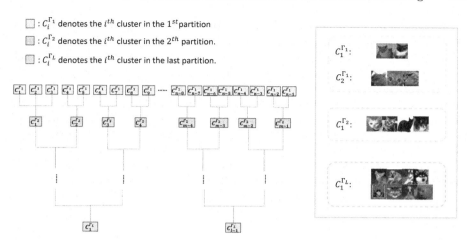

Fig. 1. Illustration of partitions and clusters during the processes of hierarchical clustering. Some instances of clusters are shown at right

differentiating between positive and negative sample pairs, we avoid the need for generating prototypes for each class using K-means, and also circumvents the issue of different centroid selections potentially influencing the clustering results. Hence, although having the same purpose, our proposed pseudo-labels generation is significantly different from that used in [5,13].

In hierarchical clustering, we first assign the nearest neighbor of each sample point and the sample point itself to the same class. Specifically, hierarchical clustering entails gathering samples and their closest neighboring points in each partition. By utilizing the integer indices that indicate the nearest neighbors for each data point, an adjacency link matrix is computed as follows:

$$A(i,j) = \begin{cases} 1, & \text{if } j = k_i \text{ or } k_j = i \text{ or } k_i = k_j, \\ 0, & \text{otherwise,} \end{cases} \tag{1}$$

where i denotes the ith sample point, k_i denotes the nearest neighbor of sample point i. The (1) generates a symmetric sparse matrix that serves as a representation of a graph, where the clusters are identified by the connected components within the graph. Through hierarchical clustering, several sets of partitions are obtained: $L = \{\Gamma_1, \Gamma_2, \ldots, \Gamma_{nl}\}$. Each partition $\Gamma_i = \{C_1^{\Gamma_i}, C_2^{\Gamma_i}, \ldots, C_{tn_i}^{\Gamma_i}\}$ denotes a valid clustering of given samples, tn_i denotes the total number of clusters in partition Γ_i, nl denotes the total number of partitions.

In Fig. 1, we illustrate the clustering assignment process, and showcase several representative clusters. As shown in Fig. 1, the right part illustrates two clusters, namely $C_1^{\Gamma_1}$ and $C_2^{\Gamma_1}$, within the first partition of hierarchical clustering. These clusters exclusively contain the most similar samples. For instance, although both $C_1^{\Gamma_1}$ and $C_2^{\Gamma_1}$ contain cats, they are assigned to different clusters, because the cats in $C_1^{\Gamma_1}$ and the cats in $C_2^{\Gamma_1}$ are not each other's nearest neighbors in the feature space. In contrast, the last partition exhibits inclusivity, where the cluster $C_1^{\Gamma_L}$ contains both cats and dogs, which do not belong to the same general semantic class. This occurs because cats and dogs share several similarities, such as being animals. Consequently, we consider the first partition the purest, and the last partition of hierarchical clustering the most inclusive.

Subsequently, to better utilize the information from hierarchical clustering, we propose a pseudo-labels generation method based on the obtained purity and inclusivity of partitions in hierarchical clustering. Assuming that if samples x_i and x_p belong to the same $C_i^{\Gamma_1}$ in the first partition, then they are most likely to belong to the same class in the true label as well. Thus, we first define positive pseudo-labels based on the clustering results of the first partition in hierarchical clustering. The positive sample pairs of sample x_i are defined as follows:

$$T(x_i) = I \setminus \{x_i\}, \tag{2}$$

$$P(x_i) = \{x_p \in T(x_i) : y_p^{pos} = y_i^{pos}\}, \tag{3}$$

where I denotes the whole data in dataset, $\setminus$ denotes the computation of set difference, the set difference $A \setminus B$ is defined by $A \setminus B = \{x : x \in A \quad and \quad x \notin B\}$, y_i^{pos} denotes positive pseudo-label of x_i, and $P(x_i)$ denotes the set of indices of positives from x_i.

Similarly, we define negative pseudo-labels based on the clustering results of the last partition in hierarchical clustering. Negative sample pairs of sample x_i are defined as follows::

$$N(x_i) = \{x_n \in T(x_i) : y_n^{neg} \neq y_i^{neg}\}, \tag{4}$$

where y_i^{neg} denotes the negative pseudo-label of x_i, and $N(x_i)$ denotes the set of indices of negatives from x_i. To expand the scope of information incorporation, we compute the similarity of whole training data samples to obtain the positive and negative sample pairs, instead of computing the similarity at each batch of training epoch as in prior work [29].

Furthermore, as the training progresses, to further alleviate the potential impact of misclassified samples in pseudo-labels on feature extraction, we utilize K-means algorithm to constrain sample pairs after a certain number of epochs. Specifically, for each batch of local features, we perform K-means clustering and exclude sample pairs from the positive pairs that are not classified into the same cluster by K-means. First, we conduct a sample pairs of sample x_i produced by K-means, named K-sample pairs, as follows:

$$K(x_i) = \{x_k \in T(x_i) : y_k^{kmeans} = y_i^{kmeans}\}, \tag{5}$$

where y_i^{kmeans} denotes the label of x_i produced by K-means, and $K(x_i)$ denotes the set of indices of K-sample pairs from x_i.

And the positive sample pairs after being constrained by K-means are updated: $P(x_i)^{update} = P(x_i) \cap K(x_i)$, where $\cap$ denotes the union of sets. In simple terms, only sample pairs that are classified into the same cluster in both

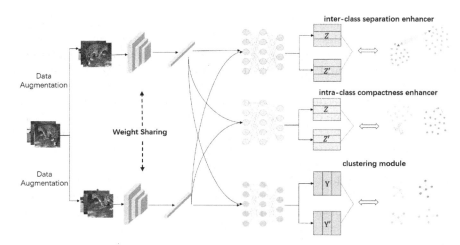

Fig. 2. Overview of the proposed CHDC method. Given data pairs, one shared DNNs is used to extract features from different augmentations. Then CHDC uses three heads with shared CNN parameters. The first head is the representation of inter-class separation enhancer, which considers the distance between samples from different semantic classes and push them further. The second head is the representation of intra-class compactness enhancer, which leads to bringing samples belonging to the same semantic class closer. The third head is the representation of the clustering module, which leads to a more compact cluster assignment

hierarchical clustering and K-means are considered as positive pairs after a certain number of epochs. Unlike we compute the similarity of whole training data samples in hierarchical clustering, we perform K-means clustering and update it at each training epoch.

3.2 Learning Clustering-Oriented Latent Space

The purpose of CL is to maximize the similarities encountered in the positive pairs and minimize the similarities encountered in the negative pairs. Previous CL-based clustering methods [25,38] consider two augmented visions of the same sample as positive pairs, and all other augmented samples as negative pairs. To illustrate this, suppose a batch has a number of B samples ($\{x_1, \ldots, x_B\}$), CL first generates $2B$ samples $\{x_1^a, \ldots, x_B^a, x_1^b, \ldots, x_B^b\}$ through image augmentation. For each sample x_i, x_i^a and x_i^b obtained through augmentation are considered positive pairs, while the other $2B - 2$ samples are considered negative pairs. However, this approach may lead to inter-class conflicts, further leading to a decline in performance. Because there might be samples in these $2B - 2$ negative pairs that belong to the same class as sample x, pushing away samples from the same class may potentially introduce conflicts within the classes during the training process [35].

To alleviate inter-class conflicts, we propose Contrastive learning using Hierarchical data similarities for Deep Clustering (CHDC). Our method consists of three modules, namely the intra-class compactness enhancer, the inter-class separation enhancer, and the clustering module. In the inter-class separation enhancer, we focus on separating samples that do not belong to the same class. While in the intra-class compactness enhancer, we aim to bring together samples that belong to the same class. Finally, the clustering module performs cluster-level contrastive learning to achieve the final cluster assignments. The framework of our method is illustrated in Fig. 2.

Inter-class Separation Enhancer. The goal of the inter-class separation enhancer is to consider increasing the distance between samples from different semantic classes and push them further. In this part, we regard samples from different semantic classes as negative sample pairs. Formally, the loss function for a given sample x_i is as follows:

$$\dot{\ell}_i^a = -\log\frac{\exp(\mathrm{sim}(z_i^a, z_i^b)/\tau_{psh})}{\sum_{k\in\{a,b\}}\sum_{n\in N(x_i)}\exp(\mathrm{sim}(z_i^a, z_n^k)/\tau_{psh})}, \tag{6}$$

where τ_{psh} denotes the temperature parameter, z_i^a denotes the mapping of x_i^a in the feature space, and $\mathrm{sim}(\cdot)$ denotes the similarity computation function, which is expressed as follows:

$$\mathrm{sim}(z_i^{k_1}, z_j^{k_2}) = \frac{(z_i^{k_1})(z_j^{k_2})^T}{||z_i^{k_1}||||z_j^{k_2}||}, \tag{7}$$

where $k_1, k_2 \in \{a, b\}$, $i, j \in \{1, \ldots, B\}$, $||\cdot||$ denotes the magnitude of a vector. In the inter-class separation enhancer, we select sample pairs with the highest probability of belonging to different classes as negative sample pairs. By doing so, we focus on pushing away samples that are more likely to belong to different classes, while better avoiding pushing operations on samples that may belong to the same class. This approach helps alleviate inter-class conflicts during the training process. For the whole samples in a batch, the inter-class separation enhancer loss is computed over every augmented vision of samples as follows:

$$L_{psh} = \frac{1}{2B}\sum_{i=1}^{N}(\dot{\ell}_i^a + \dot{\ell}_i^b). \tag{8}$$

Intra-class Compactness Enhancer. The inter-class separation enhancer primarily focuses on the distance between samples from different semantic classes, neglecting the handling of samples from the same class. To address this, we introduce the intra-class compactness enhancer, which aims to bring samples belonging to the same class closer together. Specifically, the loss function for a given sample x_i is computed as follows:

$$\ddot{\ell}_i^a = -\frac{1}{|P(x_i)|}\sum_{p\in P(x_i)}\log\frac{\sum_{k\in\{a,b\}}\exp(\mathrm{sim}(z_i^a, z_p^k)/\tau_{pl})}{\sum_{k\in\{a,b\}}\sum_{j=1}^{B}\exp(\mathrm{sim}(z_i^a, z_j^k)/\tau_{pl})}, \tag{9}$$

where τ_{pl} denotes the temperature parameter of the intra-class compactness enhancer. $|P(x_i)|$ denotes the absolute value of $P(x_i)$. The intra-class compactness enhancer loss over every augmented vision of samples is computed as follows:

$$L_{pl} = \frac{1}{2B} \sum_{i=1}^{N} (\ddot{\ell}_i^a + \ddot{\ell}_i^b). \tag{10}$$

Clustering Module. In most cases, a sample and its nearest neighbors should belong to the same semantic class, and the class assignment probabilities obtained after clustering should also be as similar as possible [38]. Therefore, we minimize the probability of dissimilar nearest neighbors in clustering module, and define the loss function of clustering module as follows:

$$\tilde{\ell}_i^a = - \sum_{g \in G(i)} \log(\mathrm{sim}'(\phi(x_i^a), \phi(x_k^a))), \tag{11}$$

$$L_{clu} = \frac{1}{|2B|} \sum_{i=1}^{N} (\tilde{\ell}_i^a + \tilde{\ell}_i^b) + \mathrm{H}(C), \tag{12}$$

where $G(i)$ denotes the set of indices of neighboring samples of sample i, ϕ denotes the softmax function, which is utilized to perform a soft assignment over the clusters C, with $\phi(x_i) \in [0,1]^C$. The probability of sample x_i being assigned to cluster c is defined as $\phi^c(x_i)$. $\mathrm{H}(C)$ denotes the entropy function, which is expressed as follows:

$$\mathrm{H}(C) = \lambda \sum_{c \in C} \phi'^c \log \phi'^c, \tag{13}$$

$$\phi'^c = \frac{1}{|B|} \sum_{i=1}^{N} \phi^c(x_i). \tag{14}$$

The entropy function is used to avoid the softmax function from assigning all samples to a single cluster [38]. In clustering module, we use the similarity measure as follows:

$$\mathrm{sim}'(\phi(x_i), \phi(x_k)) = \langle \phi(x_i), \phi(x_k) \rangle, \tag{15}$$

where $\langle \cdot \rangle$ denotes the dot product operator. We define the overall loss by combining the loss of the intra-class compactness enhancer in (8), the loss of the intra-class compactness enhancer in (10), and the loss of the clustering module in (12) as follows:

$$L_{overall} = L_{pl} + \lambda L_{psh} + \mu L_{clu}, \tag{16}$$

where λ and μ denote predefined hyperparameters to balance these three losses during training. The method's pseudocode is presented in Algorithm 1.

Algorithm 1: The Training Algorithm of CHDC.

Input: image dataset $\{x_1, x_2, \ldots, x_N\}$, training epochs E, number of clusters K, batch size N, constrain by K-means after $\hat{E}$ epochs, augmentation strategy T, temperature of intra-class compactness enhancer τ_{pl}, temperature of inter-class separation enhancer τ_{psh}, temperature of clustering component τ_{clu}

Output: model with parameters θ

1   Get $L = \{\Gamma_1, \Gamma_2, \ldots, \Gamma_{nl}\}$ through hierarchical clustering

2   Obtain positive and negative sample pairs according to Γ_1, Γ_{nl}

3   **for** *epoch = 1 to E* **do**

4   Sample a mini-batch $\{x_i\}_{i=1}^{B}$

5   **if** *epoch* $> \hat{E}$ **then**

6   Compute $P(x_i)^{update}$

7   Update $P(x_i) \leftarrow P(x_i)^{update}$;

8   **end**

9   Get the augmentations for the sampled images and their neighbors according to augmentation strategy T

10   Compute L_{psh} according to (6) and (8)

11   Compute L_{pl} according to (9) and (10)

12   Compute L_{clu} according to (11) and (12)

13   Compute overroll loss $L_{overall}$ according to (16)

14   Update θ with gradient descent by minimizing the overall loss $L_{overall}$

15 **end**

---

4 Experimental Result

This section presents a comprehensive evaluation of our proposed method. We first introduce the datasets and the implementation details. Next, we compare the proposed CHDC with 20 baseline methods. Then, we conduct a series of ablation studies to analyze the effectiveness of our method. Finally, we carry out qualitative studies to analyze the positive and negative sample pairs generated in hierarchical clustering, the confusion matrices, and the high predicted probability instances.

4.1 Datasets

In experiments, we use five widely-adopted benchmark datasets to assess the performance of CHDC, including CIFAR-10 [22], CIFAR-20 [22], ImageNet-10 [7], ImageNet-Dogs [7], and Tiny-ImageNet. CIFAR-10 and CIFAR-20 are both datasets of images with the size of 32×32. ImageNet-10, ImageNet-Dogs, and Tiny-ImageNet are all subsets of ImageNet. In ImageNet-10, there are 10 classes of images. ImageNet-Dogs contains 15 dog breeds. For ImageNet-10 and ImageNet-Dogs, we crop the images to the size of 96×96. Tiny-ImageNet is a very challenging dataset with 200 classes. Due to the limitation in computing, the images of Tiny-ImageNet are resized to 64×64.

4.2   Implementation Details

We use ResNet18 as the backbone following the prior study [38]. In addition, we adopt the SGD optimizer, and the learning rate, weight decay, and momentum are set to 0.4, 1e-4, and 0.9, respectively. The learning rate is decayed using a cosine scheduler with a decay rate of 0.1. We use a smaller batch size of 256 and train the model from scratch for 1000 epochs to compensate for the performance loss caused by small batch sizes, as recommended by [25]. We use the same data augmentation method as [8]. For temperature settings, we set the temperature of the intra-class compactness enhancer and the temperature of the inter-class separation enhancer to 0.07 and 0.1, respectively. We used NVIDIA Geforce RTX 3090 GPU for model training and testing.

Table 1. Performance comparison of different algorithms

Methods	ImageNet-10			CIFAR-10			CIFAR-20			ImageNet-Dogs			Tiny-ImageNet		
	NMI	ACC	ARI	NMI	ACC	ARI	NMI	ACC	ARI	NMI	ACC	ARI	NMI	ACC	ARI
K-means (1967) [27]	0.119	0.241	0.057	0.087	0.229	0.049	0.084	0.130	0.028	0.055	0.105	0.020	0.065	0.025	0.005
AC (1978) [10]	0.138	0.242	0.067	0.105	0.228	0.065	0.098	0.138	0.034	0.037	0.139	0.021	0.069	0.027	0.005
SC (2001) [30]	0.151	0.274	0.076	0.103	0.247	0.085	0.090	0.136	0.022	0.038	0.111	0.013	0.063	0.022	0.004
AE (2006) [2]	0.210	0.317	0.152	0.239	0.314	0.169	0.100	0.165	0.048	0.104	0.185	0.073	0.131	0.041	0.007
NMF (2009) [4]	0.132	0.230	0.065	0.081	0.190	0.034	0.079	0.118	0.026	0.044	0.118	0.016	0.072	0.029	0.005
DAE (2010) [39]	0.206	0.304	0.138	0.251	0.297	0.163	0.111	0.151	0.046	0.104	0.190	0.078	0.127	0.039	0.007
VAE (2014) [20]	0.193	0.334	0.168	0.245	0.291	0.167	0.108	0.152	0.040	0.107	0.179	0.079	0.113	0.036	0.006
JULE (2016) [45]	0.175	0.300	0.138	0.192	0.272	0.138	0.103	0.137	0.033	0.054	0.138	0.028	0.102	0.033	0.006
DEC (2016) [44]	0.282	0.381	0.203	0.257	0.301	0.161	0.136	0.185	0.050	0.122	0.195	0.079	0.115	0.037	0.007
DAC (2017) [7]	0.394	0.527	0.302	0.396	0.522	0.306	0.185	0.238	0.088	0.219	0.275	0.111	0.190	0.066	0.017
DCGAN (2018) [28]	0.206	0.304	0.138	0.265	0.315	0.176	0.120	0.151	0.045	0.121	0.174	0.078	0.135	0.041	0.007
ADC (2019) [14]	-	-	-	-	0.325	-	-	0.160	-	-	-	-	-	-	-
DDC (2019) [6]	0.433	0.577	0.345	0.424	0.524	0.329	-	-	-	-	-	-	-	-	-
DCCM (2019) [41]	0.608	0.710	0.555	0.496	0.623	0.408	0.285	0.327	0.173	0.321	0.383	0.182	0.224	0.108	0.038
IIC (2019) [17]	-	-	-	-	0.617	-	-	0.257	-	-	-	-	-	-	-
PICA (2020) [16]	0.802	0.870	0.761	0.591	0.696	0.512	0.310	0.337	0.171	0.352	0.352	0.201	0.277	0.098	0.040
CC (2021) [25]	0.850	0.893	0.811	0.705	0.790	0.637	0.431	0.429	0.266	0.445	0.429	0.274	0.340†	**0.140**	0.071†
ConCURL (2021) [31]	**0.882**	0.931†	0.869†	0.667	0.785	0.614	0.390	0.409	0.232	0.450	0.447	0.298	-	-	-
GCC (2021) [46]	0.842	0.901	0.822	**0.764**	**0.856**	**0.728**	0.472†	0.472†	0.305†	0.490†	0.526†	0.362†	**0.347**	0.138†	**0.075**
LNSCC (2022) [26]	-	-	-	0.713	0.820	-	0.446	0.439	-	-	-	-	-	-	-
CHDC (ours)	0.877†	**0.946**	**0.887**	0.759†	0.853†	0.724†	**0.477**	**0.490**	**0.315**	**0.593**	**0.626**	**0.473**	0.327	0.130	0.061

bold and † are used to denote the best result and second best result, respectively.

4.3   Performance Comparison

We evaluate the proposed method with twenty representative methods, including four traditional clustering methods [4,10,27,30], twelve methods based on classic Deep Learning [2,6,7,14,16,17,20,28,39,41,44,45], and four methods based on self-supervised contrastive frame [25,26,31,46]. For a fair comparison, the results of other methods are taken from [46]. In addition, we use three standard evaluation metrics like prior studies do [25], namely Normalized Mutual Information (NMI), Accuracy (ACC), and Adjusted Rand Index (ARI). It is worth mentioning that higher values of these metrics indicate better clustering performance.

In Table 1, we present the comparison of the performance of CHDC with that of the baseline model on CIFAR-10, CIFAR-20, ImageNet-10, ImageNet-Dogs, and Tiny-ImageNet. Compared to the baseline methods, CHDC achieves the best result on CIFAR-20, ImageNet-10, and ImageNet-Dogs. For the rest datasets, including CIFAR-10 and Tiny-ImageNet, our method is also competitive. Worthy of attention is that our CHDC significantly surpasses all traditional methods and classic DL-based methods by a large margin on most benchmarks under three different evaluation metrics. In terms of ACC, we achieve the best result on three datasets, those results show that CHDC has high accuracy for clustering. In terms of NMI, CHDC also performs well, which shows that CHDC-generated clustering results can achieve a high degree of similarity with the true labels. Different from NMI, ARI also adjusts for the size and number of categories in the dataset. The CHDC's observed improvement in ARI indicates that the algorithm not only demonstrates effective discovery of the underlying category structure in the dataset but also exhibits robustness against noisy data. CHDC's superior performance can be attributed to its ability to effectively induct the latent class information. Furthermore, compared to GCC, which also introduce the latent class information into clustering, CHDC also achieves better performance on three datasets. It can be attributed to the ability of pseudo-label generation to reduce the probability of introducing erroneous information, which is also observed when we conduct ablation studies (see Sect. 4.4). In conclusion, our method has the powerful ability of clustering.

4.4   Ablation Studies

In this subsection, we conduct a series of ablation studies to demonstrate the effectiveness of data augmentation, pseudo-label generation, each component of loss function, and global information.

Data Augmentation. We test our model on CIFAR-10 and CIFAR-20 by using two kinds of strategies of data augmentation. One from [8], named general data augmentation, is adopted in our method. And the other data augmentation proposed in [9], is named enhanced data augmentation. To verify the effectiveness of adopted data augmentation, we remove one and both of the two augmentations, and the results are presented in Table 2.

Table 2. Ablation studies on data augmentation choices

Datasets	x		x'		NMI	ACC	ARI
	T	T'	T	T'			
CIFAR-10	✓		✓		**0.759**	**0.853**	**0.724**
	✓			✓	0.731	0.824	0.672
	✓				0.701	0.797	0.645
		✓	✓		0.731	0.824	0.672
		✓		✓	0.533	0.660	0.468
					0.030	0.128	0.011
CIFAR-20	✓		✓		**0.477**	**0.490**	**0.315**
	✓			✓	0.476	0.461	0.314
	✓				0.443	0.414	0.269
		✓	✓		0.476	0.461	0.314
		✓		✓	0.212	0.268	0.101
					0.032	0.100	0.011

x and x denote sample pairs. T denotes the general data augmentation, T denotes the enhanced data augmentation.

As shown in Table 2, when data augmentation is not applied, the ACCs are only 0.128 and 0.100 on CIFAR-10 and CIFAR-20, significantly lower than the ACCs achieved in experiments with data augmentation. This finding underscores the indispensability of data augmentation in contrastive clustering. Furthermore, when both data augmentations are enhanced data augmentation, suboptimal performance, and performance degradation may occur. Conversely, when enhanced data augmentation and general data augmentation are combined, the performance is improved compared to using two enhanced data augmentations alone, but still falls short of using both general data augmentation. Notably, the best result is obtained when both data augmentations are general data augmentations, thereby validating the effectiveness of our adopted data augmentation approach.

Table 3. Ablation studies on pseudo-label generation method choices

pseudo-label generation	Datasets	NMI	ACC	ARI
by final result	CIFAR-10	0.753	0.846	0.708
	CIFAR-20	0.456	0.448	0.287
	ImageNet-10	0.876	0.944	0.881
by first partition and last partition	CIFAR-10	**0.759**	**0.853**	**0.724**
	CIFAR-20	**0.477**	**0.490**	**0.315**
	ImageNet-10	**0.877**	**0.946**	**0.887**

Table 4. Ablation study on loss function

w/ L_{pl}	w/ L_{psh}	w/ L_{clu}	CIFAR-10 ACC	NMI
✓	✓	✓	**0.855**	**0.763**
✓		✓	0.801	0.717
	✓	✓	0.816	0.729
✓	✓		0.814	0.704
✓			0.784	0.659
	✓		0.781	0.658
		✓	0.739	0.604

Table 5. Ablation study on operation choices of similarity

Dataset	Operation	NMI	ACC	ARI
CIFAR-10	Global	**0.759**	**0.853**	**0.724**
	Local	0.687	0.790	0.632
CIFAR-20	Global	**0.477**	**0.490**	**0.315**
	Local	0.427	0.426	0.262

Pseudo-label Generation. To demonstrate the effectiveness of our proposed method for generating pseudo-labels using hierarchical clustering, we conduct two experiments on three datasets. The first experiment follows our proposed method for pseudo-label generation, which uses the first and last partition. In the second experiment, we select the partition with the number of clusters closest to the ground truth as the final result of hierarchical clustering, and pseudo-labels are generated based on this final result. Specifically, we define samples belonging to the same cluster and different clusters in the final result of hierarchical clustering as positive sample pairs and negative sample pairs, respectively.

The results presented in Table 3 reveal that, in the experiments conducted on CIFAR-10, CIFAR-20, and ImageNet-10, utilizing the first partition and last partition for generating pseudo-labels yields superior performance compared to using the final result. This finding highlights the efficacy of our method in effectively leveraging the inherent semantic labels of the samples and mitigating the risk of introducing erroneous labels.

Effect of Parts of Loss Function. To verify the design of each parts of loss function, we conduct ablation studies on CIFAR-10 by removing one or two components.

As shown in Table 4, the removal of the inter-class separation enhancer leads to a decrease in ACC to 0.801, and NMI to 0.717. Similarly, when the intra-class compactness enhancer is removed, ACC decreases to 0.816 and NMI decreases to 0.726. These findings highlight the significant contributions of these two components to the learning of representation features, emphasizing that necessity inter-class separation enhancer and intra-class compactness enhancer. Furthermore, the removal of the clustering component results in an ACC decrease of 0.814, underscoring the importance of this component. It is worth noting that removing two out of the three components causes the ACC to decrease to 0.784, 0.781, and 0.739, which show more severe performance degradation. Additionally, the clustering component not only enhances performance but also allows for the representation of clustering within an end-to-end model, enabling the model to obtain clustering assignments alongside the extracted representation features.

The Necessity of Global Information. To assess the effectiveness of utilizing hierarchical clustering for calculating similarity among complete data samples, we conduct two experiments on the CIFAR-10 and CIFAR-20 datasets. The first experiment focuses on computing similarity for the entire data samples as performed in our method, named Global operation. The second experiment, inspired by previous study [29], involves computing similarity at each epoch, named Local operation. In the Global operation experiment, we follow the procedure outlined in Sect. 3.2. For the Local operation experiment, positive and negative sample pairs are generated within each batch, taking into account only local information.

Table 5 presents the results of these experiments. It can be observed that the method utilizing the Global operation achieve superior performance in terms of ACC, NMI, and ARI on both the CIFAR-10 and CIFAR-20 datasets. This observation suggests that employing global information for similarity computation can provide a more comprehensive understanding and yield improved results in clustering tasks.

4.5   Qualitative Study

In this subsection, we conduct several qualitative studies. We first visualize the positive and negative sample pairs of hierarchical clustering. Then, we represent the confusion matrices and instances of high predicted probability to display the result of clustering more intuitively.

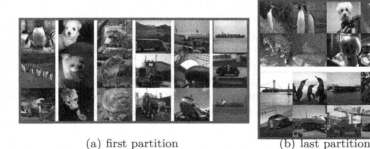

(a) first partition   (b) last partition

Fig. 3. (a) The examples of positive sample pairs. Each image in a column is positive sample of each others. (b) The examples of negative sample pairs. Images in the first two rows is negative samples of the images in the last two rows

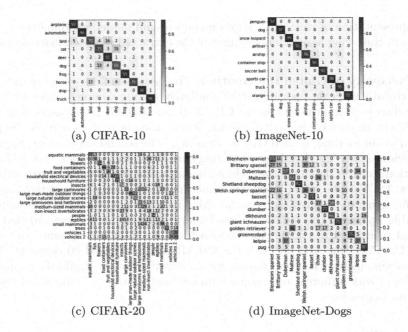

(a) CIFAR-10   (b) ImageNet-10

(c) CIFAR-20   (d) ImageNet-Dogs

Fig. 4. Confusion matrices of four datasets

Positive and Negative Sample Pairs. To provide an intuitive understanding of the pseudo-label produced by hierarchical clustering, we visualize the results when processing hierarchical clustering in Fig. 3(a). Figure 3 illustrates a set of positive sample pairs, where each column represents samples belonging to the same cluster in the initial partition of hierarchical clustering. We observe that the majority of positive sample pairs exhibit significant semantic similarities in their features. However, there are instances of misclassifications, such as the leopard in the second column and the car in the sixth column. Conversely, Fig. 3(b) displays negative sample pairs. The samples in the first and second rows belong to one cluster in the final partition, while the samples in the third and fourth rows belong to another cluster. Notably, the samples in the first cluster demonstrate distinct semantic characteristics compared to those in the second cluster.

Confusion Matrices. For a more intuitive understanding of the clustering effect, we visualize the confusion matrices in Fig. 4 of four datasets, including CIFAR-10, CIFAR-20, ImageNet-10, and ImageNet-Dog. Each confusion matrix has clear blocks, especially in CIFAR-10 and ImageNet-10, which means our method can successfully classify samples into semantic clusters. In addition, in Fig. 4(c), we can observe which clusters have poor performance and analyze the reason. For example, the 'insects' and 'non-insect invertebrates' are usually confused. That is may because both of them have similar body shapes and appear in similar living environments.

Fig. 5. The top 5 instances with the highest prediction probability in each cluster of ImageNet-10, with the top row indicating the most confidence

Instances of High Predicted Probability. To visualize the final clusters, we select the instance of high predicted probability and show the instances with the top-5 highest predicted probability in each cluster of ImageNet-10, as shown in Fig. 5. We observe that the displayed images are successfully clustered to the true semantic class.

5   Conclusion

In this work, we proposed CHDC, an algorithm for Contrastive learning using Hierarchical data similarities for Deep Clustering. This method employs a novel approach to define positive and negative sample pairs through hierarchical clustering and utilizes the information contained in these pairs to guide the training process. The experimental results demonstrate that our method outperforms existing methods on five challenging image classification datasets. In the future, we plan to delve deeper into the data and extract additional latent information. Furthermore, we also plan to extend our method to other tasks and applications, such as weakly supervised learning.

Acknowledgments. This work is supported by the National Key Research and Development Program of China (2021YFF1201200), the Jilin Provincial Department of Science and Technology Project (20230201083GX, 20220201145GX, and 20230201065GX), the National Natural Science Foundation of China (62072212, 61972174, 61972175, and 12205114), the Guangdong Universities' Innovation Team Project (2021KCXTD015), and Key Disciplines (2021ZDJS138) Projects.

References

1. Abdi, H., Williams, L.J.: Principal component analysis. Wiley Interdiscip. Rev. Comput. Stat. **2**, 433–459 (2010)
2. Bengio, Y., Lamblin, P., Popovici, D., Larochelle, H.: Greedy layer-wise training of deep networks. In: Proceedings of the Advances in Neural Information Processing System, vol. 19 (2006)
3. Bi, F., Wang, W., Chen, L.: DSCAN: density-based spatial clustering of applications with noise. J. Nanjing Univ. (Nat. Sci.) **48**, 491–498 (2012)
4. Cai, D., He, X., Wang, X., Bao, H., Han, J.: Locality preserving nonnegative matrix factorization. In: Proceedings of the International Joint Conference on Artificial Intelligence (2009)
5. Caron, M., Bojanowski, P., Joulin, A., Douze, M.: Deep clustering for unsupervised learning of visual features. In: Proceedings of the European Conference on Computer Vision, pp. 132–149 (2018)
6. Chang, J., Guo, Y., Wang, L., Meng, G., Xiang, S., Pan, C.: Deep discriminative clustering analysis (2019)
7. Chang, J., Wang, L., Meng, G., Xiang, S., Pan, C.: Deep adaptive image clustering. In: Proceedings of the IEEE/CVF International Conference on Computer Vision, pp. 5879–5887 (2017)
8. Chen, T., Kornblith, S., Norouzi, M., Hinton, G.: A simple framework for contrastive learning of visual representations. In: Proceedings of the International Conference on Machine Learning, pp. 1597–1607 (2020)
9. Cubuk, E.D., Zoph, B., Shlens, J., Le, Q.V.: Randaugment: practical automated data augmentation with a reduced search space. In: Proceedings of the IEEE/CVF Conference on Computer Vision and Pattern Recognition Workshop, pp. 702–703 (2020)
10. Gowda, K.C., Krishna, G.: Agglomerative clustering using the concept of mutual nearest neighbourhood. Pattern Recogn. **10**, 105–112 (1978)
11. Guo, X., Gao, L., Liu, X., Yin, J.: Improved deep embedded clustering with local structure preservation. In: Proceedings of the International Joint Conference on Artificial Intelligence, pp. 1753–1759 (2017)
12. Guo, X., Liu, X., Zhu, E., Yin, J.: Deep clustering with convolutional autoencoders. In: Proceedings of International Conference on Neural Information Processing, pp. 373–382 (2017)
13. Guo, Y., et al.: HCSC: hierarchical contrastive selective coding. In: Proceedings of the IEEE/CVF Conference on Computer Vision and Pattern Recognition, pp. 9706–9715 (2022)
14. Haeusser, P., Plapp, J., Golkov, V., Aljalbout, E., Cremers, D.: Associative deep clustering: training a classification network with no labels. In: Proceedings of the German Conference on Pattern Recognition, pp. 18–32 (2019)
15. Hsu, C.C., Lin, C.W.: CNN-based joint clustering and representation learning with feature drift compensation for large-scale image data. IEEE Trans. Multimedia **20**, 421–429 (2017)
16. Huang, J., Gong, S., Zhu, X.: Deep semantic clustering by partition confidence maximisation. In: Proceedings of the IEEE/CVF Conference on Computer Vision and Pattern Recognition, pp. 8849–8858 (2020)
17. Ji, X., Henriques, J.F., Vedaldi, A.: Invariant information clustering for unsupervised image classification and segmentation. In: Proceedings of the IEEE/CVF International Conference on Computer Vision, pp. 9865–9874 (2019)

18. Jiang, Z., Zheng, Y., Tan, H., Tang, B., Zhou, H.: Variational deep embedding: an unsupervised and generative approach to clustering. In: Proceedings of the International Joint Conference on Artificial Intelligence, pp. 1965–1972 (2017)
19. Khosla, P., et al.: Supervised contrastive learning. In: Proceedings of the Advances in Neural Information Processing System, vol. 33, pp. 18661–18673 (2020)
20. Kingma, D.P., Welling, M.: Auto-encoding variational Bayes. In: Proceedings of the International Conference on Learning Representations (2013)
21. Kiselev, V.Y., Andrews, T.S., Hemberg, M.: Challenges in unsupervised clustering of single-cell RNA-SEQ data. Nat. Rev. Genet. **20**, 273–282 (2019)
22. Krizhevsky, A., Hinton, G.: Learning multiple layers of features from tiny images. Handbook of Systemic Autoimmune Diseases **1** (2009)
23. Kruskal, J.B., Wish, M.: Multidimensional Scaling, vol. 11. Sage, Thousand Oaks (1978)
24. Kung, S.Y.: Kernel Methods and Machine Learning. Cambridge University Press, Cambridge (2014)
25. Li, Y., Hu, P., Liu, Z., Peng, D., Zhou, J.T., Peng, X.: Contrastive clustering. In: Proceedings of the Association for the Advancement of Artificial Intelligence, vol. 35, pp. 8547–8555 (2021)
26. Ma, X., Kim, W.H.: Locally normalized soft contrastive clustering for compact clusters. In: Proceedings of the International Joint Conference on Artificial Intelligence (2022)
27. McQueen, J.: Some methods for classification and analysis of multivariate observations. In: Proceedings of the Berkeley Symposium on Mathematical Statistics and Probability, pp. 281–297 (1967)
28. Mehralian, M., Karasfi, B.: RDCGAN: unsupervised representation learning with regularized deep convolutional generative adversarial networks. In: Proceedings of the Conference on Artificial Intelligence and Robotics and Asia-Pacific International Symposium, pp. 31–38 (2018)
29. Meng, Q., Qian, H., Liu, Y., Xu, Y., Shen, Z., Cui, L.: MHCCL: masked hierarchical cluster-wise contrastive learning for multivariate time series. In: Proceedings of the Association for the Advancement of Artificial Intelligence (2022)
30. Ng, A., Jordan, M., Weiss, Y.: On spectral clustering: analysis and an algorithm. In: Proceedings of the Advances in Neural Information Processing System, vol. 14 (2001)
31. Regatti, J.R., Deshmukh, A.A., Manavoglu, E., Dogan, U.: Consensus clustering with unsupervised representation learning. In: Proceedings of the International Joint Conference on Neural Networks, pp. 1–9 (2021)
32. Ren, Y., et al.: Deep clustering: a comprehensive survey. arXiv preprint arXiv:2210.04142 (2022)
33. Reynolds, D.A., et al.: Gaussian mixture models. Encycl. Biomet. **741**, 659–663 (2009)
34. Sarfraz, S., Sharma, V., Stiefelhagen, R.: Efficient parameter-free clustering using first neighbor relations. In: Proceedings of the IEEE/CVF Conference on Computer Vision and Pattern Recognition, pp. 8934–8943 (2019)
35. Saunshi, N., Plevrakis, O., Arora, S., Khodak, M., Khandeparkar, H.: A theoretical analysis of contrastive unsupervised representation learning. In: Proceedings of the International Conference on Machine Learning, vol. 97, pp. 5628–5637 (2019)
36. Shen, S., et al.: Structure-aware face clustering on a large-scale graph with 107 nodes. In: Proceedings of the IEEE/CVF Conference on Computer Vision and Pattern Recognition, pp. 9085–9094 (2021)

37. Springenberg, J.T.: Unsupervised and semi-supervised learning with categorical generative adversarial networks (2015)
38. Van Gansbeke, W., Vandenhende, S., Georgoulis, S., Proesmans, M., Van Gool, L.: SCAN: learning to classify images without labels. In: Proceedings of the European Conference on Computer Vision, pp. 268–285 (2020)
39. Vincent, P., Larochelle, H., Lajoie, I., Bengio, Y., Manzagol, P.A., Bottou, L.: Stacked denoising autoencoders: Learning useful representations in a deep network with a local denoising criterion. J. Mach. Learn. Res. **11**, 3371–3408 (2010)
40. Wang, L., Xiao, Y., Li, J., Feng, X., Li, Q., Yang, J.: IIRWR: internal inclined random walk with restart for lncRNA-disease association prediction. IEEE Access **7**, 54034–54041 (2019)
41. Wu, J., et al.: Deep comprehensive correlation mining for image clustering. In: Proceedings of the IEEE/CVF International Conference on Computer Vision, pp. 8150–8159 (2019)
42. Xiao, Y., et al.: Reinforcement learning-based non-autoregressive solver for traveling salesman problems. arXiv preprint arXiv:2308.00560 (2023)
43. Xiao, Y., Xiao, Z., Feng, X., Chen, Z., Kuang, L., Wang, L.: A novel computational model for predicting potential lncRNA-disease associations based on both direct and indirect features of lncRNA-disease pairs. BMC Bioinform. **21**, 1–22 (2020)
44. Xie, J., Girshick, R., Farhadi, A.: Unsupervised deep embedding for clustering analysis. In: Proceedings of the International Conference on Machine Learning, pp. 478–487 (2016)
45. Yang, J., Parikh, D., Batra, D.: Joint unsupervised learning of deep representations and image clusters. In: Proceedings of the IEEE/CVF Conference on Computer Vision and Pattern Recognition, pp. 5147–5156 (2016)
46. Zhong, H., et al.: Graph contrastive clustering. In: Proceedings of the IEEE/CVF International Conference on Computer Vision, pp. 9224–9233 (2021)

Reinforcement Learning-Based Consensus Reaching in Large-Scale Social Networks

Shijun Guo[1], Haoran Xu[2,3], Guangqiang Xie[1(✉)], Di Wen[2,3],
Yangru Huang[4], and Peixi Peng[3,4]

[1] School of Computer Science and Technological,
Guangdong University of Technology, No. 100 Waihuanxi Road,
HEMC, Guangzhou 510006, Guangdong, China
{3120005895,xiegq}@gdut.edu.cn
[2] Peng Cheng Laboratory, Xingke 1st Street, Shenzhen 518066, Guangdong, China
{xuhr,wend}@pcl.ac.cn
[3] School of Intelligent Systems Engineering, Sun Yat-sen University,
No. 66 Gongchang Road, Shenzhen 510275, Guangdong, China
{xuhr9,wend25}@mail2.sysu.edu.cn
[4] Department of Computer Science and Technology, Peking University,
Beijing 100871, China
yrhuang@stu.pku.edu.cn, pxpeng@pku.edu.cn

Abstract. Social networks in present-day industrial environments encompass a wide range of personal information that has significant research and application potential. One notable challenge in the domain of opinion dynamics of social networks is achieving convergence of opinions to a limited small number of clusters. In this context, designing the communication topology of the social network in a distributed manner is a particularly difficult. To address this problem, this paper proposes a novel perception model for agents. The proposed model, which is based on bidirectional recurrent neural networks, can adaptively reweight the influence of perceived neighbors in the convergence process of opinion dynamics. Additionally, effective differential reward functions are designed to optimize three objectives: convergence degree, connectivity, and cost of convergence. Lastly, a multi-agent exploration and exploitation algorithm based on policy gradient is designed to optimize the model. Based on the reward values in inter-agent interaction process, the agents can adaptively learn the neighbor reweighting strategy with multi-objective trade-off abilities. Extensive simulations demonstrate that the proposed method can effectively reconcile conflicting opinions among agents and accelerate convergence.

Keywords: Social network · Opinion dynamics · Reweighting perception · Reinforcement learning

S. Guo and H. Xu–Equal contribution.

1   Introduction

Recently, social network group decision-making (SNGDM) has attracted increasing attention as a valuable tool for understanding and explaining human actions [9,14]. SNGDM frameworks have excellent research and application potential in many fields, such as supplier selection [1], public opinion management [11,31], political elections [3,34], markets [2,16] and transportation [12]. To select the best alternative from a set of potential candidates, SNGDM involves a set of individuals, known as agents, who can express their opinions and communicate with their neighbors. Opinions and beliefs are crucial factors that influence our behavior, thereby defining our individuality and driving our actions [8,10,13,23]. Opinion evolution, also known as opinion dynamics, represents the process of modification of an agent's opinions by merging them with those of other agents, resulting in the formation of a stable structure. This process may involve consensus, polarization, or fragmentation [10].

A notable challenge in SNGDM is the achievement of a general and unanimous agreement among all agents [5,22]. With the rapid development of wireless communication networks and Internet-based technologies, we can now exchange opinions with a large number of people in real-time. The large-scale consensus reaching Process (LSCRP) in agents' opinions in SNGDM takes into account the opinions of agents throughout a social network [10]. The network represents the interaction rule between agents and plays a critical role in opinion dynamics [15,19]. A growing body of literature has recognized the importance of network topology in the fusion rules of opinion dynamics in the LSCRP. Lu et al. [17] allowed agents to express their trust values and relationship strengths with other agents to better reflect the actual social network and improve the efficiency of LSCRP. Chao et al. [6] constructed a two-layer network topology to address incomplete social relationships among agents on a large-scale. This framework could reconcile conflicting preferences and accelerate LSCRP at a minimal cost. Additional work on LSCRP can be found in [22,30]. Notably, the existing methods typically consider the peer-to-peer network topology, in which each agent communicates directly with all other perceived neighbors with the equal weight to update its own state.

Although consensus reaching processes have been widely studied and LSCRP investigations have achieved promising results, research on LSCRP within the context of SNGDM is still in its nascent stages [9]. The requirement of a high consensus level and presence of incomplete social relationships may render communication and opinion evolution among agents complex and challenging, especially in a large-scale scenario [6]. However, existing LSCRP methods consider only one-hop-based connections, in which agents interact with similar agents based on specific contexts, and ignore the more efficient interaction patterns that can facilitate consensus. These problems can be addressed by introducing multi-agent reinforcement learning (MARL), which has emerged as a powerful technology for accomplishing dynamic tasks online [32]. For example, Zhang et al. [33] used mean-field theory to decompose the joint action from the individual perspective in cooperative settings. Moreover, Sun et al. [21] solved the real-time Volt-Var control problem by using the multi-agent deep deterministic policy gradient method in a cooperative setting. However, the integration of MARL with LSCRP frameworks remains largely unexplored.

Considering these aspects, this study is focused on promoting consensus among a large group of people (i.e., in a multi-agent system) in a social network. We allow the agents to adaptively and discriminatively select the influence of locally perceived neighbors, thereby reconciling conflicting opinions and fostering unanimous consensus among agents at a higher rate. The main contributions of this study can be summarized as follows:

- We propose a novel agent perception model based on bidirectional long short-term Memory (LSTM), which adaptively reweights the influence of perceived local neighbors.
- We devise three types of differential reward functions within social networks to facilitate reinforcement learning.
- We design a multi-agent exploration and exploitation algorithm based on policy gradient to effectively train the agent perception model.

2   Model Formulation

This section describes the proposed model and reward functions. Section 2.1 introduces the decision-making model that adaptively adjusts the weights of neighboring states. Section 2.2 outlines the three goal-oriented differential reward functions designed to facilitate the learning process of agents.

2.1   Model Settings

The number of neighbors perceived by each agent is uncertain during the evolution of opinions, and each local neighbor needs to be evaluated in the decision-making process. Therefore, we use a recurrent neural network (RNN), which is a suitable model for time-series analysis, to address the problem of uncertain local perception input and uncertain decision length. Given that the agents must consider the overall context information of all perceived neighbors to make judgments, the perceived neighbors are encoded by extending the conventional bidirectional RNN.

As shown in Fig. 1, we encode the perceived neighbors based on the bidirectional LSTM [29] model, which is a widely used variant of RNN. The input to the proposed model is the set of state values $s_i(k) = \{x_j(k) : j \in N_i(k)\}$ of all perceived neighbors of agent i and their corresponding index $j : j \in N_i(k)$ at time k. After passing through the bidirectional LSTM network, the output is the action $a_i(k)$ for the opinion value of each neighbor, which represents the reweighting probability of each neighbor. Thus, the output layer of the LSTM model contains the SoftMax activation function, which ensures that the sum of the reweighting probabilities is 1. Therefore, the dimension of action $a_i(k)$ is consistent with that of the input state set, i.e., $[N_i(k), 1]$. Additionally, to endow the

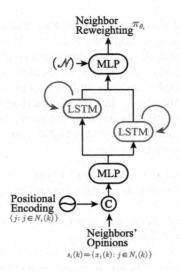

Fig. 1. Adaptively reweighting model based on LSTM

agents with exploration abilities to improve the robustness of the learning process, we add the noise $p \times \mathcal{N}$ to the output layer weight of the final action, where $\mathcal{N}$ follows the standard normal distribution, and p is a temperature parameter that controls the scale of noise $\mathcal{N}$.

The output action $a_i(k) \sim \pi_{\theta_i}$ of the model represents the aggregation of the influence weight $\pi_{\theta_i}^j \in (0,1)$ of each neighbor. The weights of the non-neighbors of agent i are set as zero. To obtain the new communication topology matrix $L(k)$ after processing through the proposed model, we aggregate the weights of the neighbors and non-neighbors of agent i in the order of agent index. Specifically, $L(k) = [l_{ij}(k)]$, where each element $l_{ij}(k)$ denotes the new communication topology relationship within the network, as defined in Eq. (1).

$$l_{ij} = \begin{cases} \pi_{\theta_i}^j, & j \in N_i(k) \\ 0, & j \notin N_i(k) \end{cases} \tag{1}$$

2.2   Reward Settings

The design of the reward function significantly affects the learning process of agents [20]. Furthermore, the reward function must be formulated considering the balance between the index coefficient and learnability. Therefore, we design the following differential reward function for the social network:

$$r_i(k) = \alpha \mathcal{G}_1(X(k+1)) + \beta \mathcal{G}_2(X(k+1)) + \gamma \mathcal{G}_3 \tag{2}$$

where $\mathcal{G}_1(\cdot)$, $\mathcal{G}_2(\cdot)$ and $\mathcal{G}_3$ represent the three objectives, with α, β, and γ representing their temperature coefficients respectively.

$\mathscr{G}_1(\cdot)$ enhances the capability of the agents in enhancing the consensus degree of the system. The convergence degree is quantified considering the standard deviation of opinion values in the multi-agent system:

$$\mathscr{G}_1 = g_1\left(X\left(k+1\right)\right) - g_1\left(X_{-i}\left(k+1\right)\right) \tag{3}$$

$$g_1\left(X\left(k+1\right)\right) = \frac{std\left(X\left(k+1\right)\right)}{n} \tag{4}$$

where $n = |V|$ represents the number of agents in the system, $X_{-i}(k+1)$ represents the state after the exclusion of agent i's state from the global state, and $std(\cdot)$ represents the standard deviation operation. The range of g_1 depends on the initial range of the system state and is typically $\left[0, \frac{|O_{max}-O_{min}|}{n}\right)$, where O_{max} and O_{min} represent the maximum and minimum values of the system state, respectively. In the Hegselmann-Krause model, (HK) [4], O_{max} is typically less than 10. A g_1 value closer to 0 corresponds to superior convergence performance of the system. The range of $\mathscr{G}_1$ is $[-\frac{|O_{max}-O_{min}|}{n}, \frac{|O_{max}-O_{min}|}{n}]$, and a value closer to $-\frac{|O_{max}-O_{min}|}{n}$ indicates a lower divergence degree of agent i with respect to the complete system.

$\mathscr{G}_2(\cdot)$ enables agents to improve the connectivity density of the system. The network topology represents the communication pattern underlying opinion dynamics and plays a key role in convergence theory [18]. Therefore, we quantify the connectivity by the density of the network topology:

$$\mathscr{G}_2 = g_2\left(X\left(k+1\right)\right) - g_2\left(X_{-i}\left(k+1\right)\right) \tag{5}$$

$$g_2\left(X\left(k+1\right)\right) = \frac{\sum_{i\in V}|N_i\left(k+1\right)|}{n^2} \tag{6}$$

where $X_{-i}(k+1)$ has the same meaning as that in Eq. (3), and $\sum_{i\in V}|N_i\left(k+1\right)|$ represents the number of connections of the system at time k. Thus, the range of g_2 is $[0,1]$, and a value is closer to 1 indicates a higher degree of connectivity in the system. The range of $\mathscr{G}_2$ is $[0,1)$, and a value closer to 1 indicates a greater influence of agent i on the density of the system distribution.

$\mathscr{G}_3$ enables agents to reduce the number of steps required to achieve a consensus. Therefore, we intuitively introduce the penalty for each convergence step:

$$\mathscr{G}_3 = -0.01 \tag{7}$$

$\mathscr{G}_3$ accumulates with the number of steps. During a round of evolution, $\mathscr{G}_3$ takes a value in $[0, -0.01 \times \hbar]$, where $\hbar$ represents the number of steps required to achieve stability. A cumulative value of $\mathscr{G}_3$ closer to 0 indicates that fewer steps are required to achieve stability.

3   Algorithm

The policy gradient-based reinforcement learning algorithm parameterizes the agent strategy and then directly optimizes it by maximizing the expected cumulative return [25]. This method can effectively enable iterative optimization during agent exploration and address the challenges associated with the continuous

---

Algorithm 1: Policy-gradient-based exploration and exploitation algorithm

---

Input: maximum training episode M, maximum time step T, learning batch size B

Output: the parameter θ of each agent's adaptively reweighting policy network

1 Initialize the policy parameter θ, the experience replay-buffer pool D, and the weights α, β, γ of differential reward function;

2 **for** $e=1$ **to** $|M|$ **do**

3   Reset and initialize the environment to obtain the global initial state $X(k)$ of the system;

4   Receive the temperature p to control the noise;

5   **for** $k=1$ **to** $|T|$ **do**

6   Each agent perceives local neighbors' state $s_i(k)$ according to Eq. (8);

7   Each agent reweights neighbors based on $a_i(k)$, and obtains $L_i(k)$ ad described in Sec. 2.1;

8   Each agent receives an instant differential reward according to Eq. (2);

9   The environment with states $X(k)$ evolves to $X(k+1)$ according to selected neighbors for all agents and Eq. (10);

10   Store the experience samples $(s_i(k), a_i(k), r_i(k))$ of all agents in the experience replay-buffer pool D;

11   **if** *done* **then**

12   Break;

13   **end**

14   **end**

15   Randomly and uniformly select trajectory samples of agents with batch size B from the experience pool D;

16   Calculate the gradient $\nabla_\theta U(\theta)$ via Eqs. (14) and (15);

17   Update policy parameters θ via Eq. (17);

18 **end**

19 **return** θ

---

action space. Based on the modeling and analysis of the state, action, and reward functions, as discussed in the previous sections, this section describes the process flow of exploration and exploitation, outlined in Algorithm 1.

For ease of reference in the following derivations, we use τ_i to denote the continuous state-action pairs $(s_i(0), a_i(0), \cdots, s_i(H), a_i(H))$ of agent i, generated through its interactions with the environment, where H denotes the length of the sequence. The proposed algorithm follows the distributed testing and centralized training framework and consists of two parts, i.e., the multi-agent exploration stage and strategy exploitation and updating stage. These stages are explained in the following sections.

Multi-agent exploration stage (lines 3–13 in Algorithm 1). In the process of exploration, each agent obtains the state value $s_i(k)$ of its neighbors within its perception range, as indicated in Eq. (8).

$$N_i(k) = \{j \in V : |x_i(k) - x_j(k)| < 1\} \tag{8}$$

Then, each agent calculates the weight of its neighbors based on the proposed model, as shown in Fig. 1. We obtain a new communication topology matrix $L(k)$ using Eq. (1). The opinions of all agents evolve synchronously as

$$x_i(k+1) = \sum_{j \in N_i(k)} x_j(k) \times l_{ij}(k) \tag{9}$$

Each agent $V_i \in V$ in the system maintains an opinion $x_i(k)$, represented by a real number, for a given issue. Let $X(k) = [x_1(k), x_2(k), \cdots, x_n(k)]^T$ be the opinion profile of all agents at time k. In this case, the model (9) can be rewritten in the following matrix form:

$$X(k+1) = L(k)X(k) \tag{10}$$

To ensure that the agents can automatically and adaptively select the exploration method according to episode, the noisy weight parameter p is decreased as the number of episodes increases.

Subsequently, to learn the policy parameters in the multi-agent exploitation stage, each agent stores the local perception state $s_i(k)$, local action $a_i(k)$ and immediate reward $r_i(k)$ at each time step in the experience buffer pool in chronological order based on the longest time span of exploration. At the end of each episode, the sequences of agents are selected for learning through uniform random sampling. If the system reaches a state of consistency, i.e., if all agents converge to a cluster, the parameter $done = true$ indicates early termination of the current round of exploration (lines 11–13).

Strategy exploitation and updating stage (lines 15–17 in Algorithm 1). At the end of each episode, the experience sequences of agents are selected for learning through uniform random sampling. To facilitate the following derivation, we use τ_i to denote the continuous state-action pairs $(s_i(0), a_i(0), \cdots, s_i(H), a_i(H))$ of agent i, where H denotes the length of the sequence. For updating the policy parameters, agents constantly explore and exploit their policies to maximize the expected cumulative return in the future:

$$U(\theta) = \mathbb{E}_{\pi_{\theta_i}} \left[\sum_k (r_i(k) | s_i(k), a_i(k)) \right] \approx \sum_{\tau_i} P(\tau_i | \theta_i) R_i(\tau_i) \tag{11}$$

$$R_i(\tau_i) = \sum_{k=0}^{H} r_i(k) \tag{12}$$

We use the gradient descent method to find the gradient $\nabla_\theta U(\theta)$ of objection function $U(\theta)$:

$$
\begin{aligned}
\nabla_\theta U(\theta) &= \nabla_\theta \sum_{\tau_i} P(\tau_i|\theta_i) R_i(\tau_i) \\
&= \sum_{\tau_i} P(\tau_i|\theta_i) \frac{\nabla_\theta P(\tau_i|\theta_i)}{P(\tau_i|\theta_i)} R_i(\tau_i) \\
&= \sum_{\tau_i} P(\tau_i|\theta_i) \nabla_\theta \log P(\tau_i|\theta_i) R_i(\tau_i)
\end{aligned}
\tag{13}
$$

According to Eq. (13), the gradient of $U(\theta)$ contains $P(\tau_i|\theta_i)$ and $\nabla_\theta \log P(\tau_i|\theta_i) R_i(\tau_i)$. Because $P(\tau_i|\theta_i)$ represents the probability of occurrence of trajectory τ_i, the gradient can be equivalent to the expectation of $\nabla_\theta \log P(\tau_i|\theta_i) R_i(\tau_i)$. Therefore, we estimate the gradient through average approximation based on the experience of the sampled trajectories:

$$
\begin{aligned}
\nabla_\theta U(\theta) &= \sum_{\tau_i} P(\tau_i|\theta_i) \nabla_\theta \log P(\tau_i|\theta_i) R_i(\tau_i) \\
&\approx \frac{1}{m} \sum_{i=1}^{m} \nabla_\theta \log P(\tau_i|\theta_i) R_i(\tau_i)
\end{aligned}
\tag{14}
$$

Furthermore, the gradient calculated using Eq. (14) can be intuitively understood as follows: The algorithm increases and decreases the probability of occurrence of trajectories with high and low reward, respectively. Then, we solve the only uncertainty $\nabla_\theta \log P(\tau_i|\theta_i)$ in Eq. (14):

$$
\begin{aligned}
&\nabla_\theta \log P(\tau_i|\theta_i) \\
&= \nabla_\theta \log \left[\prod_{k=0}^{H} \begin{array}{c} P(s_i(k+1)|s_i(k), a_i(k)) \times \\ \pi_{\theta_i}(a_i(k)|s_i(k)) \end{array} \right] \\
&= \nabla_\theta \left[\begin{array}{c} \sum_{k=0}^{H} \log P(s_i(k+1)|s_i(k), a_i(k))+ \\ \sum_{k=0}^{H} \log \pi_{\theta_i}(a_i(k)|s_i(k)) \end{array} \right] \\
&= \nabla_\theta \left[\sum_{k=0}^{H} \log \pi_{\theta_i}(a_i(k)|s_i(k)) \right] \\
&= \sum_{k=0}^{H} \nabla_\theta \log \pi_{\theta_i}(a_i(k)|s_i(k))
\end{aligned}
\tag{15}
$$

In Eq. (15), $P(s_i(k+1)|s_i(k), a_i(k))$ represents the system dynamics. Because the dynamics do not include the policy parameter θ, it can be deleted. Subsequently, we obtain the final policy gradient as:

$$
\nabla_\theta U(\theta) \approx \frac{1}{m} \sum_{i=1}^{m} \sum_{k=0}^{H} \nabla_\theta \log \pi_{\theta_i}(a_i(k)|s_i(k)) r_i(k)
\tag{16}
$$

Table 1. Parameter settings in simulations

Param	Explanation	Value
M	Maximum number of episodes	150
T	Maximum time step	20
B	Batch size	300
ζ	Learning rate	2e−3
n	Number of agents	20 to 100
ω	Density of agents	5 or 10
r_c	Perception range of agents	1
$\alpha,\ \beta,\ \gamma$	Temperature coefficients of rewards	−1, 1, 1
$interval$	Initial state range	$[0, 4]$ or $[0, 10]$
th	Convergence threshold	1e−2
$sdim$	Input dimension of state in LSTM	2
$adim$	Output dimension of action in LSTM	1
$hdim$	Hidden dimension in LSTM	36
$hlays$	Number of hidden layers in LSTM	2

Table 2. Ablation study

Exp. ID	$\mathscr{G}_1$	$\mathscr{G}_2$	$\mathscr{G}_3$	Number of clusters	Convergence step
A1	✓	%	%	4.2 ± 0.39	11.5 ± 0.49
A2	✓	%	✓	4.3 ± 0.45	9.0 ± 2.36
A3	%	✓	✓	4.4 ± 0.48	11.2 ± 2.82
A4	%	✓	✓	4.1 ± 0.51	9.9 ± 2.31
A5	✓	✓	%	$\mathbf{3.7 \pm 0.45}$	9.6 ± 1.2
A6	✓	✓	✓	3.7 ± 0.78	$\mathbf{8.4 \pm 1.35}$

Lastly, the agent's policy parameters are updated through the steepest descent method with the learning rate ζ:

$$\theta \leftarrow \theta + \zeta \nabla_\theta U(\theta) \tag{17}$$

4   Experiment

We develop a simulation environment for opinion dynamics using Python 3.7.0 and model the agent policy-gradient network with an LSTM architecture using PyTorch 1.2.0. Table 1 lists the parameters used in the experiment. To comprehensively analyze the superiority and effectiveness of our method, we use the 'convergence step' and 'number of cluster' as the evaluation metrics. The convergence step refers to the number of steps required for the system to reach a state in which the distance between any agent and its neighbors is less than the threshold th'. The achievement of system stability with fewer convergence steps and a smaller number of clusters corresponds to a superior performance [26–28] The experimental results, demonstrate that the integration of MARL with LSCRP can help reconcile conflicting opinions and promote unanimous consensus among all agents.

4.1   Ablation Study

First, we verify the effectiveness of the proposed model using the differential reward function and compare the model performance under different reward combinations, as presented in Table 2. The settings involve a system opinion range of $[0, 10]$ with $n = 100$ and $\omega = 10/r_c$. All experiments are conducted 10, and the mean and standard deviation are reported.

– **A1** vs. **A3** vs. **A5**: Compared with the scenarios in which only $\mathscr{G}_1$ (**A1**) or $\mathscr{G}_2$ (**A3**) is considered, the incorporation of both $\mathscr{G}_1$ and $\mathscr{G}_2$ (**A5**) leads to a 13.9% reduction in the average number of steps required for the system to reach stability and a 15.3% decrease in the average number of clusters at stability. Thus, using a combination of rewards $\mathscr{G}_1$ and $\mathscr{G}_2$ can help decrease the number of clusters and steps required to achieve system stability.

– **A1** vs. **A2**, and **A3** vs. **A4**: The convergence steps of **A2** and **A4** are 21.4%0 and 11.6% lower than those of **A1** and **A3**, respectively. This finding indicates that considering reward $\mathscr{G}_3$ allows agents to learn to reduce the number of steps required for system stability with an insignificant change in the number of clusters.

– **A5** vs. **A6**: By comprehensively considering rewards $\mathscr{G}_1$, $\mathscr{G}_2$ and $\mathscr{G}_3$, **A6** achieves a 12.5% reduction in convergence step compared with **A5**. This finding indicates that incorporating reward $\mathscr{G}_3$ in addition to $\mathscr{G}_1$ and $\mathscr{G}_2$ can further decrease the number of steps required to stabilize the system.

4.2   Comparative Analyses

We access the effectiveness of the model in terms of the convergence step and number of clusters when the system achieves stability. The number of convergence clusters indicates the enhancement of consistency of the model, and the convergence step indicates the number of steps required by the model to achieve stability. We compare the proposed method with the HK, common-neighbor rule (CNR) [24], group-pressure (GP) methods [7]. Because the objective is to enhance consistency, we aim to ensure convergence to fewer clusters in fewer steps. The relevant parameters in the CNR and GP models are uniformly set as $\beta = 0$, $m = 1$ and $p_i = \lambda = 0.5$.

Existing LSCRP methods consider only one-hop-based connections, in which agents interact with similar agents, and ignore the efficient interaction patterns that can facilitate consensus. We illustrate the evolution process of the three baselines and our method with an initial range of $[0, 10]$ and $n = 100$. The following observations are made:

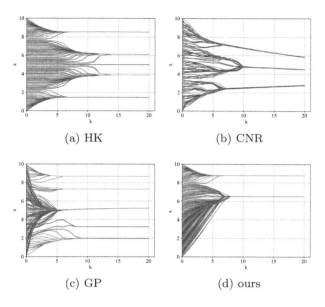

(a) HK   (b) CNR

(c) GP   (d) ours

Fig. 2. Results of the proposed and baseline methods

Table 3. Statistics of comparison simulations

ID	Init. range	n	ω	Method	Cluster number	Convergence step
M1	$[0,4]$	20	$\frac{5}{r_c}$	HK	2	5
				CNR	1	5
				GP	1	5
				ours	**1**	**5**
M2	$[0,4]$	40	$\frac{10}{r_c}$	HK	1	9
				CNR	1	4
				GP	1	4.80 ± 0.75
				ours	**1**	$\mathbf{4.45 \pm 0.50}$
M3	$[0,10]$	50	$\frac{5}{r_c}$	HK	5	11
				CNR	$\mathbf{3.33 \pm 0.94}$	19.8 ± 0.60
				GP	3.90 ± 0.70	14.90 ± 6.32
				ours	3.45 ± 0.50	$\mathbf{10.72 \pm 1.81}$
M4	$[0,10]$	100	$\frac{10}{r_c}$	HK	5	14
				CNR	8.20 ± 2.40	20.00 ± 0.00
				GP	4.50 ± 0.50	18.00 ± 3.55
				ours	$\mathbf{2.70 \pm 0.45}$	$\mathbf{9.10 \pm 1.51}$

- HK evolves by obtaining the average value of neighbors, which makes it difficult to control the numbers of steps and clusters when the system is stable (Fig. 2a).

- CNR selects local and long-range neighbors according to the confidence bounds and a common-neighbor rule, respectively. Therefore, CNR requires a large amount time to reach a consensus (Fig. 2b).
- GP takes group pressure into consideration, leading to the formation of inner opinions within the agents' bounded confidence. However, the ambient pressure reduces communication between agents, resulting in more clusters (Fig. 2c).
- The proposed method enable agents to adaptively learn the neighbor reweighting strategy with multi-objective trade-off ability. The agents required only a few steps to reach a consensus with a small number of clusters (Fig. 2d).

Table 3 summarizes the statistical results of the comprehensive comparisons. The experiments are conducted 10 times after 100 training runs, and the mean and standard deviation of the metrics are reported.

- **Horizontal analysis**: We analyze the performance of different algorithms under the same density. The proposed algorithm requires the fewest number of steps to achieve stability. With the improvement of the range (**M2, M4**), the advantage of our algorithm in terms of the number of steps is amplified.
- **Longitudinal analysis**: Under the same range, as the density increases, the numbers of steps and clusters associated with our algorithm decreases or remain stable. In high-range situations (**M3, M4**), the numbers of steps and clusters associated with the baselines increase as the density increases, whereas those of our algorithm steadily decrease.

Notably, in high situations (**M4**), the proposed method achieves an average reduction of 51% and 46.3% in the number of clusters and convergence step, respectively, compared with the baselines. These demonstrates the potential of the proposed method in resolving disagreements among agents and accelerating the consensus-building process.

5   Conclusions

With the objective of enhancing the consensus of opinion dynamics in the field of social networks, an intelligent perception model based on MARL is developed. For the convergence process, we first design an adaptive reweighting model based on bidirectional LSTM to capture the perception capability. Then, we formulate the corresponding differential reward function based on three types of goals in the opinion dynamics scenario. Finally, through the multi-agent exploration and strategy exploitation algorithm based on the policy gradient, the agents are allowed to adaptively learn an efficient neighbor reweighting strategy with multi-objective trade-off during their interaction. The experimental results verify that the proposed method can enable agents to adaptively reweight the influence of neighbors while exhibiting multi-objective trade-off abilities and effectively reconcile opinions with large differences in the social network system. Thus, the number of clusters at stability is reduced, and the convergence process is accelerated.

In future work, we will focus on the consistency enhancement method with attention mechanisms and privacy protection in social networks and verify the effectiveness and generalization ability of the proposed approach in real opinion dynamics scenarios.

Acknowledgments. This work is partially supported by National Natural Science Foundation of China, Grant Nos. 62006047 and 618760439, Guangdong Natural Science Foundation, Grant No. 2021B0101220004.

References

1. Abdel-Basset, M., Saleh, M., Gamal, A., Smarandache, F.: An approach of TOPSIS technique for developing supplier selection with group decision making under type-2 neutrosophic number. Appl. Soft Comput. **77**, 438–452 (2019)
2. Beni, S.A., Sheikh-El-Eslami, M.K.: Market power assessment in electricity markets based on social network analysis. Comput. Electr. Eng. **94**, 107302 (2021)
3. Biswas, K., Biswas, S., Sen, P.: Block size dependence of coarse graining in discrete opinion dynamics model: application to the us presidential elections. Physica A **566**, 125639 (2021)
4. Blondel, V.D., Hendrickx, J.M., Tsitsiklis, J.N.: On Krause's multi-agent consensus model with state-dependent connectivity. IEEE Trans. Autom. Control **54**(11), 2586–2597 (2009)
5. Cabrerizo, F.J., Al-Hmouz, R., Morfeq, A., Balamash, A.S., Martínez, M.Á., Herrera-Viedma, E.: Soft consensus measures in group decision making using unbalanced fuzzy linguistic information. Soft. Comput. **21**(11), 3037–3050 (2017)
6. Chao, X., Kou, G., Peng, Y., Herrera-Viedma, E., Herrera, F.: An efficient consensus reaching framework for large-scale social network group decision making and its application in urban resettlement. Inf. Sci. **575**, 499–527 (2021)
7. Cheng, C., Yu, C.: Opinion dynamics with bounded confidence and group pressure. Physica A **532**, 121900 (2019)
8. Bros, D.C.N.W.: Inception. United States (2010)
9. Dong, Y., et al.: Consensus reaching in social network group decision making: research paradigms and challenges. Knowl. Based Syst. **162**, 3–13 (2018)
10. Dong, Y., Zhan, M., Kou, G., Ding, Z., Liang, H.: A survey on the fusion process in opinion dynamics. Inf. Fusion **43**, 57–65 (2018)
11. Douven, I., Hegselmann, R.: Mis- and disinformation in a bounded confidence model. Artif. Intell. **291**, 103415 (2021)
12. Hashemi, E., Pirani, M., Khajepour, A., Fidan, B., Kasaiezadeh, A., Chen, S.: Opinion dynamics-based vehicle velocity estimation and diagnosis. IEEE Trans. Intell. Transp. Syst. **19**(7), 2142–2148 (2018)
13. Hassani, H., Razavi-Far, R., Saif, M., Chiclana, F., Krejcar, O., Herrera-Viedma, E.: Classical dynamic consensus and opinion dynamics models: a survey of recent trends and methodologies. Inf. Fusion **88**, 22–40 (2022)
14. Hua, Z., Jing, X., Martínez, L.: Consensus reaching for social network group decision making with ELICIT information: a perspective from the complex network. Inf. Sci. **627**, 71–96 (2023)

15. Huang, D.W., Yu, Z.G.: Dynamic-sensitive centrality of nodes in temporal networks. Sci. Rep. **7**(1), 1–11 (2017)
16. Kawasaki, T., Wada, R., Todo, T., Yokoo, M.: Mechanism design for housing markets over social networks. In: Dignum, F., Lomuscio, A., Endriss, U., Nowé, A. (eds.) AAMAS 2021: 20th International Conference on Autonomous Agents and Multiagent Systems, Virtual Event, United Kingdom, 3–7 May 2021, pp. 692–700. ACM (2021)
17. Lu, Y., Xu, Y., Herrera-Viedma, E., Han, Y.: Consensus of large-scale group decision making in social network: the minimum cost model based on robust optimization. Inf. Sci. **547**, 910–930 (2021)
18. Ma, Q., Qin, J., Anderson, B.D., Wang, L.: Exponential consensus of multiple agents over dynamic network topology: controllability, connectivity, and compactness. IEEE Trans. Automatic Control, 1–16 (2023). https://doi.org/10.1109/TAC. 2023.3245021
19. Nedić, A., Olshevsky, A., Rabbat, M.G.: Network topology and communication-computation tradeoffs in decentralized optimization. Proc. IEEE **106**(5), 953–976 (2018)
20. Silver, D., Singh, S., Precup, D., Sutton, R.S.: Reward is enough. Artif. Intell. **299**, 103535 (2021)
21. Sun, X., Qiu, J.: Two-stage Volt/Var control in active distribution networks with multi-agent deep reinforcement learning method. IEEE Trans. Smart Grid **12**(4), 2903–2912 (2021)
22. Ureña, R., Chiclana, F., Melançon, G., Herrera-Viedma, E.: A social network based approach for consensus achievement in multiperson decision making. Inf. Fusion **47**, 72–87 (2019)
23. Ureña, R., Kou, G., Dong, Y., Chiclana, F., Herrera-Viedma, E.: A review on trust propagation and opinion dynamics in social networks and group decision making frameworks. Inf. Sci. **478**, 461–475 (2019)
24. Wang, H., Shang, L.: Opinion dynamics in networks with common-neighbors-based connections. Physica A **421**, 180–186 (2015)
25. Weng, T., et al.: Toward evaluating robustness of deep reinforcement learning with continuous control. In: 8th International Conference on Learning Representations, ICLR 2020, Addis Ababa, Ethiopia, 26–30 April 2020. OpenReview.net (2020)
26. Xie, G., Chen, J., Li, Y.: Hybrid-order network consensus for distributed multi-agent systems. J. Artif. Intell. Res. **70**, 389–407 (2021)
27. Xie, G., Xu, H., Li, Y., Hu, X., Wang, C.D.: Fast distributed consensus seeking in large-scale and high-density multi-agent systems with connectivity maintenance. Inf. Sci. **608**, 1010–1028 (2022)
28. Xie, G., Xu, H., Li, Y., Wang, C.D., Zhong, B., Hu, X.: Consensus seeking in large-scale multiagent systems with hierarchical switching-backbone topology. IEEE Trans. Neural Networks Learn. Syst., 1–15 (2023). https://doi.org/10.1109/ TNNLS.2023.3290015
29. Yu, Y., Si, X., Hu, C., Zhang, J.: A review of recurrent neural networks: LSTM cells and network architectures. Neural Comput. **31**(7), 1235–1270 (2019)
30. Zhang, H., Dong, Y., Herrera-Viedma, E.: Consensus building for the heterogeneous large-scale GDM with the individual concerns and satisfactions. IEEE Trans. Fuzzy Syst. **26**(2), 884–898 (2018)
31. Zhang, H., Dong, Y., Xiao, J., Chiclana, F., Herrera-Viedma, E.: Consensus and opinion evolution-based failure mode and effect analysis approach for reliability management in social network and uncertainty contexts. Reliab. Eng. Syst. Saf. **208**, 107425 (2021)

32. Zhang, K., Yang, Z., Başar, T.: Multi-agent reinforcement learning: a selective overview of theories and algorithms. In: Vamvoudakis, K.G., Wan, Y., Lewis, F.L., Cansever, D. (eds.) Handbook of Reinforcement Learning and Control. SSDC, vol. 325, pp. 321–384. Springer, Cham (2021). https://doi.org/10.1007/978-3-030-60990-0_12

33. Zhang, T., Ye, Q., Bian, J., Xie, G., Liu, T.: MFVFD: a multi-agent Q-learning approach to cooperative and non-cooperative tasks. In: Zhou, Z. (ed.) Proceedings of the Thirtieth International Joint Conference on Artificial Intelligence, IJCAI 2021, Virtual Event/Montreal, Canada, 19–27 August 2021, pp. 500–506. ijcai.org (2021)

34. Zhu, L., He, Y., Zhou, D.: Neural opinion dynamics model for the prediction of user-level stance dynamics. Inf. Process. Manag. **57**(2), 102031 (2020)

Lateral Interactions Spiking Actor Network for Reinforcement Learning

Xiangyu Chen[1], Rong Xiao[1,2(✉)], Qirui Yang[3], and Jiancheng Lv[1,2]

[1] College of Computer Science, Sichuan University, Chengdu, China
chenxiangyu@stu.scu.edu.cn, lvjiancheng@scu.edu.cn
[2] Engineering Research Center of Machine Learning and Industry Intelligence,
Ministry of Education, Chengdu, China
rxiao@scu.edu.cn
[3] Arizona State University, Tempe, USA
qyang30@asu.edu

Abstract. Spiking neural network (SNN) has been shown to be a biologically plausible and energy efficient alternative to Deep Neural Network in Reinforcement Learning (RL). In prevailing SNN models for RL, fully-connected architectures with inter-layer connections are commonly employed. However, the incorporation of intra-layer connections is neglected, which impedes the feature representation and information processing capacities of SNNs in the context of reinforcement learning. To address these limitations, we propose Lateral Interactions Spiking Actor Network (LISAN) to improve decision-making in reinforcement learning tasks with high performance. LISAN integrates lateral interactions between neighboring neurons into the spiking neuron membrane potential equation. Moreover, we incorporate soft reset mechanism to enhance model's functionality recognizing the significance of residual potentials in preserving valuable information within biological neurons. To verify the effectiveness of our proposed framework, LISAN is evaluated using four continuous control tasks from OpenAI gym as well as different encoding methods. The results show that LISAN substantially improves the performance compared to state-of-the-art models. We hope that our work will contribute to a deeper understanding of the mechanisms involved in information capturing and processing in the brain.

Keywords: Reinforcement Learning · Spiking Neural Networks · Lateral Interactions

1 Introduction

Deep learning has a significant impact on machine learning, particularly in reinforcement learning (RL), leading to the development of Deep Reinforcement Learning (DRL) [1]. Recent advancements in DRL have pushed its performance beyond human-level capabilities across various reinforcement learning tasks [8,11,16]. However, the resource-intensive nature of DRL combined with

B. Luo et al. (Eds.): ICONIP 2023, CCIS 1962, pp. 184–195, 2024.
https://doi.org/10.1007/978-981-99-8132-8_14

Deep Neural Networks (DNNs) presents challenges in applications such as mobile robots which require power-efficient and low-latency processing systems [9,12]. As a result, there is a growing need to explore energy-efficient and low-latency alternative networks for DRL.

Compared to DNNs, Spiking Neural Networks (SNNs) show great potential in simulating brain-inspired topology and functions due to the complex dynamics. By integrating spiking neurons with biologically plausible plasticity principles, complex cognitive functions can be generated. The biological brain achieves efficient computation through cell assembly which prioritizes spatial-temporal coding for memory over decision-making readout [7]. In recent works, there has been a growing interest in integrating SNN into reinforcement learning algorithms [4,5,19]. While these approaches frequently depend on reward-modulated local plasticity rules which have shown success in simple control tasks, they often face challenges when they are applied to complex robotic control tasks due to their limited optimization capabilities. To address this limitation, several approaches have emerged that integrate SNNs with DRL optimization. One notable approach involves a hybrid learning framework proposed by [17], which introduces a population coded spiking actor network (PopSAN) trained alongside a deep critic network using DRL. This approach has shown impressive performance and energy efficiency in continuous control tasks. Another approach [20] presents a multi-scale dynamic coding improved spiking actor network (MDC-SAN) for reinforcement learning, aiming to achieve effective decision-making. It combines population coding at the network scale with dynamic neurons coding and incorporates 2nd-order neuronal dynamics to enable a powerful spatial-temporal state representation.

Although SNNs have shown promising results in the field of RL, there are still significant avenues for exploration regarding the neuronal behavior and dynamical equations of SNNs. In recent researches, Residual Membrane Potential (RMP) spiking neurons based on soft reset [6] have been proposed to preserve the high dynamics of biological neurons. These RMP neurons have demonstrated near lossless conversion from Artificial Neural Networks (ANN) to SNN, showcasing their effectiveness on challenging datasets. Existing SNN models for reinforcement learning predominantly utilize the leaky-integrate-and-fire (LIF) neuron model [10], which effectively extracts object features. While these models incorporate inter-layer connections for feedforward processing, they often overlook the importance of intra-layer connections which is a critical mechanism in biological neural systems for object recognition. Neuroscientists have observed that lateral interactions among retina neurons can enhance the perception of visual object edges [13]. In the fields of computational neuroscience and cognitive science, Dynamic Neuronal Field (DNF) models have gained popularity as recurrent neural networks with attractor dynamics. In DNF, a neuron excites nearby neurons while inhibiting others [14,15]. This mechanism effectively enhances important input regions, suppresses areas with noise, and preserves valuable information within the neuron population [3]. Therefore, further exploration of dynamical improvements in SNNs in the context of RL remains valuable and worthwhile.

In this paper, we introduce a novel Lateral Interactions Spiking Actor Network(LISAN) to enhance the state representation capabilities of our model in solving complex RL tasks. Our model integrates lateral interactions between adjacent neurons into the equation which governs the spiking neuron's membrane potential. Additionally, considering the significance of residual potentials observed in biological neurons, we incorporate soft reset mechanism to enhance our model's overall functionality. This comprehensive approach enables a robust and effective information processing capability within the network. Our approach leverages a hybrid Actor-Critic network by combining the strengths of SNN and DNN. To address the coding challenges introduced by multi-dimensional state inputs in continuous RL tasks, we employ population coding techniques [17]. The gradient loss of each output action is computed by a deep critic network, which is integrated with the Twin Delayed Deep Deterministic policy gradient algorithm (TD3). Experiment results conducted on the continuous control tasks from the OpenAI gym benchmark demonstrate the superior performance of our method in terms of rewards gained compared to state-of-the-art models. Furthermore, we conduct evaluations of our model's performance using different encoding methods. The results consistently demonstrate the robust performance of LISAN.

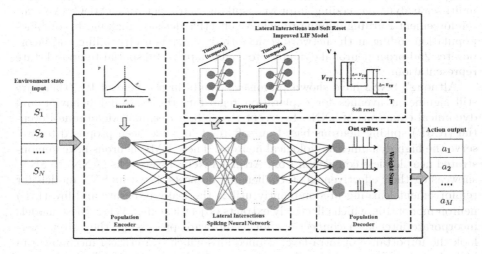

Fig. 1. The overall architecture of proposed LISAN

2   Method

In this section, we introduce the implementation details of our proposed model Lateral Interactions Spiking Actor Network (LISAN). The overall structure of LISAN is shown in Fig. 1.

2.1   Population Encoder

Population encoder refers to the encoding of information using the joint activity of multiple neurons within a population [17]. In our work, we utilize the population and deterministic coding method [20] to translate state information in RL task into spike trains that can be used as input to SNN.

For each dimension of the state S, we adopt a neuron population to perform encoding. Within each neuron population, the state information is initially transformed into stimulation strength through a Gaussian receptive field. Subsequently, the obtained stimulation strength represents the output of the presynaptic neuron, which is added to the membrane potential of the post-synaptic neuron. The sum is then compared to the threshold to determine whether a spike is fired. The overall process can be mathematically formulated as follows:

$$\begin{cases} P_{i,j} = EXP(-\frac{(S_i - \mu_{i,j})^2}{2\sigma_{i,j}^2}) \\ V_{i,j}(t) = V_{i,j}(t-1) + P_{i,j} \\ O_{i,j}(t) = \begin{cases} 1, \text{ if } V_{i,j}(t) > V_{th} \\ 0, \text{ otherwise} \end{cases} \end{cases} \tag{1}$$

where S_i is the i-th dimension of the state S, μ and σ denote the mean and standard deviation within the Gaussian receptive field, $P_{i,j}$ represents the stimulation strength of the j-th neuron within the neuron population which is responsible for encoding the i-th dimension of the state, $V_{i,j}(t)$ and $O_{i,j}(t)$ correspond to the membrane potential and spike activity of the neuron at time t, V_{th} is the firing threshold. $V_{i,j}(t)$ is reset to 0 if $O_{i,j}(t)$ is 1.

The μ and σ of the Gaussian receptive field are adjustable parameters, allowing the population encoder to progressively enhance its capability in representing the state information during the iterative training of the network.

2.2   Lateral Interactions Spiking Neuron Network

This section first provides an introduction to the conventional Leaky Integrate-and-Fire (LIF) neuron model, followed by a comprehensive definition and description of our proposed improved Lateral Interactions Spiking Neuron (LISN) framework.

LIF Neuron Model. The LIF model stands as a widely adopted mathematical framework for simulating physiological processes. The orignal LIF model can be described by:

$$\tau \frac{dV(t)}{dt} = -V(t) + RI(t) \tag{2}$$

where τ is the time constant, $I(t)$ denotes the input current accumulated from synapses and integrated into $V(t)$, which represents the membrane potential.

To capture the temporal and spatial relationship among neurons, we adopt a discretized time-step approach and partition the forward propagation within spiking neurons. This iterative model is formulated as follows:

$$I_i(t) = \alpha I_i(t-1) + \sum_j W_{ij} O_j(t-1) \tag{3}$$

$$V_i(t) = \beta V_i(t-1) + I_i(t) \tag{4}$$

$$O_i(t) = \Theta(V_i(t) - V_{th}) \tag{5}$$

where $V_i(t)$ and $I_i(t)$ represent the i-th neuron's current and voltage at time t, respectively. α, β are the current and voltage decay factor, and $\sum_j W_{ij} O_j(t-1)$ is the weighted sum of the incoming spikes from the previous layer j. $O_i(t)$ is the output spike and Θ is the Heaviside step function which is described by:

$$\Theta(x) = \begin{cases} 1, \text{ if } x > V_{th} \\ 0, \text{ otherwise} \end{cases} \tag{6}$$

When the membrane potential exceeds the firing threshold V_{th}, the neuron fires a spike and resets its membrane potential to zero. This mechanism is known as "Hard Reset".

Lateral Interactions and Soft Reset Improved Model

Soft Reset. Unlike hard reset which immediately reset the membrane potential to its initial value, soft reset allows a partial reset of the membrane potential after crossing a threshold. We content that the preservation of residual voltage which follows neuron firing confers enhanced efficacy upon SNN in complex spatiotemporal representation. Hence, to implement the soft reset mechanism in our model, we make the following modification to Eq. 4:

$$V_i(t) = \beta(V_i(t-1) - O_i(t-1) * V_{th}) + I_i(t) \tag{7}$$

Lateral Interactions. Currently, a plethora of studies shows that incorporating internal connections within neural networks enhance computational performance in many fields [2,18]. Consequently, we propose merging the internal connections among neurons into the construction of SNN. We aim to investigate the potential enhancement of processing performance for RL tasks by introducing lateral interactions among neurons within the same layer in the spiking actor network.

Furthermore, we make the following enhancements to Eq. 7 in our study:

$$V_i(t) = \beta(V_i(t-1) - O_i(t-1) * V_{th}) + I_i(t) + (O_j(t-1) * W_N) \tag{8}$$

where $O_j(t-1)$ is the spiking output of the neighbors in the same layer at time $t-1$, W_N is a learnable matrix which characterizes the interconnections among neurons within the same layer to increase network flexibility.

In conclusion, we refer to the constructed model as Lateral Interactions Spiking Neuron (LISN). To observe the difference in information representation ability between LISN model and LIF model, We present sinusoidal wave signals to simulate the voltage inputs (weighted sum of pre-synaptic neuron outputs) to the neurons. We also simulate four neighboring neurons in the LISN model. The lateral interaction weight matrix W_N is initialized with random values from a normal distribution with mean 0 and standard deviation 0.05. Under identical inputs, the contrasting dynamics between LISN and traditional LIF neuron are depicted in Fig. 2. It is evident that under identical voltage inputs, the LISN and LIF neurons exhibit distinctive voltage dynamics, leading to disparate spike patterns. Specifically, the LISN neuron model produces a total of 6 output spikes, while the LIF model generates 9 spikes (see Fig. 2b). This observation highlights that the LISN neuron, with the incorporation of soft reset and interconnection, is less susceptible to continuous fluctuations in input voltages. Furthermore, for

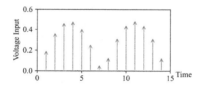

(a) Simulation of voltage input with sinusoidal wave variations

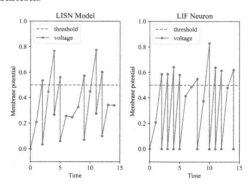

(b) Temporal Evolution of Membrane Potential. (Left) LISN Neuron (Right) LIF Neuron

Fig. 2. Contrasting dynamics between LISN and traditional LIF neurons with the same voltage inputs and threshold

the sake of clarity in presentation, we maintain fixed parameters for the W_N. However, it is important to note that in actual experimental settings, W_N is subject to learning during the training iterations which enables LISN to robustly process intricate input information.

2.3 Surrogate Gradient

The non-differentiable nature of the firing function (Eq. 8) poses challenges in training SNNs by backpropagation. To overcome the issue of gradient vanishing caused by the non-differentiable property of firing function, researchers have introduced the concept of surrogate gradient to facilitate gradient propagation in deep SNNs. In our work, we use the rectangular function equation to approximate the gradient of a spike.

$$d(x) = \begin{cases} 1, \text{ if } |x - V_{th}| < 0.5 \\ 0, \text{ otherwise} \end{cases} \tag{9}$$

where d is the pseudo-gradient, x is membrane voltage, and V_{th} is the firing threshold.

2.4 Population Decoder

In the population decoding process, spikes generated at the output layer are accumulated within a predefined time window to calculate the average firing rate. Subsequently, the output action is obtained by applying a weighted sum to the computed average firing rate.

3 Experiment

In this section, we adopt LISAN as the model for the actor network with DNN as the critic networks. We evaluate LISAN in different environments and compare it with the mainstream models.

3.1 Benchmarking LISAN Against Mainstream Models

Environment. We choose four classic MuJoCo continuous robot control tasks from the OpenAI gym (see Fig. 3) as our RL environment, the dimension information of each environment is shown in Table 1.

Benchmarking. We first compare LISAN to existing models DAN and Pop-SAN [17]. To investigate how soft reset and lateral interactions works on LIF neurons, We integrate them to PopSAN for comparison: PopSAN+SR(soft reset), PopSAN+LI(Lateral interactions). All models are trained using the TD3 algorithm combined with deep critic networks of the same structure. The hyperparameter configurations of these models are as follows:

- *DAN*: Actor network (256, relu, 256, relu, tanh), learning rate $= 1e - 3$; critic network (256, relu, 256, relu, linear), learning rate $= 1e - 3$; mini-batch size $n = 100$; reward discount factor $\gamma = 0.99$; soft target update factor $\mu = 0.005$; policy delay factor $d = 2$; maximum size of replay buffer is $1M$.
- *PopSAN*: Actor network (Population Encoder, 256, LIF, 256, LIF, Population Decoder); input population size for single state dimension is 10; output population size for single action dimension is 10; firing threshold $V_{th} = 0.5$; current decay factor $\alpha = 0.5$; voltage decay factor $\beta = 0.5$; SNN time window $T = 5$; the remaining parameters are the same as DAN.
- *LISAN*: Actor network (Population Encoder, 256, LISN, 256, LISN, Population Decoder); the remaining parameters are the same as PopSAN.
- *PopSAN+SR*: PopSAN improved with Soft Reset.
- *PopSAN+LI*: PopSAN improved with Lateral Interactions.

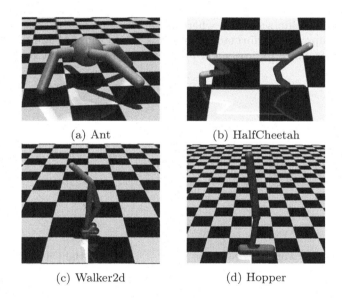

(a) Ant   (b) HalfCheetah

(c) Walker2d   (d) Hopper

Fig. 3. The overview of four OpenAI gym tasks in the simulation environment: (a) Ant: make a quadruped crawling robot run as fast as possible; (b) HalfCheetah: make a biped robot walk as fast as possible; (c) Walker2d: make a biped robot walk quickly; (d) Hopper: make a 2D robot hop forward

Training. In our experiment, we perform ten independent trainings on each model, using the same 10 random seeds to ensure consistency. During our training procedure, we train the models in each environment for one million steps, and we test every 10K steps. During each test, we calculate the average reward across 10 episodes with each episode capped at 1K steps.

Table 1. Dimension information of four OpenAI gym environments

Environment	State Dimension	Action Dimension
Ant	111	8
HalfCheetah	17	6
Walker2d	17	6
Hopper	11	3

Result. The experimental results are shown in Fig. 4 and Table 2. LISAN achieves the best performance in both simple and complex continuous control tasks. On the other had, DAN only performs well in simple environments, and fails to perform well with reinforcement learning tasks that involve high-dimensional environments and actions. Furthermore, we notice that the enhancement in model performance resulting from the soft reset mechanism is inconsistent. The residual voltage generated by soft reset proves advantageous to the model only in specific scenarios where it captures more dynamic information and leads to improved performance. And it is also obvious that LISAN exhibits a notable performance improvement even in the absence of soft reset.

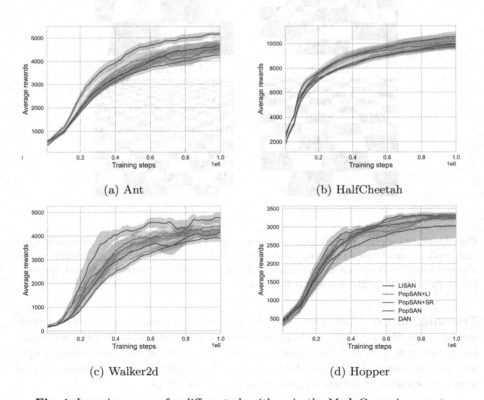

(a) Ant

(b) HalfCheetah

(c) Walker2d

(d) Hopper

Fig. 4. Learning curves for different algorithms in the MuJoCo environment

Table 2. The maximum average return over 10 random seeds. LISAN achieves the best performance marked in bold

Actor Network	Ant	HalfCheetah	Walker2d	Hopper
DAN	5038	10588	4745	3685
PopSAN	5736	10899	5605	3772
PopSAN+SR	5799	11295	5379	3590
PopSAN+LI	5811	11170	5816	3675
LISAN	**5965**	**11893**	**5901**	**3781**

3.2   Discussion of Different Input Codings for LISAN

We employ four different population coding methods [20] to encode the input state information, namely pure population coding (C_{pop}), population and Poisson coding ($C_{pop} + C_{poi}$), population and deterministic coding ($C_{pop} + C_{det}$), and population and uniform coding ($C_{pop} + C_{uni}$). We apply these coding methods to LISAN to test the performance of the Hopper task. Figure 5 shows the four integrated population-based coding methods with LISAN. All of them achieve good performance in the Hopper environment. Notably, the population and deterministic coding achieves the most rewards thanks to its faster convergence speed while the pure population coding achieves the fewest rewards. These observations highlight the robust adaptability and processing capabilities of our LISAN model across diverse coding methods.

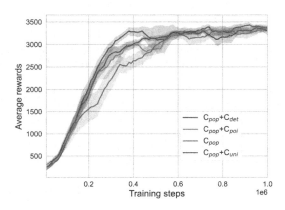

Fig. 5. Comprehensive comparison of the impact of various input coding methods on Hopper task

4   Conclusion

In this paper, we propose the Lateral Interactions Spiking Actor Network (LISAN) for deep reinforcement learning. LISAN utilizes the lateral interaction neuron model and incorporates soft reset mechanism for efficient training, which enables it to handle complex information while maintaining energy efficiency. Through extensive experiments conducted on four continuous control tasks from the OpenAI Gym, we demonstrate that LISAN outperforms the state-of-the-art deep neuron model as well as the same hybrid architecture SNN model. Additionally, we evaluate the performance of LISAN under different encoding methods, and show that LISAN achieves promising results across all of them. We hope that our work can serve as a fundamental building block for the emerging field of Spiking-RL and can be extended to a wide range of tasks in future research.

Acknowledgments. This work was supported by the National Natural Science Foundation of China (Grant No. 62206188), the China Postdoctoral Science Foundation (Grant No. 2022M712237), Sichuan Province Innovative Talent Funding Project for Postdoctoral Fellows and the 111 Project under grant B21044.

References

1. Arulkumaran, K., Deisenroth, M.P., Brundage, M., Bharath, A.A.: A brief survey of deep reinforcement learning. arXiv preprint arXiv:1708.05866 (2017)
2. Cheng, X., Hao, Y., Xu, J., Xu, B.: Lisnn: Improving spiking neural networks with lateral interactions for robust object recognition. In: IJCAI, pp. 1519–1525 (2020)
3. Evanusa, M., Sandamirskaya, Y., et al.: Event-based attention and tracking on neuromorphic hardware. In: Proceedings of the IEEE/CVF Conference on Computer Vision and Pattern Recognition Workshops (2019)
4. Florian, R.V.: Reinforcement learning through modulation of spike-timing-dependent synaptic plasticity. Neural Comput. **19**(6), 1468–1502 (2007)
5. Frémaux, N., Sprekeler, H., Gerstner, W.: Reinforcement learning using a continuous time actor-critic framework with spiking neurons. PLoS Comput. Biol. **9**(4), e1003024 (2013)
6. Han, B., Srinivasan, G., Roy, K.: RMP-SNN: residual membrane potential neuron for enabling deeper high-accuracy and low-latency spiking neural network. In: Proceedings of the IEEE/CVF Conference on Computer Vision and Pattern Recognition, pp. 13558–13567 (2020)
7. Harris, K.D., Csicsvari, J., Hirase, H., Dragoi, G., Buzsáki, G.: Organization of cell assemblies in the hippocampus. Nature **424**(6948), 552–556 (2003)
8. Hu, S., Zhu, F., Chang, X., Liang, X.: UPDeT: universal multi-agent reinforcement learning via policy decoupling with transformers. arXiv preprint arXiv:2101.08001 (2021)
9. Lahijanian, M., et al.: Resource-performance tradeoff analysis for mobile robots. IEEE Robot. Autom. Lett. **3**(3), 1840–1847 (2018)
10. Memmesheimer, R.M., Rubin, R., Ölveczky, B.P., Sompolinsky, H.: Learning precisely timed spikes. Neuron **82**(4), 925–938 (2014)
11. Mnih, V., et al.: Human-level control through deep reinforcement learning. Nature **518**(7540), 529–533 (2015)

12. Niroui, F., Zhang, K., Kashino, Z., Nejat, G.: Deep reinforcement learning robot for search and rescue applications: exploration in unknown cluttered environments. IEEE Robot. Autom. Lett. **4**(2), 610–617 (2019)
13. Ratliff, F., Hartline, H.K., Lange, D.: The dynamics of lateral inhibition in the compound eye of limulus. I. Studies on Excitation and Inhibition in the Retina: A Collection of Papers from the Laboratories of H. Keffer Hartline, p. 463 (1974)
14. Sandamirskaya, Y.: Dynamic neural fields as a step toward cognitive neuromorphic architectures. Front. Neurosci. **7**, 276 (2014)
15. Schöner, G., Spencer, J.P.: Dynamic Thinking: A Primer on Dynamic Field Theory. Oxford University Press, Oxford (2016)
16. Silver, D., et al.: Mastering the game of go with deep neural networks and tree search. Nature **529**(7587), 484–489 (2016)
17. Tang, G., Kumar, N., Yoo, R., Michmizos, K.: Deep reinforcement learning with population-coded spiking neural network for continuous control. In: Conference on Robot Learning, pp. 2016–2029. PMLR (2021)
18. Wu, Z., et al.: Modeling learnable electrical synapse for high precision spatio-temporal recognition. Neural Netw. **149**, 184–194 (2022)
19. Yuan, M., Wu, X., Yan, R., Tang, H.: Reinforcement learning in spiking neural networks with stochastic and deterministic synapses. Neural Comput. **31**(12), 2368–2389 (2019)
20. Zhang, D., Zhang, T., Jia, S., Xu, B.: Multi-sacle dynamic coding improved spiking actor network for reinforcement learning. In: Proceedings of the AAAI Conference on Artificial Intelligence, vol. 36, pp. 59–67 (2022)

An Improved YOLOv5s for Detecting Glass Tube Defects

Zhibo Wei and Liying Zheng$^{(\boxtimes)}$

School of Computer Science and Technology, Harbin Engineering, Harbin 150001, China
zhengliying@hrbeu.edu.cn

Abstract. Existing algorithms for detecting glass tube defects suffer from low recognition accuracy and huge scales. This paper proposes an improved YOLOv5s to solve these problems. Specifically, the Convolutional Block Attention Module (CBAM) is used in the improved YOLOv5s to enhance the feature extraction. Then, the Content-Aware ReAssembly of FEatures (CARAFE) is used to replace the original upsampling method, which is beneficial for feature fusion. Next, the Efficient Intersection over Union (EIoU) loss is substitute to the original Complete Intersection over Union (CIoU) loss of YOLOv5s, which improves the regression accuracy. Finally, we adopt Cosine Warmup to accelerate the convergence as well as improve the detection performance. The experimental results show that, compared with the original YOLOv5s, our improved YOLOv5s increases the mean Average Precision (mAP) by 8% while decreasing the amount of model parameters by 5.8%. Moreover, the improved detector reaches 98 Frames Per Second (FPS) that meets the requirement of real-time detection.

Keywords: YOLOv5s · Defect Detection · Model Scale Reduction · Attention Module · Loss · Cosine Warmup

1 Introduction

The manufacture of glass tubes is susceptible to the objective factors and the surrounding environment, resulting in many types of defects such as shatter, scratch and blemish. Thus, glass tube manufacturing severely depends on various defect detection methods. So far, detection for glass tube defects is still a hot and open problem.

Generally speaking, the traditional manual detection method is greatly limited by the experience of inspectors, suffering from slow speed, poor consistency, and high labor costs. Though mechanical contact detection meets the industrial production in terms of quality, they are generally expensive, inflexible and slow.

Deep learning has currently been applied to various fields of defect detection. Li et al. [1] proposed a detector based on region of interest. With the Multi-Layer Perceptron (MLP) [2] and deep learning, their detector effectively improves the detection accuracy. Jin et al. [3] proposed a multi-channel automatically encoded convolutional network model to address detection errors caused by small differences in feature space in glass detection, improving recognition accuracy and reducing training time.

B. Luo et al. (Eds.): ICONIP 2023, CCIS 1962, pp. 196–206, 2024.
https://doi.org/10.1007/978-981-99-8132-8_15

You Only Look Once (YOLO) model [4, 5] is a representative one-stage detector. Compared with the two-stage one such as Region-CNN (R-CNN) [6], YOLO is based on region extraction, with shorter training time, smaller space occupation, relatively less computational resources and faster detection speed. The first version of YOLO, i.e. YOLOv1 [7] limits the number of predictable objects, and the accuracy is not satisfied. Later, the second version, YOLOv2 [8] uses tricks such as global average pooling, leading to more stable and faster training, as well as better performance for detecting small targets. Then, YOLOv3 [9] uses multi-scale prediction and DarkNet53, which greatly improves accuracy. YOLOv4 [10] integrates many tricks that can improve the accuracy to further enhance the performance of the mode. YOLOX [11] incorporates mechanisms such as data augmentation and anchor-free. YOLOv6 [12] optimizes the model by improving the Cross Stage Partial Network (CSPNet) [13]. YOLOv7 [14] proposes a planned model structural re-parameterization and Coarse-to-Fine guided label assignment. YOLOv8 [15] mainly uses the Context-aware Cross-level Fusion network (C2f) to replace the Concentrated-Comprehensive Convolution block (C3) [16] to make the model more lightweight. Thanks to the excellent detection performance of above-mentioned YOLOs, it is widely applied to defect detection tasks including glass tube defect detection.

This paper aims to improve the detection performance of glass tube defects. We improve YOLOv5s [17, 18] to meet the requirement of glass tube manufacturing. Firstly, the Convolutional Block Attention Module (CBAM) [19] is used to improve feature extraction of the model. Secondly, the lightweight upsampling operator, i.e. Content-Aware ReAssembly of FEatures (CARAFE) [20] is used to replace the original nearest neighbor interpolation, which results in less degradation of image quality and improved detection accuracy. Then, the Efficient Intersection over Union (EIoU) [21] is chosen to optimize the calculation of the loss for the bounding box regression. Moreover, the width-to-height ratio regression of Complete Intersection over Union (CIoU) [22] is split into separate regressions of width-to-height values to improve the regression accuracy. Finally, Cosine Warmup [23] strategy is used to preheat and adjust the learning rate, making the model more stable, faster convergence and higher accuracy.

The remainder of this paper is as follows. Section 2 describes the structure of the original YOLOv5s and the problems in using it for defect detection in glass tubes. Section 3 describes improved YOLOv5s through CBAM, CARAFE, EIoU and Cosine Warmup. Section 4 presents experimental results and analysis, and Sect. 5 gives some conclusions and future work.

2 Analysis YOLOv5s for Glass Tube Defect Detection

YOLOv5 is a one-stage target detection algorithm. It divides the image into grids, and each grid predicts detection boxes for classification and localization. Compared with the previous four versions, YOLOv5 is more lightweight and faster. Among the variants of YOLOv5, the model depth of YOLOv5s is the least, and its feature map width is the smallest. Furthermore, YOLOv5s is the fastest one in YOLOv5 family.

As shown in Fig. 1, YOLOv5s consists of three modules, i.e. feature extraction, feature fusion, and prediction. Here, the feature extraction consists of Convolutional

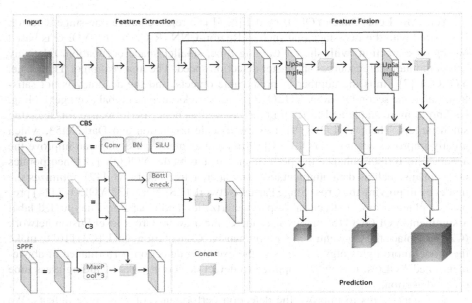

Fig. 1. YOLOv5s

Neural Network (Conv), C3 and Spatial Pyramid Pooling-Fast (SPPF) [24, 25]. It extracts features and performs pooling connection to transform an input image into feature maps. The feature fusion uses Feature Pyramid Networks (FPN) [26] and performs multi-scale fusion of feature map. With sampling, concatenating (Concat), and Conv, the feature fusion obtains quite rich features. The prediction module includes convolutional layers, pooling layers, and fully connected layers. It uses K-Means clustering, CIoU loss, and Non-Maximum Suppression (NMS) [27] to get final regression results.

As shown in Fig. 2, defects of glass tubes are divided into three categories. They are shatters, scratches, and blemishes. Shatter defects are relatively easy to identify as they have clear boundaries and the large area of a damage. Most of the scratches are long and frictional lines. Blemishes are essentially small and black dotted targets. Since the lighter blemishes not differing much from the surrounding pixels, making them difficult to detect during detection. In addition, in some cases, the same defect type has different appearances. For example, most tube scratches are continuous and clearly

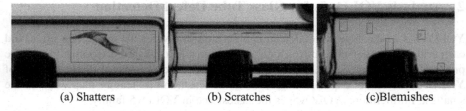

(a) Shatters   (b) Scratches   (c)Blemishes

Fig. 2. Types of glass tube defects

defined friction black lines, but there are also discontinuous that resemble tiny shatters or small blemishes.

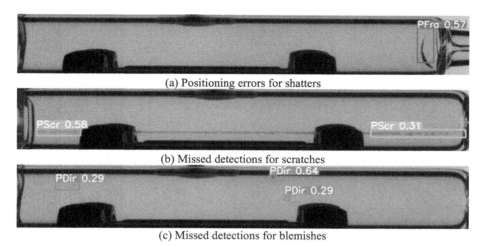

(a) Positioning errors for shatters

(b) Missed detections for scratches

(c) Missed detections for blemishes

Fig. 3. Examples for positioning errors and missed detection.

Firstly, we directly use the original YOLOv5s shown in Fig. 1 to detect glass tube defects. The experimental results show that its accuracy is relatively low. As shown in Fig. 3, there are frequently occurred missed detections for scratches and blemishes, and sometimes there exists positioning errors for shatters. Due to the obvious features in appearance, the detection accuracy for shatters is higher. But as shown in Fig. 3(a), the target location deviations for shatters are severe compared to the other two defects. When a glass tube is rotated, the scratches may overlap with the boundary of the body. In this case, YOLOv5s cannot learn the complete feature, leading to missed detections. Further, blemishes are usually small, and YOLOv5s is weak for detecting small targets, resulting in poor performances for blemishes using the original YOLOv5s. In addition, other problems such as huge training time cost and low detection efficiency need to be improved for real applications.

In summary, the results of the original YOLOv5s show the following issues: low accuracy, false detection, miss detection, and target box deviation. Therefore, this paper improves YOLOv5s to alleviate these problems.

3 Improved YOLOv5s

To solve the abovementioned problems of YOLOv5s for glass tube defect detection, we modify its structure, and proposed an improved YOLOv5s that is shown in Fig. 4. Compared with the original one, the improved model uses CBAM hybrid attention mechanism to enhance the feature extraction of the backbone of YOLOv5s. Besides, we substitute lightweight upsampling operator, CARAFE, for the original nearest neighbor interpolation to improve the feature fusion of the neck of YOLOv5s. Further, we use

EIoU to optimize the bounding box regression. Finally, the Cosine Warmup is used to adjust the learning rate decay and enhance the model performance.

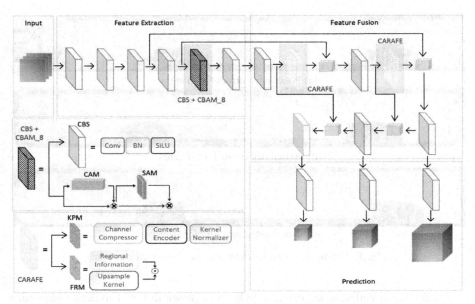

Fig. 4. Improved YOLOv5s network structure

3.1 CBAM

YOLOv5s does not pay enough attention to the key content of an input image. To solve this problem, we use the attention mechanism, i.e. CBAM to YOLOv5s. Specifically, we only replace the C3 module at the 8th layer of the backbone to CBAM.

We Use CBAM instead of other attention mechanisms, such as SE [28] and CA [29], because CBAM connects the Channel Attention Modules (CAM) and Spatial Attention Modules (SAM) in the structure, which combines the advantages of SE and CA in terms of function. Such strategy is more helpful for extracting characteristic information of glass tube defects. When inputting an image to YOLOv5s with the CBAM, our attention extraction forces the model to concentrate in key areas for detecting, discarding unrelated information, which can enhance the performance of small target detection.

3.2 CARAFE

The original YOLOv5s uses the nearest neighbor interpolation for upsampling, and taking the gray value of the nearest pixels relative to the position of the sample point to be measured as the gray value of this point. Such method causes jagged and mosaic phenomena and degrades image quality. We replace the nearest neighbor interpolation of the 11th and 15th layers of YOLOv5s to CARAFE, setting them to [512, 3, 2] and

[256, 3, 2]. Besides, their upsampling factor is 2, and the feature map size is 40 × 40 and 80 × 80. The improved feature fusion module uses a bigger receptive field to aggregate a larger range of contextual information when upsampling, and the dynamically generated adaptive kernel samples based on the content of the features. The information around the sampling point in the local area can get more attention than before, thereby reducing image loss, enhancing the characteristics of feature fusion.

3.3 Loss Function and Learning Rate

In training and reasoning, YOLOv5s uses CIoU given by (1) as the bounding box loss function.

$$L_{CIoU} = 1 - IoU + \frac{\rho^2\left(b, b^{gt}\right)}{c^2} + \alpha v \tag{1}$$

CIoU only uses the width-to-height ratio as an influence factor and does not use the width and height themselves, resulting in the width and height of the prediction frame regression cannot be increased at the same time. So does the case of decreasing. The EIoU well solves this problem by regressing the width and height values separately. Thus, we substitute EIoU given in (2) for CIoU.

$$L_{EIoU} = 1 - IoU + \frac{\rho^2\left(b, b^{gt}\right)}{c^2} + \frac{\rho^2\left(w, w^{gt}\right)}{c_w^2} + \frac{\rho^2\left(h, h^{gt}\right)}{c_h^2} \tag{2}$$

where IoU is the ratio of the intersection and the concatenation of the prediction frame and the real frame. $\rho(.)$ is the Euclidean distance between the center points of the prediction frame and the real frame. c is the diagonal distance of the smallest external rectangle of the prediction frame and the real frame. c_w and c_h are the width and height of the smallest external frame covering the prediction frame and the real frame.

Finally, because the weights of the model are randomly initialized at the beginning of the training, a larger learning rate at beginning may lead to model oscillation and non-optimal solution. Therefore, in this paper, we choose Cosine Warmup to preheat and dynamically adjust the learning rate decay. We choose a small learning rate at begin. Then we gradually increase the initial learning rate to a pre-set maximum value via cosine annealing.

4   Experimental Results and Analysis

The experimental environment is configured with Inter Core i7-11800H + NVIDIA GeForce RTX 3060 + CUDA11.6 + Python3.10.1. We use mAP to evaluate the mean average accuracy of all detected defects and select Frames Per Second (FPS) to evaluate the detection speed. Further, we use Params to indicate the amount of model parameters, and select the Floating Point Operations (FLOPs) to measure the computational complexity of the model.

Our glass tube defect dataset consists of 1200 images. The train set, test set, and validation set are divided into 8:1:1. The size of the input images is 640 × 640. We use the Mosaic data augmentation and the Early-stopping while training. In addition, the batch size, the initial learning rate, and the coefficient of EIoU loss are 8, 0.01 and 0.05, respectively.

4.1 Visual Results of YOLOv5s and Our Improved Model

Some visual results of YOLOv5s and our model are shown in Figs. 5, 6 and 7. The Confidence is indicated above the bounding box in the figures.

It can be seen from Fig. 5 that the improved model corrects the problem of bounding box deviation from the targets in the detection results of the original YOLOv5s, and the Confidence has improved significantly. As shown in Fig. 6, the improved algorithm in this paper largely solves the problem of missed detections and increases the Confidence. As shown in Fig. 7, the number of successful detections of small blemishes of our model are greatly improved. Also, its Confidence is increased.

In summary, the improved YOLOv5s outperforms the original one in terms of detecting three defect types. The problems of false detection, missed detection and positioning error are alleviated.

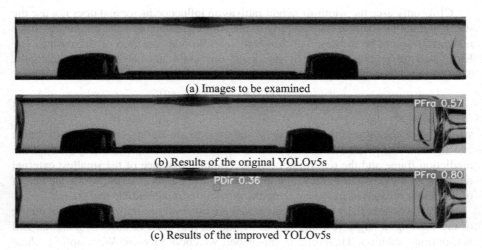

(a) Images to be examined

(b) Results of the original YOLOv5s

(c) Results of the improved YOLOv5s

Fig. 5. Detection results of shatters.

4.2 Ablation Experiments

Five ablation experiments are conducted, and results are listed in Table 1.

We can see from Table 1 that after the feature extraction network uses CBAM instead of C3, the Params are reduced by 7.6%, while the mAP being increased by 2.9%. Using CARAFE for kernel prediction and feature reorganization, increasing the Params by 0.13M, while the mAP increased by 2.5%. Using EIoU does not affect Params and FLOPs, but the mAP is increased by 1.6%. Finally, after using the Cosine Warmup to preheat and adjust the learning rate decay strategy, the mAP is increased by 1%.

Table 1 also shows that the YOLOv5s had 7.03M of Params, 16.0G of FLOPs, and 66.9% of mAP for the three types of defects. After improvement, the Params of our model are reduced by 5.8%, the FLOPs are reduced by1.9% and the mAP is increased by 8%.

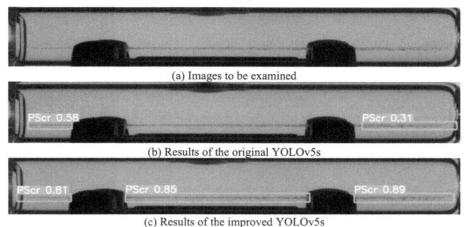

(a) Images to be examined

(b) Results of the original YOLOv5s

(c) Results of the improved YOLOv5s

Fig. 6. Detection results of scratches.

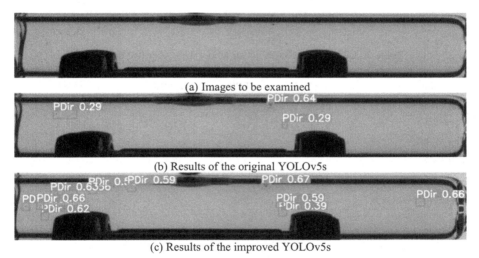

(a) Images to be examined

(b) Results of the original YOLOv5s

(c) Results of the improved YOLOv5s

Fig. 7. Detection results of blemishes

4.3 Comparative Experiments

We compare our improved YOLOv5s with the original one, the Single Shot Multi-Box Detector (SSD) [31], YOLOv3, YOLOv7 and Faster R-CNN [32]. The results are listed in Table 2.

From Table 2 one can see that the Params of SSD is 9.29 times larger than that of ours. Its mAP and FPS are 17.6% and 45.9% lower than ours. Moreover, the Params of YOLOv3 is 61.52M, and the mAP is 64.3% which is even worse than the original YOLOv5s. The Params of YOLOv7 is 5.51 times higher than ours, and the mAP is 1.2%

Table 1. Ablation results

Experiment number	CBAM	CARAFE	EIoU	Cosine Warmup	Params (M)	FLOPs (G)	mAP@0.5 (%)
N1	-	-	-	-	7.03	16.0	66.9
N2	✓	-	-	-	6.49	15.4	69.8
N3	✓	✓	-	-	6.62	15.7	72.3
N4	✓	✓	✓	-	6.62	15.7	73.9
N5	✓	✓	✓	✓	6.62	15.7	74.9

lower than ours. The Params of Faster R-CNN is 20.7 times larger than that of ours. Its FPS is very small and the mAP is 11.1% lower than ours.

Table 2 illustrates that the mAP, FPS and Params of the improved YOLOv5s are the best performers among the six algorithms. The mAP reaches 74.9%. The FPS increases by 3.1% compared with the original YOLOv5s, and it reaches 98 FPS that satisfies the actual detection requirements. The Params is the smallest one of them all.

Table 2. Comparison results of our model with other five detectors

Algorithms	mAP(%)	FPS	FLOPs(G)	Params(M)
YOLOv5s	66.9	95	16.0	7.03
SSD	57.3	62	1.1	22.20
YOLOv3	64.3	53	193.9	61.52
YOLOv7	73.7	86	103.5	36.50
Faster R-CNN	63.8	6	370.3	137.00
Improved YOLOv5s	**74.9**	**98**	15.7	**6.62**

5   Conclusion

In this paper, we have proposed an improved YOLOv5s for glass tube defect detection by introducing CBAM, CARAFE, EIoU, and Cosine Warmup to the original YOLOv5s detector. We have confirmed the effectiveness of each trick in the scheme for improving the algorithm through ablation experiments. Compared with the original model, the Params of improved Yolov5s reduced by 5.8%, the FLOPs decreased by 1.9%, and the mAP increased by 8%. The FPS reaches 98, which meets the requirements of the actual glass tube defect detection.

Our Future research work will focus on optimizing the detection accuracy and the amounts of parameters of the model.

Acknowledgments. This work is supported by the National Key R&D Program of China [grant number 2021YFF0603904] and the National Natural Science Foundation of China [grant number 61771155].

References

1. Li, C., et al.: A novel algorithm for defect extraction and classification of mobile phone screen based on machine vision. Comput. Ind. Eng. **146**, 106530 (2020)
2. Taud, H., Mas, J.F.: Multilayer perceptron (MLP). In: Olmedo, M.T.C., Paegelow, M., Mas, J.-F., Escobar, F. (eds.) Geomatic approaches for modeling land change scenarios. LNGC, pp. 451–455. Springer, Cham (2018). https://doi.org/10.1007/978-3-319-60801-3_27
3. Jin, Y., et al.: A fuzzy support vector machine-enhanced convolutional neural network for recognition of glass defects. Int. J. Fuzzy Syst. **21**, 1870–1881 (2019)
4. Redmon, J. et al.: You only look once: unified, real-time object detection. In: 2016 IEEE Conference on Computer Vision and Pattern Recognition (CVPR), pp. 779–788. IEEE, Las Vegas, Nevada, USA (2016)
5. Terven, J., Cordova-Esparza, D.: A comprehensive review of YOLO: From YOLOv1 to YOLOv8 and beyond. arXiv: 2304.00501v2, 2023, Accessed 19 May 2020
6. Girshick, R. et al.: Rich feature hierarchies for accurate object detection and semantic segmentation. In: 2014 IEEE Conference on Computer Vision and Pattern Recognition (CVPR), pp. 580–587. Columbus, OH, USA (2014)
7. Lu, J., et al.: A vehicle detection method for aerial image based on YOLO. J. Comput. Commun. **6**(11), 98–107 (2018)
8. Redmon, J., Farhadi, A.: YOLO9000: better, faster, stronger. In: 2017 IEEE Conference on Computer Vision and Pattern Recognition (CVPR), pp. 6517–6525. IEEE, Honolulu, HI, USA (2017)
9. Redmon, J., Farhadi, A.: YOLOv3: An incremental improvement. arXiv: 1804.02767v1, Accessed 8 Apr 2018
10. Bochkovskiy, A., Wang, C.: Liao H. YOLOv4: Optimal Speed and Accuracy of Object Detection. arXiv: 2004.10934v1, Accessed 23 Apr 2020
11. Ge, Z. et al.: Yolox: Exceeding yolo series in 2021. arXiv: 2107.08430v2, Accessed 6 Aug 2021
12. Li, C. et al.: YOLOv6: A single-stage object detection framework for industrial applications. arXiv: 2209.02976v1, Accessed 7 Sep 2022
13. Wang, C. et al.: CSPNet: a new backbone that can enhance learning capability of CNN. In: 2020 IEEE/CVF Conference on Computer Vision and Pattern Recognition Workshops (CVPRW), pp. 1571–1580. IEEE, Seattle, WA, USA, 2020
14. Wang, C., Bochkovskiy, A., Liao, M.: YOLOv7: Trainable bag-of-freebies sets new state -of-the-art for real-time object detectors. arXiv: 2207.02696v1, Accessed 6 July 2022
15. Kim, J., Kim, N., Won, C.: High-Speed drone detection based on Yolo-V8. In: 2023–2023 IEEE International Conference on Acoustics, Speech and Signal Processing (ICASSP), pp. 1–2. IEEE, Rhodes Island, Greece (2023)
16. Park, H. et al.: C3: Concentrated-comprehensive convolution and its application to semantic segmentation. arXiv: 1812.04920v3, Accessed 28 July 2019
17. Yang, G. et al.: Face mask recognition system with YOLOV5 based on image recognition. In: 6th International Conference on Computer and Communications (ICCC), pp. 1398–1404. IEEE, Chengdu, China (2020)

18. Zheng, L. et al.: A fabric defect detection method based on improved YOLOv5. In: 7th International Conference on Computer and Communications (ICCC), pp. 620–624. IEEE, Chengdu, China (2021)
19. WOO, S. et al.: CBAM: convolutional block attention module. In: 16th European Conference on Computer Vision (ECCV), pp. 3–19. Springer, Munich, Germany (2018). https://doi.org/10.1007/978-3-030-01234-2_1
20. Wang, J. et al.: Carafe: content-aware reassembly of features. In: 2019 IEEE/CVF International Conference on Computer Vision (ICCV), pp. 3007–3016. IEEE, Seoul, Korea (South) (2019)
21. Zhang, Y., et al.: Focal and efficient IOU loss for accurate bounding box regression. Neurocomputing **506**, 146–157 (2022)
22. Števuliáková, P., Hurtik, P.: Intersection over Union with smoothing for bounding box regression. arXiv: 2303.15067v2, Accessed 28 Mar 2023
23. Loshchilov, I., Hutter, F.: Stochastic gradient descent with warm restarts. arXiv: 1608.03983v5, Accessed 3 May 2017
24. Xia, K., et al.: Mixed receptive fields augmented YOLO with multi-path spatial pyramid pooling for steel surface defect detection. Sensors **23**(11), 5114 (2023)
25. Liu, P., et al.: A lightweight object detection algorithm for remote sensing images based on attention mechanism and YOLOv5s. Remote Sens. **15**(9), 2429 (2023)
26. Lin, T. et al.: Feature pyramid networks for object detection. In: 2017 IEEE Conference on Computer Vision and Pattern Recognition (CVPR), pp. 936–944. IEEE, Honolulu, HI, USA (2017)
27. Hosang, J., Benenson, R., Schiele, B.: Learning non-maximum suppression. In: 2017 IEEE Conference on Computer Vision and Pattern Recognition (CVPR), pp. 6469–6477. IEEE, Honolulu, HI, USA (2017)
28. Hu, J., Shen, L., Sun, G.: Squeeze-and-excitation networks. In: 2018 IEEE Conference on Computer Vision and Pattern Recognition (CVPR), pp. 7132–7141. IEEE, Salt Lake City, USA (2018)
29. Hou, Q., Zhou, D., Feng, J.: Coordinate attention for efficient mobile network design. In: 2021 IEEE/CVF Conference on Computer Vision and Pattern Recognition (CVPR), pp. 13708–13717. IEEE, Nashville, TN, USA (2021)
30. Zheng, Z. et al.: Distance-IoU loss: faster and better learning for bounding box regression. In: Proceedings of the AAAI Conference on Artificial Intelligence, pp. 12993–13000. AAAI, New York, USA (2020)
31. Liu, W. et al.: SSD: Single Shot MultiBox Detector. In: 14th European Conference on Computer Vision (ECCV), pp. 21–37. Springer, Amsterdam, The Netherlands (2016). https://doi.org/10.1007/978-3-319-46448-0_2
32. Ren, S., et al.: Faster R-CNN: towards real-time object detection with region proposal networks. Adv. Neural. Inf. Process. Syst. **28**, 91–99 (2015)

Label Selection Approach to Learning from Crowds

Kosuke Yoshimura$^{(\boxtimes)}$ and Hisashi Kashima

Kyoto University, Kyoto, Japan
yoshimura.kosuke.42e@st.kyoto-u.ac.jp, kashima@i.kyoto-u.ac.jp

Abstract. Supervised learning, especially supervised deep learning, requires large amounts of labeled data. One approach to collect large amounts of labeled data is by using a crowdsourcing platform where numerous workers perform the annotation tasks. However, the annotation results often contain label noise, as the annotation skills vary depending on the crowd workers and their ability to complete the task correctly. Learning from Crowds is a framework which directly trains the models using noisy labeled data from crowd workers. In this study, we propose a novel Learning from Crowds model, inspired by SelectiveNet proposed for the selective prediction problem. The proposed method called Label Selection Layer trains a prediction model by automatically determining whether to use a worker's label for training using a selector network. A major advantage of the proposed method is that it can be applied to almost all variants of supervised learning problems by simply adding a selector network and changing the objective function for existing models, without explicitly assuming a model of the noise in crowd annotations. The experimental results show that the performance of the proposed method is almost equivalent to or better than the Crowd Layer, which is one of the state-of-the-art methods for Deep Learning from Crowds, except for the regression problem case.

Keywords: Crowdsourcing · Human Computation · Human-in-the-Loop Machine Learning

1 Introduction

Supervised learning, especially deep supervised learning, requires large amounts of labeled data. One popular way to collect large amounts of labeled data is to use human annotators through crowdsourcing platforms human-intelligence tasks such as image classification [10] and medical image classification [1]. To achieve an accurate prediction model, we require high-quality labeled data; however, since the annotation skills vary depending on crowd workers, collecting properly labeled data at the required scale has limitations. Therefore, numerous studies were conducted for obtaining higher-quality labeled data based on the annotation results collected from crowd workers [2,9,17,31,32]. A simple strategy is to collect responses from multiple crowd workers for each data instance, and to

B. Luo et al. (Eds.): ICONIP 2023, CCIS 1962, pp. 207–221, 2024.
https://doi.org/10.1007/978-981-99-8132-8_16

aggregate the answers using majority voting or averaging. However, the majority voting fails to consider natural human differences in the skills and abilities of crowd workers. Dawid and Skene proposed a method to aggregate responses that considers workers' abilities [9], which was originally aimed at integrating doctors' examination results. Besides this, Whitehill et al. proposed GLAD, which further introduces the task's difficulty [32] into the model in addition to the workers' abilities.

A more direct approach is to train machine learning prediction models directly from crowdsourced labeled data rather than from high-quality labeled data, which is Learning from Crowds [21]. Raykar et al. proposed a method that alternates between parameter updating and EM algorithm-based answer integration within a single model using crowdsourced answers [21]. Additionally, they experimentally demonstrated that their proposed method performs better than a two-stage approach that first integrates crowdsourced labeled data by majority voting and then trains the model with this data as input [21].

In addition to classical machine learning methods, Learning from Crowds has also been studied for deep learning models [1,5,6,24]. Rodrigues and Pereira stated that the EM algorithm-based Deep Learning from Crowds method poses a new challenge: how to schedule the estimation of true labels by the EM algorithm for updates of learnable parameters [24]. The reason is that the label estimation for the whole training dataset using the EM algorithm may be computationally expensive. Moreover, using the EM algorithm in each mini-batch may not contain sufficient information to correctly estimate the worker's abilities. Rodrigues and Pereira proposed the Crowd Layer to solve this problem [24]. However, a problem with Crowd Layer and other existing methods is that they require explicitly designed models of the annotation noise given by crowd annotators. As the appropriate form of the noise model is task-dependent and is not necessarily obvious.

In this study, we propose a Label Selection Layer for Deep Learning from Crowds. The proposed method automatically selects training data from the crowdsourced annotations. It is inspired by SelectiveNet [13], an architecture proposed in selective prediction. The Label Selection Layer can be added to any DNN model and can be applied simply by rewriting the loss function. The significant feature of the Label Selection Layer is that it does not require a generative model of worker annotation. This feature will allow us to easily apply deep learning from crowds to any machine learning problem setting. This advantage is more pertinent in complex tasks such as structured output prediction rather than in simple tasks such as classification and regression problems, where annotation generation models are easier to represent in confusion matrices or additive noises. Another feature of the proposed method is that it does not use the EM algorithm. Therefore, the issue of using the EM algorithm stated in the previous study [24] does not arise.

The contributions of this study are fourfold:

– We propose a Label Selection Layer inspired by SelectiveNet as a novel Deep Learning from Crowds method.

- We proposed four variations of the Label Selection Layer: Simple, Class-wise, Target-wise, and Feature-based Label Selection Layer.
- We performed an experimental comparison of the proposed method with existing methods using real-world datasets to demonstrate the performance of the proposed method.
- We discussed the advantages and disadvantages of the proposed method based on the experimental results.

2   Related Work

This section summarizes related work from two aspects; we first review the existing research related to Learning from Crowds, which learns prediction models from noisy crowdsourced annotations. We next introduce selective prediction and related problems, which allow prediction models to neglect "too hard" instances. In particular, we explain in detail SelectiveNet [13], which is one of the typical solutions to the selective prediction problem, because this is the direct foundation of our proposed method.

2.1   Learning from Crowdsourced Annotations

Extensive research was conducted for obtaining high-quality labels from noisy labels collected in crowdsourcing platforms [2–4,17,26,31,32]. As the quality of answers obtained from crowd workers is not always high, the most common method is to collect answers with redundancy [28]. The simplest method is to collect responses from multiple workers for each task and merge these responses by majority voting. Dawid and Skene proposed a method to estimate the true label based on the estimated latent variables of each worker's ability using an EM algorithm [9]. Their original goal was to estimate the correct medical examination result when there were multiple doctors' examinations for patients. Whitehill et al. proposed a label aggregation method by simultaneously modeling the worker's abilities and task difficulties as latent variables [32]. Welinder et al. proposed a response aggregation method that models worker skills and object attributes using multidimensional vectors [31]. Oyama et al. proposed a response aggregation method to collect the confidence scores of crowd workers for their own answers and to estimate both the confidence scores and the worker's ability [17]. Baba and Kashima proposed a quality control method for more general tasks such as those requiring unstructured responses [2]. Their approach is to have a worker's answers scored by another worker and to model respondent ability and scorer bias with a graded response model. For complex tasks such as translation and ranking, Braylan and Lease proposed three modeling methods focusing on the distance between labels rather than the labels themselves. Their methods apply to tasks where requesters can appropriately define the distance function between labels [3]. Braylan and Lease also proposed a framework for response integration by decomposing and merging responses for several complex tasks, such as text span labeling and bounding box annotation [4]. This framework can handle multi-object annotation tasks that distance-based methods cannot. As a specialized method for sequential labeling, Sabetpour et al. proposed AggSLC, which considers worker's responses, confidence, prediction results by machine learning model, and the characteristics of the target sequential labeling task [26]. Wang and Dang proposed a method for generating appropriate sentences for a sentence-level task using a transformer-based generative model [30]. This generative model inputs sentences that workers answered and outputs an integrated sentence. Wang and Dang also proposed a method to create pseudo-sentences as training data for the model [30].

Extensive research has been conducted to obtain better-performing predictors using crowdsourced labeled data [1,5,6,21,24]. One intuitive approach is aggregating labels collected from multiple workers and then training them using the aggregated labels. By contrast, Raykar et al. proposed a framework called *Learning from Crowds*, which directly trains a machine learning model end-to-end using the answers obtained from the crowd workers as input [21]. Their method alternately aggregates labels based on the EM algorithm and updates the model's parameters in a single model. Kajino et al. stated that many learning-from-crowds approaches have non-convex optimization problems and proposed a new learning-from-crowds approach to make the problem convex by using individual classifiers [14]. Chu et al. assumed that the causes of label noise can be decomposed into common worker factors and individual worker factors and proposed CoNAL, which models these factors [8]. Takeoka et al. successfully improved the classifier's performance by using the information that workers answered that they did not know in the context of learning from crowds [29]. Albarqouni et al. proposed a framework for estimating the true label from crowdsourced responses based on an EM algorithm and using it to train a deep learning model, in the context of a medical image classification problem [1].

Rodrigues and Pereira investigated the Learning from Crowds method based on the EM algorithm in case of deep learning models [24]. The EM algorithm-based method had drawbacks such as it required adjustment of the timing of updating latent variables and learning parameters, and it was computationally expensive. Rodrigues and Pereira proposed *Crowd Layer* that is directly connected to the output layer of the DNNs and acts as a layer that transforms the output results into the answers of each annotator [24]. Although the nature of the Crowd Layer limits it learning to predict the labels correctly, their experiments have shown that DNN models with the Crowd Layer can produce a high-performance predictor. Chen et al. proposed SpeeLFC, which can represent each annotator's answer in the form of conditional probability, given the prediction of the output layer, because of the lack of interpretability of the Crowd Layer parameters compared to the probabilistic approaches [6]. They focused on the security aspect of learning-from-crowds systems. They stated that learning-from-crowds models are vulnerable to adversarial attacks. Therefore, they proposed A-LFC, which uses adversarial examples in the training step for this issue [5].

2.2   Selective Prediction

Machine learning models do not always return correct outputs. Therefore, selecting only highly accurate predictions is one way to obtain a more reliable predictor. This problem setup is called Selective Prediction [13], Reject Option [7], and Learning to defer [16], with minor differences in definition among them. Research in selective prediction has been conducted since 1970, initially as machine learning with a reject option [7]. There is a wide range of research related to selective prediction. Here, we describe research on selective prediction in the case of deep learning models.

One simple and effective method is to set a threshold on the confidence score obtained from the prediction model [7]. Geifman and El-Yaniv determined the application of this framework to deep learning [12]. They also proposed SelectiveNet, which allows end-to-end selective prediction in a single model [13].

We explain the details of SelectiveNet [13], which is the inspiration for our proposed method. SelectiveNet is a neural network model for selective prediction, and it consists of a base model whose role is to extract features and three final output layers connected to it. The three final output layers are the prediction head, selective head, and auxiliary head. The prediction head predicts corresponding to the target variable (denoted as $f(\mathbf{x})$). The selective head selects whether to adopt the output of the prediction head (denoted as $g(\mathbf{x})$). The auxiliary head (denoted as $h(\mathbf{x})$) is a structure used to stabilize the training and, similar to the prediction head, it predicts corresponding to the target variable. However, SelectiveNet only uses the auxiliary head in the training step.

SelectiveNet aims to train the model so that for any given set $S = \{(\mathbf{x}_n, y_n)\}_{n=1}^{m}$, where $\mathbf{x}_n$ is the input feature for n-th instance, and y_n is its true label, the empirical selection risk is minimal while keeping the empirical coverage at a predefined value. The empirical selective risk is defined by the following

$$\hat{r}_\ell(f, g|S) = \frac{\frac{1}{m}\sum_{n=1}^{m} \ell(f(\mathbf{x}_n), y_n) \cdot g(\mathbf{x}_n)}{\hat{\phi}(g|S)}, \tag{1}$$

where $\ell(\cdot, \cdot)$ is any loss function. The empirical coverage is defined by the following

$$\hat{\phi}(g|S) = \frac{1}{m}\sum_{n=1}^{m} g(\mathbf{x}_n). \tag{2}$$

Next, we show how SelectiveNet is trained. Given c as the target coverage, the main loss function of SelectiveNet is as follows

$$\mathcal{L}_{(f,g)} = \hat{r}_\ell(f, g|S) + \lambda \Psi(c - \hat{\phi}(g|S)), \tag{3}$$

where $\lambda > 0$ is a hyperparameter that balances the two terms, and $\Psi(a) = \max(0, a)^2$. Here, the loss function for the auxiliary head is defined as

$$\mathcal{L}_h = \frac{1}{m}\sum_{n=1}^{m} \ell(h(\mathbf{x}_n), y_n), \tag{4}$$

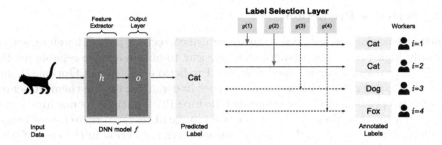

Fig. 1. The structure using Simple Label Selection Layer in the case of four workers. The ground truth label of this input data is 'Cat.' Because the annotated labels by workers $i = 3, 4$ are incorrect, label selector $g(3)$ and $g(4)$ inactivate loss for labels by workers $i = 3, 4$, 'Dog' and 'Fox,' respectively

and the overall objective function of SelectiveNet is

$$\mathcal{L} = \alpha \mathcal{L}_{(f,g)} + (1 - \alpha)\mathcal{L}_h, \tag{5}$$

where α is a hyperparameter.

3   Label Selection Layer

We propose a novel Learning from Crowds method inspired by the SelectiveNet [13], called *Label Selection Layer*. The proposed method only selects reliable labels for training. It can be added to any DNN model and applied simply by rewriting the loss function.

3.1   Problem Settings

The goal of learning from crowds scenario is to train an accurate model using labels given by annotators. We define $f \ (= o \circ h)$ as a DNN model, where o is an output layer, and h is a feature extractor. Let $\{(\mathbf{x}_n, \mathbf{y}_n)\}_{n=1}^N$ be a training data set, where $\mathbf{x}_n \in \mathbb{R}^d$ is a d-dimension feature vector, and $\mathbf{y}_n = \{y_n^i\}_{i \in \mathcal{I}}$ is an annotation vector provided by $|\mathcal{I}|$ annotators, with y_n^i representing the label annotated to the n-th sample by the i-th annotator. Note that all annotators do not necessarily annotate all of the samples. Assume that each sample has a latent ground truth label.

In the K-class classification setting, for example, y_n^i is an element of $\{1, 2, \ldots, K\}$, where each integer value denotes one of the candidate labels.

3.2   Label Selection Layer

The loss function of the proposed method reflects only the loss for the label selected by the Label Selection Layer. Let $g(i, l, \mathbf{x}, h)$ be the label selector function corresponding to the label l to a sample $\mathbf{x}$ from the i-th worker using the feature extractor h of the DNN model f. In this case, we can write the adoption rate ϕ of the training data for all annotation set A from crowd workers as

$$\phi(g|A) = \frac{1}{|A|} \sum_n \sum_{y_n^i \in A_n} g(i, y_n^i, \mathbf{x}_n, h), \tag{6}$$

where A_n means all annotation set for n-th sample.

From this equation, the empirical risk of a DNN model f in learning to use annotated labels selectively is

$$\hat{r}(f, g|A) = \frac{\frac{1}{|A|} \sum_n \sum_{y_n^i \in A_n} \ell(f(\mathbf{x}_n), y_n^i) \cdot g(i, y_n^i, \mathbf{x}_n, h)}{\phi(g|A)}, \tag{7}$$

where $\ell(\cdot, \cdot)$ is a loss function, which is defined depending on the target task. For example, we can use cross-entropy loss for the classification setting, mean square loss for the regression setting, and so on as $\ell(\cdot, \cdot)$. The numerator of Equation (7) is the loss function value for each annotation label weighted by the value of the corresponding label selection layer. To restrict the adoption rate ϕ to the training data using the hyperparameter c, the final loss function used in training can be expressed as

$$\mathcal{L}(f, g) = \hat{r}(f, g|A) + \lambda \Psi(c - \phi(g|A)), \tag{8}$$

where $\lambda > 0$ is a hyperparameter that balances the two terms, and $\Psi(a) = (\max(0, a))^2$.

In the inference phase, we use only the predicted value $f(\mathbf{x})$ as usual, and the label selection layer only selects useful training data in the training phase.

Here, we describe the differences between the proposed method and SelectiveNet. The most significant difference between the two is based on what is selected by the selection mechanism. The selective head g of SelectiveNet is a weight for the loss $\ell(f(\mathbf{x}_n), y_n)$ between the ground truth and the prediction result; therefore it takes a higher value for correct predictions. However, the label selection layer g of the proposed method is a weight for the loss $\ell(f(\mathbf{x}_n), y_n^i)$ between each worker's answer and the prediction. Therefore, it takes a higher value for workers' answers that are the same as (or close to) the prediction.

We discuss possible choices of the label selector function g. We propose four variations of the label selection layer: Simple, Class-wise, Target-wise, and Feature-based Label Selection Layer.

Simple Label Selection Layer. Simple Label Selection Layer defines the label selector g dependent only on each worker's one learnable weight scalar parameter w_i. Thus, we defined $g(i, l, \mathbf{x}, h) = g(i) = \sigma(w_i)$, where $\sigma(\cdot)$ is a sigmoid function. Figure 1 illustrates Simple Label Selection Layer. This approach assumes that each worker's annotation skills are constant for any given sample. Although the expressive power is not as significant owing to the small number of parameters, it has the advantage that it can be combined with any DNN model independent of the problem setup.

Class-Wise Label Selection Layer. Class-wise Label Selection Layer performs label selection in classification problems. It defined the label selector g that is dependent on the learnable scalar weight parameters $w_{(i,l)}$ corresponding to each combination of worker and annotated label values. Thus, we defined $g(i, l, \mathbf{x}, h) = g(i, l) = \sigma(w_{(i,l)})$. The method assumes that the annotation skills of each worker are dependent only on the label assigned by the worker and not on each sample. Although this method is used only for classification problems, it is more expressive than the Simple Label Selection Layer.

Target-Wise Label Selection Layer. Target-wise Label Selection Layer performs label selection in regression problems. It defined the label selector g that is dependent on the learnable scalar weight parameters w_i and scalar bias parameters b_i corresponding to each worker, and target (continuous) variable l. Thus, we defined $g(i, l, \mathbf{x}, h) = g(i, l) = \sigma((l \cdot w_i + b_i)^{d_0})$, where d_0 is a hyperparameter. The method assumes that the annotation skills of each worker are dependent only on the continuous label assigned by the worker and not on each sample. Although this method is used only for regression problems, it is more expressive than the Simple Label Selection Layer.

Feature-Based Label Selection Layer. Feature-based Label Selection Layer defined the label selector g that is dependent on the feature extractor output $h(\mathbf{x})$ of the DNN model f and learnable vector weight parameters W_i. Thus, we defined $g(i, l, \mathbf{x}, h) = g(i, \mathbf{x}, h) = \sigma(W_i^\top h(\mathbf{x}))$. This method assumes that the correctness of each worker's annotation depends on who labels which sample.

4   Experiments

A comparative experiment between the proposed methods and existing methods is described. We conducted the experiments on three real datasets in classification, regression, and named entity recognition problems. Table 1 shows the base model architectures used in each experimental setting. We used Adam [15] to train all the models. All the models used in the experiments were implemented in Pytorch [19] and Pytorch-lightning [11]. The codes are available at https://github.com/ssatsuki/label-selection-layer.

4.1   Image Classification

We first address a multi-class classification problem by using the LabelMe dataset, which was also used in the previous study [24]. This dataset consists of 11,688 images and each image is assigned one of the eight candidate classes. The dataset was divided into three parts: training, validation, and test data with 1,000, 500, and 1,180 labels, respectively. The validation and test data were assigned ground truth labels, and the validation data were used to determine hyperparameters and when to stop the training. Only labels annotated by 59

Table 1. The base model architecture used for the experiments with the LabelMe, the Movie Reviews, and CoNLL-2003 NER datasets. Descriptions of each layer and function are in the form of "Input Size, Details." The number in parentheses indicates the kernel size

LabelMe		Movie Reviews	
Flatten Layer	$4 \times 4 \times 512$	Embedding	1,000, 300 dim
Flatten Layer	$4 \times 4 \times 512$	Conv1d	$300 \times 1{,}000$, 128 Filters (3)
FC Layer	8,192	ReLU	128×998
ReLU	128	MaxPool1d	128×998, (5)
Dropout	128, 50%	Dropout	128×199, 50%
FC Layer	128, 8units	Conv1d	128×199, 128 Filters (5)
		MaxPool1d	128×195, (5)
CoNLL-2003		Flatten Layer	128×39
Embedding	109, 300 dim	FC Layer	4992
Conv1d	300×109, 512 Filters (5)	ReLU	32
Dropout	109×512, 50%	FC Layer	32, 1 unit
GRU	109×512, 50 time steps		
FC Layer	109, 50×10 units		

crowd workers were available as the training data, when training the proposed and conventional methods. The annotation results were collected from Amazon Mechanical Turk [25]. We used the LabelMe data with the same preprocessing as in the previous work [24].

We set the maximum number of epochs to 50, and the mini-batch size to 64. The evaluation metrics of this experiment are accuracy, one-vs-rest macro AUC, macro precision, and macro recall.

For comparison with the proposed methods, we used the base model trained using ground truth and the Crowd Layer combined with the output layer of the base model. As variations of the Crowd Layer, we selected MW, VW, VB, and VW+B proposed in [24]. MW, VW, VB, and VW+B use $\mathbf{W}^i \sigma$, $\mathbf{w}^i \odot \sigma$, $\mathbf{w}^i + \mathbf{b}$, and $\mathbf{w}^i \odot \sigma + \mathbf{b}$ as annotator-specific functions of Crowd Layer respectively, where $\mathbf{W}^i$ is a $K \times K$ trainable weight matrix, $\mathbf{w}^i$ is a K dimension trainable weight vector, $\mathbf{b}^i$ is a K dimension trainable bias vector, and σ is a K dimension output vector. We used the symbol $\odot$ as the Hadamard product operator. About the hyperparameters of the proposed method, we fixed $\lambda = 32$, and we selected c with the best performance using the validation data.

We ran each method 100 times, and the average performance is shown in Table 2. Comparison experiments using LabelMe dataset confirmed that one of the proposed methods, the Feature-based Label Selection Layer, performed better than the Crowd Layer in all evaluation metrics. However, the performance of the remaining proposed methods is worse than that of the Crowd Layer. Based

Table 2. The results of accuracy comparison for the image classification dataset: LabelMe. Higher is better in all evaluation metrics. **Bold** indicates the best average value of the results of 100 trials, excluding those trained using Ground Truth Label

Method	Acc	OvR Macro AUC	Macro Prec	Macro Rec
w/ Ground Truth	.907 (±.0056)	.994 (±.0005)	.911 (±.0053)	.911 (±.0046)
Crowd Layer [24]				
MW	.824 (±.0266)	.979 (±.0262)	.832 (±.0419)	.833 (±.0295)
VB	.824 (±.0153)	.983 (±.0105)	.834 (±.0230)	.832 (±.0157)
VW	.816 (±.0096)	.983 (±.0012)	.828 (±.0079)	.825 (±.0091)
VW+B	.824 (±.0089)	.984 (±.0010)	.834 (±.0069)	.832 (±.0083)
(Proposed) Label Selection Layer				
Simple	.796 (±.0091)	.978 (±.0012)	.811 (±.0080)	.804 (±.0091)
Class-wise	.796 (±.0092)	.979 (±.0012)	.813 (±.0074)	.804 (±.0094)
Feature-based	**.839 (±.0114)**	**.985 (±.0014)**	**.849 (±.0096)**	**.848 (±.0110)**

Table 3. The results of comparison for the regression dataset: Movie Reviews. **Bold** indicates the best average value of the results of 100 trials, excluding those trained using Ground Truth Label

Method	MAE ($\downarrow$)	RMSE ($\downarrow$)	R^2 ($\uparrow$)
w/ Ground Truth	1.061 (±.050)	1.346 (±.066)	0.447 (±.055)
Crowd Layer [24]			
S	1.188 (±.077)	1.465 (±.079)	0.344 (±.073)
B	**1.120 (±.039)**	1.465 (±.047)	**0.393 (±.041)**
S+B	1.140 (±.049)	**1.419 (±.049)**	0.386 (±.043)
(Proposed) Label Selection Layer			
Simple	1.162 (±.093)	1.442 (±.095)	0.364 (±.088)
Target-wise	1.204 (±.050)	1.494 (±.055)	0.318 (±.053)
Feature-based	1.173 (±.063)	1.460 (±.068)	0.349 (±.061)

on these results, it is possible that in the classification problem setting, the use of features for label selection can improve performance.

4.2   Rating Regression

We use the Movie Reviews dataset [18,22] for rating regression. This dataset consists of English review comments collected from the Internet in [18], and the crowd workers on Amazon Mechanical Turk predicted the writers' review scores from the comments [22]. Each annotator score takes a value between 0 and 10. However, they are not necessarily integer values. There were 5,006 reviews in English, divided into 1,498 and 3,508 review comments in training and test sets, respectively. All 135 workers predicted scores for this training set. We normalized each score y_n^i given by worker i for review comments n using the mean $\bar{y}$ and standard deviation s of the ground truth score of the training data as $\frac{y_n^i - \bar{y}}{s}$. We converted each review comment by the tokenizer into a sequence of up to 1,000 tokens.

Table 4. The results of comparison for the named entity recognition dataset: CoNLL-2003 NER. **Bold** indicates the best average value of the results of 100 trials, excluding those trained using Ground Truth Label

Method	Precision ($\uparrow$)	Recall ($\uparrow$)	F_1 score ($\uparrow$)
w/ Ground Truth	65.04 ($\pm$2.24)	66.33 ($\pm$1.08)	65.66 ($\pm$1.35)
Crowd Layer [24]			
MW (pretrained w/ MV)	54.95 ($\pm$2.14)	**47.79 ($\pm$1.61)**	**51.11 ($\pm$1.61)**
VB (pretrained w/ MV)	61.33 ($\pm$1.85)	33.71 ($\pm$2.07)	43.47 ($\pm$1.84)
VW (pretrained w/ MV)	62.42 ($\pm$2.13)	36.81 ($\pm$1.75)	46.28 ($\pm$1.70)
VW+B (pretrained w/ MV)	61.31 ($\pm$2.08)	36.32 ($\pm$1.63)	45.59 ($\pm$1.50)
(Proposed) Label Selection Layer			
Simple (pretrained w/ MV)	**64.03 ($\pm$1.53)**	33.59 ($\pm$1.88)	44.03 ($\pm$1.78)
Class-wise (pretrained w/ MV)	63.84 ($\pm$1.41)	35.33 ($\pm$1.84)	45.46 ($\pm$1.71)
Feature-based (pretrained w/ MV)	63.49 ($\pm$1.71)	31.60 ($\pm$2.07)	42.17 ($\pm$2.03)

We used the pre-trained GloVe [20] as an Embedding Layer and froze the embedding parameters. We set the maximum number of epochs to 100 and the mini-batch size to 128. The evaluation metrics of this experiment are mean absolute error (MAE), root mean squared error (RMSE), and R^2 score.

For comparison with the proposed method, we used a base model trained using ground truth and a Crowd Layer combined with the output layer of the base model. As variations of the Crowd Layer, we used S, B, and S+B proposed in [24]. S, B, and S+B use $s^i\mu$, $\mu + b^i$, and $s^i\mu + b^i$ respectively, where s^i is a trainable scaling scalar, b^i is a trainable bias scalar, and μ is an output scalar.

About the hyperparameters of the proposed methods, we set $\lambda = 32$ and selected the best c from $\{0.1, 0.2, \ldots, 0.9\}$. The hyperparameter of the Target-wise Selective Layer is fixed $d_0 = 3$.

We ran each method 100 times, and the average performance is shown in Table 3. All of the proposed variants performed worse than the best Crowd Layer variant. This is probably because the proposed method does not support learning with scaled or shifted annotation results, which is possible with the Crowd Layer.

4.3 Named Entity Recognition

As a learning task with a more complex output form, we address named entity recognition (NER). We used a dataset consisting of CoNLL 2003 shared tasks [27] and newly collected annotations based on it in [22]. We call it the CoNLL-2003 NER dataset in this study. The CoNLL-2003 NER dataset consists of 9,235 text sequences, split into training, validation, and test data with 4,788, 1,197, and 3,250 sequences, respectively.[1] For the sequences in the training data, annotation results are collected from 49 workers on the Amazon Mechanical Turk in [23]. The validation data were used to determine hyperparameters and when to stop the training.

---

[1] Note that the experimental setup in the previous study [24] differs from our split setup because they only splits the data into train and test.

We used the pre-trained GloVe [20] as an Embedding Layer. The parameters of this Embedding Layer are trainable. We set the maximum number of epochs to 30, and the mini-batch size to 64. The evaluation metrics of this experiment are recall, precision, and F_1 score. For comparison with the proposed method, we used the base model trained using ground truth and the Crowd Layer combined with the output layer of the base model. As variations of the Crowd Layer, we selected MW, VW, and VW+B, the same as in the LabelMe experiments. In the CoNLL-2003 NLP dataset, we did not successfully train the Crowd Layer model when using the worker annotation set. Therefore, referring the published experimental code of Crowd Layer [24], we first pre-trained the DNN model in 5 epochs using the results of integrating the annotator's answers with Majority Voting. We connected the Crowd Layer to the DNN model and additionally trained it with worker annotations. We also trained the proposed method in the same way as we did with CrowdLayer. About the hyperparameters of the proposed method, we fixed $\lambda = 32$, and we selected c with the best performance using the validation data.

We ran each method 100 times, and the average performance is shown in Table 4. The results show that all the proposed methods performed better than any variations of Crowd Layer on precision. Because the same approach as binary classification, which uses thresholds to adjust the Precision/Recall tradeoff, is not available, the proposed method, which can achieve high precision without adjustment, is helpful when mistakes are challenging to tolerate.

A significant feature of the Label Selection Layer is that no generative model of (errors of) worker annotation is required. This advantage is particularly demonstrated in rather complex tasks such as structured output prediction, just like this NER task, in contrast with the Crowd Layer that requires correct assumptions about the mutation from true label sequences to ones given by the workers.

5 Analysis

Finally, we analyzed how well the proposed methods capture the quality of each worker's response.

Table 5 shows the Pearson correlation coefficients between the mean label selection score over training examples corresponding to each worker and the actual worker's accuracy or RMSE. Since a higher selection score has a more significant impact on model learning, workers with high accuracy or low RMSE should have a higher selection score to learn a high-performance model. For LabelMe, all of the proposed methods show a relatively high positive correlation, which suggests that all of the proposed methods can estimate the quality of workers' responses. For CoNLL 2003, the result of Simple Label Selection Layer shows a slightly positive correlation, while the result of the Feature-based Label Selection Layer shows almost no correlation on average. This suggests that the NER task has a complex structure, and therefore a simple model is more likely to learn well than a complex model. The fact that the Class-wise Label Selection

Table 5. For each training dataset, the mean and standard deviation over 30 trials of Pearson's correlation coefficient between the metric score based on each worker's answers and the mean of the outputs of the selection layer. N/A means that the proposed method cannot apply to the target dataset. Except when using the feature-based Label Selection Layer in CoNLL-2023, because the proposed method shows a positive correlation when using accuracy and a negative correlation when using RMES, the results indicate that the proposed method estimates the quality of the worker's responses correctly

Dataset (Metric)	Simple	Class-wise	Target-wise	Feature-based
LabelMe (Acc.)	0.54 (±0.04)	0.34 (±0.00)	N/A	0.44 (±0.05)
MovieReviw (RMSE)	−0.68 (±0.11)	N/A	−0.15 (±0.05)	−0.61 (±0.10)
CoNLL-2003 (Acc.)	0.19 (±0.18)	0.90 (±0.00)	N/A	−0.03 (±0.15)

Layer shows a very high correlation may be due to the dataset's characteristics. The correlations might be overestimated due to the imbalance of labels, with most tokens having an O tag which means it is not a named entity.

For MovieReview, the method with the stronger negative correlation is considered to be better at capturing the quality of the worker's actual responses. All of the proposed methods successfully capture the quality of the worker's actual responses, as they show negative correlations to the RMSE of the MovieReview dataset. Since the absolute value of the correlation coefficient is smaller Target-wise than for the other two methods, we considered that it does not capture the quality of each worker's response well.

6   Conclusion

We proposed a new approach to deep learning from crowds: the Label Selection Layer (LSL). The proposed method is inspired by SelectiveNet [13], which was proposed in the context of selective prediction, and trains models using selective labels given by the workers. We can apply the LSL to any deep learning model by simply adding and rewriting the loss function. This flexibility is possible because the LSL does not assume an internal generative model. As variations of the LSL, we proposed the Simple, the Class-wise, the Target-wise, and the Feature-based LSL. In the regression setting, the proposed method was worse than the Crowd Layer, but in the classification and the named entity recognition setting, the proposed method performed as well as or better than the Crowd Layer. In particular, the experiments in the named entity recognition extraction setting showed that all proposed method variations outperform any Crowd Layer variation on precision. We demonstrated that the proposed method could be easily applied to complex tasks such as structured output prediction and shows high precision. The proposed method's performance and the analysis results suggest that the Feature-based Label Selection model is superior for classification problems; also, the Simple Label Selection model is superior for complex structure problems like NER setting.

In future work, we will confirm whether the proposed method shows a higher performance in the Learning from Crowds problem setting for various structured data, so that it can be applied to different problem settings without significant changes. Additionally, we have to measure the performance improvement with different selector variations as it is likely that a more fine-grained selection of labels is possible using the labels by workers and the features provided as input to the model.

References

1. Albarqouni, S., Baur, C., Achilles, F., Belagiannis, V., Demirci, S., Navab, N.: AggNet: deep learning from crowds for mitosis detection in breast cancer histology images. IEEE Trans. Med. Imaging **35**, 1313–1321 (2016)
2. Baba, Y., Kashima, H.: Statistical quality estimation for general crowdsourcing tasks. In: Proceedings of the 19th ACM SIGKDD International Conference on Knowledge Discovery and Data Mining (2013)
3. Braylan, A., Lease, M.: Modeling and aggregation of complex annotations via annotation distances. In: Proceedings of The Web Conference 2020 (2020)
4. Braylan, A., Lease, M.: Aggregating complex annotations via merging and matching. In: Proceedings of the 27th ACM SIGKDD Conference on Knowledge Discovery and Data Mining (2021)
5. Chen, P., Sun, H., Yang, Y., Chen, Z.: Adversarial learning from crowds. In: Proceedings of the AAAI Conference on Artificial Intelligence (2022)
6. Chen, Z., et al.: Structured probabilistic end-to-end learning from crowds. In: Proceedings of the 29th International Joint Conference on Artificial Intelligence (2020)
7. Chow, C.: On optimum recognition error and reject tradeoff. IEEE Trans. Inf. Theory **16**, 41–46 (1970)
8. Chu, Z., Ma, J., Wang, H.: Learning from crowds by modeling common confusions. In: Proceedings of the AAAI Conference on Artificial Intelligence (2021)
9. Dawid, A.P., Skene, A.M.: Maximum likelihood estimation of observer error-rates using the EM algorithm. J. Roy. Stat. Soc. Series C (Appl. Stat.) **28**, 20–28 (1979)
10. Deng, J., et al.: Imagenet: a large-scale hierarchical image database. In: IEEE Conference on Computer Vision and Pattern Recognition (2009)
11. Falcon, W.: The PyTorch Lightning team: PyTorch Lightning (2019)
12. Geifman, Y., El-Yaniv, R.: Selective classification for deep neural networks. In: Proceedings of the 31st International Conference on Neural Information Processing Systems (2017)
13. Geifman, Y., El-Yaniv, R.: SelectiveNet: a deep neural network with an integrated reject option. In: Proceedings of the 36th International Conference on Machine Learning (2019)
14. Kajino, H., Tsuboi, Y., Kashima, H.: A convex formulation for learning from crowds. In: Proceedings of the AAAI Conference on Artificial Intelligence (2012)
15. Kingma, D.P., Ba, J.: Adam: a method for stochastic optimization. In: 3rd International Conference on Learning Representations (2015)
16. Mozannar, H., Sontag, D.: Consistent estimators for learning to defer to an expert. In: Proceedings of the 37th International Conference on Machine Learning (2020)
17. Oyama, S., Baba, Y., Sakurai, Y., Kashima, H.: Accurate integration of crowdsourced labels using workers' self-reported confidence scores. In: Proceedings of the Twenty-Third International Joint Conference on Artificial Intelligence (2013)

18. Pang, B., Lee, L.: Seeing stars: exploiting class relationships for sentiment categorization with respect to rating scales. In: Proceedings of the 43rd Annual Meeting of the Association for Computational Linguistics (2005)
19. Paszke, A., et al.: Pytorch: an imperative style, high-performance deep learning library. In: Advances in Neural Information Processing Systems, vol. 32 (2019)
20. Pennington, J., Socher, R., Manning, C.: GloVe: global vectors for word representation. In: Proceedings of the 2014 Conference on Empirical Methods in Natural Language Processing (2014)
21. Raykar, V.C., et al.: Learning from crowds. J. Mach. Learn. Res. **11**, 1297–1322 (2010)
22. Rodrigues, F., Lourenço, M., Ribeiro, B., Pereira, F.C.: Learning supervised topic models for classification and regression from crowds. IEEE Trans. Pattern Anal. Mach. Intell. **39**, 2409–2422 (2017)
23. Rodrigues, F., Pereira, F., Ribeiro, B.: Sequence labeling with multiple annotators. Mach. Learn. **95**, 165–181 (2014)
24. Rodrigues, F., Pereira, F.C.: Deep learning from crowds. In: Proceedings of the 32nd AAAI Conference on Artificial Intelligence (2018)
25. Russell, B.C., Torralba, A., Murphy, K.P., Freeman, W.T.: LabelMe: a database and web-based tool for image annotation. Int. J. Comput. Vis. **44**, 157–173 (2008)
26. Sabetpour, N., Kulkarni, A., Xie, S., Li, Q.: Truth discovery in sequence labels from crowds. In: 2021 IEEE International Conference on Data Mining (2021)
27. Sang, T.K., F, E., De Meulder, F.: Introduction to the CoNLL-2003 shared task: language-independent named entity recognition. In: Proceedings of the Seventh Conference on Natural Language Learning at HLT-NAACL 2003 (2003)
28. Sheng, V.S., Provost, F., Ipeirotis, P.G.: Get another label? Improving data quality and data mining using multiple, noisy labelers. In: Proceedings of the 14th ACM SIGKDD International Conference on Knowledge Discovery and Data Mining (2008)
29. Takeoka, K., Dong, Y., Oyamada, M.: Learning with unsure responses. In: Proceedings of the 34th AAAI Conference on Artificial Intelligence (2020)
30. Wang, S., Dang, D.: A generative answer aggregation model for sentence-level crowdsourcing task. IEEE Transactions on Knowledge and Data Engineering 34, 3299–3312 (2022)
31. Welinder, P., Branson, S., Perona, P., Belongie, S.: The multidimensional wisdom of crowds. In: Advances in Neural Information Processing Systems (2010)
32. Whitehill, J., Wu, T.F., Bergsma, J., Movellan, J., Ruvolo, P.: Whose vote should count more: Optimal integration of labels from labelers of unknown expertise. In: Advances in Neural Information Processing Systems (2009)

New Stability Criteria for Markov Jump Systems Under DoS Attacks and Packet Loss via Dynamic Event-Triggered Control

Huizhen Chen[1], Haiyang Zhang[1,2](✉) (iD), Lianglin Xiong[1,3](✉) (iD),
and Shanshan Zhao[1]

[1] School of Mathematics and Computer science, Yunnan Minzu University, Kunming
650500, China
[2] Faculty of Mechanical and Electrical Engineering, Kunming University of Science
and Technology, Kunming 650500, China
haiya287@126.com
[3] School of Media and Information Engineering, Yunnan Open University, Kunming
650504, China
lianglin_5318@126.com

Abstract. In this paper, the exponential mean square stability of Markov jump systems under packet loss and denial-of-service (DoS) attacks is studied by the dynamic event-triggered mechanism(DETM). Different from the existing results, this paper not only considers the impacts of periodic DoS attacks on the system, but also considers random packet loss during the sleeping-period of DoS attacks. Firstly, the Bernoulli distribution is used to model the phenomenon of random packet loss, and the zero-input strategy is used to combat the impacts of DoS attacks and random packet loss on the system. Then, based on the DETM, different controllers are designed during two stages of DoS attacks, and a new switched Markov jump system model is obtained. Different from the previous piecewise Lyapunov-Krasovskii functional approach, this paper obtains the less conservative stability criteria by constructing a common Lyapunov-Krasovskii functional. Finally, the authenticity of the proposed method is illustrated through a simulation experiment.

Keywords: DoS attacks · Packet loss · Dynamic event-triggered mechanism · Markov jump systems · Exponential mean square stability · Common Lyapunov-Krasovskii functional

1 Introduction

As a special kind of hybrid system, the Markov jump systems can well simulate the structural changes of dynamic systems. Due to this characteristic, Markov jump systems have attracted the attention of many scholars [1, 2]. For instance, a switched Markov jump model is constructed for the case of network data failure caused by DoS attacks in [3].

In recent years, wireless local area network is widely used due to its easy transmission and high speed. But at the same time, it has also produced a series of problems such as cyber-attacks, packet loss and network delay. DoS attacks mainly affects the operation of the system by preventing the transmission of the data. In addition, with the increasing tension of network communication resources, data packet loss often occurs. In the existing research, the DoS attacks and packet loss are mainly characterized through the Markov process [4,5] and the Bernoulli distribution [6,7], and the hold-input strategy [8,9] and zero-input strategy [10] are adopted to combat the impacts of DoS attacks and packet loss on the system. However, recently, there is almost no research on both DoS attacks and packet loss in continuous systems.

DETM is an effective approach to improve network resource utilization. It introduces an internal dynamic variable on the basis of static event-triggered mechanism, which effectively increases the time interval between two consecutive triggered moments. Recently, DETM has been widely applied in multiple fields [11,12]. For instance, a DETM is proposed for the Markov jump systems to relieve network communication pressure in [13,14]. Inspired by these, in the case of DoS attacks and packet loss, this paper considers using DETM to design the controller to ensure the system to run smoothly and efficiently.

The contributions are as follows: (1) This article simultaneously considers the impacts of DoS attacks and random packet loss on Markov jump systems. (2) The Bernoulli stochastic process is used to model packet loss, and zero-input strategy is used to combat the impacts of DoS attacks and packet loss. (3) Different controllers are designed based on DETM at different stages of DoS attacks. (4) Different from the piecewise Lyapunov-Krasovskii functional method used for DoS attacks in the past, this paper constructs a common Lyapunov-Krasovskii functional during the action-period and sleeping-period of DoS attacks. The new exponential mean square stability criteria are obtained, which is less conservative. (5) The effectiveness of the method is verified by a simulation.

Notation: In this article, $\mathbb{R}$ and $\mathbb{N}$ represent the set of real numbers and natural numbers, respectively. $P > 0$ means P is a positive definite matrix. $\mathbb{R}^{s \times m}$ is the set of $s \times m$ real matrices. $\mathbb{R}^n$ is the $n-$ dimensional Euclidean space. $*$ denotes the symmetric entry in the symmetric matrix. $sym\{T\} = T + T^T$.

2   Problem Formulation

2.1   System Description

Consider the Markov jump systems as follows:

$$\begin{cases} \dot{x}(t) = A(\iota_t) x(t) + B(\iota_t) u(t) + f(t, x), \\ x(s) = \theta(s), s \in [-l, 0], \end{cases} \tag{1}$$

where $x(t) \in \mathbb{R}^n$ is the state vector of the system, $u(t) \in \mathbb{R}^m$ is the control input of the system. $A(\iota_t) \in \mathbb{R}^{n \times n}$ and $B(\iota_t) \in \mathbb{R}^{n \times m}$ are known constant matrices. $f(t, x)$ is the nonlinear function, which satisfies the following Assumption 1.

Assumption 1 *[15]. i) For $\hat{x}_1, \hat{x}_2 \in D$, $f(t, x)$ is one-sided Lipschitz, if $\langle f(t, \hat{x}_1) - f(t, \hat{x}_2), \hat{x}_1 - \hat{x}_2 \rangle \leqslant \rho_0 \|\hat{x}_1 - \hat{x}_2\|^2$ holds, where $\rho_0 \in \mathbb{R}$ is one-sided Lipschitz constant. And for $\Psi, \Omega \in \mathbb{R}^n$, $\langle \Psi, \Omega \rangle = \Psi^T \Omega$ represents the inner product in the space. ii) For $\hat{x}_1, \hat{x}_2 \in D$, $f(t, x)$ is quadratically inter-bounded, if $\|f(t, \hat{x}_1) - f(t, \hat{x}_2)\|^2 \leqslant \beta_0 \|\hat{x}_1 - \hat{x}_2\|^2 + \alpha_0 \langle f(t, \hat{x}_1) - f(t, \hat{x}_2), \hat{x}_1 - \hat{x}_2 \rangle$ holds, where $\beta_0, \alpha_0 \in \mathbb{R}$ are known constants.*

Remark 1. The area D is our operational area. The one-sided Lipschitz condition has the inherent advantages of the Lipschitz condition and is more general.

$\{\iota_t, t \geqslant 0\}$ is the Markov jump process in which values are taken in the finite set H. $\ell_{ij} \geqslant 0 \, (i, j \in H, i \neq j)$ represents the change rate of the system from the mode i at time t to the mode j at time $t + \pi$, and $\ell_{ii} = -\sum_{j=1, j \neq i}^{L} \ell_{ij}$ is satisfied. $M = \{\ell_{ij}\} \, (i, j \in H = \{1, 2, \ldots, L\})$ is transition rate matrix, which satisfies:

$$\Pr\{\iota_{t+\pi} = j | \iota_t = i\} = \begin{cases} \ell_{ij}(t) \pi + o(\pi), & i \neq j \\ 1 + \ell_{ij}(t) \pi + o(\pi), & i = j \end{cases}$$

where $\pi > 0$ and $\lim_{\pi \to 0} \frac{o(\pi)}{\pi} = 0$.

For $\iota_t = i, i \in H$, we denote $A(\iota_t) = A_i, B(\iota_t) = B_i$, others and so on.

Definition 1. *[16]. For $\forall t \geqslant 0$, if there exist constants $\varepsilon > 0$ and $\varpi > 0$, $E\{\|x(t)\|^2\} \leqslant \varepsilon e^{-\varpi t} \sup_{-\tau \leqslant s \leqslant 0} E\{\|\theta(s)\|^2\}$ holds, then the Markov jump systems (1) are said to be exponentially mean-square stable.*

2.2 DoS Attacks

In this paper, the random packet loss and periodic DoS attacks are considered in the control channel between the zero-order-holder (ZOH) and the actuator. It's assumed that packet loss and DoS attacks are two independent processes. And random packet loss within the action-period is not considered. The periodic DoS attacks can be expressed as:

$$\varsigma_{DoS}(t) = \begin{cases} 0, & t \in [nF, nF + F_{off}), \\ 1, & t \in [nF + F_{off}, (n+1)F), \end{cases}$$

where F is a cycle of DoS attacks. The $(n+1)th \, (n \in \mathbb{N})$ cycle of DoS attacks is divided into two stages: the sleeping-period $[nF, nF + F_{off})$ and the action-period $[nF + F_{off}, (n+1)F)$. During the action-period of DoS attacks, the data cannot be transmitted. While during the sleeping-period of DoS attacks, data can be transmitted. And at this stage, we consider the influence of random packet loss on the system. A random variable $\varsigma(t)$ that satisfies the Bernoulli distribution is introduced to model the phenomenon of random packet loss:

$$\varsigma(t) = \begin{cases} 0, & control \ packet \ loss \ occurs, \\ 1, & control \ packet \ loss \ doesn't \ occur. \end{cases}$$

And $\Pr ob\{\varsigma(t) = 1\} = \varsigma, \Pr ob\{\varsigma(t) = 0\} = 1 - \varsigma, 0 < \varsigma < 1$ is a known constant.

2.3   Dynamic Event-Triggered Mechanism

In order to reduce the frequency of data transmission, the following DETM based on sampled data is proposed. The sampler samples the system state with a fixed period h $(h > 0)$, and then calls the information of the sampling point ph $(p \in \mathbb{N})$ that meets the triggered condition through the following DETM. For $t \in [nF, nF + F)$, $t_k^{n+1}h$ $\left(\{t_k^{n+1}h \,|\, k \in \mathbb{N}, t_k^{n+1} \in \mathbb{N}\} \subset \{ph \,|\, p \in \mathbb{N}\}\right)$ denotes the latest event-triggered moment, $\left(t_k^{n+1} + j\right)h$ is the current sampled time, and the next event-triggered moment $t_{k+1}^{n+1}h$ is decided by the following rule:

$$t_{k+1}^{n+1}h = t_k^{n+1}h + \min_{j \geqslant 1, j \in \mathbb{N}} \left\{ t_k^{n+1}h \in [nF, nF + F_{off})| - e^T \left(t_k^{n+1}h + jh\right) H \right.$$
$$\left. e\left(t_k^{n+1}h + jh\right) + \sigma x^T \left(t_k^{n+1}h\right) H x \left(t_k^{n+1}h\right) + \delta\beta \left(t_k^{n+1}h + jh\right) \leqslant 0 \right\}, \quad (2)$$

where $t_{1,0} = 0, k \in \tilde{K}_n, \tilde{K}_n = \{0, 1, \ldots, \bar{k}_n\}, \bar{k}_n \triangleq \sup\{k \in \mathbb{N} \,|\, t_k^{n+1}h \leqslant nF + F_{off}\}$. $\sigma \in (0, 1)$ is a given constant. $H > 0$ is the weight matrix to be determined. $e\left(t_k^{n+1}h + jh\right) = x\left(t_k^{n+1}h\right) - x\left(t_k^{n+1}h + jh\right)$ is the error between the latest transmitted state and the current sampled state. And the dynamic variable $\beta(t)$ decided by the following dynamic rule:

$$\dot{\beta}(t) = -2\kappa_1 \beta(t) - \delta\beta(ph) + x^T(ph)\Pi x(ph), \ t \in [ph, (p+1)h), \quad (3)$$

where $\kappa_1 > 0$ and $\delta > 0$ are given constants, $\Pi > 0$ is the weight matrix to be determined. And the initial condition is $\beta(0) \geqslant 0$.

Lemma 1. *[17] For $t \in [0, \infty)$, given constants $\beta(0) \geqslant 0, \kappa_1 > 0, h > 0$ and matrix $\Pi > 0$, if $0 < \delta \leqslant -2\kappa_1 + \frac{2\kappa_1}{1 - e^{-2\kappa_1 h}}$, the dynamic variable $\beta(t)$ defined in (3) satisfies $\beta(t) \geqslant 0$.*

Remark 2. The triggered rule (2) is designed on the basis of periodic sampling, so the Zeno phenomenon will not occur.

Compared with the hold-input strategy in references [8, 18], the zero-input strategy can save energy and reduce the complexity of the equation. In this paper, when the control packet is lost due to random packet loss and DoS attacks, the zero-input strategy is adopted, that is the input of the actuator is set to zero. Based on this, the controller can be designed as follows:

$$u(t) = \begin{cases} \varsigma(t) K_i x\left(t_k^{n+1}h\right), \ t \in [t_k^{n+1}h, t_{k+1}^{n+1}h) \cap \Pi_1^n, \\ 0, \ t \in \Pi_2^n, \end{cases} \quad (4)$$

where $\Pi_1^n \triangleq [nF, nF + F_{off})$ and $\Pi_2^n \triangleq [nF + F_{off}, (n+1)F)$. And K_i is the controller gain matrix.

2.4   Switched Markov Jump System Model

Denote $q_{\bar{k}_n}^n \triangleq \sup\{j \in \mathbb{N} | t_{\bar{k}_n}^{n+1}h + jh < nF + F_{off}\}, q_k^n \triangleq \sup\{j \in \mathbb{N} | t_{k+1}^{n+1}h > t_k^{n+1}h + jh\}$, where $k \in \{0, 1, \ldots, \bar{k}_n - 1\}$. We have $\mathrm{T}_{k,n}^j =$

$\left[t_k^{n+1}h+(j-1)h,\,t_k^{n+1}h+jh\right),j=1,2,\cdots,q_k^n;\,\mathrm{T}_{k,n}^{q_k^n+1}=\left[t_k^{n+1}h+q_k^nh,t_{k+1}^{n+1}h\right);$
$\mathrm{T}_{\bar{k}_n,n}^{j}=\left[t_{\bar{k}_n}^{n+1}h+(j-1)h,t_{\bar{k}_n}^{n+1}h+jh\right),j=1,2,\cdots,q_{\bar{k}_n}^n;\,\mathrm{T}_{\bar{k}_n,n}^{q_{\bar{k}_n}^n+1}=$
$\left[t_{\bar{k}_n}^{n+1}h+q_{\bar{k}_n}^nh,nF+F_{off}\right).$

Based on the above sampling intervals, two segmented functions are defined:

(1) For $k\in\{0,1,\ldots,\bar{k}_n-1\}$, $\hat{q}_k^n=q_k^n$. For $k=\bar{k}_n$, $\hat{q}_{\bar{k}_n}^n\triangleq$
$\sup\left\{j\in\mathbb{N}\,\middle|\,\left(t_{\bar{k}_n}^{n+1}+j\right)h<nF+F\right\}$. And $0\leqslant l_{kn}(t)<l,l=h$.

$$l_{kn}(t)=\begin{cases} t-t_k^{n+1}h,\ t\in\mathrm{T}_{k,n}^1,\\ t-t_k^{n+1}h-h,\ t\in\mathrm{T}_{k,n}^2,\\ \vdots\\ t-t_k^{n+1}h-\hat{q}_k^nh,\ t\in\mathrm{T}_{k,n}^{\hat{q}_k^n+1}. \end{cases}\tag{5}$$

(2) For $k\in\tilde{K}_n$,

$$e_{kn}(t)=\begin{cases} 0,\ t\in\mathrm{T}_{k,n}^1,\\ x\left(t_k^{n+1}h\right)-x\left(t_k^{n+1}h+h\right),\ t\in\mathrm{T}_{k,n}^2,\\ \vdots\\ x\left(t_k^{n+1}h\right)-x\left(t_k^{n+1}h+q_k^nh\right),\ t\in\mathrm{T}_{k,n}^{q_k^n+1}. \end{cases}\tag{6}$$

From (5) and (6) , for $t\in\left[t_k^{n+1}h,t_{k+1}^{n+1}h\right)\cap\Pi_1^n$, we have

$$x\left(t_k^{n+1}h\right)=x\left(t-l_{kn}(t)\right)+e_{kn}(t).\tag{7}$$

Denote $\tilde{\Pi}_1^n=\left[\tilde{t}_1^n,\tilde{t}_2^n\right)=\cup_{k=0}^{\bar{k}_n}\cup_{i=1}^{q_k^n+1}\mathrm{T}_{k,n}^i\cap\Pi_1^n,\tilde{\Pi}_2^n=\left[\tilde{t}_2^n,\tilde{t}_1^{n+1}\right)=\Pi_2^n.$
Combining (1), (4) and (7), the switched Markov jump system is as follows:

$$\begin{cases} \dot{x}(t)=A_ix(t)+\varsigma(t)B_iK_i\left(x(t-l_{kn}(t))+e_{kn}(t)\right)+f(t,x),\ t\in\tilde{\Pi}_1^n,\\ \dot{x}(t)=A_ix(t)+f(t,x),\ t\in\tilde{\Pi}_2^n. \end{cases}\tag{8}$$

3   Main Results

Next, a common Lyapunov functional is constructed, and then the sufficient conditions for the exponential mean square stability of the system (8) are obtained, and the controller gain matrix and event-triggered parameters are obtained.

Theorem 1. *For positive scalars $F,F_{off},l,\sigma,\delta,\kappa_1,\kappa_2,\varepsilon_1,\varepsilon_2,\nu$ and $\kappa_1+\kappa_2>0$, satisfying Lemma 1 and $\lambda:=-\kappa_2F+(\kappa_1+\kappa_2)F_{off}>0$, if there exist symmetric matrices $\Pi_i,H_i,Q_i>0,R_i>0,\bar{P}_i>0\,(i\in1,\ldots,L)$ and matrices $N_{11i},N_{12i},N_{21i},N_{22i},M_{11i},M_{12i},M_{21i},M_{22i}\,(i\in1,\ldots,L)$ with appropriate dimensions, such that for $m=1,2$ and $q=1,2,$*

$$\Delta_{m(q)i} = \begin{bmatrix} \Phi_{m1i} & * & * & * & * & * \\ l\hat{\Gamma}_{m1} & -\left(2\nu\bar{P}_i - \nu^2 R_i\right) & * & * & * & * \\ \Phi_{m3(q)i} & 0_{n\times n} & -e^{2\alpha l}R_i & * & * & * \\ \Phi_{m4(q)i} & 0_{n\times n} & 0_{n\times n} & -3e^{2\alpha l}R_i & * & * \\ \hat{\Phi}_{m5} & 0_{n\times n} & 0_{n\times n} & 0_{n\times n} & \Phi_{m6} & * \\ \hat{\Phi}_{m7} & 0_{n\times n} & 0_{n\times n} & 0_{n\times n} & 0_{n\times n} & \Phi_{m8} \end{bmatrix} \leqslant 0, \quad (9)$$

where

$\Phi_{11i} = sym\left\{\bar{e}_{11}^T\hat{\Gamma}_{11} + (\varepsilon_2\alpha_0 - \varepsilon_1)\,\bar{e}_{11}^T\bar{P}_i\bar{e}_{16} - \varepsilon_2\bar{e}_{16}^T\bar{e}_{16} + \varXi_{11i} + \varXi_{12i}\right\}$

$+\,\bar{e}_{11}^T\left(2\kappa_1\bar{P}_i + Q_i + \ell_{ii}\bar{P}_i\right)\bar{e}_{11} - e^{-2\kappa_1 l}\bar{e}_{13}^TQ_i\bar{e}_{13} + \bar{e}_{12}^T\Pi_i\bar{e}_{12} - \bar{e}_{17}^TH_i\bar{e}_{17}$

$+\,\sigma(\bar{e}_{12} + \bar{e}_{17})^T H_i\,(\bar{e}_{12} + \bar{e}_{17})\,,\, \varXi_{11i} = le^{-2\kappa_1 l}\left(E_{15}^TN_{11i}^TE_{11} + E_{15}^TN_{12i}^TE_{12}\right),$

$\Phi_{21i} = sym\left\{\bar{e}_{21}^T\hat{\Gamma}_{21} + (\varepsilon_2\alpha_0 - \varepsilon_1)\,\bar{e}_{21}^T\bar{P}_i\bar{e}_{26} - \varepsilon_2\bar{e}_{26}^T\bar{e}_{26} + \varXi_{21i} + \varXi_{22i}\right\}$

$+\,\bar{e}_{21}^T\left(-2\kappa_2\bar{P}_i + Q_i + \ell_{ii}\bar{P}_i\right)\bar{e}_{21} + \bar{e}_{22}^T\Pi_i\bar{e}_{22} - e^{-2\kappa_1 l}\bar{e}_{23}^TQ_i\bar{e}_{23},$

$\varXi_{12i} = le^{-2\kappa_1 l}\left(E_{16}^TN_{21i}^TE_{13} + E_{16}^TN_{22i}^TE_{14}\right), \varXi_{21i} = le^{-2\kappa_1 l}\left(E_{25}^TM_{11i}^TE_{21} + E_{25}^TM_{12i}^TE_{22}\right), \varXi_{22i} = le^{-2\kappa_1 l}\left(E_{26}^TM_{21i}^TE_{23} + E_{26}^TM_{22i}^TE_{24}\right),$

$\Phi_{m6} = diag\left\{-\bar{P}_1, \cdots, -\bar{P}_{i-1}, -\bar{P}_{i+1}, \cdots, -\bar{P}_L\right\}, \Phi_{13(1)i} = lN_{11i}E_{15}, \Phi_{13(2)i} = lN_{21i}E_{16}, \Phi_{14(1)i} = lN_{12i}E_{15}, \Phi_{14(2)i} = lN_{22i}E_{16}, \Phi_{23(1)i} = lM_{11i}E_{25}, \Phi_{23(2)i} = lM_{21i}E_{26}, \Phi_{24(1)i} = lM_{12i}E_{25}, \Phi_{24(2)i} = lM_{22i}E_{26}, \Upsilon_{11} = l\Sigma_{11}, \Upsilon_{12} = l\Sigma_{12},$

$\varXi_{11} = le^{-2\kappa_1 l}\left(E_{15}^TN_{11}^TE_{11} + E_{15}^TN_{12}^TE_{12}\right), \varXi_{12} = le^{-2\kappa_1 l}\left(E_{16}^TN_{21}^TE_{13} + E_{16}^TN_{22}^TE_{14}\right), \varXi_{21} = le^{-2\kappa_1 l}\left(E_{25}^TM_{11}^TE_{21} + E_{25}^TM_{12}^TE_{22}\right), \varXi_{22} = le^{-2\kappa_1 l}\left(E_{26}^TM_{21}^TE_{23} + E_{26}^TM_{22}^TE_{24}\right), \Upsilon_{21} = l\Sigma_{21}, \Upsilon_{22} = l\Sigma_{22}, \Sigma_{11} = le^{-2\kappa_1 l}E_{15}^T$

$\left(N_{11}^TR^{-1}N_{11} + \frac{1}{3}N_{12}^TR^{-1}N_{12}\right)E_{15}, \Sigma_{12} = le^{-2\kappa_1 l}E_{16}^T\left(N_{21}^TR^{-1}N_{21}\right.$

$\left.+\frac{1}{3}N_{22}^TR^{-1}N_{22}\right)E_{16}, \Sigma_{21} = le^{-2\kappa_1 l}E_{25}^T\left(M_{11}^TR^{-1}M_{11} + \frac{1}{3}M_{12}^TR^{-1}M_{12}\right)E_{25},$

$\Sigma_{22} = le^{-2\kappa_1 l}E_{26}^T\left(M_{21}^TR^{-1}M_{21} + \frac{1}{3}M_{22}^TR^{-1}M_{22}\right)E_{26}, \Phi_{11} = sym\left\{\bar{e}_{11}^TP_i\Gamma_{11}\right.$

$+\,(\varepsilon_2\alpha_0 - \varepsilon_1)\,\bar{e}_{11}^T\bar{e}_{16} - \varepsilon_2\bar{e}_{16}^T\bar{e}_{16} + \varXi_{11} + \varXi_{12}\left.\right\} - e^{-2\kappa_1 l}\bar{e}_{13}^TQ\bar{e}_{13} + \bar{e}_{12}^T\Pi\bar{e}_{12}$

$-\,\bar{e}_{17}^TH\bar{e}_{17} + \bar{e}_{11}^T\left(2\kappa_1P_i + Q + \ell_{ii}P_i\right)\bar{e}_{11} + \sigma(\bar{e}_{12} + \bar{e}_{17})^T H\,(\bar{e}_{12} + \bar{e}_{17}),$

$\Phi_{21} = sym\left\{\bar{e}_{21}^TP_i\Gamma_{21} + (\varepsilon_2\alpha_0 - \varepsilon_1)\,\bar{e}_{21}^T\bar{e}_{26} - \varepsilon_2\bar{e}_{26}^T\bar{e}_{26} + \varXi_{21} + \varXi_{22}\right\} + \bar{e}_{22}^T\Pi\bar{e}_{22}$

$-\,e^{-2\kappa_1 l}\bar{e}_{23}^TQ\bar{e}_{23} + \bar{e}_{21}^T\left(-2\kappa_2P_i + Q + \ell_{ii}P_i\right)\bar{e}_{21}, \Gamma_{21} = A_i\bar{e}_{21} + \bar{e}_{26}, \Gamma_{11} = A_i\bar{e}_{11} + \varsigma B_iK_i(\bar{e}_{12} + \bar{e}_{17}) + \bar{e}_{16}, \Theta_m = l^2\Gamma_{m1}^TR\Gamma_{m1}, \hat{\Phi}_{17} = \Phi_{17}\bar{P}_i, \hat{\Phi}_{27} = \Phi_{27}\bar{P}_i,$

$\Phi_{17} = \left[\varepsilon_1\rho_0 + \varepsilon_2\beta_0\ 0_{n\times 6n}\right], \Phi_{27} = \left[\varepsilon_1\rho_0 + \varepsilon_2\beta_0\ 0_{n\times 5n}\right], \hat{\Gamma}_{21} = A_i\bar{P}_i\bar{e}_{21} + \bar{e}_{26},$

$\hat{\Gamma}_{11} = A_i\bar{P}_i\bar{e}_{11} + \varsigma B_iY_i(\bar{e}_{12} + \bar{e}_{17}) + \bar{e}_{16}, \Phi_{18} = \Phi_{28} = -\frac{1}{2}\left(\varepsilon_1\rho_0 + \varepsilon_2\beta_0\right),$

$$\Phi_{15} = \begin{bmatrix} \sqrt{\ell_{i1}} & \cdots & \sqrt{\ell_{i(i-1)}} & \sqrt{\ell_{i(i+1)}} & \cdots & \sqrt{\ell_{iL}} \\ 0_{n\times 6n} & \cdots & 0_{n\times 6n} & 0_{n\times 6n} & \cdots & 0_{n\times 6n} \end{bmatrix}^T, \hat{\Phi}_{15} = \Phi_{15}\bar{P}_i, E_{11} = \bar{e}_{11} - \bar{e}_{12},$$

$$\Phi_{25} = \begin{bmatrix} \sqrt{\ell_{i1}} & \cdots & \sqrt{\ell_{i(i-1)}} & \sqrt{\ell_{i(i+1)}} & \cdots & \sqrt{\ell_{iL}} \\ 0_{n\times 5n} & \cdots & 0_{n\times 5n} & 0_{n\times 5n} & \cdots & 0_{n\times 5n} \end{bmatrix}^T, \hat{\Phi}_{25} = \Phi_{25}\bar{P}_i, E_{21} = \bar{e}_{21} - \bar{e}_{22},$$

$$\Phi_{13(1)} = lN_{11}E_{15}, \Phi_{13(2)} = lN_{21}E_{16}, E_{23} = \bar{e}_{22} - \bar{e}_{23}, E_{22} = \bar{e}_{21} + \bar{e}_{22} - 2\bar{e}_{24},$$

$$\Phi_{23(1)} = lM_{11}E_{25}, \Phi_{23(2)} = lM_{21}E_{26}, E_{13} = \bar{e}_{12} - \bar{e}_{13}, E_{24} = \bar{e}_{22} + \bar{e}_{23} - 2\bar{e}_{25},$$

$$\Phi_{14(1)} = lN_{12i}E_{15}, \Phi_{14(2)} = lN_{22}E_{16}, \Phi_{24(1)} = lM_{12}E_{25}, E_{12} = \bar{e}_{11} + \bar{e}_{12} - 2\bar{e}_{14},$$

$$\Phi_{24(2)} = lM_{22}E_{26}, E_{14} = \bar{e}_{12} + \bar{e}_{13} - 2\bar{e}_{15}, E_{15} = col\{\bar{e}_{11}, \bar{e}_{12}, \bar{e}_{14}\},$$

$$E_{16} = col\{\bar{e}_{12}, \bar{e}_{13}, \bar{e}_{15}\}, E_{25} = col\{\bar{e}_{21}, \bar{e}_{22}, \bar{e}_{24}\}, E_{26} = col\{\bar{e}_{22}, \bar{e}_{23}, \bar{e}_{25}\},$$

$$\bar{e}_{1u} = \begin{bmatrix} 0_{n\times(u-1)n} & I_n & 0_{n\times(7-u)n} \end{bmatrix}, u = 1, 2, \cdots, 7,$$

$$\bar{e}_{2j} = \begin{bmatrix} 0_{n\times(j-1)n} & I_n & 0_{n\times(6-j)n} \end{bmatrix}, j = 1, 2, \cdots, 6.$$

Then, the switched Markov jump system (8) is exponentially mean-square stable. And the controller gain matrix and event-triggered parameters are $K_i = Y_i\bar{P}_i^{-1}$ and $H_i = \bar{P}_iH\bar{P}_i, \Pi_i = \bar{P}_i\Pi\bar{P}_i$, respectively.

Proof. Construct the common Lyapunov-Krasovskii functional:

$$W(t) = W(x(t), \iota_t, t) = W_1(x(t), \iota_t, t) + \beta(t), \tag{10}$$

where $\beta(t)$ is defined in (3). $Q > 0, R > 0, P_i > 0$. And $W_1(x(t), \iota_t, t) = x^T(t) P(\iota_t)x(t) + \int_{t-l}^t e^{-2\kappa_1(t-\eta)}x^T(\eta)Qx(\eta)d\eta + l\int_{t-l}^t \int_\theta^t e^{-2\kappa_1(t-\eta)}\dot{x}^T(\eta)R\dot{x}(\eta)d\eta d\theta$. The infinitesimal operator [19] of $W_1(x(t), \iota_t, t)$ is

$$LW_1(x(t), \iota_t, t) = 2x^T(t)P_i\dot{x}(t) + x^T(t)\left(\sum_{j=1}^L \ell_{ij}P_j\right)x(t) - e^{-2\kappa_1 l}x^T(t-l)Q$$
$$x(t-l) + x^T(t)Qx(t) + l^2\dot{x}^T(t)R\dot{x}(t) - 2\kappa_1\int_{t-l}^t e^{-2\kappa_1(t-\eta)}x^T(\eta)Qx(\eta)d\eta$$
$$-l\int_{t-l}^t e^{-2\kappa_1(t-\eta)}\dot{x}^T(\eta)R\dot{x}(\eta)d\eta - 2\kappa_1 l\int_{t-l}^t \int_\theta^t e^{-2\kappa_1(t-\eta)}\dot{x}^T(\eta)R\dot{x}(\eta)d\eta d\theta. \tag{11}$$

Case 1: When $t \in [\tilde{t}_1^n, \tilde{t}_2^n)$, $\varsigma_{DoS}(t) = 0$. When the event-triggered condition is not met at time t, it can be obtained according to (2) and (7):

$$-\delta\beta(t - l_{kn}(t)) < \sigma[x(t - l_{kn}(t)) + e_{kn}(t)]^T H[x(t - l_{kn}(t)) + e_{kn}(t)]$$
$$-e_{kn}^T(t)He_{kn}(t), \tag{12}$$

Combining (3) and (12), we have

$$\dot{\beta}(t) \leqslant -2\kappa_1\beta(t) + x^T(t - l_{kn}(t))\Pi x(t - l_{kn}(t)) - e_{kn}^T(t)He_{kn}(t)$$
$$+ \sigma[x(t - l_{kn}(t)) + e_{kn}(t)]^T H[x(t - l_{kn}(t)) + e_{kn}(t)], \tag{13}$$

According to (10), (11) and (13), we have

$$LW(t) \leqslant -2\kappa_1 W(t) + 2\kappa_1 x^T(t)P_ix(t) + x^T(t)Qx(t) + 2x^T(t)P_i\dot{x}(t)$$
$$+x^T(t)\left(\sum_{j=1}^L \ell_{ij}P_j\right)x(t) + x^T(t - l_{kn}(t))\Pi x(t - l_{kn}(t)) + l^2\dot{x}^T(t)R$$
$$\dot{x}(t) + \sigma[x(t - l_{kn}(t)) + e_{kn}(t)]^T H[x(t - l_{kn}(t)) + e_{kn}(t)] - e_{kn}^T(t)H$$
$$e_{kn}(t) - e^{-2\kappa_1 l}x^T(t-l)Qx(t-l) - l\int_{t-l}^t e^{-2\kappa_1(t-\eta)}\dot{x}^T(\eta)R\dot{x}(\eta)d\eta. \tag{14}$$

Let $\xi_2(t) = col\left\{x(t), x(t - l_{kn}(t)), x(t - l), \frac{1}{l_{kn}(t)}\int_{t-l_{kn}(t)}^{t} x(\eta)\,d\eta, \frac{1}{l-l_{kn}(t)}\right.$
$\left.\int_{t-l}^{t-l_{kn}(t)} x(\eta)\,d\eta, f(t,x)\right\}$, $\xi_1(t) = col\{\xi_2(t), e_{kn}(t)\}$. From Lemma 1 in [20],

$$- l\int_{t-l}^{t} e^{-2\kappa_1(t-\eta)}\dot{x}^T(\eta) R\dot{x}(\eta)\,d\eta \leqslant \xi_1^T(t)\{l_{kn}(t)\Sigma_{11} + (l - l_{kn}(t))\Sigma_{12}$$
$$+ sym\{\Xi_{11} + \Xi_{12}\}\}\xi_1(t). \tag{15}$$

From Assumption 1, it exists scalars $\varepsilon_1 > 0$ and $\varepsilon_2 > 0$,

$$\begin{aligned}
2\varepsilon_1\rho_0 x^T(t) x(t) - \varepsilon_1 x^T(t) f(t,x) - \varepsilon_1 f^T(t,x) x(t) &\geqslant 0, \\
2\varepsilon_2\beta_0 x^T(t) x(t) + 2\varepsilon_2\alpha_0 x^T(t) f(t,x) - 2\varepsilon_2 f^T(t,x) f(t,x) &\geqslant 0.
\end{aligned} \tag{16}$$

Combining (14), (15) and (16), we have

$$LW(t) \leqslant -2\kappa_1 W(t) + \xi_1^T(t)\Delta_1\xi_1(t), \tag{17}$$

where $\Delta_1 = \Phi_1 + \Theta_1 + l_{kn}(t)\Sigma_{11} + (l - l_{kn}(t))\Sigma_{12}$, $\Phi_1 = \Phi_{11} + \bar{e}_{11}^T\left(\sum_{j\neq i}^{L}\ell_{ij}P_j\right)\bar{e}_{11} + sym\left\{(\varepsilon_1\rho_0 + \varepsilon_2\beta_0)\bar{e}_{11}^T\bar{e}_{11}\right\}$.

Case 2: When $t \in [\tilde{t}_2^n, \tilde{t}_1^{n+1})$, $\varsigma_{DoS}(t) = 1$. Let $\kappa_1 + \kappa_2 > 0$, then $-2\kappa_1 < 2\kappa_2$. From Lemma 1, we have $\dot{\beta}(t) \leqslant 2\kappa_2\beta(t) + x^T(t - l_{kn}(t))\Pi x(t - l_{kn}(t))$.
Similar to the case 1, we have

$$LW(t) \leqslant 2\kappa_2 W(t) + \xi_2^T(t)\Delta_2\xi_2(t), \tag{18}$$

where $\Delta_2 = \Phi_2 + \Theta_2 + l_{kn}(t)\Sigma_{21} + (l - l_{kn}(t))\Sigma_{22}$, $\Phi_2 = \Phi_{21} + \bar{e}_{21}^T\left(\sum_{j\neq i}^{L}\ell_{ij}P_j\right)\bar{e}_{21} + sym\left\{(\varepsilon_1\rho_0 + \varepsilon_2\beta_0)\bar{e}_{21}^T\bar{e}_{21}\right\}$.

Define $\bar{P}_i = P_i^{-1}$, $\varpi \in \{Q, R, H, \Pi, N_{11}, N_{12}, N_{21}, N_{22}, M_{11}, M_{12}, M_{21}, M_{22}\}$, $\varpi_i = \bar{P}_i\varpi\bar{P}_i$, $O_{(1)} = diag\{\bar{P}_i, \bar{P}_i, \bar{P}_i, \bar{P}_i, \bar{P}_i, I_n, \bar{P}_i, I_n, \bar{P}_i, \bar{P}_i, J_i, I_n\}$, $O_{(2)} = diag\{\bar{P}_i, \bar{P}_i, \bar{P}_i, \bar{P}_i, \bar{P}_i, I_n, I_n, \bar{P}_i, \bar{P}_i, J_i, I_n\}$, $J_i = diag\underbrace{\{I_n, \cdots, I_n\}}_{L-1}$, $Y_i = K_i\bar{P}_i$. Using inequality $-R^{-1} = -\bar{P}_i R_i^{-1}\bar{P}_i \leqslant -2\nu\bar{P}_i + \nu^2 R_i$ and Schur complement, for $q = 1, 2$, $\Delta_m \leqslant 0$ $(m = 1, 2)$ is equivalent to

$$\Delta_{m(q)} = \begin{bmatrix} \Phi_{m1} & * & * & * & * & * \\ l\Gamma_{m1} & -(2\nu\bar{P}_i - \nu^2 R_i) & * & * & * & * \\ \Phi_{m3(q)} & 0_{n\times n} & -e^{2\alpha l}R & * & * & * \\ \Phi_{m4(q)} & 0_{n\times n} & 0_{n\times n} & -3e^{2\alpha l}R & * & * \\ \Phi_{m5} & 0_{n\times n} & 0_{n\times n} & 0_{n\times n} & \Phi_{m6} & * \\ \Phi_{m7} & 0_{n\times n} & 0_{n\times n} & 0_{n\times n} & 0_{n\times n} & \Phi_{m8} \end{bmatrix} \leqslant 0, \tag{19}$$

Then pre- and post-multiplying (19) by $O_{(l)}$, respectively. We can get (9). It can be further deduced from (9), (17) and (18)

$$\begin{cases} LW(t) \leqslant -2\kappa_1 W(t), & t \in [\tilde{t}_1^n, \tilde{t}_2^n), \\ LW(t) \leqslant 2\kappa_2 W(t), & t \in [\tilde{t}_2^n, \tilde{t}_1^{n+1}). \end{cases} \tag{20}$$

From (20), we have

$$EW(t) \leqslant \begin{cases} Ee^{-2\kappa_1\left(t-\tilde{t}_1^n\right)}W\left(\tilde{t}_1^n\right), & t \in \left[\tilde{t}_1^n, \tilde{t}_2^n\right), \\ Ee^{2\kappa_2\left(t-\tilde{t}_2^n\right)}W\left(\tilde{t}_2^n\right), & t \in \left[\tilde{t}_2^n, \tilde{t}_1^{n+1}\right). \end{cases} \tag{21}$$

Case1: When $t \in \left[\tilde{t}_1^n, \tilde{t}_2^n\right)$, from (21) and $\tilde{t}_1^0 = 0$, we have $EW(t) \leqslant$ $e^{-2\kappa_1\left(t-\tilde{t}_1^n\right)} EW\left(\tilde{t}_1^n\right) \leqslant e^{2(\kappa_1+\kappa_2)(F-F_{off})-2\kappa_1\left(t-\tilde{t}_2^{n-1}\right)} EW\left(\tilde{t}_2^{n-1}\right) \leqslant EW\left(\tilde{t}_1^{n-1}\right)$ $e^{-2\kappa_1\left(t-\tilde{t}_1^{n-1}\right)}e^{2(\kappa_1+\kappa_2)(F-F_{off})} \leqslant \cdots \leqslant e^{-2\kappa_1 t+2n(\kappa_1+\kappa_2)(F-F_{off})}W(0)$.

From equation $\lambda := -\kappa_2 F + (\kappa_1 + \kappa_2), F_{off} > 0, t \geqslant \tilde{t}_1^n \geqslant nF$ and $t \leqslant \tilde{t}_2^n = nF + F_{off}$, one has

$$EW(t) \leqslant e^{-2\lambda n}W(0) \leqslant e^{2\lambda\frac{F_{off}}{F}}e^{-\frac{2\lambda}{F}t}W(0). \tag{22}$$

Case2: When $t \in \left[\tilde{t}_2^n, \tilde{t}_1^{n+1}\right)$, similar to case 1

$$EW(t) \leqslant e^{-\frac{2\lambda}{F}t}W(0). \tag{23}$$

Denote $q_1 = \min_{i\in\{1,\ldots,L\}} \{\lambda_{\min}(P_i)\}, q_2 = \min_{i\in\{1,\ldots,L\}} \{\lambda_{\max}(P_i)\}, q_3 = q_2+l\lambda_{\max}$ $(Q)+l^2\lambda_{\max}(R)$. As known by the formula (10), there is a scalar $\gamma > 1$ satisfies

$$EW(t) \geqslant q_1 E\|x(t)\|^2, \quad W(0) \leqslant \gamma q_3 E\left\{\sup_{-l\leqslant s\leqslant 0} \|\theta(s)\|_l^2\right\} + \|\beta(0)\|. \tag{24}$$

For given θ and $\beta(0)$, there exists a scalar $q_4 > 0$ that satisfies $\|\beta(0)\| \leqslant$ $q_4 E\left\{\sup_{-l\leqslant s\leqslant 0} \|\theta(s)\|_l^2\right\}$. From (22), (23) and (24), we have $E\left\{\|x(t)\|^2\right\} \leqslant \frac{\vartheta q_5}{q_1}$ $e^{-\frac{\lambda}{F}t}E\left\{\sup_{-l\leqslant s\leqslant 0} \|\theta(s)\|_l^2\right\}$, where $\vartheta = \max\left\{e^{2\lambda\frac{F_{off}}{F}}, 1\right\}$ and $q_5 = \gamma q_3+q_4$. Therefore, the system (8) is exponentially stable in mean square.

4   Numerical Example

For system (1), the parameters are given: $A_1 = \begin{bmatrix} -1.4 & 0 \\ 1 & -1 \end{bmatrix}, A_2 = \begin{bmatrix} 1 & 0.5 \\ 0 & -1.5 \end{bmatrix}, B_1$ $= \begin{bmatrix} 0.8 \\ 0.5 \end{bmatrix}, B_2 = \begin{bmatrix} 0.5 \\ 0.8 \end{bmatrix}, F = 2s, F_{off} = 1.6667s, h = 0.05s, \kappa_1 = 0.1, \kappa_2 = 0.5, \sigma = 0.1, \delta = 0.3, m = 0.1, \varepsilon_1 = 0.1, \varepsilon_2 = 0.7, \alpha_0 = -0.3, \beta_0 = 0.1, \rho_0 = -0.3, f(t,x) = 6\sin(0.1x(t))$. The initial conditions are $\beta(0) = 20$ and $x(0) = [0.3 \ -0.5]^T$. The transition rate matrix is $M = \begin{bmatrix} -3 & 3 \\ 5 & -5 \end{bmatrix}$. When $\varsigma = 0.6$, the following controller gain matrix and weight matrices are obtained:

$$K_1 = [-13.0201 \ -6.5045], H_1 = \begin{bmatrix} 220.01 & 123.41 \\ 123.41 & 103.03 \end{bmatrix}, \Pi_1 = \begin{bmatrix} 1.44 & -0.16 \\ -0.16 & 1.86 \end{bmatrix},$$

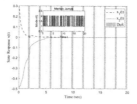

Fig. 1. The state response of the system under DETM

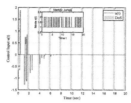

Fig. 2. The control input of the system under DETM

$$K_2 = [-14.0542 \quad -7.9044], H_2 = \begin{bmatrix} 89.82 & 119.63 \\ 119.63 & 215.10 \end{bmatrix}, \Pi_2 = \begin{bmatrix} 1.16 & -0.13 \\ -0.13 & 2.32 \end{bmatrix}.$$

Figures 1 and 2 show the state response and control input of system (8) under DETM, respectively. The control input is always zero during the action-period of the DoS attacks. Since the zero-input strategy is used when packet loss occurs, before the system reaches the stable state, the control input is zero in some periods during the sleeping-period of DoS attacks, which is due to packet loss. Figure 3 is the state response of the system (8) without control. According to Figs. 1-3, the DETM (2) can enable the system (8) to achieve exponential mean square stability under the dual influence of packet dropouts and DoS attacks. Figure 4 shows the relationship between the release instants and release intervals. And the maximum triggering interval is about 1.16s, which shows that our method can effectively alleviate the network congestion problem.

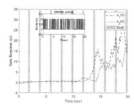

Fig. 3. The state response of the system without control

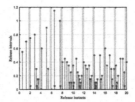

Fig. 4. The release instants and release intervals of the system under DETM

5   Conclusion

In this paper, the exponential stability in mean square of Markov jump systems with packet loss and DoS attacks is studied by DETM. Considering the influence of packet loss and DoS attacks on the system, the controller is designed based on DETM. Then by constructing common Lyapunov-Krasovskii functional, the new stability criteria of the switched Markov jump system under DoS attacks and packet loss are obtained. Finally, the practicability of the method in this paper is verified through a simulation example. In the future, our work will be developed in the following aspect: For continuous Markov jump systems, random packet loss is considered during the sleeping-period of DoS attacks, and the success rate of DoS attacks is considered during the action period of DoS attacks.

Acknowledgements. The research of the first corresponding author was supported in part by the Basic Research Youth Fund Project of Yunnan Science and Technology Department under Grant 202101AU070050, and in part by the Scientific Research Fund Project of Yunnan Provincial Department of Education under Grant 2022J0447. The research of the second corresponding author was supported by the National Nature Science Foundation of China under Grant 12061088. The research of the fourth author was supported by Yunnan Provincial Science and Technology Department under Grants 2023Y0566.

Statements and Declarations. Competing interests: The authors declare that there is no competing financial interest or personal relationship that could influenced the work reported in this paper.

References

1. Yang, T., Wang, Z., Huang, X., et al.: Aperiodic sampled-data controller design for stochastic Markovian jump systems and it application. Int. J. Robust Nonlinear Control **31**(14), 6721–6739 (2021)
2. Chen, G., Xia, J., Park, J.H., et al.: Sampled-data synchronization of stochastic Markovian jump neural networks with time-varying delay. IEEE Trans. Neural Netw. Learn. Syst. **33**(8), 3829–3841 (2021)
3. Zeng, P., Deng, F., Liu, X., et al.: Event-triggered H_∞ control for network-based uncertain Markov jump systems under DoS attacks. J. Franklin Inst. **358**(6), 2895–2914 (2021)

4. Qiu, L., Yao, F., Zhong, X.: Stability analysis of networked control systems with random time delays and packet dropouts modeled by Markov chains. J. Appl. Math. **2013**, 715072 (2013)
5. Zhang, Y., Xie, S., Ren, L., et al.: A new predictive sliding mode control approach for networked control systems with time delay and packet dropout. IEEE Access **7**, 134280–134292 (2019)
6. Su, L., Ye, D.: Observer-based output feedback H_∞ control for cyber-physical systems under randomly occurring packet dropout and periodic DoS attacks. ISA Trans. **95**, 58–67 (2019)
7. Li, L., Zhang, G., Ou, M.: A new method to non-fragile piecewise H_∞ control for networked nonlinear systems with packet dropouts. IEEE Access **8**, 196102–196111 (2020)
8. Huang, L., Guo, J., Li, B.: Observer-based dynamic event-triggered robust H_∞ control of networked control systems under DoS attacks. IEEE Access **9**, 145626–145637 (2021)
9. Liu, C., Xiong, R., Xu, J., et al.: On iterative learning control for remote control systems with packet losses. J. Appl. Math. **2013**, 245072 (2013)
10. Lu, R., Shi, P., Su, H., et al.: Synchronization of general chaotic neural networks with nonuniform sampling and packet missing: a switched system approach. IEEE Trans. Neural Netw. Learn. Syst. **29**(3), 523–533 (2016)
11. Xu, Y., Sun, J., Wang, G., et al.: Dynamic triggering mechanisms for distributed adaptive synchronization control and its application to circuit systems. IEEE Trans. Circuits Syst. I Regul. Pap. **68**(5), 2246–2256 (2021)
12. Hu, S., Yue, D., Cheng, Z., et al.: Co-design of dynamic event-triggered communication scheme and resilient observer-based control under aperiodic DoS attacks. IEEE Trans. Cybern. **51**(9), 4591–4601 (2020)
13. Wang, Y., Chen, F., Zhuang, G., et al.: Dynamic event-based mixed H_∞ and dissipative asynchronous control for Markov jump singularly perturbed systems. Appl. Math. Comput. **386**, 125443 (2020)
14. Wang, Y., Chen, F., Zhuang, G.: Dynamic event-based reliable dissipative asynchronous control for stochastic Markov jump systems with general conditional probabilities. Nonlinear Dyn. **101**(1), 465–485 (2020). https://doi.org/10.1007/s11071-020-05786-1
15. Huong, D.C., Huynh, V.T., Trinh, H.: Dynamic event-triggered state observers for a class of nonlinear systems with time delays and disturbances. IEEE Trans. Circ. Syst. II Express Briefs **67**, 3457–3461 (2020)
16. Wang, R., Luo, H.: Mean-square exponential stability analysis for uncertain stochastic neutral systems with nonlinear perturbations and distributed delays. In: 15th International Conference on Computational Intelligence and Security (CIS) Macao, China, pp. 353–357. IEEE (2019)
17. Zhao, N., Shi, P., Xing, W.: Dynamic event-triggered approach for networked control systems under denial of service attacks. Int. J. Robust Nonlinear Control **31**(5), 1774–1795 (2021)
18. Li, H., Li, X., Zhang, H.: Optimal control for discrete-time NCSs with input delay and Markovian packet losses: hold-input case. Automatica **132**, 109806 (2021)
19. Li, H., Shi, P., Yao, D.: Adaptive sliding-mode control of Markov jump nonlinear systems with actuator faults. IEEE Trans. Autom. Control **62**(4), 1933–1939 (2016)
20. Zeng, H.B., He, Y., Wu, M., et al.: New results on stability analysis for systems with discrete distributed delay. Automatica **60**, 189–192 (2015)

Image Blending Algorithm
with Automatic Mask Generation

Haochen Xue[1], Mingyu Jin[2], Chong Zhang[1], Yuxuan Huang[1], Qian Weng[1],
and Xiaobo Jin[1(✉)]

[1] Department of Intelligent Science, School of Advanced Technology,
Xi'an Jiaotong-Liverpool University, Suzhou, China
{Haochen.Xue20,Chong.zhang19,Yuxuan.Huang2002,
Qian.Weng22}@student.xjtlu.edu.cn, Xiaobo.Jin@xjtlu.edu.cn
[2] Electrical and Computer Engineering Northwestern University, Evanston, IL, USA
u9o2n2@u.northwestern.edu

Abstract. In recent years, image blending has gained popularity for its
ability to create visually stunning content. However, the current image
blending algorithms mainly have the following problems: manually creating image blending masks requires a lot of manpower and material
resources; image blending algorithms cannot effectively solve the problems of brightness distortion and low resolution. To this end, we propose a new image blending method with automatic mask generation: it
combines semantic object detection and segmentation with mask generation to achieve deep blended images, while based on our proposed
new saturation loss and two-stage iteration of the PAN algorithm to fix
brightness distortion and low-resolution issues. Results on publicly available datasets show that our method outperforms other classical image
blending algorithms on various performance metrics including PSNR and
SSIM.

Keywords: Image Blending · Mask Generation · Image
Segmentation · Object Detection

1 Introduction

Image blending is a versatile technique used in various applications [19,20] where
different images must be combined to create a unified and visually appealing
final image. It involves taking a selected part of an image (usually an object)
and seamlessly integrating it into another image at a specified location. The
ultimate goal of image fusion is to obtain a uniform and natural composite image.
This task poses two significant challenges: relatively low localization accuracy in

---

H. Xue and M. Jin—Equal contribution.

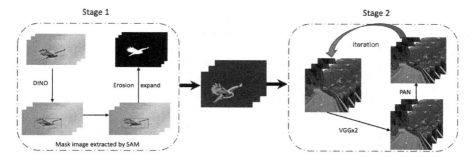

Fig. 1. Image blending process based on automatic mask generation: the first stage generates basic blending results through SAM; the second stage generates more detailed blending results after fusing PAN

cropped regions of objects and consistency issues between cropped objects and their surroundings.

GP-GAN and Poisson image editing is currently popular image blending methods [16]. In this method, the user selects an object in the source image with an associated mask to generate a high-resolution image and uses GP-GAN or Poisson to generate high-quality versions of the source and target images. However, the images generated by GP-GAN and Poisson image editing are not too realistic, where the composite image suffers from brightness distortion and often exhibits excessive brightness in small pixel clusters, thus compromising the overall realism of the image.

All image fusion algorithms require a mask as input to cut out the object to be fused from the source image, but the mask images of the previous algorithms are all handmade. These handcrafted mask images are insufficient to accurately represent the location of the foreground, which may lead to poor image fusion. Traditional segmentation methods for automatically generating masks mainly include RCNN [5], which was subsequently replaced by more powerful methods, such as the Segment Anything Model (SAM) method proposed by Meta [8]. However, SAM has its limitations in image blending, as it tends to capture all objects [17] in a particular picture, whereas image blending requires a mask for one specific object in an image.

In our work, we reconstruct the deep image blending algorithm [20] using Pixel Aggregation Network (PAN) and a new loss function, which iteratively improves the image blending process [11,15]. To address the limitations of artificial clipping masks, we apply DINO and use target text to distinguish our desired objects, resulting in better image blending. However, there remains a potential problem here that other researchers may not have noticed, namely that precise segmentation of objects may not always yield the best results. The blended image may lose important details if the mask image does not contain relevant information about the original image. We apply a classic erosion-dilation step to address this challenge, which helps preserve important details in the original image for better-blending results. Evaluation metrics including PSNR, SSIM, and MSE on multiple image datasets show that our hybrid image can outperform previous models GP-GAN, Poisson Image, etc. Our experiments show that combining DINO (DETR with improved denoising anchor boxes) [18] and SAM can generate more accurate masks than RCNN. The images generated by the hybrid algorithm have consistent brightness, higher resolution rate and smoother gradients. The whole process can be simple as shown in Fig. 1.

Our work has mainly contributed to the following three aspects:

- We propose an automatic mask generation method based on object detection and SAM segmentation, where erosion and dilation operations are used to manipulate the resulting mask for better image blending.
- We propose a new loss function, called saturation loss, for deep image blending algorithms to account for sudden changes in contrast at the seams of blended images.
- We use PAN to process blended images, solving the problem of low image resolution and distortion of individual pixel grey values of blended images.

The remainder of this paper is organized as follows: in Sect. 2, we introduce previous related work on image segmentation and detection and image blending; Sect. 3 gives a detailed introduction to our method; in Sect. 4, our algorithm will be compared with other algorithms. Subsequently, we summarize our algorithm and possible future research directions.

2   Related Work

2.1   Image Blending

The simplest approach to image blending (Copy-and-paste) is to directly copy pixels from the source image and paste them onto the destination image. Still, this technique can lead to noticeable artefacts due to sudden intensity changes at the composition's boundaries. Therefore, researchers have proposed advanced methods that use complex mathematical models to integrate source and destination images and improve the overall aesthetics of the blended image.

A traditional approach to image blending is Poisson image editing, first cited by Perez et al. [13], which exploits the concept of solving Poisson's equation to blend images seamlessly and naturally. This method transforms the source and target images into the gradient domain, thus obtaining the gradient of the target image. Another image blending technique is the gradient domain blending algorithm, proposed by Perez et al. [10]. The basic idea is to decompose the source and target images into gradient and Laplacian domains and combine them using weighted averaging. Deep Image Blending [20] refers to the gradient in Poisson image editing, and turns the gradient into a loss function, plus the loss of texture and content fidelity, resulting in higher image quality than Position image editing. Our work further optimizes Deep Image Blending to generate more realistic images.

Image inpainting is a technique [9] that uses learned semantic information and real image statistics to fill in missing pixels in an image, done by deep learning models trained on large datasets. However, this technique does not perform well in large-scale image mixing. Besides image blending, several other popular image editing tools include image denoising, image super-resolution, image inpainting, image harmonization, style transfer, etc. With the rise of generative adversarial networks (GANs), these editing tasks have become increasingly important in improving the quality of generated results, such as GP-GAN [1,2,7]. Image super-resolution involves using deep learning models to learn image texture patterns and upsampling low-resolution images to high-resolution images. PAN is often used to continuously refine and enhance the details in the image through a series of iterations. This process involves leveraging the power of neural networks to make predictions and complement the image's resolution. This trained model then predicts high-resolution details missing from the low-resolution input. We use PAN to increase the image's resolution and make the blending image more realistic.

2.2 Image Segmentation and Detection

In the past, Regions with CNN Features (RCNN) [5] was the best region-based method for semantic segmentation based on object detection. RCNN can be used on top of various CNN structures and shows significant performance improvements over traditional CNN structures. Fast R-CNN is an improved version of RCNN that improves speed and accuracy by sharing the feature extraction process. The YOLO [14] algorithm treats the target detection task as a regression problem and realizes real-time target detection by dividing the image into grids and predicting the bounding box and category in each grid cell. Still, the accuracy of its detection target localization is relatively low.

DINO is an advanced object detection and segmentation framework used by our pipeline to identify the most important objects from segmented images via SAM [4]. DINO introduces improved anchor box and mask prediction branches to implement a unified framework to support all image segmentation tasks, including instance, bloom and semantic segmentation. Mask DINO is an extension of DINO that leverages this architecture to support more general image segmentation tasks. The model is trained end-to-end on a large-scale dataset and can accurately detect and segment objects in complex scenes. Mask DINO extends DINO's architecture and training process to support image segmentation tasks, making it an effective tool for segmentation applications.

Essentially, an ideal image segmentation algorithm should be able to recognize unknown or new objects by segmenting them from the rest of the image. SegNet [3] achieves pixel-level image segmentation through an encoder-decoder structure, which is difficult to handle small objects and requires a lot of computation. The core idea of CRFasRNN [12] is to combine conditional random fields (CRF) and recurrent neural networks (RNN). Compared with SegNet, CRFasRNN has less calculation and finer segmentation results. Facebook Meta AI has developed a new advanced artificial intelligence model called Segmented Anything Model (SAM) [8] that can extract any object in an image with a single click. SAM leverages cutting-edge deep learning techniques and computer vision algorithms to accurately segment any object in an image. SAM can efficiently cut out objects from any type of image, making the segmentation process faster and more precise. This new technology is a breakthrough in computer vision and image processing because it can save a lot of time and effort when editing images and videos.

3   Proposed Approach

In this section, we first describe how to automatically generate a synthetic mask; then how to blend the source and target images to produce the initial result; finally, we propose a new saturation loss to refine the blended result image. The overall framework is shown in Fig. 2.

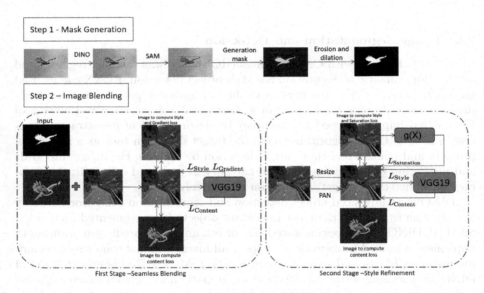

Fig. 2. The overall framework of our method: 1) generate the mask of the object through DINO and SAM; 2) use VGG19 to optimize the content loss, gradient loss and style loss to obtain an initial blending image; 3) integrate the PAN algorithm, and replace the gradient loss at the same time into saturation loss for more realistic blending image

3.1   Mask Generation

Below we describe in detail how to automatically generate high-quality masks. First, we use DINO to detect a specific region in an image based on a textual description and generate a box around that object, as shown in Fig. 4 (the input word is "bird"). We then feed the frame into the SAM and extract the mask of the region. Combining DINO and SAM, we solve the problem that SAM can only segment objects but cannot select specific objects. In Fig. 5, we can observe that our algorithm can precisely identify the desired object in the image and generate an accurate mask. This method saves time and effort compared to traditional manual editing methods. It is worth noting that after obtaining the yellow-purple Mask, the Mask image needs to be converted into a black-and-white image.

In object detection, DINO has better performance than RCNN due to its self-supervised learning method and the advantages of the Transformer network, which enables it to better capture global features. For semantic segmentation, SAM can better capture key information in images and achieve more accurate pixel-level object segmentation through its multi-scale attention mechanism and attention to spatial features. Therefore, the mask generation achieved by the combination of these two methods outperforms traditional convolutional neural networks, as shown in Fig. 3. We use IOU to measure the quality of the mask. It can be seen from the figure that the combination of DINO and SAM not only has a better segmentation effect visually, but also outperforms RCNN in terms of IOU.

Finally, an erosion and dilation operation [6] is performed on the mask to better refine the mask. The overall process of the mask operation is as follows: First, an etch operation is applied to shrink the sharp and edge areas of the mask. Second, a dilation operation is performed on the eroded mask to expand its edges to ensure that the mask completely covers the target object and maintains a smooth boundary. Through the above operations on the mask, we can improve the coverage of the mask and make the mask input of the image blending algorithm more accurate. In the image fusion stage, the processed mask can also carry part of the source image information, making the final fused image more natural.

3.2   Seamless Blending

Seamless blending is the first stage of deep image mixing [20], the style loss L_{style} is used to transfer the style information of the background image to the resulting image, it can calculate the texture difference between the generated image and the background, making the generated image style unified, more harmonious and authentic. The content loss L_{cont} measures the pixel difference between the object in the fusion image and the object in the source image, which is used to ensure the fidelity of the content in the image, and avoid the content smearing caused by the style transfer and cause the loss of details. Gradient loss L_{grad} is used to compute the pixel-wise difference between the source image and the target image on the edge for smooth blending of edges. At this stage, through continuous iteration, the fusion edge will gradually become smoother, and the texture of the fusion object will gradually resemble the background image without losing any details. Specially, we will optimize the following loss function in the first stage

$$\mathcal{L}_1 = \lambda_{grad}L_{grad} + \lambda_{cont}L_{cont} + \lambda_{style}L_{style}, \tag{1}$$

where λ_{grad}, λ_{cont} and λ_{style} respectively represent the weight of each loss.

After the first stage of processing, the blended edge of the object in the blending image is very smooth, but there are still significant differences between the blended object and the background in terms of similarity and illuminance, which may affect the quality and realism of the image. To solve this problem, we need to continue to optimize the image in terms of style refinement.

3.3   Style Refinement

In the second stage, although we use the same network architecture, it achieves a different task: to achieve the consistency of the background and source image in the blended image and to ensure that the generated image is more realistic, so we propose the new Saturation loss to replace the previous gradient loss, which computes the difference in saturation gradient between the background image and the blended image to account for blended images with unrealistic lighting and large contrast differences in the medium. The texture of the object in the final generated result image is consistent with the source image.

Fig. 3. Comparison of traditional RCNN algorithm and DINO+SAM algorithm in segmentation tasks

Since the basic brightness of the mixed background image is different from that of the target image, there will be a certain contrast difference after mixing, so that the naked eye can perceive the existence of the mixing operation. At different colour coordinates, each layer of the fused image behaves differently. We observed obvious differences at the fusion seams of R, G, and B channels under RGB colour coordinates. However, after converting the fused image to the HSV colour model, the Saturation channel of the blended image will have a sudden change in the saturation value at the edge where the source image and the target image are blended, as shown in Fig. 6.

To solve the above-mentioned sudden saturation problem of the blending boundary, we propose a new saturation loss to make the saturation change of

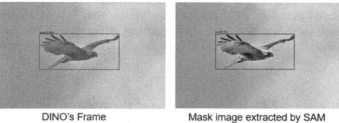

DINO's Frame   Mask image extracted by SAM

Fig. 4. Object detection by DINO and segmentation by SAM

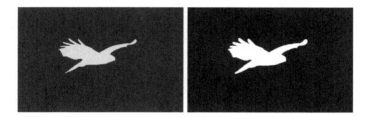

Fig. 5. Mask generation by SAM: 1) left: before dilation-corrosion operation; 2) right: after dilation-corrosion operation

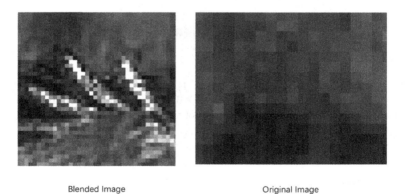

Blended Image   Original Image

Fig. 6. Pixel comparison of the S channel of both the fusion image and the original image on the blending boundary

the blended image more realistic and natural. The detailed process is shown in the Fig. 7: First, we convert the RGB colour coordinates of the background image and the blended image to HSV colour coordinates, and extract their saturation channels; then calculate the saturation gradient difference of the mixed image and the background image (H and W are height and width respectively)

$$\mathcal{L}_{sat} = \frac{g(M) - g(B)}{HW}, \tag{2}$$

where the function $g(X)$ represents the sum of the gradient values along the row and column directions on the saturation channel X_s of the image X

$$g(X) = \sum_{i=1}^{H} \sum_{j=1}^{W} |X_s(i+1,j) - X_s(i,j)| + |X_s(i,j+1) - X_s(i,j)|. \tag{3}$$

. Finally, we optimize the following loss under the original framework

$$\mathcal{L}_2 = \lambda_{sat} L_{sat} + \lambda_{cont} L_{cont} + \lambda_{style} L_{style}, \tag{4}$$

where λ_{sat}, λ_{cont}, λ_{style} are the weight coefficients of each item.

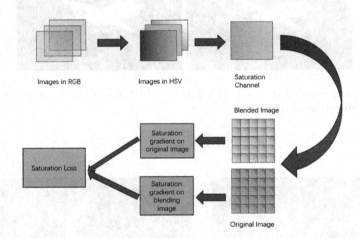

Fig. 7. Calculation steps of saturation loss: 1) convert the image from RGB space to HSV space and extract the S channel; 2) calculate the saturation gradient on the original image and the blending image respectively; 3) solve the average difference of the two saturation gradients value

4   Experiments

In this section, we describe the experimental setup, compare our method with other classical methods, and conduct ablation experiments on our method. The SSIM and PSNR scores of each image are tested multiple times to ensure data stability.

4.1 Experiment Setup

The experimental settings are shown in Table 1 below. In the loss function, the weight of the gradient loss in the first stage is set higher, because the main goal of the first stage is to solve the gradient problem and make the blending edge smoother. In the second stage, in order to solve the lighting and texture problems of the fused image, the weight of the style loss and saturation loss functions is set to 10^5.

Table 1. Experimental hyperparameter settings

Parameter	λ_{grad}	λ_{style}	λ_{cont}	λ_{sat}
Stage 1	10^4	10^3	1	0
Stage 2	0	10^5	1	10^5

4.2 Result Comparison

Fig. 8. Comparison of the effect of image blending between our method and other methods on the same input

As shown in Fig. 8, the size of all the images is 512×512, the copy-paste method copies the source image to the corresponding location on the destination. Deep reconciliation requires training a neural network to learn visual patterns from a set of images and use them to create realistic compositions or remove unwanted elements. Poisson blending computes the gradients of two images and minimizes their difference to seamlessly blend two images. The technique preserves the overall structure of an image while describing the flow of colours from one image to another. Finally, GP-GAN is a generative adversarial network (GAN) that uses a pre-trained generator network to generate high-quality images similar to the training data. Generator networks are pretrained on large image datasets and then fine-tuned on smaller datasets to generate higher-quality images. Unfortunately, these methods produce results with unrealistic borders and lighting. As can be seen from the qualitative results in Fig. 8, our algorithm produces the most visually appealing results for mixing borders, textures, and colour lighting.

Copy-paste methods for image fusion require precise control over the alignment of the target image to the background image, otherwise, it will lead to obvious incongruity and artefacts, while the inconsistency of image style will make the result have obvious artificial boundaries. GP-GAN produces worse visual results under mixed boundary and coloured lighting, where overall colours are darker and edges are not handled well. While it brings rich colour to the raw edges of an image, it introduces inconsistencies in style and texture. Compared with these algorithms, our method erodes and dilates edges, which prevents jagged edges or visible artefacts, and the two-stage algorithm adds more style and texture to the blended image.

Furthermore, we refer to the experimental protocol of the deep hybrid image algorithm and conduct comparative experiments on 20 sets of data. We compared the results of these methods on indicators such as PSNR (peak signal-to-noise ratio), SSIM (structural similarity) and MSE (mean square error), as shown in Table 2. It can be seen that the average performance of our method achieves the best results on PNSR and SSIM, while MSE is slightly worse than Poisson Blending. This is mainly because our method does not simply migrate the source image to the target image, but further refines the blended result of style and saturation consistency to make the generated picture more realistic, resulting in a slightly larger fitting error MSE. Compared with images from Deep Image Blending, images generated by our optimized model perform better in terms of PSNR, SSIM, and MSE, which also illustrates the superiority of our model.

Table 2. Quantitative comparison of average results between our method and other methods on PSNR, SSIM and MSE metrics

Method	PNSR	SSIM	MSE
GP-GAN	19.94	0.73	833.81
Poisson Blending	21.77	0.71	**472.46**
Deep Image Blending	22.11	0.72	712.64
Copy and Paste	17.98	0.57	866.41
Ours	**23.59**	**0.78**	617.51

4.3   Ablation Study

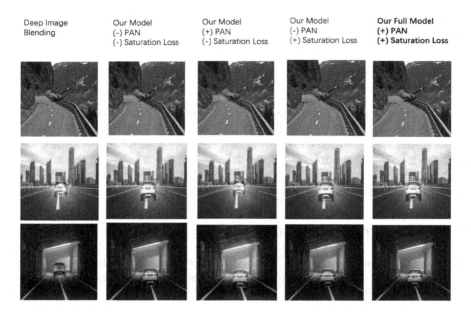

Fig. 9. The results (512×512) of the ablation experiment: (+) and (-) respectively indicate that a certain part of the algorithm participates or does not participate

We take three images as an example to conduct ablation experiments to analyze the role of PAN composition and saturation loss in our method. From Fig. 9, we can observe:

PAN in blending refinement We will keep only the saturation loss and remove the PAN component from the model. Experimental results show that the image resolution will be obviously reduced, resulting in a loss of clarity.

Saturation loss in the second stage If you remove the saturation loss and keep the PAN module, you will find that the original image still maintains the original saturation in the resulting image, which is not consistent with the background.

Both PAN and saturation loss When using both components at the same time, you will find that the generated results are more realistic, and the style of the source and target images is more consistent, especially in the blending edge area.

Table 3. Quantitative results of our algorithm's ablation experiments on 3 sets of images, where the three values in each cell represent the results on different images

Metrics	PSNR	SSIM	MSE
Baseline	17.85/18.32/17.99	0.57/0.61/0.67	718.20/661.94/594.32
+PAN	20.77/20.16/22.09	**0.74**/0.69/0.79	**543.98**/721.69/**401.49**
+Saturation Loss	19.79/19.94/22.29	0.6/0.62/0.81	681.91/658.20/383.38
+PAN+Saturation Loss	**29.57/24.58/23.64**	0.72/**0.83/0.79**	612.95/**568.47**/402.93

Table 3 further gives the quantitative results of the ablation experiments. The three numbers in the grid represent the experimental results for 3 images. As far as PNSR is concerned, the combination of PAN+Saturation performed best. Regarding SSIM, just using PAN results may be better. Finally, MSE using both PAN and Saturation does not always perform best, since our method does not simply fit a merge of source and target images.

5   Conclusion and Future Work

In our work, we address the low accuracy and low efficiency of manually cutting masks by generating masks through object detection and segmentation algorithms. Specifically, we combine DINO and SAM algorithms to generate masks. Compared with the traditional RCNN algorithm, the mask of this algorithm can cover objects better and has a stronger generalization ability. We perform erosion and dilation operations on the mask to avoid sharp protrusions in the mask. However, there may be a limitation with this part. If one object A overlaps with another object B, performing image blending after corroding and expanding the mask of A may introduce B's information into the blending process, which may lead to unreal results. Finally, we also propose a new loss, called saturation loss, to address brightness distortion in generated images. Results on multiple image datasets show that our method can outperform previous image fusion methods GP-GAN, Poisson Image, etc. Future work includes proposing new evaluation criteria to better reflect human perception and aesthetics to improve the objectivity and accuracy of the model. Another potential research direction is how to deal with object occlusion in image fusion.

6   Authorship Statement

Haochen Xue proposed the idea of automatic mask generation and constructed the overall framework of the image fusion project. Mingyu Jin uses PAN to process images, resulting in higher-quality images. Chong Zhang proposed a new loss and used erosive dilation to optimize the mask. Yuxuan Huang completed the comparison experiment and ablation experiment. Qian Weng participated in the revision of the article and undertook the polishing of the article. Xiaobo Jin supervised this work and made a comprehensive revision and reconstruction.

Acknowledgments. This work was partially supported by the "Qing Lan Project" in Jiangsu universities, National Natural Science Foundation of China under No. U1804159 and Research Development Fund with No. RDF-22-01-020.

References

1. Alsaiari, A., Rustagi, R., Thomas, M.M., Forbes, A.G., et al.: Image denoising using a generative adversarial network. In: 2019 IEEE 2nd International Conference on Information and Computer Technologies (ICICT), pp. 126–132. IEEE (2019)
2. Arjovsky, M., Bottou, L.: Towards principled methods for training generative adversarial networks. arXiv:1701.04862 (2017)
3. Badrinarayanan, V., Kendall, A., Cipolla, R.: Segnet: a deep convolutional encoder-decoder architecture for image segmentation. IEEE Trans. Pattern Anal. Mach. Intell. **39**(12), 2481–2495 (2017)
4. Chen, J., Yang, Z., Zhang, L.: Semantic segment anything (2023). www.github.com/fudan-zvg/Semantic-Segment-Anything
5. Girshick, R., Donahue, J., Darrell, T., Malik, J.: Rich feature hierarchies for accurate object detection and semantic segmentation. In Proceedings of the IEEE Conference on Computer Vision and Pattern Recognition, pp. 580–587 (2014)
6. Gonzalez, R.C.: Digital image processing. Pearson Education, India (2009)
7. Goodfellow, I., et al.: Generative adversarial networks. Commun. ACM **63**(11), 139–144 (2020)
8. Kirillov, A., et al.: Segment anything. arXiv:2304.02643 (2023)
9. Ledig, C., et al.: Photo-realistic single image super-resolution using a generative adversarial network. In: Proceedings of the IEEE Conference On Computer Vision and Pattern Recognition, pp. 4681–4690 (2017)
10. Leventhal, D., Gordon, B., Sibley, P.G.: Poisson image editing extended. In: Finnegan, J.W., McGrath, M. (eds.) International Conference on Computer Graphics and Interactive Techniques, SIGGRAPH 2006, Boston, Massachusetts, USA, July 30 - August 3, 2006, Research Posters, p. 78. ACM (2006)
11. Liu, S., Qi, L., Qin, H., Shi, J., Jia, J.: Path aggregation network for instance segmentation. In: Proceedings of the IEEE Conference on Computer Vision and Pattern Recognition, pp. 8759–8768 (2018)
12. Monteiro, M., Figueiredo, M.A.T., Oliveira, A.L.: Conditional random fields as recurrent neural networks for 3d medical imaging segmentation. arXiv:1807.07464 (2018)
13. Pérez, P., Gangnet, M., Blake, A.: Poisson image editing. ACM Trans. Graph. **22**(3), 313–318 (2003)
14. Redmon, J., Divvala, S., Girshick, R., Farhadi, A.: You only look once: unified, real-time object detection. In: Proceedings of the IEEE Conference on Computer Vision and Pattern Recognition, pp. 779–788 (2016)
15. Wang, K., Liew, J.H., Zou, Y., Zhou, D., Feng, J.: Panet: few-shot image semantic segmentation with prototype alignment. In proceedings of the IEEE/CVF International Conference on Computer Vision, pp. 9197–9206 (2019)
16. Wu, H., Zheng, S., Zhang, J., Huang, K.: GP-GAN: towards realistic high-resolution image blending. In: Proceedings of the 27th ACM International Conference on Multimedia, pp. 2487–2495 (2019)
17. Xie, G., et al.: SAM: self-attention based deep learning method for online traffic classification. In: Proceedings of the Workshop on Network Meets AI & ML, pp. 14–20 (2020)

18. Zhang, H., et al.: DINO: DETR with improved denoising anchor boxes for end-to-end object detection. arXiv:2203.03605 (2022)
19. Zhang, H., Han, X., Tian, X., Jiang, J., Ma, J.: Image fusion meets deep learning: a survey and perspective. Inf. Fusion **76**, 323–336 (2021)
20. Zhang, L., Wen, T., Shi, J.: Deep image blending. In: Proceedings of the IEEE/CVF Winter Conference on Applications of Computer Vision, pp. 231–240 (2020)

Design of Memristor-Based Binarized Multi-layer Neural Network with High Robustness

Xiaoyang Liu[1], Zhigang Zeng[2], and Rusheng Ju[1(✉)]

[1] College of Systems Engineering, National University of Defense Technology, Changsha, China
liuxiaoyang@nudt.edu.cn, jrscy@sina.com
[2] School of Artificial Intelligence and Automation, Huazhong University of Science and Technology, Wuhan, China
zgzeng@hust.edu.cn

Abstract. Memristor-based neural networks are promising to alleviate the bottleneck of neuromorphic computing devices based on the von Neumann architecture. Various memristor-based neural networks, which are built with different memristor-based layers, have been proposed in recent years. But the memristor-based neural networks with full precision weight values are affected by memristor conductance variations which have negative impacts on networks' performance. However, binarized neural networks only have two kinds of weight states, so the binarized neural networks built by memristors suffer little from the conductance variations. In this paper, a memristor-based binarized fully connected layer and a memristor-based batch normalization layer are designed. Then based on the proposed layers, the memristor-based binarized multi-layer neural network is built. The effectiveness of the network is substantiated through simulation experiments on pattern classification tasks. The robustness of the network is also explored and the results show that the network has high robustness to conductance variations.

Keywords: Artificial intelligence · Neural networks · Binarized neural network · Memristor

1 Introduction

Memristor-based neural networks (MNNs) have the ability of in-memory computing, alleviating the "memory wall" problem existing in traditional von Neu-

This work was supported by the Natural Science Foundation of China under Grants 62206306, 62103425, U1913602, and 61936004, the Natural Science Foundation of Hunan Province under Grants 2022JJ40559, the National Key R&D Program of China under Grant 2021ZD0201300, the Innovation Group Project of the National Natural Science Foundation of China under Grant 61821003, and the 111 Project on Computational Intelligence and Intelligent Control under Grant B18024.

B. Luo et al. (Eds.): ICONIP 2023, CCIS 1962, pp. 249–259, 2024.
https://doi.org/10.1007/978-981-99-8132-8_19

mann architecture-based computing devices [1–10]. They also have the characteristic of parallel computing, which could also accelerate the computing process of neural networks. But subject to manufacturing processes, memristors have conductance variations which means that the conductance could not be precisely adjusted, that is, the weight values could not be written precisely to the conductance value of memristors, resulting in conductance variations. The conductance variations cause inaccurate weight expression, leading to calculation errors. While binarized neural networks (BNN) [11] only have two kinds of weight states, and the two weight states could be represented by the high and low conductance states of the memristor, respectively. It is much easier to adjust the memristor conductance to the high and low states than to full precision values which are needed in traditional MNNs. So memristor-based BNNs suffer little from the conductance variations.

There have been some works on memristor-based BNNs [12–16]. In [12], a kind of MNN is designed based on memristor binary synapses. A memristor crossbar array is proposed in [13] to implement binarized neural networks. A binary memristor crossbar is tested in application of MNIST recognition in [14]. In [15], a BNN accelerator is built based on binary memristors, and the accelerator shows high robustness. A memristor-based BNN proposed in [16] shows high parallel computing capability. However, previous works have mostly focused on concept validations, rather than a design of memristor-based binarized networks.

The main contributions of this paper are summarized as follows

1. A memristor-based binarized fully connected (FC) layer and a memristor-based batch normalization (BN) layer are designed.
2. Based on these layers, a memristor-based binarized multi-layer neural network (MBMNN) is built.
3. The effectiveness of MBMNN is substantiated through simulation experiments on pattern classification tasks.

MBMNN is robust to conductance variations because only two states are used. The adjustment of the memristor conductance of MBMNN is also convenient compared to MNNs with full precision conductance values for MBMNN does not need precise writing voltages to adjust the conductance. MBMNN aims to accelerate the inference process of neural networks.

2   Memristor-Based Layers

In this section, the adopted memristor model is introduced, and based on the memristor model, the memristor-based binarized FC layer and the BN layer are designed. The binary activation function is also introduced.

2.1   Memristor Model

The memristor is an electronic device whose resistance or conductance could be adjusted by applying external stimuli [17,18]. The memristor model used here is the AIST memristor model depicting the behavior of AIST memristor [19,20]. Other memristor with voltage threshold and characteristic of fast-switching is also appropriate. In the AIST model

$$v\left(t\right) = i\left(t\right) R\left(t\right) \tag{1}$$

$$R\left(t\right) = R_{on} x\left(t\right) + R_{off}\left(1 - x\left(t\right)\right), \tag{2}$$

where t is the time variable, $v(t)$ is the voltage applied to the memristor, $i(t)$ is the current flowing through the memristor, $R(t)$ is the resistance of the memristor, $x(t) = \frac{w(t)}{D}$ where $w(t)$ is the internal state variable and D is the thickness, and R_{on} and R_{off} are the minimum and the maximum resistance values, respectively.

$$\frac{\mathrm{d}x(t)}{\mathrm{d}t} = \begin{cases} \mu_v \frac{R_{on}}{D^2} \frac{i_{off}}{i(t) - i_0} f\left(x(t)\right), & 0 < V_{on} < V(t) \\ 0, & V_{on} \le v(t) \le V_{off} \\ \mu_v \frac{R_{on}}{D^2} \frac{i(t)}{i_{on}} f\left(x(t)\right), & V(t) < V_{off} < 0 \end{cases} \tag{3}$$

where μ_v is the average ion mobility, i_{on}, i_{off}, and i_0 are constants, and $f\left(x(t)\right)$ is the window function. V_{on} is the positive threshold voltage and voltages larger than V_{on} will increase the conductance, and V_{off} is the negative threshold voltage and voltages smaller than V_{off} will decrease the conductance.

2.2 Memristor-Based Binarized FC Layer

The memristor-based binarized FC layer is shown in Fig. 1. In the inference phase, S_1 is closed and input voltage V_m^i ($m = 1, 2, \ldots, M$) is input to the layer. V_b is the bias voltage and V_n^o ($n = 1, 2, \ldots, N$) is the output voltage which is the input voltage to the activation function or the next layer. In the forward propagation

$$V_f = -R_f \left(\sum_{m=1}^{M} \frac{V_m^i}{R_s} + \frac{V_b}{R_s} \right), \tag{4}$$

and the current of the each column is

$$I_n = \sum_{m=1}^{M} \frac{V_m^i}{R_{mn}} + \frac{V_b}{R_{bn}} + \frac{V_f}{R_f}$$

$$= \sum_{m=1}^{M} V_m^i \cdot (G_{mn} - G_s) + V_b (G_{bn} - G_s), \tag{5}$$

where $n = 1, 2, \ldots, N$, $V_m^i = V_r \cdot x_m$ where V_r is the read voltage and x_m is the real input value, R_{mn} and G_{mn} is the resistance and the conductance values of the memristor-based synapse in the cross point of the mth row and the nth column, respectively. The weight is represented by $G_{mn} - G_s$, where $G_s = \frac{G_{on} + G_{off}}{2}$. When $G_{mn} = G_{on}$, $G_{mn} - G_s = \frac{G_{on} - G_{off}}{2}$ represents the weight value of $+1$, and when $G_{mn} = G_{off}$, $G_{mn} - G_s = \frac{G_{off} - G_{on}}{2}$ represents the weight value of -1. The output voltage of each column is

$$V_n = -I_n R_a. \tag{6}$$

When adjusting the conductance of memristors in the nth column, S_1 is open and S_2 is closed, and row writing voltage V_{rm}^w and column writing voltages V_{cn}^w and V_b^w are applied to the memristor crossbar. Writing voltages are now applied to the corresponding memristor to update its conductance to the high or the low state.

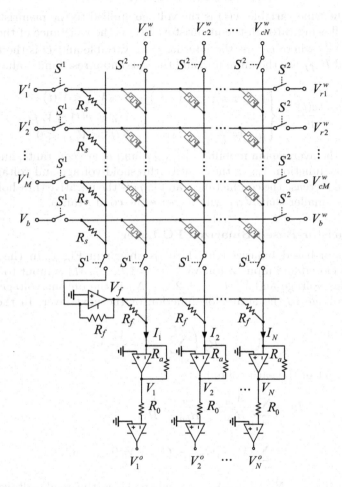

Fig. 1. The memristor-based binarized FC layer. V_m^i ($m = 1, 2, \ldots, M$) is the input and V_n^o ($n = 1, 2, \ldots, N$) is the output of the layer. V_{rm}^w, V_{cn}^w, and V_b^w are writing voltages that update the memristor conductance.

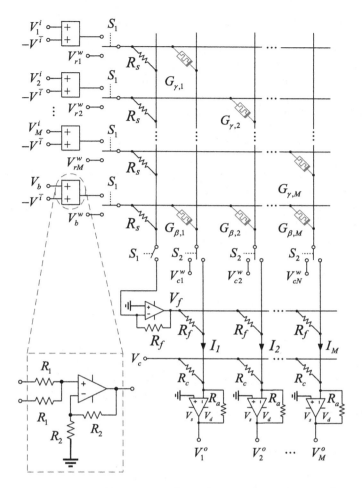

Fig. 2. The memristor-based BN layer. V_1^i to V_M^i are input voltages, $V^{\bar{i}}$ is the average value of input voltages of one mini-batch. $G_{\gamma,m}$ ($m = 1, 2, \ldots, M$) is conductance of the memristor representing the mth value of parameter vector γ and $G_{\beta,m}$ represents the value of the parameter β. V_{rm}^w, V_{cn}^w, and V_b^w are writing voltages that update the memristor conductance.

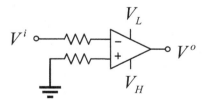

Fig. 3. The binary activation function.

2.3   Memristor-Based Ba1tch Normalization Layer

Batch normalization operation is used to normalize outputs of middle layers. Parameters in the BN layer of BNN are full precision values so the memristor conductance here is not binary and middle conductance states are also used. The memristor-based BN layer is shown in Fig. 2.

The output current of each column is

$$I_m = \left(V_m^i - V^{\bar{i}} \right) \left(G_{\gamma,m} - G_s \right) + V_b \left(G_{\beta,m} - G_s \right), \tag{7}$$

where $m = 1, 2, \ldots, M$, V_m^i is the mth input voltage, V_b is the bias voltage, $G_{\gamma,m}$ and $G_{\beta,m}$ are the memristor conductance value representing the value of parameters of γ and β, respectively, and $V^{\bar{i}}$ is the average value of input voltages of one mini-batch which is

$$V^{\bar{i}} = \frac{V_r}{Z} \sum_{z=1}^{Z} x_z, \tag{8}$$

where Z is the number of input patterns in one mini-batch. The output voltage is

$$V_m^o = -I_m R_a, \tag{9}$$

where Z is the number of samples in one mini-batch.

The memristor conductance is updated by opening S_1 and closing S_2. The row writing voltage V_{rm}^w and the column writing voltages V_{cm}^w and V_b^w are applied to the memristor in the mth row and the mth column. The pulse width of writing voltages can be determined by the lookup table which records the relation of conductance values and pulse widths of writing voltages.

2.4   Binary Activation Function

The binary activation function (BAF) can be realized by the comparer [21]. BAF circuit is shown in Fig. 3. V^i receives the output of the previous layer. Because the sign of the output voltage of the previous layer is the opposite of the actual value, V^i is input to the negative port of the comparer. The output is

$$V^o = \begin{cases} V_H, & V^i \leq 0 \\ V_L, & V^i > 0 \end{cases} \tag{10}$$

where V_H and V_L is the high and low voltage levels, respectively. BAF circuit can be placed at output ports of layers as needed.

3   Simulation Experiments

Because only two kinds of conductance states are used in the MBMNN, MBMNN suffers little from variations of memristor conductance and ex-situ training method is appropriate. MBMNN is built by the proposed FC and BN layers. The network structure is shown in Table 1. *FC* and *BN* are the FC layer and the BN layer, respectively, and the content represents the number of output units (FC layer and BAF), or the output channels (BN layer). The network model is first trained by software, and then the trained weights are downloaded to the memristors to perform inference tasks. The robustness of MBMNN to memristor conductance variations is also discussed.

Table 1. THE STRUCTURE OF MBMNN

Layer	Units
FC1	1024
BN2	1024
BAF	1024
FC2	10
BN2	10

3.1   Pattern Classification Result

The pattern classification simulation experiment is carried out on MNIST (Modified National Institute of Standards and Technology) [22] dataset. Image pixel values are first converted to input voltages through

$$V^i = xV_r, \tag{11}$$

where x is the pixel value.

The training dateset is divided into train set and validation set. Test error rate is obtained by the network with the best validation error rate. For clear illustration, weight values of every training epoch are wrote to MBMNN to obtain train and test errors. Results is shown in Fig. 4. The test error rate is about 1.98%. It can be seen that MBMNN has satisfying classification performance in classification tasks.

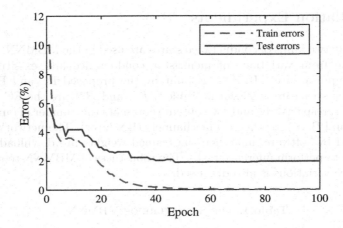

Fig. 4. Curves of train and test errors of MBMNN with training epochs.

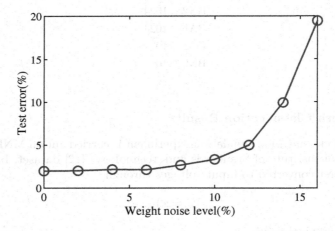

Fig. 5. The impact of different weight noise levels on test error rates of MBMNN.

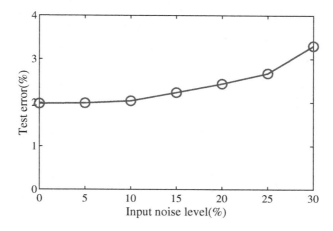

Fig. 6. The impact of different input noise levels on test error rates of MBMNN.

3.2   Impact of Noises

In practice, there are variations in the resistance or conductance adjustment of memristor, and other factors, such as the voltage supply and the change of temperature, also cause the conductance variations. The variations lead to weight noises. The supply voltages also may exist errors which lead to input noises. These noises result in the imprecise computing of MBMNN. To study the impact of the noises on MBMNN, Gaussian noises with expectation value of 0 and standard deviation of 2% to 16% with the interval of 2% of current resistance value are added as weight noises in the FC layer and the BN layer, and Gaussian noises expectation value of 0 and standard deviation of 5% to 30% with the interval of 5% of input voltages are added to the supply voltages as input noises. The conductance value and input voltage after adding noises are

$$G_{aft} = 1/\left(1/G_{bef} \times (1 + s)\right) = \frac{G_{bef}}{1 + s} \tag{12}$$

$$V_{aft}^i = V_{bef}^i (1 + s) \tag{13}$$

where G_{bef} and G_{aft} are conductance values before and after adding noises, respectively, and V_{bef}^i and V_{aft}^i are input voltages before and after adding noises, respectively. Impacts of weight noises and input noises are shown in Fig. 6 and Fig. 5, respectively.

To substantiate the robustness of MBMNN to conductance variations, comparison of test errors of MBMNN with binary conductance states and MNN with full precision conductance states are shown in Fig. 7. The two kinds of networks have the same network structure. It is seen that when the noise level is low, the latter has better performance. But with the noise level increases, its test error rises fast and exceeds MBMNN when the noise level exceeds about 12%.

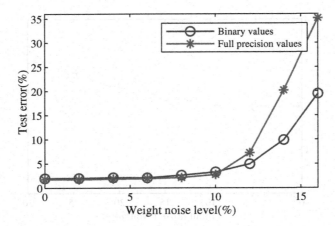

Fig. 7. Test errors of MNNs with binary conductance values and full precision conductance values under different conductance variation levels.

So MBMNN is more robustness to conductance variations. In another way, writing full precision weight values to memristor conductance is more complex than writing binary weight values.

4   Conclusions

The in-memory and parallel computing abilities of memristors devices make it promising in building acceleration devices for artificial neural networks. Memristor-based BN layer and binarized FC layer are presented in the paper, and base on these layers MBMNN is constructed. MBMNN suffers little from the conductance variations of memristors and is also easier to adjust conductance values compared with MNNs with floating full precision weight values. MBMNN is validated through pattern classification tasks. The simulation experiments also show that MBMNN has high robustness to memristor conductance variations, so it is applicable for hardware realization. Future works will focus on realizing more complex MNNs such as memristor-based convolutional neural networks.

References

1. Yakopcic, C., Hasan, R., Taha, T.: Memristor based neuromorphic circuit for ex-situ training of multi-layer neural network algorithms. In: International Joint Conference on Neural Networks, Killarney, Ireland, pp. 1–7. IEEE (2015)
2. Tanaka, G., Nakane, R., Yamane, T., et al.: Waveform classification by memristive reservoir computing. In: Liu, D., Xie, S., Li, Y., Zhao, D., El-Alfy, E.S. (eds.) ICONIP 2017. LNCS, vol. 10637, pp. 457–465. Springer, Cham (2017)
3. Li, C., Belkin, D., Li, Y., et al.: Efficient and self-adaptive in-situ learning in multilayer memristor neural networks. Nature Commun. **9**(1), 2385 (2018)

4. Yao, P., Wu, H., Gao, B., et al.: Fully hardware-implemented memristor convolutional neural network. Nature **577**(7792), 641–646 (2020)
5. Cao, Z., Sun, B., Zhou, G., et al.: Memristor-based neural networks: a bridge from device to artificial intelligence. Nanoscale Horizons **8**(6), 716–745 (2023)
6. Yi, S., Kendall, J., Williams, R., et al.: Activity-difference training of deep neural networks using memristor crossbars. Nat. Electron. **6**(1), 45–51 (2023)
7. Li, Y., Su, K., Zou, X., et al.: Research progress of neural synapses based on memristors. Electronics **12**(15), 3298 (2023)
8. Bak, S., Park, J., Lee, J., et al.: Memristor-based CNNs for detecting stress using brain imaging signals. IEEE Trans. Emerg. Topics Comput. Intell. 1–10 (2023). https://doi.org/10.1109/TETCI.2023.3297841
9. Liu, X., Zeng, Z., Wunsch, D., II.: Memristor-based LSTM network with in situ training and its applications. Neural Netw. **131**, 300–311 (2020)
10. Liu, X., Zeng, Z., Wunsch, D., II.: Memristor-based HTM spatial pooler with on-device learning for pattern recognition. IEEE Trans. Syst. Man Cybern. Syst. **52**(3), 1901–1915 (2022)
11. Courbariaux, M., Hubara, I., Soudry, D., et al.: Binarized neural networks: Training deep neural networks with weights and activations constrained to +1 or -1. arXiv preprint arXiv:1602.02830 1–11 (2016)
12. Secco, J., Poggio, M., Corinto, F.: Supervised neural networks with memristor binary synapses. Int. J. Circuit Theory Appl. **46**(1), 221–233 (2018)
13. Kim, Y., Jeong, W., Tran, S., et al.: Memristor crossbar array for binarized neural networks. AIP Adv. **9**(4), 045131 (2019)
14. Pham, K., Tran, S., Nguyen, T., Min, K.: Asymmetrical training scheme of binary-memristor-crossbar-based neural networks for energy-efficient edge-computing nanoscale systems. Micromachines **10**(2), 141 (2019)
15. Qin, Y., Kuang, R., Huang, X., Li, Y., Chen, J., Miao, X.: Design of high robustness BNN inference accelerator based on binary memristors. IEEE Trans. Electron Dev. **67**(8), 3435–3441 (2020)
16. Chen, J., Wen, S., Shi, K., Yang, Y.: Highly parallelized memristive binary neural network. Neural Netw. **144**, 565–572 (2021)
17. Strukov, D., Snider, G., Stewart, D., et al.: The missing memristor found. Nature **453**(7191), 80 (2008)
18. Dai, Yu., Li, C.: An expanded HP memristor model for memristive neural network. In: Huang, T., Zeng, Z., Li, C., Leung, C.S. (eds.) ICONIP 2012. LNCS, vol. 7667, pp. 647–653. Springer, Heidelberg (2012). https://doi.org/10.1007/978-3-642-34500-5_76
19. Li, Y., Zhong, Y., Zhang, J., et al.: Activity-dependent synaptic plasticity of a chalcogenide electronic synapse for neuromorphic systems. Sci. Rep. **4**, 4906 (2014)
20. Zhang, Y., Li, Y., Wang, X., et al.: Synaptic characteristics of Ag/AgInSbTe/Ta-based memristor for pattern recognition applications. IEEE Tran. Electron Dev. **64**(4), 1806–1811 (2017)
21. Zhang, Y., Wang, X., Friedman, E.: Memristor-based circuit design for multilayer neural networks. IEEE Trans. Circuits Syst. I Regul. Pap. **65**(2), 677–686 (2018)
22. Yann, L., Léon, B., Yoshua, B., et al.: Gradient-based learning applied to document recognition. Proc. IEEE **86**(11), 2278–2324 (1998)

FEDEL: Frequency Enhanced Decomposition and Expansion Learning for Long-Term Time Series Forecasting

Rui Chen, Wei Cui, Haitao Zhang[✉], and Qilong Han

College of Computer Science and Technology, Harbin Engineering University,
Harbin 150006, China
{zhanghaitao,hanqilong}@hrbeu.edu.cn

Abstract. Long-term Time Series Forecasting (LTSF) is widely used in various fields, for example, power planning. LTSF requires models to capture subtle long-term dependencies in time series effectively. However, several challenges hinder the predictive performance of existing models, including the inability to exploit the correlation dependencies in time series fully, the difficulty in decoupling the complex cycles of time series in the time domain, and the error accumulation of iterative multi-step prediction. To address these issues, we design a Frequency Enhanced Decomposition and Expansion Learning (FEDEL) model for LTSF. The model has a linear complexity with three distinguishing features: (i) an extensive capacity depth regime that can effectively capture complex dependencies in long-term time series, (ii) decoupling of complex cycles using sparse representations of time series in the frequency domain, (iii) a direct multi-step prediction strategy to generate the prediction series, which can improve the prediction speed and avoid error accumulation. We have conducted extensive experiments on eight real-world large-scale datasets. The experimental results demonstrate that the FEDEL model performs significantly better than traditional methods and outperforms the current SOTA model in the field of LTSF in most cases.

Keywords: LTSF · Series Decomposition · Frequency Enhancement

1 Introduction

Time series are widely available in energy, transportation, healthcare, finance, etc. Time series forecasting plays a crucial role in these areas, such as early warning, advanced planning, resource scheduling, etc. There is an urgent need to extend the prediction time period further in these practical applications. Over the past decades, solutions to time series forecasting have evolved from traditional statistical methods (e.g., ARIMA [1]) and machine learning techniques

This work was supported by the National Key R&D Program of China under Grant No. 2020YFB1710200 and Heilongjiang Key R & D Program of China under Grant No. GA23A915.

(e.g., GBRT [5]) to deep learning-based solutions (e.g., LSTM [9]), achieving better and better prediction performance. However, most existing models are only designed for short-term series forecasting, and their performance decreases sharply with increasing prediction length.

In the field of Long-term Time Series Forecasting (LTSF), three major challenges have hindered the performance of existing models. First, long-term time series have complex correlations, yet most existing studies process time series directly without considering how to tap into those unique correlations [13]. Following simple assumptions, some models based on series decomposition can only capture certain coarse and common features [15]. They need to analyze complex dependencies comprehensively. Second, a typical real-world time series is usually composed of various periods. For example, Fig. 1 shows a region's 3-year time series of hourly electricity loads. Stripping out different highly coupled cycles in the time domain is difficult. Third, traditional autoregressive or iterative multi-step (IMS) forecasting models produce significant error accumulation effects in forecasting. This inherent drawback makes most traditional models impossible to be applied to the field of LTSF.

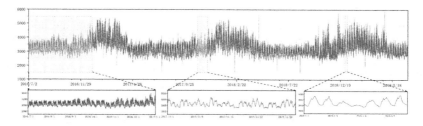

Fig. 1. We visualize the electricity load time series in a region of California to show diversified periods. In the upper part, we depict the whole time series with the length of three years, and in the bottom part, we plot three segments with the lengths of half year, one month, and one week, respectively.

To address the three aforementioned challenges, we propose a novel Frequency Enhanced Decomposition and Expansion Learning (FEDEL) model for LTSF. Inspired by the seasonal-trend decomposition method [4], FEDEL's core idea is to build a deep neural network that processes the trend and seasonal parts separately. For the trend part, we capture short-term local and long-term global dependencies. For the seasonal part, we perform a layer-by-layer extension enhancement to analyze the complex dependencies. Finally, the trend and seasonal parts are combined for end-to-end learning.

The contributions of this paper are summarized as follows:

- For complex correlations in long-term time series, we propose an extension architecture based on residual learning [6] and build a large-capacity depth system to model complex dependencies in long-term time series.

- For the difficulty of decoupling the cycles in the time domain, we transform the series to the frequency domain with the help of Fourier analysis. We decouple the series using a Frequency Enhanced Module (FEM) and enhance the appropriate cycles using learnable parameters.
- To address the error accumulation problem of IMS prediction, our model adopts a direct multi-step (DMS) prediction strategy in the LTSF process and performs vector output to avoid error accumulation.

We have conducted extensive experiments on publicly available real-world datasets, including univariate and multivariate time series forecasting. FEDEL demonstrates superior performance in handling LTSF. Compared to traditional models (LSTM, TCN), the prediction error can be reduced by more than 50% and even up to 80% in some datasets. FEDEL can also achieve better results on most datasets than the current SOTA LTSF models.

2   Related Work

Time series forecasting is a topic that has been studied for a long time, and various models have been well-developed over the decades. In the early stages, researchers developed simple and effective statistical modeling methods, such as Exponential Moving Average (EMA) [7], Autoregressive And Moving Average (ARMA) [16], Autoregressive Integrated Moving Average (ARIMA) [3], the unified state-space modeling approach as well as other various extensions. However, these statistically based methods only consider the linear dependencies of the future time series signal on past observations. Researchers use hybrid designs combining statistically-based methods with deep learning to deal with more complex dependencies. DeepAR [13] uses RNNs combined with autoregressive techniques to model the probability distribution of future series. LSTNet [9] introduces a CNN with a Recurrent and Recurrent-skip layer to capture both short-term and long-term temporal patterns. N-Beats [12] propose a deep neural architecture based on backward and forward residual links. Many studies based on Temporal Convolutional Network (TCN) [2,11] have attempted to model temporal causality using causal convolution. These models above mainly focus on modeling temporal relationships through circular linkage, temporal attention, or causal convolution.

Since its introduction, Transformer [14] has demonstrated excellent performance in sequence modeling architecture, such as Natural Language Processing (NLP) and Computer Vision (CV). However, applying Transformer to LTSF is computationally prohibitive because of the quadratic complexity of sequence length L in both memory and time. LogTrans [10] introduces the local convolution and LogSparse attention to Transformer, which reduces the complexity to $\mathcal{O}(L(logL)^2)$. Reformer [8] uses the Local-Sensitive Hashing (LSH) attention to reduce the complexity to $\mathcal{O}(LlogL)$. Informer [18] extends Transformer with KL-divergence-based ProbSparse attention and achieves $\mathcal{O}(LlogL)$ complexity. Autoformer [17] uses Auto-Correlation instead of self-attention mechanism to achieve efficient, sub-series connectivity and reduce the complexity to $\mathcal{O}(L log L)$. Fedformer [19] proposes Fourier-enhanced and Wavelet-enhanced blocks that can capture essential structures in time series through frequency domain mapping. Randomly selecting a fixed number of Fourier components reduces the model complexity to $\mathcal{O}(L)$.

3   Definition of LTSF

The problem of LTSF has been studied for a long time, but there has yet to be a clear definition for the concept of LTSF. The LTSF problem in this paper is defined as follows:

(i) With fixed size lookback window W.

(ii) Suppose that the input at moment t is $\mathcal{X}^t = \{x_1^t, x_2^t, ..., x_W^t \mid x_i^t \in \mathbb{R}^{d_x}\}$, the corresponding predicted output is $\mathcal{Y}^t = \{y_1^t, y_2^t, ..., y_L^t \mid y_i^t \in \mathbb{R}^{d_y}\}$. The output length L is at least equal to the input length W and can be much larger than the input length W.

(iii) The output can be either univariate($d_x = 1, d_y = 1$ or $d_x > 1, d_y = 1$) or multivariate($d_x > 1, d_y > 1$).

4   FEDEL Module

The general structure of FEDEL is shown in Fig. 2. All stacks have the same structure. Each stack has 3 output branches: the trend part $Trend_i$ (Eq. 1), the seasonal part $Season_i$ (Eq. 2) and the input to the next stack part X_i (Eq. 3)

$$Trend_i = SeriesDecomp\,(X_{i-1}) \tag{1}$$

$$Season_i = FrequencyEnhanced\,(X_{i-1} - Trend_i) \tag{2}$$

$$X_i = FrequencyEnhanced\,(X_{i-1} - Trend_i) + (X_{i-1} - Trend_i) \tag{3}$$

For the 0th stack, only $Trend_0$ and $Season_0$ are kept for subsequent processing, and X_1 is discarded. For the following stacks, only the X_i part is kept as input to $stack_{i+1}$, $Trend_i$, and $Season_i$ are discarded.

This section continues with the following:

(i) Time Feature Embedding Module (TFEM) embeds local positional information and global temporal features and provides a uniform representation.

(ii) Frequency Enhanced Module (FEM) that converts time series to the frequency domain and performs period enhancement using the Fourier transform.

(iii) Dependency Capture Module (DCM) captures both short-term local dependencies and long-term global dependencies of the time series.

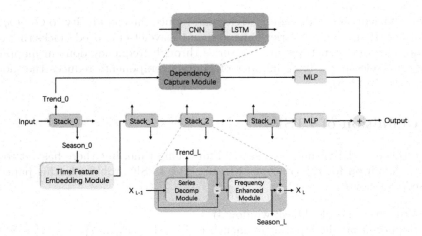

Fig. 2. FEDEL model architecture. The central part of FEDEL consists of n stacks. Stacks are stacked based on the residual network. Each stack contains Series Decomp Module (SDM) and Frequency Enhanced Module (FEB). The output of the layer 0 stack consists of two parts: the trend and seasonal. The trend part (Trend_0) will be processed by Dependency Capture Module (DCM). The seasonal part (Season_0) will go through Time Feature Embedding Module (TFEM) and then be processed by the following stacks. The processed trend and seasonal parts are passed through Multi-Layer Perceptron (MLP) separately and summed up as output.

4.1   Time Feature Embedding Module

The CNN and RNN models deal with the trend part and can capture temporal dependencies through the convolution or the recurrent structure without relying on embedding timestamps. For the processing of the seasonal part, the model itself cannot adequately capture temporal dependencies, so we artificially embed the timestamps. In the LSTF problem, capturing long-term dependencies requires global information like hierarchical time stamps (week, month, and year) and agnostic time stamps (holidays, events). We use a uniform representation to mitigate the issue. Figure 3 gives an intuitive overview.

Suppose we have t-th time series $\mathcal{X}^t$ and p types of time stamps. The feature dimension after input representation is d_{model}. We first preserve the local context by using a fixed position embedding:

$$\mathrm{PE}_{(pos,2j)} = \sin\left(pos/\left(2L_x\right)^{2j/d_{model}}\right), \tag{4}$$

$$\mathrm{PE}_{(pos,2j+1)} = \cos\left(pos/\left(2L_x\right)^{2j/d_{model}}\right), \tag{5}$$

where $j \in \{1, 2, \cdots, \lfloor d_{model}/2 \rfloor\}$. Each global timestamp is represented by a learnable embedding $SE_{(pos)}$. Each $SE_{(pos)}$ has a fixed size, up to 60, when minutes are used as the finest granularity. To align the dimension, we project the scalar context x_i^t into d_{model} vector u_i^t with 1-D convolutional filters (kernel width=3, stride=1).

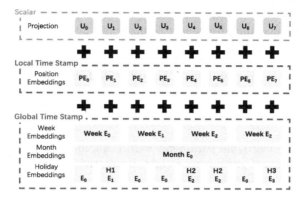

Fig. 3. Time Feature Embedding Module. The inputs' embedding consists of three separate parts: a scalar projection, the local timestamp (position), and global time stamp embeddings (minute, hour, week, month, holiday, etc.).

4.2   Frequency Enhanced Module

We use the Discrete Fourier Transform (DFT) to transform the time series from the time domain to the frequency domain, and then the result is converted back to the time domain by the Inverse Discrete Fourier Transform (IDFT).

Assuming that the length of the series in the time domain is N. The relationship between the samples in the time and frequency domains is shown in Eq. 6:

$$X_k = \sum_{n=0}^{N-1} x_n e^{-j2\pi kn/N} \quad k = 0, 1, 2, \ldots, N-1 \tag{6}$$

where X_k is the frequency domain representation of x_n, e is natural constant, j is imaginary unit.

Correspondingly, the IDFT is defined as in Eq. 7:

$$x_n = \sum_{k=0}^{N-1} X_k e^{j2\pi kn/N} \quad k = 0, 1, 2, \ldots, N-1 \tag{7}$$

where x_n is the time domain representation of X_k.

The complexity of DFT is $\mathcal{O}(N^2)$. With Fast Fourier Transform (FFT), the computation complexity can be reduced to $\mathcal{O}(NlogN)$.

The model is prone to overfitting for time series if all frequency components are simply retained, as many high-frequency components are unpredictable noise. And the model tends to be under-fitted if only the low-frequency part is simply retained, as some sharp changes in trends indicate the occurrence of important events. Therefore, we choose to uniformly sample the frequency components to take both low and high-frequency components into account and reduce the computation complexity.

The input of FEM is $x_{in} \in \mathbb{T}^{N \times D}$, and the Fourier transform of x_{in} is denoted as $Q \in \mathbb{F}^{N \times D}$. In frequency domain, only the uniformly sampled M frequency components are retained that denoted as $\tilde{Q} \in \mathbb{F}^{M \times D}$. As shown in Eq. 8:

$$\tilde{Q} = Select(Q) = Select(FFT(x_{in})) \tag{8}$$

$R \in \mathbb{F}^{M \times D \times D}$ is a learnable parameter with random initialization. Certain frequency components are given more weight during the training process. As shown in Eq. 9:

$$\tilde{Y} = \tilde{Q} \odot R \tag{9}$$

The operation of $\odot$ is defined as Eq. 10:

$$\tilde{Y}_{m,d_o} = \sum_{d_i=0}^{D} \tilde{Q}_{m,d_i} \cdot R_{d_i,d_o,m} \tag{10}$$

where $d_i = 1, 2, \ldots D$ is the input channel and $d_o = 1, 2, \ldots D$ is the output channel.

To keep the series dimension unchanged, $\tilde{Y}$ is then zero-padded to $\mathbb{F}^{N \times D}$. The final output of FEM is Eq. 11 (Fig. 4):

$$X_{out} = IFFT(Padding(\tilde{Q} \odot R)) \tag{11}$$

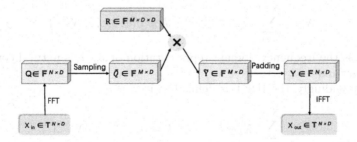

Fig. 4. Frequency Enhanced Module

4.3   Dependency Capture Module

CNN has strong local feature extraction ability because of convolution kernels. However, convolution kernels also limit the ability of CNN to handle long-term dependence on time series.

When applied to LSTF, the gradient of RNN cannot always be maintained in a reasonable range and will inevitably suffer from gradient disappearance or gradient explosion. The forget gate in LSTM will remove useless information to avoid the problems above. As shown in Fig. 5, We extract deep local dependencies from the original series by convolutional and pooling layers to obtain an effective representation. The representation is then fed into the LSTM to capture the long-term dependencies of the series. The DCM combines the advantages of CNN and LSTM and can reduce the arithmetic power and improve the model prediction accuracy.

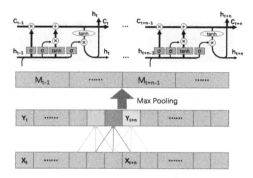

Fig. 5. Dependency Capture Module

5   Experiment

We extensively evaluate the proposed FEDEL on eight real-world datasets, covering four mainstream time series forecasting applications: energy, traffic, economics, and weather. The introduction of datasets is as follows:

ETT Electricity Transformer Temperature [18]. The dataset contains the data collected from electricity transformers, including load, oil temperature and other characteristics from July 2016 to July 2018. ETTm1 and ETTm2 recorded every 15 min, ETTh1 and ETTh2 recorded every hour.

Traffic[1] is a collection of hourly data from California Department of Transportation, which describes the road occupancy rates measured by different sensors on San Francisco Bay area freeways.

Electricity[2] dataset contains the hourly electricity consumption of 321 customers from 2012 to 2014.

Exchange records the daily exchange rates of eight different countries ranging from 1990 to 2016 [9].

Weather[3] is recorded every 10 min for 2020 whole year, which contains 21 meteorological indicators, such as air temperature, humidity,etc.

The overview of the datasets is shown in Table 1.

Table 1. Overview of dataset information

Datasets	ETTh(1, 2)	ETTm(1, 2)	Traffic	Electricity	Exchange	Weather
Variates	7	7	862	321	8	21
Timesteps	17,420	69,680	17,544	26,304	7,588	52,696
Granularity	1 h	15 min	1 h	1 h	1 day	10 min

We split all datasets into training, validation and test set in chronological order by the ratio of 6:2:2 for the ETT dataset and 7:1:2 for the other datasets.

5.1   Implementation Details

Our method is trained with the L2 loss, using the ADAM optimizer with an initial learning rate of 10^{-4}. The batch size is set to 32. The training process is early stopped within ten epochs. All experiments are repeated three times, implemented in PyTorch, and conducted on a single NVIDIA Quadro RTX 8000 (48GB) GPU. The number of stack layers in FEDEL is set to 3.

5.2   Evaluation Metrics

We use $\hat{y}$ to denote the predicted value of y, and use Mean Square Error (MSE = $\frac{1}{n}\sum_{i=1}^{n}(\mathbf{y}-\hat{\mathbf{y}})^2$) and Mean Absolute Error (MAE = $\frac{1}{n}\sum_{i=1}^{n}|\mathbf{y}-\hat{\mathbf{y}}|$) to evaluate model performance.

5.3   Main Results

To compare performances under different future horizons, we fix the input length I to 96 and define prediction lengths $O \in \{96, 192, 336, 720\}$. This setting precisely meets the definition of LTSF.

---

[1] http://pems.dot.ca.gov.
[2] https://archive.ics.uci.edu/ml/datasets/ElectricityLoadDiagrams20112014.
[3] https://www.bgc-jena.mpg.de/wetter/.

FEDEL is compared with FEDformer, Autoformer, Informe, Pyraformer, LogTrans, LSTM, and TCN. Here are the results in both the multivariate and univariate settings. Optimal results are highlighted in **bold**, and suboptimal results are highlighted by underlining.

Table 2. Multivariate LTSF results on eight datasets.

Methods		FEDEL		FEDformer		Autoformer		Informer		Pyraformer		LogTrans		LSTM		TCN	
Metric		MSE	MAE	MSE	MAE	MSE	MAE	MSE	MAE	MSE	MAE	MSE	MAE	MSE	MAE	MSE	MAE
Electricity	96	**0.179**	**0.294**	0.193	0.308	0.201	0.317	0.274	0.368	0.386	0.449	0.258	0.357	0.375	0.437	0.985	0.813
	192	**0.195**	**0.308**	0.201	0.315	0.222	0.334	0.296	0.386	0.386	0.443	0.266	0.368	0.442	0.473	0.996	0.821
	336	**0.213**	**0.322**	0.214	0.329	0.231	0.338	0.300	0.394	0.378	0.443	0.280	0.380	0.439	0.473	1.000	0.824
	720	**0.245**	**0.344**	0.246	0.355	0.254	0.361	0.373	0.439	0.376	0.445	0.283	0.376	0.980	0.814	1.438	0.784
Exchange	96	**0.122**	**0.256**	0.148	0.278	0.197	0.323	0.847	0.752	0.376	1.105	0.968	0.812	1.453	1.049	3.004	1.432
	192	**0.211**	**0.349**	0.271	0.380	0.300	0.369	1.204	0.895	1.748	1.151	1.040	0.851	1.846	1.179	3.048	1.444
	336	**0.341**	**0.453**	0.460	0.500	0.509	0.524	1.672	1.036	1.874	1.172	1.659	1.081	2.136	1.231	3.113	1.459
	720	**0.847**	**0.713**	1.195	0.841	1.447	0.941	2.478	1.310	1.943	1.206	1.941	1.127	2.984	1.427	3.150	1.458
Traffic	96	**0.568**	**0.360**	0.587	0.366	0.613	0.388	0.719	0.391	2.085	0.468	0.684	0.384	0.843	0.453	1.438	0.784
	192	**0.599**	**0.355**	0.604	0.373	0.616	0.382	0.696	0.379	0.867	0.467	0.685	0.390	0.847	0.453	1.463	0.794
	336	**0.601**	**0.326**	0.621	0.383	0.622	0.337	0.777	0.420	0.869	0.469	0.734	0.408	0.853	0.455	1.479	0.799
	720	**0.607**	**0.358**	0.626	0.382	0.660	0.408	0.864	0.472	0.881	0.473	0.717	0.396	1.500	0.805	1.499	0.804
Weather	96	0.266	0.366	**0.217**	**0.296**	0.266	0.336	0.300	0.384	0.896	0.556	0.458	0.490	0.369	0.406	0.615	0.589
	192	0.408	0.466	**0.276**	**0.336**	0.307	0.367	0.598	0.544	0.622	0.624	0.658	0.589	0.416	0.435	0.629	0.600
	336	0.377	0.418	**0.339**	**0.380**	0.359	0.395	0.578	0.523	0.739	0.753	0.797	0.652	0.455	0.454	0.639	0.608
	720	0.576	0.536	**0.403**	**0.428**	0.419	0.428	1.059	0.741	1.004	0.934	0.869	0.675	0.535	0.520	0.639	0.610
ETTh1	96	**0.363**	**0.410**	0.376	0.419	0.449	0.459	0.865	0.713	0.664	0.612	0.878	0.740	2.341	1.175	3.241	1.536
	192	**0.409**	**0.431**	0.420	0.448	0.500	0.482	1.008	0.792	0.790	0.681	1.037	0.824	2.410	1.253	2.274	1.547
	336	**0.458**	**0.461**	0.459	0.465	0.521	0.496	1.107	0.809	0.891	0.738	1.238	0.932	2.768	1.374	3.287	1.584
	720	0.651	0.603	**0.506**	**0.507**	0.514	0.512	1.181	0.865	0.963	0.782	1.135	0.852	2.820	1.399	3.313	1.591
ETTh2	96	**0.321**	**0.380**	0.346	0.388	0.358	0.397	3.755	1.525	0.645	0.597	2.116	1.197	2.430	1.164	3.201	1.511
	192	**0.411**	**0.424**	0.429	0.439	0.456	0.452	5.602	1.931	0.788	0.683	4.315	1.635	2.479	1.231	3.273	1.539
	336	**0.434**	**0.475**	0.496	0.487	0.482	0.486	4.721	1.835	0.907	0.747	1.124	1.604	2.696	1.367	3.292	1.554
	720	**0.401**	**0.441**	0.463	0.474	0.515	0.511	3.647	1.625	0.963	0.783	3.188	1.540	2.817	1.402	3.302	1.588
ETTm1	96	**0.364**	**0.406**	0.379	0.419	0.505	0.475	0.672	0.571	0.543	0.510	0.600	0.546	1.892	1.042	2.898	1.260
	192	**0.410**	**0.431**	0.426	0.441	0.553	0.496	0.795	0.669	0.557	0.537	0.837	0.700	2.031	1.113	2.989	1.312
	336	**0.438**	**0.453**	0.445	0.459	0.621	0.537	1.212	0.871	0.754	0.655	1.124	0.832	2.273	1.196	3.073	1.312
	720	**0.538**	**0.479**	0.543	0.490	0.671	0.561	1.166	0.823	0.908	0.724	1.153	0.820	2.533	1.208	3.114	1.341
ETTm2	96	**0.184**	**0.280**	0.203	0.287	0.255	0.339	0.365	0.453	0.435	0.507	0.768	0.642	1.997	1.075	2.872	1.167
	192	**0.244**	**0.322**	0.269	0.328	0.281	0.340	0.533	0.563	0.730	0.673	0.989	0.757	2.111	1.107	2.995	1.307
	336	**0.319**	**0.357**	0.325	0.366	0.339	0.372	1.363	0.887	1.201	0.845	1.334	0.872	2.365	1.186	3.082	1.341
	720	0.433	0.500	**0.421**	**0.415**	0.433	0.432	3.379	1.338	3.625	1.451	3.048	1.328	2.453	1.214	3.109	1.353
Count		52		12		0		0		0		0		0		0	

5.4   Results Analysis

Tables 2 and 3 summarize the performance of our model and SOTA model in the field of LTSF and traditional models on different datasets.

Table 3. Univariate LTSF results on four industrial datasets.

Methods		FEDEL		FEDformer		Autoformer		Informer	
Metric		MSE	MAE	MSE	MAE	MSE	MAE	MSE	MAE
ETTh1	96	**0.065**	**0.200**	0.079	0.215	<u>0.071</u>	<u>0.206</u>	0.193	0.377
	192	**0.102**	**0.238**	<u>0.104</u>	<u>0.245</u>	0.114	0.262	0.217	0.395
	336	**0.104**	**0.243**	0.119	0.270	<u>0.107</u>	<u>0.258</u>	0.202	0.381
	720	<u>0.140</u>	**0.288**	0.142	<u>0.299</u>	**0.126**	<u>0.283</u>	0.183	0.355
ETTh2	96	**0.127**	**0.265**	<u>0.128</u>	<u>0.271</u>	0.153	0.306	0.213	0.373
	192	**0.180**	**0.327**	<u>0.185</u>	<u>0.330</u>	0.204	0.351	0.227	0.387
	336	**0.223**	**0.370**	<u>0.231</u>	<u>0.378</u>	0.246	0.389	0.242	0.401
	720	**0.216**	**0.387**	0.278	0.420	<u>0.268</u>	<u>0.409</u>	0.291	0.439
ETTm1	96	**0.030**	**0.139**	<u>0.033</u>	<u>0.140</u>	0.056	0.183	0.109	0.277
	192	**0.156**	**0.176**	<u>0.158</u>	<u>0.186</u>	0.081	0.216	0.151	0.310
	336	**0.074**	**0.213**	0.084	0.231	<u>0.076</u>	<u>0.218</u>	0.427	0.591
	720	**0.095**	**0.243**	<u>0.102</u>	<u>0.250</u>	0.110	0.267	0.438	0.586
ETTm2	96	**0.065**	**0.174**	0.067	0.198	<u>0.065</u>	<u>0.189</u>	0.088	0.225
	192	**0.100**	**0.241**	<u>0.102</u>	<u>0.245</u>	0.118	0.256	0.132	0.283
	336	**0.127**	<u>0.287</u>	<u>0.130</u>	**0.279**	0.154	0.305	0.180	0.336
	720	<u>0.181</u>	0.360	**0.178**	**0.325**	0.182	<u>0.335</u>	0.300	0.435
Count		28		3		1		0	

From Table 2, we can observe that:

(i) FEDEL achieved optimal results in 52 out of 64 experiments (8 datasets, 2 evaluation metrics per dataset, 4 different prediction lengths), far exceeding the 12 of the current SOTA model FEDformer.

(ii) Compared with the traditional LSTM and TCN models, FEDEL improves significantly. For example, in the Electricity dataset, input 96 predicts 96, compared with LSTM, FEDEL reduces MSE by 52.3% and MAE by 32.7%. And compared with TCN, FEDEL reduces MSE by 81.8% and MAE by 63.8%. LSTM and TCN are typical representatives of RNN models and CNN models, respectively. The performance of FEDEL demonstrates the effectiveness of CNN combined with LSTM and corroborates the effectiveness of the DMS prediction strategy.

(iii) FEDEL achieves the best results for different prediction lengths on Electricity, Exchange, Traffic, ETTh2, and ETTm1 datasets. But the performance could be better in the Weather dataset. We partially visualize the Electricity and Weather dataset in Fig. 6. It can be seen that the Electricity dataset has apparent periodicity, while the Weather dataset has weak periodicity. It indicates that FEDEL can capture the periodicity of time series in the frequency domain effectively, and has a better prediction effect for the series with strong periodicity, but poor performance for the series without periodicity. This suggests that it is effective to transfer the time series from the time domain to the frequency domain in order to capture its seasonal characteristics.

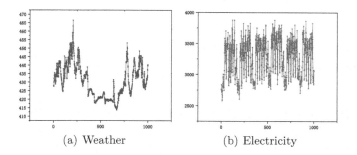

(a) Weather   (b) Electricity

Fig. 6. Electricity and Weather datas visualization

For univariate forecasting, we exclude traditional models that underperform in LTSF. FEDEL achieved optimal results in 28 and suboptimal results in 3 out of 32 experiments (4 datasets, 2 evaluation metrics per dataset, 4 different prediction lengths). We visualize the performance of four different models on four different datasets for the input length 96 and prediction length 96 conditions in Fig. 7.

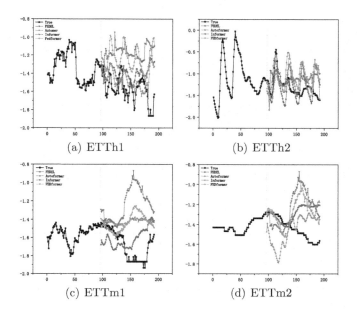

(a) ETTh1   (b) ETTh2

(c) ETTm1   (d) ETTm2

Fig. 7. Different models performance on ETT datasets

As can be seen, the prediction of FEDEL is the closest to the real results compared to the three most advanced models in the field of LTSF. FEDEL and FEDformer have a better ability to capture complex cycles than Autoformer and Informer, confirming the correctness of decoupling complex processes in the frequency domain. FEDformer has too many parameters and ignores the processing of the trend part, so the prediction accuracy is lower than FEDEL.

6   Conclusion

In this paper, we investigate the problem of LTSF and propose a Frequency Enhanced Decomposition and Expansion Learning model called FEDEL. Specifically, we design a deep extension architecture to capture complex correlation dependencies in time series, use a Frequency Enhanced Module to decouple complex cycles of series, and apply direct multi-step prediction to avoid error accumulation. Extensive experiments on real datasets have demonstrated the superiority of FEDEL in the field of LTSF. However, FEDEL's ability to handle the trend part of time series can be further enhanced, and this is an area that deserves more future investigation.

References

1. Ariyo, A.A., Adewumi, A.O., Ayo, C.K.: Stock price prediction using the ARIMA model. In: 2014 UKSim-AMSS 16th International Conference on Computer Modelling and Simulation, pp. 106–112. IEEE (2014)
2. Borovykh, A., Bohte, S., Oosterlee, C.W.: Conditional time series forecasting with convolutional neural networks. stat 1050, 17 (2018)
3. Box, G.E., Jenkins, G.M., MacGregor, J.F.: Some recent advances in forecasting and control. J. Roy. Stat. Soc.: Ser. C (Appl. Stat.) 23(2), 158–179 (1974)
4. Cleveland, R.B., Cleveland, W.S., McRae, J.E., Terpenning, I.: STL: a seasonal-trend decomposition. J. Off. Stat 6(1), 3–73 (1990)
5. Friedman, J.H.: Greedy function approximation: a gradient boosting machine. Ann. Stat. 29(5), 1189–1232 (2001)
6. He, K., Zhang, X., Ren, S., Sun, J.: Deep residual learning for image recognition. In: Proceedings of the IEEE Conference on Computer Vision and Pattern Recognition, pp. 770–778 (2016)
7. Holt, C.C.: Forecasting trends and seasonals by exponentially weighted moving averages. ONR Memorandum 52(52), 5–10 (1957)
8. Kitaev, N., Kaiser, L., Levskaya, A.: Reformer: the efficient transformer. In: International Conference on Learning Representations (2020)
9. Lai, G., Chang, W.C., Yang, Y., Liu, H.: Modeling long-and short-term temporal patterns with deep neural networks. In: The 41st International ACM SIGIR Conference on Research & Development in Information Retrieval, pp. 95–104 (2018)
10. Li, S., et al.: Enhancing the locality and breaking the memory bottleneck of transformer on time series forecasting. In: Advances in Neural Information Processing Systems, vol. 32 (2019)
11. Van den Oord, A., et al.: WaveNet: a generative model for raw audio. In: 9th ISCA Speech Synthesis Workshop, pp. 125–125 (2016)

12. Oreshkin, B.N., Carpov, D., Chapados, N., Bengio, Y.: N-beats: neural basis expansion analysis for interpretable time series forecasting. In: International Conference on Learning Representations (2020)
13. Salinas, D., Flunkert, V., Gasthaus, J., Januschowski, T.: DeepAR: probabilistic forecasting with autoregressive recurrent networks. Int. J. Forecast. **36**(3), 1181–1191 (2020)
14. Vaswani, A., et al.: Attention is all you need. In: Advances in Neural Information Processing Systems, vol. 30 (2017)
15. Vecchia, A.: Periodic autoregressive-moving average (PARMA) modeling with applications to water resources 1. JAWRA J. Am. Water Resour. Assoc. **21**(5), 721–730 (1985)
16. Whittle, P.: Hypothesis Testing in Time Series Analysis, vol. 4. Almqvist & Wiksells boktr. (1951)
17. Wu, H., Xu, J., Wang, J., Long, M.: Autoformer: decomposition transformers with auto-correlation for long-term series forecasting. Adv. Neural. Inf. Process. Syst. **34**, 22419–22430 (2021)
18. Zhou, H., et al.: Informer: beyond efficient transformer for long sequence time-series forecasting. In: Proceedings of the AAAI Conference on Artificial Intelligence, vol. 35, pp. 11106–11115 (2021)
19. Zhou, T., Ma, Z., Wen, Q., Wang, X., Sun, L., Jin, R.: FEDformer: frequency enhanced decomposed transformer for long-term series forecasting. In: International Conference on Machine Learning, pp. 27268–27286. PMLR (2022)

On the Use of Persistent Homology to Control the Generalization Capacity of a Neural Network

Abir Barbara[1,2(✉)] ⓘ, Younés Bennani[1] ⓘ, and Joseph Karkazan[2] ⓘ

[1] Université Sorbonne Paris Nord, LIPN, CNRS UMR 7030 La Maison des Sciences Numeriques, Villetaneuse, France
{Abir.Barbara,Younes.Bennani}@sorbonne-paris-nord.fr
[2] Deep Knowledge - APHP-Hôpital Avicenne, Bobigny, France
{Abir.Barbara,Joseph.Karkazan}@deepknowledge.fr

Abstract. Analyzing neural network (NN) generalization is vital for ensuring effective performance on new, unseen data, beyond the training set. Traditional methods involve evaluating NN across multiple testing datasets, a resource-intensive process involving data acquisition, processing, and labeling. The primary challenge is determining the optimal capacity for training observations, requiring adaptable adjustments based on the task and available data information. This paper leverages Algebraic Topology and relevance measures to investigate NN behavior during learning. We define NN on a topological space as a functional topology graph and compute topological summaries to estimate generalization gaps. Simultaneously, we assess the relevance of NN units, progressively pruning network units. The generalization gap estimation helps identify overfitting, enabling timely early-stopping decisions and identifying the architecture with optimal generalization. This approach offers a comprehensive insight into NN generalization and supports the exploration of NN extensibility and interpretability.

Keywords: Topological Data Analysis · Machine Learning · Deep Neural Networks · Generalization capacity

1 Introduction

Algebraic topology is a branch of mathematics, devoted to the study of properties of topological spaces, it has been employed in the data analysis field, specifically through the application of Topological Data Analysis (TDA) for data representation [7]. In recent years, it has been utilized as an analytical tool to investigate the generalization capability of deep learning models by considering an entire model as graphs and analyzing it using TDA, in order to determine its ability to perform well on previously unseen data.

A prominent method for incorporating algebraic topology in the realm of deep learning is through the use of persistent homology [2], that allows the

B. Luo et al. (Eds.): ICONIP 2023, CCIS 1962, pp. 274–286, 2024.
https://doi.org/10.1007/978-981-99-8132-8_21

examination of topological features of a space by analyzing how they change as the space is filtered at various scales. This approach can be utilized to analyze the structure of a deep learning model and to gain a deeper understanding of the relationship between this structure and the model's generalization performance. Furthermore, the utilization of algebraic topology in the field of deep learning is an active area of research, with many ongoing studies and developments aimed at further harnessing the power of these mathematical tools to improve the performance and interpretability of deep learning models.

Neural Networks, including deep ones, address complex problems without needing specific algorithms or explicit problem-solving instructions. They excel at identifying data patterns, correlations, and relationships, with their output providing solutions, a revolutionary departure from precise programming. However, a challenge arises from the "Black Box" nature of these models, where their inner workings remain largely opaque. This lack of interpretability and explainability can lead to issues such as understanding the learning process, potential under-fitting, over-fitting, or achieving good generalization. In this paper, our approach rely on the topological characteristics of the neural network by examining the nodes persistence [12]. Our method provides a novel avenue to estimate the generalization gap and evaluate the network's performance without relying on many testing or validation sets. In this regard, we were inspired by previous research to determine the relationship between the topological summaries of the neural network's structure and its generalization gap. However, instead of representing the neurons by their activations, which are related to the training examples, we have employed saliency measurements [12], to more accurately define the contribution of each node in the neural network, and to reduce the network's size by eliminating irrelevant nodes. Moreover, by definition salienty measurements, in contrast to activations, take into consideration the architecture of the consecutive layers of the neural network. Additionally, To define the distance between every two nodes, instead of relying on a computationally expensive and complex distance [4] or non-distance measure [1], we utilize the euclidean distance, which is not only a distance metric in line with the TDA theory but also easier to calculate, which means computationally inexpensive.

Our approach, built upon existing research, has obtained state-of-the-art performance when predicting the generalization gap and has the advantage of being less computationally expensive and not requiring a thorough understanding of the training process or dataset at hand. The rest of this paper will be divided into the following sections: Sect. 2 will provide a theoretical overview of generalization, Sect. 3 will present the state of the art, Sect. 4 will delve into the fundamental background of topological data analysis and relevance measurements, along with our proposed approach. Section 5 will present the experiments and results, followed by a final conclusion and perspectives.

2 Motivation

A crucial problem when learning a NN is overfitting, one possibility to remedy this is to use regularization methods to prevent this problem. These methods generally consist of adding constraint on the research space of the solution F_w. One can group the majority of these methods in two main groups; methods based on implicit regularization, and explicit regularization. In the first group mainly fluids approaches based on the use of a prior knowledge of the problem. The second group consists of techniques which introduce penalization terms in the cost function. In this paper we propose mainly an approach belonging to the first group. It describes a method which can help to determine an optimal NN architecture, based on analysis of the behavior of the functional graph associated with the NN. This graph encodes the functional behavior of the NN. The key idea of our approach is to derive a set of topological summaries measuring the topological behavior of the NN. Topological features of the network have been demonstrated to be efficient in explaining and interpreting NN [1], therefore in this research study, we propose an approach using topological approaches that enable us to analyze the generalization gap behavior of a NN during learning. The Fig. 1 illustrates the processing pipeline of the proposed approach.

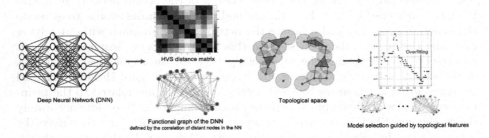

Fig. 1. Pipeline of our approach using TDA and HVS measure

We mainly present in this paper the following contributions:

- An approach for extracting topological information from NN based on persistent homology.
- Demonstrating that not the entire network but just persistent nodes define accurately the topology of the NN and correlates with generalization gap.
- Improvement in the time complexity of calculating the topological properties of a NN using relevance measure allowing progressive NN pruning.
- Enhancement of the correlation between a NN's structure and its generalization capacity as a result of improved feature extraction method and nodes selection.

3   Related Works

Analyzing the generalization gap is a critical task in machine learning that determines the performance of a model on an unseen dataset. However, **the critical**

questions we seek to answer in this paper are whether there is a relationship between the neural network structure and its generalization capacity, and whether it is possible to predict the generalization gap without relying on a testing set. This has the potential to save significant resources and time when developing and optimizing machine learning models. If we can study the generalization capacity of a neural network just by analyzing its topology or if we can accurately predict the generalization gap early on, we can focus our efforts on improving the model in specific areas where it fails, rather than testing it with different data sets multiple times.

The paper [8] describes a competition to conduct a performance evaluation of several complexity measures in predicting deep learning models' generalization capacity. The best methods used by participants do not take into account the complete architecture of the network, which may be important for accurately predicting the generalization gap, instead, they rely on internal partial representations [9]. There are numerous other methods available to estimate the generalization gap, each with its strengths and limitations, such as the generation of a meta-dataset for testing [5]. However, this method has the disadvantage of taking a significant amount of time and resources, as well as potentially not discovering overfitting. Additionally, it is uncertain whether the model truly learns using this method. Other methods are based on the Vapnik-Chervonenkis dimension (VC_{dim}) [10], but they have several limitations and drawbacks. First of all, they can be challenging to compute for complex models, then it is not always a tight bound, meaning the actual generalization gap may be smaller than the VC_{dim} suggests. An alternative approach is the use of Topological Data Analysis (TDA) to predict the generalization gap. TDA has been applied to various machine learning tasks [2], such as studying the input of neural networks to improve the model's performance, or analyzing its architecture [1,4,11].

4 Proposed Approach

4.1 Fundamental Background

Persistent homology [3] is a technique of algebraic topology that can be used to extract topological features from high-dimensional data, it is a powerful tool for understanding the topology of neural networks. It uses a concept called filtration, which involves building a sequence of simplicial complexes from a given input, which is the structures of networks in our case.

A simplicial complex is a mathematical structure that is used to represent a collection of simplices, which are geometric shapes that are formed by a set of points in a space. These simplices can be of any dimension, such as points, lines, triangles, and higher-dimensional shapes. In a simplicial complex, the simplices are connected together in a way that is consistent with the concept of a face, where a face is a subset of the points that form a simplex. This means that any subset of points that forms a face of one of the simplices in the collection is also a simplex within the collection. Each simplicial complex captures different topological features at different scales, and those that persist for the most number of simplicial complexes are considered to be the "true" topological features. The resulting persistence diagrams can then be used to study the topology of the data set, such as the number and distribution of connected components, loops, and voids. The main advantage of using persistence homology is that it provides a robust and efficient way of analyzing the topology of the network. Also the topological properties extracted using persistence homology are stable and robust to noise and small perturbations in the network [3]. One of the most widely used algorithms for computing the persistent homology of a data set is called the Vietoris-Rips algorithm (VR). This algorithm constructs a sequence of simplicial complexes by connecting points in the data set that are within a certain distance of each other, as shown in Fig. 2. The distance threshold is gradually increased, resulting in a filtration of simplicial complexes that capture the topology of the data set at different scales.

Expanding upon this notion, given a finite set of points X, which we refer to as the vertex set, and an r greater than zero, the Vietoris-Rips complex $V(X, r)$ is defined as the collection of simplexes sigma in X whose diameter is smaller than $2r$, i.e., the distance between two points of a sigma element is always less than $2r$.

$$V(X, r) = \{\sigma \in X/diam(\sigma) < 2r\}$$

The basic idea is to study topological invariants (the number of connected component, cycles, voids and n-dimentionnel voids), which are topological properties that remain invariant under a continuous transformation of the form of the data (i.e. homeomorphisms). Moreover, the topological invariants encode the topological information of the data on a multi-scale representation and are robust under noise [3]. The kth Betti number, which is equal to the dimension of $H_k(T)$ (the kth homology group), $\beta_k = dim(H_k)$, is also the number of the k-dimensional topological invariants for the space T. This justifies the usage of homology groups.

Interpretation:
-$\beta_0(T)$ is the number of connected components
-$\beta_1(T)$ is the number of "holes"
-$\beta_2(T)$ is the number of "cavities": "2-dimensionnel holes"
- ......
-$\beta_n(T)$ is the number of "n-dimensionnel holes"
A space must be first represented as a simplicial complex in order to establish its persistent homology; this process is known as triangulation.

The persistence diagram (PD) is an effective tool for representing topological features, (PD) is a collection of points, where each point represents a topological feature of the data set. Specifically, each point in the persistence diagram corresponds to a topological feature that appears and disappears at different scales in the data. The location of a point (b, d) in the diagram is determined by its birth and death scales, and those that persist longer are considered to be the "true" topological features.

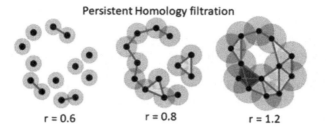

Fig. 2. Filtrations of simplicial complexes

The x-coordinate b of a point in the persistence diagram corresponds to the scale at which the topological feature appears (referred to as the birth scale), and the y-coordinate d corresponds to the scale at which the feature disappears (referred to as the death scale). To succinctly encapsulate the information presented in the persistence diagram, various numerical or vector-based persistence summaries or descriptors have been proposed, such as persistence entropy and persistence landscape. However, in this study, the following topological feature will be utilized:

The Average Life: The average life of a point or a feature in the persistence diagram is the difference between d and b, $d - b$. Therefore the average life of a persistence diagram is:

$$avg = \frac{1}{C} \sum_{i=1}^{C} (d^i - b^i)$$

where C is the number of the topological properties in the persistence diagram.

4.2   Pruning Based on the Relevance of NN's Units

We use HVS (Heuristic for variable selection) [12] as a measure for quantifying the relevance of each unit in a NN, by representing the contribution of each unit in comparison to the other units. It is used to identify the most informative or relevant units in the neural network.

The architecture of a simple feed-forward NN, denoted as $W(I, H, O)$, is comprised of three distinct layers: an input layer (I), a hidden layer (H), and an output layer (O). The connection weight value, represented by w_{ij} between two units i and j serves as an indicator of the significance of their link. This value can assume either a positive or negative value depending on the nature of the connection weight, if it is an excitatory or an inhibitory synapse. In order to focus solely on the strength of the connection weights, the absolute value $|w_{ij}|$ of the associated weight value is utilized to quantify the strength of each connection weight. In the present context, the partial contribution of a unit j, which is connected to a unit i through a connection weight w_{ij}, is defined as:

$\pi_{i,j} = \dfrac{|w_{i,j}|}{\sum\limits_{k \in fan-in(i)} |w_{i,k}|}$ where $fan-in(i)$ is the set of nodes that have a connection
to the node i, so $\pi_{i,j}$ measures the relation strength between node j and node i.

The relevance measure is the relative contribution of unit j to the output decision of the network and is defined as the weighted sum of its partial contributions, which are calculated by taking into account the relative contributions of the units to which it sends connections. This leads to the following mathematical expression:

$$S_j = \sum_{i \in fan-out(j)} \pi_{i,j} \delta_i$$

$$\text{with}: \delta_i = \begin{cases} 1 & if \quad unit \quad i \in O \\ S_i & if \quad unit \quad i \in H \end{cases}$$

4.3   Topological Functional Graph and Model Selection

In this study, we investigate the topology of a neural network at each iteration during the learning process, aiming to understand and visualize the network's behavior during training, and relating this behavior to the network's generalization capacity.

To accomplish this, we take the set of vertices in our analysis as the collection of the neural network units or nodes; $X = \{node/\ node \in N\}$, where N is our neural network.

In our analysis, each node is represented by its saliency measurement:

$$S_{node} = \sum_{i \in fan-out(node)} \pi_{i,node} \delta_i$$

which accurately characterize the relevance of the node within the neural network. Moreover, these measures are specific to the model and provide an accurate representation of the neural network, while activation values are sensitive to individual examples and iterations, and provide a more comprehensive characterization of the data rather than the model.

Subsequently, upon computing the saliency assessment, we proceed to exclude the nodes comprising the lowest 10 percent in terms of saliency measurements. This decision is grounded in their lack of persistence and inability to effectively characterize the model's underlying structure.

The metric space (X,d) is employed to define the Vietoris-Rips filtration, with d denoting the euclidean distance defined on the salienty measurements of each two nodes.

We investigate β_0, which represents the number of connected components, all the other Betti numbers are null because the set of points has no other dimensional holes. As a result, instead of dealing with the set of points, one may generate a family of simplicial complexes given a range of $i \in I$ values. This family of simplicial complexes is known as filtration.

A filtration in Algebraic Topology is an increasing sequence $(K_i)_{i \in I}$ of simplicial complexes which are ordered by inclusion. In other words, if K_1 and K_2 are two complexes where K_1 appears before K_2 in the filtration, then: K_1 is a subcomplex of K_2. The index set I denotes distance (This is how the Vietoris-Rips V complex is constructed).

Through this process, we compute the appearance and disappearance of k-dimensional homology features, which enables us to generate the persistence diagram of a k-dimensional homology group of the network at each iteration using the average life of each persistence diagram, we demonstrate experimentally that there is a correlation between those topological summaries and the generalization gap, thus enabling the prediction of the testing error without the need for a testing set.

5 Experimental Settings and Results

5.1 Datasets

Our method was tested and compared with the state-of-the-art method [1] on several public classification datasets: Dermatology, Credit default, Ionosphere, Website Phishing, Mice Protein Expression, Car Evaluation, Seeds, Haberman's Survival, Wholesale Customer Region, Iris, Waveform, from the UCI Machine Learning Repository [6].

In order to effectively evaluate and compare our proposed method with other methods, we constructed small neural networks with the aim of facilitating manipulation and allowing for a sufficient amount of experimentation time with various datasets. However, it is important to note that the proposed method should not be limited by the architectures utilized in these initial experiments. To this end, we employed a sequential neural network architecture, consisting of an input layer with a number of neurons equal to the number of features in the dataset. This was followed by two hidden layers, and an output layer with the number of neurons corresponding to the number of classes in the classification task. We conducted 100 epochs for each experiment to train these networks, which means for each dataset. During the training process, we computed various measures of saliency, for each training step. Then we eliminated 10 percent of units that have the smallest saliency measurements and from the rest of nodes, we generated persistence diagrams to capture persistent homology. Then, we summarized those diagrams in one topological feature for each iteration (the average life). We then proceeded to produce plots in which the x-axis represented the topological characterization of the network at each epoch, and the y-axis represented the generalization gap at each epoch and through these graphical representations, we were able to observe the correlation, subsequently, we employed linear regression to model these relationships and to extract quantitative insights.

5.2 Results and Discussion

The outcomes of our experiments have led us to understand that we can investigate the explainability and the interpretability of NN throughout their topological features and based on their relevance measurements. Particularly when it comes to the generalization gap.

In this part of experimentation and evaluation of our approach, we will first examine the impact of our choice of measure of correlation between the units of an NN (HVS measure) on the quality of the estimation of the gap of generalization. This impact will be compared to an approach used in other works using the correlation of the activations of the NN units thus requiring the use of the training examples to estimate the correlations of the activations of the different NN units [1], [4]. Secondly, we will show the results obtained by our approach allowing to control and adjust the generalization capacity of an NN. This adjustment will allow to select the best architecture of an NN adapted to the studied problem.

We observe a dependency between the topology of the NN and its generalization capacity, as shown in Fig. 3. We notice this connection also for the other data-sets used in this study. We can approximate this dependency by a linear correlation.

As demonstrated in Table 1, the high values of R squared indicate a strong correlation between the topology of the neural network (through the average life topological characterization of the network) and its generalization capacity. Additionally, the Table 1 illustrate that this correlation is even more significant when utilizing our approach compared to the state of the art approaches [1]. This is mainly due to the quality of salienty measurements in representing neurons; in fact, a neuron's salienty measurement summarizes its importance in the entire graph, its placement in the neural network's architecture, and its pertinence, so calculating the evolution of the topology of the neuron graph during learning throughout the salienty measurements is more representative of the neural network than using only activations, which are highly representative of examples and data but lack information about the architecture and the pertinence of each neuron in the graph.

We also explored other alternative network characterizations (midlife, entropy), yielding similar and improved results compared to state-of-the-art. However, for brevity, we present only the average life characterization in this paper, as all of those characterizations encapsulate the essence of the persistent diagram.

In this study, we approached this correlation by a linear regression which we considered to be a reasonable approach. However, in future studies, we may delve deeper into this relationship by seeking to derive an equation that more explicitly links the generalization gap to the topological features based on saliency measurements. Also, we would like to highlight that our algorithm is flexible and can be applied to any DNN architecture.

Estimating the generalization gap based on the network's topological properties could improve our understanding of network behavior and make it easier to build robust and effective machine learning models. However, it should be highlighted that these correlation parameters show model variability. This suggests that to develop a non-model-specific strategy that can precisely predict the generalization gap using these topological properties, more research is needed.

Table 1. R Squared Comparison: Evaluating the Performance of Our Regression Model Against State of the Art, for average life

datasets	state-of-the-art method	our method
Dermatology	0.95	**0.97**
Credit default	0.73	**0.75**
ionosphere	0.90	0.90
Website Phishing	0.86	**0.87**
Mice Protein Expression	0.60	**0.62**
Car	0.75	0.75
Seeds	0.88	**0.89**
Herbman's Breast Cancer	0.83	**0.89**
The Wholesale Customer	0.90	**0.91**
Iris	0.75	**0.79**
Waveform	0.83	**0.84**

To illustrate how our approach works, we'll present the results on the well-known database: Iris.

Figure 3 shows the evolution of the generalization gap as a function of the evolution of the NN's topological characteristics during learning. It can be seen from this graph that the generalization gap at the start of training is quite small, but the network has not yet reached its good performance. As the learning process progresses and the NN parameters are adapted, the gap increases and reaches a maximum, before gradually decreasing. This decrease in the gap shows that the NN is on the right learning path. After this phase, the generalization gap begins to increase, indicating that the network is just beginning to memorize the learning data: the start of overfitting. Parallel to this learning process, our approach estimates the relevance of each NN unit, and proceeds to prune the least relevant units for the problem under study. This pruning makes it possible to adjust the architecture of the NN according to the quality and quantity of information provided by the training data. This pruning is stopped as soon as overfitting is detected, by analyzing the generalization gap.

The graphs below do not directly represent the evolution of the NN architecture, but rather the correlation between the various NN units. These graph encodes the functional behavior of the NN. It can be seen that, at the start of learning, the NN units are not very correlated, resulting in a less dense graph of connections. In these graphs, two units are connected when the difference

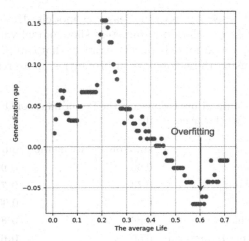

Fig. 3. Generalization gap estimation as a function of topological characteristics of the NN for Iris dataset

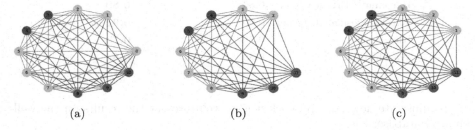

       (a)   (b)   (c)

Fig. 4. Evolution of the functional graph of the NN for Iris dataset. On the left, the functional graph at the beginning of the training, followed by that of the middle of the training and finally that of the right represents the graph at the end of the training and indicates the best network.

between their HVS is small compared with a predefined threshold. As learning iterations progress, the correlation network becomes denser and the selection of relevant features stabilizes. This process of adjusting the NN architecture is guided by the NN's topological information, which gives us an estimate of the generalization gap. By analyzing the evolution of the NN functional graph presented in Fig. 3, we can monitor the impact of the pruning of units deemed irrelevant, and identify the units correlated with each class at the end of the learning process. In other words, after pruning, the functional graph can be used to find the set of units associated with each class: the list of input variables and the list of hidden neurons linked to that class. This provides a simple explanation and characterization of each class. The starting architecture for Iris data is <4-2-3-3>, and at the end of training it becomes <3-2-3-3>. Neurons are numbered from 0 to the total number of neurons in the network minus 1. Input neurons (0,1,2,3) are in pink, the first hidden layer (4.5) in green, the second layer (6,7,8) in light blue and those of the output layer (9,10,11) in gray. We can notice that in the case of this simple example our approach to delete an entry (0) deemed irrelevant.

6   Conclusion and Perspectives

In conclusion, this study proposes an innovative approach that utilizes the generalization gap and relevance measure during neural network training to tailor the architecture to problem complexity and data characteristics. It employs persistent homology to analyze the network topology at each training iteration, revealing how neurons cooperate and aiding in visualizing network development. This approach excels in predicting generalization gaps using the data-independent HVS relevance measure, offering computational efficiency and not relying on dataset usage at every training stage. It contributes to ongoing efforts to enhance neural network interpretability, explainability, and generalization. However, a key limitation lies in the computational complexity, especially for deep networks with many layers and neurons, necessitating suitable sampling techniques and automated overfitting detection. Further research explores the fundamental connection between our approach and structural risk minimization, enabling generalization bounds without $VC - dim$ estimation.

References

1. Ballester, R., Clemente, X.A., Casacuberta, C., Madadi, M., Corneanu, C.A., Escalera, S.: Towards explaining the generalization gap in neural networks using topological data analysis (2022). https://doi.org/10.48550/ARXIV.2203.12330, https://arxiv.org/abs/2203.12330
2. Chazal, F., Michel, B.: An introduction to topological data analysis: fundamental and practical aspects for data scientists (2017). https://doi.org/10.48550/ARXIV.1710.04019, https://arxiv.org/abs/1710.04019
3. Cohen-Steiner, D., Edelsbrunner, H., Harer, J., Mileyko, Y.: Lipschitz functions have l p-stable persistence. Found. Comput. Math. **10**(2), 127–139 (2010)
4. Corneanu, C., Madadi, M., Escalera, S., Martinez, A.: Computing the testing error without a testing set (2020). https://doi.org/10.48550/ARXIV.2005.00450, https://arxiv.org/abs/2005.00450
5. Deng, W., Zheng, L.: Are labels always necessary for classifier accuracy evaluation? In: Proceedings of the CVPR (2021)
6. Dua, D., Graff, C.: UCI machine learning repository (2017). https://archive.ics.uci.edu/ml
7. Hensel, F., Moor, M., Rieck, B.: A survey of topological machine learning methods. Front. Artif. Intell. **4**, 681108 (2021). https://doi.org/10.3389/frai.2021.681108
8. Jiang, Y., et al.: NeurIPS 2020 competition: predicting generalization in deep learning. arXiv preprint arXiv:2012.07976 (2020)
9. Lassance, C., Béthune, L., Bontonou, M., Hamidouche, M., Gripon, V.: Ranking deep learning generalization using label variation in latent geometry graphs. CoRR abs/2011.12737 (2020), https://arxiv.org/abs/2011.12737

10. Vapnik, V.N., Chervonenkis, A.Y.: On the uniform convergence of relative frequencies of events to their probabilities. Theor. Probab. Appl. **16**(2), 264–280 (1971)
11. Watanabe, S., Yamana, H.: Topological measurement of deep neural networks using persistent homology. CoRR abs/2106.03016 (2021), https://arxiv.org/abs/2106.03016
12. Yacoub, M., Bennani, Y.: HVS: a heuristic for variable selection in multilayer artificial neural network classifier. In: Dagli, C., Akay, M., Ersoy, O., Fernandez, B., Smith, A. (eds.) Intelligent Engineering Systems Through Artificial Neural Networks, vol. 7, pp. 527–532 (1997)

Genetic Programming Symbolic Regression with Simplification-Pruning Operator for Solving Differential Equations

Lulu Cao[1], Zimo Zheng[1], Chenwen Ding[2], Jinkai Cai[3], and Min Jiang[1(✉)]

[1] School of Informatics, Xiamen University, Fujian, China
lulucao@stu.xmu.edu.cn, minjiang@xmu.edu.cn
[2] College of Engineering, Shantou University, Guangdong, China
[3] School of Information Science and Technology, Hainan Normal University, Hainan, China

Abstract. Differential equations (DEs) are important mathematical models for describing natural phenomena and engineering problems. Finding analytical solutions for DEs has theoretical and practical benefits. However, traditional methods for finding analytical solutions only work for some special forms of DEs, such as separable variables or transformable to ordinary differential equations. For general nonlinear DEs, analytical solutions are often hard to obtain. The current popular method based on neural networks requires a lot of data to train the network and only gives approximate solutions with errors and instability. It is also a black-box model that is not interpretable. To obtain analytical solutions for DEs, this paper proposes a symbolic regression algorithm based on genetic programming with the simplification-pruning operator (SP-GPSR). This method introduces a new operator that can simplify the individual expressions in the population and randomly remove some structures in the formulas. Moreover, this method uses multiple fitness functions that consider the accuracy of the analytic solution satisfying the sampled data and the differential equations. In addition, this algorithm also uses a hybrid optimization technique to improve search efficiency and convergence speed. This paper conducts experiments on two typical classes of DEs. The results show that the proposed method can effectively find analytical solutions for DEs with high accuracy and simplicity.

Keywords: Differential equations · Symbolic regression · Genetic programming · Genetic operator

---

This work was partly supported by the National Natural Science Foundation of China under Grant No. 62276222.

B. Luo et al. (Eds.): ICONIP 2023, CCIS 1962, pp. 287–298, 2024.
https://doi.org/10.1007/978-981-99-8132-8_22

1   Introduction

A differential equation (DE) is an equation that involves an unknown function and its derivatives, usually describing the relationship between multiple variables, such as time, space, temperature, pressure, etc. By solving a DE, researchers can obtain an expression that describes the unknown function, which is called an analytical solution of the DE. Researchers can use the analytical solution of a DE to simulate various phenomena in natural sciences and engineering, such as heat conduction, wave propagation, quantum mechanics, etc. The analytical solution of a DE can provide insights into the potential physical mechanisms and mathematical properties of the problem. The analytical solution can also serve as a benchmark for validating numerical methods and experimental results [9].

However, traditional methods for finding analytical solutions for DEs are only applicable to some special forms of DEs, such as those that are separable or reducible to ordinary differential equations. For general nonlinear DEs, it is often impossible to find analytical solutions. The solution process requires some advanced mathematical techniques and methods, such as Laplace transform, Fourier transform, Green's function, etc., which are challenging for beginners. Therefore, people have developed numerical methods to simulate the process and results of solving DEs with computers [7].

The commonly used numerical methods for DEs are the following four types: finite difference, finite element [19], finite volume [8], and spectral methods [4]. The finite difference is simple and easy to use. The finite element is popular in engineering and fluid dynamics and works well for complex geometries. The finite volume uses algebraic equations and meshless methods are useful for difficult problems but need more time and work. Spectral methods use fast Fourier transform and basis functions to obtain accurate solutions if they are smooth. Finite element and spectral methods are similar but differ in the domain of basis functions.

These numerical methods have their own advantages and disadvantages and need to be selected and compared according to specific problems and requirements. Numerical methods can handle more general and complex DEs but have some errors and instabilities. Numerical methods also require a lot of computational resources and time so there needs to be a balance between efficiency and accuracy.

Data-driven machine learning methods are a new research direction for solving DEs by improving numerical methods or learning solutions from data. The machine learning methods for DEs are the following three types: **Data-driven discretization methods** [1] use machine learning to optimize discretization in numerical methods, such as neural networks for spatial derivatives or deep reinforcement learning for grid partitioning [3]. They improve accuracy and stability on low-resolution grids. **Data-driven dimensionality reduction methods** [15] use machine learning to find coordinate systems or models to simplify DEs, such as autoencoders or manifold learning for low-dimensional features, or neural networks or support vector machines for nonlinear transformations or kernels.

They reduce computation time and resources. **Data-driven solver methods** [17] use machine learning to approximate solutions or solvers of DEs, such as neural networks or Gaussian process regression for functional forms, or neural networks or convolutional neural networks for iterative updates. They avoid discretizing DEs and reduce errors and instabilities.

Data-driven machine learning methods are usually numerical and fit data by optimizing parameters to get numerical solutions of DEs. These methods have some drawbacks, such as being black-box models that are hard to explain and do not give a deep understanding of physical or mathematical mechanisms behind data, having limited generalization ability that depends on specific data sets and parameter settings and may be inaccurate or unstable if data changes or has noise, and requiring a lot of time and resources for computational efficiency, especially for high-dimensional or complex equations. Therefore, this paper uses Genetic Programming for Symbolic Regression (GPSR) to solve DEs. This method has some advantages, such as being interpretable with a clear and structured mathematical formula that gives a deep understanding of physical or mathematical mechanisms behind data, having high generalization ability with a universal and concise mathematical formula that adapts to different data sets and parameter settings and resists noise, and being computationally efficient with a heuristic search and adaptive mutation process that finds good solutions in limited time and resources. GPSR has been successfully applied to physical and mechanical problems due to its inherent interpretability [2,10,14].

The existing GPSR can only obtain approximate expressions and are very slow when directly applied to solve differential equations analytically. To improve the speed and accuracy of generating analytical solutions for specific differential equations, we introduce a new operator in the algorithm, called the simplification-pruning operator. This operator can merge the like terms of the expressions represented by the individuals, simplify them, and prune away the redundant terms by setting some constants to zero. Meanwhile, we design multiple fitness functions that take into account both the accuracy of the solutions fitting the sample data and satisfying the differential equations. In addition, we adopt a hybrid optimization strategy that combines genetic programming with gradient descent algorithms to improve search efficiency and convergence speed. We conduct experiments on two typical classes of differential equations and the results show that our proposed method can quickly find analytical solutions for differential equations with high accuracy and conciseness.

2 Preliminaries

2.1 Genetic Programming for Symbolic Regression (GPSR)

Genetic Programming is a type of Evolutionary Computation(EC), and EC has made a lot of progress [5,11–13,18]. Genetic Programming for Symbolic Regression is a method that uses Genetic Programming (GP) to solve Symbolic Regression (SR) problems. Symbolic Regression is a problem that tries to discover a mathematical formula that can best fit a given dataset. Genetic Programming is an optimization and search method that simulates the natural evolutionary process, and it can automatically generate and evolve computer programs to solve a given problem.

The basic steps of Genetic Programming for Symbolic Regression are as Algorithm 1. Firstly, a primitive set that contains all the possible functions and terminals that can be used to construct a mathematical formula is defined, such as addition (+), subtraction (−), multiplication (×), *sin*, variables, constants, etc. Then a fitness function that measures how well each individual fits the given dataset is defined, such as mean absolute error (MAE), etc. Next, an initial population P_0 is randomly generated, each individual represented by a tree structure with functions as internal nodes and terminals as leaf nodes (Step 1). After that, the fitness value of each individual in the population is evaluated (Step 3) and some good individuals are selected as parent population P'_t (Step 4). Genetic operations such as crossover and mutation are then performed on the P'_t to produce new offspring individuals (Step 5). This process is repeated until a termination condition is met, such as reaching the maximum number of iterations or finding a satisfactory solution.

Algorithm 1. Genetic Programming for Symbolic Regression

Require: P (A primitive set of functions and terminals), f (Fitness function), N
   (population size), T (Maximum number of iterations)
1: Generate an initial population P_0;
2: **for** t=0 to T-1 **do**
3:    Compute $f(p)$ for each $p \in P_t$;
4:    Select P'_t as parent population from P_t based on f;
5:    Apply crossover and mutation to P'_t to generate P_{t+1};
6: **end for**
7: **Retrun** $\mathrm{argmin}_{p \in P_T} f(p)$;

2.2   Physics-Regularized Fitness Function

To make the GPSR algorithm more accurate in evolving solutions for differential equations, Oh *et al.* [16] introduce physics-regularized fitness function in the standard GPSR to constrain the generated solutions to satisfy the preset differential equations.

Specifically, suppose we want to solve a set of l differential equations as Eq. 1:

$$\left\{ L^{(k)}(f(X)) = 0 \right\}_{k=1}^{l} \tag{1}$$

where X is a p-dimensional vector variables, $X = x^{(1)}, x^{(2)}, \ldots, x^{(p)}$, $L^{(k)}$ is an arbitrary differential operator and $f(X)$ is the unknown function to be discovered. The physics-regularized fitness function is defined as Eq. 2:

$$F_k^{Pr} = L^{(k)}(\tilde{f}(X))^2, k = 1, \cdot, l \tag{2}$$

where $\tilde{f}(X)$ is the model proposed by GPSR.

3   Proposed Algorithm

In this section, we describe the details of the proposed SP-GPSR method. First, we present the overall framework of SP-GPSR in Sect. 3.1. Then, we introduce the simplification-pruning operator (Sect. 3.2), the hybrid optimization strategy (Sect. 3.3), and the fitness function (Sect. 3.4), which are the key steps of SP-GPSR.

3.1   Overall Framework

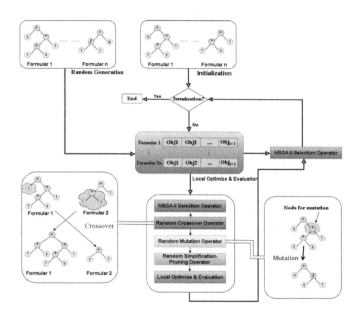

Fig. 1. The framework of SP-GPSR

In this work, we propose a simplification-pruning operator for GPSR, called SP-GPSR, to make the existing Genetic Programming for Symbolic Regression faster and more accurate in generating analytical solutions for specific differential equations. SP-GPSR follows the basic framework of GPSR.

Figure 1 shows the overall framework proposed in this work. The details of the evolution process are shown in Algorithm 2. During the evolutionary loop, we first randomly generate a population Q_t of size n (Step 4). The purpose of this step is to increase population diversity. Secondly, Optimize the individuals in the populations by fine-tuning the constants using gradient descent (Sect. 3.3). And evaluate individual by the fitness function (Sect. 3.4) (Step 5). Thirdly, select some good individuals as parent population M_t. Apply crossover operator, mutation operator and simplification-pruning operator on the population

---

Algorithm 2. SP-GPSR Framework

Require: n (Population size), T (Maximum iteration number), DE (Differential Equations), D (Observation datasets), p_c (crossover probability), p_m (mutation probability), p_s (simplification-pruning probability)

Ensure: Sol (A mathematical expression that fits the specific Differential Equations)

1: Generate an initial population P_0 of size n, where each individual is a tree representing a mathematical expression;
2: $t \leftarrow 0$
3: **while** $t < T$ **do**
4:     Random generate a population Q_t of size n;
5:     Perform local optimization and evaluation on each individual in P_t and Q_t
6:     Select parent population M_t from P_t and Q_t using NSGA-II;
7:     Apply crossover operator and mutation operator to M_t;
8:     Apply simplification-pruning operator to M_t;
9:     Perform local optimization and evaluation on M_t;
10:     Select n individual from $P_t \cup Q_t \cup M_t$ as P_{t+1} according to NSGA-II;
11:     $t \leftarrow t + 1$
12: **end while**
13: **return** The optimal solution Sol

M_t (Step 6–8). Finally, select individuals as the next generation P_{t+1} using the NSGA-II algorithm [6] (Step 9–10).

Crossover operator refers to randomly selecting the subtrees of two parent individuals and exchange them to produce two new child individuals. Mutation operation refers to randomly selecting an internal node of an individual as a mutation point, then randomly generating a new subtree, replacing the subtree with the mutation point as the root in the individual, and generating a new individual. The simplification-pruning operator, which is the core of our work, will be introduced in Sect. 3.2. The hybrid optimization strategy and the fitness evaluating function, which are also key technologies that make our work better than previous works, will be introduced in Sects. 3.3 and Sect. 3.4 respectively.

3.2   Simplification-Pruning Operator

Genetic programming (GP) often generates redundant expressions that contain components that can be eliminated. These redundant components increase the complexity and computational cost of the expressions and reduce their interpretability and generalization ability. For example, suppose we use GP to find a function that fits the data points as Eq. 3:

$$x = [0, 1, 2, 3, 4] \quad y = [0, 2, 4, 6, 8]. \tag{3}$$

GP may generate the symbolic expression as Eq. 4:

$$f(x) = (x + 1) \times (x + 1) - x \times x - 1. \tag{4}$$

This expression can perfectly fit the data points, but it contains some redundant components, such as Eq. 4 which can be simplified to $2x$. These redundant components make the expression hard to understand and optimize.

Moreover, the constant coefficients in the symbolic expressions generated by GP are sometimes epistatic, meaning that they interact with other variables or operators, making it difficult to perform local optimization. For example, GP may generate symbolic expression as Eq. 5:

$$f(x) = (x + 0.5)^2 + x \times x - x. \tag{5}$$

Equation 5 contains an epistatic constant coefficient 0.5, which is added to x and then multiplied by itself. This leads to the problem that if we want to perform local optimization on 0.5, we have to consider its interaction effects with both x and itself. Expanding $(x + 0.5)^2$ in Eq. 5, we get $x \times x + 2 \times 0.5 \times x + 0.5 \times 0.5$. The value of 0.5 can affect the coefficient of x as well as the constant coefficient. If the true expression has a positive coefficient for x and a negative constant coefficient, then we cannot determine whether to increase or decrease the value of 0.5 during local optimisation. In other words, we cannot simply adjust its value independently.

Therefore, the symbolic expressions generated by GP are rather redundant, and their constant coefficients are often epistatic, making them hard to optimize locally. To solve this problem, we propose a new operator: the simplification-pruning operator. The operator can simplify individual expressions in the population to solve the above problem. And by setting some constants to zero, randomly remove certain parts of the expression to increase the diversity of the population.

Here is an example to illustrate the simplification-pruning operator. Suppose there is an individual expression as Eq. 6:

$$u(x) = (x + 2)^2 + \sin(x) - 3. \tag{6}$$

We can simplify Eq. 6 to get Eq. 7:

$$u(x) = x^2 + 4x + \sin(x) + 1. \tag{7}$$

Then randomly select some constant terms and set them to zero, such as Eq. 8:

$$u(x) = x^2 + 0x + \sin(x) + 0. \tag{8}$$

Simplify the zeroed expression again, removing redundant zero terms and parentheses. Then obtain a new individual Eq. 9 as the result of the simplification-pruning operator. Besides reducing the redundancy and complexity of the formulas in genetic programming, this operator also increases the diversity of the population without increasing the complexity.

$$u(x) = x^2 + \sin(x). \tag{9}$$

3.3 Hybrid Optimization Strategy

Suppose we have an observation dataset $D(X_i, y_i)$ where $i \in 1, 2, \ldots, q$ based on the differential equation to be solved. The dataset D contains q observation points.

After obtaining the expressions from GP, we perform local optimization on the individuals in the population, that is, use the gradient descent method based dataset D to fine-tune the constants in each individual to make them closer to the true solution. This is the hybrid optimization strategy. The hybrid optimization strategy can improve the search efficiency and convergence speed.

3.4 Fitness Evaluating Function

In the SP-GPSR, a multi-objective function can be used to optimize both the data fitting error and the physics-regularized fitness term. It can be defined as Eq. 13:

$$F(X) = (f_1(X), f_2(X), \ldots, f_{l+1}(X)) \tag{10}$$

where $f_1(X)$ is the data fitting error, as shown in Eq. 14. $f_{k+1}(X)$ is the physical regularization term of the k-th differential equation, as shown in Eq. 15.

$$f_1(X) = \frac{1}{q} \sum_{i=1}^{q} (f(X_i) - y_i)^2 \tag{11}$$

$$f_{k+1}(X) = F_k^{Pr} \tag{12}$$

where the definition of F_k^{Pr} is shown in Eq. 2.

In the SP-GPSR, a multi-objective function can be used to optimize both the data fitting error and the physics-regularized fitness term. It can be defined as Eq. 13:

$$F(X) = (f_1(X), f_2(X), \ldots, f_{l+1}(X)) \tag{13}$$

where $f_1(X)$ is the data fitting error, as shown in Eq. 14. $f_{k+1}(X)$ is the physical regularization term of the k-th differential equation, as shown in Eq. 15.

$$f_1(X) = \frac{1}{q} \sum_{i=1}^{q} (f(X_i) - y_i)^2 \tag{14}$$

$$f_{k+1}(X) = F_k^{Pr} \tag{15}$$

where the definition of F_k^{Pr} is shown in Eq. 2.

4 Experimental Design and Result Analysis

4.1 Detailed Problem Description

Euler-Bernouli Equation. Euler-Bernouli equation is a differential equation that describes the deflection of a beam. It has the form as Eq. 16:

$$EI \frac{\partial^4 w(x)}{\partial x^4} = q(x) \tag{16}$$

where $w(x)$ is the deflection of the beam, E is the Young's modulus, I is the moment of inertia, and $q(x)$ is the distributed load on the beam.

The analytical solution of the Euler-Bernoulli equation depends on the boundary conditions and the load distribution. Here is a simple example. Suppose we have a simply supported beam with length l. The boundary conditions are such that the $w(0) = w(l) = 0$ and $w''(0) = w''(l) = 0$. The $w(x)$ is taken to be a uniform load. Then we can rewrite the constant term of Eq. 16 as Eq. 18.

$$\frac{\partial^4 w(x)}{\partial x^4} = c \tag{17}$$

The solution of Eq. 18 is shown in Eq. 18.

$$u(x) = \frac{c}{24} \left(x^4 - 2lx^3 + l^3 x \right) \tag{18}$$

Poisson's Equation. Poisson's equation is an elliptic partial differential equation of broad utility in theoretical physics. For example, the solution to Poisson's equation is the potential field caused by a given electric charge or mass density distribution; with the potential field known, one can then calculate electrostatic or gravitational fields. Poisson's equation has the form as Eq. 19:

$$\Delta \varphi = f(x) \tag{19}$$

where Δ is the Laplace operator. Suppose $f(x)$ is a source term defined by $-d\pi^2 \prod_{i=1}^{d} \sin(\pi x_i)$, where d refers to the dimension. Then the solution of Eq. 19 is shown in Eq. 20:

$$\varphi(x) = \prod_{i=1}^{d} \sin(\pi x_i). \tag{20}$$

4.2 Performance Comparison

In this section, we demonstrate the ability of SP-GPSR to solve a differential equation and a partial differential equation problem. We also complete the GPSR with a hybrid optimization strategy (HGPSR), and the GPSR with a hybrid optimization strategy and physics-regularized fitness function (PHGPSR). We compare the proposed SP-GPSR with HGPSR and PHGPSR under the same experimental settings (as shown in Table 1), to verify the effectiveness of the proposed method. Table 2 and Table 3 show the simplified optimal solutions of all algorithms, with three significant digits retained. It can be seen that our method obtained the exact analytical solution within a limited number of iterations, while the other methods only obtained approximate solutions that can fit the observed data, but not the exact solution. This demonstrates that our method can accurately and efficiently solve differential equations.

We noticed that sometimes the best solutions of the other methods, after simplification, had the correct functional structure, but the constant coefficients were not consistent with the true solution. This is because the GP-generated symbolic expressions are redundant, and their constant coefficients interact with multiple variables or constants, making it difficult to optimize them by the gradient descent method. This further illustrates the effectiveness of the simplification-pruning operator.

Table 1. GPSR hyperparameters

Equation	Population size	Generations	p_c	p_m	p_s	Operator
Bernouli	100	20	0.6	0.6	0.8	$+, -, \times$
Poisson	50	20	0.6	0.6	0.8	$+, \times, sin$

Table 2. Optimal Solution of SP-GPSR, HGPSR and PHGPSR in Euler-Bernouli Equation

c	l	SP-GPSR	HGPSR	PHGPSR
1	2	$0.041 \times x^4 - 0.166 \times x^3 + 0.333 \times x$	$-0.326 \times x^2 + 0.658 \times x - 0.009$	$0.008 \times x^4 - 0.017 \times x^3 + 0.008 \times x$
1	4	$0.041 \times x^4 - 0.333 \times x^3 + 2.66 \times x$	$8.54 \times x^2 - 14.2 \times x$	$6.57e - 6 \times x^4 + 9.86e - 6 \times x^3 + 4.32e - 11 \times x^2 - 3.00 \times x - 3.28e - 6$
3	2	$0.124 \times x^4 - 0.499 \times x^3 + 0.999 \times x$	$-1.09 \times x^2 + 0.899 \times x$	$9.29e - 6 \times x^4 + 0.000193 \times x^3 + 0.000346 \times x^2 + 2.47e - 5 \times x$
3	4	$0.124 \times x^4 - 0.999 \times x^3 + 7.999 \times x$	$-1.40 \times x^2 + 5.37 \times x$	$-1.49e - 8 \times x^4 - 1.00 \times x^3 + 2.00 \times x^2 + 1.00 \times x$
5	2	$0.208 \times x^4 - 0.834 \times x^3 + 1.66 \times x$	$-1.36 \times x^2 + 2.49 \times x$	$0.491 \times x^4 - 0.120 \times x^3 - 0.245 \times x^2 - 1.12 \times x - 0.752$
5	4	$0.208 \times x^4 - 1.66 \times x^3 + 13.3 \times x$	$-4.00 \times x^2 - 5.82 \times x$	$1.10 \times x^4 + 0.103 \times x^3 - 2.20 \times x^2 + 0.907 \times x - 0.0928$
0.5	2	$0.0209 \times x^4 - 0.0835 \times x^3 + 0.166 \times x$	$3.22 \times x^2 - 5.80 \times x - 1.30$	$4.01e - 7 \times x^4 + 1.67e - 5 \times x^3 + 3.14e - 5 \times x^2 + 1.02 \times x + 0.0251$
0.5	4	$0.0208 \times x^4 - 0.166 \times x^3 + 1.33 \times x$	$0.0311 \times x^2 + 2.03 \times x$	$2.00 \times x^4 - 1.00 \times x^3 - x + 1.00$

Table 3. Optimal Solution of SP-GPSR, HGPSR and PHGPSR in Poisson's Equation

d	SP-GPSR	HGPSR	PHGPSR
1	$sin(3.14 \times x1)$	$sin(3.10 \times x1)$	$sin(3.14 \times x1)$
2	$sin(3.14 \times x1) \times sin(3.14 \times x2)$	$-0.504 \times sin(0.261 \times x2)$	$0.964 \times sin(0.983 \times x1) \times sin(0.983 \times x2)$

4.3 Sampling Number and Performance

In this experiment, we investigated the effect of sampling number on the performance of SP-GPSR. We varied the sampling number from 3 to 11 with a step size of 2 and ran the SP-GPSR algorithm independently 30 times for each sampling number. We measured the performance by the success rate, which is defined as the proportion of runs that found an exact solution within a given number of generations.

Figure 2 shows the cumulative distribution of the success rate of finding analytical solutions of the Euler-Bernouli Equation at $c = 5$, $l = 2$. The x-axis represents the number of generations, and the y-axis represents the cumulative distribution of the success rate. The plot shows that the success rate is not sensitive to the choice of sampling number. This indicates that the variability of the performance is not affected by the sampling number. Therefore, we can conclude that the sampling number does not have a significant impact on the performance of SP-GPSR.

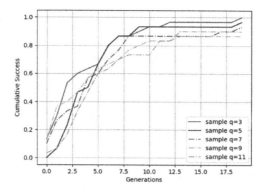

Fig. 2. Cumulative Distribution of Successful SP-GPSR Model Evolution

5   Conclusion

In this paper, we proposed a novel symbolic regression algorithm based on genetic programming with the simplification-pruning operator (SP-GPSR) to find analytical solutions to differential equations (DEs). Our method can generate simple and interpretable expressions that satisfy both the sampled data and the DEs. We conducted experiments on two typical classes of DEs. The experimental results have shown that our method can effectively find analytical solutions for DEs with high accuracy and simplicity. For future work, we plan to combine our method with transfer learning techniques, which can leverage the knowledge learned from previous problems to accelerate the solving process of new problems. Another direction is to incorporate some advanced techniques in genetic programming.

References

1. Bar-Sinai, Y., Hoyer, S., Hickey, J., Brenner, M.P.: Learning data-driven discretizations for partial differential equations. Proc. Natl. Acad. Sci. **116**(31), 15344–15349 (2019)
2. Bomarito, G., Townsend, T., Stewart, K., Esham, K., Emery, J., Hochhalter, J.: Development of interpretable, data-driven plasticity models with symbolic regression. Comput. Struct. **252**, 106557 (2021)
3. Brenner, M.: Machine learning for partial differential equations. In: APS March Meeting Abstracts, vol. 2021, pp. P61–003 (2021)
4. Bueno-Orovio, A., Pérez-García, V.M., Fenton, F.H.: Spectral methods for partial differential equations in irregular domains: the spectral smoothed boundary method. SIAM J. Sci. Comput. **28**(3), 886–900 (2006). https://doi.org/10.1137/040607575
5. Cao, L., Jiang, M., Feng, L., Lin, Q., Pan, R., Tan, K.C.: Hybrid estimation of distribution based on knowledge transfer for flexible job-shop scheduling problem. In: 2022 4th International Conference on Data-driven Optimization of Complex Systems (DOCS), pp. 1–6. IEEE (2022)

6. Deb, K., Pratap, A., Agarwal, S., Meyarivan, T.: A fast and elitist multiobjective genetic algorithm: NSGA-II. IEEE Trans. Evol. Comput. **6**(2), 182–197 (2002)
7. Denis, B.: An overview of numerical and analytical methods for solving ordinary differential equations. arXiv preprint arXiv:2012.07558 (2020)
8. Eymard, R., Gallouët, T., Herbin, R.: Finite volume methods. Handb. Numer. Anal. **7**, 713–1018 (2000)
9. Henner, V., Nepomnyashchy, A., Belozerova, T., Khenner, M.: Partial differential equations. In: Ordinary Differential Equations: Analytical Methods and Applications, pp. 479–545. Springer, Cham (2023)
10. Hernandez, A., Balasubramanian, A., Yuan, F., Mason, S.A., Mueller, T.: Fast, accurate, and transferable many-body interatomic potentials by symbolic regression. NPJ Comput. Mater. **5**(1), 112 (2019)
11. Hong, H., Jiang, M., Feng, L., Lin, Q., Tan, K.C.: Balancing exploration and exploitation for solving large-scale multiobjective optimization via attention mechanism. In: 2022 IEEE Congress on Evolutionary Computation (CEC), pp. 1–8. IEEE (2022)
12. Hong, H., Jiang, M., Yen, G.G.: Improving performance insensitivity of large-scale multiobjective optimization via Monte Carlo tree search. IEEE Trans. Cybern. (2023)
13. Jiang, M., Wang, Z., Guo, S., Gao, X., Tan, K.C.: Individual-based transfer learning for dynamic multiobjective optimization. IEEE Trans. Cybern. **51**(10), 4968–4981 (2020)
14. Karpatne, A., et al.: Theory-guided data science: a new paradigm for scientific discovery from data. IEEE Trans. Knowl. Data Eng. **29**(10), 2318–2331 (2017)
15. Meuris, B., Qadeer, S., Stinis, P.: Machine-learning-based spectral methods for partial differential equations. Sci. Rep. **13**(1), 1739 (2023)
16. Oh, H., Amici, R., Bomarito, G., Zhe, S., Kirby, R., Hochhalter, J.: Genetic programming based symbolic regression for analytical solutions to differential equations. arXiv preprint arXiv:2302.03175 (2023)
17. Sacchetti, A., Bachmann, B., Löffel, K., Künzi, U., Paoli, B.: Neural networks to solve partial differential equations: a comparison with finite elements. IEEE Access **10**, 32271–32279 (2022). https://doi.org/10.1109/ACCESS.2022.3160186
18. Wang, Z., Hong, H., Ye, K., Zhang, G.E., Jiang, M., Tan, K.C.: Manifold interpolation for large-scale multiobjective optimization via generative adversarial networks. IEEE Trans. Neural Netw. Learn. Syst. (2021)
19. Ying, L.: Partial differential equations and the finite element method. Math. Comput. **76**(259), 1693–1694 (2007). https://www.ams.org/journals/mcom/2007-76-259/S0025-5718-07-02023-6/home.html

Explainable Offensive Language Classifier

Ayushi Kohli[(✉)] and V. Susheela Devi

Indian Institute of Science, Bangalore, Karnataka, India
{ayushikohli,susheela}@iisc.ac.in

Abstract. Offensive content in social media has become a serious issue, due to which its automatic detection is a crucial task. Deep learning approaches for Natural Language Processing (or NLP) have proven to be on or even above human-level accuracy for offensive language detection tasks. Due to this, the deployment of deep learning models for these tasks is justified. However, there is one key aspect that these models lack, which is explainability, in contrast to humans. In this paper, we provide an explainable model for offensive language detection in the case of multi-task learning. Our model achieved an F1 score of 0.78 on the OLID dataset and 0.85 on the SOLID dataset. We also provide a detailed analysis of the model interpretability.

Keywords: Deep Learning · NLP · explainability and interpretability

1 Introduction

Be it social media platforms, e-commerce, or government organizations, all are in a way affected by offensive content on media. This calls for a need to look into the matter of having some automation while detecting comments or tweets as offensive or not-offensive. Using deep learning for text classification involving millions or billions of text samples is quite common. Though these NLP models achieve high performance in text classification, they act as a black box to us. Therefore there is a growing need to make these models explainable. To do so, we have two choices-first, either design inherently interpretable models. Or add explainability to the already designed model via post hoc explanations [26].

Post-hoc explanation [26] methods like LIME [1], SHAP [2], Integrated Gradients [3], etc., are quite popular. Yet there are several limitations that they suffer from, such as lack of stability, consistency, faithfulness, and robustness. One of the major concerns here is stability; that is, given the same test point, we get different explanations each time the explainability algorithm executes. This reduces trust in the model. There has been significantly less work done in designing models that are inherently interpretable. This paper is our attempt towards the same. Interpretability [25] can be classified further into the following two categories-

- **Model Interpretability:** Aims to explain what the model has learned.
- **Prediction Interpretability:** Aims to explain, for a given test data point, how the model arrived at its prediction.

B. Luo et al. (Eds.): ICONIP 2023, CCIS 1962, pp. 299–313, 2024.
https://doi.org/10.1007/978-981-99-8132-8_23

In this paper, we have first designed a model for multi-task [24] tweet classification. The task is based on a three-level hierarchy, including three sub-tasks-(i) Offensive Language Detection (Subtask A), (ii) Categorization of Offensive Language (Subtask B), (iii) Offensive Language Target Identification (Subtask C). Secondly, we talk about the interpretability of our model. First, we take the word embeddings of each word, which is then given to a CNN [4]. CNNs have been studied for interpretability in computer vision tasks. Images form a continuous space, and the spatial relationships between pixels are captured by CNNs, unlike text data, which forms a discrete space. Due to this CNNs for text is like applying brute force. They do not look for specific patterns in the text like those in the case of images; instead, they try out each and every combination of n-grams. This is analogous to computer vision, where each filter of a convolution layer specializes in detecting a particular pattern; for example, one filter detects the horizontal edges. Similarly, for text data, each filter specializes in detecting different types of n-grams; for example, one filter would capture all the positive sentiment words like good, great, fantastic, etc.

Our main contributions are as follows:

- We provide an interpretable model for hierarchical multi-level tweet classification with a high F1 score [23] for both OLID [5] and SOLID [6] datasets.
- We have calculated the threshold for each filter; that is, all the n-grams above this threshold contribute to the classification decision.
- With the help of clustering [22], we found that the filters are not homogeneous; that is, each filter activates on multiple distinct types of n-grams.
- We associated each filter to a class and analyzed to which class a particular filter relates the most.

2   Related Work

In [7], the authors discuss various methods like Naive Bayes [14], SVM [15], LSTM [16], and explainability methods like LIME [1] and LRP [17] to make offensive language detection explainable. An approach for interpreting CNNs for text by exploiting the local linear structure of ReLU-DNNs [18] has been developed in [8]. In [9], the authors talk about the interpretability aspects of ReLU-DNNs [18], that is, how a deep ReLU network can be decomposed into many local linear models, which are interpretable. The role of CNNs in text classification tasks along with interpretability aspects, is discussed in [10]. The authors in [11] have proposed a model for multi-task tweet classification. However, our model is better than theirs, as our model is designed to be inherently interpretable and simpler (Fig. 1).

3   Proposed Method

1. Model Architecture: Here, we present our multi-task [24] model for offensive tweet classification for subtasks A, B, and C. First, we take word embedding corresponding to each word of each sentence of the dataset with the help

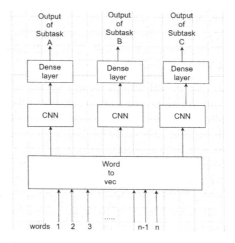

Fig. 1. Our proposed model for offensive tweet classification (subtask A), categorization of offensive language (subtask B), and offensive language target identification (subtask C).

of word2vec [12]. Then, these word embeddings are given to the convolution layer, where the convolution operation is performed between word embeddings and different filters of different sizes. This is followed by the max-pooling operation. The output of max-pooling is further given to a dense layer, which outputs its decision for subtasks A (offensive/not-offensive), B(targeted insult and threats/untargeted), or C(individual/group/other). As shown in the diagram, the word embeddings generated from word2vec are simultaneously given as input to three convolution layers each for a subtask.

2. Text Processing by CNNs: Now we look at the ways CNNs process text data. The word embeddings are given to the convolution layers. The convolution layer consists of filters of different sizes. Here, in our experiments, we have used bi-gram, tri-gram, and 4-gram filters. Filters act as n-gram detectors by giving weights to all possible n-grams and, thus, giving the highest weight to the most relevant n-gram. After this convolution operation is performed between the filter and word embeddings, max-pooling is performed. Thus, a max-pooled vector is the output of the max-pooling layer. This max-pooled vector is then given to the dense layer, which gives the final classification result. As the final classification decision depends only on the max-pooled vector, it is clear that the n-grams not in the max-pooled vector do not participate in the classification decision. But what about the n-grams in the max-pooled vector? Does the entire max-pooled vector contribute to the classification decision? We can separate the n-grams in the max-pooled vector into two categories of n-grams: n-grams contributing to the classification decision and n-grams not contributing to the classification decision. The n-grams that contribute to the classification decision are n-grams in a max-pooled vector, leading to this classification output. The n-grams which do not contribute to the classification decision are n-grams in the max-pooled vector that do not influence the classification output. They were chosen in the max-pooled vector just to choose an element. We will now separate these two types of n-grams in the max-pooled vector.

3. Determining filter thresholds: The max-pooled vector is multiplied by the weight matrix of the dense layer. The dense layer has a number of nodes equal to the number of classes. Thus, the output is the class corresponding to the highest score resulting from the multiplication between the weight matrix and the max-pooled vector. In the max-pooled vector, one value came from each filter. Thus, in a way, we are multiplying the filter scores (max-pooled vector) with the class weights (weight matrix of the dense layer) to arrive at the final classification decision. Thus each filter can be associated with a class. We associate each data point with the filter assigned class to that point. We also have a true label corresponding to every data point. This can be understood in more detail through the following toy example. Let's take the following sentence and form bi-grams from it, as shown in Table 1, **Brilliant movie must watch it!**

Table 1 shows the scores assigned to these bi-grams when passed through the filters.

Table 1. Filter scores assigned to different bi-grams.

n-grams	Brilliant, movie	movie, must	must, watch	watch, it	it, !
Filter 1	**0.8**	0.6	0.4	0.3	0.1
Filter 2	0.4	0.3	**0.7**	0.5	0.3
Filter 3	0.1	**0.6**	0.4	0.2	0.3
Filter 4	0.3	0.5	0.4	0.1	**0.6**

Table 2 shows the maximum score assigned by each filter to the n-grams (maximum of a row of Table 1), forming the max-pooled vector.

Table 2. Max-pooled vector

Filter 1	0.8
Filter 2	0.7
Filter 3	0.6
Filter 4	0.6

The first three rows of Table 3 show the layer weights (3 * 4 matrix). The last row shows the class associated with each filter, that is, the maximum weight of different classes for a filter (maximum of a column). Then this weight matrix

Table 3. Dense layer weights (first three rows) and the class associated with each filter (last row)

Filters	Filter 1	Filter 2	Filter 3	Filter 4
class 1	1	1	2	3
class 2	2	4	3	7
class 3	5	1	6	4
associated class	3	2	3	2

is multiplied with the max-pooled vector (Table 2) to obtain the classification outcome (Table 4).

$$\begin{pmatrix} 1 & 1 & 2 & 3 \\ 2 & 4 & 3 & 7 \\ 5 & 1 & 6 & 4 \end{pmatrix} * \begin{pmatrix} 0.8 \\ 0.7 \\ 0.6 \\ 0.6 \end{pmatrix} = \begin{pmatrix} 4.5 \\ 10.4 \\ 10.7 \end{pmatrix}$$

Table 4. Classification output

class 1	4.5
class 2	10.4
class 3	**10.7**

As shown in Table 3, we used the trained weight matrix to associate each filter with a class. This associated class is the maximum of each column, as seen in the table. There also exists a true class for every data point. For every filter, we have a value corresponding to the max-pooled vector score assigned by the filter to the n-gram. We also have the class associated with each filter and the actual class of every data point. We now label each point (point means max-pooled vector score corresponding to a filter) as 1 if the class associated with the filter matches the actual class and 0 if the class associated with the filter does not match the actual class. So we now have points consisting of (max-pooled vector score, 1) or (max-pooled vector score, 0). And we plot these points, resulting in a 1-dimensional scatter plot as shown in Fig. 2. We plot these plots for every filter. The point separating the two classes determines the threshold for each filter. As shown in Fig. 2, the threshold, that is, the point separating the blue and magenta points, is around 6.5. We use these filter thresholds to find whether an n-gram activates a given filter. If the n-gram score is above the filter threshold, the n-gram is relevant to the filter; else, it is irrelevant to the filter. This way, we find which n-grams were responsible for the classification decision. This can be thought of as analogous to what filters do in the case of images. For example, in the case of images, there is a filter responsible for detecting horizontal edges; similarly, in the case of text, we can say that there is a filter that activates these

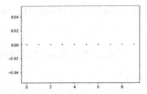

Fig. 2. n-gram scores

sets of n-grams. So, in a way, if we take all the n-grams in our corpus and pass them to a specific filter, then the filter would assign scores to these n-grams through convolution operation, and all the n-grams whose scores are above the threshold of the filter, are the n-grams which are described by the filter. So, now we know which n-grams activate the filter, but even an n-gram is composed of n words. So, can we further know which words in the n-gram are responsible for activating the filter?

The n-gram score is a sum of individual word scores. And thus can be decomposed into individual word scores by considering the inner products between every word embedding and every filter. So, we decompose n-gram scores into individual word scores.

4. Clustering: We now perform clustering [22] on these individual word scores. For clustering, we have used the Hierarchical Agglomerative Clustering [13] algorithm, which does not require prior knowledge of the number of clusters. This clustering gave us some useful insights regarding filter behavior. We found that each filter captures different semantic classes of the n-grams, and each class has some words which activate the filters, whereas some words do not activate the filter. Thus, each filter forms multiple clusters of n-grams.

5. Interpretability: Interpretability can be classified into two categories:

 i) Model Interpretability
 ii) Prediction Interpretability
   - **Model Interpretability:** Model interpretability aims to find what the model has learned during training.

 i) We have found the class each filter associates with the most.
 ii) The threshold value for each filter captures which n-grams are activated by the filter.
 iii) Through clustering, we found the linguistic patterns identified by each filter.
   - **Prediction Interpretability:** It aims to explain how the model arrived at the particular prediction for a given input. For example, in the case of sentiment classification, if for a given input the model classifies as positive, prediction interpretability aims to find which words in the input text made the model classify the input as positive. Here, instead of looking at all the words for interpretability, we have looked only at the n-grams present in the max-pooled vector. We also find which word in the n-gram influences the classification decision the most.

4 Experiments

We now present the experiments we performed to justify the above-discussed theory.

Datasets Used: We have performed experiments on two datasets for offensive language classification.

- **Offensive Language Identification Dataset (OLID)** [5]: This dataset was collected by Twitter. It comprises of 14,100 tweets, out of which 13,240 tweets are in the training set and the remaining 860 are in the test set. There are three levels of labels which are as follows:
  - i) Offensive/Not Offensive
  - ii) Targeted Insult and Threat/Untargeted
  - iii) Individual/Group/Other

  The relationship between these levels is hierarchical. For instance, if a tweet is offensive, it can be targeted or untargeted. If a tweet is offensive and targeted, it can target an individual, group, or other.
- **Semi-Supervised Offensive Language Identification Dataset (SOLID)** [6]: SOLID is similar to OLID, with the three subtasks being the same as for OLID, with a hierarchical relationship between them. However, the size of OLID is limited, whereas, in the case of SOLID, there are 9,089,140 tweets for subtask A and 1,88,973 tweets for subtasks B and C. SOLID contains average confidence (AVG_CONF) values lying between 0 to 1, which is converted to respective class by setting the threshold to 0.4. This way, SOLID is converted to a classification dataset.

Data Pre-processing Steps: First, all the characters are converted to lowercase, and all the extra spaces are removed, followed by the following pre-processing steps applied to both the datasets:

- Replacing @USER tokens - All the tweets with multiple @USER tokens were replaced by a single @USERS tokens.
- All the emojis were converted to their respective words.
- All the hashtags in the tweets were segmented by capital words.

Model Components Used

- **Word2Vec:** In order to get the word embeddings, we have used the word2vec [12] algorithm. For each word, we obtained an embedding of 300 dimensions.
- **CNN:** We have used filter sizes of {3,4,5} words, and {8,32,64} filters.
- **Loss function:** Weighted Cross-Entropy loss [19] is used, weighted over all three subtasks. The loss function,

$$L = \sum_{i \in \{A,B,C\}} w_i L_i$$

L_i is cross-entropy loss and weights w_i is loss weights for subtasks A, B, and C, found by cross-validation [20] and

$$\sum_{i \in \{A,B,C\}} w_i = 1$$

Table 5. F1 scores for subtask A

Dataset	F1 score
OLID	0.78
SOLID	0.85

Table 5 shows the macro-F1 [21] scores for subtask A for both datasets. We can see that the model performs better in the case of SOLID.

Table 6. Top tri-gram scores and word scores from a sample of eight filters for OLID dataset

Filter	n-gram	top n-gram scores	word scores for n-grams
0	black hearted troll	**14.93**	4.5, 4.76, 5.67
1	piece of s***	10.81	4.1, 0.12, 6.59
2	you are correct	6.19	1.45, 0.54, 4.2
3	he is black	5.9	0.24, 0.11, 5.55
4	so very important	3.78	0.34, 0.21, 3.23
5	drunk crazy man	9.57	3.56, 4.67, 1.34
6	all the s***	7.37	0.24, 0.35, **6.78**
7	humor or knowledge	9.14	3.56, 0.11, 5.47

Table 7. Top tri-gram scores and word scores from a sample of eight filters for SOLID dataset

Filter	n-gram	top n-gram scores	word scores for n-grams
0	who the f***	8.15	1.25, 0.12, 6.78
1	white domestic terrorist	12.35	2.67, 3.9, 5.78
2	a sunny day	6.01	0.13, 3.34, 2.54
3	hatred for america	9.04	4.9, 0.68, 3.46
4	a binary choice	4.38	0.08, 2.8, 1.5
5	f****** idiot spanish	13.92	**6.99**, 3.28, 3.65
6	f****** f****** finally	**16.87**	**6.99**, **6.99**, 2.89
7	out of season	6.34	2.4 , 0.44 , 3.5

Tables 6 and 7 shows the maximum tri-gram scores from eight filters for the OLID and SOLID dataset. We can see that filters 0,1,3,5, and 6 detect offensive tri-grams, whereas filters 2,4 and 7 detect non-offensive tri-grams, for both the datasets.

Table 8. Top tri-gram scores and word scores from a sample of eight filters for the OLID dataset by taking all possible tri-grams of the dataset.

Filter	word 1	word 2	word 3
0	gun—**7.4**	memory—5.5	sandyhook—6.5
1	manifesto—5.1	congress—3.7	commonsense—6.6
2	goldsmith—3.74	collusion—2.4	impeccable—4.5
3	polygraph—4.2	canada—3.23	black—5.55
4	hoarding—2.5	million—1.3	legalizes—3.8
5	cannabis—3.64	money—4.8	righteous—2.23
6	discretion—2.6	skeptics—2.06	s***—6.78
7	overbearing—4.56	servitude—3.89	arrogantly—6.1

Table 9. Top tri-gram scores and word scores from a sample of eight filters for the SOLID dataset by taking all possible tri-grams of the dataset.

Filter	word 1	word 2	word 3
0	tindernobody—2.3	lisp—3.45	f***—6.78
1	pussy—2.5	credit—3.5	adorable—2.67
2	dirty—2.7	concerned—4.05	emotional—6.54
3	narcissists—4.3	psychopaths—5.7	walmart—4.2
4	stupid—5.4	vibe—1.23	ugly—3.5
5	garbage—**7.5**	bro—3.3	obama—4.2
6	irrelevant—3.76	pathological—3.3	trump—4.03
7	ditch—7.23	mirror—7.01	gucci—4.21

Table 8 shows the highest tri-gram scores by eight filters when every possible tri-gram from the entire dataset is passed through the filters for the OLID dataset. The maximum n-gram score, that is, the sum of individual word scores, is higher ($7.4 + 5.5 + 6.5 = 19.4$) here as compared to that in Table 6, where the maximum n-gram score was 14.93, and the n-grams were taken as given in the dataset. We can see that each word score for a particular filter in this table is higher than the word score of the respective filter in Table 6. Table 9 shows the highest tri-gram scores by eight filters when every possible tri-gram from the entire dataset is passed through the filters for the SOLID dataset. Here the maximum tri-gram score is 18.45 (7.23 + 7.01 + 4.21 = 18.45), which is again higher than the maximum n-gram score given in Table 7, that is, 16.87. Thus, we can see that the maximum n-gram score occurs when all possible combinations of n-grams are taken from the dataset. We can see that word scores for a particular filter in this table is higher than the word scores of the respective filter in Table 7, with some exceptions. Also, we can see in Tables 8 and 9 that here no pattern is formed as in Tables 6 and 7, where some filters detected offensive tri-grams and some non-offensive tri-grams. Here, in Tables 8 and 9, the filters are just selecting the words in every tri-gram which led to its maximum activation score.

Table 10. Top-8 tri-gram scores from a filter for OLID dataset

No	n-gram	word scores for n-grams
1	piece of s***	4.1, 0.12, 6.59
2	make s*** holes	3.4, **6.59**, 5.45
3	women make s***	3.77, 3.4, **6.59**
4	a s*** hole	0.23, **6.59**, 5.44
5	all the s***	0.55, 0.12, **6.59**
6	like that s***	2.2, 0.32, **6.59**
7	with good reason	1.22, 5.43, 3.5
8	only good person	3.45, 5.43, 4.78

Table 10 shows some useful insights captured by a particular filter for the OLID dataset. This filter captures some offensive words (Numbers 1 to 6) and some non-offensive words (Numbers 7 and 8).

Table 11 shows the top eight tri-grams captured by a filter for the SOLID dataset. Tri-grams with numbers 1,2,3,4,5 and 7 are the offensive tri-grams that the filter detects, whereas numbers 6 and 8 are the non-offensive tri-grams that the filter detects. Thus we can see from Tables 10 and 11 that a filter can detect both types of tri-grams, offensive and non-offensive. Also, it is clear from this observation that a particular filter can detect multiple different types of n-grams and is not restricted to detecting a particular type of n-gram.

Tables 12 shows the maximum tri-gram scores from a particular filter for the OLID dataset, where tri-grams are formed by taking all possible combinations of words from the entire dataset. It is clear from Tables 10 and 12 that this filter focuses on words like s***, good etc. Table 13 shows the maximum tri-gram scores from a particular filter for the SOLID dataset, where tri-grams are formed by taking all possible combinations of words from the entire dataset. This filter focuses on words like f***, happy, etc.

Table 14 (Appendix) shows the result of clustering applied to tri-gram scores for the OLID dataset. One cluster focuses on offensive words like b*******,

Table 11. Top-8 tri-gram scores from a filter for SOLID dataset

No	n-gram	word scores for n-grams
1	who the f***	2.25, 0.11, **6.43**
2	what the f***	2.5, 0.11, **6.43**
3	f****** boring person	**6.5**, 3.22, 3.77
4	scared as f***	4.27, 0.66, **6.43**
5	a f****** ford	0.31, **6.5**, 5.33
6	happy everyone is	4.22, 5.34, 0.35
7	so f****** bored	0.34, **6.5**, 3.4
8	happy birthday to	4.22, 5.13, 0.41

Table 12. Top-8 tri-gram scores from a particular filter for the OLID dataset by taking all possible tri-grams of the dataset.

No	word 1	word 2	word 3
1	thank—1.48	s***—**6.59**	control—5.45
2	what—3.42	good—5.43	control—5.45
3	strict—3.2	result—6.47	audience—5.44
4	people—1.03	s***—**6.59**	selective—5.54
5	food—3.34	good—5.43	testify—5.44
6	circus—3.54	s***—**6.59**	murder—4.55
7	diamond—2.6	criminal—6.13	control—5.45
8	democracy—5.5	good—5.43	violence—4.56

whereas the other focuses on non-offensive words like **beautiful**. Table 15 (Appendix) shows the result of clustering applied to tri-gram scores for the SOLID dataset. Here, one cluster focuses on offensive words like **a****, the second cluster focuses on words like **congratulate**, while the third focuses on words like **pretty**, etc. One conclusion that can be drawn from Tables 14 and 15 (Appendix) is as follows:

Filters do not detect one particular type of n-grams; rather, they detect multiple different types of n-grams, thus forming different clusters as shown here. N-grams within a cluster are similar to one another. But as shown here, a filter forms multiple clusters and not just a single cluster, clearly stating that a filter does not just detect a particular type of n-gram. Figure 3 (Appendix) shows the results when clustering is applied to a filter for both datasets. As shown in Fig. 3 (a), two clusters are formed for OLID, showing the non-homogeneous nature of the filter. For the SOLID dataset (Fig. 3 (b)), the filter forms multiple clusters, confirming the non-homogeneous behavior of the filter.

Table 13. Top-8 tri-gram scores from a particular filter for the SOLID dataset by taking all possible tri-grams of the dataset.

No	word 1	word 2	word 3
1	happy—4.22	and—0.13	f***—**6.43**
2	s***—4.56	rapist—4.78	f***—**6.43**
3	racist—3.45	police—6.34	apologies—4.55
4	happy—4.22	drug—5.44	bored—6.41
5	racist—3.54	hypocrites—5.45	buddy—5.43
6	happy—4.22	fight—3.42	f******—6.5
7	her—1.27	embarrassment—2.32	b*******—4.52
8	spoiling—2.46	disgusting—3.67	f***—6.5

5 Conclusion

In this paper, we have presented an interpretable model for text classification for offensive language detection. We have shown a detailed analysis of the internal working for the entire model. We looked at interpretability through two lenses, model interpretability and prediction interpretability. Through model interpretability, we have shown the class each filter associates with, the threshold value for each filter, and the clusters of n-grams that activates a particular filter. Through prediction interpretability, we have shown which n-grams are responsible for predicting the class of the input sentence. Here we have only used the n-grams, which were presented in the max-pooled vector, as they only contribute to the classification task. Finally, we have a useful insight to the working of filters through·clustering; that is, a filter does not form a single cluster of n-grams, but rather a single filter forms multiple different clusters of n-grams. There has been very less work done in the area of inherently interpretable models for NLP. This paper is one such attempt. More work needs to be done in this area of research. We can try other model architectures and see the internal working of these models for interpretability in future work.

Acknowledgments. We gratefully acknowledge the financial support provided by Fujitsu Research India Private Limited. This funding played a crucial role in facilitating our research and enabling us to achieve our research objectives.

Appendix

Due to space constraints, we have added the results of clustering applied to SOLID and OLID datasets here.

Table 14. Clustering the tri-grams scores of a filter for OLID dataset

n-gram	word scores for n-gram	cluster
the b******* flag	0.15, **7.32**, 5.43	0
made up b*******	3.24, 0.80, **7.32**	0
insanely ridiculous b*******	4.56, 3.55, **7.32**	0
another b******* story	3.24, **7.32**, 3.86	0
country has b*******	4.33, 0.34, **7.32**	0
a beautiful person	0.34, 5.13, 3.44	1
the beautiful bears	0.15, 5.13, 4.43	1
she is beautiful	0.25, 0.64, 5.13	1

Table 15. Clustering the tri-grams scores of a filter for SOLID dataset

n-gram	word scores for n-gram	cluster
his weirdo a**	1.13, 3.23, **6.45**	0
by fake a**	0.23, 3.42, **6.45**	0
your crusty a**	0.56, 3.23, **6.45**	0
to congratulate others	0.22, 5.57, 4.45	1
congratulate with one	5.57, 0.23, 2.54	1
pretty good stuff	5.32, 3.46, 4.86	2
pretty easy to	5.32, 3.23, 0.22	2
please pretty please	4.23, 5.32, 4.23	2

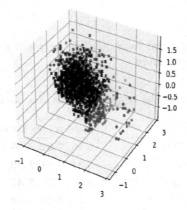

(a) OLID dataset

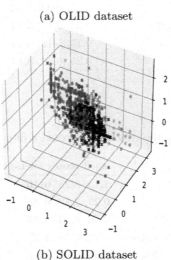

(b) SOLID dataset

Fig. 3. Clustering applied to a filter for both datasets.

References

1. Ribeiro, M.T., Singh, S., Guestrin, C.: Why Should I Trust You?: Explaining the Predictions of Any Classifier (2016)
2. Lundberg, S., Lee, S.-I.: A Unified Approach to Interpreting Model Predictions (2017)
3. Sundararajan, M., Taly, A., Yan, Q.: Axiomatic Attribution for Deep Networks (2017)
4. O'Shea, K., Nash, R.: An Introduction to Convolutional Neural Networks (2015)
5. Zampieri, M., Malmasi, S., Nakov, P., Rosenthal, S., Farra, N., Kumar, R.: Predicting the Type and Target of Offensive Posts in Social Media (2019)
6. Rosenthal, S., Atanasova, P., Karadzhov, G., Zampieri, M., Nakov, P.: SOLID: A Large-Scale Semi-Supervised Dataset for Offensive Language Identification (2020)

7. Risch, J., Ruff, R., Krestel, R.: Explaining Offensive Language Detection (2020)
8. Zhao, W., Singh, R., Joshi, T., Sudjianto, A., Nair, V.N.: Self-interpretable Convolutional Neural Networks for Text Classification (2021)
9. Sudjianto, A., Knauth, W., Singh, R., Yang, Z., Zhang, A.: Unwrapping The Black Box of Deep ReLU Networks: Interpretability, Diagnostics, and Simplification (2020)
10. Jacovi, A., Shalom, O.S., Goldberg, Y.: Understanding Convolutional Neural Networks for Text Classification (2018)
11. Dai, W., Yu, T., Liu, Z., Fung, P.: Kungfupanda at SemEval-2020 Task 12: BERT-Based Multi-Task Learning for Offensive Language Detection (2020)
12. Mikolov, T., Chen, K., Corrado, G., Dean, J.: Efficient Estimation of Word Representations in Vector Space (2013)
13. Müllner, D.: Modern hierarchical, agglomerative clustering algorithms (2011)
14. Rish, I.: An empirical study of the Naive Bayes classifier (2001)
15. Evgeniou, T., Pontil, M.: Workshop on Support Vector Machines: Theory and Applications (1999)
16. Hochreiter, S., Schmidhuber, J.: Long Short-term Memory (2010)
17. Montavon, G., Binder, A., Lapuschkin, S., Samek, W., Müller, K.-R.: Layer-wise relevance propagation: an overview. In: Samek, W., Montavon, G., Vedaldi, A., Hansen, L.K., Müller, K.-R. (eds.) Explainable AI: Interpreting, Explaining and Visualizing Deep Learning. LNCS (LNAI), vol. 11700, pp. 193–209. Springer, Cham (2019). https://doi.org/10.1007/978-3-030-28954-6_10
18. Agarap, A.F.M.: Deep Learning using Rectified Linear Units (ReLU) (2018)
19. Özdemir, Ö., Sönmez, E.B.: Weighted Cross-Entropy for Unbalanced Data with Application on COVID X-ray images (2020)
20. Berrar, D.: Cross-validation
21. Sokolova, M., Japkowicz, N., Szpakowicz, S.: Beyond accuracy, f-score and ROC: a family of discriminant measures for performance evaluation. In: Sattar, A., Kang, B. (eds.) AI 2006. LNCS (LNAI), vol. 4304, pp. 1015–1021. Springer, Heidelberg (2006). https://doi.org/10.1007/11941439_114
22. Bindra, K., Mishra, A.: A Detailed Study of Clustering Algorithms (2017)
23. Lipton, Z.C., Elkan, C., Naryanaswamy, B.: Thresholding Classifiers to Maximize F1 Score (2014)
24. Zhang, Y., Yang, Q.: A Survey on Multi-Task Learning (2021)
25. Zhang, Y., Tino, P., Leonardis, A., Tang, K.: A Survey on Neural Network Interpretability (2021)
26. Lakkaraju, H., Adebayo, J., Singh, S.: Explaining Machine Learning Predictions: State-of-the-Art, Challenges, Opportunities (2020)

Infrared Target Recognition Technology Based on Few Shot Learning

Jing Zhang[1,2], Yangyang Ma[3], and Weihong Li[2(✉)]

[1] College of Aeronautics, Northwestern Polytechnical University, Xi'an 710072, China
[2] First aircraft design and Research Institute of aviation industry, Xi'an, China
251834800@qq.com
[3] Luoyang electro optic equipment Research Institute of aviation industry,
Luoyang 471000, China

Abstract. Aiming at the problems of sparse sample data and high difficulty of embedded implementation under the constraint of limited resources in the military application of infrared target recognition, this paper proposes a lightweight target recognition technology based on few shot learning. This technology improves the structure of generator and discriminator network by designing the Cycle Generative Adversarial Network model, and realizes the migration from visible image to infrared image, so as to achieve the purpose of expanding the training data; Through the improvement of YOLOV5s network, the recognition accuracy is improved without reducing the magnitude of model parameters, and the characteristics of high real-time processing are retained. The experimental results show that the generative adversarial network model designed based on this project can process the visible image and generate the near-infrared image. After adding the training data, the model accuracy is effectively improved. The improved YOLOV5s model is 2% higher than the original model map 0.5, and is easier to be embedded.

Keywords: Generative adversarial network · Infrared target recognition · Convolutional neural network

1 Introduction

As one of the important detection sensors of aircraft, infrared sensor has the advantages of reliable operation and strong anti-interference ability, and infrared target recognition technology is one of the important technologies to improve the intelligent application of airborne infrared sensor.

At present, the target recognition technology usually adopts the deep learning method [1] to realizes the intelligent recognition of the target in the image by building the deep convolution neural network model. Target recognition technology based on deep learning is mainly divided into two stage [2,3] and one stage [4–6]. The deep learning target detection and recognition algorithm based on two stage structure is mainly divided into two steps. First, a series of candidate frames are extracted from the image through the Regional Proposal Network, and then each candidate frame is further classified and regressed through the convolution network in the second stage to obtain the recognition results. The deep learning target detection and recognition algorithm based on one stage structure does not need to extract the candidate target region in advance, but solves the target location, category and confidence by regression, without the saliency detection process, so it is faster than the former.

However, either deep learning method requires a large number of learning samples for modeling, otherwise it is difficult to obtain recognition models with high generalization and robustness. And compared with the visible light data, the infrared target database is relatively sparse, especially the military target samples. Therefore, the small sample size is one of the major problems faced by infrared target recognition. In addition, deep learning models are often complex, with hundreds of millions of parameters occupying large amounts of memory and computational resources. In military applications, due to the constraints of environmental adaptability, hardware, volume and power consumption of equipment, the resources of hardware platform required by target recognition model are limited, and it is difficult to deploy complex deep learning model.

In view of the lack of infrared image, this paper uses the data augmentation technology based on the generative adversarial network to process the visible image and generate the near-infrared image [7,8]. In the generative adversarial networks, Pix2Pix [9] and CycleGAN [10] provide a general network framework for converting visible images to infrared images. Aiming at the problems of unclear texture and missing structure in the infrared image generated by the general method, this paper takes the visible image as the input, outputs the equivalent infrared image, improves the quality of the generated infrared image by improving the generator network and discriminator network, and designs the generator network based on SENet [11] to effectively use the deep and shallow features of the extracted visible image, Through the feature statistics of the generated infrared image, the discriminator network can guide the gray level and structure information of the generated infrared image, weaken the constraint on the generator network, and release the greater potential of the generator network.

Aiming at the problem of embedded implementation caused by the complexity of deep learning model, this paper adopts lightweight network design. At present, there are two main directions of research on lightweight networks: first, constructing complex high-precision models, and then compressing the model parameters and streamlining the structure without losing accuracy through network pruning techniques [12]; second, designing lightweight models [13], and then further improving the accuracy of the models through learning training or design improvement [14]. However, the first kind of methods encounters the problem of reduced generalization in practical applications. At present, network compression methods are limited to limited data sets. However, for unlimited application scenarios, it is inevitable to encounter such situations: during pruning iterative training, the "unimportant" network connections pruned off are not really unimportant, but are only unimportant on the current data set, and often play a key role in some other scenarios, This leads to the problem that the generalization of the pruned model decreases when applied. In this paper, YOLOV5s [15] with balanced accuracy and speed is selected for improvement.

2   Infrared Image Generation Based on Generative Adversarial Network Model

The target database is a necessary for the training of deep learning algorithm. In practical application, image acquisition and target annotation require a lot of work, which makes the iteration cycle of the model longer. At present, most of the data sets are visible images, such as COCO, ImageNet, VisDrone, and so on. The training sample learning method based on generative adversarial network can simulate the real data distribution and generate sample data to improve the performance of target recognition model under the condition of few shot.

2.1   Structure Design of Generative Adversarial Network Model

The generative adversarial network constructs a confrontation game framework consisting of a generator model and a discriminator model [16]. The generator model G is trained by learning the joint probability distribution of samples and labels $P(X, Y)$ through the observation data to capture the high-order correlation of the data, and generate new data in line with the sample distribution by learning the essential characteristics of the real data; The discriminator model D samples and judges the data generated by the generation model and the real samples in the training set, predicts the probability of the data source, and identifies the authenticity of the input samples. The generator model G continuously improves the data generation ability according to the discrimination results of the discriminator model D, making the generated results closer and closer to the real samples, making it difficult for the discriminator model to distinguish; The discriminator model D improves its ability to distinguish the true from the false by constantly learning the difference between the generated data and the real samples. The generator model and the discriminator model are mutually antagonistic, game and promote each other, so that their respective performance is continuously improved iteratively until the generator model and the discriminator model converge and reach the Nash equilibrium. Finally, the generator model can generate data that is confused with the real, while the discriminator model cannot distinguish the generated data from the real samples, so as to obtain a generator model with superior performance.

In the method of generative adversarial network, the cycle generative adversarial network does not need to label data, nor does it need paired data, but it can achieve good conversion effect between image domains, and has strong practicability. Considering the lack of paired visible light and infrared images in practical applications, this paper uses the cycle generative adversarial network of unpaired style migration network as the basic architecture. By designing a bidirectional cycle learning strategy, it completes the Bi shot constraint between the source domain and the target domain, and expands the infrared data set based on visible light migration.

2.2   Generator Design of Cycle Generative Adversarial Network Model

The generator of Cycle Generative Adversarial network is composed of three parts: encoder, transformation and decoder. In the first stage, the encoder is composed of three convolution layers, which is mainly used to extract the features of the input image for encoding. In the second stage, the transformation layer is composed of several ResNet Blocks [17]. ResNet Blocks can add the input image and the image after two layers of convolution to ensure that the output image and the input image are as similar in structure as possible. The generator is shown in Fig. 1.

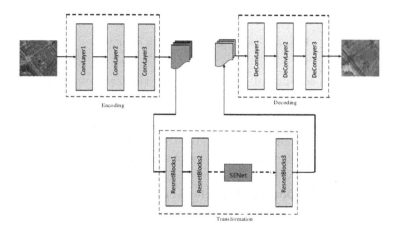

Fig. 1. Generator Network Structure Diagram

In the transformation layer of the generator, the input features are fused with the convoluted output features. The transformation layer consists of nine identical residual convolution blocks [17] and an intermediate SENet network.

In this paper, the structure of SENet is improved and designed as a channel attention module. The channel attention module can learn the relationship between the channels of the input features, and learn the multi-channel features for the input or the important parts of the images for the whole. Each feature channel can be regarded as a feature detector. After the global average pooling operation, the global compressed feature vector of the current feature graph is obtained. After activating the function Relu, it is connected to the bottleneck structure constructed by 1×1 convolution layer.

2.3 Design of Discriminator for Cycle Generative Adversarial Network Model

For the discriminator part, different from the PatchGAN [18] discriminator which only focuses on the receptive field of the image, this paper designs a multi-scale local discriminator to distinguish the true or false in the visible image and the synthetic infrared image. The structure is shown in Fig. 2.

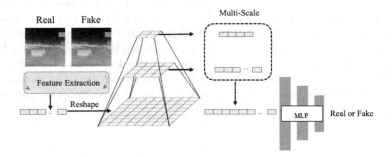

Fig. 2. Discriminator Network Structure

Specifically, the adversarial network extracts the features of the input image through the pre trained ViT model, and then reshapes the size of the obtained features. It uses the average pooling to down sample the features, extracts the features through the convolution layer with the same structure on three different scales of the features, and then calculates the statistical features. Finally, the total loss is obtained by adding the losses obtained from the linear layer.

2.4 Loss Function Design of Cycle Generative Adversarial Network Model

After the structure of generator and discriminator is determined, the loss function of the whole network model is defined as:

$$L = L_{adv} + \alpha L_{cys} + \lambda L_{percep} + \beta L_{SSIM} \tag{1}$$

where, α, λ, β are the super parameters. By changing this value, the regularization effect of the whole model can be achieved. Larger values make the network model have better reconstruction loss, but the network generalization performance is reduced. Smaller values increase the universality of the model and may introduce artificial defects. L_{adv} is the adversarial loss, the purpose is to enable the network to learn the distribution of the image in the target domain. In the process of generating the infrared image from the visible image, its generator G_X and discriminator D_X correspond to the sum respectively, which X is the real visible image, and Y is the real infrared image, then:

$$L_{adv} = \sum [\log D_X(y)] + \sum [1 - \log D_X(G_Y(x))] \tag{2}$$

L_{cys} is the cycle consistency loss, where $G_Y(x)$ is the real visible image, is the generated infrared image, $D_X(G_Y(x))$ is the visible image reconstructed by the cycle generator and the loss is defined by L_2 norm:

$$L_c(G_X, G_Y) = \sum_t ||x - G_X(G_Y(x))||^2 \tag{3}$$

L_{percep} is the perception loss, where x_r is the real visible image, x_f is the generated visible image, and ϕ is the feature value of VGG network output layer. The visual perception effect of the generated image is measured by comparing the Euclidean distances of different levels of features:

$$L_{percep} = \frac{1}{N} \sum_{k=1}^{N} (\phi_{i,j}^k(x_r) - \phi_{i,j}^k(x_f))^2 \tag{4}$$

L_{SSIM} is the loss function of structural consistency. The addition of this loss function can further improve the visual quality of the image while preventing the deformation of the generated image, where N and M are the size of the image:

$$L_{SSIM} = \frac{1}{NM} \sum_{p=1}^{P} (1 - SSIM(p)) \tag{5}$$

3   Design of Target Recognition Model Based on Lightweight Network

Based on YOLOV5s network, set the input to 1024×1024, and the following improvements are made:

The prediction branch is added on the basis of the structure, and the target recognition is carried out on the larger resolution feature map to improve the recognition ability of small targets. In addition, the SiLU activation function is changed to ReLU for subsequent embedded implementation. The network structure before and after modification is shown in Fig. 3.

As shown in Fig. 3, the red part is the added prediction branch. Through the structural design idea of PANet [19], the deep low-resolution high semantic features are upsampled and the low semantic features at the bottom are fused, so as to complement each other in semantic description and feature information. It has the following advantages:

In the shallow neural network prediction, the resolution of the feature map is large, and even small targets can have certain feature information, which is conducive to improving the recognition ability of small targets. Through the two-way feature fusion of bottom and deep layers, the ability of feature description is improved.

In addition, the activation function is piecewise linear function LeakyReLU rather than nonlinear functions such as SiLU and Mish. The change of activation function is mainly considered from the embedded implementation. The computational complexity of nonlinear activation function is higher than that of linear activation function, which is not conducive to the deployment of hardware platform. The curve diagram of the activation function is shown in Fig. 4.

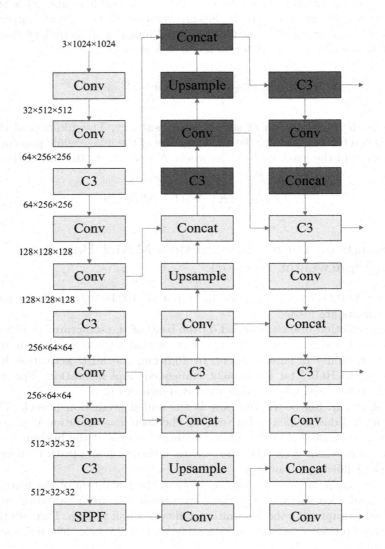

Fig. 3. Schematic Diagram Of Network Structure

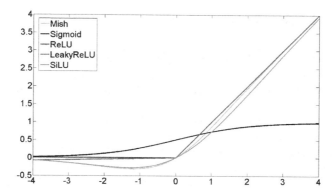

Fig. 4. Schematic Diagram Of Activation Function.

4   Simulation Test and Result Analysis

The data used in the experiment are 1683 infrared vehicle images collected by UAV. 842 images are randomly selected as the training set and other images as the test set. Before training the recognition model, the generative adversarial network model is trained based on the infrared training set images and 5265 visible images, and the infrared image is augmented.

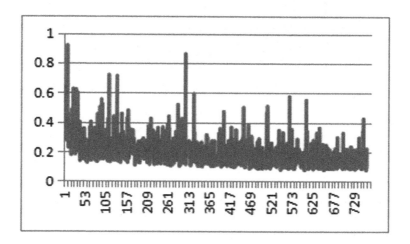

Fig. 5. Network Generation Loss Curve

4.1   Infrared Image Enhancement

In this paper, 256×256 images are obtained by randomly clipping the images during the training process as network input, and the size of each mini batch is 1. In order to improve the training stability, the generator network is first pretrained, and the loss function is only the adversarial loss function during pretraining. Adam is used to optimize. The initial learning rate is 10-4. Every two epochs, the learning rate is reduced by 10%. After the learning rate is reduced to 10-6, the training can be stopped by training one epoch. After the pretraining, the complete target loss function is used for the adversarial training. At this time, the initialization learning rate is 10-4. Adam is used for optimization. When the training is half finished, the cycle consistent loss weight is linearly halved to prevent the training from becoming unstable. When the training loss does not change, the learning rate is halved. If the loss does not change after 5 epochs, the training stops.

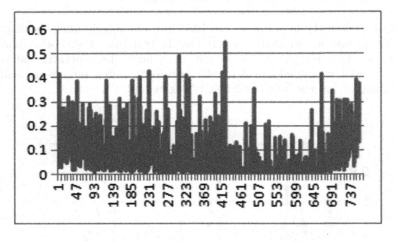

Fig. 6. Network Discrimination Loss Curve

Fig. 5 and Fig. 6 respectively show the generation loss and the change curve of discrimination loss of the network during the training process. From the figure, it can be seen that the overall generation loss function shows a gradual downward trend, and the discrimination loss curve is relatively stable, indicating the effectiveness of the network.

Through the training of generation network and discrimination network, a complete generative adversarial network model is obtained, and then the visible image is processed to generate a near-infrared image, as shown in the Fig. 7.

Fig. 7. Infrared Image Generated Based On Visible Light

4.2 Analysis of the Impact of Infrared Augmented Image on Recognition Accuracy

In order to further verify the impact of infrared data augmentation based on visible light migration on target recognition performance, based on the original YOLOV5s model, comparative tests of different data sets were set up. The experimental results on the test set are shown in Table 1.

Table 1. Target recognition based on augmented infrared image.

Group	Number of real samples	Number of augmented samples	Map 0.5	Map 0.5 lifting
1	500	0	57.5%	–
2	500	500	60.6%	(+) 3.1%
3	800	0	74.1%	–
4	800	800	80.6%	(+) 6.5%

By comparing the recognition results of experimental groups 1 and 2 and experimental groups 3 and 4, it is shown that compared with the original real data set, the target recognition performance of the data set with doubled data volume is improved by 3.1% and 6.5%, respectively. It is verified that the target recognition performance can be effectively improved by increasing the number of generated samples.

4.3 Infrared Target Recognition Test

Through the circular generation countermeasure network model designed in this paper, 5265 infrared images are generated based on visible images, and the training images are expanded to 12647 by means of rotation and translation.

The initial learning rate during training is set to $2.5 \times 10\text{-}4$, momentum is set to 0.949, and decay is set to $5 \times 10\text{-}4$, and the number of iteration rounds is set to 150. The training adopts linear extended learning rate, learning rate preheating, unbiased attenuation and other means to accelerate the convergence speed of the model. Compare the current common one stage target recognition methods, including YOLOV4CSP [20], YOLOV5X [15], YOLOv7X [21], and set the network input to 1024 for unified comparison. The training and testing were carried out on NVIDIA RTX 2080ti graphics card, and FPS was counted.

Table 2 shows the test results. Among them, "Improved YOLOV5s for three branch prediction" linearizes the activation function on the basis of YOLOV5s network structure, and "Improved YOLOV5s for four branch prediction" adds an additional prediction branch on its basis.

Table 2. Test comparison results.

Method	Map0.5	Parameter quantity (MB)	FPS
YOLOv7X	0.8909	143.1	16
YOLOv4CSP	0.8865	173.3	17
YOLOv4CSP	0.7871	146.5	21
YOLOv5s	0.8141	14.6	30
Improved YOLOv5s for three branch rediction	0.8010	14.6	31
Improved YOLOv5s for four branch prediction	0.8339	14.6	29

The experimental results show that compared with the more complex YOLOV5X and YOLOv7X, YOLOV5s and the improved model in this paper have a certain gap in map0.5, but in terms of parameters and processing speed, YOLOV5s and the improved model in this paper have more advantages.

However, from the perspective of practical application, the improved YOLOV5s in this paper has more advantages because:

Although the improved YOLOV5s based on four branch prediction is worse than YOLOV5X and YOLOv7X in map0.5, each value of precision-best, recall-best and YOLOV5X is about 3% lower. From the perspective of application, precision-best and recall-best are 3% lower than each other. In fact, the gap is not obvious. What we focus on is the amount of model parameters and running speed.

From the perspective of engineering implementation, the nonlinear activation function is far inferior to the linear operator on the embedded hardware platform, and this gap is larger than that on the server side, because the embedded implementation often needs to access external memory, resulting in a large amount of data throughput time.

In addition, compared with the original YOLOV5s, the improved four branch YOLO-V5s has an increase of about 2% in map0.5. The generalization of the model is stronger, but the size of the model is only 14.6mb, and the computational complexity is low. Figure 8 is an example of the identification result.

Fig. 8. Example Of Identification Results.

5   Conclusion

In this paper, an infrared image is generated by designing a cycle generative adversarial network model to achieve data Augmention. It is improved on the basis of YOLOV5s to achieve higher precision target recognition while maintaining the real-time processing speed. The experimental results show that the performance of the infrared target recognition model under the condition of few shot can be effectively improved through the expansion of the infrared image. In addition, the improved YOLOV5s model in this paper has higher recognition accuracy than the original model, but it is not suitable for 1024×1024. The image processing speed of 1024 can reach 29fps, which has high application value.

References

1. Girshick, R., Donahue, J., Darrell, T., Malik, J.: Rich feature hierarchies for accurate object detection and semantic segmentation. IEEE Computer Society (2014)
2. Girshick, R.: Fast r-cnn. Comput. Sci. (2015)
3. Ren, S. , He, K. , Girshick, R. , Sun, J.: Faster R-CNN: towards real-time object detection with region proposal networks. In: NIPS (2016)

4. Wang, Y. , Wang, C. , Zhang, H. , Zhang, C. , Fu, Q.: Combing single shot multibox detector with transfer learning for ship detection using Chinese Gaofen-3 images. In: 2017 Progress in Electromagnetics Research Symposium - Fall (PIERS - FALL). IEEE (2017)

5. Redmon, J. , Farhadi, A.: YOLO9000: better, faster, stronger. In: IEEE Conference on Computer Vision and Pattern Recognition, pp. 6517-6525. IEEE (2017)

6. Qiang, W., He, Y., Guo, Y., Li, B., He, L.: Exploring underwater target detection algorithm based on improved ssd. Xibei Gongye Daxue Xuebao/J. Northwestern Polytech. Univ. (4) (2020)

7. Liu, B., Tan, C., Li, S., He, J., Wang, H.: A data augmentation method based on generative adversarial networks for grape leaf disease identification. IEEE Access **PP**(99), 1-1 (2020)

8. Chen, Z., Fang, M., Chai, X., Fu, F., Yuan, L.: U-gan model for infrared and visible images fusion. Xibei Gongye Daxue Xuebao/J. Northwestern Polytech. Univ. **38**(4), 904–912 (2020)

9. Isola, P., Zhu, J.Y., Zhou, T., Efros, A.A.: Image-to-image translation with conditional adversarial networks. IEEE (2016)

10. Zhu, J.Y., Park, T., Isola, P., Efros, A.A.: Unpaired image-to-image translation using cycle-consistent adversarial networks. IEEE (2017)

11. Tan, W.R., Chan, C.S., Aguirre, H.E., Tanaka, K.: Fuzzy qualitative deep compression network. Neurocomputing (2017)

12. Yanchen, W.U., Wang, Y.: A research on underwater target recognition neural network for small samples. J. Northwestern Polytech. Univ. **40**(1), 40–46 (2022)

13. Heydari, M., Sadough, S.M.S., Chaudhry, S.A., Farash, M.S., Mahmood, K.: An improved one-to-many authentication scheme based on bilinear pairings with provable security for mobile pay-tv systems. Multimedia Tools Appli. **76**(12), 14225–14245 (2017)

14. Dai, X., Jiang, Z., Wu, Z., Bao, Y., Zhou, E.: General instance distillation for object detection (2021)

15. Li, C., Wand, M.: Precomputed real-time texture synthesis with markovian generative adversarial networks. In: Leibe, B., Matas, J., Sebe, N., Welling, M. (eds.) ECCV 2016. LNCS, vol. 9907, pp. 702–716. Springer, Cham (2016). https://doi.org/10.1007/978-3-319-46487-9_43

16. Goodfellow, I.J., Pouget Abadie, J., Mirza, M., et al.: generative adaptive nets. In: Procedures of the 27th International Conference on Neural Information Processing Systems, Montreal, 8-13 Dec 2014, pp. 2672-2680, MIT Press, Cambridge (2014)

17. He, K., Zhang, X., Ren, S., Sun, J.: Deep residual learning for image recognition. IEEE (2016)

18. Jocher, G.: Ultralytics/YOLOV5: V5.0- YOLOV5-p6 1280 models, AWS, supervise.ly and Youtube integrations (2021)

19. Liu, S., Qi, L., Qin, H., Shi, J., Jia, J.: Path aggregation network for instance segmentation. IEEE (2018)

20. Wang, C.Y., Bochkovskiy, A., Liao, H.Y.M.: Scaled-YOLOv4: scaling cross stage partial network. In: Computer Vision and Pattern Recognition. IEEE (2021)

21. Wang, C.Y., Bochkovskiy, A., Liao, H.Y.M.: Yolov7: trainable bag-of-freebies sets new state-of-the-art for real-time object detectors. arXiv e-prints (2022)

An Approach to Mongolian Neural Machine Translation Based on RWKV Language Model and Contrastive Learning

Xu Liu, Yila Su$^{(\boxtimes)}$, Wu Nier, Yatu Ji, Ren Qing Dao Er Ji, and Min Lu

Inner Mongolia University of Technology, Inner Mongolia Hohhot 49 Aimin Street,
Mongolia, China
suyl@imut.edu.cn

Abstract. Low-resource machine translation (LMT) is a challenging task, especially for languages with limited resources like Mongolian. In this paper, we propose a novel Mongolian-to-Chinese machine translation approach based on the RWKV language model and augmented with contrastive learning. Traditional methods that perturb the embedding layer often suffer from issues such as semantic distortion and excessive perturbation, leading to training instability. To address these problems, we introduce a contrastive learning approach combined with adversarial perturbation. Additionally, the RWKV language model, as a new architecture, has shown to be more efficient in terms of training and inference time compared to traditional transformer models in various natural language processing tasks. In this work, we employ the RWKV language model as the core of our machine translation model. We evaluate our approach on a benchmark dataset of Mongolian-to-Chinese parallel sentences. The experimental results demonstrate that our method outperforms the state-of-the-art approaches in Mongolian machine translation. Furthermore, our research indicates that the proposed approach significantly mitigates the training instability caused by adversarial perturbation and demonstrates the effectiveness of employing the RWKV language model in improving translation performance.

Keywords: Low-resource machine translation · RWKV language model · Contrastive learning · Adversarial perturbation

1 Introduction

In the context of globalization and rapid advancements in information technology, machine translation plays an increasingly important role in facilitating cross-language communication. However, low-resource machine translation (LMT) tasks, particularly for languages like Mongolian with limited resources, still face significant challenges [3]. To overcome this issue, researchers have

B. Luo et al. (Eds.): ICONIP 2023, CCIS 1962, pp. 327–340, 2024.
https://doi.org/10.1007/978-981-99-8132-8_25

been exploring innovative approaches to improve the quality and efficiency of Mongolian -to-Chinese machine translation.

Traditional data augmentation methods have achieved some success in machine translation and other NLP (natural language processing) tasks, but they still have some limitations [2]. Conventional methods often rely on rule-based or model-based expansions, such as synonym replacement or syntactic transformation. However, these methods fail to capture the true distribution of the data accurately, resulting in noticeable discrepancies between generated samples and real samples, introducing the problem of semantic distortion [1]. On the other hand, adversarial perturbation, as another commonly used data augmentation method, generates new samples by perturbing the original samples [12]. However, excessive perturbation can lead to significant differences between generated and original samples, introducing unnecessary noise that reduces translation quality and causes training instability [8]. In Fig. 1, the x-axis represents epochs, and the y-axis represents the changes in BLEU scores. On the left, we can see the translation results after applying traditional data augmentation to the Mongolian language. It becomes apparent that traditional data augmentation falls short of addressing the existing overfitting phenomenon. However, on the right, when incorporating adversarial perturbations, we observe a reduction in overfitting. Nonetheless, this improvement comes at the expense of introducing training instability.

To overcome the limitations of traditional data augmentation methods, this work proposes an improved data augmentation approach called contrastive learning. Contrastive learning constrains the level of perturbation by comparing the similarity between original and perturbed samples, thereby reducing the problems of semantic distortion and excessive perturbation. Specifically, we treat the original samples as positive samples and the perturbed samples as negative samples and optimize the similarity measure between samples during the training process. This approach effectively balances the level of perturbation, reducing semantic distortion while avoiding excessive perturbation, thus improving translation quality and accuracy. Compared to traditional data augmentation methods and adversarial perturbation, contrastive learning has several advantages: it does not rely on additional training and computational resources, making it easily applicable to existing model frameworks; it better captures the true distribution of data, generating samples that are closer to real samples and enhancing model generalization [7].

Furthermore, this work introduces a new language model architecture called the RWKV language model [13] as the backbone model for improving Mongolian-to-Chinese machine translation. The RWKV language model combines the strengths of traditional Transformers [18] and RNN [10] models and captures long-range dependencies in input sequences better by introducing a Transformer-inspired attention mechanism called TimeMix. Compared to traditional RNN models, the RWKV language model improves translation accuracy and fluency [19]. Additionally, the RWKV language model is more efficient in terms of training and inference time compared to traditional Transformer models, reducing computational costs and speeding up the training and inference process.

In summary, this work aims to improve the translation quality and efficiency of Mongolian-to-Chinese machine translation. By introducing contrastive learning as an improved data augmentation method and incorporating the use of the RWKV language model as the backbone model, we hope to address the issues associated with traditional methods and achieve better translation performance.

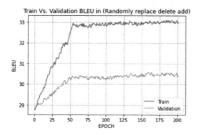
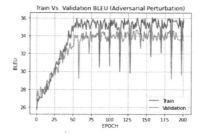

Fig. 1. On the left, the translation results after applying traditional data augmentation in Mongolian language are displayed. On the right, the translation results when adversarial perturbations are added.

Finally, the main contributions of this work include:

- We introduce a novel data augmentation approach for low-resource machine translation using contrastive learning
- We apply the RWKV model to Mongolian-to-Chinese machine translation for the first time.
- We demonstrate a synergistic enhancement in Mongolian-to-Chinese machine translation by combining the contrastive learning data augmentation with the RWKV language model

2   Relate Work

2.1   Data Augmentation

In reference [9], the authors introduced data augmentation techniques such as synonym replacement and random noise addition for text data. This augmentation method involved adding random noise to the text and replacing words in the sentence with synonyms to enhance the data. However, the effectiveness of synonym replacement methods was limited for Mongolian due to a lack of sufficient synonym resources. Moreover, Mongolian possessed unique grammar rules and features, and randomly adding noise might not have adhered to the language's rules, resulting in unnatural or incomprehensible sentences. In reference [4], the authors applied the back-translation (also known as round-trip translation) method in the field of neural machine translation. This technique involved translating text into another language and then translating it back. This app-roach helped capture the meaning of the text more accurately and improved the model's performance in downstream tasks. Amane [15] utilized back-translation in machine translation. However, due to the substantial differences between Mon-golian and Chinese, this approach might have led to inaccurate translations and loss of the original sentence meanings. Adversarial perturbation [11] was initially used in the field of computer vision and showed significant results. In reference [12], adversarial perturbation was applied to text by perturbing the embeddings at the embedding layer to achieve data augmentation, and it was proven to be effective in NLP tasks. However, directly applying adversarial perturbation to NLP tasks might have led to the problem of high variance [8]. Therefore, in this work, contrastive learning was proposed to address the issues associated with random perturbations. By training a model through contrastive learning, the aim was to guide the random perturbation and generate higher-quality pertur-bations, thus reducing excessive perturbation and semantic distortion.

2.2 Transformer Improve

Transformer has achieved high performance in the field of NLP. However, its self-attention mechanism, which has a time complexity of $O(N^2)$, lacks com-putational efficiency. To address this issue, various improved models based on Transformer's self-attention have emerged, such as Reformer [5] and Linformer [20] which aim to linearize self-attention. Although these models improve the efficiency of self-attention, they have not outperformed the original Transformer in terms of overall performance. Another approach to improving the time com-plexity of self-attention is replacing it with alternative components. Models like MLP-Mixer [17], Synthesizer [16], gMLP [22], aMLP [6], and AFT [21] have been proposed. However, these methods have their own limitations. For exam-ple, while AFT is faster than Transformer, it has a design flaw that prevents it from effectively handling very long texts.

In this neural machine translation (NMT) task, the RWKV model is based on the Transformer block, but it replaces self-attention with Position Encoding and Time-Mixing, and the Feed-Forward Network (FFN) is replaced with Channel-Mixing. Experimental results have shown that this approach can not only handle longer texts but also reduce the computational time of the model.

3   Method

3.1   Contrastive Learning Data Augmentation

The strategy for improving embedding perturbation data augmentation using contrastive learning can be described using the following formula: Let's assume we have an input sample x, and we perturb it using the embedding layer to obtain the perturbed sample x'. Our goal is to train a model using contrastive learning so that the perturbed sample becomes more similar to the original sample while preserving semantic information and being robust to perturbations. First, we use an embedding function $f(x)$ to map the input sample x to an embedding vector in a vector space, denoted as $f(x) = v$, where v represents the embedding vector of the input sample x. Next, we apply the embedding operation to both the input sample x and the perturbed sample x', resulting in their embedding vectors v and v', respectively, i.e., $v = f(x)$ and $v' = f(x')$. Then, we introduce the strategy of contrastive learning to measure the similarity between samples. One commonly used method in contrastive learning is training using positive and negative sample pairs. For a positive sample pair (x, x'), we want their embedding vectors to be close to each other in the vector space, i.e., $||v - v'||$ should be minimized.

To improve the model's generalization capability, we introduce negative sample pairs. For each positive sample pair (x, x'), we randomly select a sample x_{neg} that is unrelated to x and perturb it to obtain the perturbed sample x'_{neg}. Then, we aim to maximize the distance between the embedding vectors of the negative sample pair (x, x'_{neg}) in the vector space, i.e., $||v - v'_{neg}||$ should be maximized. Taking these objectives into account, we can define a contrastive loss function, such as the triplet loss:

$$L_{triplet} = max(0, ||v - v'|| - ||v - v'_{neg}|| + margin) \qquad (1)$$

Here, $||.||$ represents the Euclidean distance, and $margin$ is a predefined boundary value used to adjust the distance between positive and negative sample pairs.

By minimizing the contrastive loss function, our model will learn a representation that better preserves semantic similarity in the embedding space [14]. Therefore, even with the introduction of variations during the embedding perturbation, the model can maintain semantic consistency through contrastive learning and alleviate the impact of perturbations.

This approach can more effectively enhance the effectiveness of data augmentation, reducing semantic distortion and excessive perturbation, and improving the model's generalization ability and translation quality.

3.2   RWKV Language Model

Figure 3 provides a visual representation of the RWKV architecture. The RWKV architecture consists of a series of stacked residual blocks, A residual block is composed of a Channel-Mixing block and a Time-Mixing block, and is connected through residuals as shown on the left side of the Fig. 3, and each residual block contains cyclic structures of Time-Mixing and Channel-Mixing sub-blocks.

The entire RWKV architecture is composed of stacked residual blocks, with each block containing Time-Mixing and Channel-Mixing sub-blocks. This architecture is designed to improve the original Transformer model by incorporating

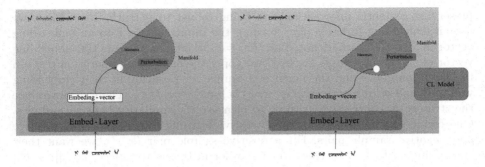

Fig. 2. On the left, the traditional approach of adversarial perturbation is depicted, which suffers from the drawback of introducing random perturbation terms, leading to training instability and ultimately unstable translation results. On the right, the use of a contrastive learning model to guide adversarial perturbations is illustrated. This approach has the advantage of minimizing the variance of samples generated by adversarial perturbations.

Time-Mixing and Channel-Mixing, aiming to enhance the computational efficiency and performance of the model.

Time-Mixing. The Time-Mixing Block is a key component of the RWKV architecture shown in Fig. 4 right. It is used to model the temporal dependencies between tokens in a sequence, which is important for many natural language processing tasks. Time-Mixing Block can be used to model complex temporal dependencies between tokens in a sequence, while also maintaining numerical stability and avoiding vanishing gradients. This makes it well-suited for language modeling tasks, where accurate predictions for next-token prediction are critical.

Channel-Mixing. Channel-Mixing is illustrated in Fig. 4 on the left side. The purpose of Channel-Mixing is to allow the model to learn more complex interactions between different channels (i.e., feature dimensions) in the input data. This can be useful for tasks where there are complex dependencies between different features. In Channel-Mixing, the input is linearly projected and divided into two parts: R (Representation) and K (Key). R is the representation of the current time step, while K is the representation of the previous time step. This design allows the model to use past time step information to assist in the calculation of the current time step. Next, the Channel-Mixing subblock uses a loop structure to mix R and K. Specifically, through Linear interpolation, the R of the current time step is mixed with the K of the previous time step. This interpolation operation can be adjusted independently for each linear projection as needed.

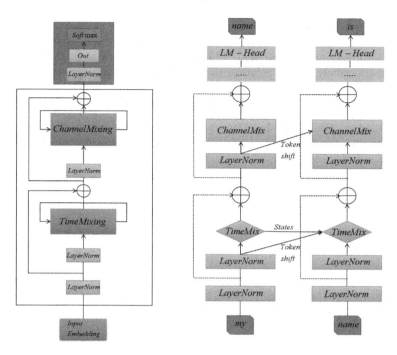

Fig. 3. RWKV residual block with a final head for language modeling architectures(left) and the RWKV language model architectures show in right

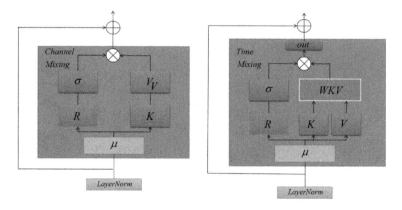

Fig. 4. The Channel-Mixing block is shown on the left, while the Time-Mixing is illustrated on the right

In summary, by combining Channel-Mixing and Time-Mixing, the RWKV architecture can significantly improve computational efficiency while retaining its effectiveness in handling complex NLP tasks. Channel-Mixing and Time-Mixing are two complementary components that address input data information at different levels. Channel-Mixing focuses on interactions across the channel dimensions of input data, while Time-Mixing emphasizes interactions along the temporal sequence dimension. This hierarchical approach allows the model to more effectively learn relationships within different dimensions, thereby enhancing its expressive capabilities. In terms of computational efficiency, both Channel-Mixing and Time-Mixing employ several optimization strategies:

- **Parameter Sharing**: In Time-Mixing, as each time step's calculation involves the previous time step's input, parameter sharing can reduce redundant computations. Similarly, for each channel in Channel-Mixing, parameter sharing helps decrease the computational load.
- **Linear Operations**: Most computations in both Channel-Mixing and Time-Mixing involve linear weighted summation operations, which can be efficiently performed.
- **Element-wise Operations**: Channel-Mixing utilizes element-wise ReLU activation and sigmoid gating, which have efficient implementations on modern hardware.
- **Locality**: Many computations involve localized information, considering only a small portion of the input. This enhances computation locality, leading to improved cache hit rates.

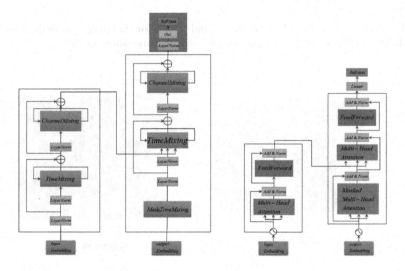

Fig. 5. The RWKV model is shown on the left, and the Transformer model is shown on the right

Compared to the traditional Transformer architecture in Fig. 5, RWKV (Recurrent Weighted Key-Value) introduces several improvements that enhance its performance in tackling complex NLP tasks. Here are the key enhancements of RWKV over the traditional Transformer:

- **Hybrid Structure**: RWKV incorporates a hybrid structure of Channel-Mixing and Time-Mixing. While the traditional Transformer heavily relies on the Self-Attention mechanism, RWKV extends this by introducing interactions in both the temporal and channel dimensions. This combined approach allows RWKV to capture correlations comprehensively within data.
- **Recurrent Structure**: In Time-Mixing, RWKV introduces a recurrent structure by modeling time dependencies through linear interpolation between the current input and the input at the previous time step. This enables the model to effectively capture long-range dependencies in time-series data.
- **Parameter Sharing and Locality**: RWKV employs parameter sharing to reduce redundant computations. Additionally, due to the majority of calculations being localized, the model's computational efficiency is improved. This contributes significantly to both training and inference speed.

Figure 5 illustrates the comparison between RWKV and the Transformer architecture.

4 Experiment

4.1 Experiment Setup

In this experiment, we use the Adam optimizer, set the weight decay and warm-up learning rate scheduling to 0.001, and we set the batch size to 64 and the max sequence length for tokenizer is set to 128 or 64. We trained the model for 200 epochs and evaluated each epoch to calculate its BLEU value.

Introduction to the Dataset. We will use two datasets for this experiment

- English-French (EN- FR): 22.5 million sentences in French and English, We allocated 30% of the data for validation and the remaining 70% for training.
- Mongolia-Chinese (MN-CH):1.7 million sentences in Mongolia and Chinese, We allocated 30% of the data for validation and the remaining 70% for training.

Introduce Method

- TR: Traditional data augmentation strategies include random word replacement, deletion, and addition.
- AP: Add adversarial random disturbance strategy.
- CLAP: Use a contrastive learning strategy to optimize adversarial perturbations.

4.2 Model Selection

For this experiment, the selection of Transformer, RWKV, LSTM, AFT, Performer, AdaMRA, Synthesizer, and Linformer models was driven by the intention to thoroughly explore a spectrum of attention mechanism enhancements and classic architectures in diverse contexts. This ensemble of models was chosen to comprehensively analyze their respective contributions and effectiveness in various scenarios, shedding light on the efficacy of attention-related strategies and their impact on the task at hand.

- Transformer: A standard Transformer.
- Performer: A strategy to improve the attention mechanism of Transformer through nonlinear mapping.
- Linformer: An improved strategy for Transformer's Attention Mechanism through Low-Rank Matrix Approximation.
- RWKV: An improved strategy for combining Transformer and RNN.
- LSTM: LSTM stands for long short-term memory networks that use RNN.
- AFT: A strategy to improve the attention mechanism of Transformer through AFT layer.

4.3   Results

The presented results in Tables 1 illustrate the effectiveness of three different methods (TR, AP, CLAP) in the MN-CH task. These methods, namely traditional data augmentation, adversarial perturbation, and contrastive learning guided perturbation, were evaluated in terms of their impact on performance. The RWKV model consistently achieved a relatively stable score of 38.1 when the token length was either 128 or 64. Notably, AFT performed exceptionally well with a score of 38.1 when the token length was 64, although there was a slight decrease in performance when the token length was set to 128. Conversely, the LSTM model exhibited the lowest performance among the evaluated methods.

Moving on to the EN-FR task, Tables 2 present the performance of the same three methods (TR, AP, CLAP). While these methods demonstrated some improvement in translation performance on the EN-FR dataset, the observed enhancement was not as substantial as in the MN-CH task. Similarly to the MN-CH task, AFT exhibited better performance when the token length was 64 compared to when it was set to 128. In both tasks, the RWKV model consistently showcased good performance across different token lengths.

Base on Fig. 6, The left side displays the translation results for Mongolian language using TR, AP, and CLAP. We observed that the traditional data augmentation method yielded slightly lower results. Although there was some improvement when using adversarial perturbation, it failed to address the issue of unstable training. However, when CLAP was employed, we discovered that it effectively alleviated overfitting. On the right side, we have the validation translation results when CLAP was incorporated. We found that using this approach not only mitigated overfitting but also helped alleviate training instability.

In summary, the results indicate that traditional data augmentation alone may not yield significant improvements, while adversarial perturbation alone cannot fully address training instability. However, incorporating contrastive learning guided perturbation (CLAP) proved effective in alleviating overfitting and training instability. This suggests that a combination of these techniques could lead to more robust and stable translation performance.

Table 1. MN-CH

model	TR	AP	CLAP
Transformer(128 tokens)	33.1	35.1	37.1
AFT(128 tokens)	33.1	35.1	37.7
RWKV(128 tokens)	33.1	36.1	**38.1**
Reformer(128 tokens)	33.1	35.1	36.1
LSTM(128 tokens)	28.1	29.1	33.1
Linformer(128 tokens)	32.1	34.5	36.1
Transformer(64 tokens)	33.1	35.1	37.1
AFT(64 tokens)	33.1	35.1	**38.1**
RWKV(64 tokens)	33.1	36.1	**38.1**
Reformer(64 tokens)	33.1	35.1	36.1
LSTM(64 tokens)	28.1	29.1	33.1
Linformer(64 tokens)	32.1	34.5	36.1

Table 2. EN-FR

model	TR	AP	CLAP
Transformer(128 tokens)	41.5	42.1	43.8
AFT(128 tokens)	40.5	44.3	**44.6**
RWKV(128 tokens)	42.4	44.3	**44.6**
Reformer(128 tokens)	42.1	44.3	45.4
LSTM(128 tokens)	34.8	35.3	36.8
Linformer(128 tokens)	39.2	41.3	42.8
Transformer(64 tokens)	41.5	42.3	43.8
AFT(64 tokens)	42.1	42.3	42.9
RWKV(64 tokens)	42.4	42.8	**43.1**
Reformer(64 tokens)	41.7	41.9	42.9
LSTM(64 tokens)	34.8	36.2	38.4
Linformer(64 tokens)	40.2	41.3	41.8

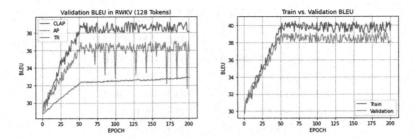

Fig. 6. The left side displays the translation results for Mongolian language using TR (traditional data augmentation), AP (adversarial perturbation), and CLAP (contrastive learning guided adversarial perturbation). On the right side, we have the validation translation results when CLAP was incorporated.

5   Conclusion

In summary, this work introduces a novel approach to Mongolian-to-Chinese machine translation by incorporating the RWKV language model and contrastive learning-based data augmentation. Our aim was to address challenges posed by traditional data augmentation methods and enhance translation quality. By leveraging the strengths of the RWKV model and innovative data augmentation, we sought to achieve more accurate and fluent translations in a low-resource language translation scenario.

The primary contribution of this work lies in the successful combination of the RWKV language model and contrastive learning-based data augmentation. This approach not only mitigates the issues of semantic distortion and excessive perturbation common in traditional augmentation methods but also significantly improves translation quality. The RWKV language model, a hybrid of Transformer and RNN architectures, enables better capturing of long-range dependencies, thus elevating translation accuracy and fluency. Furthermore, the model's computational efficiency optimizes training and inference times compared to conventional Transformer models.

Looking forward, several research directions emerge. Expanding the application of this approach to other language pairs and domains can validate its generalizability. Further optimizing the RWKV model, exploring hybrid architectures, and developing fine-tuning strategies tailored to this approach hold potential for even more substantial improvements. Additionally, investigating interpretability, scalability, and real-time translation aspects can address specific challenges and enhance the applicability of this approach in practical scenarios. These future endeavors will undoubtedly contribute to advancing the field of machine translation and fostering more effective and efficient translation systems.

References

1. Chen, J., Tam, D., Raffel, C., Bansal, M., Yang, D.: An empirical survey of data augmentation for limited data learning in nlp. Trans. Assoc. Comput. Ling. **11**, 191–211 (2023)
2. Feng, S.Y., et al.: A survey of data augmentation approaches for nlp. arXiv preprint arXiv:2105.03075 (2021)
3. Haddow, B., Bawden, R., Barone, A.V.M., Helcl, J., Birch, A.: Survey of low-resource machine translation. Comput. Linguist. **48**(3), 673–732 (2022)
4. Hayashi, T., et al.: Back-translation-style data augmentation for end-to-end asr. In: 2018 IEEE Spoken Language Technology Workshop (SLT), pp. 426–433. IEEE (2018)
5. Kitaev, N., Kaiser, Ł., Levskaya, A.: Reformer: The efficient transformer. arXiv preprint arXiv:2001.04451 (2020)
6. Lalis, J.T., Maravillas, E.: Dynamic forecasting of electric load consumption using adaptive multilayer perceptron (amlp). In: 2014 International Conference on Humanoid, Nanotechnology, Information Technology, Communication and Control, Environment and Management (HNICEM), pp. 1–7. IEEE (2014)
7. Le-Khac, P.H., Healy, G., Smeaton, A.F.: Contrastive representation learning: A framework and review. IEEE Access **8**, 193907–193934 (2020)
8. Lee, S., Lee, D.B., Hwang, S.J.: Contrastive learning with adversarial perturbations for conditional text generation. arXiv preprint arXiv:2012.07280 (2020)
9. Liu, P., Wang, X., Xiang, C., Meng, W.: A survey of text data augmentation. In: 2020 International Conference on Computer Communication and Network Security (CCNS), pp. 191–195. IEEE (2020)
10. Medsker, L.R., Jain, L.: Recurrent neural networks. Design Appli. **5**, 64–67 (2001)
11. Moosavi-Dezfooli, S.M., Fawzi, A., Fawzi, O., Frossard, P.: Universal adversarial perturbations. In: Proceedings of the IEEE Conference on Computer Vision and Pattern Recognition, pp. 1765–1773 (2017)
12. Morris, J.X., Lifland, E., Yoo, J.Y., Grigsby, J., Jin, D., Qi, Y.: Textattack: a framework for adversarial attacks, data augmentation, and adversarial training in nlp. arXiv preprint arXiv:2005.05909 (2020)
13. Peng, B., et al.: Rwkv: reinventing rnns for the transformer era. arXiv preprint arXiv:2305.13048 (2023)
14. Robey, A., Chamon, L., Pappas, G.J., Hassani, H., Ribeiro, A.: Adversarial robustness with semi-infinite constrained learning. Adv. Neural. Inf. Process. Syst. **34**, 6198–6215 (2021)
15. Sugiyama, A., Yoshinaga, N.: Data augmentation using back-translation for context-aware neural machine translation. In: Proceedings of the Fourth Workshop on Discourse in Machine Translation (DiscoMT 2019), pp. 35–44 (2019)
16. Tay, Y., Bahri, D., Metzler, D., Juan, D.C., Zhao, Z., Zheng, C.: Synthesizer: Rethinking self-attention for transformer models. In: International Conference on Machine Learning, pp. 10183–10192. PMLR (2021)
17. Tolstikhin, I.O., Houlsby, N., Kolesnikov, A., Beyer, L., Zhai, X., Unterthiner, T., Yung, J., Steiner, A., Keysers, D., Uszkoreit, J., et al.: Mlp-mixer: An all-mlp architecture for vision. Adv. Neural. Inf. Process. Syst. **34**, 24261–24272 (2021)
18. Vaswani, Aet al.: Attention is all you need. In: Advances in Neural Information Processing Systems 30 (2017)
19. Wang, L.: Rrwkv: capturing long-range dependencies in rwkv. arXiv preprint arXiv:2306.05176 (2023)

20. Wang, S., Li, B.Z., Khabsa, M., Fang, H., Ma, H.: Linformer: self-attention with linear complexity. arXiv preprint arXiv:2006.04768 (2020)
21. Zhai, S., et al.: An attention free transformer. arXiv preprint arXiv:2105.14103 (2021)
22. Zhang, W., et al.: Gmlp: Building scalable and flexible graph neural networks with feature-message passing. arXiv preprint arXiv:2104.09880 (2021)

Event-Triggered Constrained H_∞ Control Using Concurrent Learning and ADP

Shan Xue[1], Biao Luo[2(✉)], Derong Liu[3], and Dongsheng Guo[1]

[1] School of Information and Communication Engineering, Hainan University, Haikou 570100, China
shan.xue0807@foxmail.com, gdongsh2022@hainanu.edu.cn
[2] School of Automation, Central South University, Changsha 410083, China
biao.luo@hotmail.com
[3] School of System Design and Intelligent Manufacturing, Southern University of Science and Technology, Shenzhen 518055, China
liudr@sustech.edu.cn

Abstract. In this paper, an optimal control algorithm based on concurrent learning and adaptive dynamic programming for event-triggered constrained H_∞ control is developed. First, the H_∞ control system under consideration is based on event-triggered constrained input and time-triggered external disturbance, which saves resources and reduces the network bandwidth burden. Second, in the implementation of the control scheme, a critic neural network is designed to approximate unknown value function. Moreover, concurrent learning techniques participate in weight training, making the implementation process simple and effective. Lastly, the stability of the system and the effectiveness of the algorithm are demonstrated through theorem proofs and simulation results.

Keywords: Adaptive dynamic programming · event-triggered control · neural networks · H_∞ control · concurrent learning

1 Introduction

In the early 1950 s, when American mathematician R. Bellman and others studied the optimization problem of multi-stage decision-making process, they put forward the famous optimization principle, thus creating dynamic programming [1]. The effect of dynamic programming on solving multi-stage decision-making problems is obvious, but it also has certain limitations. On the one hand, it does not have a unified processing method, and it must be handled according to various properties of the problem and combined with certain skills. On the other hand, when the dimension of the variable increases, the total calculation and storage increases sharply, leading to the dilemma of "curse of dimensionality" [2,3]. Fortunately, in the era of rapid development of machine learning, an algorithm for approximately solving dynamic programming problems was proposed. It usually uses an approximator such as neural networks (NNs) to approximate

© The Author(s), under exclusive license to Springer Nature Singapore Pte Ltd. 2024
B. Luo et al. (Eds.): ICONIP 2023, CCIS 1962, pp. 341–352, 2024.
https://doi.org/10.1007/978-981-99-8132-8_26

unknown components in the dynamic programming problem. Typical approximators include critic NN and actor NN, which play the role of agents. The agent is learning in a trial-and-error manner, and continuously adjusts its strategy based on the rewards or punishments obtained through interaction with the environment [4]. Eventually the goal of optimizing system performance is reached. Such algorithms are called reinforcement learning, adaptive critic designs [5], approximate dynamic programming, neural dynamic programming, adaptive dynamic programming [6], neuro dynamic programming [7], etc., by the academic community. In this article, we use the acronym "ADP" to uniformly represent such algorithms.

However, ADP usually solves dynamic programming problems in a time-triggered manner. For example, [8–13] solve the optimization control problem of continuous-time and discrete-time systems, respectively. It should be pointed out that this widely used time-triggered ADP algorithm is effective, but it does not achieve the purpose of saving resources. In order to save resources and reduce the burden on the network, an event-triggering mechanism has attracted attention [14]. Compared to the time-triggered mode, its main idea is that it only executes strategy learning when the environment requires it. Whether the environment is needed depends on the design requirements and control objectives. In the field of ADP, related event-triggered work already exists, and they have solved basic dynamic programming problems [15–19].

Because of the constrained input and external disturbances, they are common and cannot be ignored in engineering control [20,21]. Therefore, it is meaningful to solve the optimal control strategy for this kind of control problem. In addition, it is considered that in the traditional dynamic programming problem, persistence of excitation (PE) conditions are usually used to ensure the full exploration and learning of the environment. However, traditional PE conditions are difficult to verify. In order to relax the demand for PE conditions, a simple and effective technique, namely concurrent learning, is proposed [22]. In recent years, this concurrent learning technique has been applied to solve problems such as input regulation, output regulation [23], gaming [24], etc. Therefore, in this paper, an algorithm using concurrent learning and ADP is proposed to solve the constrained event-triggered H_∞ control problem. The concurrent learning technique is applied to the weight training of the critic NN. The learning of the control system is based on event-triggered control input and time-triggered disturbance. This method not only solves the optimal control problem of the H_∞ system, but also has the benefits of saving resources and reducing the burden of network bandwidth. And the employment of concurrent learning techniques makes the implementation process simple and effective.

The following content is arranged as follows. In Sect. 2, the problem description of the constrained H_∞ control system is presented. In Sect. 3, the event-triggering mechanism, the implementation method using concurrent learning and ADP, and stability analysis are given. The simulation results are shown in Sect. 4. Finally, the conclusion is described in Sect. 5.

Notation: $\mathbb{N}$, $\mathbb{R}^n$ and $\mathbb{R}^{n \times m}$ describe the set of non-negative integers, the n-dimensional Euclidean space and $n \times m$ dimensional matrix space, respectively. I_p is an p-dimensional identity matrix. The superscript T, $\lambda_{\min}(\cdot)$ and $\|\cdot\|$ denote take a transpose, the minimum eigenvalue and the 2-norm of a vector or matrix, respectively. $\nabla V \triangleq \frac{\partial V}{\partial x}$ and $(\cdot)^+$ are the gradient operator and right continuity of a function.

2   Problem Description

Let us consider the following nonlinear systems

$$\dot{x}(t) = f(x) + g(x)u(t) + h(x)w(t), \tag{1}$$

where $x \in \mathbb{R}^m$, $u(t) \in \mathbb{R}^n$ with the positive upper bound b_u and $w(t) \in \mathbb{R}^p$ denote the state, control input and external disturbance, respectively, and $w(t) \in L_2[0, \infty)$. $f(x) \in \mathbb{R}^m$, $g(x) \in \mathbb{R}^{m \times n}$, and $h(x) \in \mathbb{R}^{m \times p}$.

The aim of this paper is to design $u(x)$ to make the system (1) with $w(t) \equiv 0$ asymptotically stable and the following L_2-gain is satisfied [25],

$$\int_0^\infty \left(x^\mathsf{T} Q x + Y(u) \right) \mathrm{d}t \le \rho^2 \int_0^\infty w^\mathsf{T} w \mathrm{d}t, \tag{2}$$

where $Q \in \mathbb{R}^{m \times m}$ is a positive definite matrix,

$$Y(u) = 2 \int_0^u b_u \zeta^{-1} \left(\frac{\nu}{b_u} \right)^\mathsf{T} \mathrm{d}\nu = 2 \sum_{j=1}^n \int_0^{u_j} b_u \zeta^{-1} \left(\frac{\nu_j}{b_u} \right) \mathrm{d}\nu_j, \tag{3}$$

$\zeta(\cdot)$ represents a monotone bounded odd function and its first derivative has an upper bound, and the positive constant ρ denotes the prescribed level of disturbance-attenuation. Without loss of generality, the function $\zeta(\cdot)$ is selected as hyperbolic tangent function in this paper. According to (2), define the following value function

$$V(x(t)) = \int_t^\infty (x^\mathsf{T} Q x + Y(u) - \rho^2 w^\mathsf{T} w) \mathrm{d}\nu, \tag{4}$$

where $t \ge 0$. For the admissible control input, differentiating (4), and the following Lyapunov equation is obtained

$$x^\mathsf{T} Q x + Y(u) - \rho^2 w^\mathsf{T} w + (\nabla V(x))^\mathsf{T} (f + gu + hw) = 0. \tag{5}$$

Based on (5), define the following Hamiltonian

$$H(x) = x^\mathsf{T} Q x + Y(u) - \rho^2 w^\mathsf{T} w + (\nabla V(x))^\mathsf{T} (f + gu + hw). \tag{6}$$

The H_∞ control problem is usually regarded as a two-player zero-sum game, in which two players, u and w, try to minimize and maximize the value function, respectively. According to Bellman's optimality principle, the optimal value function $V^*(x)$ satisfies

$$\min_u \max_d [x^\mathsf{T} Q x + Y(u) - \rho^2 w^\mathsf{T} w + (\nabla V^*(x))^\mathsf{T} (f + gu + hw)] = 0, \tag{7}$$

where $V^*(0) = 0$ and (7) is called the Hamilton-Jacobi-Isaacs (HJI) equation. According to the HJI equation (7), the optimal control input $u^*(x)$ and the worst disturbance $w^*(x)$ satisfy

$$u^*(x) = -b_u \zeta \left(\frac{1}{2b_u} g^\mathsf{T} \nabla V^*(x) \right), \tag{8}$$

$$w^*(x) = \frac{1}{2\rho^2} h^\mathsf{T} \nabla V^*(x), \tag{9}$$

respectively, where $(u^*(x), w^*(x))$ is the saddle point solution. Using $u^*(x)$ and $w^*(x)$ in (7), we have

$$x^\mathsf{T} Q x + 2 \int_0^{-b_u \zeta \left(\frac{1}{2b_u} g^\mathsf{T} \nabla V^*(x) \right)} b_u \zeta^{-1} \left(\frac{\nu}{b_u} \right) d\nu + (\nabla V^*(x))^\mathsf{T} f$$

$$- (\nabla V^*(x))^\mathsf{T} g b_u \zeta \left(\frac{1}{2b_u} g^\mathsf{T} \nabla V^*(x) \right) + \frac{1}{4\rho^2} (\nabla V^*(x))^\mathsf{T} h h^\mathsf{T} \nabla V^*(x) = 0. \tag{10}$$

It is noted that the HJI equation is a nonlinear partial differential equation and it is difficult to obtain its analytic solution. Next, we develop an event-triggered constrained H_∞ control method to solve the HJI equation.

3   Event-Triggered Constrained H_∞ Control

3.1   Event-Triggering Mechanism of the H_∞ Control

Define the time sequences of the event-triggering mechanism as

$$\tau_0, ..., \tau_k, \tau_{k+1}, ..., \tag{11}$$

where $k \in \mathbb{N}$, $\tau_0 = 0$, and τ_k is the k-th sampling instant. Based on (11), the state sequence is

$$x(\tau_0), ..., x(\tau_k), x(\tau_{k+1}), ... \tag{12}$$

Thus, the control sequence is defined as

$$u(x(\tau_0)), ..., u(x(\tau_k)), u(x(\tau_{k+1})), ... \tag{13}$$

The H_∞ control system (1) based on (13) becomes

$$\dot{x}(t) = f(x) + g(x)u(x(\tau_k)) + h(x)w(t). \tag{14}$$

Then, the optimal control input (8) based on event-triggering mechanism is

$$u^*(x(\tau_k)) = -b_u \zeta \left(\frac{1}{2b_u} g^\mathsf{T}(x(\tau_k)) \nabla V^*(x(\tau_k)) \right), \tag{15}$$

For all $t \in [\tau_k, \tau_{k+1})$, define the following triggering error $e_\tau(t)$,

$$e_\tau(t) = x(\tau_k) - x(t). \tag{16}$$

The sampling instant under the event-triggering mechanism is aperiodic and it depends on the defined triggering rule. Generally, when the triggering error exceeds a given threshold, the triggering rule is said to be violated. When the triggering rule is violated, an event is triggered and the triggering error is reset to zero.

3.2 Implementation Using Concurrent Learning and ADP

Given the powerful nonlinear approximation ability of neural networks (NNs), the following NN with i hidden neurons is employed to approximate the optimal value function $V^*(x)$,

$$V^*(x) = \mathcal{P}^\mathsf{T}\theta(x) + \varepsilon(x), \tag{17}$$

where $\mathcal{P} \in \mathbb{R}^i$, $\theta(x) \in \mathbb{R}^i$ and $\varepsilon(x)$ are the ideal weights, the activation function and the error of the NN, respectively, and $i > 0$. Using $\hat{\mathcal{P}}$ to approximate the unknown $\mathcal{P}$, then, the estimate of $V^*(x)$ is

$$\hat{V}(x) = \hat{\mathcal{P}}^\mathsf{T}\theta(x), \tag{18}$$

where (18) is the critic NN. Based on (17) and (18), we have

$$\widetilde{\mathcal{P}} = \mathcal{P} - \hat{\mathcal{P}}, \tag{19}$$

where $\widetilde{\mathcal{P}}$ is the weight estimation error. Based on (18), the Hamiltonian (6) becomes

$$\hat{H}(x) = x^\mathsf{T}Qx + Y(u) - \rho^2 w^\mathsf{T} w + \hat{\mathcal{P}}^\mathsf{T}\nabla\theta(x)(f + gu + hw). \tag{20}$$

In the process of critic NN training, $\hat{\mathcal{P}}$ should be designed to minimize $\hat{H}(x)$. Note that, based on (18), the following event-triggered control input and time-triggered disturbance are employed

$$\hat{u}(x(\tau_k)) = -b_u\zeta\left(\frac{1}{2b_u}g^\mathsf{T}(x(\tau_k))\hat{\mathcal{P}}^\mathsf{T}\nabla\theta(x(\tau_k))\right), \tag{21}$$

$$\hat{w}(x) = \frac{1}{2\rho^2}h^\mathsf{T}(x)\hat{\mathcal{P}}^\mathsf{T}\nabla\theta(x). \tag{22}$$

Considering concurrent learning, for all $t_c \in [\tau_k, \tau_{k+1})$, the following data set is stored

$$\mathcal{C} \triangleq \{x(t)|_{t=t_c}\}_{c=1}^l, \tag{23}$$

where $l \in \mathbb{N}$. For $x(t_c)$, the Hamiltonian is

$$\hat{H}(x(t_c)) = x^\mathsf{T}(t_c)Qx(t_c) + Y(u) - \rho^2 w^\mathsf{T} w + \hat{\mathcal{P}}^\mathsf{T}\nabla\theta(x(t_c))(f + gu + hw). \tag{24}$$

According to (20) and (24), define the following objective function

$$E(x, x(t_c)) = \frac{1}{2}\hat{H}^\mathsf{T}(x)\hat{H}(x) + \frac{1}{2}\hat{H}^\mathsf{T}(x(t_c))\hat{H}(x(t_c)). \tag{25}$$

Using

$$\frac{\partial E(x, x(t_c))}{\partial\hat{\mathcal{P}}} = \nabla\theta(x)(f + gu + hw)\hat{H}(x)$$

$$+ \sum_{c=1}^l \nabla\theta(x(t_c))(f + gu + hw)\hat{H}(x(t_c)),$$

the following gradient descent algorithm is designed

$$\dot{\hat{P}} = -r\frac{\pi\hat{H}(x)}{(1+\pi^\mathsf{T}\pi)^2} - \sum_{c=1}^{l} r\frac{\pi(t_c)\hat{H}(x(t_c))}{(1+(\pi(t_c))^\mathsf{T}\pi(t_c))^2}, \tag{26}$$

where

$$\pi = \nabla\theta(x)(f + g\hat{u}(x(\tau_k)) + h\hat{w}(x)),$$
$$\pi(t_c) = \nabla\theta(x(t_c))(f + g\hat{u}(x(\tau_k)) + h\hat{w}(x(t_c))),$$

and $r > 0$ is the learning rate. In this paper, $\{\pi(t_c)\}_{c=1}^{l}$ needs to satisfy rank$(\varXi) = i$, where $\varXi = [\pi(t_1), ..., \pi(t_c)]$, to replace the traditional PE condition. According to [26], the linearly independent of $\{\theta(x_l)\}_{c=1}^{l}$ ensures that $\{\pi(t_c)\}_{c=1}^{l}$ is linearly independent. According to (26), we have

$$\begin{aligned}
\dot{\tilde{P}} &= r\frac{\pi\hat{H}(x)}{(1+\pi^\mathsf{T}\pi)^2} + \sum_{c=1}^{l} r\frac{\pi(t_c)\hat{H}(x(t_c))}{(1+(\pi(t_c))^\mathsf{T}\pi(t_c))^2} \\
&= -r\frac{\pi\pi^\mathsf{T}\tilde{P}}{(1+\pi^\mathsf{T}\pi)^2} - \sum_{c=1}^{l} r\frac{\pi(t_c)\pi(t_c)^\mathsf{T}\tilde{P}}{(1+(\pi(t_c))^\mathsf{T}\pi(t_c))^2} \\
&\quad + r\frac{\pi\eta}{(1+\pi^\mathsf{T}\pi)^2} + \sum_{c=1}^{l} r\frac{\pi(t_c)\eta(t_c)}{(1+(\pi(t_c))^\mathsf{T}\pi(t_c))^2},
\end{aligned} \tag{27}$$

where

$$\eta = \nabla\varepsilon(x)(f + gu + hw), \eta(t_c) = \nabla\varepsilon(x(t_c))(f + gu + hw).$$

3.3   Stability Analysis

The following assumption is proposed for the analysis of the theorem [27–29].

Assumption 1. $\nabla\theta(x)$ and $g(x)$ are Lipschitz continuous with $\mathcal{C}_\sigma > 0$ and $\mathcal{C}_g > 0$. $\|g(x)\|$, $\|\nabla\theta(x)\|$, $\nabla\varepsilon(x)$ and η have positive upper bounds b_g, b_θ, b_ε and b_η, respectively.

Theorem 1. Consider the nonlinear system (1) with the event-triggered control input (21) and time-triggered disturbance input (22). Let Assumption 1 holds. The threshold satisfies

$$\|E_\tau(t)\|^2 = \frac{\lambda_{\min}(Q)(1-\varPhi_3^2)\|x\|^2 + Y(\hat{u}(x(\tau_k))) - \rho^2\hat{w}^\mathsf{T}\hat{w}}{\|\hat{P}\|^2\varPhi_4^2}, \tag{28}$$

where $\varPhi_3$ and $\varPhi_4$ are defined in the following proof. When the critic network is trained using (26), x and $\tilde{P}$ are uniformly ultimately bounded.

Proof. Select the following Lyapunov function

$$\mathcal{L} = \frac{1}{2}\widetilde{\mathcal{P}}^\mathsf{T}\widetilde{\mathcal{P}} + V^*(x) + V^*(x(\tau_k)). \tag{29}$$

The introduction of event-triggering mechanism makes the system become an impulse system, so the proof is divided into two cases.

Case 1: For the flow dynamics, i.e., $t \in [\tau_k, \tau_{k+1})$. Denote

$$\mathcal{L}_p \triangleq \frac{1}{2}\widetilde{\mathcal{P}}^\mathsf{T}\widetilde{\mathcal{P}}. \tag{30}$$

Based on (27), we have

$$
\begin{aligned}
\dot{\mathcal{L}}_p &= -\frac{r\widetilde{\mathcal{P}}^\mathsf{T}\pi\pi^\mathsf{T}\widetilde{\mathcal{P}}}{(1 + \pi^\mathsf{T}\pi)^2} - \sum_{c=1}^{l} \frac{r\widetilde{\mathcal{P}}^\mathsf{T}\pi(t_c)(\pi(t_c))^\mathsf{T}\widetilde{\mathcal{P}}}{(1 + (\pi(t_c))^\mathsf{T}\pi(t_c))^2} \\
&\quad + \frac{r\widetilde{\mathcal{P}}^\mathsf{T}\pi\eta}{(1 + \pi^\mathsf{T}\pi)^2} + \sum_{c=1}^{l} \frac{r\widetilde{\mathcal{P}}^\mathsf{T}\pi(t_c)\eta(t_c)}{(1 + (\pi(t_c))^\mathsf{T}\pi(t_c))^2} \\
&\leq -\frac{1}{2}r\lambda_{\min}(\varPhi_1)\|\widetilde{\mathcal{P}}\|^2 + \frac{1}{2}r\varPhi_2,
\end{aligned}
\tag{31}
$$

where

$$\varPhi_1 = \frac{\pi\pi^\mathsf{T}}{(1 + \pi^\mathsf{T}\pi)^2} + \sum_{c=1}^{l} \frac{\pi(t_c)(\pi(t_c))^\mathsf{T}}{(1 + (\pi(t_c))^\mathsf{T}\pi(t_c))^2}, \varPhi_2 = \eta^\mathsf{T}\eta + \sum_{c=1}^{l}(\eta(t_c))^\mathsf{T}\eta(t_c).$$

For the second term $\dot{V}^*(x)$, it satisfies

$$\dot{V}^*(x) = \nabla V^{*\mathsf{T}}(f(x) + g(x)\hat{u}(x(\tau_k)) + h(x)\hat{w}(x)). \tag{32}$$

Based on the time-triggered HJI equation, we obtain

$$(\nabla V^*(x))^\mathsf{T} f(x) = -x^\mathsf{T} Q x - Y(u^*) - \rho^2 w^{*\mathsf{T}} w^* - (\nabla V^*(x))^\mathsf{T} g u^*. \tag{33}$$

From (8) and (9), we have

$$(\nabla V^*(x))^\mathsf{T} g = -2b_u\zeta^{-1}\left(\frac{u^*}{b_u}\right)^\mathsf{T}, \tag{34}$$

$$(\nabla V^*(x))^\mathsf{T} h = 2\rho^2 w^{*\mathsf{T}}. \tag{35}$$

Considering (33)–(35), (32) becomes

$$
\begin{aligned}
\dot{V}^*(x) &= -x^\mathsf{T} Q x - Y(u^*) - \rho^2 w^{*\mathsf{T}} w^* \\
&\quad + 2b_u\zeta^{-1}\left(\frac{u^*}{b_u}\right)^\mathsf{T}(u^* - \hat{u}(x(\tau_k))) + 2\rho^2 w^{*\mathsf{T}}\hat{w} \\
&= 2b_u \int_{\hat{u}(x(\tau_k))}^{u^*} \left(\zeta^{-1}\left(\frac{u^*}{b_u}\right) - \zeta^{-1}\left(\frac{\nu}{b_u}\right)\right)^\mathsf{T} d\nu \\
&\quad - x^\mathsf{T} Q x - Y(\hat{u}(x(\tau_k))) + \rho^2(2w^{*\mathsf{T}}\hat{w} - w^{*\mathsf{T}}w^*).
\end{aligned}
\tag{36}
$$

Denote

$$v^*(x) \triangleq \zeta^{-1}\left(\frac{u^*}{b_u}\right) = -\frac{1}{2b_u}g^\mathsf{T}(x)(\mathcal{P}^\mathsf{T}\nabla\theta(x) + \nabla\varepsilon(x)),$$

$$\hat{v}(x(\tau_k)) \triangleq \zeta^{-1}\left(\frac{\hat{u}(x(\tau_k))}{b_u}\right) = -\frac{1}{2b_u}g^\mathsf{T}(x(\tau_k))\hat{\mathcal{P}}^\mathsf{T}\nabla\theta(x(\tau_k)).$$

Then, the first term of (36) satisfies

$$2b_u\int_{\hat{u}(x(\tau_k))}^{u^*}\left(\zeta^{-1}\left(\frac{u^*}{b_u}\right) - \zeta^{-1}\left(\frac{\nu}{b_u}\right)\right)^\mathsf{T}\mathrm{d}\nu$$
$$\leq 2b_u\|u^* - \hat{u}(x(\tau_k))\|\|v^*(x) - \hat{v}(x(\tau_k))\|. \tag{37}$$

Since the upper bound of the first derivative of the hyperbolic tangent function is 1, we have

$$\|u^* - \hat{u}(x(\tau_k))\| \leq b_u\|v^*(x) - \hat{v}(x(\tau_k))\|,$$

thus, (37) satisfies

$$2b_u\int_{\hat{u}(x(\tau_k))}^{u^*}\left(\zeta^{-1}\left(\frac{u^*}{b_u}\right) - \zeta^{-1}\left(\frac{\nu}{b_u}\right)\right)^\mathsf{T}\mathrm{d}\nu$$
$$\leq 2b_u^2\|v^*(x) - \hat{v}(x(\tau_k))\|^2$$
$$\leq 0.5\|g^\mathsf{T}(\mathcal{P}^\mathsf{T}\nabla\theta + \nabla\varepsilon) - g^\mathsf{T}(x(\tau_k))\hat{\mathcal{P}}^\mathsf{T}\nabla\theta(x(\tau_k))\|$$
$$\leq \|\nabla\theta g - \nabla\theta(x(\tau_k))g(x(\tau_k))\|^2\|\hat{\mathcal{P}}\|^2 + \|g^\mathsf{T}\widetilde{P}^\mathsf{T}\nabla\theta + g^\mathsf{T}\nabla\varepsilon\|$$
$$\leq \|(\nabla\theta - \nabla\theta(x(\tau_k)))g + \nabla\theta(x(\tau_k))(g - g(x(\tau_k)))\|^2\|\hat{\mathcal{P}}\|^2$$
$$\quad + 2\|g^\mathsf{T}\widetilde{P}^\mathsf{T}\nabla\theta\|^2 + 2\|g^\mathsf{T}\nabla\varepsilon\|^2$$
$$\leq 2(C_\sigma^2 b_g^2 + b_\theta^2 C_g^2)\|e_\tau(t)\|^2\|\hat{\mathcal{P}}\|^2 + 2b_g^2 b_\theta^2\|\widetilde{P}\|^2 + 2b_g^2 b_\varepsilon^2 \tag{38}$$

For $\rho^2(2w^{*\mathsf{T}}\hat{w} - w^{*\mathsf{T}}w^*)$ in (36), we have

$$\rho^2(2w^{*\mathsf{T}}\hat{w} - w^{*\mathsf{T}}w^*) \leq \rho^2\hat{w}^\mathsf{T}\hat{w}. \tag{39}$$

Considering (38) and (39), (36) becomes

$$\dot{V}^*(x) \leq 2(C_\sigma^2 b_g^2 + b_\theta^2 C_g^2)\|e_\tau(t)\|^2\|\hat{\mathcal{P}}\|^2 + 2b_g^2 b_\theta^2\|\widetilde{P}\|^2$$
$$\quad + 2b_g^2 b_\varepsilon^2 - \lambda_{\min}(Q)\|x\|^2 - Y(\hat{u}(x(\tau_k))) + \rho^2\hat{w}^\mathsf{T}\hat{w}. \tag{40}$$

Since $\dot{V}^*(x(\tau_k)) = 0$, we have

$$\dot{\mathcal{L}} \leq -(\frac{1}{2}r\lambda_{\min}(\Phi) - 2b_g^2 b_\theta^2)\|\widetilde{\mathcal{P}}\|^2 + \frac{1}{2}r\Phi_2 + 2b_g^2 b_\varepsilon^2 + 2(C_\sigma^2 b_g^2 + b_\theta^2 C_g^2)\|e_\tau(t)\|^2\|\hat{\mathcal{P}}\|^2$$
$$\quad - \lambda_{\min}(Q)\|x\|^2 - Y(\hat{u}(x(\tau_k))) + \rho^2\hat{w}^\mathsf{T}\hat{w}. \tag{41}$$

Therefore, if the triggering rule is satisfied, i.e.,

$$\|e_\tau(t)\|^2 < \frac{\lambda_{\min}(Q)(1 - \Phi_3^2)\|x\|^2 + Y(\hat{u}(x(\tau_k))) - \rho^2 \hat{w}^\mathsf{T}\hat{w}}{\|\hat{\mathcal{P}}\|^2 \Phi_4^2}$$

where $\Phi_3 \in [0, 1]$ and $\Phi_4^2 = \mathcal{C}_\sigma^2 b_g^2 + b_\theta^2 \mathcal{C}_g^2$, (41) satisfies

$$\dot{\mathcal{L}} < -\Phi_5 \|\widetilde{\mathcal{P}}\|^2 + \Phi_6 - \Phi_3^2 \lambda_{\min}(Q)\|x\|^2, \tag{42}$$

where $\Phi_5 = \frac{1}{2} r \lambda_{\min}(\Phi) - 2 b_g^2 b_\theta^2$ is positive by adjusting the learning rate r and $\Phi_6 = \frac{1}{2} r \Phi_2 + 2 b_g^2 b_\varepsilon^2$. According to the standard Lyapunov extension theorem, x and $\widetilde{\mathcal{P}}$ are uniformly ultimately bounded when $\|\widetilde{\mathcal{P}}\|$ is outside the boundary $\sqrt{\Phi_6/\Phi_5}$ or $\|x\|$ is outside the boundary $\sqrt{\Phi_6/[\Phi_3^2 \lambda_{\min}(Q)]}$.

Case 2: For the jump dynamics, i.e., $t = \tau_k$. Let's consider the following differential form of (29)

$$\Delta\mathcal{L} = \Delta\mathcal{L}_p + \Delta V^*(x) + \Delta V^*(x(\tau_k)), \tag{43}$$

where

$$\Delta\mathcal{L}_p = \frac{1}{2}(\widetilde{\mathcal{P}}^+)^\mathsf{T}\widetilde{\mathcal{P}}^+ - \frac{1}{2}\widetilde{\mathcal{P}}^\mathsf{T}\widetilde{\mathcal{P}}, \Delta V^*(x) = V^*(x^+) - V^*(x),$$
$$\Delta V^*(x(\tau_k)) = V^*(x_k^+) - V^*(x(\tau_k)).$$

$\Delta\mathcal{L}_p$ and $\Delta V^*(x)$ less than zero can be guaranteed since $\widetilde{\mathcal{P}}$ and x are continuous and they are uniformly ultimately bounded as demonstrated in Case 1. In addition, $x_k^+ = x_{k+1}$, $\Delta V^*(x(\tau_k)) = V^*(x_{k+1}) - V^*(x(\tau_k))$ less than zero can also be guaranteed. Thus, $\Delta\mathcal{L} < 0$. Considering Case 1 and 2, the conclusion of Theorem 1 can be proved. $\qquad\square$

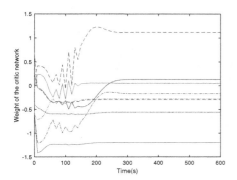

Fig. 1. The evolution of $\hat{\mathcal{P}}$.

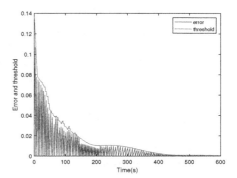

Fig. 2. The evolution of the event-triggering rule.

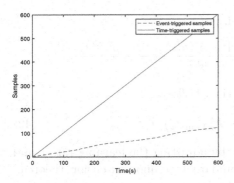

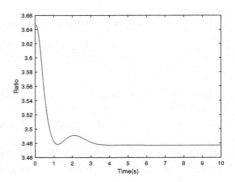

Fig. 3. The evolution of the sample.

Fig. 4. The evolution of the disturbance-attenuation.

4   Simulation Results

Consider the following system dynamics

$$f(x) = [-x_1 + x_2 + 0.5\sin(x_1)^3 \cos(x_2)^2, -2x_1 - x_2^3)]^\mathsf{T},$$
$$g(x) = [\sin(x_2), 1]^\mathsf{T}, h(x) = [0, \cos(x_1)]^\mathsf{T}.$$

where $x = [x_1, x_2]^\mathsf{T}$. The parameters are selected as $x_0 = [1, -1]^\mathsf{T}$, $Q = I_2$, $b_u = 1$, $\rho = 4$, $l = 10$, $r = 10$, $\Phi_3 = 0.01$, and $\Phi_4 = 10$. The initial value of $\hat{\mathcal{P}}$ is randomly selected between $[-1, 1]$. $\theta(x)$ is selected as $[x_1^2, x_2^2, x_1x_2, x_1^4, x_2^4, x_1^3x_2, x_1^2x_2^2, x_1x_2^3]^\mathsf{T}$. The evolution of $\hat{\mathcal{P}}$ is shown in Fig. 1, which eventually converges to

$$[0.1300, 1.1157, -0.1691, -0.5596, 0.0524, -1.1988, -0.2956, -0.2804]^\mathsf{T}.$$

The evolution of the event-triggering rule is shown in Fig. 2. It can be found that $e_\tau(t)$ does not exceed the threshold, thus ensuring the stability of the system. The evolution of the sample is shown in Fig. 3, where the sample size of the event-triggered method is greatly reduced compared with that of the time-triggered method. An external disturbance signal $0.5e^{-t}\cos(t)$ is used to verify performance. As shown in Fig. 4, ρ converges to 3.4772, which is less than 4. Therefore, the validity of the algorithm is verified.

5   Conclusions

An event-triggered constrained H_∞ optimal control algorithm based on concurrent learning and ADP is proposed in this paper. The event-triggering mechanism is introduced into the H_∞ control system, so that the learning of the system is based on the event-triggered constrained input and time-triggered external disturbance. This achieves the purpose of saving resources and reducing the burden of network bandwidth. Moreover, a critic NN is designed for the implementation of control scheme. Note that the gradient descent algorithm based on concurrent learning is used to update the NN weights, which makes the implementation process simple and effective. Finally, the effectiveness of the algorithm is shown in the simulation results.

Acknowledgments. This paper was supported in part by the National Natural Science Foundation of China under Grants 62022094, 62073085, 62373375, and the Zhejiang Lab (No. 2021NB0AB01), in part by Scientific Research Fund of Hainan University under Grant KYQD(ZR)23025.

References

1. Bellman, R.E.: Dynamic programming. Princeton University Press (1957)
2. Liu, D., Xue, S., Zhao, B., Luo, B., Wei, Q.: Adaptive dynamic programming for control: a survey and recent advances. IEEE Trans. Syst. Man Cybern. Syst. **51**(1), 142–160 (2021)
3. Liu, D., Wei, Q., Wang, D., Yang, X., Li, H.: Adaptive Dynamic Programming with Applications in Optimal Control. AIC, Springer, Cham (2017). https://doi.org/10.1007/978-3-319-50815-3
4. Lewis, F.L., Vrabie, D.: Reinforcement learning and adaptive dynamic programming for feedback control. IEEE Circuits Syst. Mag. **9**(3), 32–50 (2009)
5. Wang D., Li X., Zhao M., Qiao J.: Adaptive critic control design with knowledge transfer for wastewater treatment applications. IEEE Trans. Industrial Inform. https://doi.org/10.1109/TII.2023.3278875
6. Wang, F.-Y., Zhang, H., Liu, D.: Adaptive dynamic programming: an introduction. IEEE Comput. Intell. Mag. **4**(2), 39–47 (2009)
7. Bertsekas, D.P., Homer, M.L., Logan, D.A., Patek, S.D., Sandell, N.R.: Missile defense and interceptor allocation by neuro-dynamic programming. IEEE Trans. Syst. Man Cybern.-Part A: Syst. Hum. **30**(1), 42–51 (2000)
8. Zhang, H., Cui, L., Luo, Y.: Near-optimal control for nonzero-sum differential games of continuous-time nonlinear systems using single network ADP. IEEE Trans. Cybern. **43**(1), 206–216 (2013)
9. Luo, B., Wu, H.-N., Huang, T., Liu, D.: Data-based approximate policy iteration for affine nonlinear continuous-time optimal control design. Automatica **50**(12), 3281–3290 (2014)
10. Jiang, Y., Jiang, Z.-P.: Computational adaptive optimal control for continuous-time linear systems with completely unknown dynamics. Automatica **48**(10), 2699–2704 (2012)
11. Luo, B., Liu, D., Wu, H.-N.: Adaptive constrained optimal control design for data-based nonlinear discrete-time systems with critic-only structure. IEEE Trans. Neural Netw. Learn. Syst. **29**(6), 2099–2111 (2018)
12. Liu, D., Wei, Q.: Finite-approximation-error-based optimal control approach for discrete-time nonlinear systems. IEEE Trans. Cybern. **43**(2), 779–789 (2013)
13. Wang, D., Liu, D., Wei, Q., Zhao, D., Jin, N.: Optimal control of unknown nonaffine nonlinear discrete-time systems based on adaptive dynamic programming. Automatica **48**(8), 1825–1832 (2012)
14. An, T., Wang, Y., Liu, G., Li, Y., Dong, B.: Cooperative game-based approximate optimal control of modular robot manipulators for human-robot collaboration. IEEE Trans. Cybern. **53**(7), 4691–4703 (2023)

15. Vamvoudakis, K.G., Mojoodi, A., Ferraz, H.: Event-triggered optimal tracking control of nonlinear systems. Int. J. Robust Nonlinear Control **27**(4), 598–619 (2017)
16. Yang, X., He, H.: Adaptive critic designs for event-triggered robust control of nonlinear systems with unknown dynamics. IEEE Trans. Cybern. **49**(6), 2255–2267 (2019)
17. Zhang, Q., Zhao, D.: Data-based reinforcement learning for non-zero-sum games with unknown drift dynamics. IEEE Trans. Cybern. **49**(8), 2874–2885 (2019)
18. Luo, B., Huang, T., Liu, D.: Periodic event-triggered suboptimal control with sampling period and performance analysis. IEEE Trans. Cybern. **51**(3), 1253–1261 (2021)
19. Yang, Y., Vamvoudakis, K.G., Modares, H., Yin, Y., Wunsch, D.C.: Hamiltonian-driven hybrid adaptive dynamic programming. IEEE Trans. Syst. Man Cybern. Syst. **51**(10), 6423–6434 (2021)
20. Xue, S., Luo, B., Liu, D.: Event-triggered adaptive dynamic programming for zero-sum game of partially unknown continuous-time nonlinear systems. IEEE Trans. Syst. Man Cybern. Syst. **50**(9), 3189–3199 (2020)
21. Wang, D., Hu, L., Zhao, M., Qiao, J.: Dual event-triggered constrained control through adaptive critic for discrete-time zero-sum games. IEEE Trans. Syst. Man Cybern. Syst. **53**(3), 1584–1595 (2023)
22. Kamalapurkar, R., Klotz, J.R., Dixon, W.E.: Concurrent learning based approximate feedback-nash equilibrium solution of N-player nonzero-sum differential games. IEEE/CAA J. Autom. Sinica **1**(3), 239–247 (2014)
23. Luo, B., Yang, Y., Liu, D.: Adaptive Q-learning for data-based optimal output regulation with experience replay. IEEE Trans. Cybern. **48**(12), 3337–3348 (2018)
24. Zhao, D., Zhang, Q., Wang, D., Zhu, Y.: Experience replay for optimal control of nonzero-sum game systems with unknown dynamics. IEEE Trans. Cybern. **46**(3), 854–865 (2016)
25. Yang, X., Xu, M., Wei, Q.: Approximate dynamic programming for event-driven H_∞ constrained control. IEEE Trans. Syst. Man Cybern. Syst. doi: https://doi.org/10.1109/TSMC.2023.3277737
26. Beard, R.W., Saridis, G.N., Wen, J.T.: Galerkin approximations of the generalized Hamilton-Jacobi-Bellman equation. Automatica **33**(12), 2159–2177 (1997)
27. Zhao, B., Liu, D.: Event-triggered decentralized tracking control of modular reconfigurable robots through adaptive dynamic programming. IEEE Trans. Industr. Electron. **67**(4), 3054–3064 (2020)
28. Bai, W., Li, T., Long, Y., Chen, C.L.P.: Event-triggered multigradient recursive reinforcement learning tracking control for multiagent systems. IEEE Trans. Neural Netw. Learn. Syst. **34**(1), 366–379 (2023)
29. Liu, J., Wu, Y., Sun, M., Sun, C.: Fixed-time cooperative tracking for delayed disturbed multi-agent systems under dynamic event-triggered control. IEEE/CAA J. Autom. Sinica **9**(5), 930–933 (2022)

Three-Dimensional Rotation Knowledge Representation Learning Based on Graph Context

Xiaoyu Chen[1,2], Yushui Geng[1,2(✉)], and Hu Liang[1,2]

[1] Key Laboratory of Computing Power Network and Information Security, Ministry of Education, Shandong Computer Science Center, Qilu University of Technology (Shandong Academy of Sciences), Jinan, China
gys@qlu.edu.cn

[2] Shandong Engineering Research Center of Big Data Applied Technology, Faculty of Computer Science and Technology, Qilu University of Technology (Shandong Academy of Sciences), Jinan, China

Abstract. The goal of knowledge graph representation learning is to map entities and relations into low-dimensional continuous vector Spaces in order to learn their semantic information representation. However, most existing models often struggle to model the basic features of knowledge graphs effectively, such as symmetric/antisymmetric, inverse, and combinatorial relational patterns. In addition, many models ignored the information about the neighborhood of entities in the triples in the graph. In order to solve these problems, this paper proposes a learning model of three-dimensional rotating knowledge graph representation based on graph context. The model first uses the quaternion mathematical framework to represent the entity as a set of vectors in three-dimensional space, and interprets the relationship as a three-dimensional rotation transformation between the entities. Then, by calculating the semantic similarity between entities and relations, the graph context information is fused into the vector representation. Experiments on public data sets FB15K-237 and WN18RR demonstrate the effectiveness of the proposed model. The experimental results show that the model can capture the relational pattern of knowledge graph better and make full use of the neighborhood information of entities in the graph.

Keywords: Knowledge graph · Representation learning · Three-dimensional rotation · Quaternion · Graph context · Semantic similarity

1 Introduction

Knowledge graphs utilize directed graph structures to represent knowledge and capture connections between different entities in the world. In this representation,

This work was supported by Natural Science Foundation of Shandong Province (No. ZR2022LZH008) and The 20 Planned Projects in Jinan (Nos. 2021GXRC046, 202228120).

nodes represent entities and edges represent relationships between entities. In recent years, a number of large-scale knowledge graphs have emerged on the Internet, such as Freebase, DBpedia, YAGO, and WordNet. However, these large-scale knowledge graphs suffer from two major limitations in practical application. Firstly, knowledge representation based on network symbols has low computational efficiency and sparse data [1,2]. Secondly, the constructed knowledge graphs are often incomplete and lack a large number of factual relationships [3]. To overcome the challenges of low computational efficiency and sparse data in knowledge graphs, a knowledge representation learning model is proposed. These models are designed to map knowledge graphs into low-dimensional continuous vector Spaces. This method improves the computational efficiency of knowledge graph processing and alleviates the problem of sparse data. Starting with the simplest and most efficient TransE model [4], researchers have proposed many knowledge graph representation learning models, including TransH [5], TransR [6], DisMult [7], ComplEx [8], RotatE [9] and so on. These models enrich the repertoire of methods for learning representations in knowledge graphs, enabling a more effective utilization of the semantic information within these graphs.

Recent research has shown that there are multiple patterns of relationships between nodes (entities) and edges (relationships) in the knowledge graph. Finding a way to model and infer all relational patterns is very important for knowledge representation learning. Existing knowledge representation learning models have some limitations in dealing with different relational patterns. For example, the TransE model treats relationships as a translation from head to tail entity and can model inverse and combinatorial relationships. ComplEx models introduce complex embedding to model symmetric and antisymmetric patterns, but cannot reason combinatorial relationships. Most existing models can only learn and reason about some of these relational patterns. To solve this problem, Sun et al. proposed the RotatE model, which models the three relational patterns using complex space, that is, each relationship is defined as the rotation of the head entity to the tail entity in a complex vector space. On the other hand, when learning the embedded representation of entities and relations, many knowledge representation learning models usually only focus on the structural information of a single triplet, ignoring the graph structure and context information between entities and relations in the knowledge graph.

Based on the above two problems, this paper proposes a 3D rotation knowledge representation learning model (Context_3D) based on graph context. The core idea of the model is to introduce a Graph Neural Network (GNN) on the basis of the RotatE model. To capture graph structure and context information between entities and relationships in the knowledge graph. Specifically, the model first uses the RotatE model for representation learning of triples in knowledge graph, representing the relationship as rotation in complex vector space. Then, GNN module is used to model the graph structure of knowledge graph, and the embedded representation of entities and relationships is integrated with the graph context information. By combining the advantages of rotational models (such as RotatE) with the capabilities of graph neural networks, the Context_3D model provides a more comprehensive approach to learning and reasoning about various relational patterns. It enhances the representation of knowledge graphs and improves performance in dealing with the complexities of relationships.

2   Related Work

Many existing knowledge representation learning models adopt different approaches to model relational patterns in knowledge graphs [9]. For instance, the TransE model treats relationships as a transfer process from the head entity to the tail entity. However, TransE has limitations in modeling and reasoning about symmetrical relational patterns. To address this limitation, improved models such as TransH and TransR have been proposed. By restricting the representations of head and tail entities to the same hyperplane, TransH can better capture the symmetry and antisymmetry properties of relationships. By transforming the head entity into the semantic space where the tail entity resides through the relation's projection matrix, TransR models the translation relationship between the head and tail entities. This approach enables TransR to effectively model complex relational patterns, including symmetric, antisymmetric, and combinatorial relationships.

The DisMult model [7] i uses subtraction operators to capture linear relationships between entities, and uses matrix multiplication to extract semantic information from multiple relationships. DisMult effectively models symmetrical relational patterns. However, it lacks the ability to support the antisymmetric relational model. The ComplEx model [8] extends the representation space of DisMult to the complex space. By representing entities and relationships as complex numbers, complex can effectively model antisymmetric relational patterns. However, ComplEx models are still difficult to model complex combinatorial relationship patterns. The RotatE model [9] introduces complex vectors to represent entities and relationships, enabling it to capture the semantics of relationships through rotation operations. This representation enables the model to effectively model symmetric, antisymmetric, and combinatorial relational patterns. However, the RotatE model does not preserve the order of the relationships when encoding the combinatorial relationships. Therefore, the change of relation order will significantly affect the semantics of relation combination.

Parcollet et al. [10] proposed quaternion recurrent neural networks (QRNN and QLSTM) and applied them to speech recognition tasks. These quaternion-based models outperformed traditional recurrent neural networks (RNN and LSTM), demonstrating the enhanced expressive power and performance achieved by introducing quaternions in tasks like speech recognition. This indicates that the incorporation of quaternions can improve the capabilities and efficiency of neural networks. In the realm of knowledge graph representation learning, Zhang et al. [11] introduced the QuatE model, which maps entities and relations in the knowledge graph to the quaternion space. By utilizing quaternion multiplication, QuatE enhances the interaction between entities and relationships, thereby capturing potential linking relationships more effectively.

Some representation learning models pay more attention to the encoding of a single triplet while ignoring the structure of the directed graph and the neighborhood information of the entity. In order to solve this problem, some methods put forward the following strategies: PTransE [12] introduces the multi-hop relational paths mined from the knowledge graph, and integrates these path information into the TransE model to enrich the modeling ability of the model for the graph structure. GAKE [13] is a general knowledge graph representation learning model, which introduces the context of neighbors, edges and paths, describes the structure information of graphs from different perspectives, and integrates these context information into the model. In addition, some researchers have introduced graph neural networks into knowledge graph representation learning, such as R-GCN [14], CompGCN [15]. These methods [14–16] first use graph neural networks to encode graph structure data, and then pass entity and relationship embeddings with contextual information to subsequent representation learning models for tasks such as link prediction.

In this paper, we propose a novel approach that utilizes quaternion vectors to represent entities and relations in knowledge graphs. We represent entities as sets of vectors in a three-dimensional vector space, while relationships are interpreted as three-dimensional rotational transformations between entities. Drawing inspiration from graph neural network models, our method integrates graph context information based on the TransE model. By leveraging the benefits of quaternion vectors and graph context information, our proposed method offers can improve the ability of representation learning model to model knowledge graph.

3   Methodology

3.1   Relational 3D Rotation Modeling Based on Quaternion Multiplication

In our model, entities are represented as collections of vectors in a three-dimensional space, while relationships are interpreted as three-dimensional rotational transformations between head and tail entities. What sets our approach apart from traditional methods for entity and relation embedding is the utilization of quaternion vectors to represent entities, relations, and their associations.

$$h = x_h i + y_h j + z_h k \tag{1}$$

$$t = x_t i + y_t j + z_t k \tag{2}$$

$$r = sin\frac{\alpha}{2} \circ u + cos\frac{\alpha}{2}, u = u_x i + u_y j + u_z k \tag{3}$$

where, $h, t, u \in T^k$ said pure quaternion function vector, vector $\circ$ said the hadamard product between x_h, y_h, useful, x_t, y_t, z_t, u_x, u_y, offers, $\alpha \in$ fairly R^k, quaternion vector $r \in H^k$ said.

For each triplet (h, r, t), the model expects the following equations to hold in vector space after model training:

$$t_i \approx r_i \cdot h_i \cdot r_i^{-1}, i \in \{0, 1, ..., k-1\} \tag{4}$$

where, $h_i = x_{h,i}i + y_{h,i}j + z_{h,i}k + 0$

$$t_i = x_{t,i}i + y_{t,i}j + z_{t,i}k + 0 \tag{5}$$

$$r_i = sin\frac{\alpha_i}{2} \cdot (u_{x,i}i + u_{y,i}j + u_{z,i}k) + cos\frac{\alpha_i}{2} \tag{6}$$

The ith element,r_i represents the ithelement of the relationship embedded; The element h_i in the head entity embedding will be rotated in three dimensions around the rotation axis u_i and rotation Angle α_i determined by the relation element r_i to obtain the corresponding element t_i in the tail entity embedding.

For the convenience of subsequent expression, formula 4 is denoted as the mapping between the head entity and the relation to the tail entity, and rewritten as follows:

$$t = Rot^*(h, r) \tag{7}$$

Then, the following distance scoring function is defined for the above process:

$$d((h, r), t) = \sum_{i=1}^{k} \left\| r_i \cdot h_i \cdot r_i^{-1} - t_i \right\| \tag{8}$$

Using quaternion vectors as representations of entities and relationships offers several advantages. Firstly, the unique structure of quaternions allows them to accurately represent complex rotation transformations, enabling the representation learning models to capture the intricate relationships between entities more effectively. Secondly, quaternion vectors preserve the properties of uniqueness and reversibility of rotations. This property is crucial in knowledge graph representation learning as it ensures that the relationships between entities are accurately described.

3.2 Fuses Graph Context Information

The context information of node e in the knowledge graph is taken as the model input as shown in Fig. 1, and an embedding vector containing the context information of the entity is obtained, that is, the potential entity is embedded in v_{ec}, and then the v_{ec} is fused into the initial embedding vector of the entity to obtain the output of the model, that is, the entity is embedded in v_e. This paper presents a model method expression for embedding potential entities into v_{ec}, expressed as:

$$v_{ec} = f(g(M_{ht}, M_r))M_{ht} \tag{9}$$

where, $M_{ht} \in R^{(x+y) \times n}$ and $M_r \in R^{(x+y) \times n}$ are the context matrix and context relation matrix of graph context information of node e. M_{ht} and M_r Together constitute the complete graph context information of e. In the process of defining the above two variables, this paper adopts the model assumption of TransE, that is, for the real triplet, there is a hypothesis: $h+r-t \approx 0$. Based on this condition, M_{ht} and M_r are defined, expressed as:

$$M_{ht} = [h_1 + r_{h1}, h_2 + r_{h2}, ..., h_x + r_{hx}, t_1 - r_{t1}, t2 - r_{t2}, ..., t_y - r_{ty}] \tag{10}$$

$$M_r = [r_{h1}, r_{h2}, ..., r_{hx}, r_{t1}, r_{t2}, ..., r_{ty}] \tag{11}$$

where, $\{h_1, h_2, ..., h_x\}$, $\{t_1, t_2, ..., t_y\}$ and $\{r_{h1}, r_{h2}, ..., r_{hx}, r_{t1}, r_{t2}, ...r_{ty}\}$ is the set of head entities with e as tail entity triples in the graph context of node e, the set of tail entities with e as head entity triples, and the set of associated edges, respectively. On the basis of obtaining the representation of M_{ht} and M_r In the graph context, the similarity between each context vector in M_{ht} and the corresponding relation vector in M_r can be calculated. In this paper, cosine similarity is used to calculate the similarity. The similarity function is defined as:

$$g(M_{ht}, M_r)_{[i]} = \frac{(M_{ht})_{[i]}(M_r)_{[i]}}{\sqrt{\sum_{j=1}^{n}(M_{ht})^2_{[i,j]}}\sqrt{\sum_{j=1}^{n}(M_r)^2_{[i,j]}}} \tag{12}$$

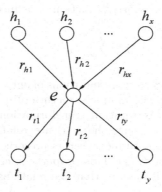

Fig. 1. Location of entity e and its graph context information

The similarity between the ith context vector in the context matrix M_{ht} and the ith relation vector in the context relation matrix M_r is obtained from the above formula. Through the normalization of each similarity, the weight of the graph context is obtained, that is, the weight function $f(\cdot)$ in Eq. 9. The definition is expressed as:

$$f(g(M_{ht}, M_r))_{[i]} = \frac{exp(g(M_{ht}, M_r)_{[i]})}{\sum_{j=1}^{x+y} exp(g(M_{ht}, M_r)_{[j]})} \tag{13}$$

The weights of each context vector in the context matrix M_{ht} of node e(entity e) can be obtained from the above formula. According to Eq. 9, the weighted summation of the context vectors in M_{ht} is performed to calculate the potential entity embedding v_{ec}. Finally, the initial embedding vector of e is fused with v_{ec}. You can get the output entity of the model embedded v. The fusion process is expressed as:

$$(v_e)_{[k]} = (v_e)_{[k-1]} \oplus v_{ec} \tag{14}$$

where, $(v_e)_{[k]}$ represents the embedded vector representation obtained by node e(entity e) after the kth generation selection; $\oplus$ means additive fusion.

For a certain edge r(that is, relation r) in the knowledge graph, there are z triples whose relation is r. In this paper, the relation is embedded in v_r, and the relationship model is established as follows:

$$v_r = \frac{1}{z}\sum_{i=1}^{z}(W_{ht})_{[i]} \tag{15}$$

where, $W_{ht} \in R^{z \times n}$ represents the graph context matrix of edge r. W_{ht} expressed as:

$$W_{ht} = [t_{r1} - h_{r1}, t_{r2} - h_{[r2]}, ..., t_{rz} - h_{rz}] \tag{16}$$

where, $\{(h_{r1}, t_{r1}), (h_{r2}, t_{r2}), ..., (h_{rz}, t_{rz})\}$ represents the set of all entity pairs associated with edge r, that is, the graph context information for r. In the relational model, the initial embedding vector of r is ignored because it not only prevents information overlap but also improves the efficiency of the model while satisfying the assumptions of the model.

To get better training results, define the objective function as follows. First, the Sigmoid function $\sigma(\cdot)$ is used to calculate the probability of each candidate entity, and the calculation formula is expressed as:

$$f_o(v_{ec}, v_o) = b - \|v_{ec} - v_e\| \tag{17}$$

$$P_{[i]} = \sigma(f_o(v_{ec}, v_o)_{[i]}) \tag{18}$$

where, v_o represents the embedding vector of a candidate entity, b is the paranoid parameter; The function $f_o(v_{ec}, v_o)$ calculates the distance between the current entity's potential entity embedding v_{ec} and the candidate entity v_e. Formula 18 represents the probability P that all candidate entities are target entities under the transformation of the function $\sigma(\cdot)$. The cross-entropy loss function, the objective function, is defined based on the probability of the candidate entity, expressed as:

$$L(P,t) = min(\frac{-1}{N}\sum_{i=1}^{N} t_{[i]}(log(P_{[i]}) + (1 - t_{[i]})log(1 - P_{[i]}))) \tag{19}$$

where, N represents the number of candidate entities; $t_{[i]}$ represents the label of the ith candidate entity, which takes the value 0,1. is 1 when the ith candidate entity is the target entity, and 0 otherwise.

In the optimization of the model, the stochastic gradient descent method was used in this paper. In order to prevent overfitting of the data during the experiment, $\|v_e\| \leq 1$ and $\|v_r\| \leq 1$ were set.

3.3   Model Training

In order to obtain the final distance scoring function for knowledge reasoning and model training, the above distance scoring function formulas are combined as follows:

$$d_{final}(h,r,t) = \frac{1}{4}(d((h,r),t) + d(h,(r,t)) + d_c((h,r),t) + d_c(h,(r,t))) \quad (20)$$

where, $dist1(h,r,t)$ represents the distance measure based on the header entity, relationship, and tail entity, and $dist2(h,r,t)$ represents the distance measure based on the graph context representation of the header entity and the graph context representation of the relationship. α and β are weight parameters that balance the importance of the two distance measures.

In order to optimize the model, the self-adversarial negative sampling loss function in literature [9] is adopted as the optimization objective of model training. The expression of the loss function is as follows:

$$L = -\sum_{i=1}^{n_c} p(h_i', r, t_i') log\sigma(d_{final}(h_i', r, t_i') - \gamma) - log\sigma(\gamma - d_{final}(h,r,t)) \quad (21)$$

where σ is the sigmoid function, γ is the fixed margin, (h_i', r, t_i') represents the negative case triplet, and n_c is the number of negative cases. The goal of optimizing the loss function is to separate the positive case triples from the negative case triples. By optimizing the loss function, the score of the positive case triples is as low as possible and the score of the negative case triples is as high as possible.

The sampling weights of negative triples are defined as follows:

$$p(h_i', r, t_i') = \frac{exp(-a \cdot d(h_i', r, t_i'))}{\sum_{i=1}^{n_c} exp(-a \cdot d(h_i', r, t_i'))} \quad (22)$$

where, $a \geq 0$ indicates the sampling temperature. When $a = 0$, the above sampling degenerates to uniform sampling. When $a > 0$, the lower the score of the negative case triplet, the greater the corresponding sampling weight.

4   Experiment and Result Analysis

4.1   Data Set

In this experiment, we use two well-known reference datasets in the field of knowledge graphs: WN18RR and FB15K-237. The WN18RR dataset contains a variety of common sense knowledge across different domains. FB15k-237 is the result of rearranging relationships and resampling negative triples from the original FB15K dataset. Using these datasets, we can evaluate the performance and effectiveness of our model in capturing and understanding complex relationships in the knowledge graph (Table 1).

4.2   Description of Evaluation Methods and Experimental Settings

The link prediction task is to find a missing item in a triple, To evaluate the model, several evaluation metrics are used: Mean Rank (MR): It calculates the

Table 1. Statistics of Datasets

Dataset	Entities	Relations	Training	Validation	Test	Indegree
WN18RR	40943	11	86835	3034	3134	2.12
FB15K-237	14541	237	272115	17535	20466	18.7

average ranking of all the predicted entities. Lower MR values indicate better performance. Mean Reciprocal Rank (MRR): It calculates the average of the reciprocal ranks of all predicted entities. MRR emphasizes the ranking of the correct entity higher in the list and provides a measure of overall accuracy. Hits@N: It calculates the hit ratio of the predicted entity among the top N candidate entities. It measures the proportion of correct predictions within the top N ranks. Common choices for N are 1, 3, or 10. MRR, which stands for mean-reverse rank, is a widely used evaluation method in search ranking algorithms that provides a better reflection of the model's accuracy. MRR is a commonly used evaluation method in search ranking algorithms, which reflects the accuracy of the proposed model is higher than the comparisons. MRR is defined as inverse of the ranking of the first correct answer in the list of all possible candidates. The formula for calculating MRR is:

$$MRR = \frac{1}{n} \sum_{i=1}^{n} \frac{1}{rank(i)} \tag{23}$$

where n represents the number of test set triples and $rank(i)$ represents the rank of the ratings of the original correct triples among the ratings of all triples. A higher MRR indicates that the model is better at ranking the correct missing entity in a triple.

Hits@n measures the proportion of correctly predicted entities or relationships within the top N scores of the test set triples. In other words, it measures the percentage of correct predictions among the top $N\%$ of all entities or relationships.

The hyperparameters are set as follows: self-adjoint negative sampling temperature $\alpha \in \{0.5,1,1.5\}$, embedding dimension of entities and relationships $k \in \{128,256,512\}$, fixed interval $\gamma \in \{6,12,15,18,21\}$, batch size batch_size $\in \{128,256,512\}$, the number of negative cases n in the training phase is fixed at 256, and d_n is set to 3k.

4.3   Experimental Results and Analysis

To validate the effectiveness of the proposed Context_3D model, it is compared with several widely used knowledge graph representation learning models: TransE, DisMult, ComplEx, RotatE, ConvE, and R-GCN, which incorporates graph context information. The comparison is conducted on WN18RR and FB15k-237 datasets, and the findings are summarized as follows (Table 2):

Table 2. Link prediction results of different models on the WN18RR and FB15K-237 datasets

Model	WN18RR					FB15K-237				
	MR	MRR	Hits@1	Hits@3	Hits@10	MR	MRR	Hits@1	Hits@3	Hits@10
TransE	3382	0.293	-	-	0.465	357	0.226	-	-	0.465
RotatE	3341	0.475	0.427	0.491	0.572	177	0.338	0.240	0.381	0.532
DisMult	5111	0.420	0.390	0.440	0.481	254	0.240	0.155	0.262	0.418
ComplEx	5260	0.430	0.420	0.460	0.501	338	0.246	0.158	0.276	0.428
ConvE	4185	0.431	0.402	0.441	0.521	237	0.331	0.241	0.364	0.501
R-GCN	-	-	-	-	-	-	0.249	0.151	0.264	0.417
Context_3D	2854	0.472	0.429	0.489	0.568	162	0.352	0.267	0.402	0.548

1) Context_3D outperforms basic models (TransE, DisMult, and ComplEx) in various evaluation metrics. This demonstrates that Context_3D exhibits superior performance and capability in knowledge graph representation learning tasks. Compared to R-GCN, Context_3D shows certain improvements. Although it may not achieve optimal results on all metrics, it consistently achieves sub-optimal performance, confirming the effectiveness and enhancement of the proposed method. On the FB15k-237 dataset, Context_3D achieves optimal results on most metrics, except for sub-optimal performance in Hits@1. This further validates the superiority of Context_3D in capturing relational patterns in knowledge graphs. On the WN18RR dataset, Context_3D achieves optimal results in MR, MRR, and Hits@3, and sub-optimal results in Hits@1 and Hits@10, although not optimal. This indicates that the Context_3D model can deliver relatively good performance across different datasets.

2) The WN18RR and FB15k-237 datasets consist of three relational modes: symmetry, antisymmetry, and combination, which are prevalent in the data. Compared to the RotatE model, Context_3D demonstrates improvements in various evaluation metrics. This improvement can be attributed to Context_3D's ability to maintain the order of relationships in composite relational patterns.

3) Compared with the WN18RR dataset, the implementation of the Context_3D model combined with the entity graph context is more significantly improved on the FB15k-237 dataset. This difference is mainly due to the fact that entities in the FB15k-237 dataset have richer graph context information. In knowledge graphs with more neighborhood relationships, Context_3D can make more efficient use of entity graph context information, thus improving the performance of representation learning tasks. This further emphasizes the importance of integrating the context of the entity diagram and provides guidance for model and dataset selection.

To assess the capability of the Context_3D model in handling diverse complex relationships, experiments were conducted on the FB-15K237 dataset, and different relationship categories were analyzed. The results are presented in Fig. 2, and the following observations can be made: For 1-1 relationships, the Con-

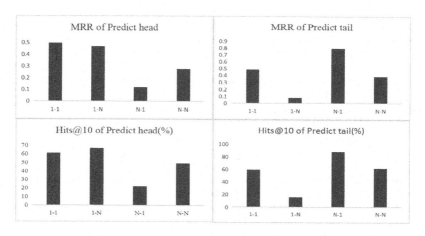

Fig. 2. Relationship Category Testing on FB15K-237 Dataset

text_3D model demonstrates superior performance in predicting both head and tail entities. This indicates that incorporating the graph context information of entities enhances the accuracy of entity embeddings. In the case of complex relationships, particularly those falling into N-1, 1-N, and N-N, the Context_3D model outperforms other models by a significant margin. This suggests that the fusion of graph context information enables the model to effectively differentiate between different entities within these complex relationship categories.

5   Conclusion

In this paper, a 3D rotating knowledge graph representation learning model Context_3D is proposed, which includes the context of the synthesis graph. Entities are represented as vectors in three-dimensional space to better capture various relational patterns. The model learns the representation of the entity graph context to enhance the embedded entity representation. Experimental results show that compared with existing representation learning models, the performance of Context_3D model is improved.

Acknowledgments. This work was supported by Natural Science Foundation of Shandong Province (No. ZR2022LZH008) and The 20 Planned Projects in Jinan (Nos. 2021GXRC046, 202228120).

References

1. Zhiyuan, L., Maosong, S., Yankai, L.: Knowledge representation learning: a review. Comput. Res. Dev. **53**(2), 247–261 (2016)
2. Wang, Q., Mao, Z., Wang, B.: Knowledge graph embedding: as survey of approaches and applications. IEEE Trans. Knowl. Data Eng. **29**(12), 2724–2743 (2017)

3. Dong, X., Gabrilovich, E., Heitz, G.: Knowledge vault: a web scale approach to probabilistic knowledge fusion. In: Proceedings of the 20th ACM SIGKDD International Conference on Knowledge Discovery and Data Mining, pp. 601–610 (2014)
4. Bordes, A., Usunier, N., Garcia-Duran, A.: Translating embeddings for modeling multi-relational data. In: International Conference on Neural Information Processing Systems (2013)
5. Wang, Z., Zhang, J., Feng, J.: Knowledge graph embedding by translating on hyper planes. In: Proceedings of the AAAI Conference on Artificial Intelligence, vol. 28, no. 1, pp. 1112–1119 (2014)
6. Ji, G., He, S., Xu, L.: Knowledge graph embedding via dynamic mapping matrix. In: Proceedings of the 53rd Annual Meeting of the Association for Computational Linguistics and the 7th International Joint Conference on Natural Language Processing (Volume 1; Long Papers), pp. 687–696 (2015)
7. Yang, B., Yih, W., He, X.: Embedding entities and relations for learning and inference in knowledge bases. arXiv preprint arXiv:1412.6575 (2014)
8. Trouillon, T., Welbl, J., Riedel, S.: Complex embeddings for simple link prediction. In: International Conference on Machine Learning, pp. 2071–2080 (2016)
9. Sun, Z., Deng, Z.H., Nie, J.Y.: RotatE; knowledge graph embedding by relational rotation in complex space. In: International Conference on Learning Representations (2018)
10. Parcollet, T., Ravanelli, M., Morchid, M.: Quaternion recurrent neural networks. In: International Conference on Learning Representations (2018)
11. Zhang, S., Tay, Y., Yao, L.: Quaternion knowledge graph embed-dings. In: Conference on Neural Information Processing Systems, pp. 2735–2745 (2019)
12. Lin, Y., Liu, Z., Luan, H.: Modeling relation paths for representation learning of knowledge bases. In: Proceedings of the 2015 Conference on Empirical Methods in Natural Language Processing, pp. 705–714 (2015)
13. Feng, J., Huang, M., Yang, Y.: GAKE: graph aware knowledge embedding. In: Proceedings of COLING 2016, the 26th International Conference on Computational Linguistics: Technical Papers, pp. 641–651 (2016)
14. Schlichtkrull, M., Kipf, T.N., Bloem, P., van den Berg, R., Titov, I., Welling, M.: Modeling relational data with graph convolutional networks. In: Gangemi, A., et al. (eds.) ESWC 2018. LNCS, vol. 10843, pp. 593–607. Springer, Cham (2018). https://doi.org/10.1007/978-3-319-93417-4_38
15. Vashishth, S., Sanyal, S., Nitin, V.: Composition-based multi-relational graph convolutional networks. In: International Conference on Learning Representations (2019)
16. Yao, S.Y., Zhao, T.Z., Wang, R.J.: Rule-guided joint embedding learning of knowledge graphs. Comput. Res. Dev. **57**(12), 2514–2522 (2020)
17. Cariow, A., Cariowa, G., Majorkowska-Mech, D.: An algorithm for quaternion-based 3D rotation. Int. J. Appl. Math. Comput. Sci. **30**(1), 149–160 (2020)

TextBFA: Arbitrary Shape Text Detection with Bidirectional Feature Aggregation

Hui Xu[1,2]($\boxtimes$), Qiu-Feng Wang[3], Zhenghao Li[1], Yu Shi[1], and Xiang-Dong Zhou[1]

[1] Chongqing Institute of Green and Intelligent Technology, Chinese Academy of Sciences, Chongqing, China
{xuhui,lizh,shiyu,zhouxiangdong}@cigit.ac.cn

[2] Chongqing School, University of Chinese Academy of Sciences (UCAS Chongqing), Chongqing, China

[3] Department of Intelligent Science, School of Advanced Technology, Xi'an Jiaotong-Liverpool University (XJTLU), Suzhou, China
qiufeng.wang@xjtlu.edu.cn

Abstract. Scene text detection has achieved great progress recently, however, it is challenging to detect arbitrary shaped text in the scene images with complex background, especially for those unobvious and long texts. To tackle this issue, we propose an effective text detection network, termed TextBFA, strengthening the text feature by aggregating high-level semantic features. Specifically, we first adopt a bidirectional feature aggregation network to propagate and collect information on feature maps. Then, we exploit a bilateral decoder with lateral connection to recover the low-resolution feature maps for pixel-wise prediction. Extensive experiments demonstrate the detection effectiveness of the proposed method on several benchmark datasets, especially on inconspicuous text detection.

Keywords: Scene text detection · inconspicuous text · feature aggregation

1 Introduction

Scene text detection plays an important role aiming to locate text regions in the natural images. Recently, the regular shape text detection methods have achieved great progress, however, it is challenging to detect arbitrary shape texts such as examples shown in Fig. 5.

Nowadays, researchers [1,9,13,28] have started to focus on arbitrary shape text detection for practical applications, which can be roughly classified into two categories: regression-based [19,21,22] and segmentation-based methods

B. Luo et al. (Eds.): ICONIP 2023, CCIS 1962, pp. 365–377, 2024.
https://doi.org/10.1007/978-981-99-8132-8_28

[12,18,23]. Segmentation-based methods obviously have advantages in the geo-metric representation of arbitrary shape texts. However, existing segmentation-based methods [9,30] heavily rely on the prediction accuracy of Feature Pyra-mid Networks (FPN) [10] framework, neglecting high correlation between pix-els in the text. Inconspicuous text, including text with complex illumination, small scale, obstacles and other text with inconspicuous features as shown in Fig. 1 and Fig. 5, is still challenging to detect in the wild. If we think of FPN as an encode-decoder structure, such text detection methods can be divided into encoder, decoder and postprocessing stages, as shown in the top pipeline in Fig. 1. Most methods improve the model in the part of post-processing, includ-ing the belonging-relation learning between pixels [28] and boundary fine-tuning [27], etc. The difference is that we add a bidirectional feature aggregation module between the encoder and the decoder to optimize the model in feature extraction.

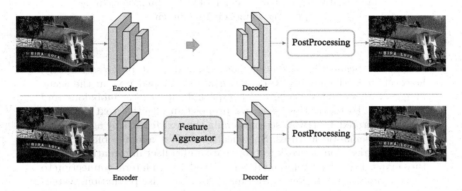

Fig. 1. Feature aggregation illustration. The top pipeline represents most current segmentation-based text detection methods, while the bottom pipeline illustrates our detector incorporating feature aggregation.

In this paper, we propose a novel text detector for arbitrary shape text detec-tion, which increases the cohesion between pixels within the text by feature aggregation. Our text detection network is mainly composed of a CNN-based encoder, a bidirectional feature aggregation module and a bilateral up-sampling decoder, as shown in Fig. 2. Inspired by RESA [29], we design the bidirec-tional feature aggregation network to collect information within the features and propagate spatial information along both horizontal and vertical directions. The aggregated feature $\mathcal{F}_s$ is combined from three branches: horizontal aggre-gator, vertical aggregator and the direct link. Each aggregator first slices the feature map along the current direction, and then aggregates each sliced feature with another sliced feature adjacent to a certain stride. Each spatial location is updated synchronously through multiple steps and finally can gather infor-mation in the whole space. In this way, the relationship between pixels inside the text is better mined, and the background information is also strengthened.

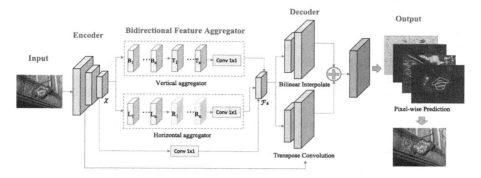

Fig. 2. The overview framework of TextBFA, which is composed by encoder, bidirectional feature aggregator and decoder. 'T_n' and 'B_n' represent "top-to-bottom" and "bottom-to-top" respectively at the n-th layer in vertical aggregator, and 'L_n' and 'R_n' represent "left-to-right" and "right-to-left" respectively in horizontal aggregator.

After that, we design a bilateral decoder with lateral connection to up-sample the low-resolution feature maps into pixel-wise prediction.

In summary, the main contributions of this paper are three-fold: 1) We propose a novel unified end-to-end trainable framework for arbitrary shape text detection, which can strengthen the cohesion between the pixels inside the text. 2) To the best of our knowledge, our work presents one of the very first attempts to perform feature aggregation for inconspicuous text detection. 3) The proposed method is demonstrated to be effective for inconspicuous text detection and achieves competitive performance on benchmark datasets.

2   Related Work

Scene text detection methods based on deep learning can be roughly divided into two categories: top-down methods and bottom-up methods.

Top-Down Methods. Methods of this type explicitly encode text instances with contours of text region. Different from generic objects, texts are often presented in irregular shapes with various aspect ratios. MOST [7] focused on text instances of extreme aspect ratios and apply Position-Aware Non-Maximum Suppression (PA-NMS) to produce the final predictions. TextRay [19] formulated the text contours in the polar coordinates. TextBPN [27] adopted a GCN-based adaptive boundary deformation model to generate accurate text instance. Although these methods have achieved good performance in quadrilateral text detection, The constrained representation capability of point sequence limits the detection performance of arbitrary shape text. In addition, they always need some hand-designed processes, such anchor generation and NMS.

Bottom-Up Methods. Methods of this type firstly detect fundamental components (e.g., pixels and segments), then combine these components to produce the detection results. DB [9] performed an adaptive binarization process in a segmentation network, which simplifies the post-processing and enhances the detection performance. Seglink++ [17] took small segments of a text instance as the fundamental elements and link them together to form the bounding boxes. CRAFT [1] detected text regions by exploring affinities between characters. Zhang *et al.* [28] applied GCN to infer the linkages between different text components. TCM [25] focused on turning the CLIP Model directly for text detection without pretraining process. These bottom-up methods have less constraints than the top-down ones on the representation of arbitrary shapes, but the their performance are still affected by the quality of instance segmentation accuracy.

3   Methodology

3.1   Overview

The pipeline of the proposed method is shown in Fig. 2. We use the backbone network to extract shared features. The bidirectional feature aggregator consists of vertical aggregator and horizontal aggregator. We exploit the bilateral decoder with lateral connection to perform feature concatenation and upsampling based on the aggregated feature $\mathcal{F}_s$. After that, text boundaries can be obtained under the guidance of pixel-wise predictions.

3.2   Bidirectional Feature Aggregator

The bidirectional feature aggregator consists of a veritical aggregator and a horizontal aggregator. The vertical aggregator is composed by "top-to-bottom" and "bottom-to-up" layers, and the horizontal aggregator is composed by "right-to-left" and "left-to-right" layers. As examples, we have only presented the structure of feature aggregation in two transfer directions, which are "bottom-to-top" and "left-to-right" as shown in Fig. 3. In "bottom-to-up" layers, the feature is passed up from the bottom piece of features with a certain stride. Similarly, in "top-to-bottom" layers, the feature is passed down from the top piece of features. The same is true of "left-to-right" and "right-to-left" layers.

Feature map $\mathcal{X}$ is split into H slices in the horizontal direction or W slices in the vertical direction. In each transfer direction, we totally utilize several layers with different shift strides to perform feature aggregation. Given a 3D feature map tensor $\mathcal{X}$ of size $C \times H \times W$, where C, H and W denote the number of channels, rows and columns respectively, the aggregated feature map is calculated as follows,

$$\mathcal{F}_s = Concat(\boldsymbol{K}_v \otimes \mathcal{F}_v, \boldsymbol{K}_h \otimes \mathcal{F}_h, \boldsymbol{K}_x \otimes \mathcal{X}), \qquad (1)$$

where $\boldsymbol{K}_v, \boldsymbol{K}_h$ and $\boldsymbol{K}_x$ are 1×1 convolution kernels to change the number of feature channels. $\mathcal{F}_v$ and $\mathcal{F}_h$ are feature maps output by vertical aggregator and horizontal aggregator respectively. Let $L = \{L_1, L_2, ..., L_n\}, R =$

$\{R_1, R_2, ..., R_n\}, T = \{T_1, T_2, ..., T_n\}$ and $B = \{B_1, B_2, ..., B_n\}$ represent the aggregation layers of "left-to-right", "right-to-left", "top-to-bottom" and "bottom-to-top" respectively, where n is the number of layers. The feature $\mathcal{F}_v$ and $\mathcal{F}_h$ are as follows,

$$\mathcal{F}_v = T_n(T_{n-1}...(T_1(B_n(B_{n-1}...(B_1(\mathcal{X})))))) \qquad (2)$$
$$\mathcal{F}_h = R_n(R_{n-1}...(R_1(L_n(L_{n-1}...(L_1(\mathcal{X})))))),$$

In the current vertical aggregator, we perform "bottom-to-top" layers first and then "top-to-bottom" layers. In fact, it doesn't matter if the order of the

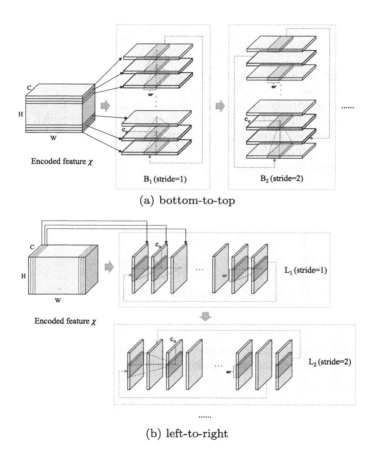

(a) bottom-to-top

(b) left-to-right

Fig. 3. Illustration of Bidirectional Feature Aggregator. Similarly, the feature propagation in "top-to-bottom" layers is sequential downward from the top feature slice by stride, in the opposite direction to "bottom-to-top" layers. The same is true for "right-to-left" layers.

two are switched. The same goes for the horizontal aggregator. The process of feature aggregation in four types of layers can be represented as follows,

$$\mathcal{X}_{c_o,i,j}^{t'} = \mathcal{X}_{c_o,i,j}^{t} + \beta f(Z_{c_o,i,j}^{t}), \tag{3}$$

where $\mathcal{X}^t$ denotes the feature at the t-th layer (e.g. L_t, R_t, T_t, B_t), c_o, i and j represent indexes of channel, row and column respectively, Z^t is the intermediate results for information passing. After that, $\mathcal{X}^t$ is updated to $\mathcal{X}^{t'}$ by a feature aggregation. $f(\cdot)$ is a nolinear activation function such as ReLU, and β is a hyper-parameter to control the degree of feature aggregation. The information to be transmitted is calculated as follows,

$$Z_{c_o,i,j}^{t} = \sum_{n_{in},w_k} K_{n_{in},c_o,w_k} \otimes \mathcal{X}_{n_{in},p_h,p_w}^{t}, \tag{4}$$

where K is a 1D convolution kernel of size $N_{in} \times N_{out} \times w$, where N_{in}, N_{out} and w represent the number of input channels, the number of output channels and kernel width, respectively. Both N_{in} and N_{out} are equal to C. n_{in}, c_o and w_k denote the index of input channel, output channel and kernel width respectively. p_h and p_w are the index values corresponding to the feature map when calculating the information passing. They are different in vertical and horizontal aggregator, which are designed as follows,

$$\begin{cases} p_h = (i + s_t) \bmod H \\ p_w = j + w_k - \lfloor w/2 \rfloor \\ s_t = \dfrac{W}{2^{\lfloor log_2 W \rfloor - t + 1}}, \quad t = 1, ..., \lfloor log_2 W \rfloor \end{cases} \tag{5}$$

$$\begin{cases} p_h = i + w_k - \lfloor w/2 \rfloor \\ p_w = (j + s_t) \bmod W \\ s_t = \dfrac{H}{2^{\lfloor log_2 H \rfloor - t + 1}}, \quad t = 1, ..., \lfloor log_2 H \rfloor \end{cases}, \tag{6}$$

where (5) is for vertical aggregator and (6) is for horizontal aggregator. s_t is the shift stride to determine the information passing distance dynamically.

3.3   Bilateral Decoder with Lateral Connection (BDLC)

The bilateral up-sampling decoder in RESA [29] directly upsamples the output of the last layer, which only considers features at a single scale. However, multi-scale detection still performs better especially for small objects, and the text is usually small target in a scene image. Inspired by FPN [10], we design a bilateral decoder with lateral connection to associate low-level feature maps across resolutions and semantic levels as shown in Fig. 4.

Each lateral connection merges feature maps of the same spatial size between encoder and decoder. Though bottom feature maps in encoder are of low-level

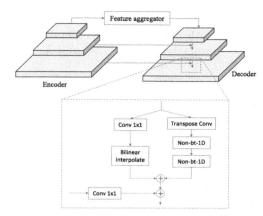

Fig. 4. A building block illustrating the bilateral upsampling and lateral connection, which constructs our high-resolution feature maps. Non-bt-1D block is composed by two 3×1 convolution layers and two 1×3 convolution layers as described in [15].

semantics, their activations are more accurately localized as they were subsampled fewer times. With a low-resolution feature map, we double the spatial resolution by up-sampling in two pathways: bilinear interpolation based and transpose convolution based. The upsampled feature map is then merged with the corresponding feature map in the encoder by element-wise addition. This process is iterated until the finest resolution map is generated.

The bilinear interpolation based pathway utilize the bilinear upsampling to generate a relatively rough feature map of $2\times$ size. A 1×1 convolution is applied to reduce the number of channels, which follows a BN and a ReLU. This pathway can quickly recover the obvious contour features, although some details may be lost. By contrast, the transpose convolution based pathway can get more refined features due to more learnable kernel parameters. We combine the two passways by adding their outputs together.

3.4 Training

The proposed model is trained in an end-to-end manner, with the following objective function,

$$\mathcal{L} = \lambda_{tr}\mathcal{L}_{tr} + \lambda_{tcl}\mathcal{L}_{tcl} + \lambda_s\mathcal{L}_{scale} + \lambda_o\mathcal{L}_{ori}, \tag{7}$$

where $\mathcal{L}_{tr}$ and $\mathcal{L}_{tcl}$ are the cross-entropy loss for text region and text center region. To solve the sample imbalance problem, OHEM [16] is adopted for $\mathcal{L}_{tr}$ with the ratio between negative and positive samples being $3 : 1$. $\mathcal{L}_{scale}$ and $\mathcal{L}_{ori}$ are the Smoothed-L1 loss for text height maps and orientation. The loss function can be formulated as

$$\begin{pmatrix} \mathcal{L}_{scale} \\ \mathcal{L}_{ori} \end{pmatrix} = SmoothedL1 \begin{pmatrix} \frac{\hat{s}-s}{s} \\ \hat{\theta} - \theta \end{pmatrix}, \tag{8}$$

where $\hat{s}$ and $\hat{\theta}$ are the predicted values of the scale and the orientation, while s and θ are their groundtruth correspondingly. Details of $\mathcal{L}_{ori}$ and θ can be found in [24]. The hyper-parameters, $\lambda_{tr}, \lambda_{tcl}, \lambda_s$ and λ_o are all set to 1 in our experiments.

4   Experiments

We varify the effectiveness of our model on three representative benchmarks: TD500 [3], Total-Text [2] and CTW1500 [11]. We pre-train the network on Synth-Text [5] and MLT2017 [14] datasets, then fine-tune it on the official training set of the corresponding dataset.

4.1   Datasets

The datasets used for the experiments in this paper are briefly introduced below:
   SynthText [5] is a large scale dataset that contains about $800K$ synthetic images, which is only used to pre-train our model.
   TD500 [3] consists of 300 training images and 200 test images, including multi-oriented and long text. The texts in the images are either English or Chinese scripts, annotated by rotate rectangles.
   Total-Text [2] is one of the most important arbitrarily-shaped scene text benchmarks, which contains 1255 training images and 300 test images. The images were collected from various scenes, including text-like background clutter and low-contrast texts, with word-level polygon annotations.
   CTW1500 [11] contains both Englist and Chinese texts with text-line level annotations, where 1000 images for training, and 500 images for testing. It also contains horizontal, multi-orientated and curved texts.

4.2   Implementation Details

We use the ResNet50 [6] pre-trained on ImageNet as our backbone. The model is implemented in Pytorch and trained on 8 NVIDIA GTX GeForce 1080T GPUs using Adam optimizer [8]. We pre-train the network on Synth-Text and MLT2017 dataset and then fine-tune it on the official training set of the target dataset. We adopt data augmentation strategies, e.g., randomly rotating the images with degree in range of $[-15°, 15°]$, random brightness and contrast on input images, and random crop, where we make sure that no text has been cut. In the pre-training stage, the randomly cropped images are resized to 512×512. In the fine-tuning stage, multi-scale training is used. The short side of the image is resized to $\{640, 800, 960\}$, and the long sizd of the image is maintained to 1280. Blurred texts labeled as DO NOT CARE are ignored during training. The initial learning rate for pre-training and fine-tuning is $1e^{-3}$ and $1e^{-4}$, respectively. Both of them are reduced $0.8\times$ every 50 epoches.

4.3 Ablation Study

We conduct the ablation study to verify the effectiveness of bidirectional feature aggregator and BDLC.

Effect of Each Component: To verify the performance gain of the proposed bidirectional feature aggregator and BDLC, we conduct several ablation experiments on CTW1500. We summarize the detection performance of each module in Table 1. As we can see, both modules can improve text detection performance, which verifies the capabilities of proposed modules. It is worth noting that the version without BDLC uses transposed convolution for upsampling.

Table 1. Ablation experiments of each module on CTW1500.

Baseline	Aggregator	BDLC	F-measure (%)
✓			82.0
	✓		84.1
		✓	82.5
	✓	✓	**84.8**

Effectiveness of Lateral Connection in Decoder: To evaluate the effectiveness of the proposed BDLC, we conduct servel experiments on CTW1500. We compare the performance of decoders with and without lateral connection. In addition, we compare the performance of BDLC with unilateral decoders based on bilinear upsampling and transpose convolution upsampling. The results in Table 2 shows that our decoder outperforms others.

Table 2. The comparison between different upsampling decoders. † refers to the removal of lateral connection.

Method	Precision (%)	Recall (%)	F-measure (%)
bilinear based	87.3	81.3	84.2
transpose convolution based	87.5	81.8	84.5
BDLC†	87.2	81.5	84.3
BDLC	**87.9**	**82.0**	**84.8**

4.4 Comparisons with State-of-the-Art Methods

We compare our method with recent state-of-the-art methods on TD500, Total-Text and CTW1500. Before this, we conducted qualitative analysis on the detection results of some inconspicuous texts, taking the results of TextBPN [27] as a comparison.

Qualitative Analysis. As shown in Fig. 5, some texts with unobvious features are not detected or detected incompletely by TextBPN. In the first row of Fig. 5, the shaded part of the text "GEMBIRA" is not completely detected in TextBPN. The text "the" and "BREWING CO" in the second and third rows are also ignored by TextBPN due to their small scale and surrounded by larger scale text. The long text in the detection results of TextBPN are easily cut off by gaps or icons, such as the baseball representing the letter "O" in the fourth row and the bottle icon in the fifth row.

Evaluation on Text Benchmarks. We evaluate the proposed method onTD500, Total-Text and CTW1500 to test its performance for multi-oriented texts, curved texts and long texts. As shown in Table 3, the proposed method

Fig. 5. Detection results of inconspicuous text on Total-Text and CTW1500 dataset.

Table 3. Comparison with related methods on TD500, CTW1500 and Total-Text.

Method	CTW1500			Total-Text			TD500		
	P(%)	R(%)	F(%)	P(%)	R(%)	F(%)	P(%)	R(%)	F(%)
TextSnake [13]	69.7	**85.3**	75.6	82.7	74.5	78.4	83.2	73.9	78.3
TextDragon [4]	84.5	82.8	83.6	85.6	75.7	80.3	-	-	-
TextField [23]	83.0	79.8	81.4	81.2	79.9	80.6	87.4	75.9	81.3
CARFT [1]	86.0	81.1	83.5	87.6	79.9	83.6	88.2	78.2	82.9
LOMO [26]	85.7	76.5	80.8	87.6	79.3	83.3	-	-	-
ContourNet [22]	83.7	84.1	83.9	86.9	83.9	85.4	-	-	-
Boundary [20]	-	-	-	88.9	85.0	87.0	-	-	-
DRRG [28]	85.9	83.0	84.5	86.5	84.9	85.7	88.1	82.3	85.1
FCENet [30]	87.6	83.4	**85.5**	89.3	82.5	85.8	-	-	-
TextBPN [27]	86.5	83.6	85.0	**90.7**	**85.2**	**87.9**	86.6	**84.5**	**85.6**
Ours	**87.9**	82.0	84.8	88.9	83.7	86.2	**88.3**	82.9	85.5

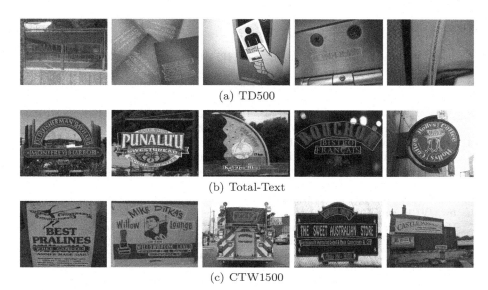

(a) TD500

(b) Total-Text

(c) CTW1500

Fig. 6. Detection results on different benchmarks.

achieves a competitive performance. TextBPN [27] has the best overall performance, which adopts graph convolutional network (GCN) to model the relationship between the pixels and group them into fine representations. In other words, it adds a complex GCN-based post-processing process after pixel-wise prediction. In contrast, our approach has no additional post-processing, but achieves a competitive performance by deeply mining the information of the features themselves. The visualization of the detection results are shown in Fig. 6.

Through the analysis of experimental results, we find that some texts with small intervals are easily to be integrated into one long text in feature aggregation, which is what we will focus on in the future.

5   Conclusion

This paper focuses on the detection of visually inconspicuous text in natural scene images. We propose a novel arbitrary-shaped text detection method, which uses spatial aggregation of features to enhance the cohesion between the pixels inside the text. Different from previous text detection methods that directly use FPN [10] for pixel-level prediction, we add the bidirectional feature aggregation network before up-sampling to make full use of higher-level semantic informations. After that, bilateral upsampling decoder with lateral connection is adopted to obtain more accurate pixel-level prediction. The experimental results demonstrate that our method have a good performance on inconspicuous texts.

Acknowledgments. This research is supported by the National Natural Science Foundation of China under No. 62106247.

References

1. Baek, Y., Lee, B., Han, D., Yun, S., Lee, H.: Character region awareness for text detection. In: CVPR, pp. 9365–9374 (2019)
2. Ch'Ng, C.K., Chan, C.S., Liu, C.L.: Total-text: toward orientation robustness in scene text detection. Doc. Anal. Recognit. **23**(1), 31–52 (2019)
3. Cong, Y., Xiang, B., Liu, W., Yi, M., Tu, Z.: Detecting texts of arbitrary orientations in natural images. In: CVPR, pp. 1083–1090 (2012)
4. Feng, W., He, W., Yin, F., Zhang, X.Y., Liu, C.L.: Textdragon: an end-to-end framework for arbitrary shaped text spotting. In: ICCV, pp. 9076–9085 (2019)
5. Gupta, A., Vedaldi, A., Zisserman, A.: Synthetic data for text localisation in natural images. In: CVPR (2016)
6. He, K., Zhang, X., Ren, S., Sun, J.: Deep residual learning for image recognition. In: CVPR, pp. 770–778 (2016)
7. He, M., Liao, M., Yang, Z., Zhong, H., Bai, X.: Most: a multi-oriented scene text detector with localization refinement. In: CVPR, pp. 8813–8822 (2021)
8. Kingma, D.P., Ba, J.: Adam: a method for stochastic optimization. arXiv preprint arXiv:1412.6980 (2014)
9. Liao, M., Wan, Z., Yao, C., Chen, K., Bai, X.: Real-time scene text detection with differentiable binarization. In: AAAI, vol. 34, no. 07, pp. 11474–11481 (2020)
10. Lin, T.Y., Dollar, P., Girshick, R., He, K., Hariharan, B., Belongie, S.: Feature pyramid networks for object detection. In: CVPR, pp. 2117–2125 (2017)
11. Liu, Y., Jin, L., Zhang, S., Luo, C., Zhang, S.: Curved scene text detection via transverse and longitudinal sequence connection. PR **90**, 337–345 (2019)
12. Long, S., Qin, S., Panteleev, D., et al.: Towards end-to-end unified scene text detection and layout analysis. In: CVPR, pp. 1049–1059 (2022)
13. Long, S., Ruan, J., Zhang, W., He, X., Wu, W., Yao, C.: Textsnake: a flexible representation for detecting text of arbitrary shapes. In: ECCV (2018)
14. Nayef, N., Yin, F., Bizid, I., et al.: ICDAR2017 robust reading challenge on multilingual scene text detection and script identification-RRC-MLT. In: ICDAR (2017)
15. Romera, E., Alvarez, J.M., Bergasa, L.M., Arroyo, R.: Erfnet: efficient residual factorized convnet for real-time semantic segmentation. T-ITS **19**(1), 1–10 (2017)
16. Shrivastava, A., Gupta, A., Girshick, R.: Training region-based object detectors with online hard example mining. In: CVPR, pp. 761–769 (2016)
17. Tang, J., Yang, Z., Wang, Y., Zheng, Q., Bai, X.: Detecting dense and arbitrary-shaped scene text by instance-aware component grouping. PR **96**, 106954 (2019)
18. Tang, J., Zhang, W., Liu, H., et al.: Few could be better than all: Feature sampling and grouping for scene text detection. In: CVPR (2022)
19. Wang, F., Chen, Y., Wu, F., Li, X.: Textray: contour-based geometric modeling for arbitrary-shaped scene text detection. In: ACM MM, pp. 111–119 (2020)
20. Wang, H., Lu, P., Zhang, H., et al.: All you need is boundary: toward arbitrary-shaped text spotting. In: AAAI (2020)
21. Wang, X., Jiang, Y., Luo, Z., et at.: Arbitrary shape scene text detection with adaptive text region representation. In: CVPR, pp. 6449–6458 (2019)
22. Wang, Y., Xie, H., Zha, Z.J., Xing, M., Fu, Z., Zhang, Y.: Contournet: taking a further step toward accurate arbitrary-shaped scene text detection. In: CVPR, pp. 11753–11762 (2020)
23. Xu, Y., Wang, Y., Zhou, W., et al.: Textfield: learning a deep direction field for irregular scene text detection. TIP **28**(11), 5566–5579 (2019)

24. Yang, M., Guan, Y., Liao, M., He, X., Bai, X.: Symmetry-constrained rectification network for scene text recognition. In: ICCV (2019)
25. Yu, W., Liu, Y., Hua, W., Jiang, D., Ren, B., Bai, X.: Turning a clip model into a scene text detector. In: CVPR (2023)
26. Zhang, C., Liang, B., Huang, Z., En, M., Han, J., Ding, E., Ding, X.: Look more than once: an accurate detector for text of arbitrary shapes. In: CVPR, pp. 10552–10561 (2019)
27. Zhang, S., Zhu, X., Yang, C., Wang, H., Yin, X.: Adaptive boundary proposal network for arbitrary shape text detection. In: ICCV (2021)
28. Zhang, S.X., Zhu, X., Hou, J.B., Liu, C., Yang, C., Wang, H., Yin, X.C.: Deep relational reasoning graph network for arbitrary shape text detection. In: CVPR, pp. 9699–9708 (2020)
29. Zheng, T., Fang, H., Zhang, Y., Tang, W., Cai, D.: Resa: recurrent feature-shift aggregator for lane detection. In: AAAI, vol. 35, no. 4, pp. 3547–3554 (2021)
30. Zhu, Y., Chen, J., Liang, L., Kuang, Z., Jin, L., Zhang, W.: Fourier contour embedding for arbitrary-shaped text detection. In: CVPR, pp. 3123–3131 (2021)

Bias Reduced Methods to Q-learning

Patigül Abliz$^{(\boxtimes)}$ and Shi Ying

School of Computer Science, Wuhan University, Wuhan 430072, China
patigvl@whu.edu.cn

Abstract. It is well known that Q-learning (QL) suffers from overestimation bias, which is caused by using the maximum action value to approximate the maximum expected action value. To solve overestimation issue, overestimation property of Q-learning is well studied theoretically and practically. In general, most work on reducing overestimation bias is to find different estimators to replace maximum estimator in order to mitigate the effect of overestimation bias. These works have achieved some improvement on Q-learning. In this work, we still focus on overestimation bias reduced methods. In these methods, we focus on M samples action values, and one of these samples estimated by remaining samples' maximum actions. We select median and max members from these new samples which are estimated by maximum actions. We call these max and median members as Bias Reduced Max Q-learning (BRMQL) and Bias Reduced Median Q-learning (BRMeQL). We first theoretically prove that BRMQL and BRMeQL suffer from underestimation bias and analyze the effect of number of M Q-functions on the performance of our algorithms. Then we evaluate the BRMQL and BRMeQL on benchmark game environments. At last, we show that BRMQL, and BRMeQL less underestimate the Q-value than Double Q-learning (DQL) and perform better than several other algorithms on some benchmark game environments.

Keywords: Bias · Q-learning · Underestimation Bias ·
Overestimation Bias · Action Value Samples · Double Q-learning ·
Cross-Validation Estimator

1 Introduction

Reinforcement learning (RL) concerns sequential problems which are formed as markow decision process (MDP) [1]. Its goal is to maximize the future cumulative reward signal [2] through mapping states to actions. Q-learning (QL) as a useful practical method has been introduced by [3] to address this problem. Q-learning convergence to optimal action value has been depicted in [4–6] under the condition all policies end up with a zero reward absorbing state. Q-learning has been used to solve wide various practical RL problems such as Meal delivery system [7], Mobile robot navigation [8], soccer games [9].

Even if Q-learning is a useful method and simple to update [10], it suffers from overestimation bias [11–13]. Overestimation bias causes poor performance

B. Luo et al. (Eds.): ICONIP 2023, CCIS 1962, pp. 378–395, 2024.
https://doi.org/10.1007/978-981-99-8132-8_29

in the learning process especially in the stochastic environments, including non-optimal convergence [11,14], unstable in learning process [15,16]. To solve over-estimation bias, double Q-learning (DQL) [13] uses two action value estimators $Q_1(s_t, a_t), Q_2(s_t, a_t)$, and $Q_1(s_{t+1}, argmax_a Q_2(s_{t+1}, a))$ is used to update Q-learning and vice versa. In this way, they could use different estimators to evaluate maximum action and its value, thus DQL mitigates the overestimation bias problem in some settings. Twin delayed deep deterministic policy gradient (TD3) [17] uses two action value estimators. One of them is used to evaluate maximum action, and min of these two action values which are estimated by maximum action used to update Q-function. TD3 suffers from underestimation bias because of using min operation. Maxmin Q-learning (Maxmin QL) [18] selects the minimum of M action value functions, then this min estimator is used to update Q-function. Maxmin QL gains underestimation problem as TD3. Quasi-Median Q-learning (QMQ) [19] selects the median value of M action value estimators instead of min value which is proposed in Maxmin QL. QMQ reduces the overestimation bias in Q-learning and mitigates the underestimation issue in DQL, TD3, and Maxmin QL, because median estimator is a mild estimator. Random Ensembled Q-learning (REDQ) [20] also uses M action value estimators as Maxmin QL. However, the differences between REDQ and Maxmin Q-learning are that REDQ minimizes over a random subset of M action value estimators. REDQ controls overestimation bias and variance of the Q estimate in more flexible way. Averaged DQN [21] is based on the average of the M action value estimators' means, which leads to a stable training procedure and reducing approximation error variance in the target values. Adaptive Ensemble Q-learning (AdaEQ) [22] uses ensemble size adaptation in ensemble methods such as Random Ensembled Q-learning (REDQ), Maxmin Q-learning, and Averaged DQN towards minimizing the estimation bias. AdaEQ adapts the ensemble size by driving the bias to be nearly zero, thereby coping with the impact of the time-varying approximation errors accordingly. AdaEQ solves the either over-estimation or underestimation bias by not fixing the ensemble size M. Monte Carlo Bias Correction Q-learning (MCBCQL) [23] subtracts bias estimate from action value estimates. MCBCQL tackles the overestimation bias without using multiple estimator scheme and hyperparameter tuning. Heterogeneous Update Strategy Q-learning (HetUp QL) [24] finds the optimal action by enlarging the normalized gap. HetUp QL controls overestimation or underestimation bias by inputting optimal action to decide whether a state action pair should be overestimated or underestimated, thus HetUp can be a flexible bias control method.

Most previous works are lack of consideration about underestimation bias. Besides, both the overestimation bias and underestimation bias may improve learning performance empirically depending on the environment they meet [18]. Thus, reducing overestimation bias through underestimation bias methods is desirable sometime. Most previous works aim to solve overestimation bias through M action value estimators' mean or min, or through single estimator or through two estimators. Moreover, previous methods directly evaluate maximum action through Q estimator itself or another Q estimator, which causes overestimation bias or underestimation bias. Thus, how about evaluating one Q estimator's maximum action through remaining $M - 1$ Q estimators? or evaluating $M - 1$ Q estimators' maximum action on one remaining Q estimator? In this way, the new methods update the Q-function through selecting one member from the new $M - 1$ Q estimators. Thus, new methods are benefited from exploring among new $M - 1$ Q estimators, then it is better than exploring among $M = 2$ Q estimators. As a result, the new methods maybe can avoid serious underestimation or overestimation bias problems.

Thus motivated, in this work we focus on reducing overestimation bias through underestimation bias methods, especially on M action value estimators. We depict Bias Reduced Max Q-learning (BRMQL) and Bias Reduced Median Q-learning based on M action-value estimators. In these methods, we randomly choose one estimator that is evaluated by other M-1 estimators' maximum actions. These maximum actions evaluate this one estimator and we get new M-1 estimators. We choose the max and median value of these new M-1 estimators to estimate maximum action value in Q-learning. Theoretically, BRMeQL and BRMQL suffer from underestimation bias and thus reduce overestimation bias in Q-learning. To visualize bias properties of our methods, we test the median value of these M-1 estimators on Meta-Chain MDP example (Fig. 2), and test max and median values on Taxi-v3 toy environment (Fig. 4). The test results show that BRMQL less underestimates Q-value than BRMeQL, and BRMeQL less underestimates Q-value than DQL. At last, BRMQL, and BRMeQL perform better than some of compared algorithms on some of the current benchmark game environments.

Contributions of this work are as follows:

1. We theoretically analyze the overestimation problem of maximum estimator over i.i.d random variables. And we depict our estimators and prove their underestimation problem over i.i.d random variables. Our estimators reduce overestimation bias through underestimation bias method, and our estimators are benefited from M sample size by exploring among M-1 new constructed action value estimators, thus our new estimators can reduce both underestimation and overestimation bias. Our methods' bias properties are different from each other' and depend on sample size, and specific condition they meet.
2. We incorporate Q-learning into above i.i.d random variables. Then, we introduce BRMeQL and BRMQL and theoretically prove that they less underestimate the Q-value than DQL. Also, BRMQL less underestimates the Q-value than BRMeQL. We visualize bias property of these algorithms on Meta-Chain MDP and Taxi-v3 toy environment.
3. We empirically investigate their performance on four benchmark game environments. Finally, we show that BRMDQN, BMReDQN perform better than some of other methods on some of the current environments.

Paper Outline: Next, we organized the paper as follows. Section 2 introduces backgrounds that are used in following sections. Section 3 depicts our methods' theoretic proof on i.i.d random variables and introduces our methods to Q-learning. Section 4 evaluates our methods on several discrete control problems. Section 5 concludes the paper with some future works.

2   Background

In this section, we first introduce the maximum estimator and its bias [12], then we introduce its application of Q-learning. Then, we introduce the cross-validation estimator [25], which is useful in next part.

2.1   Maximum Estimator Bias in Maximum Estimator

Consider a set of random variables (RVs) $U = \{U_1, ...U_M\}$, $M \geq 2$. $\mu_1, ..., \mu_M$ is finite means of these RVs. The maximum value $\mu_*(U)$ defined as

$$\mu_*(U) \equiv \arg\max_i \mu_i \equiv \arg\max_i E\{U_i\} \tag{1}$$

Most often the distribution of U_i is unknown, so we get a set of samples S. Now, we consider how best to estimate $\mu_*(U)$. The maximum estimator (ME) is a common estimator which is used to estimate $\hat{\mu}(S) \approx \mu_*(U)$. When $S_i \in S$, $\hat{\mu}_i$ is the average of S_i. Then ME overestimates the μ_* averagely [25,26]. Now we depict the reason of overestimation in ME. Let i^* denote the maximal estimated value $\hat{\mu}_{i^*} = max\{\hat{\mu}_1, ..., \hat{\mu}_M\}$. Let j^* denote the maximal estimated value $\mu_{j^*} = max\{\mu_1, ..., \mu_M\}$. Then we have $\mu_{i^*} - \hat{\mu}_{i^*} \leq \mu_{j^*} - \hat{\mu}_{i^*} \leq \mu_{j^*} - \hat{\mu}_{j^*}$. Then we take expectations on condition μ, we have $E[\mu_{i^*} - \hat{\mu}_{i^*}|\mu] \leq E[\mu_{j^*} - \hat{\mu}_{j^*}|\mu] = 0$. Integrating on uncertain μ, we get that $E[\hat{\mu}_i|\mu_1, ..., \mu_M] = \mu_i$, namely conditionally unbiased. At last we have $E[\mu_{i^*} - \hat{\mu}_{i^*}] \leq 0$. Then ME overestimates the μ_* even if the value estimates are conditionally unbiased.

Next part, we introduce Maximum estimator application of Q-learning.

Q-learning. Markov Decision process is defined with four tuple (S, A, P, R). S is finite state space including $|S|$ sates, A is finite action space including $|A|$ actions. The state transition function is $P : S \times A \times S \leftarrow [0, 1]$ and its values is $P(s, a, s^{'}) = Prob[s_{t+1} = s^{'}|s_t = s, a_t = a]$. The stochastic reward function is described as $R : S \times A \rightarrow R$. The reward evaluates the immediate effect of action a_t, namely the transition from s_t to s_{t+1}, but does not imply anything about its long-term effects. The agent goal is to maximize the object $E[\sum_{t=1}^{\infty} \gamma^{t-1} R(s_t, a_t)]$, where $\gamma \in (0, 1)$ is a discount factor. γ gives a trade-off to the importance of delayed and immediate rewards.

A similar optimal state-action value function is defined as $Q^* : S \times A \rightarrow R$ for all $(s, a) \in S \times A$ and given as follows:

$$Q^*(s_t, a_t) = E[R(s_t, a_t) + \gamma \max_{a^{'}} Q^*(s_{t+1}, a^{'})] \tag{2}$$

Where s_{t+1} is the next state drawn from distribution $P(s_t, a_t, .)$. The Eq. (1) is called Bellaman optimality function.

2.2   Cross-Validation Estimator

We show some definitions which will be useful in the follow. We still use the random variables stated in maximum estimator section. The definition of optimal indices set of the RVs U is:

$$\mathcal{O}(U) \equiv \{i | \mu_i = \mu_*\} \tag{3}$$

The definition of maximal indices set of samples S is:

$$\mathcal{M}(S) \equiv \left\{ i | \hat{\mu}_i(S) = \max_j \hat{\mu}_j(S) \right\} \tag{4}$$

Here we note that above two set may be different. Actually, we notice that μ_* can be a weighted average of the means of all optimal variables:

$$\mu_* = \frac{1}{|\mathcal{O}(U)|} \sum_{i=1}^{M} \mathcal{I}(i \in \mathcal{O}(U)) \mu_i \tag{5}$$

Where $\mathcal{I}$ means the indicator function. Although we know that $\mathcal{O}(U)$ and μ_i can not be known, we can approximate these with $\mathcal{M}(C)$ and $\mathcal{M}(D)$ with sample sets C, D and is obtained by:

$$\mu_* \approx \hat{\mu}_* \equiv \frac{1}{|\mathcal{M}(C)|} \sum_{i=1}^{M} \mathcal{I}(i \in \mathcal{M}(C)) \hat{\mu}_i(D) \tag{6}$$

If $C = D = S$, this is the same as $\hat{\mu}_*^{ME}$. This is called two-fold cross-validation estimator, and we have K-fold cross-validation estimator. CVE incurs underestimation bias [27].

3   Bias Reduced Methods to Q-learning

In this section we first depict our estimators suffer from underestimation bias, then we give its theoretic analysis. Next part, we introduce our estimators to Q-learning.

3.1   Underestimation Bias of Our Estimators

In this part, we first introduce some notations which are useful in the following. Each sample S is split into K disjoint sets S^k. We define $\hat{\mu}_i^k \equiv \hat{\mu}_i(S^k)$. We construct an indices set $\hat{a}^k$ and its value set $\hat{v}^k$ for each $k \in [K]$. Let $K - 1$ set S is used to determine the indices set $\hat{a}^k$, and the remaining one set is used to build its value: $\hat{a}_i^k \equiv \hat{\mu}_i(S \backslash S^k)$ and $\hat{v}_i^k \equiv \hat{\mu}(S^k) \equiv \hat{\mu}_i^k$. It is obvious that $E\left\{ \hat{a}_i^k \right\} = E\left\{ \hat{v}_i^k \right\} = \mu_i$. From 2.3 expression (4) we know that $\mathcal{M}^k = \mathcal{M}(S \backslash S^k)$ is the maximal indices set. Then the value of these indices is:

$$\hat{\mu}_*^k \equiv \frac{1}{|\mathcal{M}^k|} \sum_{i \in \mathcal{M}^k} \hat{v}_i^k \tag{7}$$

All sets are averaged on K and get the K-fold cross validation estimator:

$$\hat{\mu}_*^{CV} \equiv \frac{1}{K} \sum_{k=1}^{K} \hat{\mu}_*^k = \frac{1}{K} \sum_{k=1}^{K} \frac{1}{|\mathcal{M}^k|} \sum_{i \in \mathcal{M}^k} \hat{v}_i^k \qquad (8)$$

Lemma 1. *If $E\left\{\hat{\mu}_i^k|U\right\} = \mu_i$ is unbiased, then we have $E\left\{\hat{\mu}_*^{CV}|U\right\}$ is negatively biased, namely $E\left\{\hat{\mu}_*^{CV}|U\right\} \leq \mu_*$.*

CV is negatively biased. Now we give Theorem 1. In Theorem 1, we give a theoretic foundation to show our estimators suffer from underestimation bias.

Theorem 1. *we have K disjoint estimators sets $\hat{\mu}^{K_1} = \left\{\hat{\mu}_1^{K_1}, .., \hat{\mu}_M^{K_1}\right\}, ...,$ $\hat{\mu}^{K_K} = \left\{\hat{\mu}_1^{K_K}, .., \hat{\mu}_M^{K_K}\right\}$. $\forall K_i \in [K]$, the remaining $K-1$ sets are to build maximal indices set, namely $\left\{j_p| \max_j \hat{\mu}_j^{K_p}, K_p \in [K] \setminus K_i\right\}$, for each K_p, if there multiple j_p, we randomly choose one. We evaluate each indices value on $\hat{\mu}^{K_i}$, then we have $\bigcup\limits_{j_p \in \left\{\arg\max_j \hat{\mu}_j^{K_p}\right\}} \hat{\mu}_{j_p}^{K_i}$ is negatively biased.*

The Theorem 1 implies that our estimators suffer from underestimation bias theoretically. Now, we prove the Theorem 1.

Proof of Theorem 1. For $\forall K_p \in [K] \setminus K_i, j_p \in \mathcal{M}^{K_i}$ where $\mathcal{M}^{K_i} = \mathcal{M}(S \setminus S^{K_i})$ we have

$$\begin{aligned}
E\left\{\hat{\mu}_{j_p}^{K_i}\right\} &= P(j_p \in \mathcal{O})E\left\{\hat{\mu}_{j_p}^{K_i}|j_p \in \mathcal{O}\right\} + P(j_p \notin \mathcal{O})E\left\{\hat{\mu}_{j_p}^{K_i}|j_p \notin \mathcal{O}\right\} \\
&= P(j_p \in \mathcal{O}) \max_{j_p} E\left\{U_{j_p}\right\} + P(j_p \notin \mathcal{O})E\left\{\hat{\mu}_{j_p}^{K_i}|j_p \notin \mathcal{O}\right\} \\
&\leq P(j_p \in \mathcal{O}) \max_{j_p} E\left\{U_{j_p}\right\} + P(j_p \notin \mathcal{O}) \max_{j_p} E\left\{U_{j_p}\right\} \\
&= \max_{j_p} E\left\{U_{j_p}\right\}
\end{aligned} \qquad (9)$$

Thus $E\left\{\hat{\mu}_{j_p}^{K_i}\right\} \leq \max_{j_p} E\left\{U_{j_p}\right\} = \mu_*$, with inequality is strict if and only if there exists a $j_p \notin \mathcal{O}$ such that $P(j_p \in \mathcal{M}^{K_i}) > 0$. The results are true for all $K_p \neq K_i$ and $K_p \in [K]$, then obviously we have

$$E\left\{\max_{K_i} \hat{\mu}_{j_p}^{K_i}\right\} \leq \max_{j_p} E\left\{U_i\right\} = \mu_* \qquad (10)$$

3.2 Our Proposed Methods to Q-learning

Now we introduce our methods. Firstly, we store M Q-functions estimators: $Q^1, Q^2, .., Q^M$ and any i^{th} member is randomly, uniformly selected from these M Q-functions estimators. Different strategy from DQL, we use each of remaining $M-1$ estimators to evaluate next-sate maximum actions, namely

$\{a_j^* | \arg\max_a Q^j(s_{t+1}, a), j \in [M] \setminus i\}$, if there are multiple maximum actions for each Q-function, we randomly choose one. We recall that N actions available at state s_{t+1} for $M-1$ Q-functions estimators. We evaluate each of these maximum actions values through $\{Q^i(s_{t+1}, a_j^*) | j \in [M] \setminus i\}$, thus we have $M-1$ action values in state s_{t+1}. We select the max or median of these $M-1$ next-state action values $\{Q^i(s_{t+1}, a_j^*), j \in [M] \setminus i\}$. Next we mathematically prove our method reduces underestimation bias than Double Estimator (DE). We use Double Estimator as a basic estimator, which our method aims to approximate. This is because the bias between our method and DE is easy to calculate here, and we know that DE underestimates the Q-value. If the bias between our method and DE is greater than 0, our method less underestimates Q-value than DE, vice versa, our method underestimates Q-value more than DE. Here, we visualize the bias of median action values (BRMeQL) on simple Meta-chain MDP games.

The bias between $Q^i(s_{t+1}, a_j^*)$ estimator and DE is defined as following

$$B_j = Q^i(s_{t+1}, a_j^*) - Q^{DE}(s_{t+1}, a_j^*), \tag{11}$$

Where $B_j, j \in [M] \setminus i$, and we assume that B_j satisfies different distributions. Here we use uniform, independent random variables in the interval $[-\epsilon, \epsilon](\epsilon > 0)$. We borrow the uniform, independent random assumption from [11]. $Q^{DE}(s_{t+1}, a_j^*)$ refers to Double Estimator. In the following parts, we mathematically explain max of $B_j, j \in [M] \setminus i$ reduces overestimation in the Q value function and underestimates the Q-value less than DE. The reason why we choose max of $B_j, j \in [M] \setminus i$ and calculate the bias between the DE and max of $B_j, j \in [M] \setminus i$ are that other members of $B_j, j \in [M] \setminus i$ values are less than max of $B_j, j \in [M] \setminus i$ value, thus we can indirectly infer that other members of $B_j, j \in [M] \setminus i$ underestimate Q-value more or less than max of $B_j, j \in [M] \setminus i$.

Theorem 2. *When $B_j, j \in [M] \setminus i$ satisfy the uniform independently variables in $[-\epsilon, \epsilon](\epsilon > 0)$ then,*

$$E[\max_{j \in [M] \setminus i} B_j] = \frac{[N(M-1) - 1]\epsilon}{N(M-1) + 1} \tag{12}$$

Here, we introduce Lemma 2 as a tool to prove Theorem 2.

Lemma 2. *Assume $S_1, \ldots, S_N$ is M i.i.d. random variables. Let $S_1, \ldots, S_M$ be an absolutely continuous distribution and its cumulative distribution function (CDF) is $F(S)$ and probability density function (PDF) is $f(s)$. Denote $\mu \overset{def}{=} E[S_i]$ and $\sigma^2 \overset{def}{=} Var[S_i] < +\infty$. We have $S_{(1)} \leq S_{(2)} \leq \ldots \leq S_{(M)}$, if we arrange $S_1, \ldots, S_M$ with not decreasing order. Obviously, $S_{(M)}$ is max member of $\{S_1, \ldots, S_M\}$, denote $S_{(M)}$ PDF and CDF as $f_{s_{(M)}}(s)$ and $F_{s_{(M)}}(s)$ respectively, then we have*

$$F_{s_{(M)}}(s) = F(s)^M \tag{13}$$

$$f_{s_{(M)}}(s) = Mf(s)F(s)^{M-1} \tag{14}$$

Proof of Lemma 2. Consider the CDF of $S_{(M)}$. $F_{s_{(M)}} = P(S_{(M)} \leq s) = P(S_1 \leq s, ..., S_M \leq s) = P(S_1 \leq s)......P(S_M \leq s) = F(s)^M$. Now the PDF of $S_{(M)}$ is $f_{s_{(M)}}(s) = \frac{dF_{s_{(M)}}}{ds} = Mf(s)F(s)^{M-1}$.

Now we proof Theorem 2.

Proof of Theorem 2. We denote B(x) and b(x) are the CDF and PDF of B_j, respectively. Similarly, then $\max_{j \in [N] \setminus i} B_j$ PDF and CDF are $b_{x_{(M)}}(x)$ and $B_{x_{(M)}}(x)$. Since we are sampling B_j from *Uniform* $(-\epsilon, +\epsilon)$, we are easy to get $b(x) = \frac{1}{2\epsilon}$ and $B(x) = \frac{1}{2} + \frac{x}{2\epsilon}$. By using Lemma 2, we have $B_{(M)}(x) = [\frac{1}{2} + \frac{x}{2\epsilon}]^{M-1}$ and $b_{(M)}(x) = (M-1)b(x)B(X)^{M-2} = \frac{(M-1)}{2\epsilon}[\frac{1}{2} + \frac{x}{2\epsilon}]^{(M-2)}$. Then the expectation bias is

$$
\begin{aligned}
E[\max_{j \in [M] \setminus i} B_j] &= E[\max_{j \in [M] \setminus i} Q^i(s_{t+1}, a_j^*) - Q^{DE}(s_{t+1}, a_j^*)] \\
&= \int_{-\epsilon}^{+\epsilon} (M-1)xb(x)B(X)^{(M-2)}dx \\
&= \int_{-\epsilon}^{+\epsilon} N(M-1)x\frac{1}{2\epsilon}[\frac{1}{2} + \frac{x}{2\epsilon}]^{(N-1)}((\frac{1}{2} + \frac{x}{2\epsilon})^N)^{M-2}dx \\
&= \frac{\epsilon[N(M-1)-1]}{N(M-1)+1}
\end{aligned}
\tag{15}
$$

In Theorem 2, $\frac{\epsilon[N(M-1)-1]}{N(M-1)+1} \geq 0$, which implies that the bias between BRMQL and DE is greater than 0. Thus BRMQL's value is greater than DE's value. Even if BRMQL suffers from underestimation bias theoretically, it does not underestimate Q-value as much as DE. In this way, BRMQL can be a less underestimation method and thus can come to first when reducing overestimation bias in Q-learning. When the available action numbers N are fixed, the bias between BRMQL and DE is increasing with increasing M, which implies that BRMQL's underestimation bias is decreasing with increasing sample size M.

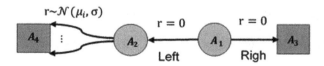

Fig. 1. A simple Meta chain MDP game. It includes four states. This MDP usually is used to illustrate different estimators' over or under-estimation bias.

We visualize the bias of BRMeQL on MDP example and show it in Fig. 2. The MDP is shown in Fig. 1. In this example, the simple game has four states, which includes two non-terminal state A_1 and A_2 and two terminal states A_3, A_4. A starting state always can be A_1 for each Episode with a choice between right action and left action. The right action results the terminal state A_3 with a zero return. The left action results in A_2 with a zero return. Besides from A_2, the possible left actions results in immediate termination with a return, which are drawn from a Gaussian $\mathcal{N}(-0.1, 1)$. Thus, in any trajectory, left actions bring about the expected return -0.1. Then the optimal policy is selecting the action left 5% from A_1 for this simple game.

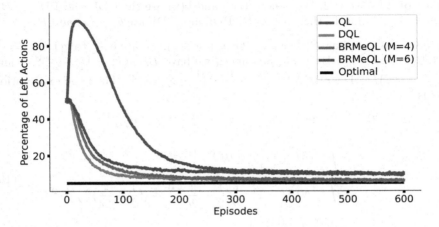

Fig. 2. Maximization bias example. Percent of left actions taken by QL, DQL, BRMeQL with parameter settings $\epsilon = 0.1$, $\alpha = 0.1$, and $\gamma = 1$.

From Theorem 1, 2, and from above MDP example, we get following Theorem 3, 4.

Theorem 3. *Under the condition stated above we have:*

$$bias(Q^{DE}(s_{t+1}, \hat{a}_j^*)) \leq bias(\max_{j \in [M] \backslash i} Q^i(s_{t+1}, \hat{a}_j^*)) \leq bias(\max_a Q(s_{t+1}, a)) \quad (16)$$

Theorem 4. *Under the condition stated above we have:*

$$bias(Q^{DE}(s_{t+1}, \hat{a}_j^*)) \leq bias(\underset{j \in [M] \backslash i}{median} Q^i(s_{t+1}, \hat{a}_j^*)) \leq bias(\max_{j \in [M] \backslash i} Q^i(s_{t+1}, \hat{a}_j^*)) \quad (17)$$

The Fig. 2 indicates that on different M, BRMeQL bias is different, and BRMeQL underestimation bias is decreasing with increasing M. In this example on small M, BRMeQL converge to an optimal policy quickly, which implies it mitigates overestimation efficiently. Thus, BRMeQL leads different bias to Q-value function on different M.

Algorithm 1. BRMQL algorithm

Input: step-size α, exploration parameter $\epsilon > 0$.
Initialize M action-value functions $\{Q^1, Q^2, ..., Q^M\}$ randomly. Observe initial state s_t

for $t = 0, ...T \in$ each episode **do**

From s_t, using the policy ϵ-greedy in $\sum_{i=1}^{M} Q^i(s_t, a)$ to choose action a_t
Take action a_t, observe s_{t+1}, r_t
From $\{Q^1, Q^2, ..., Q^M\}$, uniformly select an estimator Q^i
Note $\{a_j^* | \arg\max_a Q^j(s_{t+1}, a), j \in [M] \setminus i\}$
Note $\hat{Q}(s_{t+1}, a_j^*) = \max_i Q^i(s_{t+1}, a_j^*), j \in [M] \setminus i$
Update: We randomly choose $\forall \tilde{k} \in [M]$
$Q^{\tilde{k}}(s_t, a_t) \leftarrow (1 - \alpha_t)Q^{\tilde{k}}(s_t, a_t) + \alpha_t[r_t + \gamma\hat{Q}(s_{t+1}, a_j^*)]$
$s_t \leftarrow s_{t+1}$

End For
Return $\{Q^1, Q^2, ..., Q^M\}$

From the results of Fig. 2 and Theorem 2, 3, 4, we conclude that BRMQL less underestimates the Q-value than DE, and BRMeQL. BRMQL can be a less underestimation and overestimation bias reducing method to Q-learning. BRMQL and BRMeQL biases are between DQL's and QL's theoretically, and their biases are related to sample size M, therefore they can be flexible bias control methods to Q-learning.

Next, we give our algorithm, and show it in Algorithm 1. In Algorithm 1, we refers to BRMQL, the algorithm of BRMeQL can be achieved by replacing max operator with median operator in Algorithm 1. Note as $\hat{Q}(s_{t+1}, a_j^*) = \max_i Q^i(s_{t+1}, a_j^*), j \in [M] \setminus i$. We incorporate the $\hat{Q}(s_{t+1}, a_j^*)$ into Q-learning. And the update rule is as follows:

$$\hat{Q}(s_{t+1}, a_j^*) = \max_i Q^i(s_{t+1}, a_j^*), j \in [M] \setminus i \tag{18}$$

We randomly choose $\forall \tilde{k} \in [M]$ to update the algorithm.

$$Q^{\tilde{k}}(s_t, a_t) \leftarrow (1 - \alpha_t)Q^{\tilde{k}}(s_t, a_t) + \alpha_t[r_t + \gamma\hat{Q}(s_{t+1}, a_j^*)] \tag{19}$$

4 Experiment

In this section, we first evaluate our alogorithms on benchmark environments including Catcher, Pixelcopter, Lunarlander, SpaceInvaders [28–30]. We exploit the open source code from (MaxMin Q-learning, Qingfeng Lan &Pan [30]). We reused the hyper-parameters and settings of neural networks in (MaxMin Q-learning, Qingfeng Lan & Pan [30]). We reused the update, train, and test rule in (MaxMin Q-learning, Qingfeng Lan & Pan [30]). In games Catcher and Pixelcopter, learning rate is $[10^{-3}, 3*10^{-4}, 10^{-4}, 3*10^{-5}, 10^{-5}]$, in games LunarLander, and SpaceInvaders, the learning rate is $[3*10^{-3}, 10^{-3}, 3*10^{-4}, 10^{-4},$

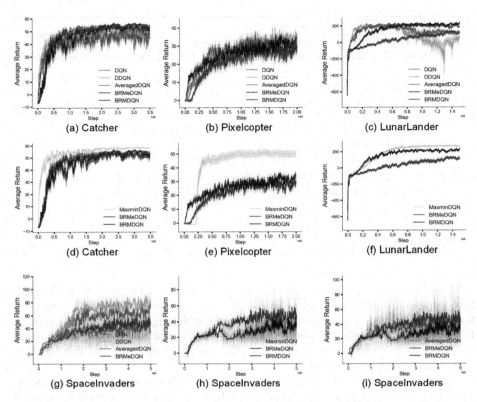

Fig. 3. The performance investigation of DQN, DDQN, MaxminDQN, AveragedDQN, BRMeDQN, BRMDQN on Catcher, Pixelcopter, LunarLander, and SpaceInvaders. The results were averaged over 20 runs on games Catcher, pixelcopter, and LunarLander. In the first row, plots (a,b,c) show the performance of DQN, DDQN, AveragedDQN, BRMeDQN, BRMDQN on Catcher, Pixelcopter, LunarLander. In the second row, plots (d,e,f) show the performance of MaxminDQN, BRMeDQN, BRMDQN on Catcher, Pixelcopter, LunarLander. In the third row, plot (g) shows the performance of DQN, DDQN, AveragedDQN, BRMeDQN on SpaceInvaders. In the third row, plot (h) shows the performance of MaxminDQN, BRMeDQN, BRMDQN on SpaceInvaders. In the third row, plot (i) shows the performance of AveragedDQN, BRMeDQN, BRMDQN on SpaceInvaders.

$3 * 10^{-5}$]. We compare BRMDQN and BRMeDQN with DQN, DDQN, MaxminDQN, and AverageDQN. In these four algorithms, M can be selected from $[2, 3, 4, 5, 6, 7, 8, 9]$. In BRMDQN, and BRMeDQN M can be selected from $[3, 4, 5, 6, 7, 8]$. In Fig. 3 we show the results.

Figure 3 (a,b,c,d,e,f) show that BRMDQN performs better than DQN, DDQN, AveragedDQN, BRMeDQN on games Catcher, Pixelcopter, LunarLander. Also, Fig. 3 (a,b,c,d,e,f) show that MaxminDQN performance is better than BRMDQN on games Catcher, Pixelcopter, LunarLander. When environments become more complicated, for example, on game SpaceInvaders, Fig. 3 (g,h,i) show that the classic algorithm DDQN performs better than DQN, and DQN performs better than BRMDQN, BRMeDQN, AveragedDQN, MaxminDQN. Figure 3 (h,i) show that BRMeDQN, and BRMDQN perform better than MaxminDQN, and AveragedDQN on SpaceInvaders. MaxminDQN suffers from underestimation bias (MaxminDQN contains min operator, which suffers from underestimation bias.), so with the increasing complexity of the environments, MaxminDQN's performance is not very well. Also, BRMeDQN performs better than BRMDQN on game SpaceInvaders. BRMDQN, and BRMeDQN's convergences are fast as well as other algorithms', especially on game Copter. Our algorithms' performances are stable on these four games and perform better than other algorithms on most of four games. One better explanation for this is that BRMDQN, and BRMeDQN reduce both underestimation and overestimation bias, which is also better explained by the Theorem 1, 2. At last, we get that BRMDQN, and BRMeDQN can be good estimators for Q-learning on most of four games.

Next, In Table 1, we give the final step averaged return of BRMDQN, and BRMeDQN on copter and catcher, to show how M effects their performances. In Table 2, we show the best parameters for different algorithms on benchmark environments, to further confirm that M is related to algorithms' performances. In Table 3, we show the memories and times consumed by different algorithms on benchmark environments. In Table 4, we show how M effects bias properties of BRMQL, and BRMeQL on MDP, and Taxi-v3.

In Table 1, we show the final step averaged return of BRMDQN, and BRMeDQN on different M on games Catcher and Copter. Table 1 shows that on the last step of the games, BRMDQN, and BRMeDQN reach different final averaged return. Besides, the order of final averaged return is not linear correlation to the order of M. At last, we would get that the performances of BRMDQN, and BRMeDQN depend on M and different environments they interact.

Table 1. Different algorithms performance on different M

		Catcher		Copter
	M	Final Averaged Return	M	Final Averaged Return
BRMDQN	3	47.355	4	17.316
	5	51.184	6	16.895
	7	43.615	8	17.790
BRMeDQN	4	45.575	3	6.755
	6	38.940	5	8.545
	8	45.750	7	9.810

In below Table 2, we show the best parameters for different algorithms. Table 2 shows that the performances of different algorithms depend on action-value function number M, and learning rate. Also, all the algorithms' performances can get best on certain M, and the best parameter M is different for each algorithm. Finally, the results of Table 2 and Fig. 3 confirm our previous claim that performances of BRMDQN, and BRMeDQN on benchmark environments depend on conditions they meet and are related to M.

Table 2. Best parameters for different algorithms

	Catcher		Copter		LunarLander		SpaceInvaders	
	M	Learning rate	M	Learning rate	M	Learning rate	M	Learning rate
DQN	\	10^{-4}	\	3×10^{-5}	\	3×10^{-4}	\	3×10^{-4}
DDQN	\	10^{-4}	\	3×10^{-5}	\	3×10^{-4}	\	3×10^{-4}
AveragedDQN	2	10^{-4}	7	3×10^{-5}	4	3×10^{-4}	5	3×10^{-4}
MaxminDQN	6	3×10^{-4}	6	3×10^{-4}	3	10^{-3}	4	3×10^{-4}
BRMeDQN	8	3×10^{-5}	8	3×10^{-4}	6	3×10^{-4}	3	3×10^{-4}
BRMDQN	5	3×10^{-5}	3	10^{-4}	5	10^{-3}	3	3×10^{-4}

In below Table 3, we show the memories and times consumed by each algorithm on total episode on each game (Each algorithm runs 20 times on each game, and memories and times are averaged over 20 runs.). Table 3 shows that BRMDQN memory is almost the same with the DQN's, DDQN's, and AveragedDQN's on cather, copter, and spaceInvaders, while its memory is less than DQN's, DDQN's, AveragedDQN's on lunarlander. BRMeDQN memory is almost the same with the DQN's, DDQN's, and AveragedDQN's on catcher, and copter, and its memory is better than DQN's, DDQN's, and AveragedDQN's on lunarlander. MaxminDQN memory is more than DQN's, DDQN's, AveragedDQN's, and BRMDQN's on copter, lunarlander, and spaceInvaders. MaxminDQN memory is more than BRMeDQN's on copter, and lunarlander. BRMDQN, and BRMeDQN's time have no advantages over DQN's, DDQN's, and AveragedDQN's time. However, BRMDQN's time is less than MaxminDQN's on these four games. BRMeDQN's time is less than MaxminDQN on copter, and spaceInvaders. Even if MaxminDQN performance is the best on previous three games, however, it is a memory and time consuming on these four games. Besides, our methods' time and memory are almost same with the DQN's, DDQN's, and AveragedDQN's, and our method performances are increasing with the increasing complexity of the environments. All in all, each algorithm's memory and time depend on different environments they meet, and each algorithm's performances are not linear related to each algorithm's memory and time they consumed.

Table 3. Memories and Times consumed by Different algorithms

	Catcher					
	DQN	DDQN	AveragedDQN	MaxminDQN	BRMDQN	BRMeDQN
Memory	274.60	274.41	274.77	273.22	277.13	278.85
Time	117.46	123.10	119.81	197.77	178.89	261.79
	Copter					
	DQN	DDQN	AveragedDQN	MaxminDQN	BRMDQN	BRMeDQN
Memory	289.23	288.80	288.84	330.89	291.03	293.05
Time	78.34	83.00	93.65	174.41	97.17	165.89
	Lunarlander					
	DQN	DDQN	AveragedDQN	MaxminDQN	BRMDQN	BRMeDQN
Memory	426.65	426.72	421.24	458.32	396.05	397.76
Time	82.62	87.92	84.88	97.21	106.15	159.29
	SpaceInvaders					
	DQN	DDQN	AveragedDQN	MaxminDQN	BRMDQN	BRMeDQN
Memory	1024.95	1031.01	1032.03	1061.27	1040.16	1064.25
Time	1030.51	1123.96	1459.4	2602.45	2036.95	2002.73

Then we test our algorithms bias on Taxi-v3 toy environment [31]. Taxi-v3 environment (Taxi Cab Navigation given inset in Fig. 4) has 4 labeled by different letters locations. Agent's goal is to pick up and drop passenger off in different locations. Agent get one penalty for every timestep and receive 20 points for dropping off successfully. Agent also lose 10 points for illegal drop-off and pick-up actions. In Taxi-v3 environment, the learning rate is $\alpha_t = 0.1, \epsilon = 0.1$, and discount factor is $\gamma = 1$.

In Fig. 4, we show the bias comparison of different algorithms. In Fig. 4(a), (b), (c), BRMeQL bias is between QL and DQL, in Fig. 4(d), (e) QL bias is between BRMeQL and BRMQL. This implies that BRMeQL doesn't suffer from overestimation bias, however BRMQL does. Then we get from Fig. 4 that BRMeQL and BRMQL reduce underestimation bias theoretically and empirically. However, the inherent property of max operation and its cooperation with sample size, BRMQL suffer from overestimation bias. Even if BRMQL suffers overestimation bias, its performance is good in Fig. 3 on most games. Besides, BRMeQL's bias is mild, and it can reduce overestimation and underestimation bias effectively, but BRMeDQN's performance is not better than BRMDQN on most games. At last, we conclude that in practice it is hard to find a perfect estimator for Q-learning. The different algorithms' properties depend on sample size, environments, and algorithms themselves. As a result, we need consider different aspects to find a proper estimator for Q-learning.

In the below Table 4 we show how bias properties of our algorithms are changing on different M. Table 4 shows that BRMQL suffers greater overestimation

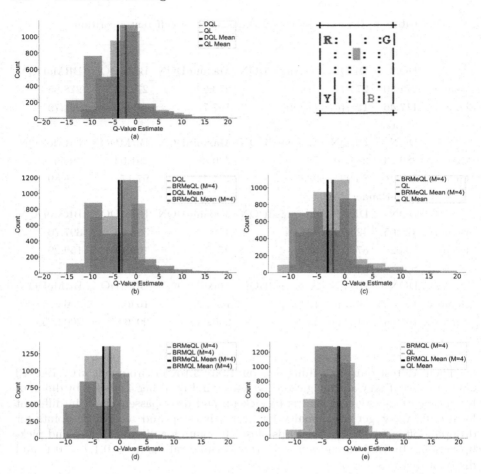

Fig. 4. Frequency distribution histograms of Q value estimates and mean of them with different methods on total $40k$ episodes. Q value estimate is final Q-value of each method on every episode. In each method, final Q-value is calculated by averaging over the learned Q-values. (a) Compared QL with DQL. (b) Compared BRMeQL ($M = 4$) with DQL. (c) Compared BRMeQL ($M = 4$) with QL. (d) Compared BRMeQL ($M = 4$) with BRMQL ($M = 4$). (e) Compared BRMQL ($M = 4$) with QL.

bias than QL, and its overestimation bias is increasing with the increasing M on both MDP and Taxi-v3, however, when M gets bigger, the BRMQL's bias gets stable, which means current M's bias has no big difference from previous M's bias, e.g. M = 6 and M = 8. BRMeQL's bias is between DQL and QL, and with increasing M, its bias is increasing on Taxi-v3, but BRMeQL doesn't suffer from overestimation bias. On MDP, BRMeQL's bias is still between DQL and QL, and with increasing M, its bias is decreasing. At last, we get that bias properties of different algorithms depend on environments they meet.

Table 4. Bias properties of BRMeQL and BRMQL

Number of Q-function	MDP	
	BRMQL's bias	BRMeQL's bias
M = 4	$> QL$	$> DQL, < QL$
M = 6	$> BRMQL(M = 4)$	$> BRMeQL(M = 4), \leq QL$
M = 8	$\geq BRMQL(M = 6)$	$\geq BRMeQL(M = 4), < BRMeQL(M = 6)$
Number of Q-function	Taxi-v3	
	BRMQL's bias	BRMeQL's bias
M = 4	$> QL$	$> DQL, < QL$
M = 6	$> BRMQL(M = 4)$	$> BRMeQL(M = 4), < QL$
M = 8	$\geq BRMQL(M = 6)$	$> BRMeQL(M = 6), < QL$

5   Conclusion and Future Work

Overestimation bias in Q-learning leads bad influence on learning process, especially when the agent interacts with stochastic environments. Various algorithms have been proposed to settle this issue. In this paper, we introduce less underestimation methods BRMeQL and BRMQL to help Q-learning reduce overestimation bias. Our methods are related to number of M Q-functions. We mathematically prove our methods suffer from underestimation bias, and they thus help Q-learning reduce overestimation bias. On meta-chain MDP and toy environment Taxi-v3, we visualize bias properties of BRMeDQL and BRMQL, and the results better explain the related theorems. We show how the different algorithms' performances depend on different parameters in table form. Also, in table form, we show that the algorithms performances are not linear related to time and memory consumed by each algorithm, and their times and memories are different from each other's on different environments they meet. At last, we evaluate the performance of BRMDQN and BRMeDQN on four benchmark environments, and the results show that BRMDQN, and BRMeDQN performances are more stable than other algorithms. Besides, the results show that the algorithms' performances not only depend on inherent properties of algorithms itself but also depend on the complexity of environments they interact, so it is hard to find a perfect, best estimator for Q-learning.

As future work, extend our work to continuous reinforcement problems such as mujoco environments. Another future work is to study the bias properties of a member other than max and median members.

Acknowledgments. The authors wish to thank the referees for some helpful comments.

References

1. Szepesvt'ari, C.: Reinforcement Learning Algorithms for MDPs. Wiley Encyclopedia of Operations Research and Management Science (2011)
2. Sutton, R.S., Barto, A.G.: Reinforcement learning: an introduction (2018)
3. Watkins, C.J., Dayan, P.: Q-learning. Mach. Learn. 8(3–4), 279–292 (1992)
4. Tsitsiklis, J.N.: Asynchronous stochastic approximation and Q-learning. Mach. Learn. 16(3), 185–202 (1994)
5. Jaakkola, T., Jordan, M.I., Singh, S.P.: On the convergence of stochastic iterative dynamic programming algorithms. Neural Comput. 6(6), 1185–1201 (2014)
6. Zhang, Y., Qu, G., Xu, P., Lin, Y., Chen, Z., Wierman, A.: Global convergence of localized policy iteration in networked multi-agent reinforcement learning. Proc. ACM Meas. Anal. Comput. Syst. 7(1), 13:1–13:51 (2023). https://doi.org/10.1145/3579443
7. Jahanshahi, H., et al.: A deep reinforcement learning approach for the meal delivery problem. Knowl.-Based Syst. 243, 108489 (2022)
8. de Moraes, L.D., et al.: Double deep reinforcement learning techniques for low dimensional sensing mapless navigation of terrestrial mobile robots 2023, CoRR abs/2301.11173. arXiv:2301.11173. https://doi.org/10.48550/arXiv.2301.11173
9. Brandao, B., de Lima, T.W., Soares, A., Melo, L.C., Mt'aximo, M.R.O.A.: Multi-agent reinforcement learning for strategic decision making and control in robotic soccer through self-play. IEEE Access 10, 72628–72642 (2022). https://doi.org/10.1109/ACCESS.2022.3189021
10. Sutisna, N., Ilmy, A.M.R., Syafalni, I., Mulyawan, R., Adiono, T.: FARANE-Q: fast parallel and pipeline Q-learning accelerator for configurable reinforcement learning SOC. IEEE Access 11, 144–161 (2023). https://doi.org/10.1109/ACCESS.2022.3232853
11. Thrun, S., Schwartz, A.: Issues in using function approximation for reinforcement learning. In: Proceedings of the Fourth Connectionist Models Summer School, Hillsdale, NJ, pp. 255–263 (1993)
12. Smith, J.E., Winkler, R.L.: The optimizer's curse: skepticism and post-decision surprise in decision analysis. Manage. Sci. 52(3), 311–322 (2006)
13. Hasselt, H.: Double Q-learning. Adv. Neural. Inf. Process. Syst. 23, 2613–2621 (2010)
14. Devraj, A.M., Meyn, S.P.: Zap Q-learning. In: Guyon, I., et al. (eds.) Advances in Neural Information Processing Systems 30: Annual Conference on Neural Information Processing Systems, 4–9 December 2017, Long Beach, CA, USA, pp. 2235–2244 (2017). https://proceedings.neurips.cc/paper/2017/hash/4671aeaf49c792689533b00664a5c3ef-Abstract.html
15. Cini, A., D'Eramo, C., Peters, J., Alippi, C.: Deep reinforcement learning with weighted Q-learning, CoRR abs/2003.09280 (2020). arXiv:2003.09280
16. Lale, S., Shi, Y., Qu, G., Azizzadenesheli, K., Wierman, A., Anandkumar, A.: KCRL: krasovskii-constrained reinforcement learning with guaranteed stability in nonlinear dynamical systems, CoRR abs/2206.01704 (2022). arXiv:2206.01704. https://doi.org/10.48550/arXiv.2206.01704
17. Fujimoto, S., Hoof, H., Meger, D.: Addressing function approximation error in actor-critic methods. In: International Conference on Machine Learning, PMLR, pp. 1587–1596 (2018)

18. Lan, Q., Pan, Y., Fyshe, A., White, M.: Maxmin Q-learning: controlling the estimation bias of Q-learning. In: 8th International Conference on Learning Representations, ICLR 2020, Addis Ababa, Ethiopia, 26–30 April 2020. OpenReview.net (2020). https://openreview.net/forum?id=Bkg0u3Etwr

19. Wei, W., Zhang, Y., Liang, J., Li, L., Li, Y.: Controlling underestimation bias in reinforcement learning via quasi-median operation. In: Proceedings of the 36th Conference on Innovative Applications of Artificial Intelligence, AAAI 2022, Virtual, 2–9 February 2002 (2002)

20. Chen, X., Wang, C., Zhou, Z., Ross, K.W.: Randomized ensembled double Q-learning: learning fast without a model. In: 9th International Conference on Learning Representations, ICLR 2021, Virtual Event, Austria, 3–7 May 2021. OpenReview.net (2021). https://openreview.net/forum?id=AY8zfZm0tDd

21. Anschel, O., Baram, N., Shimkin, N.: Averaged-DQN: variance reduction and stabilization for deep reinforcement learning. In: International Conference on Machine Learning, pp. 176–185. PMLR (2017)

22. Wang, H., Lin, S., Zhang, J.: Adaptive ensemble Q-learning: minimizing estimation bias via error feedback. Adv. Neural. Inf. Process. Syst. **34**, 24778–24790 (2021)

23. Papadimitriou, D.: Monte Carlo bias correction in Q-learning. In: Goertzel, B., Iklé, M., Potapov, A., Ponomaryov, D. (eds.) AGI 2022. LNCS, vol. 13539, pp. 343–352. Springer, Cham (2023). https://doi.org/10.1007/978-3-031-19907-3_33

24. Tan, T., Hong, X., Feng, L.: Q-learning with heterogeneous update strategy. Available at SSRN 4341963 (2023)

25. Van Hasselt, H.: Estimating the maximum expected value: an analysis of (nested) cross validation and the maximum sample average, arXiv preprint arXiv:1302.7175 (2013)

26. Bengio, Y., Grandvalet, Y.: No unbiased estimator of the variance of k-fold cross-validation. J. Mach. Learn. Res. **5**(Sep), 1089–1105 (2004)

27. Berrar, D.: Cross-validation (2019)

28. Brockman, G., et al.: Openai gym, arXiv preprint arXiv:1606.01540 (2016)

29. Tasfi, N.: Pygame learning environment, GitHub repository (2016)

30. Lan, Q.: A pytorch reinforcement learning framework for exploring new ideas (2019). https://github.com/qlan3/Explorer

31. Dieterich, T.G.: An overview of MAXQ hierarchical reinforcement learning. In: Choueiry, B.Y., Walsh, T. (eds.) SARA 2000. LNCS (LNAI), vol. 1864, pp. 26–44. Springer, Heidelberg (2000). https://doi.org/10.1007/3-540-44914-0_2

A Neural Network Architecture for Accurate 4D Vehicle Pose Estimation from Monocular Images with Uncertainty Assessment[*]

Tomasz Nowak[(✉)] [iD] and Piotr Skrzypczyński [iD]

Poznan University of Technology, Institute of Robotics and Machine Intelligence, ul. Piotrowo 3A, 60-965 Poznań, Poland
tomasz.nowak@doctorate.put.poznan.pl, piotr.skrzypczynski@put.poznan.pl

Abstract. This paper proposes a new neural network architecture for estimating the four degrees of freedom poses of vehicles from monocular images in an uncontrolled environment. The neural network learns how to reconstruct 3D characteristic points of vehicles from image crops and coordinates of 2D keypoints estimated from these images. The 3D and 2D points are used to compute the vehicle pose solving the Perspective-n-Point problem, while the uncertainty is propagated by applying the Unscented Transform. Our network is trained and tested on the ApolloCar3D dataset, and we introduce a novel method to automatically obtain approximate labels for 3D points in this dataset. Our system outperforms state-of-the-art pose estimation methods on the ApolloCar3D dataset, and unlike competitors, it implements a full pipeline of uncertainty propagation.

Keywords: Vehicle pose estimation · 3D scene understanding · Deep learning

1 Introduction

It is essential for autonomous cars to be able to predict the poses of different objects in the environment, as it allows these agents to localise relatively to other vehicles or infrastructure on the road. Although the pose of another vehicle can be accurately measured using 3D laser scanner data [12], it is practical to obtain good accurate estimates of the pose of a vehicle from a single-camera image [18].

Song *et al.* [25] demonstrated that it is difficult to estimate vehicle poses from monocular images in a traffic scenario, because the problem is ill-posed due to the lack of obvious constraints and because of unavoidable occlusions. In our recent research [21], we showed how a deep neural network derived from the HRNet [29], originally created for human pose estimation, can be used to

[*] Research Supported by Poznań University of Technology Internal Grant 0214/SBAD/0242

accurately position a single camera in relation to a known road infrastructure object. In this paper, we introduce a novel deep learning architecture, also based on HRNet, which estimates 4D poses (3D Cartesian position and yaw angle) of vehicles in a realistic traffic scenario. This architecture estimates 2D keypoints on RGB image crops and then leverages these point features to estimate 3D coordinates of the characteristic points of a vehicle CAD model. The vehicle pose is computed by solving the Perspective-n-Point problem, given a set of n estimated 3D points of the vehicle model and their corresponding 2D keypoints in the image.

Unfortunately, the dataset we use, ApolloCar3D [25], does not contain associations between the 3D coordinates of the characteristic points of different car models and the annotated 2D keypoints in the images. To overcome this problem we propose also a new approach to automatically pre-process the dataset in order to obtain these associations leveraging the known poses and 3D CAD models of the vehicles seen in the images.

Knowing an estimate of the uncertainty in object pose estimates is crucial for the safe and reliable operation of autonomous cars. Therefore, we propose a two-step uncertainty estimation process. In the first step, we augment the architecture of the proposed neural network to estimate the uncertainty of the obtained keypoints. This uncertainty is represented by covariance matrices and then is propagated in the second step through the PnP algorithm using the unscented transform technique [22], which takes into account the non-linearity of the pose estimation algorithm being applied. Pose estimates with compatible covariance matrices are a unique feature among pose estimation algorithms that employ deep learning.

We thoroughly evaluate our approach on the ApolloD3 benchmark, proposed in the same paper as the dataset we use for training, demonstrating that the results outperform the state-of-the-art methods in terms of accuracy, while they are accompanied by human-interpretable estimates of spatial uncertainty.

The remainder of this paper is organised as follows: The most relevant related works on pose, keypoints, and uncertainty estimation are briefly reviewed in Sect. 2. The structure of the proposed pose estimation system is detailed in Sect. 3, while the approach to supervised training of our neural network, including pre-processing of the dataset, is explained in Sect. 4. Next, techniques applied to estimate and propagate the uncertainty are presented in Sect. 5. Finally, experiments on the ApoloCar3D dataset and the obtained results are presented in Sect. 6, followed by the conclusions given in Sect. 7.

2   Related Work

2.1   Vehicle Pose Estimation

Vehicle pose estimation solutions are of interest to researchers mainly due to the development of autonomous cars. Many solutions to the problem of determining the 3D pose of a vehicle based on observations from depth sensors such as LiDAR have been developed [30, 32]. The pose of a vehicle can also be accurately estimated from scene depth data obtained using stereo vision [14]. Estimating the pose of a vehicle from a single monocular camera image is most difficult, due to the lack of depth data and the difficulty in extracting pose constraints imposed by the environment from the image [25]. Nowadays, the state-of-the-art in estimating vehicle pose from monocular images involves the use of deep learning techniques combined with geometric priors [5]. Two basic types of approaches can be distinguished: direct, without detection of feature points, and indirect, using the detection of vehicle feature points in images.

Older direct pose estimation systems often relied on external neural network models that generated partial solutions, such as 2D proposals in [20], which were then cropped and processed in another network to estimate 3D bounding boxes and their orientations. Also, Xu and Chen [31] used separated neural networks to predict a depth image of the scene converted then to a point cloud and to generate 2D bounding boxes used to sample from this point cloud. More recent approaches, like the M3D-RPN [1] utilise a standalone 3D region proposal network leveraging the geometric relationship of 2D and 3D perspectives and allowing 3D bounding boxes to use features generated in the image-space. Yet differently, [9] exploits class-specific shape priors by learning a low dimensional shape-space from CAD models. This work uses a differentiable render-and-compare loss function which allows learning of the 3D shapes and poses with 2D supervision.

The indirect approaches typically align 3D car pose using 2D-3D matching with 2D keypoints detected on images and the provided CAD models of vehicles [2], exploiting also geometric constraints, e.g. the co-planar constraints between neighbouring cars in [2]. The RTM3D [15] predicts the nine keypoints of a 3D bounding box in the image space, then applies the geometric relationship of 2D and 3D perspectives to recover the parameters of the bounding box, including its orientation. In indirect approaches to vehicle pose estimation an important problem is to localise the occluded keypoints. This problem is addressed by the Occlusion-Net [23], which predicts 2D and 3D locations of occluded keypoints for objects. One of the recent developments, the BAAM system [10], leverages various 2D primitives to reconstruct 3D object shapes considering the relevance between detected objects and vehicle shape priors.

The approach in [17] is similar to ours in using estimated 2D and 3D points, but it directly regresses 20 semantic keypoints that define the 3D pose of a vehicle, whereas we extract more keypoints and compute poses indirectly, solving the PnP problem.

In this work, fewer keypoints are detected and used by the PnP algorithm, whereas the evaluation is done only on the KITTI dataset. The same problem is considered in [24] and, like us, the authors aim to select the best points for pose estimation, but the results they show use ground truth data – either about depth or 2D points.

Similarly to our solution, reprojection loss is used to train the model in the indirect GSNet [7] system, which is also evaluated on the ApolloCar3D dataset. However, the authors focus more on vehicle shape estimation, which is out of the scope of our work.

2.2   Keypoints and Uncertainty

In terms of accurate and robust detection of keypoints in images, significant progress has recently been made in human pose estimation applications [26]. The High Resolution Network (HRNet) [29] is a leading solution for keypoint detection in human pose estimation. As this architecture ensures high resolution of the feature maps through the entire network processing pipeline, which results in accurate location of the keypoints, we have selected the HRNet as a backbone of the neural part of our vehicle pose estimation system. In our previous work [21], we showed that it is possible to adapt HRNet to the task of estimating the feature points of an object with a known CAD model from RGB images. In the same work, we also introduced a method for estimating the spatial uncertainty of feature points in the form of a covariance matrix, inspired by the algorithm for estimating the uncertainty of keypoints in a face recognition task [8]. While there are classical methods for the propagation of uncertainty in computer vision [4], the problem of estimating feature points uncertainty or object pose uncertainty in deep learning based systems has been addressed in relatively few research papers. Several versions of a camera pose estimation method that consider the uncertainty of point and line features while computing the pose are proposed in [27], but these methods do not produce a covariance matrix for the final pose estimate. The geometric (spatial) uncertainty of features in the PnP problem is also considered in [3]. Recently, approaches to pose estimation that consider spatial uncertainty were presented for deep learning-based human pose recognition [13] and general object pose determination from images [16,33]. As far as we know, there are no publications tackling the estimation of the keypoints uncertainty and propagation of this uncertainty to the final vehicle pose for monocular pose estimation in the context of automotive applications.

3   Structure of the Pose Estimation System

The diagram of the whole processing pipeline is shown in Fig. 1. The details of heads architecture are presented in Fig. 2. It accepts as input only an image crop containing the considered vehicle. The processing pipeline consists of several modules: The 2D Keypoint Estimation Head, the 3D Keypoint Estimation Head, the Keypoint Score Head (KSH) for evaluating the accuracy of individual point estimates, and the Uncertainty Estimation Head for estimating the uncertainty of 2D and 3D points.

3.1   Estimation of 2D Keypoints

The first module is a deep neural network that estimates the 2D coordinates of the car's characteristic points on the image. Our approach combines the HRNet48 [29] as a backbone with a head for the feature extraction and heatmap estimation of 66 distinctive points on the image. We utilise input image crops of 256×192 pixels, providing sufficient detail for precise and fast feature extraction. The output consists of 48 feature maps of size $w \times h$, in our models $w=64$ and $h=48$. During training, we compute the loss function only for visible points,

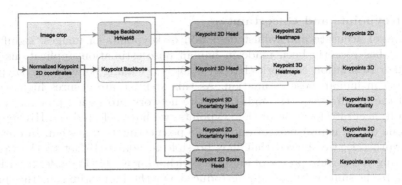

Fig. 1. Architecture of the proposed vehicle pose estimation system

applying a visibility mask that defines which points are evident for each instance. This process ensures that the model focuses on relevant data and is not influenced by obscured or irrelevant points. We employ unbiased encoding and decoding methods described in [6] for the preparation of ground-truth heatmaps for training and the subsequent decoding coordinates from the estimated heatmaps. This technique enhances prediction accuracy by reducing the quantization error.

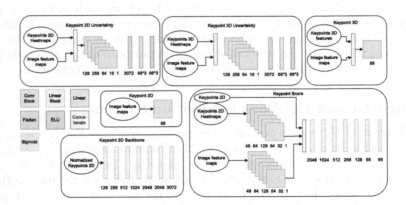

Fig. 2. Architecture details of the implemented network heads. Conv Block means Convolution layer followed by a batch normalization and ReLU activation. Linear Block means Linear layer followed by a batch normalization and ReLU activation

3.2   Estimation of 3D Keypoints

Estimation of 3D points requires the feature maps derived from the image crops initially obtained for the 2D point estimation network and the features extracted from the estimated 2D points by the Keypoint 2D Backbone. For the formation of features from the 2D points, a Multilayer Perceptron (MLP) consisting of seven layers is employed. This MLP takes the normalised 2D coordinates with respect to the bounding box as input. The feature maps derived from the image and the 2D points are concatenated to create an input for the 3D point estimation head. This module estimates the 3D points in the canonical pose, which is constant, regardless of the observed pose of the vehicle. The canonical pose implies that the coordinate system's origin is situated in the vehicle's geometric center, and the vehicle's front always faces the same direction. This consistent approach to 3D point estimation simplifies the problem, reducing computational complexity while increasing estimation accuracy. The 3D Keypoint Head estimates two separate feature maps. One of these corresponds to the $X - Y$ plane, from which the x and y coordinates are estimated. The second feature map corresponds to the $X - Z$ plane and serves to estimate the z coordinate. By independently addressing two orthogonal planes, this dual-map approach allows us to use a similar processing pipeline that is applied for the estimation of 2D keypoints. The 3D point estimation module's design allows it to seamlessly integrate with the 2D point estimation network by reusing feature maps extracted from the image. Note that our method does not need a mask from pre-trained Mask R-CNN, which is used by the baseline methods considered in [25].

3.3   Selection of the Best Estimated Keypoints

In this section, we introduce a uniquely designed module that focuses on assessing the precision of the point estimations. From any point of view, a significant portion of characteristic points is obscured, making their precise localisation on the image challenging. Such inaccurately estimated points can easily distort pose estimation. To counteract this phenomenon, we have designed an additional head dedicated to evaluating the accuracy of each estimated point. The accuracy evaluation head accepts as input the image feature map, the estimated heatmaps, and the normalised coordinates of the estimated points.

The image feature maps and heatmaps are processed through a sequence of six convolutional layers that outputs two feature vectors, each having length equal to 3072. These feature vectors, along with the estimated point coordinates, are then concatenated and processed through a sequence of six linear layers. The last layer is the sigmoid activation layer, which limits the output values to a range from zero to one. The output of the head is a set of 66 values, each corresponding to the estimation precision score of a given 2D point on the image, hereinafter called KSH scores. During the training phase, as learning targets we used only two values: '1' for a correctly estimated point, and '0' for the opposite case. A point is considered correctly estimated (has a '1' label) if the estimation error normalised to the bounding box is less than 0.04. We also conducted experiments using directly the normalised estimation errors as the learning targets, but the achieved results were worse compared to binary targets generated by thresholding the error.

3.4   Training Process

The training process is organized into distinct stages, each stage focusing on a specific aspect of the network. Initially, during the first stage, we directed our focus towards the 2D and 3D point heads and the backbone network. Following this, we select the best performing model from this initial training phase. The model's performance is evaluated using the Percentage of Correct Keypoints (PCK) metric for keypoints 2D and Mean Per Joint Position Error (MPJPE) for 3D point estimations. In the PCK metric: $PCK = \frac{P_{correct}}{P} \cdot 100$, where P is the total number of predicted points, we sum up to the number of correct points $P_{correct}$ only the keypoints with the coordinates estimation error normalised to the bounding box smaller than 0.05. The MPJPE metric is defined as a mean of the Euclidean distances between the estimated points and the corresponding ground truth points.

Additionally, for the training of 2D and 3D keypoint heads, we used a reprojection loss function. This function is an extension of an approach presented in [21] that maintains the geometric consistency of the estimated coordinates. It works by comparing the projected 3D points using ground truth transformation with the predictions of the 2D points, ensuring that the network's estimations of 2D and 3D points are consistent with each other and with real-world geometries. This loss is defined as:

$$\mathcal{L}_{\text{reprojection}} = \sum_{i=1}^{n} \left(\left\| \pi \left(\mathbf{T}, \hat{\mathbf{p}}_i^{3d}, \mathbf{K} \right) - \hat{\mathbf{p}}_i^{2d} \right\|_2 \right)^2, \tag{1}$$

where π is the projection function, $\mathbf{T}$ is a ground truth pose, $\hat{\mathbf{p}}_i^{3d}$ are the estimated 3D coordinates of the i-th characteristic point, $\mathbf{K}$ is the camera intrinsics matrix, and $\hat{\mathbf{p}}_i^{2d}$ are the estimated 2D coordinates of the i-th keypoint on image. The application of this loss function significantly improves the accuracy of the point estimation heads.

Once the best model is selected, its weights are frozen to preserve the best results for keypoint estimation. In the second stage, the focus shifts to the training of the Uncertainty Estimation Head and the Keypoint Score Head. By training the UEH and KSH components after the Keypoint 2D and 3D Heads, we ensure that the training process can leverage the well-tuned features provided by the backbone network and the point-generating heads. The loss functions used during training are described by Eq. 2, 3

$$\mathcal{L}_{stage1} = w_{repr} \cdot \mathcal{L}_{reprojection} + \mathcal{L}_{heatmap3Dxy} + \mathcal{L}_{heatmap3Dxz} + \mathcal{L}_{heatmap2D} \tag{2}$$

$$\mathcal{L}_{stage2} = \mathcal{L}_{uncertainty2D} + \mathcal{L}_{uncertainty3D} + \mathcal{L}_{keypoint\_score}, \tag{3}$$

where $w_{repr} = 1e^{-7}$ and $\mathcal{L}_{heatmap3Dxy}, \mathcal{L}_{heatmap3Dxz}, \mathcal{L}_{heatmap2D}, \mathcal{L}_{keypoint\_score}$ are defined as the Mean Squared Error loss function.

3.5   Pose Estimation

Given 2D points in the image, their corresponding 3D points in the model, and the camera matrix $\mathbf{K}$, various PnP algorithms can be applied to derive the pose. The 2D car characteristic points and corresponding 3D points are estimated by our network. During our research, we used two approaches: the Efficient

Perspective-n-Point (EPNP) algorithm [11], and a dedicated procedure called Solve-PnP-BFGS implemented using the SciPy library [28]. For input to the PnP algorithm, we select a subset of N best-estimated points according to KSH scores. In our experiments, we used $N=17$ as it gives the best results. The Solve-PnP-BFGS procedure works by minimizing the reprojection error using the BFGS optimization algorithm. The reprojection error is defined similarly to (1), but the ground truth transformation $\mathbf{T}$ is replaced by the optimized transformation. The optimisation search space in the BFGS algorithm is constrained by bounds applied to the estimated translation. The optimisation process is carried out five times, with a different starting point randomly selected from the search space each time. The selected solution is the one with the lowest value of the cost function. This repetitive process reduces the possibility of optimisation being stuck at the local minimum. To estimate the shape of a vehicle, we employ a predefined library consisting of all CAD models included in the dataset and a set of 3D characteristic points in canonical form for each model. This library serves as a reference during the shape estimation process, allowing us to capture the full mesh of the vehicles.

During the inference stage, we compare the estimated 3D points with those in the library and select the CAD model that yields the smallest MPJPE. To get the most reliable results, to calculate the MPJPE, we remove firstly the points that are supposed to be the least accurately estimated according to the KSH score value. The best results were achieved by removing points with a KSH score below the threshold $s = 0.19$.

4   Dataset Preparation for Supervised Learning

For our experiments, we use the ApolloCar3D dataset introduced in [25]. This dataset comprises 5,277 images derived from traffic scenes, containing over 60,000 instances of vehicles. Each vehicle instance is defined by a set of 66 characteristic points, with annotations of only visible points. The dataset also provides a set of 34 CAD models of the vehicles appearing in the images. Ground truth pose data, relative to the camera coordinates, are provided for each vehicle. A significant challenge with respect to our pose estimation task is the absence of a mapping between the 2D points in the images and the corresponding 3D points from the CAD models. Such a mapping (provided as labels) is necessary for supervised learning of 3D points detection.

To overcome this problem, we employ a procedure illustrated in Fig. 3. This process begins by transforming the CAD model using the given translation and rotation. Subsequently, the parameters of the ray encompassing all 3D points, which project onto the image at the annotated keypoints for the considered vehicle, are established. For each face of the CAD model, we check if this ray intersects it using the Möller-Trumbore algorithm [19], subsequently determining the coordinates of the intersection point.

In cases where multiple intersection points are discovered, the point closest to the camera is chosen. This strategy is in line with the dataset's approach of marking only non-occluded points. The final stage involves applying the inverse of the ground truth rotation and translation to the intersection point, yielding the coordinates with respect to the car's canonical pose. The point coordinates derived from this process are then fine-tuned to achieve a more precise match. For this fine-tuning, we employ the Nelder-Mead optimization algorithm on all instances of a particular car type in the training set. This algorithm adjusts the 3D coordinates to minimize the translation error, defined as the square of the distance between the ground truth translation and the translation estimated by the EPnP [11] method, taking into account the given camera parameters and the 2D point coordinates in the image.

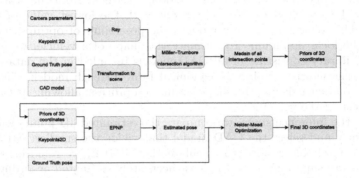

Fig. 3. Pipeline for dataset pre-processing in order to obtain labelled 3D points for supervised learning

5   Uncertainty Estimation

5.1   Estimation of Keypoints Uncertainty

In this section, we discuss a module designed to estimate the uncertainty of characteristic point estimations. The keypoints uncertainty estimation process is an extension of the approach presented in our previous paper [21]. However, our new method estimates both 2D and 3D point uncertainties, contributing to a comprehensive understanding of the estimation process's precision. For each of the 2D points, a 2×2 covariance matrix, Σ_{2D}, is estimated for x and y coordinates. The Uncertainty Estimation Head (UEH) consists of a stack of 5 blocks built by convolutional layer, ReLU, and batch normalization. Convolutional blocks are followed by one linear layer. The UEH takes as input the keypoint heatmaps and image feature maps and calculates three positive numbers, filling in the lower triangle of the 2D Cholesky factor matrix, $\mathbf{L}_{2D}$. We ensure the positivity of the estimated numbers through the application of an Exponential Linear Unit (ELU) activation function and the addition of a constant to the output values. Then, by multiplying the matrix $\mathbf{L}_{2D}$ by $\mathbf{L}_{2D}^T$, we acquire a covariance matrix Σ_{2D}. This approach ensures the generation of a positive semi-definite covariance matrix, a necessary characteristic for valid uncertainty estimation. During the training phase, we employ the Gaussian Log-Likelihood Loss function, as suggested in [8].

When it comes to 3D points, we estimate the values found on the main diagonal of the 3×3 covariance matrix. Only the main diagonal elements are filled in the 3D Cholesky factor matrix, $\mathbf{L}_{3D}$. The subsequent processing steps are analogous to those used in the 2D point uncertainty estimation, reinforcing consistency across the module's operation.

5.2 Propagation Using Unscented Transform

The estimated uncertainties of both 2D and 3D points can be propagated to the uncertainty of the estimated vehicle pose. To accomplish this, we employ the Unscented Transform method, as presented in [22]. The propagation of uncertainty of the n-dimensionsal input is carried out using $2n+1$ sigma points χ_i, $i = 0, ..., n$, while the coordinates of these points are determined by formulas 4:

$$\chi_0 = \mathbf{m_x}, \; \chi_{2i-1} = \mathbf{m_x} + \sqrt{n + \lambda}\left[\sqrt{\mathbf{C_x}}\right], \; \chi_{2i} = \mathbf{m_x} - \sqrt{n + \lambda}\left[\sqrt{\mathbf{C_x}}\right], \quad (4)$$

for $i = 1, ..., n$, where $\mathbf{m_x}$ and $\mathbf{C_x}$ are the mean and variance of the estimated points, $\lambda = \alpha^2(n + k) - n$ is a scaling factor, and α and k are the parameters influencing how far the sigma points are away from the mean. Then, by applying a nonlinear transformation, which is the PnP algorithm, we obtain a set of points from which the mean and covariance of the estimation after the nonlinear transformation are estimated. In our implementation, the Unscented Transform parameters α and k [22] are set to 0.9 and 50, respectively.

The propagation process for 2D and 3D points is executed independently. After the uncertainties have been individually propagated, the resulting covariance matrices of equal dimension are summed up, assuming statistical independence between the coordinates of the 3D and 2D keypoints.

6 Evaluation Results

Model evaluation was carried out on the ApoloCar3D dataset validation set. It contains 200 images according to the split made by the authors of the Apolo-Car3D dataset. On a Nvidia 1080Ti GPU the proposed pipeline is capable to run at 20 FPS for the EPNP variant without uncertainty propagation and at 18 FPS with full uncertainty propagation.

6.1 Keypoints 2D

For the evaluation of the 2D keypoints estimation network, we used the Percentage of Correct Keypoints metric, assuming three thresholds: 5 pixels, 10 pixels, and 15 pixels.

The first two rows in Table 1 were calculated using as reference only visible vehicle points, i.e., those located in the ground truth annotations of the dataset. We provide metric values for all marked points and for those points that the network identified as having the highest KSH score. The results demonstrated that the network correctly assesses the accuracy of the point estimation and is capable of improving the results achieved on the PCK metric.

The last two rows in Table 1 present analogous PCK metric values. The difference is that we used the projections of 3D points as ground truth annotations, using the ground truth translation and rotation of a given vehicle. This approach allows for an evaluation of the invisible points that is more relevant to real-world conditions, where we do not have a priori knowledge about the visibilities of points. The presented results, especially those acquired for the points selected by the KSH score provide sufficient accuracy for the 3D pose estimation using PnP algorithms.

Table 1. PCK metric of the 2D keypoints estimations comparing to manually labeled visible points (top) and all 66 points acquired by projection of 3D points (bottom)

Threshold	5 px	10 px	15 px
Visible points PCK	55.0	81.0	89.0
Selected visible points PCK	60.0	84.0	91.0
All points PCK	13.1	19.7	22.2
Selected points PCK	40.5	56.0	60.9

6.2   Keypoints 3D

In this section, we present the results related to the estimation of 3D points on vehicles using our proposed method. Module performance is evaluated using the Mean Per Joint Position Error (MPJPE) metric, which measures the average distance error of the estimated points. Our results reveal an averaged MPJPE value of 0.119 m for all points estimated by the network. When we consider only the N points with the highest KSH score, the error decreases to 0.105 m, which confirms that the KSH score promotes better keypoints. The most accurately estimated points are on the rear corner of the car handle of the right front car door, with a mean error of 0.073 m, while the points of lowest accuracy (0.235 m on average) are located on the left corner of the rear bumper.

6.3   A3DP Metrics

We utilised the Absolute 3D Pose Error (A3DP-Abs) metric introduced in [25] to evaluate our results. This metric focuses on the absolute distances to objects, considering three key components: the estimated shape of the car, its position, and its rotation. The translation error metric is defined as:

$$c_{\text{trans}} = \left| \mathbf{t}_{gt} - \hat{\mathbf{t}} \right|_2 \leq \delta_t, \tag{5}$$

where $\mathbf{t}_{gt}$ denotes ground truth translation, $\hat{\mathbf{t}}$ denotes estimated translation and δ_t is an acceptance threshold. The rotation error metric is defined as:

$$c_{\text{rot}} = \arccos\left(|\mathbf{q}_{gt} \cdot \hat{\mathbf{q}}|\right) \leq \delta_{rot}, \tag{6}$$

where $\mathbf{q}_{gt}$ denotes ground truth rotation quaternion, $\hat{\mathbf{q}}$ denotes estimated rotation quaternion, and δ_{rot} is an acceptance threshold. Similarly to metrics proposed for the COCO dataset, the authors of ApolloCar3D proposed a set of metric thresholds from strict to loose. The translation thresholds were established from 2.8 m to 0.1 m in increments of 0.3 m, while the rotation thresholds range from $\pi/6$ to $\pi/60$ with steps of $\pi/60$. Beyond the 'mean' metric, which averages results across all thresholds, two single-threshold metrics were defined. The loose criterion (denoted as $c - l$) utilises $[2.8, \pi/6]$ thresholds for translation and rotation, whereas the strict criterion (denoted as $c - s$) employs $[1.4, \pi/12]$ as thresholds for these respective parameters. To evaluate the 3D shape reconstruction, a predicted mesh is rendered from 100 different perspectives. Intersection over Union (IoU) is computed between these renderings and the ground truth masks. The average of these IoU calculations is then used to evaluate the shape reconstruction.

Table 2 shows our results and compares them against state-of-the-art methods. Note that for the algorithms proposed in [2,9] the implementations provided in [25] as baselines were used for a fair comparison.

Table 2. Comparison of results with the state-of-the-art methods on A3DP-Abs metrics

algorithm	mean	$c - l$	$c - s$
3D-RCNN [9]	16.4	29.7	19.8
DeepMANTA [2]	20.1	30.7	23.8
GSNet [7]	18.9	37.4	18.4
BAAM-Res2Net [10]	25.2	47.3	23.1
Ours EPnP	23.4	44.6	31.7
Ours BFGS	**25.6**	**47.7**	**34.6**

The results of the A3DP-Abs mean metric indicate that our system's performance is superior to the very recent BAAM [10] method, which demonstrates the proficiency of our solution in estimating 3D points accurately. A significant advantage of our network is its performance on the strict criterion $c - s$. This metric measures how well our module estimates 3D points under stringent conditions, a challenging task that demands a high degree of precision and reliability. Our system outperforms all existing state-of-the-art solutions on this metric by a large margin, highlighting the solution's excellence in providing highly accurate 3D characteristic points.

6.4   Uncertainty Assessment

In this section, we present results related to the uncertainty estimation of the 2D and 3D characteristic points, as well as to the uncertainty of the vehicle's pose. The first part of our evaluation process involved examining the percentage of point and vehicle translation estimations that fall within the ranges of 1, 2, and 3 standard deviations (σ). The results of this examination are shown in the

Table 3. Percentage of estimation that fall within the ranges of one, two and three σ

	Keypoints 2D		Keypoints 3D			Pose translation		
	x	y	x	y	z	x	y	z
1 σ	79.0	79.8	82.5	80.3	64.8	84.5	85.0	81.3
2 σ	93.1	93.8	94.2	94.1	80.9	94.9	95.6	93.7
3 σ	97.2	97.2	97.5	97.3	87.5	98.6	98.6	97.5

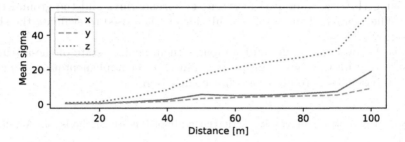

Fig. 4. Plot of the mean σ (standard deviation) depending on the observation distance

Table 3. We further examine the relationship between the mean value of σ and the vehicle's distance in Fig. 4.

These findings suggest that as distance increases, the uncertainty in pose estimation also increases. This trend is to be expected, as distant objects inherently have the lower resolution in the image, which tends to lead to higher uncertainties in the estimation. An additional noteworthy observation is that the uncertainty along the z-axis (depth direction) is greater than along the remaining two axes. This finding is justified, as it is often more challenging to determine the position along the axis perpendicular to the image plane. Due to the nature of monocular vision and image projection, depth information is less reliable and can result in higher uncertainties. Figure 5 presents six examples of predictions and the uncertainty of the estimated pose from a bird's eye view. The uncertainty of position (x, y) for each vehicle is represented by its uncertainty ellipse. Observe that cars that are partially occluded or are located further from the camera have larger uncertainty ellipses than the closer, fully visible cars.

Fig. 5. Visualizations of estimated pose and uncertainty. Circles show the estimated positions of cars, squares are ground truth positions of these cars, colour rays show ground truth yaw angles, grey fans show the estimated one-sigma yaw range. Ellipses visualise one sigma position uncertainty (Color figure online)

7   Conclusions

In this paper, we have introduced a comprehensive processing pipeline capable of estimating vehicle pose solely based on data from a single RGB camera. This innovative pipeline comprises a network for estimating 2D and 3D points, an evaluation of the credibility of individual point estimations, and an optimisation algorithm estimating pose based on information derived from previous steps.

Our system outperforms all previously reported models on the ApolloCar3D dataset and is capable of estimating the uncertainty of pose estimates. The unique ability to handle uncertainty has significant implications for real-world applications, offering an additional layer of validation and error mitigation in pose estimation and further decision making.

We also introduce an unsupervised method for preparing a training dataset containing 3D point coordinates. This approach leverages 2D point annotations, CAD models, and vehicle poses, offering a novel methodology to generate reliable training data without the need for laborious and error-prone manual annotation.

In conclusion, our approach represents a significant advancement in the field of vehicle pose estimation, providing a powerful, robust, and accurate pipeline for this complex task. Furthermore, the ability to estimate the uncertainty of pose estimations and the introduction of an unsupervised method for training dataset preparation pave the way for future advancements in this field. Further work will include the integration of geometric priors stemming from the scene structure and semantics, such as the coplanarity constraint applied in [25].

References

1. Brazil, G., Liu, X.: M3D-RPN: monocular 3D region proposal network for object detection. In: IEEE/CVF International Conference on Computer Vision (ICCV), pp. 9286–9295 (2019)
2. Chabot, F., Chaouch, M., Rabarisoa, J., Teuliere, C., Chateau, T.: Deep MANTA: A coarse-to-fine many-task network for joint 2D and 3D vehicle analysis from monocular image. In: IEEE/CVF Conference on Computer Vision and Pattern Recognition (CVPR), pp. 2040–2049 (2017)
3. Ferraz, L., Binefa, X., Moreno-Noguer, F.: Leveraging feature uncertainty in the PNP problem. In: Proceedings of the British Machine Vision Conference (2014)
4. Haralick, R.M.: Propagating covariance in computer vision. Int. J. Pattern Recogn. Artif. Intell. **10**(5), 561–572 (1996)
5. Hoque, S., Xu, S., Maiti, A., Wei, Y., Arafat, M.Y.: Deep learning for 6d pose estimation of objects - a case study for autonomous driving. Expert Syst. Appl. **223**, 119838 (2023)
6. Huang, J., Zhu, Z., Guo, F.: The devil is in the details: delving into unbiased data processing for human pose estimation. arXiv:2008.07139 (2020)
7. Ke, L., Li, S., Sun, Y., Tai, Y.-W., Tang, C.-K.: GSNet: joint vehicle pose and shape reconstruction with geometrical and scene-aware supervision. In: Vedaldi, A., Bischof, H., Brox, T., Frahm, J.-M. (eds.) ECCV 2020. LNCS, vol. 12360, pp. 515–532. Springer, Cham (2020). https://doi.org/10.1007/978-3-030-58555-6_31
8. Kumar, A., Marks, T.K., Mou, W., Feng, C., Liu, X.: UGLLI face alignment: Estimating uncertainty with gaussian log-likelihood loss. In: IEEE/CVF International Conference on Computer Vision Workshop (ICCVW), pp. 778–782 (2019)
9. Kundu, A., Li, Y., Rehg, J.M.: 3D-RCNN: instance-level 3D object reconstruction via render-and-compare. In: IEEE/CVF Conference on Computer Vision and Pattern Recognition (CVPR), pp. 3559–3568 (2018)

10. Lee, H.J., Kim, H., Choi, S.M., Jeong, S.G., Koh, Y.J.: BAAM: monocular 3d pose and shape reconstruction with bi-contextual attention module and attention-guided modeling. In: Proceedings of the IEEE/CVF Conference on Computer Vision and Pattern Recognition (CVPR), pp. 9011–9020, June 2023

11. Lepetit, V., Moreno-Noguer, F., Fua, P.: EPnP: an accurate O(n) solution to the PnP problem. Int. J. Comput. Vision **81**, 155–166 (2009)

12. Li, B., Zhang, T., Xia, T.: Vehicle detection from 3d lidar using fully convolutional network. In: Hsu, D., Amato, N.M., Berman, S., Jacobs, S.A. (eds.) Robotics: Science and Systems XII. University of Michigan, Ann Arbor (2016)

13. Li, H., et al.: Pose-oriented transformer with uncertainty-guided refinement for 2d-to-3d human pose estimation (2023)

14. Li, P., Chen, X., Shen, S.: Stereo R-CNN based 3D object detection for autonomous driving. In: IEEE/CVF Conference on Computer Vision and Pattern Recognition (CVPR), pp. 7636–7644 (2019)

15. Li, P., Zhao, H., Liu, P., Cao, F.: RTM3D: Real-time monocular 3d detection from object keypoints for autonomous driving. In: Vedaldi, A., Bischof, H., Brox, T., Frahm, J.-M. (eds.) ECCV 2020. LNCS, vol. 12348, pp. 644–660. Springer, Cham (2020). https://doi.org/10.1007/978-3-030-58580-8_38

16. Liu, F., Hu, Y., Salzmann, M.: Linear-covariance loss for end-to-end learning of 6d pose estimation. CoRR abs/2303.11516 (2023)

17. LÃṣpez, J.G., Agudo, A., Moreno-Noguer, F.: Vehicle pose estimation via regression of semantic points of interest. In: 11th International Symposium on Image and Signal Processing and Analysis (ISPA), pp. 209–214 (2019)

18. Marti, E., de Miguel, M.A., Garcia, F., Perez, J.: A review of sensor technologies for perception in automated driving. IEEE Intell. Transp. Syst. Mag. **11**(4), 94–108 (2019)

19. Möller, T., Trumbore, B.: Fast, minimum storage ray-triangle intersection. J. Graph. Tools **2**(1), 21–28 (1997)

20. Mousavian, A., Anguelov, D., Flynn, J., Kosecka, J.: 3d bounding box estimation using deep learning and geometry. In: IEEE Conference on Computer Vision and Pattern Recognition, CVPR. pp. 5632–5640 (2017)

21. Nowak, T., Skrzypczyński, P.: Geometry-aware keypoint network: accurate prediction of point features in challenging scenario. In: 17th Conference on Computer Science and Intelligence Systems (FedCSIS), pp. 191–200 (2022)

22. PÃl'rez, D.A., Gietler, H., Zangl, H.: Automatic uncertainty propagation based on the unscented transform. In: IEEE International Instrumentation and Measurement Technology Conference (I2MTC), pp. 1–6 (2020)

23. Reddy, N.D., Vo, M., Narasimhan, S.G.: Occlusion-Net: 2D/3D occluded keypoint localization using graph networks. In: IEEE/CVF Conference on Computer Vision and Pattern Recognition (CVPR), pp. 7318–7327 (2019)

24. Shi, J., Yang, H., Carlone, L.: Optimal pose and shape estimation for category-level 3d object perception. arXiv:2104.08383 (2021)

25. Song, X., et al.: ApolloCar3D: a large 3D car instance understanding benchmark for autonomous driving. In: IEEE/CVF Conference on Computer Vision and Pattern Recognition (CVPR), pp. 5447–5457 (2019)

26. Toshpulatov, M., Lee, W., Lee, S., Haghighian Roudsari, A.: Human pose, hand and mesh estimation using deep learning: a survey. J. Supercomput. **78**(6), 7616–7654 (2022)

27. Vakhitov, A., Colomina, L.F., Agudo, A., Moreno-Noguer, F.: Uncertainty-aware camera pose estimation from points and lines. In: IEEE/CVF Conference on Computer Vision and Pattern Recognition (CVPR), pp. 4657–4666 (2021)

28. Virtanen, P., et al.: SciPy 1.0 contributors: SciPy 1.0: fundamental algorithms for scientific computing in python. Nat. Methods **17**, 261–272 (2020). https://doi.org/10.1038/s41592-019-0686-2
29. Wang, J., et al.: Deep high-resolution representation learning for visual recognition. IEEE Trans. Pattern Anal. Mach. Intell. **43**, 3349–3364 (2021)
30. Wang, Q., Chen, J., Deng, J., Zhang, X.: 3D-CenterNet: 3D object detection network for point clouds with center estimation priority. Pattern Recogn. **115**, 107884 (2021)
31. Xu, B., Chen, Z.: Multi-level fusion based 3D object detection from monocular images. In: IEEE/CVF Conference on Computer Vision and Pattern Recognition, pp. 2345–2353 (2018)
32. Yang, B., Luo, W., Urtasun, R.: PIXOR: real-time 3D object detection from point clouds. In: IEEE/CVF Conference on Computer Vision and Pattern Recognition, pp. 7652–7660 (2018)
33. Yang, H., Pavone, M.: Object pose estimation with statistical guarantees: conformal keypoint detection and geometric uncertainty propagation. CoRR abs/2303.12246 (2023)

PSO-Enabled Federated Learning for Detecting Ships in Supply Chain Management

Y Supriya[1] , Gautam Srivastava[2(✉)] , K Dasaradharami Reddy[1] ,
Gokul Yenduri[1], Nancy Victor[1] , S Anusha[3] ,
and Thippa Reddy Gadekallu[1,4,5,6,7]

[1] School of Information Technology and Engineering, Vellore Institute of Technology,
Vellore, Tamil Nadu, India
supriya.2020@vitstudent.ac.in, dasaradharami.k@vit.ac.in,
gokul.yenduri@vit.ac.in, drnancyvictor@ieee.org, thippareddy.g@vit.ac.in
[2] Department of Math and Computer Science, Brandon University, Brandon, Canada
srivastavag@brandonu.ca
[3] Department of Computer Science and Engineering, NBKR Institute of Science and
Technology, Vidyanagar, Tirupathi, Andhra Pradesh, India
anusha.dasaradh@gmail.com
[4] Zhongda Group, Haiyan County, Jiaxing City, Zhejiang Province, China
[5] Department of Electrical and Computer Engineering, Lebanese American
University, Byblos, Lebanon
[6] Research and Development, Lovely Professional University, Phagwara, India
[7] College of Information Science and Engineering, Jiaxing University, Jiaxing 314002,
China

Abstract. Supply chain management plays a vital role in the efficient and reliable movement of goods across various platforms, which involves several entities and processes. Ships in the supply chain are very important for the global economy to be connected. Detecting ships and their related activities is of paramount importance to ensure successful logistics and security. To improve logistics planning, security, and risk management, a strong framework is required that offers an efficient and privacy-preserving solution for identifying ships in supply chain management. In this paper, we propose a novel approach called Particle Swarm Optimization-enabled Federated Learning (PSO-FL) for ship detection in supply chain management. The proposed PSO-FL framework leverages the advantages of both Federated Learning (FL) and Particle Swarm Optimization (PSO) to address the challenges of ship detection in supply chain management. We can train a ship identification model cooperatively using data from several supply chain stakeholders, including port authorities, shipping firms, and customs agencies, thanks to the distributed nature of FL. By improving the choice of appropriate participants for model training, the PSO algorithm improves FL performance. We conduct extensive experiments using real-world ship data that is gathered from various sources to evaluate the effectiveness of our PSO-FL approach. The results demonstrate that our framework achieves

B. Luo et al. (Eds.): ICONIP 2023, CCIS 1962, pp. 413–424, 2024.
https://doi.org/10.1007/978-981-99-8132-8_31

superior ship detection accuracy of 94.88% compared to traditional centralized learning approaches and standalone FL methods. Furthermore, the PSO-FL framework demonstrates robustness, scalability, and privacy preservation, making it suitable for large-scale deployment in complex supply chain management scenarios.

Keywords: Federated Learning · Particle Swarm Optimization · Ships · Supply Chain Management · Industry 5.0

1 Introduction

Supply chain management is the coordination and management of the entire process of producing, acquiring, transforming, and distributing commodities and services from their place of origin to their point of consumption. Ensuring the efficient movement of goods and information along the supply chain includes the integration and management of numerous functions, such as production, inventory management, transportation, and warehousing [13].

In recent years, the emergence of Industry 5.0 has transformed the manufacturing and supply chain management landscape, with a focus on integrating advanced technologies such as artificial intelligence (AI), the Internet of Things (IoT), and robotics into industrial processes [11]. One of the key challenges in Industry 5.0 is the need to ensure safety, security, and efficiency in supply chain management, particularly about the detection of ships [8].

The detection of ships is a crucial aspect of transportation in supply chain management. Ships are one of the primary means of transporting goods across long distances, and their timely detection is essential for ensuring that goods are delivered on time [22].

Several Machine Learning (ML) and Deep Learning (DL) techniques have been used widely in ship detection tasks as they can learn from large amounts of data and extract complex features. ML algorithms such as Support Vector Machines (SVM), Random Forests, or Logistic Regression and other techniques like Convolutional Neural Networks (CNN), Transfer Learning, and Recurrent Neural Networks (RNN) have been used in ship detection [24].

However, while ML and DL techniques showed good potential in ship detection, there also several limitations and challenges like imbalanced data availability, large computational requirements, and lack of a good framework that provides an effective and privacy-preserving solution for ship detection [20].

FL has emerged as a promising approach for ship detection addressing the limitations of traditional methods due to its ability to address privacy concerns and leverage distributed data sources [2]. FL in ship detection also faces certain challenges, including communication bandwidth limitations, heterogeneity of data sources, model convergence issues, and coordination among participants [1,14,15,18].

PSO-enabled FL [10] in ship detection combines the optimization capabilities of PSO with the collaborative and privacy-preserving nature of FL, enabling improved ship detection accuracy, faster convergence, and better utilization of distributed data sources [12]. It provides a promising approach for developing robust ship detection models in scenarios where data privacy and collaboration among multiple entities are essential.

1.1 Key Contributions

The key contributions of our work include the following:

- Proposal of a novel PSO-enabled FL approach: The paper introduces a unique approach that combines the advantages of FL and PSO for ship detection in supply chain management.
- Enhancement of ship detection performance: The proposed approach significantly improves ship detection performance compared to traditional centralized methods. The paper provides experimental results and performance metrics to demonstrate the effectiveness of the PSO-enabled FL approach.
- Consideration of data privacy and security: The paper addresses the security and privacy concerns by employing an FL framework that allows local models to be trained on edge devices without sharing sensitive data with a centralized server.

2 Related Work

The following section briefs about related work. Ship detection in supply chain management plays a critical role in ensuring efficient and secure global trade operations. With the advent of advanced technologies and data-driven decision-making, researchers and practitioners have explored various techniques for ship detection, including ML and FL. In this section, we review the related work that focuses on how the state-of-the-art techniques are used for ship detection in supply chain management.

Ship Detection in Supply Chain Management: El Mekkaoui et al. [3] focused on the problem of ship speed prediction within the context of Maritime Vessel Services operating in the Saint Lawrence Seaway area. The primary challenge lies in developing a real-time predictive model that can effectively account for varying routes and vessel types. To address this challenge, the study proposes a data-driven solution that leverages DL sequence methods and historical ship trip data to forecast ship speeds at different stages of a voyage.

Faustina et al. [4] addressed the challenge of ship detection in synthetic aperture radar (SAR) imagery. Ship detection is a critical task in maritime surveillance and security, and the utilization of multi-temporal information can significantly enhance detection accuracy. A novel approach was proposed that leverages transfer learning techniques to adapt multi-temporal information for optimized ship detection. SAR image dataset was used to train a DL model.

FL for Ship Detection: Zhang et al. [23] present an interesting and valuable contribution to the field of fault diagnosis in the IoT domain, specifically for ships. The paper addresses an important challenge of privacy preservation in FL, and the adaptive nature of the framework enhances its practical applicability. Conducting empirical evaluations and extending the discussion on related works would further enhance the paper's contribution.

Wang et al. in [19] present an innovative solution to address the challenge of privacy protection and efficient incumbent detection in spectrum-sharing scenarios. The paper introduces a well-designed FL framework and provides a comprehensive explanation of the proposed approach. Further empirical evaluations and comparisons would enhance the paper's contribution and validate its effectiveness in practical spectrum-sharing applications.

Teng et al. [17] introduced an innovative FL framework that incorporates prior knowledge and a bilateral segmentation network for accurate and efficient edge extraction. The paper presents a detailed description of the framework, supported by mathematical formulations and experimental evaluations. The proposed approach demonstrates superior performance compared to existing methods, making it a valuable contribution to the field of FL for image analysis tasks.

PSO-enabled FL: Supriya et al. [16] introduced a novel framework that combines FL and PSO for efficient and accurate early detection of forest fires. The paper presents a comprehensive explanation of the proposed approach, supported by experimental evaluations and discussions on practical implications.

Kandati et al. [7] proposed an FPS optimization algorithm that employs PSO to increase FL's effectiveness and decrease the average amount of data the client communicates to the server. The suggested approach distributes the model's score value from the server where it was trained.

Based on the above literature, although ship detection in supply chain management has been extensively studied, the integration of FL and PSO in this domain remains relatively unexplored. In our research, we aim to bridge this gap by developing a PSO-enabled FL approach for ship detection in supply chain management. Our work contributes to the advancement of ship detection techniques in the context of Industry 5.0 and provides valuable insights for optimizing supply chain operations and ensuring secure maritime logistics.

3   Proposed Work

3.1   Dataset Collection and Preparation

This work uses an image dataset from Kaggle[1] as was also used in [21]. Using this dataset, ships can be detected in the test images by classifying the images. The dataset contains 4000 images. Figure 1 depicts the sample images from the satellite imagery of the ship dataset. This dataset consists of two directories, no-ship, and ship. The no-ship directory has 3000 images, and the ship directory has 1000 images.

---

[1] https://www.kaggle.com/datasets/apollo2506/satellite-imagery-of-ships.

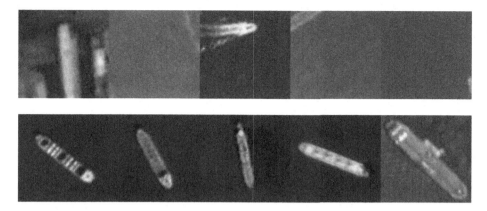

Fig. 1. Sample images from the satellite imagery of ships dataset. The images are of two classes namely no-ship and ship.

3.2 Dataset Pre-processing

It is necessary to perform some data cleaning to improve the quality of the training and the outcomes. As a consequence of this, we cropped and scaled to a constant size of 150×150 pixels.

3.3 Dataset Partitioning

The train_test_split function splits the dataset. The test_size=0.2 specifies that 20% of the data is allocated to the test set and 80% of the data is allocated to the train set.

3.4 CNN

Our CNN model consists of four primary layers: two convolutional layers followed by max-pooling layers. We utilize the ReLU activation function for its efficiency in dealing with the vanishing gradient problem. The model is compiled using the Adam optimizer with a learning rate of 0.01, and binary cross-entropy is used as the loss function. Table 1 refers to the architecture of the CNN model.

3.5 Methodology

The primary objective of this study is to propose a method for the early detection of ships in supply chain management. The combination of FL and PSO makes such a model feasible. Adding more layers to CNN models is a common procedure to enhance their accuracy. This is known as a deep neural network. Training time for weight parameters increases as the number of layers increases. The cost of transmitting the client-learned model to the server increases. Thus, we propose a unified framework that uses PSO characteristics to relocate the trained model, irrespective of its size, with the highest possible score (such as accuracy or loss).

Table 1. CNN Model Architecture

Layer	Type	Filters	Filter Size	Stride	Padding	Activation
Conv Layer 1	Conv2D	32	3×3	1	Same	ReLU
Pool Layer 1	MaxPool2D	N/A	2×2	2	Valid	N/A
Conv Layer 2	Conv2D	64	3×3	1	Same	ReLU
Pool Layer 2	MaxPool2D	N/A	2×2	2	Valid	N/A
Flatten	Flatten	N/A	N/A	N/A	N/A	N/A
FC Layer	Dense	64	N/A	N/A	N/A	ReLU
Output Layer	Dense	1	N/A	N/A	N/A	Sigmoid

Initially, we will start by evaluating the proposed PSO-enabled FL. PSO-enabled FL, the recommended model, only allows model weights from the top-scoring client rather than from all clients. Figure 2 shows how the process works. The client's potential loss value after training is a key determinant of the most significant attainable score. In this instance, the loss value is just 4 bytes long. PSO-enabled FL finds the finest model by applying the *pb* and *gb* variables, and it then changes the value of S for each weighted array member that represents the finest model.

PSO-enabled FL's weights were updated using Eq. 1. The weight update appears as follows:

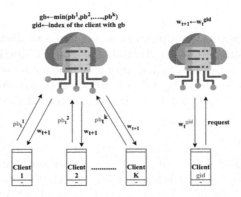

Fig. 2. Particle swarm optimization-enabled Federated Learning

$$S_i^a = \alpha.S_i^{a-1} + d_1.rd_1.(pb - Si^{a-1}) + d_2.rd_2.(gb - S_i^{a-1})$$
$$w_i^a = w_i^{a-1} + S^a \tag{1}$$

Table 2. Constants

Name of the constant	Value
Clients	10
Epochs	30
Client-Epochs	5
Batch size	16
Learning rate	0.01

For each layer with weight w, S in CNN may be calculated using Eq. 1. The current step weight w_{a-1} is calculated by adding the previous step weight w_a. According to 1, the acceleration constants for pb and gb are, respectively, d_1 and d_2. Any random number between 0 and 1 can be substituted for the values of rd_1 and rd_2.

Algorithm 1 presents the basic idea of PSO-enabled FL. The PSO-based Algorithm 1 is an enhancement of the FedAvg Algorithm. Looking at Line 5, the client only sends the pb variables to the function ServerAggregation but not the weight w. Lines 6–8 carry out the process of finding the client with the lowest pb value in the collected data. After that, CNN will start using the UpdateClient function to implement the PSO. Lines 13 and 14 estimates the previously used variable S, the clients desired w^{pb} value, and the w^{gb} value transmitted to the server. This procedure is carried out again for each layer weight. In Line 15, we add S to the w we retrieved from the previous round to determine the w we will use in this round. Lines 16–18 are iterated as many times as client epoch E. Lines 20–23 of GetAptModel requested the server for the client model with the highest possible score. Table 2 depicts the hyper-parameter values used in the paper.

4   Results and Analysis

A summary of the tests performed to estimate and analyze the PSO-enabled FL architecture is provided in this section. In the following area, the performance of PSO-enabled FL on the ship dataset is studied.

4.1   Performance of the Proposed Approach

In this paragraph, the training and performance evaluation of the model is evaluated. Using samples from the ship dataset, the server model is initially trained. Clients are then given access to the server model. Typically, we test the model's performance on 10 clients. The learning rate value that we chose was 0.01. The observations are selected at random for every client device in the dataset. Figure 3 shows the model's successful performance in each round and is compared with the performance of the simple FL algorithm.

4.2 Evaluation Metrics

Several metrics are used to evaluate the performance of different ship detection models on the dataset, including recall rate (R), precision rate (P), F1-score and specificity. Eqs. 2, 3, 4, and 5 are the formulas used to calculate precision, recall, F1-score and specificity, respectively.

$$\text{Precision} = \frac{\text{True Positives}}{\text{True Positives} + \text{False Positives}} \times 100 \tag{2}$$

$$\text{Recall} = \frac{\text{True Positives}}{\text{True Positives} + \text{False Negatives}} \times 100 \tag{3}$$

$$F\text{1-score} = \frac{2 \times \text{Precision} \times \text{Recall}}{\text{Precision} + \text{Recall}} \tag{4}$$

Algorithm 1. PSO-enabled FL algorithm

1: **function** SERVERAGGREGATION(η_N)
2:     initialize w_0, $gbid$, pb, gb,
3:     **for** every iteration $a = 1, 2, \ldots$ **do**
4:         **for** every parallel client t **do**
5:             $pb \leftarrow$ UPDATECLIENT(t, w_t^{gbid})
6:             **if** $gb > pb$ **then**
7:                 $gb \leftarrow pb$
8:                 $gbid \leftarrow t$
9:         $w_{a+1} \leftarrow$ GETAPTMODEL($gbid$)
10: **function** UPDATECLIENT(t, w_a^{gbid})
11:     initialize S, w,w^{pb}, α, d_1, d_2
12:     $\beta \leftarrow$ (divide p_t into batches each of size B)
13:     **for** every layer with weight $r = 1, 2, \ldots$ **do**
14:         $S_r \leftarrow \alpha.S_r + d_r.rd.(w^{pb} - S_r) + d_2.rd.(w_a^{gb} - S_r)$
15:     $w \leftarrow w + S$
16:     **for** every epoch of client i from 1 to G **do**
17:         **for** batch $b \in B$ **do**
18:             $w \leftarrow w - \eta \delta l(w; b)$
19:     **return** pb to the server
20: **function** GETAPTMODEL(gid)
21:     sends a request to the client (gbid)
22:     receive w from Client
23:     **return** w to the server

$$\text{Specificity} = \frac{\text{True Negatives}}{\text{True Negatives} + \text{False Positives}} \times 100 \qquad (5)$$

where TP denotes the number of positive samples that are correctly identified, FP denotes the number of false positive negative samples, and FN denotes the number of false positive samples. Table 3 refers to the evaluation metrics values.

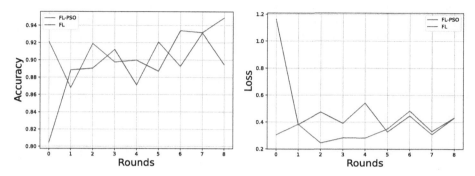

Fig. 3. Accuracy and Loss values of our approach and FL algorithm

4.3 Performance Comparison with Other Models

To evaluate the accuracy of the proposed model, we compare it with the recent state-of-the-art algorithms in prior works. These include the work in [6] which proposes a CNN model to classify ships and achieve an accuracy of 86%. The authors in [5] propose automated ship detection and category recognition from high-resolution aerial images. The SLC-HRoI Detection (SHD) model showed an accuracy of 81.5% while the Rotation Dense Feature Pyramid Network (RDFPN) and Sequence Local Context-HRoI Detection (SHD-HBB) had an accuracy of 76.2% and 69.5% respectively. The work in [9] used a micro aerial vehicle (MAV)

Table 3. Evaluation metrics

Round Number	F1 Score	Precision	Recall	Recall	Accuracy of FL
1	0.927	0.921	0.932	0.923	0.804
2	0.886	0.825	0.956	0.875	0.888
3	0.923	0.938	0.908	0.913	0.890
4	0.899	0.951	0.853	0.895	0.912
5	0.909	0.886	0.932	0.903	0.871
6	0.892	0.909	0.876	0.885	0.920
7	0.937	0.951	0.924	0.937	0.892
8	0.935	0.947	0.924	0.931	0.931
9	0.951	0.967	0.936	0.954	0.894

for the same purpose. In comparison to the earlier methods, our method achieved an accuracy of 94.88%.

PSO can optimize the decision-making process. In the context of detecting ships, PSO helps FL in making decisions while selecting the clients for aggregation. PSO also can deal with ambiguous or uncertain data like unclear satellite images or inaccurate location data. This hybrid method can provide a more robust solution that can work well under different conditions and variations. PSO-FL also can provide more accurate results because of its optimization capabilities and the way it handles uncertainties.

PSO is chosen over lightweight ML algorithms due to its robustness and adaptability. As a global optimization technique, PSO excels at handling problems with numerous local optima or those that are not differentiable or continuous, often stumbling blocks for traditional ML methods (Table 4).

Table 4. Accuracy comparison of previous ship detection models

Model	Accuracy
Sequence Local Context-HRoI Detection(SHD) [5]	69.5%
Rotation Dense Feature Pyramid Network(RDFPN) [5]	76.2%
SLC-HRoI Detection(SHD-HBB) [5]	81.5%
CNN [5]	86%
Micro aerial vehicle(MAV) [9]	92%
Federated Learning	93.1%
Proposed work	94.88%

5   Conclusion

This work presents a novel approach, PSO-FL, for ship detection in supply chain management. By using the benefits of FL and PSO, our proposed work addresses the challenges of ship detection in complex supply chain scenarios. Our framework achieves superior ship detection accuracy of 94.88% when compared to the traditional centralized learning approach and regular FL methods. By using the collaborative nature of FL and the optimization capabilities of PSO, the PSO-FL approach optimizes client selection and model aggregation. This duo helps the framework to identify the most relevant and informative data sources for model training which results in higher accuracy. The PSO-FL framework is efficient in handling the uncertainty and ambiguity in ship detection tasks, particularly when dealing with unclear satellite images or imprecise location data. By offering a practical and private ship detection method, this research contributes to the field of supply chain management. The supply chain's logistical planning, security, and risk management may all be improved with the use of the suggested PSO-FL framework. This work sets the path for further improvements in intelligent supply chain management systems by using distributed learning approaches and metaheuristic algorithms. Though PSO is effective in increasing the accuracy of prediction of ships, PSO has many disadvantages as well. PSO takes longer to calculate and suffers from early convergence in the beginning. When compared to larger datasets, PSO works effectively for smaller datasets. By substituting other naturally inspired algorithms for the PSO technique, our work can be made more effective. The future research objectives include investigating the use of PSO-FL in additional facets of supply chain management, such as cargo monitoring, anomaly detection, and predictive analytics. Additionally, examining the integration of other metaheuristic algorithms with FL can improve the effectiveness and performance of ship identification and associated supply chain operations. Overall, the PSO-FL framework presented in this paper demonstrates significant potential for improving ship detection capabilities and enhancing supply chain management processes. The utilization of distributed learning techniques combined with metaheuristic optimization contributes to the advancement of intelligent and efficient supply chain operations in an increasingly complex and interconnected world.

References

1. AbdulRahman, S., Tout, H., Mourad, A., Talhi, C.: Fedmccs: multicriteria client selection model for optimal IoT federated learning. IEEE Internet Things J. **8**(6), 4723–4735 (2020)
2. AbdulRahman, S., Tout, H., Ould-Slimane, H., Mourad, A., Talhi, C., Guizani, M.: A survey on federated learning: the journey from centralized to distributed on-site learning and beyond. IEEE Internet Things J. **8**(7), 5476–5497 (2020)
3. El Mekkaoui, S., Benabbou, L., Caron, S., Berrado, A.: Deep learning-based ship speed prediction for intelligent maritime traffic management. J. Marine Sci. Eng. **11**(1), 191 (2023)
4. Faustina, A., Aravindhan, E., et al.: Adapting multi-temporal information for optimized ship detection from SAR image dataset using transfer learning application. In: Handbook of Research on Advanced Practical Approaches to Deepfake Detection and Applications, pp. 275–287. IGI Global (2023)
5. Feng, Y., Diao, W., Sun, X., Yan, M., Gao, X.: Towards automated ship detection and category recognition from high-resolution aerial images. Remote Sens. **11**(16), 1901 (2019)
6. Gallego, A.J., Pertusa, A., Gil, P.: Automatic ship classification from optical aerial images with convolutional neural networks. Remote Sens. **10**(4), 511 (2018)
7. Kandati, D.R., Gadekallu, T.R.: Federated learning approach for early detection of chest lesion caused by Covid-19 infection using particle swarm optimization. Electronics **12**(3), 710 (2023)
8. Karmaker, C.L., et al.: Industry 5.0 challenges for post-pandemic supply chain sustainability in an emerging economy. Int. J. Prod. Econ. **258**, 108806 (2023)
9. Lan, J., Wan, L.: Automatic ship target classification based on aerial images. In: International Conference on Optical Instruments and Technology: Optical Systems and Optoelectronic Instruments. vol. 7156, pp. 316–325. SPIE (2009)
10. Li, Y., Wang, X., Zeng, R., Donta, P.K., Murturi, I., Huang, M., Dustdar, S.: Federated domain generalization: a survey. arXiv preprint arXiv:2306.01334 (2023)

11. Maddikunta, P.K.R., et al.: Industry 5.0: a survey on enabling technologies and potential applications. J. Indust. Inform. Integr. **26**, 100257 (2022)
12. Minyard, E., Kolawole, S., Saxena, N.: Evolutionary federated learning using particle swarm optimization (2023)
13. Nayeri, S., Sazvar, Z., Heydari, J.: Towards a responsive supply chain based on the industry 5.0 dimensions: a novel decision-making method. Expert Syst. Appl. **213**, 119267 (2023)
14. Rahman, S.A., Tout, H., Talhi, C., Mourad, A.: Internet of things intrusion detection: centralized, on-device, or federated learning? IEEE Netw. **34**(6), 310–317 (2020)
15. Song, Y., Dong, G.: Federated target recognition for multi-radar sensor data security. IEEE Trans. Geosci. Remote Sens. (2023)
16. Supriya, Y., Gadekallu, T.R.: Particle swarm-based federated learning approach for early detection of forest fires. Sustainability **15**(2), 964 (2023)
17. Teng, L., et al.: Flpk-bisenet: federated learning based on priori knowledge and bilateral segmentation network for image edge extraction. IEEE Trans. Netw. Serv. Manage. (2023)
18. Wahab, O.A., Mourad, A., Otrok, H., Taleb, T.: Federated machine learning: survey, multi-level classification, desirable criteria and future directions in communication and networking systems. IEEE Commun. Surv. Tutorials **23**(2), 1342–1397 (2021)
19. Wang, N., Le, J., Li, W., Jiao, L., Li, Z., Zeng, K.: Privacy protection and efficient incumbent detection in spectrum sharing based on federated learning. In: 2020 IEEE Conference on Communications and Network Security (CNS), pp. 1–9. IEEE (2020)
20. Wang, Z., Wang, R., Ai, J., Zou, H., Li, J.: Global and local context-aware ship detector for high-resolution SAR images. IEEE Trans. Aerospace Electron. Syst. (2023)
21. Xie, X., Li, B., Wei, X.: Ship detection in multispectral satellite images under complex environment. Remote Sens. **12**(5) (2020). https://doi.org/10.3390/rs12050792,https://www.mdpi.com/2072-4292/12/5/792
22. Yasir, M., et al.: Ship detection based on deep learning using SAR imagery: a systematic literature review. Soft. Comput. **27**(1), 63–84 (2023)
23. Zhang, Z., Guan, C., Chen, H., Yang, X., Gong, W., Yang, A.: Adaptive privacy-preserving federated learning for fault diagnosis in internet of ships. IEEE Internet Things J. **9**(9), 6844–6854 (2021)
24. Zhao, P., Ren, Y., Xiao, H.: An image-based ship detector with deep learning algorithms. In: Encyclopedia of Data Science and Machine Learning, pp. 2540–2552. IGI Global (2023)

Federated Learning Using the Particle Swarm Optimization Model for the Early Detection of COVID-19

K. Dasaradharami Reddy[1] , Gautam Srivastava[2(✉)] , Yaodong Zhu[7] ,
Y. Supriya[1] , Gokul Yenduri[1] , Nancy Victor[1] , S. Anusha[3] ,
and Thippa Reddy Gadekallu[1,4,5,6,7]

[1] School of Information Technology and Engineering, Vellore Institute of Technology,
Vellore, Tamil Nadu, India
{dasaradharami.k,gokul.yenduri,thippareddy.g}@vit.ac.in,
supriya.2020@vitstudent.ac.in, drnancyvictor@ieee.org
[2] Department of Mathematics and Computer Science, Brandon University,
Brandon, Canada
srivastavag@brandonu.ca
[3] Department of Computer Science and Engineering, NBKR Institute of Science and
Technology, Vidyanagar, Tirupathi, Andhra Pradesh, India
[4] Zhongda Group, Haiyan County, Jiaxing, Zhejiang, China
[5] Department of Electrical and Computer Engineering,
Lebanese American University, Byblos, Lebanon
[6] Research and Development, Lovely Professional University, Phagwara, India
[7] College of Information Science and Engineering, Jiaxing University,
Jiaxing 314001, China

Abstract. The COVID-19 pandemic has created significant global
health and socioeconomic challenges, which creates the need for efficient and effective early detection methods. Several traditional machine
learning (ML) and deep learning (DL) approaches have been used in
the detection of COVID-19. However, ML and DL strategies face challenges like transmission delays, a lack of computing power, communication delays, and privacy concerns. Federated Learning (FL), has emerged
as a promising method for training models on decentralized data while
ensuring privacy. In this paper, we present a novel FL framework for early
detection of COVID-19 using the Particle Swarm Optimization (PSO)
model. The proposed framework uses advantages of both FL and PSO.
By employing the PSO technique the model aims to achieve faster convergence and improved performance. In order to validate the effectiveness
of the proposed approach, we performed experiments using a COVID-19
image dataset which was collected from different healthcare institutions.
The results indicate that our approach is more effective as it achieves a
higher accuracy rate of 94.36% which is higher when compared to traditional centralized learning approaches. Furthermore, the FL framework

B. Luo et al. (Eds.): ICONIP 2023, CCIS 1962, pp. 425–436, 2024.
https://doi.org/10.1007/978-981-99-8132-8_32

ensures data privacy and security by keeping sensitive patient information decentralized and only sharing aggregated model updates during the training process.

Keywords: Federated Learning · Particle Swarm Optimization · COVID-19 detection

1   Introduction

The rapid spread of COVID-19 has a significant impact on human health and life. The COVID-19 pandemic, now affecting millions of people around the world, has been ranked among the most catastrophic health problems in the last decade [13]. The COVID-19 virus produces respiratory particles that enter the air when an infected person sneezes, coughs, or speaks. The earliest possible detection of COVID-19 is critical to limiting its potential spread. Detecting COVID-19 outbreaks rapidly and accurately is becoming increasingly important. The World Health Organization (WHO) declared COVID-19 an acute public health emergency due to the severity of its outbreak. Many people were affected by this deadly outbreak in different ways. The COVID-19 virus has been detected globally through several tests [6]. Current methods for detecting COVID-19 infection in the human body are as follows:

- 3D images are captured by computed tomography scans (CT scans) to detect COVID-19.
- Reverse transcription polymerase chain reaction (RT-PCR) is a useful tool for detecting infectious Ribonucleic acid (RNA) from nasal swabs.
- CT scan machines require more equipment, while chest X-rays (CXR) require less and are easier to transport. Furthermore, CXR tests are fast, taking about 15 seconds per individual.

However, CT scans are not available in hospitals or many health centers. Furthermore, RT-PCR tests cannot be used in most hospitals to diagnose COVID-19 infections and this requires a lot of time. Due to its ease of availability, reliability, and accuracy, we used the COVID-19 infection chest X-ray image dataset in this study.

Medical professionals can diagnose COVID-19 infection as early as possible if they possess enough epidemiological data, including protein composition (mainly RNA), serological findings, and diagnostic tests. Imaging data can also be used to determine the clinical significance of COVID-19 infections [26]. A drawback of healthcare and epidemic management is the difficulty in diagnosing COVID-19 infection early in clinical specimens. Early detection of diseases can prevent their spread [5].

Data scientists and artificial intelligence (AI) engineers can detect COVID-19 infections. COVID-19 infection can be predicted based on chest X-rays using

Deep Learning (DL) models [32]. A prediction model based on past instances can be developed using AI and DL which prevent the spread of disease further [30]. Therefore, Machine Learning (ML) models are required for detecting patients with COVID-19 infection or predicting their future transmission. Developing an accurate model would be impossible without sufficient patient data, but patient information is considered to be sensitive [19]. Therefore, it is necessary to create a model that reliably predicts outcomes without disclosing individual patient information.

Federated Learning (FL) was first introduced to the world by Google in 2016. By using FL, we can build ML models from different datasets without disclosing any personal information [22]. Moreover, FL enables the client and server to communicate more efficiently, thus reducing communication costs [20]. There is less time spent waiting for the information to be delivered and received since client-side data is not forwarded to the server for training [8]. In addition to making use of enormous quantities of information locally, FL is also capable of doing precisely the same thing virtually [18]. However, in FL, time spent communicating takes longer than time spent computing. Reduced network connection time is necessary to increase FL's efficiency. Research in recent years [24,25] has shown that FL may be used to reliably detect COVID-19 infection from chest X-ray images. However, earlier investigations used FL's default setup, which proves unsuccessful when client input is unclear and requires a lot of processing power to distribute updated models.

The recommended research uses Particle Swarm Optimization (PSO), a technique that quickly updates models by using a distributed algorithm to determine the optimal solution [9]. Because PSO finds the best solution through a random process, many repetitions are necessary. PSO succeeds in adaptable and complicated environments like FL [14]. This motivates us to develop a novel approach, that involves the PSO in FL.

1.1 Key Contributions

Our work has made the following key contributions:

- Proposal of a novel FPS optimization approach: Using FL and PSO together for chest X-ray (COVID-19 and Pneumonia) detection, this paper introduces a novel technique.
- Enhancement of chest X-rays (COVID-19 and Pneumonia) detection performance: Comparing the proposed approach with traditional centralized methods, it significantly improves the detection of COVID-19 and Pneumonia on chest X-rays. In this paper, experimental results and performance metrics are presented as evidence of the effectiveness of the FPS optimization approach.
- Consideration of data privacy and security: The paper addresses issues related to privacy and security through the use of an FL approach that enables training of local models on peripheral devices without sharing sensitive data.

2   Related Work

The following section briefly describes related work. In recent research, FL has been shown to improve client communication to a great extent. FL's unstable network environment causes many problems for mobile devices. There are several issues with this, including regular collapses, modifications to node groups, extra work for a centralized server, and significant latency improvements as the number of nodes grows in size. Communication between client and server in terms of data volume should be taken into account in specific circumstances. Recent studies [10,12,31] jointly select devices and design beamforming algorithms to solve the bandwidth bottleneck problem. Conventional FL methods encounter numerous difficulties, including limited capacity and resources in Internet of Things (IoT) networks [29], high communication overhead when transferring local updates to the server in more extensive networks [2], and occasionally being vulnerable to certain attacks like Byzantine attacks [1].

Recently, a significant amount of effort has been made to build algorithms using DL for the diagnosis of COVID-19 disease from chest X-ray images [3, 17]. Chowdhury et al. [7] and Narin et al. [21] developed convolutional neural networks (CNN)-based models for detecting COVID-19 patients from chest X-ray images.

The PSO algorithm simulates the behavior of animals swarming to solve convex and differentiable optimization problems [9]. Some recent efforts to improve ML performance have made use of PSO concepts. The PSO technique is used when Convolutional Neural Networks (CNN) are used to improve image recognition and categorization accuracy [27]. PSO and FL can be integrated to improve FL's performance [23]. In FL, [23] data is learned, and in the PSO, the optimal hyperparameters are determined.

A client model is selected using PSO for each round of global model updates. Fine-tuning FL parameters using the PSO algorithm maximizes its performance.

Supriya et al. [28] present an extensive approach based on FL and a PSO algorithm that enables researchers to respond promptly to forest fires. Kandati et al. [14] suggested an FPS optimization algorithm that employs PSO to improve FL's performance and decrease the quantity of data the client transmits to the server. Previous efforts to improve FL's efficiency have focused on increasing communication with clients and optimizing its global operations. Our primary goal is to improve FL's performance through the usage of PSO-based client-server communication.

3   Proposed Work

There is a need to detect COVID-19 infection as early as possible. COVID-19 infection can be detected through different AI and DL models. However, we need a model which detects infection as early as possible without compromising privacy of patient data. This type of model is possible through the combination of FL and PSO. Adding more layers to a CNN model can improve its accuracy. Deep neural networks are used for this purpose. Weight parameters are influenced by layer depth in terms of training effort. The cost of transmitting the client-learned model to the server rises as the network distance between the devices increases. Consequently, we offer an integrated approach that uses PSO characteristics to relocate the trained model with the best score (for instance, accuracy or loss), no matter the model's size.

FPS Optimization, the proposed model, only allows model weights from the top-scoring client rather than from all clients. The process is illustrated in Fig. 1. Client training is targeted in such a way that concludes the client attaining the lowest possible loss value, resulting in the highest score. Only 4 bytes are used in this case to represent the loss value. By applying pb and gb to a weighted array, FPS Optimization determines the optimal model and then sets S for each of its weighted array members.

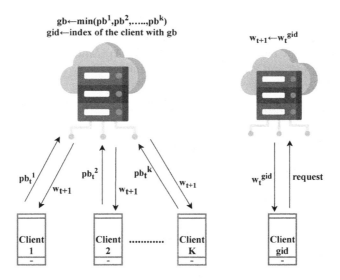

Fig. 1. FPS Optimization Process

The FPS optimization weights are updated using Eq. 1. The weight update appears as follows:

$$S_l^t = \alpha.S_l^{t-1} + c_1.rand_1.(pb - Sl^{t-1}) + c_2.rand_2.(gb - S_l^{t-1})$$
$$w_i^t = w_i^{t-1} + S^t \qquad (1)$$

Each layer of CNN has a weight S, which is expressed by Eq. 1. The current step weight w^t is determined by adding S to the w^{t-1} prior step weight. In Eq. 1, α represents the inertia weight, c_1 denotes the constant of acceleration for pb, and c_2 denotes the constant of acceleration for gb. $Rand_1$ and $Rand_2$ are randomly generated numbers between 0 and 1.

Based on Algorithm 1, the proposed framework aims to solve the problem. The FPS optimization given in Algorithm 1, is an enhancement of the FedAvg Algorithm. According to ServerAggregation, client pb values are only accepted on line 5, in contrast to traditional methods. Initializing variables w_0, gid, pb, gb is part of the ServerAggregation function. The pb value is updated using the UpdateClient function. It compares w^{gb} with pb, and w^{gb} is updated with the new pb value, while gid is updated with K, which is the client id. Among the available samples, lines 6–8 select the client with the lowest loss value (pb). The UpdateClient function is performed by CNN through the PSO. The variables S, w,w^{pb}, α, c_1, c_2 are initialized in the UpdateClient function. The batch sizes are divided by a constant called β. The particle velocity is updated for each layer. As the batch size increases, the weights are also updated. Lines 13–14 compute variable S along with the user-stored best value of w^{pb} and the server-received w^{gb} value. The procedure is repeated for each successive layer weight. To calculate the value of w for the current iteration, Line 15 adds variable S to w in the last iteration. Training should be continued through lines 16–18 until the client epoch E is reached. In lines 20–23, GetAptModel inquires about the server to retrieve the most effective client model. GetAptModel sends the request to the client using gid and receives w as confirmation.

Algorithm 1. Federated PSO algorithm

1: **function** SERVERAGGREGATION(η_N)
2:       initialize w_0, gid, pb, gb,
3:       **for** every iteration $t = 1, 2, \ldots$ **do**
4:             **for** every parallel client k **do**
5:                   $pb \leftarrow$ UPDATECLIENT(k, w_t^{gid})
6:                   **if** $gb > pb$ **then**
7:                         $gb \leftarrow pb$
8:                         $gid \leftarrow k$
9:                   $w_{t+1} \leftarrow$ GETAPTMODEL(gid)
10: **function** UPDATECLIENT(k, w_t^{gid})
11:       initialize S, w,w^{pb}, α, c_1, c_2
12:       $\beta \leftarrow$ (divide p_k into batches each of size B)
13:       **for** every layer with weight $l = 1, 2, \ldots$ **do**
14:             $S_l \leftarrow \alpha.S_l + c_1.rand.(w^{pb} - S_l) + c_2.rand.(w_t^{gb} - S_l)$
15:       $w \leftarrow w + S$
16:       **for** every epoch of client i from 1 to E **do**
17:             **for** batch $b \in B$ **do**
18:                   $w \leftarrow w - \eta \delta l(w; b)$
19:       **return** pb to the server
20: **function** GETAPTMODEL(gid)
21:       request Client(gid)
22:       will be acknowledged w from Client
23:       **return** w to the server

3.1   Dataset Collection and Preparation

The dataset used in this work is taken from the Kaggle repository[1].

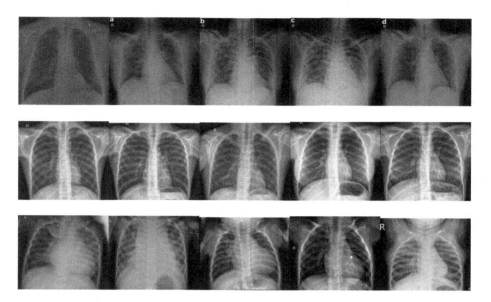

Fig. 2. Sample images from the chest X-ray (covid-19 and Pneumonia) dataset. The images are divided into three categories: Covid, Normal, and Pneumonia

The dataset contains 6432 images. The training dataset folder has 5144 images and the testing dataset folder has 1288 images.

The images in each folder belong to three categories namely:

– Covid19
– Normal
– Pneumonia

The purpose of this study is to use PSO to detect the onset of a chest X-ray (COVID-19 and Pneumonia). Figure 2 shows the sample images from the chest X-ray (COVID-19 and Pneumonia) dataset.

3.2   Dataset Pre-processing

Images from various perspectives are included in the dataset. A dataset of this kind can be used to fine-tune a model to distinguish between images associated with COVID-19, Normal, and Pneumonia. To improve training and results, it is necessary to clean the dataset. Therefore, we pre-processed the images so

---

[1] https://www.kaggle.com/datasets/prashant268/chest-xray-covid19-Pneumonia.

that only content related to the identified problem was included in the images. Cropping and scaling were then performed on each image to maintain a constant size of 150 × 150 pixels. Chest X-rays are structured into testing and training directories. We used 80% of the data for training and 20% for testing.

4   Results and Analysis

Experimental results for FPS optimization are summarized in this section. Finally, we will evaluate the performance of FPS optimization and standard FL on the chest X-ray (COVID-19 and Pneumonia) dataset.

4.1   Performance of the Proposed Approach

The purpose of this subsection is to assess the model's effectiveness and approach. To train the server-side model, we begin with data from the chest X-ray (COVID-19 and Pneumonia) dataset. Based on the client ratio, servers are assigned to clients. The study we performed randomly allocated 10 clients. In our case, the learning rate was set at 0.01. Data accuracy is ensured by randomly selecting observations for each client device in the dataset. The model's outcome in comparison to each iteration is shown in Fig. 3. Test findings for the chest X-ray (COVID-19 and Pneumonia) dataset are shown in Fig. 3. Each of these graphs is designed based on accuracy of the test. FPS optimization outperformed in all 30 epochs with greater accuracy (94.36%). Data is transmitted from the server to the client in parallel as volume increases, which leads to higher accuracy when C is higher. However, FPS optimization takes fewer iterations to converge.

4.2   Evaluation Metrics

Several metrics are used to evaluate the performance of different COVID-19 detection models on the dataset, including recall rate (R), precision rate (P), and F1-score:

$$\text{Recall} = \frac{TP}{TP + FN}$$

$$\text{Precision} = \frac{TP}{TP + FP}$$

$$F1\text{-score} = \frac{2 \times \text{Precision} \times \text{Recall}}{\text{Precision} + \text{Recall}}$$

where TP denotes the number of positive samples that are correctly identified, FP denotes the number of false positive negative samples, and FN denotes the number of false positive samples, respectively. Table 1 refers to the evaluation metrics values.

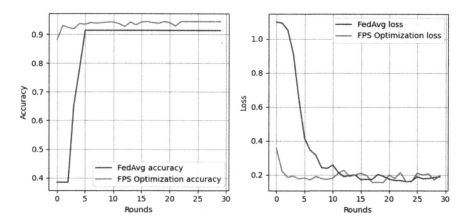

Fig. 3. Comparison of the FPS Optimization and FedAvg Accuracy and Loss Results for the chest X-ray (COVID-19 and Pneumonia) dataset

Table 1. F1-score, Precision and Recall values

Round Number	Accuracy	Precision	Recall	F1-score
5	0.9359	0.9352	0.9356	0.9355
10	0.9417	0.9410	0.9414	0.9413
15	0.9320	0.9313	0.9317	0.9316
20	0.9436	0.9429	0.9433	0.9432
25	0.9320	0.9313	0.9317	0.9316
30	0.9436	0.9429	0.9433	0.9432

Table 2. Comparison of the proposed model with recent studies

Work	Accuracy	Dataset	Method
Khaloufi et al. [16]	79%	COVID-19 symptom	ANN is an AI-enabled framework that uses a smartphone to diagnose a chest lesion caused by COVID-19 infection
Horry et al. [11]	86%	X-ray, ultrasound, CT scan	VGG19
Ayan et al. [4]	87%	Chest X-Ray images (Pneumonia)	VGG16
Kermany et al. [15]	92.8%	Chest X-Ray images (Pneumonia)	Identifying medical diagnoses and treating diseases with image-based deep learning
Proposed work	94.36%	chest X-ray (COVID-19 and Pneumonia) dataset	FPS Optimization

4.3 Performance Comparison with Other Models

Using the state-of-the-art algorithms in prior works, we evaluate the accuracy of the proposed model. The authors in [16] present a robust framework for early detection of COVID-19 using smartphone-embedded sensors that utilize advanced ML techniques and data analytics techniques. The authors in [11] demonstrate that COVID-19 can be detected by DL models by using X-ray images. Table 2 shows a comparison of the methodologies. Compared to more traditional FL methods, FPS Optimization is a significant improvement. COVID-19 hyperparameters are more easily refined using our method since it is more robust. This simplifies the diagnosis process. Our technology guarantees that FL performance can be optimized by enhancing the client model's measurements before sending them to the server.

5 Conclusion

In this work, we presented a novel FL framework that utilizes PSO for the early detection of COVID-19. The proposed framework uses the benefits of both FL and PSO. One of the significant advantages of our framework is its ability to address privacy concerns associated with centralized data storage and processing. By keeping sensitive patient information decentralized and sharing only aggregated model updates during training, our approach ensures data privacy while maintaining the benefits of collaborative learning. This aspect is crucial for maintaining public trust and facilitating large-scale deployment of intelligent healthcare systems. Our experimental results demonstrated that the recommended approach performed better than existing methods, with predictions on the COVID-19 infection dataset showing a 94.36% accuracy rate. PSO takes longer to calculate and suffers from early convergence in the beginning. When compared to bigger datasets, PSO is more suited for smaller datasets. PSO has trouble with bigger datasets because of its slow convergence time and wide search space. The PSO algorithm can be replaced with other nature-inspired algorithms to improve the effectiveness of our work.

Acknowledgements. This work is supported by Zhejiang Key Research and Development Project under Grant 2017C01043.

References

1. Agrawal, S., Chowdhuri, A., Sarkar, S., Selvanambi, R., Gadekallu, T.R.: Temporal weighted averaging for asynchronous federated intrusion detection systems. Comput. Intell. Neurosci. **2021**, 5844728 (2021)
2. Agrawal, S., et al.: Federated learning for intrusion detection system: concepts, challenges and future directions. Comput. Commun. **195**, 346–361 (2022)
3. AlMohimeed, A., Saleh, H., El-Rashidy, N., Saad, R.M., El-Sappagh, S., Mostafa, S.: Diagnosis of Covid-19 using chest X-ray images and disease symptoms based on stacking ensemble deep learning. Diagnostics **13**(11), 1968 (2023)

4. Ayan, E., Ünver, H.M.: Diagnosis of pneumonia from chest X-ray images using deep learning. In: 2019 Scientific Meeting on Electrical-Electronics & Biomedical Engineering and Computer Science (EBBT), pp. 1–5. IEEE (2019)
5. Bhapkar, H., Mahalle, P.N., Dey, N., Santosh, K.: Revisited COVID-19 mortality and recovery rates: are we missing recovery time period? J. Med. Syst. **44**(12), 1–5 (2020)
6. Chowdhury, D., et al.: Federated learning based Covid-19 detection. Expert Syst. **40**, e13173 (2022)
7. Chowdhury, M.E., et al.: Can AI help in screening viral and COVID-19 pneumonia? IEEE Access **8**, 132665–132676 (2020)
8. Dasaradharami Reddy, K., Gadekallu, T.R., et al.: A comprehensive survey on federated learning techniques for healthcare informatics. Comput. Intell. Neurosci. **2023**, 8393990 (2023)
9. Eberhart, R., Kennedy, J.: A new optimizer using particle swarm theory. In: MHS 1995. Proceedings of the Sixth International Symposium on Micro Machine and Human Science, pp. 39–43. IEEE (1995)
10. Fan, X., Wang, Y., Huo, Y., Tian, Z.: Joint optimization of communications and federated learning over the air. IEEE Trans. Wireless Commun. **21**, 4434–4449 (2021)
11. Horry, M.J., et al.: COVID-19 detection through transfer learning using multimodal imaging data. IEEE Access **8**, 149808–149824 (2020)
12. Jing, S., Xiao, C.: Federated learning via over-the-air computation with statistical channel state information. IEEE Trans. Wireless Commun. **21**, 9351–9365 (2022)
13. Kandati, D.R., Gadekallu, T.R.: Genetic clustered federated learning for COVID-19 detection. Electronics **11**(17), 2714 (2022). https://doi.org/10.3390/electronics11172714, www.mdpi.com/2079-9292/11/17/2714
14. Kandati, D.R., Gadekallu, T.R.: Federated learning approach for early detection of chest lesion caused by COVID-19 infection using particle swarm optimization. Electronics **12**(3), 710 (2023)
15. Kermany, D.S., et al.: Identifying medical diagnoses and treatable diseases by image-based deep learning. Cell **172**(5), 1122–1131 (2018)
16. Khaloufi, H., et al.: Deep learning based early detection framework for preliminary diagnosis of COVID-19 via onboard smartphone sensors. Sensors **21**(20), 6853 (2021)
17. Khan, I.U., Aslam, N.: A deep-learning-based framework for automated diagnosis of COVID-19 using X-ray images. Information **11**(9), 419 (2020)
18. Konečnỳ, J., McMahan, B., Ramage, D.: Federated optimization: distributed optimization beyond the datacenter. arXiv:1511.03575 (2015)
19. Liu, B., Yan, B., Zhou, Y., Yang, Y., Zhang, Y.: Experiments of federated learning for COVID-19 chest X-ray images. arXiv:2007.05592 (2020)
20. McMahan, B., Moore, E., Ramage, D., Hampson, S., y Arcas, B.A.: Communication-efficient learning of deep networks from decentralized data. In: Artificial Intelligence and Statistics, pp. 1273–1282. PMLR (2017)
21. Narin, A., Kaya, C., Pamuk, Z.: Automatic detection of coronavirus disease (COVID-19) using X-ray images and deep convolutional neural networks. Pattern Anal. Appl. **24**, 1207–1220 (2021)
22. Pandya, S., et al.: Federated learning for smart cities: a comprehensive survey. Sustain. Energy Technol. Assess. **55**, 102987 (2023)
23. Qolomany, B., Ahmad, K., Al-Fuqaha, A., Qadir, J.: Particle swarm optimized federated learning for industrial IoT and smart city services. In: GLOBECOM 2020–2020 IEEE Global Communications Conference, pp. 1–6. IEEE (2020)

24. Rahman, M.A., Hossain, M.S., Islam, M.S., Alrajeh, N.A., Muhammad, G.: Secure and provenance enhanced internet of health things framework: a blockchain managed federated learning approach. IEEE Access **8**, 205071–205087 (2020)
25. Rieke, N., et al.: The future of digital health with federated learning. NPJ Digit. Med. **3**(1), 1–7 (2020)
26. Santosh, K.: AI-driven tools for coronavirus outbreak: need of active learning and cross-population train/test models on multitudinal/multimodal data. J. Med. Syst. **44**(5), 1–5 (2020)
27. Serizawa, T., Fujita, H.: Optimization of convolutional neural network using the linearly decreasing weight particle swarm optimization. arXiv:2001.05670 (2020)
28. Supriya, Y., Gadekallu, T.R.: Particle swarm-based federated learning approach for early detection of forest fires. Sustainability **15**(2), 964 (2023)
29. Taheri, R., Shojafar, M., Alazab, M., Tafazolli, R.: Fed-IIoT: a robust federated malware detection architecture in industrial IoT. IEEE Trans. Industr. Inf. **17**(12), 8442–8452 (2020)
30. Wieczorek, M., Siłka, J., Woźniak, M.: Neural network powered COVID-19 spread forecasting model. Chaos, Solitons Fractals **140**, 110203 (2020)
31. Yang, K., Jiang, T., Shi, Y., Ding, Z.: Federated learning based on over-the-air computation. In: ICC 2019–2019 IEEE International Conference on Communications (ICC), pp. 1–6. IEEE (2019)
32. Zhang, W., et al.: Dynamic-fusion-based federated learning for COVID-19 detection. IEEE Internet Things J. **8**(21), 15884–15891 (2021)

Domain Generalization via Implicit Domain Augmentation

Zhijie Rao[1,3], Qi Dong[1,3], Chaoqi Chen[2], Yue Huang[1,3(✉)], and Xinghao Ding[1,3]

[1] School of Informatics, Xiamen University, Xiamen, China
huangyue05@gmail.com
[2] The University of Hong Kong, Hong Kong, China
[3] Institute of Artificial Intelligent, Xiamen University, Xiamen, China

Abstract. Deep convolutional neural networks often suffer significant performance degradation when deployed to an unknown domain. To tackle this problem, domain generalization (DG) aims to generalize the model learned from source domains to an unseen target domain. Prior work mostly focused on obtaining robust cross-domain feature representations, but neglecting the generalization ability of the classifier. In this paper, we propose a novel approach named *Implicit Domain Augmentation* (IDA) for classifier regularization. Our motivation is to prompt the classifier to see more diverse domains and thus become more knowledgeable. Specifically, the styles of samples will be transferred and re-equipped to original features. To obtain the direction of meaningful style transfer, we use the multivariate normal distribution to model the feature statistics. Then new styles are sampled from the distribution to simulate potential unknown domains. To efficiently implement IDA, we achieve domain augmentation implicitly by minimizing an upper bound of expected cross-entropy loss on the augmented feature set instead of generating new samples explicitly. As a plug-and-play technique, IDA can be easily applied to other DG methods and boost the performance, introducing negligible computational overhead. Experiments on several tasks demonstrate the effectiveness of our method.

Keywords: Domain generalization · Classifier regularization · Robust loss function · Transfer learning

1 Introduction

Despite the excellent achievement of convolutional neural networks in various vision tasks, severe performance degradation occurs when encountering domain shift phenomenon, which is caused by differences in style, lighting, texture, etc.

The work was supported in part by the National Natural Science Foundation of China under Grant 82172033, U19B2031, 61971369, 52105126, 82272071, 62271430, and the Fundamental Research Funds for the Central Universities 20720230104.

B. Luo et al. (Eds.): ICONIP 2023, CCIS 1962, pp. 437–448, 2024.
https://doi.org/10.1007/978-981-99-8132-8_33

To address this issue, domain generalization (DG) learns transferable knowledge on the source domains and thus generalizes to an unseen target domain.

Most previous DG studies have focused on extracting robust feature representations from the source domains to cope with domain perturbations. Their methods are broadly categorized as domain augmentation [1,8,9,14,15] and learning strategies [3,11,16]. The former aims at making the model automatically summarize the essential laws of the objects from a large amount of augmented data. The latter forces the model to learn domain-invariant features through special learning strategies such as adversarial learning, meta-learning and self-supervised learning.

Although their approaches achieved promising results, they did not pay enough attention to the generalization needs of the classifier. A recent study [18] pointed out that tuning the classifier can effectively improve the accuracy of out-of-distribution data in the case of well-trained features. This implies that the classifier trained only on the source domains is not sufficient for out-of-domain data. To enhance the robustness of the classifier, Li et al. [2] design multiple independent networks and introduce interactive training to eliminate the effect of domain shift. Dou et al. [12] use meta-learning and constrain label distribution consistency to strengthen the classifier. However, their methods require complex network modules and introduce significant additional computational costs, which hinder their practicality and versatility.

In this paper, we present a novel method named *Implicit Domain Augmentation* (IDA) to mitigate the aforementioned problems. The basic idea of IDA is to expand the perceptual horizon of the classifier by increasing the diversity of source domains. To instantiate this idea, we perform style transfer on the features to simulate the latent unknown domains. It has been known that feature statistics, i.e., mean and standard deviation, represent the style information of an image [19]. Therefore, style transfer is accomplished by changing the feature statistics. We model the feature statistics with multivariate Gaussian distribution to reveal the direction of meaningful change. Meanwhile, we adopt a memory bank to estimate the variance of the distribution more accurately. Then new styles are sampled from the distribution and equipped with original features to augment the source domains. To avoid explicitly generating new samples, we train the network by minimizing an upper bound of expected cross-entropy loss on the augmented feature set. IDA is essentially a more robust loss function, which works in a plug-and-play manner.

The primary contributions of this paper are:

- We investigate the DG problem explicitly focusing on generalization ability of the classifier, which has not been fully explored yet.
- A novel DG method named *Implicit Domain Augmentation* (IDA) enforces cross-domain robustness by adding domain-disturbed regularization to classifier.
- The proposed IDA is a plug-and-play method. It can be easily applied to any other DG methods, which introduces negligible extra computational cost.

– The experimental results on category classification and instance retrieval verify the effectiveness of the proposed IDA.

2   Related Work

To cope with the degradation of model performance due to domain shift, domain generalization learns a sufficiently robust model on several source domains to generalize to an invisible target domain. Over the past decade, a plethora of approaches have been developed. The core ideas of these approaches can be summarized into two categories, 1) learning domain invariant features; 2) classifier regularization.

Learning domain invariant features aims at making the model to extract features that contain more essential information about the object and less redundant information that is not useful for recognition. The existing methods are mainly divided into domain augmentation [1,7–9,14,15] and learning strategies [3,4,10,11,16,17].

Domain Augmentation. Domain augmentation is a plain method that hopes the model automatically summarize the essential features of the object from the huge amount of data. To achieve this goal, Zhou et al. [7] design a generative network and force the model to output image-level samples that are distinct from the source domains by adversarial training. Yue et al. [8] use an auxiliary dataset to randomize the visual appearance style of the source domain images. Then pyramidal learning is introduced to enhance domain consistency. In contrast, Zhou et al. [1] resorts to feature-level augmentation to explore potential domains by performing mixup on the feature statistics. Further, Li et al. [9] extends the domain diversity using Gaussian modeling.

Our approach is also based on the motivation of domain augmentation. However, the difference is that IDA is designed for classifier regularization and does not require the generation of explicit augmented samples, avoiding high additional computational costs.

Learning Strategies. Many studies resort to special learning strategies, such as adversarial learning, meta-learning, to guide the model to learn domain invariant features. For example, Matsuura et al. [4] assigns pseudo-domain labels to samples by unsupervised clustering and then utilizes adversarial learning to extract domain-invariant features. Motiian et al. [10] introduces contrast learning to shorten the intra-class distance, enabling tighter semantic feature distribution. Balaji et al. [3] proposes a new meta-learning based regularization method. Carlucci et al. [11] introduce a self-supervised learning task, jigsaw puzzles, to help the model understand the picture content.

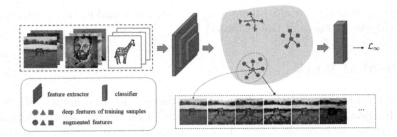

Fig. 1. An overview of IDA. The network consists of a feature extractor and a linear classifier. After the feature extractor maps images into deep feature space, we transfer the styles of the features to augment the source domains. The potential new samples form an augmented set. To avoid extra computational costs, the network is trained by minimizing an upper bound of expected cross-entropy loss on the augmented set.

Although feature learning effectively mitigates the problem of domain shift, the weak classifier may limit the performance. Classifier regularization aims to train a robust classifier, preventing it from overfitting on the source domains.

Consistency Constraints. Many studies regularize the classifier by constraining the consistency of output for different domains' samples. To name a few, Dou et al. [12] leverage the KL divergence to align the softmax output of similar samples from different source domains. Li et al. [2] design separate feature extractors and classifiers for each domain, and then reduces the sensitivity of the classifiers to domain perturbations by interactive training. In addition, other works [3,6,13] also use consistency constraint techniques in their methods.

Despite the good results of their methods, customized auxiliary network modules and complex learning processes hinder their practicality and versatility. Instead, our method is a plug-and-play regularization method that driven by domain augmentation. Simply replacing the cross entropy loss with the proposed IDA can facilitate the model performance.

3 Methodology

An overview of our method is depicted in Fig. 1. The network is decomposed into a feature extractor and a linear classifier. In the deep feature space, the styles of features are transferred to simulate unseen domains. The samples with new styles form an augmented set. Instead of explicitly generate samples, we optimize the network by minimizing an upper bound of expected cross-entropy loss on the augmented set.

3.1 Preliminaries

A deep neural network consists of a feature extractor F_Ψ and a linear classifier T_Θ, where Ψ and Θ are their parameters respectively. The $x \in \mathbb{R}^{L \times H \times W}$ is the feature extracted by F_Ψ, with L, H, W representing the dimension of the channel, height and width. The style of an image can be quantified as the feature statistics. The formula is as follows:

$$\mu(x) = \frac{1}{HW} \sum_{h=1}^{H} \sum_{w=1}^{W} x, \tag{1}$$

$$\sigma(x) = \sqrt{\frac{1}{HW} \sum_{h=1}^{H} \sum_{w=1}^{W} (x - \mu(x))^2}, \tag{2}$$

where $\mu(x), \sigma(x)$ denote channel-wise mean and standard deviation respectively. AdaIN [19] achieved arbitrary style transfer by assigning the feature statistics of instance y to instance x:

$$\text{AdaIN}(x) = \sigma(y) \frac{x - \mu(x)}{\sigma(x)} + \mu(y). \tag{3}$$

3.2 Style Augment

AdaIN [19] reveals that styles can be augmented by transferring the mean μ and the standard deviation σ of the feature channels. A key point is to find a reasonable transfer direction. Arbitrary transfer may lead to the destruction of semantic information and thus make the training process uncontrollable. To remedy this issue, we empirically use the variance to indicate the direction of transfer. Our motivation is to automatically summarize the variation pattern by counting the range of variation of style in the source domains.

If there are S samples, the range of variation in style is expressed as

$$\Sigma(\mu) = \frac{1}{S} \sum_{}^{S} (\mu(x) - \mathbb{E}[\mu(x)])^2, \tag{4}$$

$$\Sigma(\sigma) = \frac{1}{S} \sum_{}^{S} (\sigma(x) - \mathbb{E}[\sigma(x)])^2, \tag{5}$$

where $\mathbb{E}[\cdot]$ is the mean value. $\Sigma(\mu) \in \mathbb{R}^C$ and $\Sigma(\sigma) \in \mathbb{R}^C$ provide a reasonable range of variation, minimizing the risk of meaningless or harmful style transfer.

Memory Bank. A naive approach to calculate $\Sigma(\mu)$ and $\Sigma(\sigma)$ is storing feature statistics of the total training dataset. It is resource-consuming and outdated features may lead to inaccurate results. Considering the practicality and accuracy, we adopt two memory banks to cache μ and σ of the latest S samples, i.e. S is the size of memory banks. When the banks are full, the data in which will be updated on a first-in-first-out basis.

3.3 Augmented Features

To instantiate the process of sample generation, we model the style with the multivariate Gaussian distribution, while $\Sigma(\mu)$ and $\Sigma(\sigma)$ are the variance of the distribution. Then the new style can be obtained by sampling from the distribution. The formula is as follows:

$$\beta(x) \sim \mathcal{N}(\mu(x), \lambda\Sigma(\mu)), \tag{6}$$

$$\gamma(x) \sim \mathcal{N}(\sigma(x), \lambda\Sigma(\sigma)). \tag{7}$$

In the above equation, $\beta(x), \gamma(x)$ denote new style representation and λ is a coefficient to control the strength of augmentation. To alleviate the effect of deviation of variance in the early training stage. We set $\lambda = \frac{t}{T}\lambda_0$, t and T represent current training epoch and total epoch respectively. λ_0 is a constant and fixed to 1.0 in this paper.

Now we first de-stylize the original features to obtain the normalized features,

$$\tilde{x} = \frac{x - \mu(x)}{\sigma(x)}, \tag{8}$$

Then the new styles are equipped to the normalized features to get the augmented features,

$$X = \gamma(x)\tilde{x} + \beta(x). \tag{9}$$

The augmented features are fed into the classifier after being pooled and flattened.

3.4 Implicit Domain Augmentation

According to Eq. (9), the augmented samples can be obtained by repeated sampling of $\gamma(x)$ and $\beta(x)$. However, high computational costs are introduced when the number of augmented samples is large. To avoid this problem, following [5], we achieve augmentation implicitly by a novel loss function instead of explicitly generating samples. Specifically, suppose we sample M times, an augmented feature set is obtained as $\{(X_i^1, y_i), (X_i^2, y_i), ..., (X_i^M, y_i)\}_{i=1}^{N}$, y denote the labels and N is the batch size. Then we feed the augmented feature set into the classifier T_Θ and compute the cross-entropy loss:

$$\mathcal{L}_M(\Psi, W, b) = \frac{1}{N}\sum_{n=1}^{N}\frac{1}{M}\sum_{m=1}^{M} -\log\frac{e^{w_{y_n}^T X_n^m + b_{y_n}}}{\sum_{j=1}^{C} e^{w_j^T X_n^m + b_j}}, \tag{10}$$

where $W = [w_1, ..., w_C]^T \in \mathbb{R}^{C \times K}$ and $b = [b_1, ..., b_C]^T \in \mathbb{R}^C$ are the weight matrix and biases of the classifier. Note that here C represents the number of classes.

Notice that when M approach infinity, an expectation of the Eq. (10) can be computed:

$$\lim_{M \to +\infty} \mathcal{L}_M(\Psi, W, b) = \frac{1}{N} \sum_{n=1}^{N} \mathbb{E}_{X_n}[-\log \frac{e^{w_{y_n}^T X_n + b_{y_n}}}{\sum_{j=1}^{C} e^{w_j^T X_n + b_j}}]. \tag{11}$$

However, it is difficult to work out the result directly. Alternatively, we minimize an upper bound of Eq. (11) to train network. According to the Jensen's inequality $\mathbb{E}[\log(X)] \leq \log(\mathbb{E}[X])$ and the logarithmic function $\log(\cdot)$ is concave, we have:

$$\mathcal{L}_\infty = \frac{1}{N} \sum_{n=1}^{N} \mathbb{E}_{X_n}[\log \sum_{j=1}^{C} e^{(w_j^T - w_{y_n}^T)X_n + (b_j - b_{y_n})}] \tag{12}$$

$$\leq \frac{1}{N} \sum_{n=1}^{N} \log \mathbb{E}_{X_n}[\sum_{j=1}^{C} e^{(w_j^T - w_{y_n}^T)X_n + (b_j - b_{y_n})}] \tag{13}$$

$$= \frac{1}{N} \sum_{n=1}^{N} \log(\sum_{j=1}^{C} e^{\xi + \xi'}). \tag{14}$$

where $\xi = (w_j^T - w_{y_n}^T)x_n + (b_j - b_{y_n})$ and $\xi' = \frac{\lambda}{2}(w_j^T - w_{y_n}^T)(\tilde{x}_n^T \Sigma(\sigma)\tilde{x}_n + \Sigma(\mu))(w_j - w_{y_n})$. The Eq. (14) is calculated by the moment-generating function $\mathbb{E}[ta] = e^{t\mu + \frac{1}{2}\sigma^2 t^2}$, $a \sim \mathcal{N}(\mu, \sigma^2)$, as X_n in Eq. (14) can be seen as a Gaussian random variable, i.e., $X_n \sim \mathcal{N}(x_n, \tilde{x}_n^T \Sigma(\sigma)\tilde{x}_n + \Sigma(\mu))$. So we have:
$(w_j^T - w_{y_n}^T)X_n + (b_j - b_{y_n}) \sim \mathcal{N}((w_j^T - w_{y_n}^T)x_n + (b_j - b_{y_n}), \lambda(w_j^T - w_{y_n}^T)(\tilde{x}_n^T \Sigma(\sigma)\tilde{x}_n + \Sigma(\mu))(w_j - w_{y_n}))$.

We can observe that ξ is the cross-entropy loss function term and ξ' is the regularization term. So the proposed IDA is a surrogate loss of cross-entropy loss to achieve domain augmentation implicitly.

4 Experiments

4.1 Generalization for Category Classification

Dataset and Setup. We use the standard benchmark *PACS* [20] to evaluate our method. *PACS* covers 7 categories with a total of 9991 images and consists of four domains, *Photo*, *Art-painting*, *Cartoon* and *Sketch*. We train 30 epochs with ResNet-18 and ResNet-50 [21] pretrained on the ImageNet. The learning rate is 0.001 and decayed by 0.1 at 80% of the total epochs. The batch size is set to 64. The optimizer is SGD with a momentum of 0.9 and weight decay of 5e−4. Source domains are split into a training set and a validation set in a ratio of 9 to 1. The size of the memory bank is 2000. Following prior studies [11,20], we use the leave-one-domain-out protocol which means one domain is selected as the target domain while the remaining domains are source domains for training.

Results. The results are reported in Table 1 and Table 2. We can see that the proposed IDA improves over the baseline method by 1.81% (ResNet-18) and 1.33% (ResNet-50) on average. Except for *Photo* as the target domain on Resnet-18 architecture, our approach outperforms the baseline method on all of the domains. In particular, IDA significantly improves the classification accuracy for *Sketch* which has a large domain gap with source domains, which manifests the strong generalization ability of our method. Furthermore, we plugged IDA into the SOTA DG method and boost its performance. It demonstrates that our method is superior to the cross-entropy function, and combining which with other methods can yield the further improvements.

Table 1. Results of category classification task on PACS with ResNet-18. The random shuffle version of MixStyle is reported for fair comparisons.

Method	Art	Cartoon	Photo	Sketch	Avg
MASF [12]	80.29	77.17	94.99	71.69	81.04
Epi-FCR [2]	82.10	77.00	93.90	73.00	81.50
L2A-OT [7]	83.30	78.20	**96.20**	73.60	82.80
JiGen [11]	79.42	75.25	96.03	71.35	80.51
MetaReg [3]	83.70	77.20	95.50	70.30	81.70
EISNet [17]	81.89	76.44	95.93	74.33	82.15
MixStyle [1]	82.30	79.00	96.30	73.80	82.80
DSU [9]	83.60	**79.60**	95.80	77.60	84.10
SagNet [13]	83.58	77.66	95.47	76.30	83.25
SFA-A [15]	81.20	77.80	93.90	73.70	81.70
ISDA [5]	79.05	74.91	95.73	73.24	80.73
ResNet-18	78.39	75.28	95.81	71.06	80.14
+ IDA	80.71	76.61	95.71	74.77	81.95
FACT [6]	**85.37**	78.38	95.15	79.15	84.51
+ IDA	85.06	78.50	95.32	**79.64**	**84.63**

4.2   Generalization for Instance Retrieval

Dataset and Setup. Person re-identification (re-ID) aims to match people across disjoint camera views. Since each camera can be viewed as a distinct domain, person re-ID is essentially a cross-domain problem. To validate the generalization ability of the proposed IDA for Instance Retrieval, we use two common datasets *Market1501* [23] and *Duke* [22] to experiment. One dataset is used for training while the other is used for testing. In this experiment, ResNet-50 is adopted as the backbone. We use the code based on [24].

Results. The results are shown in Table 3. It can be see that IDA improves over baseline by 0.8 ($Market1501 \rightarrow Duke$) and 1.3 ($Duke \rightarrow Market1501$) on mAP. We also compare with some common regularization methods, including RandomErase [25] and LabelSmoothing [26]. From the table, we can see that RandomErase is not beneficial for cross-domain generalization. LabelSmoothing has very little improvement on $Duke$ since it does not consider the domain shift problem. Instead, our method implicitly generates samples of potential domains, thus enabling the model to generalize to unknown domains.

Table 2. Results of category classification task on PACS with ResNet-50.

Method	Art	Cartoon	Photo	Sketch	Avg
MASF [12]	82.89	80.49	95.01	72.29	82.67
MetaReg [3]	87.20	79.20	97.60	70.30	83.60
EISNet [17]	86.64	81.53	97.11	78.07	85.84
ResNet-50	86.72	78.56	97.62	76.24	84.79
+ IDA	87.11	81.30	**97.74**	78.32	86.12
FACT [6]	89.63	81.77	96.75	84.46	88.15
+ IDA	**89.73**	**82.78**	96.82	**85.55**	**88.72**

Table 3. Results of instance retrieval task with ResNet-50.

Method	$Market1501 \rightarrow Duke$			
	mAP	R1	R5	R10
ResNet-50	16.5	29.6	45.5	51.8
+ RandomErase [25]	16.7	30.1	45.7	51.9
+ LabelSmoothing [26]	16.6	30.5	45.3	52.3
+ IDA	**17.3**	**31.6**	**46.8**	**54.8**
Method	$Duke \rightarrow Market1501$			
	mAP	R1	R5	R10
ResNet-50	18.4	42.7	60.8	67.5
+ RandomErase [25]	18.4	43.2	60.9	67.5
+ LabelSmoothing [26]	19.0	44.1	61.8	69.1
+ IDA	**19.7**	**44.7**	**63.8**	**71.2**

4.3 Further Analysis

Ablation Study. We conduct an extensive ablation analysis on *PACS* to study the effectiveness of each component in IDA. The results are shown in Table 4. The meanings of main items are as follows: (1) **w/o $\Sigma(\mu)$**: $\beta(x)$ sample from $\mathcal{N}(\mu(x), 0)$ instead of $\mathcal{N}(\mu(x), \lambda\Sigma(\mu))$; (2) **w/o $\Sigma(\sigma)$**: $\gamma(x)$ sample from $\mathcal{N}(\sigma(x), 0)$ instead of $\mathcal{N}(\sigma(x), \lambda\Sigma(\sigma))$; (3) **constant λ**: λ is fixed to a constant instead of changing with epoch. It can be observed that each component plays an active role.

Effect of Memory Bank Size. We perform an analysis to study the effect of memory bank size. Figure 2 shows the results. We can see that the memory bank-based method outperforms the iterative method and the best results appear when the size is 2k or 4k. However, size and performance are not positively correlated, this is because outdated features lead to greater statistical bias when the size is too large.

Table 4. Ablation study results on PACS.

Method	Art	Cartoon	Photo	Sketch	Avg
ResNet-18	78.39	75.28	**95.81**	71.06	80.14
$w/o\ \Sigma(\mu)$	78.86	75.51	95.75	74.06	81.05
$w/o\ \Sigma(\sigma)$	78.65	75.98	95.75	72.51	80.72
$constant\ \lambda$	79.20	75.80	95.49	73.59	81.02
IDA	**80.71**	**76.61**	95.71	**74.77**	**81.95**

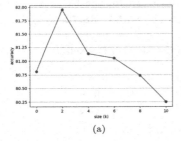
(a)

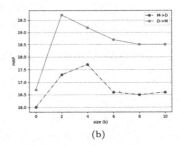
(b)

Fig. 2. Evaluation on the memory bank size. The horizontal axis represents the size of memory bank, in thousands. Size is 0 represents using iterative approach instead of memory bank. (a) The results of multi-source domain generalization task on PACS with ResNet-18. (b) The results of instance retrieval task with ResNet-50. "D" and "M" denote *Duke* and *Market1501* respectively.

5   Conclusion

In this paper, we propose a novel regularization technique called *Implicit Domain Augmentation* (IDA) for domain generalization (DG) to improve the robustness of the classifier. Our approach is based on the motivation of domain augmentation, which augments the source domain by generating valid diverse samples by performing style transfer in the feature space. To determine meaningful transfer direction, we model the feature statistics with a Gaussian distribution and introduce memory banks to estimate the distribution variance. Finally, we utilize a cross-entropy expectation upper bound to optimize the network, avoiding the additional computational cost associated with explicit augmentation. Our approach can be easily applied to vanilla models and other DG methods to boost the performance. Experimental results in several tasks demonstrate the effectiveness of the proposed IDA.

References

1. Zhou, K., Yang, Y., Qiao, Y., Xiang, T.: Domain generalization with MixStyle. In: ICLR (2021)
2. Li, D., Zhang, J., Yang, Y., Liu, C., Song, Y.-Z., Hospedales, T.M.: Episodic training for domain generalization. In: Proceedings of the IEEE/CVF International Conference on Computer Vision, pp. 1446–1455 (2019)
3. Balaji, Y., Sankaranarayanan, S., Chellappa, R.: MetaReg: towards domain generalization using meta-regularization. Adv. Neural. Inf. Process. Syst. **31**, 998–1008 (2018)
4. Matsuura, T., Harada, T.: Domain generalization using a mixture of multiple latent domains. In: Proceedings of the AAAI Conference on Artificial Intelligence, vol. 34, no. 07, pp. 11 749–11 756 (2020)
5. Wang, Y., Pan, X., Song, S., Zhang, H., Huang, G., Wu, C.: Implicit semantic data augmentation for deep networks. In: Advances in Neural Information Processing Systems, vol. 32, pp. 12 635–12 644 (2019)
6. Xu, Q., Zhang, R., Zhang, Y., Wang, Y., Tian, Q.: A Fourier-based framework for domain generalization. In: Proceedings of the IEEE/CVF Conference on Computer Vision and Pattern Recognition, pp. 14 383–14 392 (2021)
7. Zhou, K., Yang, Y., Hospedales, T., Xiang, T.: Learning to generate novel domains for domain generalization. In: Vedaldi, A., Bischof, H., Brox, T., Frahm, J.-M. (eds.) ECCV 2020. LNCS, vol. 12361, pp. 561–578. Springer, Cham (2020). https://doi.org/10.1007/978-3-030-58517-4_33
8. Yue, X., Zhang, Y., Zhao, S., Sangiovanni-Vincentelli, A., Keutzer, K., Gong, B.: Domain randomization and pyramid consistency: simulation-to-real generalization without accessing target domain data. In: Proceedings of the IEEE/CVF International Conference on Computer Vision, pp. 2100–2110 (2019)
9. Li, X., Dai, Y., Ge, Y., Liu, J., Shan, Y., Duan, L.: Uncertainty modeling for out-of-distribution generalization. In: International Conference on Learning Representations (2022). https://openreview.net/forum?id=6HN7LHyzGgC
10. Motiian, S., Piccirilli, M., Adjeroh, D.A., Doretto, G.: Unified deep supervised domain adaptation and generalization. In: Proceedings of the IEEE International Conference on Computer Vision, pp. 5715–5725 (2017)
11. Carlucci, F.M., D'Innocente, A., Bucci, S., Caputo, B., Tommasi, T.: Domain generalization by solving jigsaw puzzles. In: Proceedings of the IEEE/CVF Conference on Computer Vision and Pattern Recognition, pp. 2229–2238 (2019)
12. Dou, Q., de Castro, D.C., Kamnitsas, K., Glocker, B.: Domain generalization via model-agnostic learning of semantic features. In: Advances in Neural Information Processing Systems, vol. 32, pp. 6450–6461 (2019)
13. Nam, H., Lee, H., Park, J., Yoon, W., Yoo, D.: Reducing domain gap by reducing style bias. In: Proceedings of the IEEE/CVF Conference on Computer Vision and Pattern Recognition, pp. 8690–8699 (2021)
14. Zhou, K., Yang, Y., Hospedales, T., Xiang, T.: Deep domain-adversarial image generation for domain generalisation. In: Proceedings of the AAAI Conference on Artificial Intelligence, vol. 34, no. 07, pp. 13 025–13 032 (2020)

15. Li, P., Li, D., Li, W., Gong, S., Fu, Y., Hospedales, T.M.: A simple feature augmentation for domain generalization. In: Proceedings of the IEEE/CVF International Conference on Computer Vision, pp. 8886–8895 (2021)
16. Li, S., Song, S., Huang, G., Ding, Z., Wu, C.: Domain invariant and class discriminative feature learning for visual domain adaptation. IEEE Trans. Image Process. **27**(9), 4260–4273 (2018)
17. Wang, S., Yu, L., Li, C., Fu, C.-W., Heng, P.-A.: Learning from extrinsic and intrinsic supervisions for domain generalization. In: Vedaldi, A., Bischof, H., Brox, T., Frahm, J.-M. (eds.) ECCV 2020. LNCS, vol. 12354, pp. 159–176. Springer, Cham (2020). https://doi.org/10.1007/978-3-030-58545-7_10
18. Kumar, A., Raghunathan, A., Jones, R., Ma, T., Liang, P.: Fine-tuning can distort pretrained features and underperform out-of-distribution. arXiv:2202.10054 (2022)
19. Huang, X., Belongie, S.: Arbitrary style transfer in real-time with adaptive instance normalization. In: Proceedings of the IEEE International Conference on Computer Vision, pp. 1501–1510 (2017)
20. Li, D., Yang, Y., Song, Y.-Z., Hospedales, T.M.: Deeper, broader and artier domain generalization. In: Proceedings of the IEEE International Conference on Computer Vision, pp. 5542–5550 (2017)
21. He, K., Zhang, X., Ren, S., Sun, J.: Deep residual learning for image recognition. In: Proceedings of the IEEE Conference on Computer Vision and Pattern Recognition, pp. 770–778 (2016)
22. Ristani, E., Solera, F., Zou, R., Cucchiara, R., Tomasi, C.: Performance measures and a data set for multi-target, multi-camera tracking. In: Hua, G., Jégou, H. (eds.) ECCV 2016. LNCS, vol. 9914, pp. 17–35. Springer, Cham (2016). https://doi.org/10.1007/978-3-319-48881-3_2
23. Zheng, L., Shen, L., Tian, L., Wang, S., Wang, J., Tian, Q.: Scalable person re-identification: a benchmark. In: Proceedings of the IEEE International Conference on Computer Vision, pp. 1116–1124 (2015)
24. Luo, H., Gu, Y., Liao, X., Lai, S., Jiang, W.: Bag of tricks and a strong baseline for deep person re-identification. In: Proceedings of the IEEE/CVF Conference on Computer Vision and Pattern Recognition Workshops (2019)
25. Zhong, Z., Zheng, L., Kang, G., Li, S., Yang, Y.: Random erasing data augmentation. In: Proceedings of the AAAI Conference on Artificial Intelligence, vol. 34, no. 07, pp. 13 001–13 008 (2020)
26. Szegedy, C., Vanhoucke, V., Ioffe, S., Shlens, J., Wojna, Z.: Rethinking the inception architecture for computer vision. In: Proceedings of the IEEE Conference on Computer Vision and Pattern Recognition, pp. 2818–2826 (2016)

Unconstrained Feature Model and Its General Geometric Patterns in Federated Learning: Local Subspace Minority Collapse

Mingjia Shi⬤, Yuhao Zhou⬤, Qing Ye⬤, and Jiancheng Lv$^{(\boxtimes)}$⬤

College of Computer Science, Sichuan University, Chengdu 610065, China
{yeqing,lvjiancheng}@scu.edu.com

Abstract. Federated Learning is a decentralized approach to machine learning that enables multiple parties to collaborate in building a shared model without centralizing data, but it can face issues related to client drift and the heterogeneity of data. However, there is a noticeable absence of thorough analysis regarding the characteristics of client drift and data heterogeneity in FL within existing studies. In this paper, we reformulate FL using client-class sampling as an unconstrained feature model (UFM), and validates the soundness of UFM in FL through theoretical proofs and experiments. Based on the model, we explored the potential information loss, the source of client drifting, and general geometric patterns in FL, called *local subspace minority collapse*. Through theoretical deduction and experimental verification, we provide support for the soundness of UFM and observe its predicted phenomenon, neural collapse.

Keywords: Federated Learning · Neural Network Theory · Neural Collapse · Information Bottleneck · Local Subspace Minority Collapse

1 Introduction

Federated Learning (FL) is an approach to machine learning that enables multiple parties, each holding private data, to collaborate in building a shared model without requiring data to be centrally stored or communicated. [15] This decentralized approach to training models allows for improved privacy and security while still providing the benefits of large-scale machine learning. By aggregating locally trained models across devices or servers, FL can result in more accurate and robust models that reflect the diversity of data across different devices or user populations. However, in classic FL settings, which involve private data protection and local training, client drift and data heterogeneity are inevitable issues. [26,27]

Recently, insightful works are proposed addressing issues about client drift and heterogeneous data in FL. To address the former, FedProx [13] uses a regularization term to limit how far local clients can drift from the global model

B. Luo et al. (Eds.): ICONIP 2023, CCIS 1962, pp. 449–464, 2024.
https://doi.org/10.1007/978-981-99-8132-8_34

during local training. To further address the latter, many personalized FL methods have been proposed to find personalized models that fit local data, such as pFedMe [20,21] which, while using a regularization term, also re-models the local problem as a Moreau envelope. This provides further theoretical support compared to heuristic aggregation of personalized models. Furthermore, in large-scale training and with larger models, Vit-FL [18] has been proposed to adapt to the needs of new DNN model structures [25]. However, these works do not deeply deconstruct client drift and data heterogeneity characteristics under FL. In order to further promote the development of FL and meet its potential explainability needs, we introduce the theories of information and geometry into FL and further explore the FL specialized patterns of the client drift and data heterogeneity characteristics beyond classical deep learning (DL).

In this paper, we use client-class sampling to reformulate FL as an unconstrained feature model (UFM) [16]. Based on this modeling, we theoretically prove that there is potential sampling information loss in FL and that the source of client drift is feature drift. We also demonstrate through experiments that client-class sampling information exists and that the neural collapse phenomenon predicted through UFM does exist in FL as a similar pattern as the one in DL, called local subspace minority collapse. This is encouraging and exciting because it validates the soundness of UFM in FL, supporting the future and existing works [9,14] which prerequisite that neural collapse exists in FL.

The contributions of this paper are shown as follows:

- We propose the first UFM in FL by extracting client-class sampling information, and validate the existence of this information in classic FL through experiments.
- Theoretically, we further discuss the potential information loss, the source of client drift, and the rationality of UFM.
- Empirically, based on the proposed UFM, we observe and summarize the geometric characteristics of classical FL, called *local subspace minority collapse* and *frame drift*.

2   Preliminary and Related Work

Neural collapse (NC) is a geometric phenomenon observed in the final layer of DNNs, which occurs during the terminal phase of training (TPT) in DL. [17] For further research, we are trying to employ NC to support our analysis about FL. However, the prerequisites [10,11] for previous NC are vastly different from the assumptions and settings in FL (e.g., model aggregation, imbalanced and heterogeneous data). Therefore, it is necessary to study whether NC occurs in FL. If not, there might be a similar general pattern or phenomenon. Several geometric metrics have been introduced to identify the patterns and phenomenon. Encouragingly, these metrics characterize the relationships between features and classifiers, helping us find the general patterns. Moreover, we try to explain these observed results via Information Bottleneck (IB).

The NC phenomenon is that in the TPT, single-label supervised deep learning with a balanced dataset and sampling, the features of the last layer of each class collapse to the class-mean of the features (NC-1). Moreover, all the class-means collapse into an equiangular tight frame (ETF) with the global mean point as the center, which means that the angle between them is maximized (NC-2). For the classifier, the last layer linear classifiers match their class-means almost perfectly (NC-3) and are behaviorally equivalent to a nearest class-center decision rule (NC-4) [17].

One method motivated by NC proposes fixing a random simplex ETF geometric structure to serve as the local model for clients [9]. Another method motivated by NC proposes fixing the classifier as ETF similarly and finetuning the extractor and classifier alternatively [14].

These works motivated by NC do not focus on studying the NC phenomenon itself in FL. [9,14] These motivations presuppose that the NC phenomenon does exist in FL and performs in a similar way to how the NC-2 phenomenon occurs in FL settings. This highlights the necessity of exploring whether NC exists in FL. Additionally, we aim to investigate the geometry of FL in a broader field.

The IB was introduced by [23,24] as a fundamental principle for understanding the behavior of deep neural networks. It shows that neural networks trained with gradient descent on a given dataset tend to converge to solutions that have a small number of highly informative features (the "bottleneck"), which capture the essence of the data and enable accurate prediction. This suggests that neural networks are able to learn efficient representations of complex data by discarding redundant or irrelevant information.

IB divides the entire inference process into two parts: feature fitting and feature compression. In the fitting process, DNNs try to learn and discover the information and features hidden in the data. In the compression process, DNNs forget what they have learned. Furthermore, IB regards a DNN as a process of extracting information. Starting from the data, the amount of feature information is high in the lower layers of the network and decreases towards the higher ones. The pattern of the mutual information between features and labels during the training process of DNNs is to first increase and then decrease.

2.1 Normalized Gram Matrix and Generalized Coherence Estimate

Generalized coherence estimation (GCE) and normalized Gram matrix (NGM) are commonly used in the field of signal processing. [2,5,7,8,12,19] They contain information about the coherence between a set of vectors, where the value represents the strength of coherence between them. In this paper, we observe the class-wise and client-wise GCE statistics of classifiers, as well as the feature means in the FL process.

The normalized Gram matrix is define as: $\hat{\mathbf{G}}(\mathbf{X}) := \{\frac{\langle \mathbf{x}_i, \mathbf{x}_j \rangle}{||\mathbf{x}_i||||\mathbf{x}_j||}\}_{i,j}$ where $\mathbf{X} = [\mathbf{x}_1, ..., \mathbf{x}_N]$, each column $\mathbf{x}_i$ is vector in a Hilbert space, $\langle \rangle$ is the inner product and $|| \cdot ||^2 = \langle \cdot, \cdot \rangle$ is the norm induced by the inner product. The Gram matrix is $\mathbf{G}(\mathbf{X}) := \mathbf{X}^T \mathbf{X}$.

The GCE is define as: $gce(\mathbf{X}) := 1 - det(\hat{\mathbf{G}}(\mathbf{X}))$ The GCE captures the overall angular relationship between a set of basis vectors $\mathbf{X}$, providing insight into how they are oriented with respect to the span space of the other vectors. The larger the value of GCE, the more reasonable it is for us to speculate that there is statistical independence and white Gaussian noise in this set of vectors. In terms of geometry, intuitively, NGM reflects the squared normalized volume of the high-dimensional parallelogram formed by this set of vectors, $\mathbf{X} \in \mathbb{R}^{d \times N}$.

3   Reformulated FL: Unconstrained Feature Model

3.1   Motivation of Reformulation

Previously, NC-related work discussed one or more special cases and proposed theoretical properties and phenomenological descriptions based on an UFM with different parameters and free feature solution spaces. However, the existence of NC phenomenon in FL is questionable because FL settings with non-IID data involve special settings, unbalanced sampling within clients, different distributions of similar data sources among clients, and client drift caused by limitations in data transmission.

Traditional modeling based on client sampling cannot effectively describe the above three points. In client-side optimization, problems are often constructed using the expectation of loss on local data sampling or by directly using Empirical Risk Minimization (ERM). Meanwhile, in global optimization, problems are often constructed based on ERM or the expectation of loss on data sampling using client sampling. Moreover, even for IID FL, it is necessary to reconsider the combined effects of model aggregation solution space, client drift, and unbalanced sampling.

The modeling approach based on client sampling and data sampling within clients cannot describe the relationship between the class mean of different features and the classifier in the NC phenomenon, so we need a new FL modeling.

3.2   Categories of FL

Our approach to FL categorization differs from other methods that are based on instance-attribute categories. [26,27] Specifically, we categorize FL into two types: class-homogenous clients and class-heterogeneous clients, which means $\mathbf{P}_{x|k} = \mathbf{P}_{x|k,i}$ or not, where $\mathbf{P}_{x|k,i}$ is the probability of unlabeled data x is sampled from k^{th}-class of client i^{th} ($k \in [K]$, $i \in [N]$ for simplification.). For example, (x, k) can be used as a data sample for supervised learning, while x can be used as a data sample for unsupervised learning.

3.3   Deep Learning Formulation

General Formulation with Regularization. The population risk functional $\mathcal{R} : \hat{\mathcal{H}} \to \mathbb{R}/\mathbb{R}^-$ is represented as:

$$\mathcal{R}(h) = \int_{\mathbb{R}^d} \mathcal{D}(h(x), k_x) \mathbf{P}(\mathrm{d}x)$$

where, $\hat{\mathcal{H}}$ is the function space ($h \in \hat{\mathcal{H}}$), k. is the $\mathbf{P}$ measurable label function and $\mathcal{D} : \mathbb{R}^K \times \mathbb{R}^K \to \mathbb{R}/\mathbb{R}^-$ is a generic loss function that measuring the distance of its inputs.

To parameterize, we set $h(x; W, b, W^-) = WH(x; W^-) + b$, where $H : \mathbb{R}^D \to \mathbb{R}^F$ is the feature extractor of the final layer of a DNN parameterized by W^-. Thus, the population risk of classic classification formulation in DL is:

$$\min_{W, b, W^-} \mathbf{E}_{x,k}[\mathcal{D}(WH(x; W, b, W^-), k)] + \lambda_{\mathbf{R}} \mathbf{R}(W, b, W^-)$$

where $\lambda_{\mathbf{R}} \mathbf{R}(\cdot)$ is the regularization term of parameters. We highlight a commonly used trick here, weight decay (WD), as $\lambda_{\mathbf{R}} \mathbf{R}_{wd}(\cdot) = \frac{\lambda}{2} || \cdot ||_F^2$.

Unconstrained Feature Models. The nonlinearity within DNNs and their inter-layer interactions pose many challenges to analyzing the behavior and performance of DNNs. Fortunately, the neural networks used in recent years are typically over-parameterized and can approximate almost any continuous function in the function space involved. Additionally, the phenomenon of NC only concerns the geometric properties of the final layer of the neural network. Given these two factors, an intuitive but effective way to analyze DNNs is to view the final features as free variables and include them as part of the optimization problem's solution space (unconstrained feature model). [16]

Next, the process from issues related to using WD in DL to UFM is discussed. As the Lagrangian dual, the regularization term $\frac{\lambda_W}{2} ||W^-||^2$ is a relaxation of the constrain $||W^-||^2 \leq C_{W^-}$ (C. is any constant about $\cdot$). The DL problem with WD tunes into:

$$\min_{W, b, W^-} \mathbf{E}_{x,k}[\mathcal{D}(WH(x; W^-) + b, k)],$$
$$\text{s.t.} ||W||^2 \leq C_W, ||W^-||^2 \leq C_{W^-}, ||b||^2 \leq C_b. \tag{1}$$

To apply UFM, we equivalently write Equation (1) as:

$$\min_{W, b, H_x} \mathbf{E}_{x,k}[\mathcal{D}(WH_x + b, k)],$$
$$\text{s.t.} ||W||^2 \leq C_W, ||b||^2 \leq C_b, H_x \in \{H(x; W^-) : ||W^-||^2 \leq C_{W^-}\} \tag{2}$$

For simplification, the same ansatz in [4] is made that the range of $H(x; W^-)$ under the constraint $||W^-||^2 \leq C_{W^-}$ is approximately an ellipse in the sense that:

$$\{H(x; W^-) : ||W^-||^2 \leq C_{W^-}\} \approx \{H_x : ||H_x||^2 \leq C_H\},$$

which makes sense due to the self-duality of each classifier and feature, i.e., NC-3.

3.4   FL Modeling with Client-Class Sampling

Quantities in Client-Class Sampling. The main relevant quantities are as follows:

- Client-class sampling matrix: $P_{N,K} = \{\mathbf{P}_{i,k}\}_{i\in[N],k\in[K]} \in \mathbb{R}^{N\times K}$.
- Local class sampling matrix: $P_{K|N} = \{\mathbf{P}_{k|i}\}_{i\in[N],k\in[K]} \in \mathbb{R}^{N\times K}$.
- Client sampling vector: $P_N = [\mathbf{P}_i]_{i\in[N]} \in \mathbb{R}^N$.

Random Factor Extraction: Local Basis and Global Solution. To build a FL model that uses multi-model averaging as a global aggregation solution, we gradually extract the random factors of the problem. First, we extract the random factors of the client sampling, as shown below:

- $\mathcal{W} = \{W_i\}_{i\in[N]}$ is a set of basis (third-order tensor) and $\bar{W} = \mathbf{E}_i W_i = \mathcal{W}P_N$ is on the basis with coordinates P_N.
- $\mathcal{B} = \{b_i\}_{i\in[N]}$ is a set of basis and $\bar{b} = \mathbf{E}_i b_i = \mathcal{B}P_N$ is a point on the basis with coordinates P_N.
- $\mathcal{W}^- = \{W_i^-\}_{i\in[N]}$ is a set of basis and $\bar{W}^- = \mathbf{E}_i W_i^- = \mathcal{W}^- P_N$ is a point on the basis with coordinates P_N.
- $\mathcal{H}_{N,K}(\cdot) = \{H_{\cdot,i} := H(\cdot; W_i^-)\}_{i\in[N],k\in[K]}$, $\mathcal{H}$ for simplification.
- $\bar{\mathcal{H}}_{N,K}(\cdot) = \{\bar{H}_\cdot := H(\cdot; \bar{W}^-)\}_{i\in[N],k\in[K]}$, $\bar{\mathcal{H}}$ for simplification.

To extract random factor of WD, we have:

$$\frac{\lambda_W}{2}||\bar{W}||^2 + \frac{\lambda_W}{2}||\bar{W}^-||^2 + \frac{\lambda_b}{2}||\bar{b}||^2$$
$$\leq \mathbf{E}_i[\frac{\lambda_W}{2}||W_i||^2 + \frac{\lambda_W}{2}||W_i^-||^2 + \frac{\lambda_b}{2}||b_i||^2], \tag{3}$$

which is a upper bound of WD regularization, constructed by basis.

Next, we extract the class-wise sampling on given client. First, we define the random element of class-wise sampling and its statistics, which corresponds to the local loss function by class for specific local loss. By combining client and on-client class sampling, we obtained the population risk of FL as shown below:

$$\min_{\mathcal{W},\mathcal{B},\bar{\mathcal{H}}} L(\mathcal{W},\mathcal{B},\bar{\mathcal{H}}) := \langle P_{N,K}, L_{N,K}(\mathcal{W},\mathcal{B},\bar{\mathcal{H}})\rangle \tag{4}$$

where $L_{N,K}(\mathcal{W},\mathcal{B},\bar{\mathcal{H}}) := \{\mathbf{E}_{x|i,k}\mathcal{D}(\mathcal{W}P_N\bar{H}_x + \mathcal{B}P_N, k)\}_{i\in[N],k\in[K]}$. By sampling different classes on client side, we can return to the classical problem modeling based on client sampling in federated learning, as shown below:

$$\min_{\mathcal{W},\mathcal{B},\bar{\mathcal{H}}} \langle P_N, L_N(\mathcal{W},\mathcal{B},\bar{\mathcal{H}})\rangle$$

where $L_N := \{\langle P_{K|i}, L_{i,K}\rangle\}_{i\in[N]}$, and $Mtrx_{i,K}$ $(i \in [N])$ is the ith row of matrix $Mtrx_{N,K}$.

Client-Class Sampling FL with WD. Combining the above, UFM with client-class sampling and WD, we obtain the following optimization objective function of FL and its upper bound, as shown below (with convexity of $||\cdot||^2$).

$$\min_{\mathcal{W},\mathcal{B},\bar{\mathcal{H}}} L(\mathcal{W},\mathcal{B},\bar{\mathcal{H}}) + \frac{\lambda_H}{2}||\bar{H}||^2 + \frac{\lambda_W}{2}||\mathcal{W}P_N||^2 + \frac{\lambda_b}{2}||\mathcal{B}P_N||^2$$
$$\leq L(\mathcal{W},\mathcal{B},\bar{\mathcal{H}}) + \mathbf{E}[\frac{\lambda_H}{2}||\bar{H}_x||^2 + \frac{\lambda_W}{2}||W_i||^2 + \frac{\lambda_b}{2}||b_i||^2]$$

(5)

where $\bar{H} := \mathbf{E}_x[\bar{H}_x := H(x;\bar{W})]$ is the expectation (mean) of global feature. Note that the final layer regularization term of the global classifier $\frac{\lambda_W}{2}||\mathcal{W}P_N||^2 + \frac{\lambda_b}{2}||\mathcal{B}P_N||^2$ is upper-bounded with the expectation of client sampling $\mathbf{E}_i[\frac{\lambda_W}{2}||W_i||^2 + \frac{\lambda_b}{2}||b_i||^2]$, as part of Equation (3), which can be obtained in a local epoch. However, the feature regularization term cannot do so due to the expectation being of data sampling, thus failing to decouple the dependency of client sampling.

Sampling Noise and Empirical Risk Minimization. To reduce computational costs and further analyze the actual optimization process, a common technique is to use stochastic gradient descent and data sampling for modeling, and convert the expected loss into empirical loss. To meet above, we divide the data sampling into three parts: client sampling ($\mathbf{E}_i$), class sampling ($\mathbf{E}_{k|i}$), and client-class source sampling ($\mathbf{E}_{x|i,k}$). The first two parts involve uniform sampling and are well-known. For the latter, we perform uniform sampling on the data in the given dataset, whether it is full or mini-batch.

Note full data in FL system is d, the data on the ith client is $d_i = \{(x,k)\}$ and the data of the kth class in d_i is $d_{i,k} = \{\hat{x} : k = \hat{k}, (\hat{x},\hat{k}) \in d_i\}$. The empirical risk minimization (ERM) is as shown below:

$$\min_{\bar{W},\bar{b},\bar{H}_x} \sum_{i\in[N]} \frac{|d_i|}{|d|} \sum_{k\in[K]} \frac{|d_{i,k}|}{|d_i|} \sum_{x\in d_{i,k}} \frac{1}{|d_{i,k}|}[\mathcal{D}(\bar{W}\bar{H}_x + \bar{b}, k) + \mathcal{R}]$$

where $\mathcal{R}$ is the regularization term. This is equivalent to the classic problem formulation on data sampling, but client-class sampling and UFM are highlighted.

3.5   Observations and Insights

By emphasizing different samplings and combining Hilbert space with the norm induced by its inner product, we can predict certain phenomenon that occur in the TPT through optimization problem modeling. This also allows us to understand the properties of classifiers and unconstrained feature representations near the optimal solution. Note that we, in order to more generally and fundamentally discuss the problem, do not provide a specific loss function $\mathcal{D}$, inner product $\langle\cdot,\cdot\rangle$, or norm $||\cdot||^2 = \langle+\cdot,\cdot\rangle$.

Observation 1. *The sampled loss function may loss the sampling information.*

This is a potential pattern that should be relatively easy for people to notice, but is often overlooked and not given much attention. In practical calculations, we usually define the inner product in Euclidean space, as $< \xi, \cdot >= \cos \angle(\xi, \cdot) \times || \cdot || \times ||\xi||$. We may only want to reduce the norm $|| \cdot ||$ of the loss and do not consider reducing the impact of $\cos \angle(\xi, \cdot)$, especially in uniform sampling. In DL rather than FL, we don't consider the information loss between the sampling and our sampled loss even when the loss is small and the angle has a significant impact. We also don't care about whether sampling would cause the loss values and probability distribution to be orthogonal or if it would cause information loss leading to performance degradation. In FL, we should do.

A general consensus related to tensor geometry is that in an m-dimensional space, complete information about an n-dimensional geometry requires an m dimensional n^{th}-order tensor.

From a geometric perspective, it makes sense, because the vector of data sampling is considered as a high-dimensional first-order tensor, its high-dimensional and higher-order information is projected onto a one-dimensional space (line), thereby considering only its numerical value and not considering its angles with respect to each other, although this might imply information loss. Note that, in Equation (4), the information of $P_{N,K}$ is second-order, while classic client sampling information P_N is first-order.

However, in FL, we should consider preserving sampling information, especially when studying optimization and client sampling strategies based on local loss functions. This appears to be an unavoidably sampling-sensitive phenomenon, because in the TPT, when its loss value is small, the objective and sampling tend to be orthogonal, due to the sampling itself. The corresponding solutions are discussed in Sect. 6, and a empirical observation is proposed in Sect. 4.

Observation 2. *Client drift originates from feature drift.*

By categorizing parameters as UFM, we observe from the optimization process that the source of client drift is in the final feature extractor rather than the classifier as shown in Theorem 1.

With corrections made via the drift matrix in Theorem 1, we have:

$$\mathbf{P}_i \nabla_{\bar{H}_x} L(\mathcal{W}, \mathcal{B}, \bar{\mathcal{H}}) = (\nabla_{\bar{W}^-} \bar{H}_x)^\dagger (\nabla_{W_i^-} H_{x,i}) \nabla_{H_{x,i}} L(\mathcal{W}, \mathcal{B}, \bar{\mathcal{H}})$$

which represents the un-drifted gradient. If each local epoch is only one gradient step, and this correction is applied, averaging aggregation would not drift theoretically. However, the classic and ideal setting is to set $(\nabla_{\bar{W}^-} \bar{H}_x)^\dagger (\nabla_{W_i^-} H_{x,i}) = I_m$, the unit matrix. This leads to the first client drift, and provides a metric to evaluate the client drift instead of local loss value. However, this metric requires an additional computational cost, and is costly when the DNN is deep and the feature's dimension is high.

Observation 3. *The feature mean can be controlled by classifier.*

When weight decay is used, an upper bound on the norm of the features near the minimum is the expectation on client sampling of the norm of the local classifier.

Note that the origin of the regularizer term for features in Eq. (5) depends on the assumption that they lie within an ellipse. This still differs somewhat from actual DNN training, as we typically adjust parameters through backpropagation rather than directly adjusting features. However, with Theorem 2, near the optimum of UFM, robust optimization can be obtained by controlling only the parameters of the final classifier.

Table 1. Feature Means

Summation	client	$\leftarrow$
class	Global Mean	Local Mean
$\uparrow$	Global Class-Mean	Local Class-Mean

3.6 Related Theorems

Theorem 1. (First Drift in Local Training) *In FL in the form of Eq. (4), if averaging $\bar{\cdot} = \mathbf{E}_{i} \cdot_i$ is the client aggregation operator, the client drift due to the partially local training, originates from the loss gradient to the local feature $H_{x,i}$. The drift matrix is $(\nabla_{\bar{W}} - \bar{H}_x)^{\dagger}(\nabla_{W_i} - H_{x,i})$, where $\dagger$ means the generalized inverse operator.*

$$\nabla_{W_i} L(\mathcal{W}, \mathcal{B}, \bar{\mathcal{H}}) = \mathbf{P}_i \nabla_{\bar{W}} L(\mathcal{W}, \mathcal{B}, \bar{\mathcal{H}})$$
$$\nabla_{b_i} L(\mathcal{W}, \mathcal{B}, \bar{\mathcal{H}}) = \mathbf{P}_i \nabla_{\bar{b}} L(\mathcal{W}, \mathcal{B}, \bar{\mathcal{H}})$$
$$\nabla_{H_{x,i}} L(\mathcal{W}, \mathcal{B}, \bar{\mathcal{H}}) \neq \mathbf{P}_i \nabla_{\bar{H}_x} L(\mathcal{W}, \mathcal{B}, \bar{\mathcal{H}}).$$

Theorem 2. (Feature Mean Bound around Minimum) *The feature norm satisfies the following inequality in the vicinity of the optimal value of UFM with WD:*

$$\frac{\lambda_H}{\lambda_W} ||\bar{H}||_F^2 \leq \mathbf{E} ||W_i||_F^2$$

Theorem 3. (Feature Class-Mean Pair and Angle) *The angle between the feature class-mean of each pair of local sampling class satisfies the following:*

$$\kappa^{-1} \cos \angle (\bar{\mathcal{H}}_{i,k_1} - \mathbf{E}_{k|i}\bar{\mathcal{H}}_{i,k}, \bar{\mathcal{H}}_{i,k_2} - \mathbf{E}_{k|i}\bar{\mathcal{H}}_{i,k})$$
$$= \mathbf{P}_{\widetilde{1,2}|i} \langle \bar{\mathcal{H}}_{i,k_1}, \bar{\mathcal{H}}_{i,k_2} \rangle - \mathbf{P}_{1,2|i} \mathbf{E}_{k|1,2} ||\bar{\mathcal{H}}_{i,k}||^2$$
$$- \mathbf{P}_{\widetilde{1,2}|i} \langle \bar{\mathcal{H}}_{i,k_1} + \bar{\mathcal{H}}_{i,k_2}, \mathbf{E}_{k|\widetilde{1,2},i}\bar{\mathcal{H}}_{i,k} \rangle + ||\mathbf{E}_{k|i}\bar{\mathcal{H}}_{i,k}||^2$$

where $\mathbf{P}_{1,2|i}$ is the probability of being sampled in any class pair set $\{k_1, k_2\}$ (sequenced as 1st and 2nd respectively for distinction) on given client i, and $\mathbf{P}_{\widetilde{1,2}|i}$ is the one of not being sampled, $\kappa^{-1} = ||\bar{\mathcal{H}}_{i,k_1} - \mathbf{E}_{k|i}\bar{\mathcal{H}}_{i,k}|| \times ||\bar{\mathcal{H}}_{i,k_2} - \mathbf{E}_{k|i}\bar{\mathcal{H}}_{i,k}||$. Note that, we have $\mathbf{E}_{k|i}\bar{\mathcal{H}}_{i,k} = \mathbf{P}_{\widetilde{1,2}|i}\mathbf{E}_{k|\widetilde{1,2},i}\bar{\mathcal{H}}_{i,k} + \mathbf{P}_{1,2|i}\mathbf{E}_{k|1,2,i}\bar{\mathcal{H}}_{i,k}.$

4   Experiments

4.1   Where's Client-Class Sampling Information Found in FL?

To find the validity of the modeling proposed and to verify if the client-class sampling information can be observed in experiments, we conduct some explorations. Encouragingly, we find that many client-wise sets of vectors including client-class-sampling information, about both the classifiers and features: (Fig. 3 as gound true)

- The set of vectors from global classifier $\mathcal{W}\mathbf{P}_N$ to each basis in $\mathcal{W}$. The NGM $\hat{\mathbf{G}}(\mathcal{W} - \mathcal{W}P_N\mathbf{1}_N^T)$ is shown in Fig. 1.
- On given public class, the set of vectors from each feature mean bolded in Table 1 to local class-mean of the given 0th class. The NGMs of them are shown in Fig. 2.

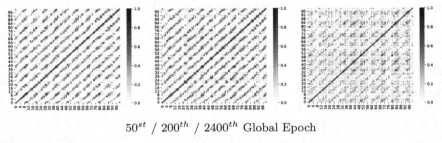

$$50^{st} \; / \; 200^{th} \; / \; 2400^{th} \text{ Global Epoch}$$

Fig. 1. The visualization of $\hat{\mathbf{G}}(\mathcal{W} - \mathcal{W}P_N\mathbf{1}_N^T)$ at the 50st, 200th and 2400th (final) global epoch on MNIST

4.2   Local Subspace Minority Collapse: The Geometric Pattern in FL

One common geometric phenomenon in DL is NC. However, the geometry of FL cannot be simply described with NC, because the classical assumptions in previous NC-related work that may not hold in FL setting. In this section, we observe the geometry in FL of features and classifiers in UFM and provide support for NC-motivated FL methods. The UFM's feature space visualization of 3 classes is shown in Fig. 5.

In DL with imbalanced data, the phenomenon of NC observed under the UFM model manifests as Minority Collapse (MC). In [4], imbalanced data is described as dividing the data into two groups based on class, with inter-group sampling imbalance and intra-group balance. During the TPT, in the group containing the minority class, the angles between classifiers for each class are equal, and decrease as the imbalance ratio increases.

In FL, MC can be more complex because local data is typically not easily divided into two balanced groups. The observations of MC with FL characteristics are shown in Fig. 4.

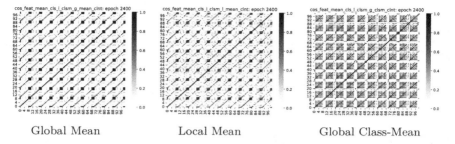

Global Mean   Local Mean   Global Class-Mean

Fig. 2. The visualization of the NGM of vector sets mentioned in Table 1 at the 2400th (final) global epoch on MNIST with given 1st public class.(From top to bottom)

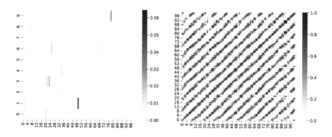

Fig. 3. The figures above are the visualization of client-class sampling matrix $P_{N,K}^T$ and its NGM $\hat{\mathbf{G}}(P_{N,K}^T)$

4.3 Settings

The DNN models mentioned in this paper:

- **DNN**: 2-layers DNN with 100-units hidden layer.
- **DNN-A-B**: 3-layers DNN with first and second hidden layers of A and B units respectively.
- **CNN**: we use CifarNet [6] as our CNN model in this paper.

The datasets mentioned in this paper:

- **FEMNIST**: The dataset in LEAF Benchmark [1], 198-client and 62-classification task with 10.1% aggregation ratio by default.
- **FMNIST & MNIST**: The dataset in [22], 100-client and 10-classification task with 20% aggregation ratio by default.
- **CIFAR-10**: The dataset in [3], 20-client and 10-classification task with 20% aggregation ratio by default.

5 Analyses

5.1 Local Subspace Minority Collapse

As shown in Fig. 4, the majority classes on $0/32/42/49^{th}$ client are respectively classes of $\{0\}/\{2,3,4,9\}/\{2,3,4\}/\{1,9\}$ and the entire parameter space of the

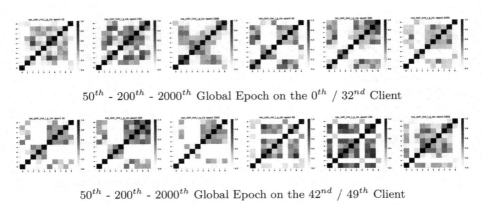

50th - 200th - 2000th Global Epoch on the 0th / 32nd Client

50th - 200th - 2000th Global Epoch on the 42nd / 49th Client

Fig. 4. The local subspace MC phenomenon in FL. The visualization of the classifier on $0/32/42/49^{th}$ client in DNN-MNIST setting

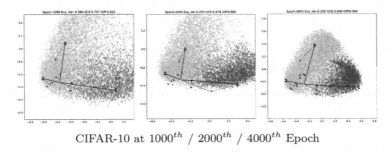

CIFAR-10 at 1000th / 2000th / 4000th Epoch

Fig. 5. Neural Collapse in FL. The figures are the UFM visualization on task CNN / CIFAR-10, generated with kernel-PCA

classifier has been divided (with blank parts) into two parts, these being the classifier majority class and minority class.

Local subspace MC: in the latter stages of training, after each local fine-tuning, the angle between the majority-class classifiers increases, as does the angle between the minority-class classifiers. Additionally, the changes in angles between the minority-class classifiers are more random compared to the majority-class classifiers and a fixed sampling matrix.

One possible explanation for this randomness could be that during local training, the proportion of intrinsic information in the data increases due to feature compression, while the proportion of sampling information decreases as it is forgotten. Hence, when adjusting the classifiers, there is less focus on the information from minority-class data, as it may contain less intrinsic information.

Another possible explanation is that the loss weight of minority classes in the local objective is smaller, leading to inadequate reduction in loss during previous training processes. As a result, they enter the feature compression stage later or slower than the majority class, which means the angle between them takes longer

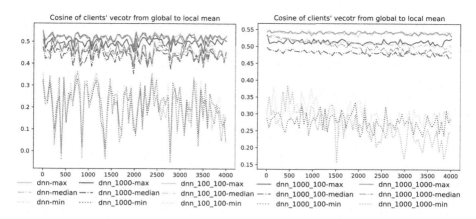

Impacts of Depth and Width of DNN on CIFAR-10 with 20% / Full Aggregation

Fig. 6. The client-wise maximum/median/minimum of cosine between each pair of the vector from global to local classifier and global to local class-mean on given i^{th} client, $\cos \angle (W_i - \mathcal{W}p_N), [\mathbf{E}_{x|i,k} - \mathbf{E}_{x|k} \bar{\mathcal{H}}_{i,k}(x)]_{k \in [K]})$

to open up. As shown in Fig. 6, it shows the impact of local iteration numbers on the global process, which means that overall global features enter the feature compression stage earlier, and the prerequisite for UFM, i.e., the enough feature representing capability of the over-parameterized DNN, which means more parameters do not affect the overall geometric properties of federated learning. The classifiers of lack classes are almost completely unpredictable, due to the fact that they have not been trained.

5.2   The Lost Sampling Information

Based on empirical observations in Fig. 1 and Fig. 2, the two sets mentioned in Sect. 4.1 both contain sampling information, the vector set from the global to local models, as well as from each features mean to each local class mean.

However, with the progress of global training, the inter-client NGM about local class feature mean increasingly highlights the client-class sampling information, while the NGM about local models loses this information more and more.

A straightforward inference is that more and more non-sampling information is being introduced, which means that more randomness is being introduced as well. Furthermore, we are inspired by the explanation in Sect. 5.1, which suggests that this randomness may come from intrinsic information in the data or from the unpredictable optimization landscape and algorithmic stochastic noise during imbalanced training.

5.3   The Frame Drift

Continuing from the conclusion drawn in Sect. 5.1, it is certain that the results of local training using a minority class framework as noise will be amplified when using client sampling probability as model aggregation weight. This aggregation strategy may be the very reason for the phenomenon of FD. It can be specifically represented as when aggregating that $\bar{W} \leftarrow Aggregate_{N,K}(\mathcal{W}) := \mathcal{W}P_N$ in classic FL with aggregation via client sampling, but it should be

$$Aggregate_{N,K}(\mathcal{W}) := (\mathcal{W}P_{N,K}), \bar{W}^k = \mathbf{E}_{i,k}W_i^k,$$

where $\cdot^k$ is the kth covariant basis or row vector of $\cdot$.

Let's take an example of a binary classification problem for two clients: $P_{i=0,K} = [9999, 1]$, $P_{i=1,K} = [1, 99]$. In this case, we can calculate that the proportion $P_{i=0} = 10000/10100 \approx 99\%$, $P_{i=1} = 100/10100 \approx 1\%$. This means that the noise for the classifier of 1th class from 0th client is amplified by a factor about 10,000, which can interfere with the aggregation of this classifiers.

6   Inspiration and Discussions

As mentioned in **Observation 1** in Sect. 3.5, at the end of FL training, due to the presence of cosine factors, the distribution of the loss value will lose the client-class sampling information because they tend to be orthogonal theoretically. If strategy based on client-class sampling information is needed, one can avoid using differences between loss functions, or switch strategies at the end of training. It should be noted that for cases with fewer local epochs or larger inter-class differences, it takes a long time to reach the "end of training" condition.

With Theorem 3, on the client side, the features of classes with a lower probability of being sampled are more likely to be influenced by the features between them, and spread out the angle. Among classes with a large difference in sampling probability, the features of classes with a higher probability of being sampled have a greater influence on the angle. We can use this to construct a strategy to adjust the MC phenomenon in FL.

As mentioned in Sect. 5, based on the phenomenon of local subspace MC, intrinsic information in the noise of the classifiers of local minority classes can be extracted, which theoretically should improve local training effectiveness. Based on the phenomenon of FD, as well as an analysis of its causes, we suggest to use client-class sampling aggregation in FL.

7   Conclusion

In this paper, we discuss the geometric characteristics of FL both theoretically and empirically, providing Fundamental insight and inspiration for future work

Theoretically, we propose a FL modeling based on client-class sampling, and an UFM based on this modeling. We introduce the theorem that the UFM model is bounded near the minimum, allowing it to return to the FL optimization pattern based on regularization, by splitting the global model into client sampling expectation, and to some extent improving the rationality of UFM in FL. Meanwhile, we explore the source of FL client drift. We rewrite the closed-form cosine of feature mean angle, decoupling the complex relationship between angles and class sampling probabilities on given client.

Empirically, we find the client-class sampling information that is inherent in classical FL supporting our modeling, which gradually disappears between classifiers on the client-side and manifests itself in the features. We propose the general local and global geometric patterns of FL, which are respectively the local subspace MC and FD.

Acknowledgements. This work is supported by the Key Program of National Science Foundation of China (Grant No. 61836006) from College of Computer Science (Sichuan University) and Engineering Research Center of Machine Learning and Industry Intelligence (Ministry of Education), Chengdu 610065, P. R. China.

References

1. Caldas, S., et al.: Leaf: A benchmark for federated settings. arXiv:1812.01097 (2018)
2. Cochran, D., Gish, H., Sinno, D.: A geometric approach to multiple-channel signal detection. IEEE Trans. Signal Process. **43**(9), 2049–2057 (1995)
3. Dinh, C.T., Vu, T.T., Tran, N.H., Dao, M.N., Zhang, H.: Fedu: a unified framework for federated multi-task learning with laplacian regularization. arXiv:2102.07148 (2021)
4. Fang, C., He, H., Long, Q., Su, W.J.: Exploring deep neural networks via layer-peeled model: Minority collapse in imbalanced training. Proc. National Acad. Sci.**118**(43), e2103091118 (2021)
5. Gish, H., Cochran, D.: Generalized coherence (signal detection). In: ICASSP-88., International Conference on Acoustics, Speech, and Signal Processing, pp. 2745–2748 vol 5 (1988)
6. Hosang, J., Omran, M., Benenson, R., Schiele, B.: Taking a deeper look at pedestrians. In: 2015 IEEE Conference on Computer Vision and Pattern Recognition (CVPR), pp. 4073–4082. IEEE, Boston, MA, USA (2015), cIFARNET
7. Howard, S.D., Sirianunpiboon, S., Cochran, D.: Invariance of the distributions of normalized gram matrices. In: 2014 IEEE Workshop on Statistical Signal Processing (SSP), pp. 352–355 (2014)
8. Howard, S.D., Sirianunpiboon, S., Cochran, D.: The geometry of coherence and its application to cyclostationary time series. In: 2018 IEEE Statistical Signal Processing Workshop (SSP), pp. 348–352 (2018)
9. Huang, C., Xie, L., Yang, Y., Wang, W., Lin, B., Cai, D.: Neural collapse inspired federated learning with non-iid data (2023)
10. Hui, L., Belkin, M., Nakkiran, P.: Limitations of neural collapse for understanding generalization in deep learning (2022)
11. Kothapalli, V.: Neural collapse: A review on modelling principles and generalization (2023)

12. Li, G., Zhu, Z., Yang, D., Chang, L., Bai, H.: On projection matrix optimization for compressive sensing systems. IEEE Trans. Signal Process. **61**(11), 2887–2898 (2013)
13. Li, T., Sahu, A.K., Zaheer, M., Sanjabi, M., Talwalkar, A., Smith, V.: Federated optimization in heterogeneous networks. Proc. Mach. Learn. Syst. **2**, 429–450 (2020)
14. Li, Z., Shang, X., He, R., Lin, T., Wu, C.: No fear of classifier biases: Neural collapse inspired federated learning with synthetic and fixed classifier (2023)
15. McMahan, B., Moore, E., Ramage, D., Hampson, S., Arcas, B.A.y.: Communication-Efficient Learning of Deep Networks from Decentralized Data. In: Singh, A., Zhu, J. (eds.) Proceedings of the 20th International Conference on Artificial Intelligence and Statistics. Proceedings of Machine Learning Research, vol. 54, pp. 1273–1282. PMLR (2017)
16. Mixon, D.G., Parshall, H., Pi, J.: Neural collapse with unconstrained features (2020)
17. Papyan, V., Han, X.Y., Donoho, D.L.: Prevalence of neural collapse during the terminal phase of deep learning training. Proc. Natl. Acad. Sci. **117**(40), 24652–24663 (2020)
18. Qu, L., et al.: Rethinking architecture design for tackling data heterogeneity in federated learning. In: Proceedings of the IEEE/CVF Conference on Computer Vision and Pattern Recognition (CVPR), pp. 10061–10071 (2022)
19. Sastry, C.S., Oore, S.: Detecting out-of-distribution examples with Gram matrices. In: III, H.D., Singh, A. (eds.) Proceedings of the 37th International Conference on Machine Learning. Proceedings of Machine Learning Research, vol. 119, pp. 8491–8501. PMLR (2020)
20. Shi, M., Zhou, Y., Ye, Q., Lv, J.: Personalized federated learning with hidden information on personalized prior (2022)
21. T Dinh, C., Tran, N., Nguyen, J.: Personalized federated learning with moreau envelopes. J. Adv. Neural Inform. Process. Syst. **33**, 21394–21405 (2020)
22. T. Dinh, C., Tran, N., Nguyen, J.: Personalized federated learning with moreau envelopes. In: Advances in Neural Information Processing Systems, vol. 33, pp. 21394–21405. Curran Associates, Inc. (2020)
23. Tishby, N., Pereira, F.C., Bialek, W.: The information bottleneck method (2000)
24. Tishby, N., Zaslavsky, N.: Deep learning and the information bottleneck principle. In: 2015 IEEE Information Theory Workshop (ITW), pp. 1–5 (2015)
25. Vaswani, A., et al.: Attention is all you need. In: Guyon, I., Luxburg, U.V., Bengio, S., Wallach, H., Fergus, R., Vishwanathan, S., Garnett, R. (eds.) Advances in Neural Information Processing Systems. vol. 30. Curran Associates, Inc. (2017)
26. Zhang, C., Xie, Y., Bai, H., Yu, B., Li, W., Gao, Y.: A survey on federated learning. Knowl.-Based Syst. **216**, 106775 (2021)
27. Zhu, H., Xu, J., Liu, S., Jin, Y.: Federated learning on non-iid data: a survey. Neurocomputing **465**, 371–390 (2021)

Mutually Guided Dendritic Neural Models

Yanzi Feng[1], Jian Wang[2(✉)], Peng Ren[1], and Sergey Ablameyko[3,4]

[1] College of Oceanography and Space Informatics, China University of Petroleum
(East China), Qingdao, China
[2] College of Science, China University of Petroleum (East China), Qingdao, China
wangjiannl@upc.edu.cn
[3] Belarusian State University, Minsk, Belarus
[4] United Institute for Informatics Problems of NAS of Belarus, Minsk, Belarus

Abstract. With the explosive growth of Internet data, it has become
quite easy to collect unlabeled training data in many practical machine
learning and its applications, but it is relatively difficult to obtain a
large amount of labeled training data. Meanwhile, based on the idea
of semi-supervised learning, we propose the mutually guided dendritic
neural models (MGDNM) framework, which can realize the expansion of
labeled datasets. MGDNM utilizes two base classifiers for data exchange,
so as to achieving complementary advantages. To simulate the problem
of insufficient labeled data, we used 20% of the dataset as the training
dataset. On this basis, we conducted experiments on three datasets (Iris,
Breast Cancer and Glass). By calculating the accuracy and confusion
matrices, the comparison shows that the classification effect of MGDNM
is significantly higher than dendritic neural model (DNM), Support Vector Machine (SVM), Gaussian Naive Bayesian (GaussianNB) and Back
Propagation Neural Network (BP). It shows that MGDNM framework
is effective and feasible.

Keywords: Mutual Guide · Dendritic Neural Model ·
Semi-Supervised Learning

1 Introduction

According to different learning models, machine learning can be divided into
supervised learning, semi-supervised learning and unsupervised learning [11].
Traditional supervised learning is used to build a model to predict new data by
learning from a large amount of labeled training data [12]. Unsupervised learning
discovers structure and patterns in the data by training with unlabeled data. To
avoid the potentially significant resource requirements for obtaining label data,
Gong et al. [5] proposed a novel adaptive autoencoder with redundancy control
(AARC) as an unsupervised feature selector. Supervised learning and unsupervised learning have their own advantages and disadvantages, the former is more
suitable for labeled data, while the latter is more suitable for unlabeled data.

B. Luo et al. (Eds.): ICONIP 2023, CCIS 1962, pp. 465–475, 2024.
https://doi.org/10.1007/978-981-99-8132-8_35

Semi-supervised learning is a combination of supervised learning and unsupervised learning. It uses a small number of labeled data and a large number of unlabeled data to train.

With the development of technology and the progress of the times, data in the internet is growing explosively. In many practical machine learning applications, collecting unlabeled training data has become quite easy, but obtaining a large amount of labeled training data has become relatively difficult. For instance, in medical image diagnosis and therapy, collecting and annotating a large number of medical images is a significant challenge, as it takes years to collect and requires a substantial workload and cost for radiologists to annotate [13]. Therefore, with limited labeled training data, semi-supervised learning attempts to use unlabeled data to help improve learning performance, which has attracted extensive attention and application in the past few years.

At present, many semi-supervised learning methods have been applied to practice. Han et al. [6] proposed a Co-teaching network, in which two networks sample their small loss instances as useful knowledge in each small batch of data, and pass these useful instances on to their peer networks for further training. The error flow propagates in a zigzag shape, allowing for robust training of deep neural networks under noise supervision. Berthelot et al. [3] proposed MixMatch algorithm, which unifies multiple semi-supervised learning methods in a framework, and obtains more advanced results in many datasets and labeled data volumes. On the basis of MixMatch, Li et al. [9] proposed the DivideMix network, which uses a Gaussian mixture model (GMM) to split the training dataset into labeled and unlabeled sets. The trained data is not only used for another classifier, but also for the current classifier for training. Tai et al. [14] proposed a mutual guide framework, which used ELMs as the basic classifier to achieve hyperspectral image classification. From the aspect of data, it uses prior knowledge to enhance training data and uses extended sample to obtain reliable model. Liu et al. [10] concentrated on distributed classification with positive and unlabeled data, and introduce a distributed positive and unlabeled learning (PU learning) algorithm based on Semi-Supervised Support Vector Machine. Wang et al. [17] proposed the Ada-FedSemi system, which enhances the performance of DL models by utilizing on-device labeled data and in-cloud unlabeled data, based on the federated semi-supervised learning (FSSL) technology. Adedigba et al. [1] proposed data augmentation methods for mammograms, as there is a restricted quantity of medical images available, and a class imbalance problem in the data. Zhang et al. [18] proposed a new hyperspectral image classification method called DPCMF to address the issue of limited training samples. It utilizes dense pyramidal convolution and multi-feature fusion to extract and exploit spatial and spectral information, achieving significant improvements in classification accuracy compared to other methods.

In this paper, we develop the mutually guided dendritic neural models (MGDNM) framework aimed at solving the problem of insufficient labeled samples. MGDNM uses two different dendritic neural models (DNM) to train. Each classifier is initialized with different parameters, and then iterate through the exchange of newly labeled data between two classifiers to obtain the final model. We conducted experiments on three universal datasets using MGDNM framework and compared it with some classic algorithms. MGDNM is superior to classical algorithms in performance.

2 Methodology

2.1 Dendritic Neural Model

Dendritic Neural Model (DNM) is a new type of single neuron model that incorporates plastic dendritic morphology, inspired by the biological behavior of neurons in vivo [8]. The neural structure of the DNM consists of four layers: the synaptic layer, the dendritic layer, the membrane layer, and the cell body. Each layer has different excitation functions that correspond to different neural functions. Theoretical and empirical evidence shows that the DNM can achieve satisfactory performance on various practical applications, such as computer-aided medical diagnosis [7], business risk [15], financial time series prediction [16] and so on.

2.2 Mutually Guided Dendritic Neural Models

We develop a mutually guided dendritic neural models (MGDNM) framework for classification, as shown in Fig. 1. First, we put the labeled training data into two base classifiers with different initialization parameters for training, enabling them to learn different aspects from the training data. Then the unlabeled data is predicted by each classifier, and each classifier selects a certain number of high confidence data for labeling. The two classifiers exchange their selected high confidence data, that is, one classifier puts the high confidence data and labels into the training data of the other. With the increase of the number of iterations, the labeled training set is gradually expanded, while the unlabeled data is gradually reduced, which makes the training effectiveness of the basic classifier continuously improved. The mutual guidance operation between the two basic classifiers terminates when the amount of guidance data is large enough to provide effective training.

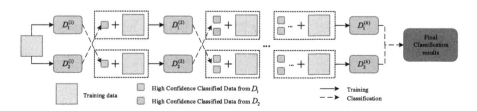

Fig. 1. Mutually guided dendritic neural models framework

Let L denote the labeled training dataset, U denote the unlabeled dataset, and Y_L denote the corresponding label of L. And let D_1 and D_2 denote two base classifiers with different initialization parameters α_1 and α_2. The classification results of the two classifiers on the unlabeled data are as follows:

$$Y_1 = D_1(U_1, \alpha_1, \beta_1) \tag{1}$$

$$Y_2 = D_2(U_2, \alpha_1, \beta_2) \tag{2}$$

where β_1 and β_2 represent the training parameters of the base classifiers D_1 and D_2, respectively.

When training to the kth iteration, it is as follows:

$$\hat{\beta}_1^{(k)} = \underset{\beta_1}{argmino}(Y_{L_1}^{(k-1)}, D_1(L_1^{(k-1)}, \alpha_1, \beta_1)) \tag{3}$$

$$\hat{\beta}_2^{(k)} = \underset{\beta_2}{argmino}(Y_{L_2}^{(k-1)}, D_2(L_2^{(k-1)}, \alpha_2, \beta_2)) \tag{4}$$

where $\hat{\beta}_1^{(k)}$ and $\hat{\beta}_2^{(k)}$ are the optimal model parameters obtained during training, and $o(\cdot)$ is the error function between the classification result and the correct result.

The unlabeled data is then classified according to the current model parameters $\hat{\beta}_1^{(k)}$ and $\hat{\beta}_2^{(k)}$. The results are shown in Eq. (5) and (6):

$$Y_{1H}^{(k)} = \Re_\gamma(D_1(U_1^{(k)}, \alpha_1, \hat{\beta}_1^{(k)})) \tag{5}$$

$$Y_{2H}^{(k)} = \Re_\gamma(D_2(U_2^{(k)}, \alpha_2, \hat{\beta}_2^{(k)})) \tag{6}$$

where $\Re_\gamma(\cdot)$ is the sorting function, which takes data with the top γ percentages of confidence, and Y_{1H} and Y_{2H} are the classification results with high confidence.

Let X_{1H} and X_{2H} denote the data in U corresponding to Y_{1H} and Y_{2H}, respectively. Finally, X_{2H} is taken out from U_2 and put into the labeled dataset of classifier D_1, so the labeled training dataset of classifier D_1 is updated. Similarly, X_{1H} is taken out from U_1 and put into the labeled dataset of classifier D_2, so the labeled training dataset of classifier D_2 is updated. The details are as follows:

$$L_1^{(k)} = L_1^{(k-1)} \cup X_{2H}^{(k)} \tag{7}$$

$$L_2^{(k)} = L_2^{(k-1)} \cup X_{1H}^{(k)} \tag{8}$$

$$Y_1^{(k)} = Y_1^{(k-1)} \cup Y_{2H}^{(k)} \tag{9}$$

$$Y_2^{(k)} = Y_2^{(k-1)} \cup Y_{1H}^{(k)} \tag{10}$$

The above completes an expansion of the labeled training dataset. In this way, the two classifiers are trained with the expanded dataset, and the mutual guidance is stopped when the labeled training dataset is expanded enough or the number of iterations reaches the specified number of rounds.

3   Experiments Setup

3.1   Datasets

In order to verify the effectiveness of MGDNM, we conducted experiments on three datasets, including Iris, Breast Cancer and Glass. Table 1 summarizes the information on these datasets.

Table 1. Summary of four classification datasets

Dataset	Classes	Features	Dataset Size
Iris	3	4	150
Breast Cancer	2	9	699
Glass	2	9	214

3.2   Evaluation Metrics

To evaluate the performance of our model, we used two metrics, namely accuracy and the confusion matrix.

In machine learning, accuracy is often used to evaluate the performance of a model. It is calculated by dividing the number of correct predictions by the total number of predictions. So it is a measure of how close a predicted value is to the actual value. In other words, it is a measure of how well a model is able to correctly predict the outcome of a given task. Overall, accuracy is an important metric for evaluating the performance of a model.

Confusion matrix is a popular tool used in the field of machine learning to evaluate the performance of a classification model. It summarizes the performance of a model by comparing its predicted outputs with the actual outputs. Confusion matrix is a square matrix with the number of rows and columns equal to the number of classes in the dataset. It contains four different categories, namely true positive, false positive, true negative, and false negative. The true positive (TP) represents the number of correctly predicted positive instances, while false positive (FP) represents the number of incorrectly predicted positive instances. Similarly, true negative (TN) represents the number of correctly predicted negative instances, and false negative (FN) represents the number of incorrectly predicted negative instances. In addition to evaluating the performance of a classification model, confusion matrix can also be used to identify areas where the model needs improvement.

3.3   Experimental Design

According to the principle of MGDNM framework, the designed experiment is shown in Fig. 2.

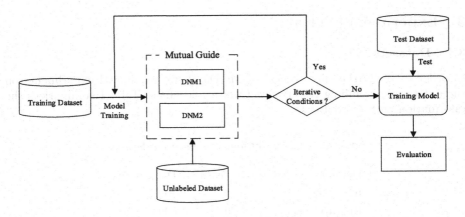

.**Fig. 2.** Experimental flow chart

To demonstrate the performance of MGDNM, we compare it with DNMs (the two base classifiers DNM1 and DNM2), Support Vector Machine (SVM) [4], Gaussian Naive Bayesian (GaussianNB) and Back Propagation Neural Network (BP) [2]. These algorithms use the same training set, and then we adjust parameters of them to achieve the best results compared to MGDNM. Each dataset is randomly divided into three parts, namely an unlabeled dataset, a test dataset, and a labeled training dataset. For each dataset, 20% of the data is used as the training dataset, 25% of the data is used as the test dataset, and 55% of the data is used as the unlabeled dataset. The performance of each method is measured by accuracy and confusion matrices.

4   Results

4.1   Iris Dataset

For Iris dataset, we split the dataset into two categories, resulting in a binary classification problem of Setosa and Non-Setosa. The evaluation metrics obtained by the six algorithms on Iris dataset are shown in Table 2. Due to the small number of training dataset, MGDNM still shows high accuracy on iris dataset. Based on DNMs, MGDNM can improve the accuracy by nearly 20%. Moreover, the accuracy of MGDNM is also higher than the other three algorithms SVM, GaussianNB and BP.

Table 2. Comparison of different methods on Iris dataset

	MGDNM	DNM1	DNM2	SVM	GaussianNB	BP
training accuracy(%)	100.00	70.00	70.00	66.67	93.33	66.67
test accuracy(%)	92.50	72.50	77.50	70.00	82.50	70.00

Observing both Table 2 and Fig. 3, it can be seen that MGDNM has improved in some misclassified instances of DNMs. As shown in Fig. 3, there is no false negative in MGDNM, but 9 false negatives in DNM1, 3 false negatives in DNM2. Correspondingly, the amount of true positive in MGDNM increased compared to DNM1 and DNM2. This validates the improvement of MGDNM in the classification of positive instances. Similarly, we can see that MGDNM also improves over DNM1 and DNM2 in negative instance classification. However, MGDNM has one more true negative and one less false positive than DNM1. This indicates that the learning ability of MGDNM in terms of negative instances is still insufficient. In addition, comparing the other three algorithms SVM, GaussianNB and BP, it can be seen that MGDNM has higher classification ability than them, which is consistent with the results in Table 2.

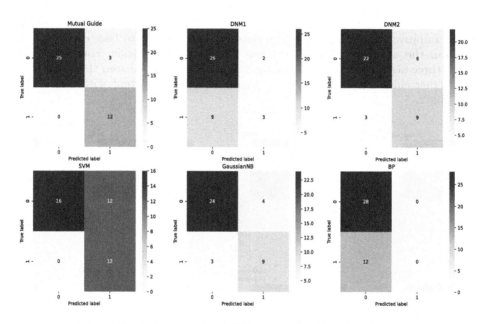

Fig. 3. Confusion matrix of different methods on Iris dataset

4.2   Breast Cancer Dataset

For Breast Cancer dataset, the data is divided into benign and malignant types. The evaluation metrics obtained by the six algorithms on Breast Cancer dataset are shown in Table 3. Although the number of training dataset is relatively small, each algorithm still has good classification results. Based on DNMs, MGDNM can improve the accuracy by nearly 2%. Moreover, the accuracy of MGDNM is also higher than the other three algorithms SVM, GaussianNB and BP.

Table 3. Comparison of different methods on Breast Cancer dataset

	MGDNM	DNM1	DNM2	SVM	GaussianNB	BP
training accuracy(%)	97.79	95.59	94.85	88.24	96.32	92.65
test accuracy(%)	96.17	93.99	92.90	86.34	93.44	95.08

Observing both Table 3 and Fig. 4, it can be seen that MGDNM has improved in some misclassified instances of DNMs. As shown in Fig. 4, there is 4 false negatives in MGDNM, but 8 false negatives in DNM1, 10 false negatives in DNM2. Correspondingly, the amount of true positives in MGDNM increased compared to DNM1 and DNM2. This validates the improvement of MGDNM in the classification of positive instances. However, MGDNM has the same number of false positives and true negatives as DNM1 and DNM2, which indicates that MGDNM is not improved in negative instance classification by the two base classifiers. In general, the accuracy of MGDNM has improved. In addition, comparing the other three algorithms SVM, GaussianNB and BP, it can be seen that MGDNM has higher classification ability than them, which is consistent with the results in Table 2.

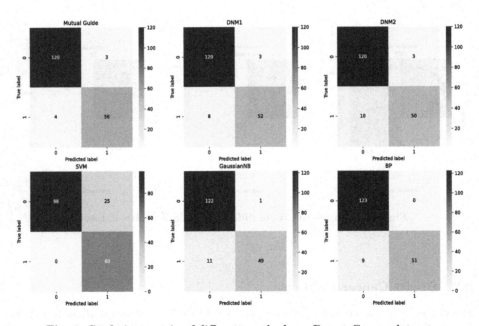

Fig. 4. Confusion matrix of different methods on Breast Cancer dataset

4.3 Glass Dataset

For Glass dataset, we will utilize nine characteristics to differentiate whether a sample is a window glass or a non-window glass. The evaluation metrics obtained by the six algorithms on Glass dataset are shown in Table 4. Although the number of training dataset is relatively small, each algorithm still has good classification results. Based on DNMs, MGDNM can improve the accuracy by nearly 2%. Moreover, the accuracy of MGDNM is also higher than the other three algorithms SVM, GaussianNB and BP. Meanwhile, the accuracy of BP on the training dataset is as high as that of MGDNM.

Table 4. Comparison of different methods on Glass dataset

	MGDNM	DNM1	DNM2	SVM	GaussianNB	BP
training accuracy(%)	100.00	97.62	97.62	92.86	90.48	100.00
test accuracy(%)	93.10	91.38	91.38	79.31	89.66	91.38

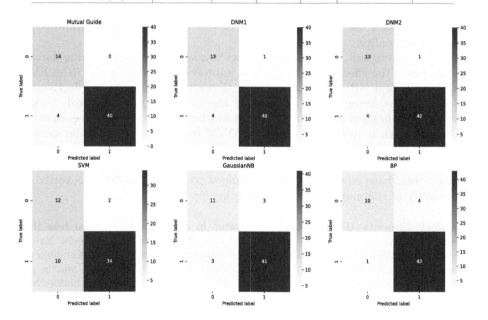

Fig. 5. Confusion matrix of different methods on Glass dataset

Observing both Table 4 and Fig. 5, it can be seen that MGDNM has improved in some misclassified instances of DNMs. As shown in Fig. 5, there is no false positive in MGDNM, but 1 false positive in DNM1, 1 false positive in DNM2. Correspondingly, the amount of true negatives in MGDNM increased compared to DNM1 and DNM2. This validates the improvement of MGDNM in the classification of negative instances. However, MGDNM has the same number of false negatives and true positives as DNM1 and DNM2, which indicates that MGDNM is not improved in positive instance classification by the two base classifiers.

In addition, comparing the other three algorithms SVM, GaussianNB and BP, it can be seen that MGDNM has higher classification ability than them, which is consistent with the results in Table 4. There is 40 true positives in MGDNM, but 41 true positives in GaussianNB, 43 true positives in BP. This indicates that MGDNM is deficient in the ability to classify positive instances. Of course, it can also be seen from Fig. 5 that MGDNM has advantages in negative instance classification. In general, the test accuracy of MGDNM is higher than that of other algorithms DNM, SVM, GaussianNB and BP.

5   Conclusion

In this paper, we proposed MGDNM framework, which is based on two base classifier DNMs with different initialization parameters. The two basic classifiers exchange the newly labeled data and labels, and the training dataset is extended after many iterations, so as to improve the classification performance. MGDNM framework makes use of the advantages of two base classifiers to improve the classification ability.

In this experiment, 20% of each dataset was selected as the training dataset. On this basis, the effect of MGDNM was improved compared with DNM, SVM, GaussianNB and BP, by 2% to 20%. It shows that MGDNM is effective and feasible. At the same time, by observing the confusion matrix, it can be seen that the classification ability of MGDNM has been improved in some aspects.

Future work is mainly to apply MGDNM framework to more datasets for experiments, such as remote sensing images and other actual scenes. At the same time, MGDNM should be improved to enhance its data exchange ability. And the two base classifiers can be changed to other classifiers to verify the universality of the mutual guide framework.

Acknowledgments. This work was supported in part by the National Key R&D Program of China under Grant 2019YFA0708700; in part by the National Natural Science Foundation of China under Grant 62173345; and in part by the Fundamental Research Funds for the Central Universities under Grant 22CX03002A; and in part by the China-CEEC Higher Education Institutions Consortium Program under Grant 2022151; and in part by the Introduction Plan for High Talent Foreign Experts under Grant G2023152012L; and in part by the "The Belt and Road" Innovative Talents Exchange Foreign Experts Project under Grant DL2023152001L.

References

1. Adedigba, A.P., Adeshina, S.A., Aibinu, A.M.: Performance evaluation of deep learning models on mammogram classification using small dataset. Bioengineering **9**(4), 161 (2022)

2. Bansal, A., Singhrova, A.: Performance analysis of supervised machine learning algorithms for diabetes and breast cancer dataset. In: 2021 International Conference on Artificial Intelligence and Smart Systems (ICAIS), pp. 137–143. IEEE (2021)

3. Berthelot, D., Carlini, N., Goodfellow, I., Papernot, N., Oliver, A., Raffel, C.A.: Mixmatch: a holistic approach to semi-supervised learning. In: Advances in Neural Information Processing Systems 32 (2019)

4. Chandra, M.A., Bedi, S.: Survey on SVM and their application in image classification. Int. J. Inf. Technol. **13**, 1–11 (2021)

5. Gong, X., Yu, L., Wang, J., Zhang, K., Bai, X., Pal, N.R.: Unsupervised feature selection via adaptive autoencoder with redundancy control. Neural Netw. **150**, 87–101 (2022)

6. Han, B., et al.: Co-teaching: Robust training of deep neural networks with extremely noisy labels. In: Advances in Neural Information Processing Systems 31 (2018)

7. Ji, J., Dong, M., Lin, Q., Tan, K.C.: Noninvasive cuffless blood pressure estimation with dendritic neural regression. IEEE Trans. Cybern. **53**(7), 4162–4174 (2023). https://doi.org/10.1109/TCYB.2022.3141380

8. Ji, J., Gao, S., Cheng, J., Tang, Z., Todo, Y.: An approximate logic neuron model with a dendritic structure. Neurocomputing **173**, 1775–1783 (2016)

9. Li, J., Socher, R., Hoi, S.C.: Dividemix: learning with noisy labels as semi-supervised learning. arXiv:2002.07394 (2020)

10. Liu, Y., Zhang, C.: Distributed semi-supervised learning with positive and unlabeled data. In: 2022 3rd International Conference on Big Data, Artificial Intelligence and Internet of Things Engineering (ICBAIE), pp. 488–492. IEEE (2022)

11. Ning, X., et al.: A review of research on co-training. Concurrency and computation: practice and experience p. e6276 (2021)

12. Saravanan, R., Sujatha, P.: A state of art techniques on machine learning algorithms: A perspective of supervised learning approaches in data classification. In: 2018 Second International Conference on Intelligent Computing and Control Systems (ICICCS), pp. 945–949 (2018). https://doi.org/10.1109/ICCONS.2018.8663155

13. Suzuki, K.: Small data deep learning for lung cancer detection in ct. In: 2022 IEEE Eighth International Conference on Big Data Computing Service and Applications (BigDataService), pp. 114–118 (2022). https://doi.org/10.1109/BigDataService55688.2022.00025

14. Tai, X., Li, M., Xiang, M., Ren, P.: A mutual guide framework for training hyperspectral image classifiers with small data. IEEE Trans. Geosci. Remote Sens. **60**, 1–17 (2021)

15. Tang, Y., Ji, J., Zhu, Y., Gao, S., Tang, Z., Todo, Y., et al.: A differential evolution-oriented pruning neural network model for bankruptcy prediction. Complexity 2019 (2019)

16. Tang, Y., Song, Z., Zhu, Y., Hou, M., Tang, C., Ji, J.: Adopting a dendritic neural model for predicting stock price index movement. Expert Syst. Appl. **205**, 117637 (2022)

17. Wang, L., Xu, Y., Xu, H., Liu, J., Wang, Z., Huang, L.: Enhancing federated learning with in-cloud unlabeled data. In: 2022 IEEE 38th International Conference on Data Engineering (ICDE), pp. 136–149. IEEE (2022)

18. Zhang, J., et al.: Hyperspectral image classification based on dense pyramidal convolution and multi-feature fusion. Remote Sens. **15**(12), 2990 (2023)

Social-CVAE: Pedestrian Trajectory Prediction Using Conditional Variational Auto-Encoder

Baowen Xu[1,2] ⓘ, Xuelei Wang[1(✉)] ⓘ, Shuo Li[1,2] ⓘ, Jingwei Li[1,2] ⓘ,
and Chengbao Liu[1] ⓘ

[1] Institute of Automation, Chinese Academy of Sciences, Beijing 100190, China
xubaowen2021@ia.ac.com
[2] School of Artificial Intelligence, University of Chinese Academy of Sciences,
Beijing 100049, China

Abstract. Pedestrian trajectory prediction is a fundamental task in applications such as autonomous driving, robot navigation, and advanced video surveillance. Since human motion behavior is inherently unpredictable, resembling a process of decision-making and intrinsic motivation, it naturally exhibits multimodality and uncertainty. Therefore, predicting multi-modal future trajectories in a reasonable manner poses challenges. The goal of multi-modal pedestrian trajectory prediction is to forecast multiple socially plausible future motion paths based on the historical motion paths of agents. In this paper, we propose a multi-modal pedestrian trajectory prediction method based on conditional variational auto-encoder. Specifically, the core of the proposed model is a conditional variational auto-encoder architecture that learns the distribution of future trajectories of agents by leveraging random latent variables conditioned on observed past trajectories. The encoder models the channel and temporal dimensions of historical agent trajectories sequentially, incorporating channel attention and self-attention to dynamically extract spatio-temporal features of observed past trajectories. The decoder is bidirectional, first estimating the future trajectory endpoints of the agents and then using the estimated trajectory endpoints as the starting position for the backward decoder to predict future trajectories from both directions, reducing cumulative errors over longer prediction ranges. The proposed model is evaluated on the widely used ETH/UCY pedestrian trajectory prediction benchmark and achieves state-of-the-art performance.

Keywords: Pedestrian trajectory prediction · Conditional Variational Auto-Encoder · Attention · RWKV · RNN

This work is supported by National Key Research and Development Program of China [Grant 2022YFB3305401] and the National Nature Science Foundation of China [Grant 62003344].

1   Introduction

Recently, pedestrian trajectory prediction has attracted widespread attention in various fields, including autonomous driving vehicles, service robots, intelligent transportation, and smart cities. By accurately predicting the positions of surrounding pedestrians, autonomous driving vehicles can avoid traffic accidents. Robots with social awareness need to predict human trajectories in order to better plan their cruising paths and provide more efficient services. Intelligent tracking and surveillance systems used in urban planning must understand how crowds interact in order to better manage city infrastructure. However, due to the complexity and uncertainty of human decision-making processes, accurately predicting the future trajectory of pedestrians still faces significant challenges.

The trajectory prediction task can be divided into Deterministic Trajectory Prediction (DTP) task and Multimodal Trajectory Prediction (MTP) task. Deterministic Trajectory Prediction refers to the model providing only one deterministic prediction for each agent. However, human behavior is naturally multimodal and uncertain, meaning that given past trajectories and surrounding environmental information, an agent can have multiple seemingly reasonable trajectories in the future. Multimodal Trajectory Prediction requires the model to provide multiple acceptable solutions, which can more realistically simulate human behavior.

The recently popular multimodal pedestrian trajectory prediction frameworks are noise-based models or deep generative models. Specifically, these models inject random noise into the model to generate multiple plausible paths. Noise-based trajectory prediction models can be categorized into frameworks such as Generative Adversarial Networks (GAN) [1], Conditional Variational Autoencoders (CVAE) [2–4], Normalizing Flows (NF) [5], and Denoising Diffusion Probabilistic Models (DDPM) [6]. The model we propose adopts Conditional Variational Autoencoder architecture and introduces a Social Conditional Variational Autoencoder (Social-CVAE) for multimodal pedestrian trajectory prediction. Specifically, Social-CVAE utilizes a conditional variational autoencoder architecture to learn the distribution of future trajectories for agents, conditioned on observed past trajectories through random latent variables. The encoder models the channel and time dimensions of the past trajectories successively to obtain latent variables. Subsequently, the endpoints of multi-modal trajectories are estimated. Finally, the multi-modal trajectories are predicted by combining the hidden states of forward propagation from the current position and the hidden states of backward propagation from the estimated trajectory endpoints.

In summary, the contributions of this work are as follows,

- We propose a novel framework based on Conditional Variational Auto-encoder (CVAE) for pedestrian multi-modal trajectory prediction. This framework learns the distribution of future trajectories of the agent by utilizing random latent variables conditioned on observed past trajectories. Extensive experiments on the ETH/UCY benchmarks demonstrate state-of-the-art performance.

- In the encoder, channel attention and self-attention are introduced to model the channel dimension and the time dimension, respectively, to dynamically extract spatio-temporal features from observable past trajectories. A bidirectional decoder is designed, which first predicts the multi-modal trajectory target points. Then, the estimated trajectory target points are used as a condition for the backward decoder, while the current position serves as the starting point for the forward decoder. The gate recurrent unit (GRU) and Receptance Weighted Key Value (RWKV) are employed as the basic modules for the backward and forward decoders, respectively.

2   Related Work

2.1   Research on Social Interaction Model

In the early days, social force [7] were used for interactions between individuals, modeling the interactions through forces (repulsion or attraction). However, social force failes to accurately capture interactions based on manually crafted features. Social-LSTM [8] represents the first application of recursive neural networks to simulate interactions between individuals in crowded scenes, modeling the social interactions and potential conflicts among pedestrians. Each trajectory is modeled as an LSTM layer, and different LSTM layers can share information through a social pooling layer to generate conflict-free trajectories. To address the deficiency of Social-LSTM in neglecting the current important intentions of neighbors, SR-LSTM [9] reevaluates the contributions of adjacent pedestrians to the target through a social perception information selection mechanism. Social-BiGAT [10] formulates pedestrian social interactions as a graph and assigns higher edge weights to more significant interactions. NSP [11] achieves multimodal trajectory prediction by incorporating an explicit physical model with learnable parameters within a deep neural network. In contrast to social force models and their variants, the key parameters of NSP's deterministic model can be learned from data instead of being manually selected and fixed.

2.2   Research on Multimodal Pedestrian Trajectory Prediction

Human behavior is inherently multimodal and uncertain, meaning that given past trajectories and surrounding environmental information, an agent can have multiple seemingly reasonable future trajectories. Multimodal trajectory prediction requires models to provide multiple acceptable solutions, which can more realistically simulate human behavior. Recently popular multimodal pedestrian trajectory prediction frameworks are based on noise models or deep generative models, specifically injecting random noise into the model to generate multiple plausible paths.

Generative models-based prediction methods utilize core architectures such as Generative Adversarial Networks (GANs), Conditional Variational Autoencoders (CVAEs), and Denoising Diffusion Probabilistic Models (DDPM). GAN-based methods utilize random noise as input to the model to generate stochastic

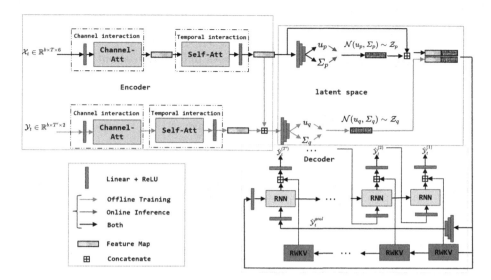

Fig. 1. Overview of Model Architecture.

results. Social-GAN [1], SoPhie [12], and Social-Ways [13] employ GAN architectures. Regarding social interactions, Social-GAN proposes a pooling network to aggregate the states of neighbors, while Social-Ways utilizes an attention mechanism to aggregate the states of neighbors. PECNet [14], Trajectron++ [15], Social-VAE [2], and BiTrap [4] use the CVAE architecture to directly predict trajectory distributions. MID [6] achieves multimodal trajectory prediction by employing a diffusion model that gradually reduces the prediction uncertainty through learning the reversal from deterministic ground truth trajectories to stochastic Gaussian noise.

3   Method

In this section, we first briefly introduce the data preparation for pedestrian trajectory prediction; Finally, the proposed model architecture and Loss function are introduced in detail.

3.1   Data Preparation

The position of agent i at timestamp t is represented by $p_i^{(t)} = (x, y) \in \mathbb{R}^2$, where x and y are 2D spatial coordinates. The trajectory of agent i from timestamp 1 to T is defined as $P_i^{1:T} = \{p_i^{(1)}, p_i^{(2)}, ..., p_i^{(T)}\} \in \mathbb{R}^{T \times 2}$. The prediction task in this paper is to infer the future trajectory distribution of each agent in the scene based on the agent's past trajectory. Specifically, based on the joint observations of T frames in the entire scene, inference is performed on the distribution of future positions $\mathcal{Y}_t = \{p^{(T+1)}, p^{(T+2)}, \dots p^{(T+T')}\}$ for T' frames. This can be represented as the predictive distribution $p(\hat{\mathcal{Y}}_t | \mathcal{X}_t)$, where $\mathcal{X}_t^i = \{p_i^{(t)}, d_i^{(t)} = p_i^{(t)} - p_i^{(t-1)}, a_i^{(t)} = d_i^{(t)} - d_i^{(t-1)}\} \in \mathbb{R}^6$ includes information about the position, velocity, and acceleration of agent i.

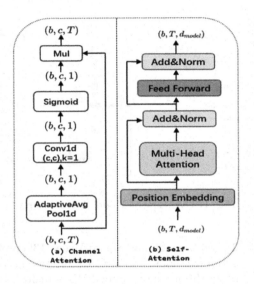

Fig. 2. Description of the attention used.

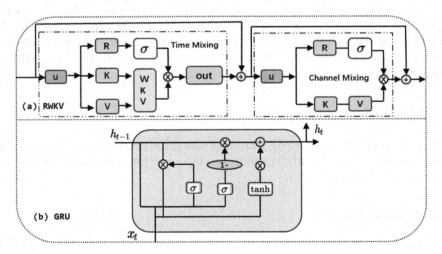

Fig. 3. Description of RWKV and GRU.

3.2   Overall Model Architecture

To generate random predictions, we employ a variational autoencoder architecture to achieve multimodal trajectory prediction by introducing random latent variables. Given the past trajectory information $\mathcal{X}_t$ of the agent, we first estimate the target endpoint $\hat{\mathcal{Y}}_t^{goal}$ and then predict the future trajectory $\hat{\mathcal{Y}}_t$ of the agent. The proposed CVAE architecture model is divided into a generative model for generating predicted trajectories and an inference model for posterior approximation. The detailed descriptions of the inference model and the generative model are presented in Sects. 3.3 and 3.4, respectively, and the overall architecture of the proposed model is illustrated in Fig. 1.

3.3   Inference Model

Let $\mathcal{Z}$ represent the latent variable, and $\mathcal{X}_t$ and $\mathcal{Y}_t$ represent the agent's historical trajectory and future trajectory, respectively. The inference model approximating the posterior q on the latent variable can be described as follows. Firstly, the feature encodings of the agent's historical and future trajectories can be described as follows,

$$\mathcal{H}_x = \psi_x(\mathcal{X}_t), \tag{1}$$

$$\mathcal{H}_y = \psi_y(\mathcal{Y}_t), \tag{2}$$

where ψ_x and ψ_y represent the networks that encode the historical and future trajectories, respectively. The posterior features can be represented as follows,

$$\mathcal{H}_{xy} = \psi_{xy}(\mathcal{H}_x, \mathcal{H}_y), \tag{3}$$

where ψ_{xy} is the embedding neural network. The distribution followed by the posterior latent variable can be represented as,

$$p(\mathcal{Z}|\mathcal{X}_t, \mathcal{Y}_t) := q_\phi(\mathcal{Z}|\mathcal{H}_{xy}), \tag{4}$$

where ϕ represents the parameters of the posterior neural network. The posterior can be obtained by sampling from $\mathcal{Z} \sim q_\phi(\mathcal{Z}|\mathcal{H}_{xy})$.

3.4   Generative Model

The generative model based on the CVAE architecture can be represented as,

$$p(\hat{\mathcal{Y}}_t|\mathcal{X}_t) = \int_{\mathcal{Z}} p(\hat{\mathcal{Y}}_t|\mathcal{X}_t, \mathcal{Z})p(\mathcal{Z}|\mathcal{X}_t)d\mathcal{Z},$$
$$= \int_{\mathcal{Z}} p(\hat{\mathcal{Y}}_t|\mathcal{X}_t, \hat{\mathcal{Y}}_t^{goal}, \mathcal{Z})p(\hat{\mathcal{Y}}_t^{goal}|\mathcal{X}_t, \mathcal{Z})p(\mathcal{Z}|\mathcal{X}_t)d\mathcal{Z}. \tag{5}$$

The encoding $\mathcal{H}_x$ of the agent's history $\mathcal{X}_t$ can be represented as shown in Eq. (1). The third term in the integral of Eq. (5) can be described as,

$$p(\mathcal{Z}|\mathcal{X}_t) := p_\theta(\mathcal{Z}|\mathcal{H}_x), \tag{6}$$

where θ represents the parameters of the prior neural network. The prior can be obtained by sampling from $\mathcal{Z} \sim p_\theta(\mathcal{Z}|\mathcal{H}_x)$. The second term in the integral of Eq. (5) can be described as,

$$\mathcal{H}_{xz} = Concat(\mathcal{H}_x, \mathcal{Z}), \ p(\hat{\mathcal{Y}}_t^{goal}|\mathcal{X}_t, \mathcal{Z}) := p_\epsilon(\hat{\mathcal{Y}}_t^{goal}|\mathcal{H}_{xz}), \tag{7}$$

where ϵ represents the network parameters of the trajectory endpoint decoder, and $Concat(\cdot, \cdot)$ represents the concatenation operation. The first term in the integral of Eq. (5) implies sampling the future trajectory from the hidden features $\mathcal{H}_{xz}$, and estimated target endpoint $\hat{\mathcal{Y}}_t^{goal}$.

$$\hat{\mathcal{Y}}_t \sim p_\xi(\cdot|\mathcal{H}_{xz}, \hat{\mathcal{Y}}_t^{goal}), \tag{8}$$

where ξ represents decoder network parameters.

3.5   Observation Encoding

Due to the fact that trajectory data belongs to spatio-temporal data and has spatio-temporal characteristics. Therefore, we model the agent's trajectory separately in the channel dimension and the temporal dimension. Specifically, the feature extraction module includes channel interaction and temporal interaction. Channel interaction is composed of feed-forward neural networks and channel attention. Given the agent's history $\mathcal{X}_t \in \mathbb{R}^{T \times 6}$, the channel interaction $\mathcal{X}_{channel} \in \mathbb{R}^{T \times d_{model}}$ can be described as follows:

$$\mathcal{X}'_{channel} = \mathcal{X}_t W_{channel} + b_{channel}, \ \mathcal{X}_{channel} = f_{channel}(\mathcal{X}'_{channel}), \tag{9}$$

where $W_{channel} \in \mathbb{R}^{6 \times d_{model}}$ and $b_{channel} \in \mathbb{R}^{d_{model}}$ represent the parameters of the feed-forward neural networks. $f_{channel}(\cdot)$ is the channel attention as shown in Fig. 2(a), and the channel attention [16] can be expressed as follows:

$$s = \sigma(W_2 \delta(W_1 GAP(\mathcal{X}'_{channel}))), \ \mathcal{X}_{channel} = s\mathcal{X}'_{channel}, \tag{10}$$

where σ and δ respectively represent Sigmoid and ReLU, W_1 and W_2 represent the fully-connected layers, and $GAP(\cdot)$ represents global average pooling. The temporal interaction is composed of multi-head attention, as shown in Fig. 2 (b). Given the output $\mathcal{X}_{channel}$ of the channel interaction, the output $\mathcal{X}_{temporal}$ of the temporal interaction can be represented as follows. First, we follow the strategy of Fourier feature positional encoding to preserve the temporal information of the historical positions.

$$\mathcal{PE}(pos, 2i) = \sin(pos \cdot e^{-\frac{4i}{d_{model}}}), \ \mathcal{PE}(pos, 2i+1) = \cos(pos \cdot e^{-\frac{4i}{d_{model}}}), \tag{11}$$

where d_{model} represents the model embedding dimension, $i \in [0, d_{model})$ specifies the position in the model embedding dimension, and $pos \in [0, T)$ specifies the position in temporal. The output $\mathcal{X}_{att} \in \mathbb{R}^{T \times d_{model}}$ of the multi-head attention can be described as follows:

$$
\begin{aligned}
\mathcal{X}'_{att} &= \mathcal{X}_{channel} + \mathcal{PE}, \\
Q_i &= \mathcal{X}'_{att} W_i^Q, K_i = \mathcal{X}'_{att} W_i^K, V_i = \mathcal{X}'_{att} W_i^V, \\
h_i &= Attention(Q_i, K_i, V_i), \\
\mathcal{X}_{att} &= Concat[h_1, h_2, ..., h_i] W^o + \mathcal{X}'_{att},
\end{aligned}
\tag{12}
$$

where W_i^Q, W_i^K, W_i^V, and W^o represent the embedding parameter matrices. The self-attention can be described as follows:

$$
Attention(Q, K, V) = softmax(\frac{QK^T}{\sqrt{d}})V, \tag{13}
$$

where Q, K and V represent query, key, and value, respectively, and d represents the embedding dimension. Finally, reshaping the dimension $\mathbb{R}^{T \times d_{model}}$ of $\mathcal{X}_{att}$ into $\mathbb{R}^{T \cdot d_{model}}$, we obtain the output $\mathcal{X}_{temporal} \in \mathbb{R}^{d_{model}}$ of the temporal interaction through a feed-forward neural network.

$$
\mathcal{X}_{temporal} = \mathcal{X}_{temporal} W_{temporal} + b_{temporal}, \tag{14}
$$

where $W_{temporal} \in \mathbb{R}^{T \cdot d_{model} \times d_{model}}$ and $b_{temporal} \in \mathbb{R}^{d_{model}}$ represent the parameters of the feed-forward neural networks.

3.6   Latent Variables Estimation

The trajectory prediction model based on CVAE achieves multimodal trajectory prediction by introducing a latent variable $\mathcal{Z}$. It assumes that $\mathcal{Z}$ follows a Gaussian distribution, with a prior network $p_\theta(\mathcal{Z}_p | \mathcal{H}_x) \sim \mathcal{N}(\mu_p, \Sigma_p)$ and an recognition network $q_\phi(\mathcal{Z}_q | \mathcal{H}_{xy}) \sim \mathcal{N}(\mu_q, \Sigma_q)$. Both the prior and recognition networks are composed of three feed-forward neural networks. Then, the re-parameterization technique is used to obtain samples of $\mathcal{Z}$ from $q_\phi(\cdot | \mathcal{H}_{xy})$ during training or from $p_\theta(\cdot | \mathcal{H}_x)$ during testing. This process can be represented as:

$$
\mathcal{Z} = \mu + \epsilon \odot \Sigma, \epsilon \sim \mathcal{N}(0, I). \tag{15}
$$

3.7   Trajectory Decoding

The trajectory decoding first relies on the hidden features $\mathcal{H}_{xz} = Concat(\mathcal{H}_x, \mathcal{Z})$ to obtain estimated trajectory endpoint $\hat{\mathcal{Y}}_t^{goal}$. The predicted trajectory endpoint is then used as inputs to the trajectory decoder to predict multimodal trajectories. Inspired by the bidirectional decoder in [4], the trajectory decoder of the proposed model is also bidirectional. The forward decoding utilizes Receptance Weighted Key Value (RWKV) [17] as the basic module, while the backward decoding employs Recurrent Neural Network (RNN) as the basic module. The above two components are shown in Fig. 3. The decoder can be described as follows:

$$\mathcal{H}_t^f = RWKV(\mathcal{H}_{t-1}^f),$$
$$\mathcal{H}_t^b = RNN(\mathcal{H}_{t+1}^b, \hat{y}_{t+1}W_b + b_b), \tag{16}$$
$$\hat{y}_t = Concat[\mathcal{H}_t^f, \mathcal{H}_t^b]W_o + b_o,$$

where $\hat{y}_{T+1} = \hat{y}_t^{goal}$. $\mathcal{H}_t^f$ and $\mathcal{H}_t^b$ represent the forward and backward hidden states, respectively. $\{W_b, b_b\}$ and $\{W_o, b_o\}$ represent the backward and output parameter matrices, respectively. The estimated trajectory endpoint can be represented as,

$$\hat{y}_t^{goal} = \phi_{goal}(\mathcal{H}_{xz}), \tag{17}$$

where ϕ_{goal} represents the trajectory endpoint decoder network composed of 3-layer feed-forward neural network. In this paper, RNN refers to GRU. The RWKV block consists of time-mixing and channel-mixing blocks with a recursive structure. The recursion is formulated as a linear interpolation between the current input and the input from the previous time step. Let x_t represent the current input, the time-mixing blocks can be described as,

$$r_t = W_r \cdot (\mu_r x_t + (1 - \mu_r) x_{t-1}), k_t = W_k \cdot (\mu_k x_t + (1 - \mu_k) x_{t-1}),$$
$$v_t = W_v \cdot (\mu_v x_t + (1 - \mu_v) x_{t-1}),$$
$$wkv_t = \frac{\sum_{i=1}^{t-1} e^{-(t-1-i)w+k_i} v_i + e^{u+k_t} v_t}{\sum_{i=1}^{t-1} e^{-(t-1-i)w+k_i} + e^{u+k_t}}, o_t = W_o \cdot (\sigma(r_t) \odot wkv_t). \tag{18}$$

The channel-mixing block is given by the following equation:

$$r_t = W_r \cdot (\mu_r x_t + (1 - \mu_r) x_{t-1}), k_t = W_k \cdot (\mu_k x_t + (1 - \mu_k) x_{t-1}),$$
$$o_t = \sigma(r_t) \odot \left(W_v \cdot \max(k_t, 0)^2\right). \tag{19}$$

3.8 Training

From Eq. (5), it can be observed that the objects to be optimized include three components: latent variables, trajectory endpoints, and predicted trajectories. In this paper, Kullback-Leibler (KL) divergence is used to optimize the prior and posterior distributions of latent variables, while residual prediction and best-of-many (BoM) l_2 loss are employed to optimize trajectory endpoints and predicted trajectories. In conclusion, Assuming there are a total of K modes in the predicted trajectory, where $\hat{y}_t^{(i)}$ represents the ith mode of the predicted trajectory, the final loss can be described as:

$$\ell_{total} = \mathbb{E}_{\substack{\hat{y}_t \sim p_\xi(\cdot|\mathcal{H}_{xz}, \hat{y}_t^{goal}) \\ \hat{y}_t^{goal} \sim p_\epsilon(\cdot|\mathcal{H}_{xz}) \\ \mathcal{Z} \sim q_\phi(\cdot|\mathcal{H}_{xy})}} [\min_{i \in K} \left\| \mathcal{Y}_t - \mathcal{X}_t - \hat{y}_t^{(i)} \right\|_2$$
$$+ \min_{i \in K} \left\| \mathcal{Y}_t^{goal} - \mathcal{X}_t - \hat{y}_t^{goal,(i)} \right\|_2] - \mathcal{D}_{KL}[q_\phi(\mathcal{Z}|\mathcal{H}_{xy}) \| p_\theta(\mathcal{Z}|\mathcal{H}_x)], \tag{20}$$

where $\hat{\mathcal{Y}}_t^{goal} \in \mathbb{R}^{K \times 2}$ and $\hat{\mathcal{Y}}_t \in \mathbb{R}^{K \times T' \times 2}$ respectively represent the estimated trajectory endpoints and predicted trajectories relative to the current position $\mathcal{X}_t$.

4 Experimental Results and Analysis

4.1 Experimental Setup

Datasets. We evaluate our method on two publicly available datasets: ETH [18] and UCY [19]. These datasets include pedestrian interactions, non-linear trajectories, collision avoidance, standing, and group pedestrian trajectory coordinates. They also contain 5 unique outdoor environments recorded from a fixed top-down view. The ETH dataset consists of two subsets, ETH and Hotel, while the UCY dataset consists of three subsets: Zara1, Zara2, and Univ.

Table 1. ADE_{20}/FDE_{20} of Pedestrian Trajectory Prediction.

Models	ETH	Hotel	Univ.	Zara01	Zara02	Average
Social-STGCNN	0.64/1.11	0.49/0.85	0.44/0.79	0.34/0.53	0.30/0.48	0.44/0.75
SGCN	0.63/1.03	0.32/0.55	0.37/0.70	0.29/0.53	0.25/0.45	0.37/0.65
Social-GAN	0.63/1.09	0.46/0.98	0.56/1.18	0.33/0.67	0.31/0.64	0.70/1.52
SoPhie	0.70/1.43	0.76/1.67	0.54/1.24	0.30/0.63	0.38/0.78	0.72/1.54
Social-Ways	0.39/0.64	0.39/0.66	0.55/1.31	0.44/0.64	0.51/0.92	0.46/0.83
PECNet	0.54/0.87	0.18/0.24	0.35/0.60	0.22/0.39	0.17/0.30	0.29/0.48
Trajectron++	0.54/0.94	0.16/0.28	0.28/0.55	0.21/0.42	0.16/0.31	0.27/0.50
Social-VAE	0.51/0.85	0.15/0.24	0.25/0.47	0.21/0.38	**0.16/0.30**	0.25/0.43
Social-DualCVAE	0.66/1.26	0.42/0.78	0.40/0.77	0.31/0.56	0.27/0.45	0.41/0.76
Bitrap	0.51/0.74	**0.13/0.20**	0.24/0.45	0.22/0.43	0.16/0.32	0.25/0.43
Social-CVAE	**0.47/0.74**	0.14/0.23	**0.24/0.43**	0.21/0.38	0.16/0.31	**0.24/0.42**
Social-CVAE + FPC	0.42/0.62	0.13/0.21	0.22/0.38	0.19/0.31	0.15/0.28	0.22/0.36

Similar to previous works [1, 2, 4], we employ leave-one-out cross-validation, training on four sets and testing on the remaining set. The observation/prediction window is set at 3.2/4.8 s (*i.e.*, 8/12 frames).

Compared Methods. To validate the effectiveness of the proposed model, we compare it with the following baselines, including previous state-of-the-art multimodal prediction methods: GAN-based methods: Social-GAN [1], SoPhie [12], SocialWays [13]. CVAE-based methods: PECNet [14], Trajectron++ [15], Social-VAE [2], SocialDualCVAE [3], and Bitrap [4]. Graph-based methods: Social-STGCNN [20], SGCN [21].

Implementation Details. All models are implemented in PyTorch and trained and tested on a single Nvidia 3090 40 GB GPU. Parameter optimization is performed using the *Adam* optimizer with a batch size of 128 and an initial learning rate of $3e^{-4}$. The total number of training epochs is 120. Metrics on the test set is based on the model with the best performance on the validation set.

The hyper-parameter of the baseline methods are the same as those in the original paper. The hyper-parameter settings for the proposed model are as follows: the multi-head attention consists of 4 attention heads and one layer, with an embedding dimension of 256. The hidden size of both the GRU and RWKV in the decoder is 256.

4.2   Evaluation Indicator

We use commonly used metrics in the literature [1,3], namely Average Displacement Error (ADE) and Final Displacement Error (FDE), to evaluate the performance of the model. ADE represents the average l_2 distance between the predicted trajectory and the ground truth, while FDE measures the l_2 distance between the estimated trajectory's final point and the ground truth's final point. Assuming there are N agents in the scene, where $\mathcal{Y}_i^{(t),(k)}$ represents the position of agent i at timestamp t of the kth trajectory, the prediction task is to forecast K plausible trajectories. Mathematically, ADE_K and FDE_K are represented as,

$$ADE_K = \min_{k \in K} \frac{\sum_{t=1}^{T'} \sum_{i=1}^{N} ||\hat{\mathcal{Y}}_i^{(t),(k)} - \mathcal{Y}_i^{(t)}||_2}{T' \cdot N} \tag{21}$$

$$FDE_K = \min_{k \in K} \frac{\sum_{i=1}^{N} ||\hat{\mathcal{Y}}_i^{(T'),(k)} - \mathcal{Y}_i^{(T')}||_2}{N} \tag{22}$$

where ADE_K represents the minimum average displacement error over K samples, and FDE_K represents the minimum final displacement error over K samples.

4.3   Results and Analyses

Table 1 displays the ADE_{20} and FDE_{20}, which represent the best sample of Average Displacement Error (ADE) and Final Displacement Error (FDE) among 20 trajectories, respectively, for different models in various scenarios. The ADE/FDE results are reported in meters. The bold black font represents the state-of-the-art without the FPC condition in Social-CVAE, while the underline indicates the state-of-the-art among all models.

In addition, we also introduce a variant of Final Position Clustering (FPC) [2] that enhances prediction diversity within a limited number of samples. Specifically, the number of predictions performed exceeds the required number of samples. K-means clustering is used to cluster the final positions of all predicted trajectories, and only the predicted trajectories of the average final position closest to the cluster are retained. Finally, K predictions are generated as the final result. In this article, the setting of FPC is as follows: the model predicts 5 times, generates 100 (5 × 20) prediction samples, and generates the final 20 predictions through K-Means clustering.

In the absence of FPC, the prediction results of the proposed Social-CVAE are on par with or even surpass the baseline predictions of the state-of-the-art. Social-CVAE demonstrates an improvement of approximately 45.5% in terms of Average Displacement Error (ADE) and a 44% improvement in terms of Final Displacement Error (FDE) compared to Social-STGCNN. With the application of FPC, Social-CVAE achieves an additional improvement of approximately 50% and 52% respectively. This indicates that the application of FPC can lead to predictions that better align with the ground truth in scenarios with a limited number of samples.

5   Conclusion

In this paper, we propose Social-CVAE, which is a novel architecture based on conditional Variational autoencoder for multi-modal pedestrian trajectory prediction. It utilizes attention-based mechanisms to extract pedestrian trajectory features from short-term observations and relies on a bidirectional decoder that uses estimated trajectory target endpoint as conditions to generate stochastic predictions for future trajectories.

Predicting the next actions of an agent based on the surrounding environment is also helpful. However, in the validation pedestrian trajectory prediction dataset used in this paper, there is a lack of detailed environmental information. Therefore, a module for extracting the surrounding environment is not considered in the model design process. If we could obtain more detailed environmental information, such as road conditions, traffic signals, pedestrian density, it would contribute to improving the performance of the prediction model.

In future work, we aim to further optimize the model's inference speed and prediction accuracy, enabling its deployment in practical applications such as intelligent cameras and robotic systems.

References

1. Gupta, A., Johnson, J., Fei-Fei, L., Savarese, S., Alahi, A.: Social GAN: socially acceptable trajectories with generative adversarial networks. In: Proceedings of the IEEE Conference on Computer Vision and Pattern Recognition, pp. 2255–2264 (2018)
2. Xu, P., Hayet, J.-B., Karamouzas, I.: SocialVAE: human trajectory prediction using timewise latents. In: Avidan, S., Brostow, G., Cissé, M., Farinella, G.M., Hassner, T. (eds.) Computer Vision-ECCV, 17th European Conference, Tel Aviv, Israel, Proceedings, Part IV, vol. 13664, pp. 511–528. Springer, Heidelberg (2022). https://doi.org/10.1007/978-3-031-19772-7_30
3. Gao, J., Shi, X., Yu, J.J.: Social-dualcvae: multimodal trajectory forecasting based on social interactions pattern aware and dual conditional variational auto-encoder. arXiv preprint arXiv:2202.03954 (2022)

4. Yao, Y., Atkins, E., Johnson-Roberson, M., Vasudevan, R., Du, X.: BiTraP: bidirectional pedestrian trajectory prediction with multi-modal goal estimation. IEEE Rob. Autom. Lett. **6**(2), 1463–1470 (2021)

5. Liang, R., Li, Y., Zhou, J., Li, X.: Stglow: a flow-based generative framework with dual graphormer for pedestrian trajectory prediction. arXiv preprint arXiv:2211.11220 (2022)

6. Gu, T., Chen, G., Li, J., Lin, C., Rao, Y., Zhou, J., Lu, J.: Stochastic trajectory prediction via motion indeterminacy diffusion. In: Proceedings of the IEEE/CVF Conference on Computer Vision and Pattern Recognition, pp. 17113–17122 (2022)

7. Helbing, D., Molnar, P.: Social force model for pedestrian dynamics. Phys. Rev. E **51**(5), 4282 (1995)

8. Alahi, A., Goel, K., Ramanathan, V., Robicquet, A., Fei-Fei, L., Savarese, S.: Social LSTM: human trajectory prediction in crowded spaces. In: Proceedings of the IEEE Conference on Computer Vision and Pattern Recognition, pp. 961–971 (2016)

9. Zhang, P., Yang, W., Zhang, P., Xue, J., Zheng, N.: SR-LSTM: state refinement for LSTM towards pedestrian trajectory prediction. In: Proceedings of the IEEE/CVF Conference on Computer Vision and Pattern Recognition, pp. 12085–12094 (2019)

10. Kosaraju, V., Sadeghian, A., Martín-Martín, R., Reid, I., Rezatofighi, H., Savarese, S.: Social-BIGAT: multimodal trajectory forecasting using bicycle-gan and graph attention networks. Adv. Neural Inf. Process. Syst. **32**, 1–10 (2019)

11. Yue, J., Manocha, D., Wang, H.: Human trajectory prediction via neural social physics. In: Avidan, S., Brostow, G., Cissé, M., Farinella, G.M., Hassner, T. (eds.) Computer Vision-ECCV,: 17th European Conference, Tel Aviv, Israel, 23–27 October 2022, Proceedings, Part XXXIV, vol. 13694, pp. 376–394. Springer, Heidelberg (2022). https://doi.org/10.1007/978-3-031-19830-4_22

12. Sadeghian, A., Kosaraju, V., Sadeghian, A., Hirose, N., Rezatofighi, H., Savarese, S.: Sophie: an attentive gan for predicting paths compliant to social and physical constraints. In: Proceedings of the IEEE/CVF Conference on Computer Vision and Pattern Recognition, pp. 1349–1358 (2019)

13. Amirian, J., Hayet, J.-B., Pettré, J.: Social ways: learning multi-modal distributions of pedestrian trajectories with gans. In: Proceedings of the IEEE/CVF Conference on Computer Vision and Pattern Recognition Workshops (2019)

14. Mangalam, K., et al.: It is not the journey but the destination: endpoint conditioned trajectory prediction. In: Vedaldi, A., Bischof, H., Brox, T., Frahm, J.-M. (eds.) ECCV 2020. LNCS, vol. 12347, pp. 759–776. Springer, Cham (2020). https://doi.org/10.1007/978-3-030-58536-5_45

15. Salzmann, T., Ivanovic, B., Chakravarty, P., Pavone, M.: Trajectron++: dynamically-feasible trajectory forecasting with heterogeneous data. In: Vedaldi, A., Bischof, H., Brox, T., Frahm, J.-M. (eds.) ECCV 2020. LNCS, vol. 12363, pp. 683–700. Springer, Cham (2020). https://doi.org/10.1007/978-3-030-58523-5_40

16. Hu, J., Shen, L., Sun, G.: Squeeze-and-excitation networks. In: Proceedings of the IEEE Conference on Computer Vision and Pattern Recognition, pp. 7132–7141 (2018)

17. Peng, B., et al.: RWKV: reinventing RNNs for the transformer era. arXiv preprint arXiv:2305.13048 (2023)

18. Pellegrini, S., Ess, A., Van Gool, L.: Improving data association by joint modeling of pedestrian trajectories and groupings. In: Daniilidis, K., Maragos, P., Paragios, N. (eds.) ECCV 2010. LNCS, vol. 6311, pp. 452–465. Springer, Heidelberg (2010). https://doi.org/10.1007/978-3-642-15549-9_33

19. Leal-Taixé, L., Fenzi, M., Kuznetsova, A., Rosenhahn, B., Savarese, S.: Learning an image-based motion context for multiple people tracking. In: Proceedings of the IEEE Conference on Computer Vision and Pattern Recognition, pp. 3542–3549 (2014)
20. Mohamed, A., Qian, K., Elhoseiny, M., Claudel, C.: Social-STGCNN: a social spatio-temporal graph convolutional neural network for human trajectory prediction. In: Proceedings of the IEEE/CVF Conference on Computer Vision and Pattern Recognition, pp. 14 424–14 432 (2020)
21. Shi, L., et al.: SGCN: sparse graph convolution network for pedestrian trajectory prediction. In: Proceedings of the IEEE/CVF Conference on Computer Vision and Pattern Recognition, pp. 8994–9003 (2021)

Distributed Training of Deep Neural Networks: Convergence and Case Study

Jacques M. Bahi[✉], Raphaël Couturier, Joseph Azar,
and Kevin Kana Nguimfack

Université de Franche-Comté, CNRS, Institut FEMTO-ST, F-25000 Besançon, France
{jacques.bahi,raphael.couturier,joseph.azar,
kevin.kana_nguimfack}@univ-fcomte.fr

Abstract. Deep neural network training on a single machine has become increasingly difficult due to a lack of computational power. Fortunately, distributed training of neural networks can be performed with model and data parallelism and sub-network training. This paper introduces a mathematical framework to study the convergence of distributed asynchronous training of deep neural networks with a focus on sub-network training. This article also studies the convergence conditions in synchronous and asynchronous modes. An asynchronous and lock-free training version of the sub-network training is proposed to validate the theoretical study. Experiments were conducted on two well-known public datasets, namely Google Speech and MaFaulDa, using the Jean Zay supercomputer of GENCI. The results indicate that the proposed asynchronous sub-network training approach, with 64 GPUs, achieves faster convergence time and better generalization than the synchronous approach.

Keywords: Distributed Deep Learning · Asynchronous convergence · Subnet Training · Parallelism

1 Introduction

The work presented in this paper concerns distributed and parallel machine learning with a focus on asynchronous deep learning model training. Model parallelism and data parallelism are the two primary approaches to distributed training of neural networks [1,2]. In data parallelism, each server in a distributed environment receives a complete model replica but only a portion of the data. Locally on each server, a replica of the model is trained on a subset of the entire dataset [3]. Model parallelism involves splitting the model across multiple servers. Each server is responsible for processing a unique neural network portion, such as a layer or subnet. Data parallelism could be deployed on top of model parallelism to parallelize model training further. In a deep convolutional neural network, for instance, the convolution layer requires a big number of calculations, but the required parameter coefficient W is small, whereas the fully connected

© The Author(s), under exclusive license to Springer Nature Singapore Pte Ltd. 2024
B. Luo et al. (Eds.): ICONIP 2023, CCIS 1962, pp. 490–503, 2024.
https://doi.org/10.1007/978-981-99-8132-8_37

layer requires a small number of calculations, and the required parameter coefficient W is large. Consequently, data parallelism is appropriate for convolutional layers, while model parallelism is appropriate for fully connected layers.

Nevertheless, data and model parallelism approaches have numerous drawbacks. Data parallel approaches suffer from limited bandwidth since they must send the whole model to every location to synchronize the computation. A huge model with many parameters cannot meet this requirement. Model parallel approaches are not viable in most distribution settings. When data are shared across sites, model parallel computing requires that different model components can only be updated to reflect the data at a specific node. Fine-grained communication is required to keep these components in sync [4].

1.1   Objective

The study has two main objectives. Given a neural network and its partition into sub-networks, the first objective is to establish the mathematical relationship between deep learning on the complete network and deep learning on the partitioned sub-networks. The second is to use this mathematical model to clarify sufficient conditions that ensure the effective convergence, after training of the sub-networks, towards the solution that would have been obtained by the complete neural network. The research question **RQ** is *"whether, by training and combining smaller parts of a full neural network, the convergence of the global solution to the original problem is obtained. In other words how to ensure that weight parameters obtained by training the partitioned parts of the full network reconstitute the correct weight parameters that tune the full network."*

The mathematical model we introduce is general; indeed, it covers both sub-network training and federated learning approaches, as well as their execution in both synchronous and asynchronous modes. To validate our model, we conduct synchronous and asynchronous implementations of sub-network training with a coordinator and several workers and compares their performance.

Our implementation aims to validate the proposed mathematical model and its necessary conditions, rather than to offer a novel deep learning architecture for distributed deep learning. For this purpose, we present an implementation based on the Independent Subnet Training (IST), an approach proposed by Yuan et al. [5]. The experiments we conducted[1] demonstrate how asynchronous distributed learning leads to a shorter waiting time for model training and a faster model updating cycle.

1.2   Methodology

We introduce a general mathematical model to describe the sequences generated[2] by sub-networks partitioning-based approaches. The model considers the

---

[1] Experiments are conducted on GENCI's Jean Zay supercomputer in France, which contains 64 GPUs.

[2] The generated sequences, or produced sequences, are in fact the sequence of iterations and therefore the computation of the weights at each update by the backpropagation algorithm.

dependencies of the unknown variables (weights parameters) that the original neural network must tune. Consequently, different techniques can be modeled, including independent sub-network training and federated learning methods. Specifically, this mathematical model is based on a fixed-point function defined on an extended product space, see [6]: the sequences generated by the extended fixed-point function describe the aggregation of iterations produced by gradient descent algorithms applied to neural sub-networks. Using the fixed point theorem, it is possible to study the numerical convergence of sub-network-based approaches running back-propagation algorithms to the actual solution that would be obtained by original neural network.

The numerical convergence of sub-network partitioning algorithms in asynchronous mode is then examined. To this aim, we apply a classical convergence theorem of asynchronous iterations developed in [7], and in other papers such as [8,9], to the aforementioned fixed-point function. The convergence conditions and cautious implementation details are then specified.

The rest of the paper is organized as follows: The related work to this paper is presented in Sect. 2. Section 3 recalls preliminaries on successive approximations, fixed-point principle and their usefulness to modelize sequential, synchronous and asynchronous iterations. The general mathematical model and the convergence study of the algorithms proposed by this paper are presented in Sect. 4. We first introduce the fixed-point mapping that models the behavior of a certain number of backpropagations on sub-networks of the original network. Then we prove thanks to Proposition 1 and 2 the convergence result of synchronous and asynchronous sub-networks training algorithms. Section 5 is devoted to the empirical validation of the results, the experimental use case is also presented. Section 6 discusses this paper's results and presents future work and perspectives and Sect. 7 concludes the paper.

2   Related Work

Stochastic gradient descent (SGD) uses a randomly chosen sample to update model parameters, providing benefits such as quick convergence and effectiveness. Nodes must communicate to update parameters during gradient calculation. However, waiting for all nodes to finish before updating can decrease computing speed if there is a significant speed difference between nodes. If the speed difference is small, synchronization algorithms work well, but if the difference is significant, the slowest node will reduce overall computing speed, negatively impacting performance. Asynchronous algorithms most effectively implement it [10].

Google's DistBelief system is an early implementation of async SGD. The DistBelief model allows for model parallelism [11]. In other words, different components of deep neural networks are processed by multiple computer processors. Each component is processed on a separate computer, and the DistBelief software combines the results. It is similar to how a computer's graphics card can be used to perform certain computations faster than the CPU alone. The processing time is decreased because each machine does not have to wait its turn to perform a portion of the computation, but can instead work on it simultaneously. Async SGD has two main advantages: (1) First, it has the potential to achieve higher throughput across the distributed system: workers can spend more time performing useful computations instead of waiting for the parameter averaging step to complete. (2) Workers can integrate information from other workers (parameter updates), which is faster than using synchronization (every N steps). However, a consequence of introducing asynchronous updates to the parameter vector, is a new issue known as the stale gradient problem. A naive implementation of async SGD would result in a high gradient staleness value. For example, Gupta et al. [12] show that the stale value of the average gradient is equal to the number of workers. If there are M workers, these gradients are M steps late when applied to the global parameter vector. The real-world impact is that high gradient staleness values can greatly slow down network convergence or even prevent convergence for some configurations altogether. Earlier async SGD implementations (such as Google's DistBelief system) did not consider this effect. Some ways to handle expired gradients include, among others, implementing "soft" synchronization rules such as in [13] and delaying the faster learner if necessary to ensure that the overall maximum delay is less than a threshold [14]. Researchers have proposed several schemes to parallelize stochastic gradient descent, but many of these computations are typically implemented using locks, leading to performance degradation. The authors of [15] propose a straightforward technique called Hogwild! to remove locks. Hogwild! is a scheme that enables locking-free parallel SGD implementations. Individual processors are granted equal access to shared memory and can update memory components at will using this method. Such a pattern of updates leads to high instruction-level parallelism on modern processors, resulting in significant performance gains over lock-based implementations. Hogwild! has been extended by the Dogwild! model in [16]. Dogwild! has extended the Hogwild! model to distributed-memory systems, where it still converges for deep learning problems. To reduce the interference effect of overwriting w at each step, the gradient ∇w from the training agents is transferred in place of w. The empirical evaluation of this paper is inspired by the Independent sub-network Training (IST) model [5], which decomposes the neural network layers into a collection of subnets for the same goal by splitting the neurons across multiple locations. Before synchronization, each of these subnets is trained for one or more local stochastic gradient descent (SGD) rounds. IST is better than model parallelism approaches in some ways. Given that subnets are trained independently during local updates, there is no need to synchronize them with each other. Still, IST has some of the same benefits as model parallelism methods. Since each machine only gets a small piece of the whole model, IST can be used to train models that are too big to fit in the RAM of a node or a device. This can be helpful when training large models on GPUs, which have less memory than CPUs. The following sections provide proof of convergence for partitioning-based networks in synchronous and asynchronous modes. The empirical evaluation section considers the IST model as a use case and proposes an asynchronous implementation.

3   Preliminaries: Successive Approximations. Synchronous and Asynchronous Algorithms

In the sequel, x_i^k denotes the kth iterate of the ith component of a multi-dimensional vector x.

A Numerical iterative method is usually described by a fixed-point application, say T. For example, given an optimization problem $\min_x f(x)$, the first challenge is to build an adequate T, such that the iterative sequence $x^{k+1} = T(x^k)$ for $k \in \mathbf{N}$ converges to a fixed point x^*, solution of $\min_x f(x)$. Successive iterations generated by the numerical method are then described by:

$$
\begin{aligned}
&Given x^0 = (x_1^0, \ldots, x_n^0)^T \in \mathbb{R}^n \\
&for\, i = 1, \ldots, n \\
&\quad for\, k = 1, 2 \ldots until\, convergence \\
&\quad x_i^{k+1} = T_i(x_1^k, \ldots, x_n^k)
\end{aligned}
\tag{1}
$$

Suppose that x is partitioned into m blocks $x_l \in \mathbb{R}^{n_l}$ so that $\Sigma_{l=1}^m n_l = n$ and that m processors are respectively in charge of a block x_l of components of x, then let us then define sets $S(k)$ describing the blocks of x updated at the iteration k. The other blocks are supposed not to be updated at iteration k. Asynchronous algorithms, in which the communications between blocks are free with no synchronization between the processors in charge of the blocks, are described by the following iterations:

$$
\begin{aligned}
&Given x^0 = (x_1^0, \ldots, x_m^0)^T \in \Pi_{l=1}^m \mathbb{R}^{n_l} \\
&for\, k = 1, 2 \ldots until\, convergence \\
&\quad for\, l = 1, \ldots, m \\
&\quad x_l^{k+1} = \begin{cases} T_l(x_1^{\rho_1(k)}, \ldots, x_m^{\rho_m(k)}) if\, l \in S(k) \\ x_l^k if\, l \notin S(k) \end{cases}
\end{aligned}
\tag{2}
$$

Here $\rho_i(k)$ is the delay due to processor i at the kth iteration. Note that with this latter formulation, synchronous per-blocks iterative algorithms are particular cases of asynchronous per-blocks iterative algorithms. Indeed, one has to simply assume that there is no delay between the processors, i.e., $\forall i, k \quad \rho_i(k) = k$, and that all the blocks are updated at each iteration, $\forall k, S(k) = \{1, \ldots, m\}$. In this paper, $T^{(\beta)}$ will denote the fixed-point mapping describing the gradient descent method or its practical version SGD with learning rate β. Under suitable conditions and construction, the fixed-point application T describing the gradient descent method converges to a desired solution. In what follows, a fixed-point mapping $\mathcal{T}$ will also be introduced, this is a key function of our study on traininig deep neural netwoks. This function involves the computation of weighted parameters associated with network training, as well as combination matrices that model the aggregation of partial computations from sub-networks.

4   Convergence Study

4.1   Mathematical Modeling

Consider the general case of a feedforward neural network of some number of layers and neurons. Suppose that the number of interconnections between neurons in the network is n, so that the network is tuned by n weight vectors.

Let us consider a training set composed of inputs and targets. Consider a loss-function f (e.g. Mean Squared Error, Cross Entropy, etc.) that should be minimized on the traing set in order to fit the inputs to their targets. To do so, let $x \in \mathbb{R}^n$ be the weight vector that the neural network must tune to minimize f. Suppose that the gradient descent algorithm is used to perform this minimization. The gradient descent algorithm performs the iterations:

$$x^{k+1} = x^k - \beta \nabla_x f(x^k), k \in \mathbb{N}, x^0 \in D(f) \tag{3}$$

Under the right conditions on the function f and the learning rate β, the function $I - \nabla_x f$ is contractive on its definition domain $D(f)$ and the successive iterations (Eq. 3) converge to x^* satisfying the solution:

$$x^* = argmin_x f(x), x^* \in D(f) \tag{4}$$

Here $\nabla_x f(x) = \left(\frac{\partial f_i}{\partial x_j}\right)_{ij}$ is the Jacobian matrix consisting of the partial derivatives of f with respect to weight vector x of the interconnections of the network. A Backpropagation with a learning rate β_i applied to the network consists in computing successive iterations associated to the Descent gradient algorithm:

$$x^0 \quad being \quad the \quad initialization \quad weights$$
$$x^{k+1} = x^k - \beta_i \nabla_x f(x^k), k \in \mathbb{N}, \tag{5}$$

Let $T^{(\beta_i)}$ denotes the fixed point application describing the iterations (Eq. 5):

$$y = T^{(\beta_i)}(x) = x - \beta_i \nabla_x f(x)$$

Consider m backpropagation algorithms on the neural network Net with learning rates β_i, $i = 1, \ldots, m$. The learning rates $\beta_1, ..., \beta_m$ are chosen in such a way that the gradient descent algorithms converge to x^*, solution of (4), i.e.

$$T^{(\beta_i)}(x^*) = x^*, \forall i \in \{1, \ldots, m\} \tag{6}$$

For a network Net with n weighted vector, and m backpropagation algorithms, let us introduce the following fixed-point mapping which is called aggregation fixed point mapping:

$$\mathcal{T} : (\mathbb{R}^n)^m \longrightarrow (\mathbb{R}^n)^m$$
$$\left(x^1, \ldots, x^m\right) \longrightarrow \left(y^1, \ldots, y^m\right)$$
$$y^l = T^{(\beta_l)} \left(\sum_{k=1}^{m} E_k x^k\right) \tag{7}$$

where E_k are diagonal matrices called weight matrices so that

$$\sum_{k=1}^{m} E_k = I \tag{8}$$

I is the identity matrix $\in \mathbb{R}^{n \times n}$. Since the weight matrices are diagonal, the aggregation fixed-point mapping modelizes the execution of m backpropagations on m sub-networks. It should be noted that condition (8) is essential for the formal convergence study.

Proposition 1. $\mathcal{T}$ is a contractive function
The aggregation fixed-point mapping $\mathcal{T}$ is contractive with constant of contraction α satisfying, $\alpha \leq \max_{l=1}^{m}(\alpha_l)$, where α_l are the constants of contraction of $T^{(\beta_l)}$

Proof. *Let $y^l = T^{(\beta_l)}(\Sigma_{k=1}^{m} E_k x^k)$*
From (6) each gradient descent converges to x^* the solution of (4), so each fixed-point mapping $I - \beta \nabla_x J$ is contractive.
For $l \in \{1, \dots, m\}$, denote by α_l the constant of contraction of $T^{(\beta_l)}$ with respect to the l_2 norm $\|x\|_2$.
By (7) we have

$$\|y^l - x^*\|_2 = \|T^{(\beta_l)}(\Sigma_{k=1}^{m} E_k x^k) - T^{(\beta_l)}(x^*)\|_2 \leq \alpha_l \|\Sigma_{k=1}^{m} E_k x^k - x^*\|_2$$

Since E_k are diagonal matrices and $\Sigma_{k=1}^{m} E_k = I$, we have

$$\|\Sigma_{k=1}^{m}(E_k x^k - x^*)\|_2 = \|\Sigma_{k=1}^{m} E_k (x^k - x^*)\|_2 \leq \max_{k=1}^{m} \|x^k - x^*\|_2$$

Thus, $\|y^l - x^*\|_2 \leq \alpha_l \max_{k=1}^{m} \|x^k - x^*\|_2$
Hence we obtain, $\max_{l=1}^{m} \|y^l - x^*\|_2 \leq (\max_{k=1}^{m} \alpha_l) \max_{k=1}^{m} \|x^k - x^*\|_2$
Let's define the norm, $\|(y^1, \dots, y^m)\|_\infty = \max_{1 \leq l \leq m} \|y^l\|_2$, then

$$\|\mathcal{T}(x^1, \dots, x^m) - (x^*, \dots, x^*)\|_\infty \leq (\max_{l=1}^{m} \alpha_l)\|(x^1, \dots, x^m) - (x^*, \dots, x^*)\|_\infty$$

This proves the claimed result.

Proposition 2. *Asynchronous convergence of $\mathcal{T}$*
The asynchronous iterations generated by the fixed-point mapping $\mathcal{T}$ converge to $(x^*, \dots, x^*)^T$.

Proof. *Proposition 1 implies that we are in the framework of contractive fixed-point mapping with respect to a maximum norm on a product space. Indeed, $\mathcal{T}$ is a contractive mapping with respect to the maximum norm $\|..\|_\infty$, on the product space $\Pi_{i=1}^{m} \mathbb{R}^n = (\mathbb{R}^n)^m$. These are the required sufficient conditions for the convergence of asynchronous algorithms associated to $\mathcal{T}$. see [7]. Thus Proposition 2 is proved.*

Remark 1. *The aggregation fixed point mapping modelizes Independent Sub-network Training, indeed, if the neurons are disjoint then $\forall i, (E_k)_{ii}$ is equal to either 0 or 1. It should be noted that we don't use the notion of compressed iter-ates and the assumptions related to the expectancy of the compression operator $\mathcal{M}(.)$ as in [5]. We prove the convergence of the iterates values instead of their convergence on expectation. To modelize federated learning like algorithms the weighted matrices become $\forall i, (E_k)_{ii} = 1/M$.*

5   Empirical Validation

5.1   The Links Between the Mathematical Model and the Implementation

- To obtain asynchronous iterations produced by the aggregation of the compu-tations of m sub-networks, it is sufficient to consider that $T = \mathcal{T}$ in (2).
- Synchronous per-blocks computations correspond to $S(k) = \{1, \ldots, m\}$, and $\rho_i{}^k = k, \forall k$ (no-delays and all the blocks are updated at each iteration).
- A sub-network l is defined by all nodes j corresponding to the non-zero entries of E_l: $(E_l)_{jj} \neq 0$, so the aggregation of the weight vectors computed by the sub-networks must be done carefully so that $\sum_{k=1}^{m} E_k = I$ is satisfied.
- Even if in the mathematical formulation $\mathcal{T}$ is defined on the product space $\Pi_{i=1}^{m} \mathbb{R}^n$, each sub-network l updates at each iteration only its own components, this is because the weight matrices E_l are diagonal matrices and $(E_l)_{jj} = 0$ if node j does not belongs into sub-network l.
- To execute the computations in asynchronous mode, it should be noted that the convergence theory requires some conditions on the block-component updates and the delays between the processors. These are listed below.

  – $\forall i \in \{1, ...m\}$ the set $\{k \in \mathbb{N}/i \in S(k)\}$ is infinite. This simply means that any sub-network i is guaranteed to update its computations, i.e. not to be permanently inactive.
  – The delays must "follow" the iterations, the mathematical formulation for that is $\forall i, \lim_{k \to \infty}(\rho_i(k)) = \infty$.

For more precision see the literature on asynchronous iterations, e.g. [7], [9, 17–19].

It can be noticed that in a practical implementation these assumptions are realistic.

5.2   Use Case: Independent Subnet Training

As a use case, let us consider the IST model for this research. Note that the proposed mathematical model is general and could also cover dependent partitioning-based networks. The publicly available Github implementation[3], the

---

[3] https://github.com/BinhangYuan/IST_Release.

clarity of the code, and the recent date of publication were among the reasons to consider this model for our use case.

In the original work, the authors suggest to regularly resample the subnets and to train them for fewer iterations between resamplings. This paper proposes that the coordinator distribute subnets only at the beginning of training to prevent blockage. Similarly to the original IST model, the following constraints are respected:

- Input and output neurons are common to all subnets.
- Hidden neurons are randomly partitioned via uniform assignment to one of n possible compute nodes.
- The complete Neural Network (NN)'s weights are partitioned based on the number of activated neurons in each subnet.
- No collisions occur because the parameter partition is disjoint.

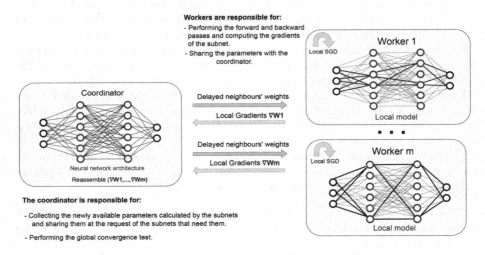

Fig. 1. The architecture of a distributed asynchronous training based on the partitioning-based framework.

The training process of the asynchronous IST training is as follows: (1) The coordinator distributes the sub-networks to the workers at the beginning of the training process. Contrary to the original IST, no new group of subnets is constructed to avoid synchronization. Note that during the training process, the coordinator trains a subnet, not the full neural network. (2) Each cluster (coordinator and workers) computes the loss of its local model based on its local data (i.e., forward pass). (3) Each cluster computes the gradients based on the loss (i.e., backward pass). (4) Each worker sends its parameters to the coordinator. In our asynchronous version, it is done using Pytorch's *isend* method that sends a tensor asynchronously. (5) Given that the partition is disjoint, the coordinator copies the parameters into the full neural network without collisions. The asynchronous version is considered using Pytorch's *irecv* method that receives a tensor asynchronously.

The architecture and training procedure considered in this research (Fig. 1) are comparable to frameworks such as the parameter server with centralized or federated learning techniques [20]. The local models, however, are subnets of the full model and not a replica. Note that Fig. 1 illustrates a general partitioning-based framework described by the proposed mathematical model and not necessarily the particular IST model. To represent the async IST implemented by this paper, one could ignore the transmission of the delayed parameters from the coordinator to the workers after the first partition, given that each worker has an independent subnet that does not require the parameters of other workers. One of the challenges of asynchronous training is that the asynchronous communication will be rendered meaningless if one continues to allow communication between workers: This issue is solved by avoiding parameters exchange between subnets and avoiding synchronization during local updates or the coordinator's updates.

5.3   Experiments

The Google Speech [21] and Mafaulda[4] [22] datasets were considered to demonstrate the advantage of asynchronous distributed training of neural networks over synchronous training. The Google Speech dataset describes an audio dataset of spoken words that can be used to train and evaluate keyword detection systems. The objective is to classify 35 labeled keywords extracted from audio waveforms. In the dataset, the training set contains roughly 76,000 waveforms, and the testing set has around 19,000 waveforms. The Mafaulda dataset is a publicly available dataset of vibration signals acquired from four different industrial sensors under normal and faulty conditions. This database is composed of 1951 multivariate time-series and comprises six different states. Similarly to [5], a three-layer MLP (multilayer perceptron) has been developed. The number of neurons in each layer is a parameter. Using subnets as mentioned in [5], the MLP is distributed across all processes and GPUs. To simplify, links between neurons are divided among all the nodes, and each link is controlled exclusively by a single node. Consequently, the coordinator node (which also participates in the training computation) and the worker node share link weights during parallel training.

In contrast to [5], where all tasks are synchronized, an implementation with asynchronous tasks and iterations has been built. In this version, there is no synchronization following the initialization. The coordinator node has a heavier workload compared to other nodes in the implementation of the IST code. It is responsible for coordinating workers, performing tests with the test dataset, sending and receiving the link weights of other nodes' neurons. Additionally,

---

[4] Dataset: https://www02.smt.ufrj.br/~offshore/mfs/page_01.html, last visited: 20-10-2022.

the coordinator is also responsible for a subnet (local gradient descent) in this implementation, leading to a higher amount of calculations.

For the experiments, the French supercomputer Jean Zay of GENCI was used. This supercomputer contains more than 2,000 GPUs (most of them are NVIDIA V100).

The experiments compare the performance of three different versions of a neural network training algorithm: the old synchronous version (the training version implemented by [5]), the synchronous version with partitioning at the first iteration, and the asynchronous version with no synchronizations between the workers. The main point that the experiments aim to prove is that the asynchronous method provides better testing accuracy in a shorter time and lower execution times compared to the synchronous methods.

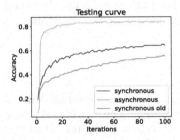
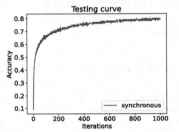

(a) Testing curve for the synchronous and the asynchronous cases (limited to 100 epochs).

(b) Testing curve for the synchronous case (limited to 1,000 epochs).

Fig. 2. Experiments with the Google Speech dataset.

Table 1. Execution times for the Google Speech dataset.

Algorithm	Number of epochs	Execution times (in s)	Accuracy	Asynchronous acceleration
Synchronous old version [5]	100	531	0.56	1.08
Synchronous version	100	550	0.67	1.11
Synchronous version	1,000	5,501	0.80	11.18
Asynchronous version	100	492	0.84	1

The training of three models was executed ten times with 64 GPUs. A supercomputer with specialized architecture was used, resulting in minimal variations in execution times. Testing accuracy was averaged and plotted to assess the convergence of the models. Results from Figs. 2a and 3a indicate that the asynchronous method outperforms the synchronous mode due to worker nodes completing more iterations than the coordinator node (workers don't wait for the coordinator to synchronize at the end of each epoch), while in synchronous mode, the number of iterations completed is equal to the number of epochs. The number of epochs was fixed at 100 for the Google Speech dataset and 200 for Mafaulda based on several experiments, as those values better show the difference in performance between synchronous and asynchronous training. Additionally, the average testing accuracy for 1,000 epochs was computed only for the synchronous model with one partitioning at the beginning to compare the performance of the asynchronous model with a smaller number of epochs against the synchronous model's accuracy after 1,000 epochs. Figures 2b and 3b demonstrate that the synchronous case has lower testing accuracy even with 1,000 epochs compared to the asynchronous case. Table 1 shows the average execution times rounded to the nearest second of three versions. The maximum accuracy is shown. The last column shows the asynchronous acceleration compared to the other versions. The asynchronous training was faster than the other synchronous approaches on both Google Speech and Mafaulda datasets. The code used for the experiments is available[5].

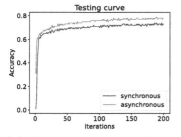
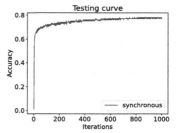

(a) Testing curve for the synchronous and the asynchronous cases (limited to 200 epochs).

(b) Testing curve for the synchronous case (limited to 1,000 epochs).

Fig. 3. Experiments with the Mafaulda dataset.

6   Discussion and Perspectives

This paper aims to address numerical convergence in deep learning frameworks for large problems. To achieve this, the neural network is partitioned into subnets and back-propagation algorithms are applied to these subnets instead of the original network. The paper also raises questions about reconstituting the global solution from partial solutions of subnets and the differences between synchronous and asynchronous algorithms during convergence.

The paper presents a mathematical model that describes the behavior of partitioning-based network methods in synchronous and asynchronous modes,

---

[5] https://github.com/rcouturier/async_mlp.

and proves the convergence of such methods. To validate this approach, the specific case of Independent Subnets Training is considered, and the asynchronous mode is implemented and compared to the standard synchronous mode. The study concludes that the asynchronous mode has better performance in terms of execution time and accuracy.

The presented mathematical model is general and describes various scenarios, including federated learning with workers possessing the same network architecture. Future works will investigate different concepts and distributed architectures to broaden the scope of the study.

7   Conclusion

This paper proposes a general model to investigate the convergence of distributed asynchronous training of deep neural networks. The model addresses the question of whether partial solutions of sub-networks can be used to reconstitute the global solution. Two experimentations based on the Independent sub-network Training model on different datasets are provided, demonstrating that using the asynchronous aggregation sub-networks model is interesting for reducing execution time and increasing accuracy, even on a supercomputer. The study expects a better performance from asynchronous aggregation algorithms in contexts with weak computation time to communication time ratios.

Acknowledgment. This work was partially supported by the EIPHI Graduate School (contract "ANR-17-EURE-0002"). This work was granted access to the AI resources of CINES under the allocation AD010613582 made by GENCI and also from the Mesocentre of Franche-Comté.

References

1. Ben-Nun, T., Hoefler, T.: Demystifying parallel and distributed deep learning: an in-depth concurrency analysis. ACM Comput. Surv. (CSUR) **52**(4), 1–43 (2019)
2. Verbraeken, J., Wolting, M., Katzy, J., Kloppenburg, J., Verbelen, T., Rellermeyer, J.S.: A survey on distributed machine learning. ACM Comput. Surv. (CSUR) **53**(2), 1–33 (2020)
3. Li, S., et al.: Pytorch distributed: experiences on accelerating data parallel training. arXiv preprint arXiv:2006.15704 (2020)
4. Podareanu, D., Codreanu, V., Sandra Aigner, T., van Leeuwen, G.C., Weinberg, V.: Best practice guide-deep learning. Partnership for Advanced Computing in Europe (PRACE), Technical Report, vol. 2 (2019)
5. Yuan, B., Wolfe, C.R., Dun, C., Tang, Y., Kyrillidis, A., Jermaine, C.: Distributed learning of fully connected neural networks using independent subnet training. Proc. VLDB Endow. **15**(8), 1581–1590 (2022)
6. Bahi, J., Miellou, J.-C., Rhofir, K.: Asynchronous multisplitting methods for nonlinear fixed point problems. Numer. Algor. **15**(3), 315–345 (1997)
7. El Tarazi, M.N.: Some convergence results for asynchronous algorithms. Numer. Math. **39**(3), 325–340 (1982)

8. Baudet, G.: Asynchronous iterative methods for multiprocessors. J. Assoc. Comput. Mach. **25**, 226–244 (1978)
9. Bahi, J.: Asynchronous iterative algorithms for nonexpansive linear systems. J. Parallel Distrib. Comput. **60**(1), 92–112 (2000)
10. Lian, X., Zhang, W., Zhang, C., Liu, J.: Asynchronous decentralized parallel stochastic gradient descent. In: International Conference on Machine Learning, pp. 3043–3052. PMLR (2018)
11. Dean, J., et al.: Large scale distributed deep networks. Adv. Neural Inf. Process. Syst. **25**, 1–9 (2012)
12. Gupta, S., Zhang, W., Wang, F.: Model accuracy and runtime tradeoff in distributed deep learning: a systematic study. In: 2016 IEEE 16th International Conference on Data Mining (ICDM), pp. 171–180. IEEE (2016)
13. Zhang, W., Gupta, S., Lian, X., Liu, J.: Staleness-aware ASYNC-SGD for distributed deep learning (2015). arXiv preprint arXiv:1511.05950
14. Ho, Q., et al.: More effective distributed ml via a stale synchronous parallel parameter server. Adv. Neural Inf. Process. Syst. **26**, 1–9 (2013)
15. Recht, B., Re, C., Wright, S., Niu, F.: Hogwild!: a lock-free approach to parallelizing stochastic gradient descent. Adv. Neural Inf. Process. Syst. **24**, 1–9 (2011)
16. Noel, C., Osindero, S.: Dogwild!-distributed hogwild for CPU & GPU. In: NIPS Workshop on Distributed Machine Learning and Matrix Computations, pp. 693–701 (2014)
17. Bertsekas, D.P., Tsitsiklis, J.N.: Parallel and distributed computation: numerical methods (2003)
18. Spiteri, P.: Parallel asynchronous algorithms: a survey. Adv. Eng. Softw. **149**, 102896 (2020)
19. Frommer, A., Szyld, D.B.: On asynchronous iterations. J. Comput. Appl. Math. **123**(12), 201–216 (2000)
20. Elbir, A.M., Coleri, S., Mishra, K.V.: Hybrid federated and centralized learning. In: 29th European Signal Processing Conference (EUSIPCO), pp. 1541–1545. IEEE (2021)
21. Warden, P.: Speech commands: a dataset for limited-vocabulary speech recognition. arxiv:1804.03209 (2018)
22. Ribeiro, F.M., Marins, M.A., Netto, S.L., da Silva, E.A.: Rotating machinery fault diagnosis using similarity-based models. In: XXXV Simpósio Brasileiro de Telecomunicações e Processamento de Sinais-sbrt2017 (2017)

Chinese Medical Intent Recognition Based on Multi-feature Fusion

Xiliang Zhang[1,2,3], Tong Zhang[1,2,3], and Rong Yan[1,2,3]($\boxtimes$)

[1] College of Computer Science, Inner Mongolia University, Hohhot 010021, China
[2] Inner Mongolia Key Laboratory of Mongolian, Information Processing Technology, Hohhot 010021, China
[3] National and Local Joint Engineering Research Center of Intelligent Information, Processing Technology for Mongolian, Hohhot 010021, China
csyanr@imu.edu.cn

Abstract. The increasing popularity of online query services has heightened the need for suitable methods to accurately understand the truth of query intention. Currently, most of the medical query intention recognition methods are deep learning-based. Because of the inadequate of corpus of the medical field in the pre-trained phase, these methods may fail to accurately extract the text feature constructed by medical domain knowledge. What's more, they rely on a single technology to extract the text information, and can't fully capture the query intention. To mitigate these issues, in this paper, we propose a novel intent recognition model called EDCGA (ERNIE-Health+D-CNN+Bi-GRU+Attention). EDCGA achieves text representation using the word vectors of the pretrained ERNIE-Health model and employs D-CNN to expand the receptive field for extracting local information features. Meanwhile, it combines Bi-GRU and attention mechanism to extract global information to enhance the understanding of the intent. Extensive experimental results on multiple datasets demonstrate that our proposed model exhibits superior recognition performance compared to the baselines.

Keywords: Deep learning · Intent recognition · Feature fusion

1 Introduction

Accurately extracting useful information from a large number of Chinese medical texts is a valuable and challenging task, and it is of great significance to effectively utilize medical information and reduce medical costs. Currently, medical search engines primarily rely on keyword-based full-text searches, which only match one or multiple keywords and fail to capture the search intent of users accurately. It may affect the quality of search results [17]. Intent recognition in information retrieval [11] involves analyzing the search texts created by users to identify their search intent. In the medical field, intent recognition models need to delve deeper into understanding the specific requirements of users [17]. By incorporating intent information from search texts, medical search engines

B. Luo et al. (Eds.): ICONIP 2023, CCIS 1962, pp. 504–515, 2024.
https://doi.org/10.1007/978-981-99-8132-8_38

can significantly improve the accuracy of search results and enhance the overall search experience for users.

In recent years, pre-trained models have gained increasing attention in NLP (Natural Language Processing) tasks in the medical domain. Zhang [23] utilizes a large volume of specialized medical corpus data to train MC-BERT, which surpassed the performance of BERT [6] in medical information retrieval and intent recognition tasks. Wang et al. [15] employ Baidu's ERNIE [13] knowledge-enhanced pre-trained model and medical knowledge enhancement techniques to train the ERNIE-Health pre-trained model. This model successfully captures professional medical knowledge and performs well in many NLP tasks in the medical field.

However, there are still many challenges and issues in the intent recognition field. Firstly, the provided Chinese search text is often short in length and lacks contextual information, which makes it hard to fully express the intention of users, so the quality of extracted semantic features is compromised [19]. Secondly, the search behaviors of users often do not adhere to strict formal language grammar structures, which leads to issues such as grammar errors, typos, and non-standard language usage in search queries [14,18]. It is difficult for search engines to understand and accurately respond to the search needs of users.

In this paper, we propose a novel intent recognition model called EDCGA (ERNIE-Health+D-CNN+Bi-GRU+Attention). It employs D-CNN (Dilated Convolutional Neural Network) [20] with an enlarged receptive field to extract local feature information. Additionally, it combines Bi-GRU (Bidirectional Gated Recurrent Unit) and attention mechanism to capture global feature information. The objective is to improve the quality of text feature representation and achieve better recognition performance. Experimental results on multiple publicly available datasets demonstrate that the proposed model outperforms existing intent recognition models in the Chinese medical domain.

2 Related Work

2.1 D-CNN

D-CNN [20] is a specialized type of convolutional neural network. It utilizes dilated convolution kernels to extract features. The key characteristic of dilated convolution is the introduction of holes or gaps within the standard convolution mapping, to expand the receptive field and capture multi-scale feature information. Currently, dilated convolutional neural networks have been successfully applied in many fields, including speech recognition and generation [3], image denoising [16], and image semantic segmentation [10]. Compared to classical convolutional approaches, they have demonstrated excellent efficiency. In the field of NLP, dilated convolutional kernels are primarily applied in the context of text convolutional neural networks, such as TextCNN [7], to expand the receptive field of the model. Strubell et al. [12] first apply dilated convolution in NLP for NER (Named Entity Recognition) tasks. The IDCNN-CRF model effectively identifies named entities by stacking dilated convolutional layers with shared parameters.

2.2 Bi-GRU

GRU [2] is a recursive neural network unit utilized for sequence prediction. It is a simplified variation of LSTM [5]. Compared to LSTM, GRU retains fewer parameters, faster training speed, and shows comparable performance in NER and text classification tasks.

Bi-GRU is a dual-directional recurrent neural network model. It is an enhanced version of the traditional GRU, where the GRU is replaced with a bidirectional GRU. It allows for capturing information from both preceding and succeeding directions in a sequence and is beneficial for better extraction of contextual semantic features.

2.3 Attention Mechanism

Attention mechanism is a commonly used technique in deep learning models. It allows the model to automatically focus on important parts, rather than considering the entire input as a whole. In the field of NLP, attention mechanism is often applied in encoder-decoder models and feature extraction modules. It computes a weight coefficient to determine the importance of each input word, thus assigning different levels of attention. Our proposed method focuses on improving the effectiveness of the model in capturing essential information from sentences while reducing sensitivity to noisy information.

3 Our Method

In this paper, we propose a novel intent recognition model called EDCGA (ERNIE-Health+D-CNN+Bi-GRU+Attention). As shown in Fig. 1, it consists of four main components: the text encoding layer, the feature extraction layer, the feature fusion layer, and the intent recognition layer. Specifically, (1) the text encoding layer utilizes the ERNIE-Health pre-trained model to map each word in the sentence to a continuous low-dimensional vector space; (2) the feature extraction layer employs both Bi-GRU with attention mechanism and the dilated convolutional neural network to extract global and local feature information, respectively; (3) the feature fusion layer combines the global and local feature information through vector concatenation; (4) the intent recognition layer classifies the concatenated feature vectors by using a fully connected layer, and the recognition results are obtained through softmax activation.

3.1 The Text Encoding Layer

The preprocessed text, denoted as $x = \{x_1, x_2, \cdots, x_n\}$, is inputted into ERNIE-Health for text encoding. Firstly, the masking mechanism of ERNIE-Health is

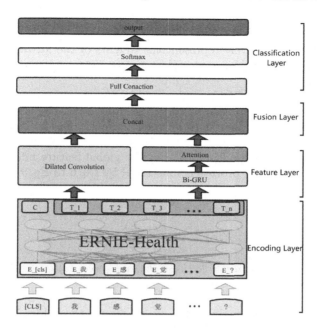

Fig. 1. The structure of EDCGA

applied to obtain word embeddings. Then, a Transformer bidirectional encoder is used for text encoding. The Transformer bidirectional encoder has the ability to capture semantic information and global relationships within the text. In the end, ERNIE-Health generates a vector representation $H_E = \{h_1, h_2, \cdots, h_n\}$ that represents the semantic information of the entire text.

3.2 The Feature Extraction Layer

The encoded word vectors H_E are separately fed into the local feature information module and the global feature information module to extract corresponding feature vectors H_D and H_{GA}.

The Bi-GRU model can effectively process sequential information in the text. The forward GRU captures preceding contextual information, and the backward GRU captures succeeding contextual information. The outputs of the forward and backward GRUs are concatenated to form the output H_G of Bi-GRU. The calculation process is illustrated by Eq. 1 to Eq. 5:

$$\overrightarrow{q_t} = GRU(h_t, \overrightarrow{q_{t-1}}) \tag{1}$$

$$\overleftarrow{q_t} = GRU(h_t, \overleftarrow{q_{t-1}}) \tag{2}$$

$$h_t' = [\overrightarrow{q_t}, \overleftarrow{q_t}] \tag{3}$$

$$BiGRU = [h_1', h_2', \cdots, h_n'] \tag{4}$$

$$H_G = BiGRU(H_E) \tag{5}$$

where w_t and v_t respectively represent the weights of the forward hidden state $\overrightarrow{q_t}$ and the backward hidden state $\overleftarrow{q_t}$ at time t, b_t represents the bias of the hidden state at time t.

The output H_G of Bi-GRU is influenced by the attention mechanism, which assigns attention scores to each word in the input text. By assigning different weights to the corresponding word vectors, the entire model can focus on the key parts of the feature and reduce the impact of non-keywords on the text. We can consider the resulting output H_{GA} as the global feature information of the sentence. The calculation process is illustrated by Eq. 6 to Eq. 9.

$$u_t = tanh(W_t h_t') \tag{6}$$

$$\alpha_t = \frac{exp(u_t)}{\sum_{k=1}^{n} exp(u_k)} \tag{7}$$

$$Attention = [\alpha_1 h_1', \alpha_2 h_2', \alpha_3 h_3', \cdots, \alpha_n h_n'] \tag{8}$$

$$H_{GA} = Attention(H_G) \tag{9}$$

where W_t represents the weight matrix, u_t represents the hidden layer calculated by h_t' through $tanh$, h_t' represents the feature vector output by Bi-GRU at time t, α_t represents the attention weight of h_t'.

The combination of Bi-GRU and attention mechanism in the global feature information extraction module enables the capture of long-distance contextual dependencies while preserving semantic coherence. Figure 2 illustrates the structure of the global feature information extraction module.

D-CNN injects holes into the standard convolutional mapping to expand the receptive field and capture multi-scale contextual information. Considering the characteristics of Chinese medical retrieval information, such as short length, fewer characters, and sparse features. In this work, we utilize D-CNN to extract local feature information. The calculation process is illustrated by Eq. 10–13. Figure 3 shows the structure of the local feature information extraction module.

$$p_i = Conv(E_H) \qquad i = 3 \times l, 4 \times l, 5 \times l \tag{10}$$

$$p_i' = Relu(p_i) \qquad i = 3 \times l, 4 \times l, 5 \times l \tag{11}$$

$$p_i'' = Maxpool(p_i') \qquad i = 3 \times l, 4 \times l, 5 \times l \tag{12}$$

$$H_D = [p_3'', p_4'', p_5''] \tag{13}$$

where $Conv()$ and $Maxpool()$ represent convolution operation and pooling operation respectively, i represents the size of convolution kernel and l represents the length of the sentence.

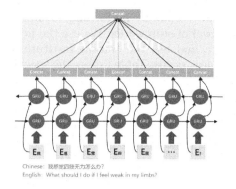 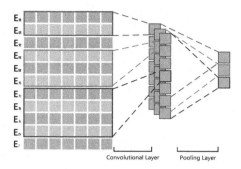

Fig. 2. Global feature information extraction module

Fig. 3. Local feature information extraction module

3.3 The Feature Fusion Layer

D-CNN extracts text features through convolutional layers and retains only locally salient features by discarding irrelevant information through pooling layers. It can effectively capture local feature information but may not preserve semantic coherence. Bi-GRU with an attention mechanism can capture the contextual dependencies of the entire sentence and maintain semantic coherence. Since the two modules have complementary advantages, we connect and fuse the text features extracted by both modules to obtain a higher-quality feature representation, denoted as H. The calculation is shown in Eq. 14.

$$H = Conact(H_D, H_{GA}) \tag{14}$$

where H_D represents the local feature information vector, H_{GA} represents the global feature information vector.

3.4 The Intent Recognition Layer

After sending the fused feature vector into the fully connected layer, use the $Softmax$ classifier to get the final output y, the calculation can be expressed as Eq. 15:

$$y = max(Softmax(wH + b)) \tag{15}$$

where w and b represent real matrix and bias term respectively.

4 Experiments

4.1 Datasets

In this paper, we use KUAKE-QIC [22], CMID [1],CMedIC [23] and XUNFEI-QIC[1] four publicly available datasets to compare and verify our proposed method.

---

[1] https://challenge.xfyun.cn/topic/info?type=medical-search&option=stsj.

In the first process, we preprocess the raw corpus by performing an initial step that includes tokenization and the removal of stop words. We use the jieba tokenization tool[2] to tokenize the text data and use the Harbin Institute of Technology stop word list to remove meaningless words. Table 1 gives the specific statistical information for each dataset.

Table 1. Information of the datasets.

dataset	category	max length	avg length	train	val	test
KUAKE-QIC	11	469	13	6,931	1,955	1,994
CMID	13	221	35	3,143	391	398
CMedIC	3	39	10	1,512	189	184
XUNFEI-QIC	10	153	21	1,699	212	213

4.2 Experimental Settings

In this paper, we utilize the ERNIE-Health pre-trained model [15] to obtain word embedding representations. Our model is trained with a dropout of 0.1, and we optimize cross-entropy loss using Adam optimizer [8]. The experimental parameter settings are provided in Table 2.

Table 2. Experimental parameter settings.

parameter	value
embedding size	768
filter sizes	3,4,5
num filters	256
dilation	2
Bi-GRU hidden size	256
learning rate	0.00002
batch size	32

4.3 Baselines

To verify the effectiveness of the proposed model, this paper conducts comparative experiments with Bert [6], Roberta [9], ERNIE [13], MC-BERT [23], Bert-CNN [4], and Bert-LSTM [21].

---

[2] https://github.com/fxsjy/jieba.

4.4   Results

Table 3 shows the comparison results of our proposed method with other baselines. Here, we utilize the commonly used Accuracy (acc) as the evaluation metric for experimental results in intent recognition.

Table 3. Comparison results of different models on all datasets.

	KUAKE-QIC	CMID	CMedIC	XUNFEI-QIC
BERT	84.40	79.65	95.65	56.81
Roberta	84.76	79.90	95.11	57.75
MC-BERT	84.81	80.40	96.20	53.05
ERNIE3.0	85.66	77.39	96.20	49.77
BERT-CNN	84.40	77.39	95.11	56.81
BERT-LSTM	83.95	73.62	94.57	57.28
EDCGA	**86.51**	**81.16**	**96.74**	**59.15**

As shown in Table 3, it shows that our proposed EDCGA outperforms baselines in terms of recognition performance and exhibits good stability on the four open-source datasets. Particularly, on XUNFEI-QIC, it achieves a 1.4% higher accuracy than the best-performing Roberta. BERT-CNN and BERT-LSTM attempt to extract deep semantic features by employing feature extraction modules after text encoding. However, their performance falls short in specialized and feature-sparse medical intent recognition tasks. We analysis that this limitation is due to the general domain bias of BERT usage and the limited diversity of feature extraction modules. However, their performance falls short in specialized and feature-sparse medical intent recognition tasks. This limitation is due to the general domain bias of BERT usage and the limited diversity of feature extraction modules.

Comparing the results of ERNIE and MC-BERT, it also can be observed that pretraining models with large-scale and domain-specific corpora can enhance recognition performance. However, due to their limited capability to extract multi-level semantic features, these methods heavily rely on specific datasets, leading to inconsistent performance and lacking robustness. In contrast, the proposed EDCGA uses a feature fusion approach with word embeddings from the ERNIE-Health model trained on extensive medical domain corpora. By incorporating multi-level text feature information, the EDCGA model achieves superior recognition performance.

4.5   Ablation Experiments

Table 4 presents the experimental results of the ablation experiments on the four datasets, where -A, -D, and -G respectively indicate the removal of the attention mechanism, D-CNN, and Bi-GRU.

Table 4. Results of ablation experiment.

	KUAKE-QIC	CMID	CMedIC	XUNFEI-QIC
EDCGA-A-D-G	86.21	76.38	96.20	57.75
EDCGA-G-A	86.06	78.89	96.20	50.23
EDCGA-A	86.21	80.40	96.20	55.87
EDCGA-G	86.32	80.38	95.65	56.18
EDCGA-D	86.26	79.15	95.65	56.34
EDCGA	**86.51**	**81.16**	**96.74**	**59.15**

As shown in Table 4, each component of the proposed EDCGA contributes to its overall performance to varying degrees. Specifically, when the local feature extraction module is removed, such as D-CNN, the model experiences a decrease in recognition accuracy ranging from 0.25% to 2.81% across the datasets. It indicates that fusing this module can effectively extract local feature information from the text and contribute to the recognition performance of EDCGA.

Similarly, when the attention mechanism is not used in the global feature extraction module, the model experiences a decrease in recognition accuracy ranging from 0.3% to 3.28% across the datasets. It suggests that incorporating the attention mechanism in the sentence-level features extracted by Bi-GRU can effectively highlight important parts of the sentences and captures deeper semantic features. In comparison to fine-tuning with the standalone ERNIE-Health pre-trained model, using a single feature extraction module alone does not significantly enhance the recognition performance of the model. In fact, it may even weaken the performance of the pre-trained model when dealing with complex input texts.

Furthermore, we conduct experiments on CMID, which contains a larger number of categories. We compare the $F1$ score of the models with the local information feature extraction module removed, the global information feature extraction module removed, and the model using only the ERNIE-health pre-training model for each intent category. The experimental results are illustrated in Fig. 4. This further validates the effectiveness of the proposed modules in this paper.

4.6   Selection of Parameters

To assess the impact of the number of hidden units in Bi-GRU and the number of convolutional kernels in D-CNN on the performance of the model, we conducted

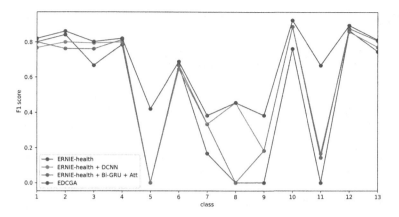

Fig. 4. The $F1$ score comparisons on CMID.

experiments on the CMID dataset. These experiments aim to determine the optimal hyperparameters, and the results are illustrated in Fig. 5. The horizontal axis represents the number of convolutional kernels in D-CNN, and the number of hidden units in Bi-GRU is also depicted. The vertical axis represents the accuracy under different parameters.

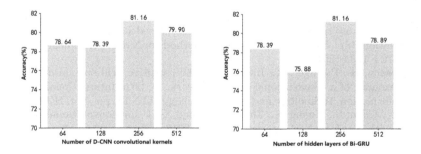

Fig. 5. Performance comparisons with different parameters.

As shown in Fig. 5, it indicates that with other parameters unchanged, the model achieves the highest accuracy when the number of convolutional kernels in D-CNN and the number of hidden units in Bi-GRU are both set to 256. Specifically, when the quantities of convolutional kernels and hidden units are too small, the model fails to capture the rich semantic features in the text, resulting in poor recognition performance. Conversely, when the quantities of convolutional kernels and hidden units are too large, the model may extract excessive feature information, leading to overfitting during training and the inclusion of redundant noise features, thereby reducing the accuracy of the recognition results.

5 Conclusion

In this paper, we propose a novel recognition method called EDCGA for the task of intent recognition in the medical field. It utilizes word embeddings generated by the ERNIE-Health pre-trained model and integrates D-CNN, Bi-GRU, and an attention mechanism as the feature extraction layer. With performance comparisons with baseline models, EDCGA shows superior performance on four open-source medical intent recognition datasets. By conducting ablation experiments and selecting optimal experimental parameters, the effectiveness of each proposed component and the selected parameters are validated. This model obtains a promising application prospects in medical information retrieval systems and medical intelligent question-answering systems, as well as contributing to the advancement of smart healthcare.

Acknowledgements. This research is jointly supported by the Natural Science Foundation of Inner Mongolia Autonomous Region (Grant No. 2023MS06023) and the Self-project Program of Engineering Research Center of Ecological Big Data, Ministry of Education.

References

1. Chen, N., Su, X., Liu, T., Hao, Q., Wei, M.: A benchmark dataset and case study for chinese medical question intent classification. BMC Med. Inform. Decis. Mak. **20**(3), 1–7 (2020)
2. Cho, K., van Merrienboer, B., Gulcehre, C., Bougares, F., Schwenk, H., Bengio, Y.: Learning phrase representations using RNN encoder-decoder for statistical machine translation. In: Conference on Empirical Methods in Natural Language Processing, pp. 1724–1734 (2014)
3. Gabbasov, R., Paringer, R.: Influence of the receptive field size on accuracy and performance of a convolutional neural network. In: 2020 International Conference on Information Technology and Nanotechnology (ITNT), pp. 1–4. IEEE (2020)
4. He, C., Chen, S., Huang, S., Zhang, J., Song, X.: Using convolutional neural network with bert for intent determination. In: 2019 International Conference on Asian Language Processing (IALP), pp. 65–70 (2019)
5. Hochreiter, S., Schmidhuber, J.: Long short-term memory. Neural Comput. **9**(8), 1735–1780 (1997)
6. Kenton, J.D.M.W.C., Toutanova, L.K.: Bert: pre-training of deep bidirectional transformers for language understanding. In: Proceedings of NAACL-HLT, pp. 4171–4186 (2019)
7. Kim, Y.: Convolutional neural networks for sentence classification. In: Proceedings of the 2014 Conference on Empirical Methods in Natural Language Processing (EMNLP), pp. 1746–1751. Association for Computational Linguistics, Doha (2014)
8. Kingma, D.P., Ba, J.: Adam: a method for stochastic optimization. In: Proceedings of the International Conference on Learning Representations (2015)
9. Liu, Y., et al.: Roberta: a robustly optimized bert pretraining approach. arXiv preprint arXiv:1907.11692 (2019)
10. Mehta, S., Rastegari, M., Caspi, A., Shapiro, L., Hajishirzi, H.: Espnet: efficient spatial pyramid of dilated convolutions for semantic segmentation. In: Proceedings of the European Conference on Computer Vision (ECCV), pp. 552–568 (2018)

11. Park, K., Jee, H., Lee, T., Jung, S., Lim, H.: Automatic extraction of user's search intention from web search logs. Multimedia Tools Appl. **61**(1), 145–162 (2012)
12. Strubell, E., Verga, P., Belanger, D., McCallum, A.: Fast and accurate entity recognition with iterated dilated convolutions. In: Proceedings of the 2017 Conference on Empirical Methods in Natural Language Processing, pp. 2670–2680 (2017)
13. Sun, Y., et al.: Ernie 3.0: large-scale knowledge enhanced pre-training for language understanding and generation. arXiv preprint arXiv:2107.02137 (2021)
14. Wang, J., Wang, Z., Zhang, D., Yan, J.: Combining knowledge with deep convolutional neural networks for short text classification. In: Proceedings of the 26th International Joint Conference on Artificial Intelligence, pp. 2915–2921 (2017)
15. Wang, Q., et al.: Building Chinese biomedical language models via multi-level text discrimination. arXiv preprint arXiv:2110.07244 (2021)
16. Wang, Y., Wang, G., Chen, C., Pan, Z.: Multi-scale dilated convolution of convolutional neural network for image denoising. Multimedia Tools Appl. **78**, 19945–19960 (2019)
17. Wang, Y., Wang, S., Li, Y., Dou, D.: Recognizing medical search query intent by few-shot learning. In: Proceedings of the 45th International ACM SIGIR Conference on Research and Development in Information Retrieval, pp. 502–512 (2022)
18. Wang, Y., Wang, S., Yao, Q., Dou, D.: Hierarchical heterogeneous graph representation learning for short text classification. In: Proceedings of the 2021 Conference on Empirical Methods in Natural Language Processing, pp. 3091–3101 (2021)
19. Wang, Y., Yao, Q., Kwok, J.T., Ni, L.M.: Generalizing from a few examples: a survey on few-shot learning. ACM Comput. Surv. **53**(3), 1–34 (2020)
20. Yu, F., Koltun, V.: Multi-scale context aggregation by dilated convolutions. arXiv preprint arXiv:1511.07122 (2015)
21. Yu, Z., Hu, K.: Study on medical information classification of BERT-ATT-BILSTM model. IEEE J. Comput. Age **3**(6), 1–4 (2020)
22. Zhang, N., et al.: CBLUE: a Chinese biomedical language understanding evaluation benchmark. In: Proceedings of the 60th Annual Meeting of the Association for Computational Linguistics (Volume 1: Long Papers), pp. 7888–7915 (2022)
23. Zhang, N., Jia, Q., Yin, K., Dong, L., Gao, F., Hua, N.: Conceptualized representation learning for Chinese biomedical text mining. arXiv preprint arXiv:2008.10813 (2020)

Multi-mobile Object Motion Coordination with Reinforcement Learning

Shanhua Yuan, Sheng Han, Xiwen Jiang, Youfang Lin, and Kai Lv[✉]

Beijing Key Laboratory of Traffic Data Analysis and Mining, School of Computer and Information Technology, Beijing Jiaotong University, Beijing 100044, China
lvkai@bjtu.edu.cn

Abstract. Multi-mobile Object Motion Coordination (MOMC) refers to the task of controlling multiple moving objects to travel from their respective starting station to the terminal station. In this task, the objects are expected to complete their travel in a shorter amount of time with no collision. Multi-mobile object motion coordination plays an important role in various application scenarios such as production, processing, warehousing and logistics. The problem can be modeled as a Markov decision process and solved by reinforcement learning. The current main research methods are suffering from long training time and poor policy stability, both of which can decrease the reliability of practical applications. To address these problems, we introduce a State with Time (ST) model and a Dynamically Update Reward (DUR) model. The results of experiments show that the ST model can enhance the stability of learned policies, and the DUR model can improve the training efficiency and stability of the strategy, and the motion coordination solutions obtained through this algorithm are superior to those of similar algorithms.

Keywords: Reinforcement learning · Motion coordination · Reward model

1 Introduction

The multi-mobile object motion planning problem is a fundamental research problem in multi-mobile object systems [1]. It refers to a specific scenario where multiple mobile objects need to move from their respective starting station to the terminal station. During the task execution, it is necessary to ensure that there are no collisions between different mobile objects while ensuring the quality and time of the entire task. This problem can be divided into path planning [7,14,18] and motion coordination [4,12,19]. An important constraint of path planning is that there should be no collisions between mobile objects during the motion, while ensuring the quality and feasibility of the path.

B. Luo et al. (Eds.): ICONIP 2023, CCIS 1962, pp. 516–527, 2024.
https://doi.org/10.1007/978-981-99-8132-8_39

Deep reinforcement learning [6, 17] is a branch of artificial intelligence that learns and adjusts strategies by interacting with the environment. The core concept of reinforcement learning is to adjust the decision-making of agent based on rewards and punishments, and continuously optimize strategies through interaction with the environment. This makes reinforcement learning widely used in fields such as automation control, robotics, and intelligent games, as it can learn through trial and error, continuously optimizing strategies and improving the performance of agent. Using reinforcement learning to solve the motion coordination problem for multiple mobile objects has also become a hot research topic. Hao et al. [5] modeled the motion coordination problem as a Markov decision process and designed a state-action space and reward function model in deep reinforcement learning to solve the problem. Zhao et al. [20] proposed a new reward function model based on this work and improved the performance of the model in some tasks. However, in current research, the existing reward function in modeling the motion coordination problem for multiple mobile objects can lead to unstable training of agent and the inability to converge to stable solutions. In this paper, we propose a State with Time (ST) model and a Dynamically Updated Reward (DUR) model to address these issues and improve the efficiency and stability of motion coordination tasks for multiple mobile objects.

2   Related Work

2.1   Traditional Solution Methods for Motion Coordination

Traditional methods for solving motion coordination for multiple mobile objects include the Longest Job First (LJF) algorithm and the mixed-integer programming coordination method [13]. Although LJF algorithm is simple, it cannot guarantee the time and quality of motion. In the mixed-integer programming method, it is necessary to calculate whether two different mobile objects will collide based on the intersection of their displacement points on their respective paths. Based on the calculated path collision constraint relationships, each path will be divided into several path segments, as shown in Fig. 1. However, as the

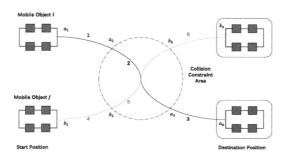

Fig. 1. The constraint relationships between different paths.

number of mobile objects increases, the computational complexity of optimizing the inequality goal in the mixed-integer programming method becomes very large and difficult to solve.

2.2 Double Deep Q-Network

The DQN algorithm [10] has achieved good results in many tasks of deep reinforcement learning, but the Q value estimation process of the DQN algorithm has been shown to sometimes result in overestimation of the Q value. In order to solve this problem, the Double Deep Q Network (DDQN) algorithm [16] improves the updating process of the Q network. Specifically, when choosing the next action with the maximum value, the DDQN algorithm finds the action with the maximum output value in the Q network, and then finds the output value of the Q target network corresponding to that action. The calculation of the target Q value in the DDQN algorithm is:

$$y_j = r_j + \gamma \hat{Q}\left(s_{j+1}, \text{argmax}_{a'} Q\left(s_{j+1}, a', \theta\right); \theta^-\right). \tag{1}$$

2.3 Reinforcement Learning Methods for Solving Motion Coordination Problems

In the problem of motion coordination for multiple mobile objects, Hao et al. [5] first proposed using reinforcement learning to solve this problem and developed the Multi-Loss Double Deep Q Network (MLDDQN) algorithm and Time-Dependent Rewards (TDR) model. Existing works in this area have used a centralized training framework [2,3,11], which requires a central controller to control all mobile objects. The central controller collects real-time state information from all mobile objects and makes unified scheduling decisions based on this information.

Zhao et al. [20] proposed a motion coordination algorithm for multiple mobile objects based on this architecture, called the Partial Tolerant Collision Double Deep Q Network (PTCDDQN) algorithm. It is an improvement on the Double Deep Q Network (DDQN) algorithm [16] and uses a dynamic priority strategy (DPS) reward function model, which performs well in some coordination tasks.

2.4 Path Chessboard Graph

In this paper, the Path Chessboard Graph [8] is used as one of the basic environments for collision detection and model training, aiming at improving the efficiency of collision detection and speed up the training process of the model. The Path Chessboard Graph is an environment built upon collision detection and path planning algorithms. The path planning algorithm used in this paper is the third-order Bezier curve algorithm [9], and the generated path graph with intersection relationships is called a path network. The collision detection algorithm used is the oriented bounding box algorithm [15]. The Path Chessboard Graph algorithm can use the computer to calculate collision and non-collision intervals for each mobile object A_i, and finally generate the Path Chessboard Graph.

3   State with Time and Reward Model Algorithms for Motion Coordination

3.1   Centralized System

In a centralized system, a central agent commands the actions of all mobile objects. At the same time, centralized architecture treats all mobile objects as a whole and abstracts them as a composite object, whose state information includes the position information o_i of each mobile object, forming a joint state s. The joint action a emitted by the composite object directs each moving object to perform an action. After all moving objects perform their actions, the environment transitions to the next global state s' and returns a reward value r to the agent. The framework of the centralized model is shown in Fig. 2.

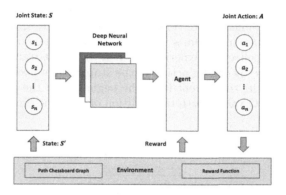

Fig. 2. The framework of the centralized system.

3.2   State Setting

In terms of state setting, each moving object has its own position o_i, and the position information of each moving object needs to be handed over to the central agent network to form a global joint state s, that is, $s = \langle o_1, o_2, \ldots, o_n \rangle$. The state o_i is composed of a quadruple $\langle ID, P, G, R \rangle$. The meaning of each variable in the quadruple will be introduced below:

- ID: The current mobile object is at the node index of the Path Chessboard Graph, that is which segment of the path it is in.
- P: The distance traveled by the mobile object on the node with index ID in the Path Chessboard Graph, initialized to 0.
- G: A flag indicating whether the mobile object has reached its destination position. If it has reached, the flag set $True$; otherwise, it set $False$.
- R: The remaining distance between the mobile object and its destination position.

3.3   Action Setting

In terms of action setting, similar to the state setting, it is necessary to combine the actions of multiple objects into a joint action a. The joint action contains the action information a_i of each mobile object. That is, the joint action space can be represented as $a_1 \times a_2 \times \ldots \times a_n$, and a_i is the executable action for each mobile object. In this problem, the agent can choose to move forward($a_i=0$) or wait($a_i=1$) for each object.

3.4   State with Time Model

The neural network used to fit the Q function $Q(s, a)$ is a function of both the state and action, and when the reward value is related to time, the neural network lacks information about the task execution time t, which leads to the problem of inconsistent target Q values. To address this issue, this paper proposes the ST model. That is, the dimension of time is added to the state transition process so that the Q value function becomes a function of the state s, action a, and time t. In the process of collecting experience, when the agent transitions from state s to the next state s' by taking action a, the stored tuple becomes $\langle s, a, t, r, s' \rangle$, where t represents the time spent from the beginning of the task until now. In this way, the target Q values with the addition of the time dimension in become as shown in Table 1.

Table 1. The target Q value function obtained with ST model.

policy	function	value
t_1	$q(s_1,\ 0,\ 0)$	$2 \cdot \gamma^2$
	$q(s_2,\ 1,\ 0)$	$2 \cdot \gamma$
	$q(s_3,\ 2,\ 0)$	2
t_2	$q(s_1,\ 0,\ 1)$	$3 \cdot \gamma$
	$q(s_3,\ 1,\ 0)$	3

After adding the time dimension to the Q value expression, it can be seen that the problem of inconsistent Q values in the above two tables has been solved. The calculation of the target Q value after adding the ST model to the DDQN algorithm is:

$$y_j = r_j + \gamma \max_{a'} \hat{Q}\left(s_{j+1}, t_{j+1}, a'; \theta^-\right). \tag{2}$$

The loss function is:

$$L_i(\theta_i) = \mathbf{E}_{s,t,a \sim \rho(s,t,a)}\left[(y_i - Q(s, t, a; \theta))^2\right]. \tag{3}$$

where $\rho(s,\ t,\ a)$ is the probability distribution on the state s, task execution time t, and action a.

Then, the effect of adding the ST model to the DDQN algorithm is tested in a validation experiment. The reward function is set as shown in formula 4 below:

$$Reward = \begin{cases} N - time, & \text{if finishing} \\ -1, & \text{if collision} \end{cases} \cdot \tag{4}$$

N is a hyperparameter, which is set to 4 in the validation experiment. The experiment simulates a simple 3 mobile objects coordination task, as shown in Fig. 4. The blue mobile object needs to reach the target position in two steps, while the yellow and red mobile objects each need to reach the target position in one step. If two vehicles pass through the intersection at the same time, a collision will occur. The results of the experiment are shown in Fig. 3.

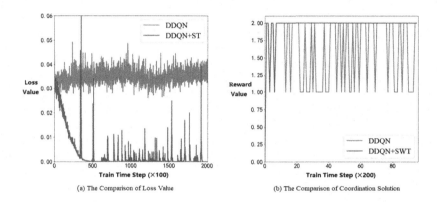

(a) The Comparison of Loss Value   (b) The Comparison of Coordination Solution

Fig. 3. The results of validation experiment. (Color figure online)

In Fig. 3(a), the horizontal axis represents the number of training steps, and the vertical axis represents the algorithm's loss function. The green curve in the figure represents the $td - loss$ of the DDQN algorithm, while the blue curve represents the $td - loss$ of the DDQN algorithm with the ST module added to the state. It can be seen that the algorithm can further converge the $td - loss$ after adding the ST module. In Fig. 3(b), the horizontal axis represents the number of testing times, with a test conducted every 200 time steps, and the vertical axis represents the reward obtained by the agent when completing the task during the test. It can be seen that the algorithm with the ST module can enable the agent to learn a stable coordination strategy, and the corresponding reward value for this strategy is 2, indicating that the coordination time is short and the coordination effect is good. The algorithm without the ST module cannot enable the agent to learn a stable coordination strategy, and the learned strategy keeps changing during the training process.

3.5 Dynamically Updated Reward Model

This paper proposes the DUR model to address the long training time and slow policy convergence issues of the DPS and TDR models. The DUR model dynamically updates the reward function based on the interaction between the agent and the environment, so as to better guide the agent's actions and improve the training efficiency and the policy convergence speed. The goal of the agent is to maximize the future cumulative expected reward. It can suppose a motion coordination task, at time t, the state of the agent is s, the optimal coordination strategy after taking action a still requires coordination time T, and the reward obtained after completing the task is r. The reward is discounted at each time step by γ, so the expected reward return $R(s, t, a)$ that the agent can expect after taking action a in state s is $\gamma^T \times r$. This means that the expected reward return decreases as the coordination time T increases and increases as the reward r increases:

$$\frac{\partial R}{\partial r} > 0. \tag{5}$$

Therefore, in order to make it easier for the agent to distinguish among the advantages and disadvantages of different strategies, the differences in the reward values r that different coordination strategies can obtain when the task is completed should be expanded. However, since there are many different coordination strategies with different coordination times, if a unique reward value is given to each coordination strategy, the range of all reward values will be very large, which undoubtedly increases the burden of agent training. To address this issue, this paper proposes a dynamically updated reward function model for motion coordination tasks, which can speed up the training process while allowing the agent to more stably learn the best strategies explored. The specific reward function is shown in formula 6:

$$\text{Reward} = \begin{cases} 1/\left(1 + e^{(-N/(\alpha \times T))}\right), & \text{Goal \& (T} <= g) \\ -1, & \text{Collision} \\ \delta, & \text{T} > g \end{cases} \tag{6}$$

- Goal&(T $<= g$): If all mobile objects safely reach the destination and the coordination time obtained by the current strategy is less than or equal to the best coordination time ever explored. In this reward function, the symbol N represents the number of successful coordination attempts, the symbol α is a hyperparameter greater than zero, and the symbol T represents the time required to complete the coordination task.
- Collision: If collisions occur between mobile objects, the agent will receive a negative reward.
- T $>$ best: If during training, it is found that the current coordination time for moving the objects is already greater than the time spent by the best explored strategy, the agent will receive a reward value δ. The value of δ is less than $1/\left(1 + e^{(-N/(\alpha \times T))}\right)$. and greater than or equal to -1, and the end flag is set to $True$, and the current training round is terminated.

4 Experiments and Result Analysis

4.1 Experiment Platform

The experimental platform used in this paper is a multi-mobile object motion coordination simulation system developed using the $PyQt5$ framework. In this simulation system, the path planning algorithm will generate Bezier curve paths for each mobile task. At the same time, it will generate a path chessboard diagram based on the collision constraint relationships among the paths. The path chessboard diagram can quickly detect the section where the mobile object is located and accelerate judgment on whether collisions occur between mobile objects. The path chessboard diagram generated by the system for the motion coordination task in Fig. 4(a) is shown in Fig. 4(b).

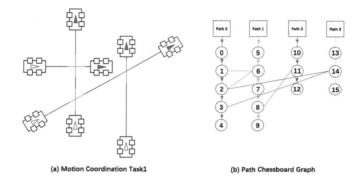

(a) Motion Coordination Task1   (b) Path Chessboard Graph

Fig. 4. The example of path chessboard graph. (Color figure online)

4.2 Evaluation Metrics

The effectiveness of motion coordination can be evaluated by the success rate, the proportion of valid solutions, the model training time, and the MakeSpan (MS) during testing. In this paper, the LFJ algorithm is used as the baseline. When the coordination solution obtained by the algorithm is better than the coordination solution obtained by the LFJ algorithm, the solution is considered valid and a score is obtained. The maximum training time step in the experiments in this chapter is 400,000, and the results are compared based on this.

In a motion coordination task, the coordination time can be represented as $max\,(T_1, T_2, \ldots, T_n)$, where T_i represents the coordination end time of mobile object i. The scoring function is shown in formula 7:

$$\text{score} = \begin{cases} MS_{LJF} - \text{T}, \text{T} <= MS_{LJF} \\ 0, \qquad\qquad \text{others} \end{cases}. \tag{7}$$

Here, MS_{LJF} represents the coordination time of the LJF algorithm. From the formula, it can be seen that when the coordination time obtained by the algorithm is not less than MS_{LJF}, the task is considered a failure and the score is 0. The higher the score, the better the quality of the solution obtained.

4.3   Experiment Results and Analysis

To verify the effectiveness of the ST method and DUR model, this paper first conducted comparative experiments in Motion Coordination Task1. The compared algorithms were the MLDDQN algorithm [5], PTC-DDQN [20] algorithm, and DDQN algorithm. In the experiments, different algorithms were trained under different reward models. The results of the success rate and valid proportion for the various experiments using the ST method are shown in Table 2.

Table 2. The experiment results of coordination task 1.

Algorithm	Model	success rate (%) ↑	the valid proportion (%) ↑	optimal time step
MLDDQN	TDR	100	70.33	12
	DUR	100	97.33	12
PTC-DDQN	DPS	100	92.33	13
	DUR	100	97.00	12
DDQN	TDR	100	59.67	12
	DPS	100	91.33	13
	DUR	100	96.33	12

From the experimental results in Table 2, it can be seen that in terms of success rate, all algorithms combined with the ST model can achieve 100%, and the agent can learn a collision-avoidance coordination strategy stably. In terms of the proportion of valid solutions, the proportion of valid solutions obtained by the MLDDQN algorithm and PTC-DDQN algorithm using the DUR reward model was 97.33% and 97.00%, respectively, which is better than the success rate of 70.33% of the MLDDQN algorithm using the TDR model and the proportion of valid solutions of 92.33% of the PTC-DDQN algorithm using the DPS reward model. In the experiments using the DDQN algorithm with the three reward models, the proportion of valid solutions using the DUR reward model was 96.33%, higher than the 59.67% using the TDR model and the 91.33% using the DPS reward model. In terms of the coordination time, the LFJ algorithm took 15 time steps, all three algorithms surpassed the greedy algorithm in different reward models. Based on the above analysis, the DUR reward model is superior to the TDR and DPS reward models in terms of the proportion of valid solutions and the optimal coordination time obtained.

The score results of the MLDDQN algorithm under the DUR and TDR reward models are shown in Fig. 5(a), where the horizontal axis represents the training time step of the model, and the vertical axis represents the average score

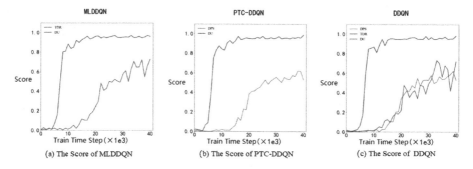

(a) The Score of MLDDQN

(b) The Score of PTC-DDQN

(c) The Score of DDQN

Fig. 5. The scores of different algorithms in coordination task 1. (Color figure online)

of the coordination solutions during testing. The red and blue lines represent the experimental results of the algorithm under the DUR and TDR reward models, respectively. From the experimental results, it can be seen that the convergence time of the MLDDQN algorithm under the DUR reward model is much shorter than that under the TDR reward model, and the obtained solution is also better than that under the TDR reward model. The score results of the PTC-DDQN algorithm under the DUR and DPS reward models are shown in Fig. 5(b), where the green line represents the experimental results of the DPS reward model. Similar to the former results, the convergence time of the PTC-DDQN algorithm under the DUR reward model is much shorter than that under the DPS reward model, and the obtained solution is also better than that under the DPS reward model. The results of the DDQN algorithm under the three reward models are shown in Fig. 5(c). From the figure, it can be seen that under the DUR reward model, the convergence time of the DDQN algorithm is still much shorter than that under the DPS and TDR reward models, and the quality of the obtained solution is also better than the other two reward models.

4.4 Validation of the Effectiveness of the State with Time Model

This paper conducted an ablation experiment on this method in Task1. The baseline algorithm for the experiment was DDQN, using the DUR reward model. The experimental results are shown in Fig. 6.

   In the experimental results graph, the orange line represents the results without using the ST model, and the red line represents the experimental results using the ST model. The success rate of Task1 is shown in Table 3. The success rate without using the ST model is 98.67%. In terms of the proportion of valid solutions, the proportion of valid solutions without the ST model is only 27.67%, which is much lower than the 96.33% with the ST model. The experimental results indicate that after solving the problem of inconsistent target Q-values with the ST model, the agent can learn a stable motion coordination strategy in an environment where the reward value is associated with time.

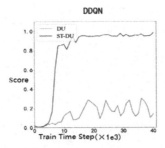

Fig. 6. The score of ablation experiment. (Color figure online)

Table 3. The results of ablation experiment.

Algorithm	Model	success rate (%) ↑	the valid proportion (%) ↑
DDQN	DUR+ST	100	96.33
	DUR	98.67	27.67

5   Conclusion

The multi-mobile object motion coordination problem is a common scheduling problem in production with significant economic value. However, this problem is difficult to obtain the optimal solution due to its high computational complexity. When using reinforcement learning to solve this problem, it belongs to the sparse reward problem in reinforcement learning, and existing reward models may have unstable solutions when solving this problem. To address this issue, this paper proposes the ST model, which aims at enabling the agent to converge to a stable coordination strategy after training. The experiments show that this improvement speeds up the learning process and allows the agent to converge to a more stable and optimal coordination strategy.

Acknowledgements. This work was supported by the National Natural Science Foundation of China under Grant 62206013.

References

1. Al-Jarrah, R., Shahzad, A., Roth, H.: Path planning and motion coordination for multi-robots system using probabilistic neuro-fuzzy. IFAC-PapersOnLine **48**(10), 46–51 (2015)
2. Busoniu, L., Babuska, R., De Schutter, B.: A comprehensive survey of multiagent reinforcement learning. IEEE Trans. Syst. Man Cybern. Part C (Appl. Rev.) **38**(2), 156–172 (2008)
3. Gronauer, S., Diepold, K.: Multi-agent deep reinforcement learning: a survey. Artif. Intell. Rev. **55**, 1–49 (2022)
4. Han, S., Lin, Y., Guo, Z., Lv, K.: A lightweight and style-robust neural network for autonomous driving in end side devices. Connect. Sci. **33**, 1–18 (2022)

5. Hao, X., Wu, Z., Zhou, H., Bai, X., Lin, Y., Han, S.: Motion coordination of multiple robots based on deep reinforcement learning. In: 2019 IEEE 31st International Conference on Tools with Artificial Intelligence (ICTAI), pp. 955–962. IEEE (2019)
6. Hernandez-Leal, P., Kartal, B., Taylor, M.E.: A survey and critique of multiagent deep reinforcement learning. Auton. Agent. Multi-Agent Syst. **33**(6), 750–797 (2019)
7. Hu, X., Wu, Z., Lv, K., Wang, S., Lin, Y.: Agent-centric relation graph for object visual navigation. arXiv preprint arXiv:2111.14422 (2021)
8. Le, T.S., Nguyen, T.P., Nguyen, H., Ngo, H.Q.T.: Integrating both routing and scheduling into motion planner for multivehicle system. IEEE Can. J. Electr. Comput. Eng. **46**(1), 56–68 (2023)
9. Lin, C.C., Chuang, W.J., Liao, Y.D.: Path planning based on bezier curve for robot swarms. In: 2012 Sixth International Conference on Genetic and Evolutionary Computing, pp. 253–256. IEEE (2012)
10. Mnih, V., et al.: Playing atari with deep reinforcement learning. arXiv preprint arXiv:1312.5602 (2013)
11. Nguyen, T.T., Nguyen, N.D., Nahavandi, S.: Deep reinforcement learning for multiagent systems: a review of challenges, solutions, and applications. IEEE Trans. Cybern. **50**(9), 3826–3839 (2020)
12. Peng, J., Akella, S.: Coordinating multiple robots with kinodynamic constraints along specified paths. Int. J. Rob. Res. **24**(4), 295–310 (2005)
13. Ravikumar, S., Quirynen, R., Bhagat, A., Zeino, E., Di Cairano, S.: Mixed-integer programming for centralized coordination of connected and automated vehicles in dynamic environment. In: 2021 IEEE Conference on Control Technology and Applications (CCTA), pp. 814–819. IEEE (2021)
14. Stern, R., et al.: Multi-agent pathfinding: definitions, variants, and benchmarks. In: Proceedings of the International Symposium on Combinatorial Search, vol. 10, pp. 151–158 (2019)
15. Tan, T., Wu, Q., Zhao, H.: OBB hierarchy bounding boxes improved constructed method. Comput. Eng. Appl. **44**(5), 79–81 (2008)
16. Van Hasselt, H., Guez, A., Silver, D.: Deep reinforcement learning with double q-learning. In: Proceedings of the AAAI Conference on Artificial Intelligence, vol. 30 (2016)
17. Wang, H., Lin, Y., Han, S., Lv, K.: Offline reinforcement learning with diffusion-based behavior cloning term. In: 16th International Conference on Knowledge Science, Engineering and Management, KSEM 2023, Korea (2023)
18. Wang, S., Wu, Z., Hu, X., Lin, Y., Lv, K.: Skill-based hierarchical reinforcement learning for target visual navigation. IEEE Trans. Multimedia (2023)
19. Zhang, H., Lin, Y., Han, S., Lv, K.: Lexicographic actor-critic deep reinforcement learning for urban autonomous driving. IEEE Trans. Veh. Technol. **72**, 4308–4319 (2022)
20. Zhao, L., Han, S., Lin, Y.: Collision-aware multi-robot motion coordination deep-RL with dynamic priority strategy. In: 2021 IEEE 33rd International Conference on Tools with Artificial Intelligence (ICTAI), pp. 65–72. IEEE (2021)

Comparative Analysis of the Linear Regions in ReLU and LeakyReLU Networks

Xuan Qi[ID], Yi Wei[ID], Xue Mei[(✉)][ID], Ryad Chellali[ID], and Shipin Yang[ID]

Nanjing Tech University, Nanjing 211816, China
{mx,spyang}@njtech.edu.cn, rchellali@hotmail.fr

Abstract. Networks with piecewise linear activation functions partition the input space into numerous linear regions. As such, the number of linear regions can serve as a metric to quantify the expressive capacity of networks employing ReLU (Rectified Linear Unit) and LeakyReLU activations. One notable drawback of the ReLU network lies in the potential occurrence of the "dying ReLU" issue during training, whereby the output and gradient remain zero when the input to a ReLU layer is negative. This results in ineffective weight updates and renders the affected neurons unresponsive, consequently impeding their contribution to network training. In this study, we perform statistical analysis on the actual number of linear regions expressed by ReLU and LeakyReLU networks, providing an intuitive explanation for the "dying ReLU" problem. Our findings indicate that, under consistent input distributions and network parameters, LeakyReLU networks generally exhibit stronger expressive capacity in terms of linear regions compared to ReLU networks. We hope that our research can provide inspiration for the design of activation functions and contribute to the exploration and analysis of the behaviors exhibited by piecewise linear activation functions in networks.

Keywords: Deep Networks · Linear regions · Piecewise Linear Activation Functions

1 Introduction

Activation functions are a fundamental component of neural networks, responsible for introducing non-linearity and enabling models to learn complex relationships in data. Among the commonly used activation functions, LeakyReLU (Leaky Rectified Linear Unit) [14] and ReLU (Rectified Linear Unit) [7,16] have gained significant popularity due to their simplicity and effectiveness. Network models employing the ReLU activation function, such as CFDM [6], CondenseNetV2 [21], and FATTN [4], as well as those utilizing the LeakyReLU activation function, such as EffNet [5], Scaled-YOLOv4 [18], and YOLObile [2], have achieved remarkable advancements in the field of deep learning tasks. ReLU maps negative inputs to zero and keeps positive inputs unchanged. By introducing a non-linearity, ReLU allows neural networks to learn complex patterns and improves their expressive power. Additionally, ReLU promotes sparsity in network activations, making computations more efficient. However, ReLU suffers from a limitation known as the "dying ReLU" problem, where neurons can become stuck at zero and cease to learn. LeakyReLU addresses the "dying ReLU" problem by modifying the activation function to allow a small non-zero output for negative inputs. Instead of mapping negative values to zero, LeakyReLU introduces a small slope for negative inputs, typically a small constant. This small slope ensures that neurons receive a gradient even for negative values, preventing them from becoming inactive. By providing a non-zero output for negative inputs, LeakyReLU promotes continuous learning and improves the stability and robustness of neural networks. Figure 1 illustrates the network unit employed in this research endeavor.

Networks harness the efficacy of piecewise linear activation functions, such as ReLU and LeakyReLU, in order to fit an extensive array of distinct linear functions. Particularly, when dealing with standard deep neural networks that employ piecewise linear activation functions, we are capable of effectuating the transformation of the input space into a diverse set of linear convex regions [9,17]. Regarding multi-classification tasks, the ultimate classification outcome relies on the hierarchical arrangement of linear functions within the networks. Figure 2 visually shows the visualization results of the linear regions in a two-dimensional input space, obtained through the multi-classification of ReLU and LeakyReLU networks. Notably, the different color blocks featured in the figure represent the distinctive linear regions formed by linear functions within the networks. Moreover, each neuron within the networks possesses the ability to partition the input space into two regions along a hyperplane. Through the collective influence of numerous neurons, a larger region can be effectively segmented into smaller, more refined subregions. Consequently, each region encompasses linear functions that symbolize the ultimate outcome of multi-classification, with the predicted label determined as the maximum value among the results of the linear functions. In ReLU and LeakyReLU networks, the neurons in subsequent layers further partition the linear regions previously established by preceding layers. As a result, the linear functions are organized in a layered fashion, leading to the generation of a substantial multitude of linear regions during network operation. Figure 3 illustrates the arrangement of internal linear convex regions within the ReLU and LeakyReLU network.

In this paper, our primary objective is to transform networks into mappings within the input space. We endeavor to explain the phenomenon of "dying ReLU" from the perspective of piecewise linear functions within the networks. Furthermore, we conduct a comparative analysis of the linear regions in both ReLU and LeakyReLU networks. The primary contributions of our study can be summarized as follows:

(1) We elucidate the evolutionary process of the linear regions in ReLU and Leaky-ReLU networks based on two-dimensional inputs.
(2) We show visualizations of the linear regions at different epochs and layer-wise in ReLU and LeakyReLU networks.

(3) Under consistent input distributions and network parameters, LeakyReLU networks generally exhibit stronger expressive capacity in terms of linear regions compared to ReLU networks. This phenomenon primarily arises from the "dying ReLU" issue, which reduces the number of piecewise linear functions within ReLU networks.

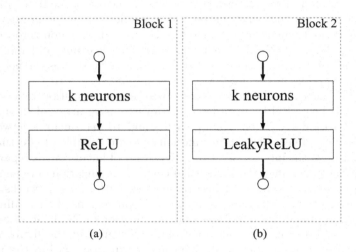

(a)   (b)

Fig. 1. The minimum network units used in this paper. (a) Block 1, a network with k neurons activated by ReLU. (b) Block 2, a network with k neurons activated by LeakyReLU.

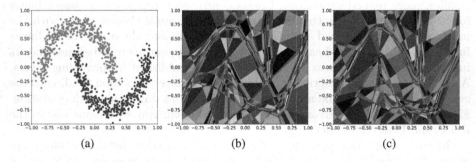

(a)   (b)   (c)

Fig. 2. The figure shows the relationship between the linear regions and input distributions of the networks under various activation states, all of which are mapped in the two-dimensional input space. Each distinct color block represents a different linear region of the networks. (a) The two-dimensional input data set *make moons*. (b) The linear regions obtained from a network utilizing architecture Fig. 3 (a) trained for 50 epochs, with each layer incorporating a Block 1 (k = 32). (c) The linear regions obtained from a network utilizing architecture Fig. 3 (a) trained for 50 epochs, with each layer incorporating a Block 2 (k = 32).

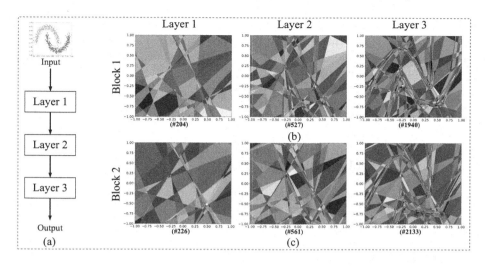

Fig. 3. During the training process with dataset *make moons* (1000 samples), we empirically observe the evolution process of linear regions and count the number of linear regions in each layer of different networks. (a) The network consists of three fully connected layers, each layer containing a minimum unit Block 1 or Block 2. (b) The evolution process of linear regions in each layer of the three-layer Block 1 network during training for 100 epochs (k = 32). (c) The evolution process of linear regions in each layer of the three-layer Block 2 network during training for 100 epochs (k = 32).

2   Related Work

In recent years, a group of scholars has made noteworthy contributions to the examination of linear regions and the operational mechanisms of networks. The research conducted in [9] presented an analysis of the correlation between the number of neurons and the count of linear regions in networks utilizing piecewise linear activation functions. Moreover, [10] established an upper bound on the number of activation regions of ReLU networks. The impact of network depth, width, and activation complexity on the quantity of linear regions was explored in [8]. Additionally, [3,12,15,19,20] investigated both upper and lower bounds of the number of linear regions of ReLU and LeakyReLU networks. By partially solving the algebraic topology problem based on ReLU networks, [11] provided an approximate analysis of how initialization affects the number of linear regions. Furthermore, [1] proposed a training approach for ReLU networks that enables the convergence of parameters to the global optimum. Notably, SplineCam [13] introduced a method for accurately calculating the geometric shape of networks.

3   Linear Regions of Networks

The ReLU [11] activation function is defined as follows:

$$ReLU(x) = \max(0, x), \tag{1}$$

when the input x is greater than 0, the output of ReLU function is x. Conversely, when x is less than or equal to 0, the output is 0 and the corresponding node is inactive and does not contribute to the output. During gradient back propagation, the derivative of ReLU is:

$$ReLU'(x) = \begin{cases} 1, & \text{if } x > 0, \\ 0, & \text{otherwise}. \end{cases} \tag{2}$$

LeakyReLU differs from ReLU by assigning a non-zero slope to all negative values. Specifically, it divides the negative portion by a value greater than 1. We define the LeakyReLU function as follows:

$$LeakyReLU(x) = \begin{cases} x, & \text{if } x >= 0, \\ x/a, & \text{if } x < 0, \end{cases} \tag{3}$$

where a is a constant greater than 1, representing the slope of the function for negative values. When the input x is greater than or equal to 0, the output of LeakyReLU function is x. Conversely, when x is less than 0, the output is x/a. During gradient back propagation, the derivative of LeakyReLU is:

$$LeakyReLU'(x) = \begin{cases} 1, & \text{if } x >= 0, \\ 1/a, & \text{otherwise}. \end{cases} \tag{4}$$

Under the activation of ReLU and LeakyReLU, let us consider a network with Z activation layers, assuming H^z is the number of neurons with z layers, where z is the number of activation layers. Therefore, the output of each H^z is

$$o^z = \{o_h^z, h \in H^z\}, \tag{5}$$

$$o_h^z = (\mathbf{w}_h^z)^T \hat{o}^{z-1} + \mathbf{b}^z, \quad h \in H^z, z \leq Z, \tag{6}$$

where $\mathbf{b}^z$ signifies the bias term pertaining to layer z, and $\mathbf{w}_h^z$ represents the weight of neurons h within layer z. Therefore, networks with ReLU and LeakyReLU activation can be seen as a connected piecewise linear function. As a result, every neuron in a network can be represented as a piecewise linear function of the input $x \in \mathbb{R}^d$:

$$f_h^z(x) = (\mathbf{S}_h^z)^T x + \mathbf{b}_h^z, \quad h \in H^z, \tag{7}$$

where d is the dimension of the input, and the linear weight and bias of the current neuron input are denoted by $\mathbf{S}_h^z$ and $\mathbf{b}_h^z$, respectively.

Thus, each linear inequality arrangement in ReLU and LeakyReLU networks forms convex regions. We define the linear regions of networks with ReLU and LeakyReLU activation as follows:

Definition 1 (adapted from [10]) Let Q be a network with input dimension d_{inp} and fix τ, a vector of trainable parameters for Q. Define

$$K_Q(\tau) := \{x \in \mathbb{R}^{d_{inp}} \mid \nabla Q(\cdot; \tau) \text{ is discontinuous at } x\}. \tag{8}$$

The linear regions of Q at τ are the connected parts of input without K_Q:

$$\text{linear regions } (Q, \tau) = \{x \mid x \in \mathbb{R}^{d_{inp}}, x \notin K_Q(\tau)\}. \tag{9}$$

4   "Dying ReLU" Based on Piecewise Linear Functions

Theorem 1. *Given that A is a ReLU network and D is a LeakyReLU network, under the conditions of consistent input distribution, initialization, network structure, and other parameters, the total number of continuous and piecewise-linear (CPWL) functions generated during training for A and D can be denoted as follows:*

$$\mathbb{E}[\#\{\ CPWL\ functions\ in\ \ A\ \}] = \beta(1), \tag{10}$$

$$\mathbb{E}[\#\{\ CPWL\ functions\ in\ \ D\ \}] = \gamma(1). \tag{11}$$

Then, the relationship between the number of CPWL functions generated during training for networks A and D can be expressed as follows:

$$|\beta(1)| \leq |\gamma(1)|. \tag{12}$$

This theorem implies that under consistent input distributions and network parameters, LeakyReLU networks generally exhibit stronger expressive capacity in terms of $CPWL$ functions compared to ReLU networks. The main reason of this phenomenon is the inability of the neurons in the ReLU layer to update their weights when encountering negative inputs, which is known as the "dying ReLU" problem. We provide the proof process for Theorem 1 as follows:

Proof. Let C represent the loss function of the network, and let $x_i^{(j)}$ denote the output of the ith neuron in the jth layer. Suppose $f(v) = \max(0, v)$ is a ReLU neuron, and $v_i^{(j)}$ is the linear input to the $(j+1)^{th}$ layer. Then, according to the chain rule, the derivative of the loss with respect to the weight $w_i^{(j)}$ connecting the jth and $(j+1)^{th}$ layers is given by:

$$\frac{\partial C}{\partial w_i^{(j)}} = \frac{\partial C}{\partial x_i^{(j+1)}} \frac{\partial x_i^{(j+1)}}{\partial w_i^{(j)}}. \tag{13}$$

Referring to (13), the first term on the right-hand side can be recursively calculated. The second term on the right-hand side is the only part directly related to $w_i^{(j)}$ and can be decomposed into:

$$\frac{\partial x_i^{(j+1)}}{\partial w_i^{(j)}} = \frac{\partial f\left(v_i^{(j)}\right)}{\partial v_i^{(j)}} \frac{\partial v_i^{(j)}}{\partial w_i^{(j)}}$$

$$= f'\left(v_i^{(j)}\right) x_i^{(j)}. \tag{14}$$

Therefore, in the case of ReLU networks, if the output of the previous layer within the network is consistently negative, the weights of the neuron will not be updated, rendering the neuron non-contributory to the learning process.

Referring to (7), assuming that the number of neurons in the $(j+1)^{th}$ layer of the ReLU network whose weights $\mathbf{S}_h^z$ are not updated is denoted as M, it follows that there are M neurons in the $(j+1)^{th}$ layer of the ReLU network that are unable to form new $CPWL$ functions. We can express the expected total number of $CPWL$ functions generated in the $(j+1)^{th}$ layer of the ReLU network as follows:

$$\mathbb{E}[\#\{ CPWL \text{ functions in the } (j+1)^{th} \text{ layer with ReLU}\}] = \lambda(1). \quad (15)$$

However, LeakyReLU does not suffer from the issue of weights not being updated. Under the condition of consistent parameters, if we replace the activation function between the jth and $(j+1)^{th}$ layers of the network with LeakyReLU, the expected total number of $CPWL$ functions generated in the $(j+1)^{th}$ layer of the LeakyReLU network can be expressed as follows:

$$\mathbb{E}[\#\{ CPWL \text{ functions in the } (j+1)^{th} \text{ layer with LeakyReLU}\}] = \eta(1). \quad (16)$$

Therefore, without considering the generation of duplicate $CPWL$ functions in the network and the special case where the jth layer of the network has no negative output, we typically have the following expression:

$$|\lambda(1)| \leq |\eta(1)|. \quad (17)$$

Referring to (10) and (11), assuming that A is a ReLU network and D is a LeakyReLU network, under the conditions of consistent input distribution, initialization, network structure, and other parameters, we easily get (12).

5   Experiments

We employed the PyTorch framework for our experiments. Each neuron in the network is represented as a linear function based on the input. The linear function of each neuron can be obtained through forward propagation. The datasets used in the experiments consist of two-dimensional data, as two-dimensional input data is easier to visualize. We used three types of two-dimensional datasets. The first dataset consists of 500 samples uniformly generated within the range of -1 to 1, with randomly generated labels dividing the data into two classes: 0 and 1. The second dataset is a binary classification dataset *make moons* with 1000 samples. The third dataset is a five-class classification dataset *make gaussian quantiles* with 1000 samples. Figure 4 illustrates the distribution of the three input data types in the network. Regarding the network parameters and configurations, we utilized network structures with different numbers of Blocks, as shown in Fig. 1, to examine the expressive capacity of the linear regions for both ReLU and LeakyReLU networks. Additionally, we employed the Adam optimizer, set the batch size to 32, used the Cross Entropy loss function, set the learning rate to 0.001, and set the value of a in (3) to 100.

Figure 5, Fig. 6, and Fig. 7 present the number of linear regions and the visualization results in ReLU and LeakyReLU networks trained on the three types of two-dimensional input samples for different epochs. Through these figures, we can visually observe a significant difference in the expressive capacity of linear regions between ReLU and LeakyReLU networks, under the same external

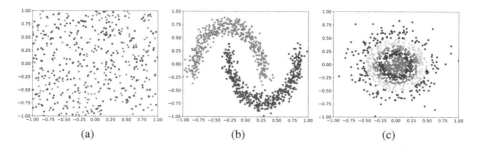

(a)   (b)   (c)

Fig. 4. (a) Two-dimensional input dataset consists of 500 samples are randomly generated within the range of -1 to +1, with randomly assigned labels for two classes. (b) Two-dimensional binary dataset *make moons* comprising 1000 samples. (c) Two-dimensional five-class dataset *make gaussian quantiles* comprising 1000 samples.

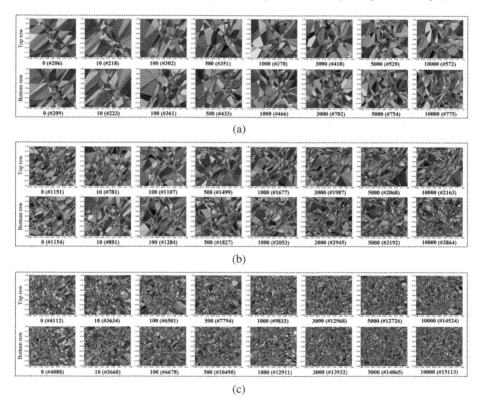

Fig. 5. The number of linear regions and the visualization results of the networks trained for r epochs with input Fig. 4 (a), for $r = 0, 10, 100, 500, 1000, 3000, 5000, 10000$. Top row (with ReLU): The network utilizes three fully connected Block 1. Bottom row (with LeakyReLU): The network employs three fully connected Block 2. (a) k = 16. (b) k = 32. (c) k = 64.

conditions. Based on our analysis of the number of linear regions and the visualization results, considering the same input distribution, number of neurons, and training parameters, in the majority of cases (while acknowledging the possibility of rare occurrences of $CPWL$ function duplication and linear region repetition during network training), LeakyReLU networks demonstrate a stronger ability to express linear regions compared to ReLU networks.

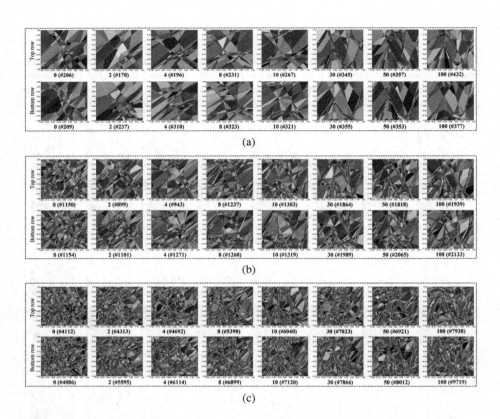

Fig. 6. The number of linear regions and the visualization results of the networks trained for r epochs with input Fig. 4 (b), for $r = 0, 2, 4, 8, 10, 30, 50, 100$. Top row (with ReLU): The network utilizes three fully connected Block 1. Bottom row (with LeakyReLU): The network employs three fully connected Block 2. (a) k = 16. (b) k = 32. (c) k = 64.

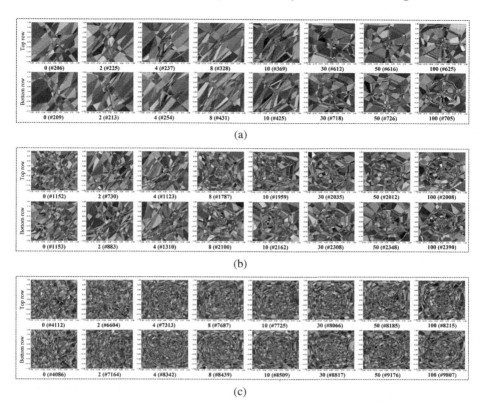

Fig. 7. The number of linear regions and the visualization results of the networks trained for r epochs with input Fig. 4 (c), for $r = 0, 2, 4, 8, 10, 30, 50, 100$. Top row (with ReLU): The network utilizes three fully connected Block 1. Bottom row (with LeakyReLU): The network employs three fully connected Block 2. (a) k = 16. (b) k = 32. (c) k = 64.

6 Conclusion

The actual expressive capacity of most $CPWL$ networks is still not well understood. In this study, we investigated a key characteristic of ReLU and LeakyReLU networks, which is their ability to express the number of linear regions. This serves as a representative measure of the complexity of functions they can effectively learn.

We conducted a series of experiments to explore how the linear regions of ReLU and LeakyReLU networks evolve during the training process. This included analyzing the evolution of linear regions layer by layer in the network and examining the evolution of linear regions across different epochs. Furthermore, we provided an intuitive explanation of the "dying ReLU" problem based on the expressive capacity of $CPWL$ functions within the network. Specifically, we found that under the same conditions of network input, network structures, and training parameters, LeakyReLU networks generally exhibit a stronger ability to express linear regions compared to ReLU networks.

We hope that our research can provide inspiration for the design of activation functions and contribute to the exploration and analysis of the behavior exhibited by $CPWL$ functions within networks.

Acknowledgements. This work was supported by the National Natural Science Foundation of China (Grant No. 61973334).

References

1. Arora, R., Basu, A., Mianjy, P., Mukherjee, A.: Understanding deep neural networks with rectified linear units. arXiv preprint arXiv:1611.01491 (2016)
2. Cai, Y., et al.: Yolobile: real-time object detection on mobile devices via compression-compilation co-design. In: Proceedings of the AAAI Conference on Artificial Intelligence, vol. 35, pp. 955–963 (2021)
3. Chen, H., Wang, Y.G., Xiong, H.: Lower and upper bounds for numbers of linear regions of graph convolutional networks. arXiv preprint arXiv:2206.00228 (2022)
4. Chen, P., Zhuang, B., Shen, C.: FATNN: fast and accurate ternary neural networks. In: Proceedings of the IEEE/CVF International Conference on Computer Vision, pp. 5219–5228 (2021)
5. Freeman, I., Roese-Koerner, L., Kummert, A.: EffNet: an efficient structure for convolutional neural networks. In: 2018 25th IEEE International Conference on Image Processing (ICIP), pp. 6–10. IEEE (2018)
6. Gao, Z., Chen, X., Xu, J., Yu, R., Zong, J.: A comprehensive vision-based model for commercial truck driver fatigue detection. In: Tanveer, M., Agarwal, S., Ozawa, S., Ekbal, A., Jatowt, A. (eds.) Neural Information Processing. ICONIP 2022. LNCS, vol. 13625, pp. 182–193. Springer, Cham (2023). https://doi.org/10.1007/978-3-031-30111-7_16
7. Glorot, X., Bordes, A., Bengio, Y.: Deep sparse rectifier neural networks. In: Proceedings of the Fourteenth International Conference on Artificial Intelligence and Statistics, pp. 315–323. JMLR Workshop and Conference Proceedings (2011)
8. Goujon, A., Etemadi, A., Unser, M.: The role of depth, width, and activation complexity in the number of linear regions of neural networks. arXiv preprint arXiv:2206.08615 (2022)
9. Hanin, B., Rolnick, D.: Complexity of linear regions in deep networks. In: International Conference on Machine Learning, pp. 2596–2604. PMLR (2019)
10. Hanin, B., Rolnick, D.: Deep ReLU networks have surprisingly few activation patterns. In: Advances in Neural Information Processing Systems, vol. 32 (2019)
11. Hinz, P.: Using activation histograms to bound the number of affine regions in ReLU feed-forward neural networks. arXiv preprint arXiv:2103.17174 (2021)
12. Hu, Q., Zhang, H., Gao, F., Xing, C., An, J.: Analysis on the number of linear regions of piecewise linear neural networks. IEEE Trans. Neural Netw. Learn. Syst. **33**(2), 644–653 (2020)
13. Humayun, A.I., Balestriero, R., Balakrishnan, G., Baraniuk, R.G.: SplineCam: exact visualization and characterization of deep network geometry and decision boundaries. In: Proceedings of the IEEE/CVF Conference on Computer Vision and Pattern Recognition, pp. 3789–3798 (2023)

14. Maas, A.L., Hannun, A.Y., Ng, A.Y., et al.: Rectifier nonlinearities improve neural network acoustic models. In: Proceedings of International Conference on Machine Learning, vol. 30, p. 3. Atlanta, Georgia, USA (2013)
15. Montúfar, G., Ren, Y., Zhang, L.: Sharp bounds for the number of regions of maxout networks and vertices of minkowski sums. SIAM J. Appl. Algebra Geom. **6**(4), 618–649 (2022)
16. Nair, V., Hinton, G.E.: Rectified linear units improve restricted boltzmann machines. In: Proceedings of the 27th International Conference on Machine Learning (ICML-10), pp. 807–814 (2010)
17. Tseran, H., Montufar, G.F.: On the expected complexity of maxout networks. Adv. Neural. Inf. Process. Syst. **34**, 28995–29008 (2021)
18. Wang, C.Y., Bochkovskiy, A., Liao, H.Y.M.: Scaled-yolov4: scaling cross stage partial network. In: Proceedings of the IEEE/CVF Conference on Computer Vision and Pattern Recognition, pp. 13029–13038 (2021)
19. Wang, Y.: Estimation and comparison of linear regions for ReLU networks. In: International Joint Conference on Artificial Intelligence (2022)
20. Xiong, H., Huang, L., Yu, M., Liu, L., Zhu, F., Shao, L.: On the number of linear regions of convolutional neural networks. In: International Conference on Machine Learning, pp. 10514–10523. PMLR (2020)
21. Yang, L., et al.: CondenseNet v2: sparse feature reactivation for deep networks. In: Proceedings of the IEEE/CVF Conference on Computer Vision and Pattern Recognition, pp. 3569–3578 (2021)

Heterogeneous Graph Prototypical Networks for Few-Shot Node Classification

Yunzhi Hao[1,2(✉)], Mengfan Wang[1], Xingen Wang[1], Tongya Zheng[1],
Xinyu Wang[1], Wenqi Huang[3], and Chun Chen[1]

[1] Zhejiang University, Hangzhou 310058, China
{ericohyz,21951202,newroot,tyzheng,wangxinyu,chenc}@zju.edu.cn
[2] Zhejiang University - China Southern Power Grid Joint Research Centre on AI,
Hangzhou 310058, China
[3] Digital Grid Research Institute, China Southern Power Grid, Guangzhou 510663,
China
huangwq@csg.cn

Abstract. The node classification task is one of the most significant applications in heterogeneous graph analysis, which is widely used for modeling multi-typed interactions. Meanwhile, Graph Neural Networks (GNNs) have aroused wide interest due to their remarkable effects on graph node classification. However, there are some challenges when applying GNNs to heterogeneous graph node classification: the cumbersome node labeling cost, and the heterogeneity of graphs. Existing GNNs require sufficient annotation while learning classifiers independently with node embeddings cannot exploit graph topology effectively. Recently, few-shot learning has achieved competitive results in homogeneous graphs to address the performance degradation in the label sparsity case. While heterogeneous graph few-shot learning is limited by the difficulties of extracting multiple semantics. To this end, we propose a novel Heterogeneous graph Prototypical Network (HPN) with two modules: Graph structural module generates node embeddings and semantics for meta-training by capturing heterogeneous structures. Meta-learning module produces prototypes with heterogeneous induced subgraphs for meta-training classes, which improves knowledge utilization compared with the traditional meta-learning. Experimental results on three real-world heterogeneous graphs demonstrate that HPN achieves outstanding performance and better stability.

Keywords: Few-shot Learning · Heterogeneous Graph Node Classification · Meta-learning · Graph Convolutional Networks

1 Introduction

Heterogeneous graphs analysis has practical significance and applications since the interactions between real-world entities are often multi-typed [22,30], such as social graphs and citation graphs. And node classification [2,35] finds its crucial applications over a wide domain such as document categorization and protein classification [4]. Graph neural networks (GNNs) [12,15,26], owing to the large power on node classification in heterogeneous graphs, have gained increasing popularity [8,13,31]. However, there are three main challenges for the heterogeneous graph node classification task: (1) The rich structural and semantic proximities are difficult to be aggregated as formatted auxiliary information, despite their importance. (2) The model performance highly relies on a large amount of annotated samples. (3) Defining node labels requires prior domain knowledge, which makes the task cumbersome to be deployed in practice [32]. For the first challenge, existing heterogeneous embedding methods [8,10] first aggregate heterogeneous structures and then apply them to downstream tasks. Such two-phase frameworks have an inherent deficiency that task-related information cannot be effectively exploited. Focusing on two latter challenges, existing GNNs [28,33,34] for the heterogeneous graph node classification usually follow a semi-supervised paradigm. However, they are prone to errors when available labeled nodes are less than 10%, especially without labels from all node classes.

Meta-learning, which aims to handle the problem of data deficiency by recognizing new categories with very few labeled samples [1], has yielded significant progress [24]. And meta-learning methods on graphs show superior performance in few-shot node classifications [6,37]. Meta-GNN [38] applies meta-learning to GNNs for the first time. GPN [7] generates effective class prototypes with high robustness by *network encoder* and *node valuator*. G-Meta [14] collects meta-gradients from local graphs to update network parameters. However, the aforementioned graph few-shot learning methods cannot deal with various semantics and can only be applied to homogeneous graphs. Very recently, several attempts have also been made towards applying meta-learning to heterogeneous graphs. MetaHIN [18] proposes a novel semantic-enhanced tasks constructor and a co-adaptation meta-learner. HG-Meta [36] builds a graph encoder to aggregate heterogeneous information with meta-paths and leverages unlabelled information to alleviate the low-data problem. However, there are three main difficulties in applying meta-learning to heterogeneous graphs:

1) *Meta-training task sampling.* Meta-learning first trains a model with sampled similar tasks and then test on new few-shot classes. These sampled training tasks are assumed to be independent of each other in traditional scenarios. However, heterogeneous graph nodes are naturally related and show diverse importance, training on far-related nodes results in a decrease in efficiency. How to effectively sample effective training tasks with global graph knowledge has a significant impact on the model performance.

2) *Structural and semantic heterogeneities.* Multiple semantics and structures in heterogeneous graphs are difficult to be captured as auxiliary knowledge in the meta-training process by existing methods. How to effectively extract information from different neighbors remains a problem.

3) *Model robustness.* Noticeably, models with low robustness show a significant decline in effectiveness once the new class is difficult to classify. The robustness

analysis in Sect. 5 verifies that the performances of previous methods differ largely across different testing sets, which illustrates their limitations.

To address the above difficulties, we propose a novel **H**eterogeneous graph **P**rototypical **N**etworks (HPN) to address the heterogeneous graph few-shot node classification task. The core of HPN is to generate effective prototypes with heterogeneous relational structures for multiple node classes. We first sample effective meta-training classes with global graph knowledge to avoid redundancy and make the model more adaptable to new classes. Then the graph structural module captures the multiple semantics to produce node representations and an attention matrix that quantifies the importance of different semantics. And the meta-learning module gets training on the sampled tasks and generates practical prototypes with the induced heterogeneous subgraphs. In this way, HPN produces representative class prototypes with various auxiliary knowledge in the meta-training process and shows higher performance on new node classification tasks with few-shot settings. We summarize the main contributions as follows:

- We propose a novel graph-based meta-learning scheme HPN to tackle the heterogeneous few-shot node classification task.
- A graph structural module and a meta-learning module are proposed for learning representative prototypes by incorporating heterogeneous relational structures into each sampled class.
- Extensive experiments on three heterogeneous graph analysis tasks across various domains prove that our HPN significantly outperforms state-of-the-art methods in heterogeneous knowledge amalgamation.

2   Related Work

2.1   Heterogeneous Graph Node Classification

Node classification, as one of the most significant graph tasks, aims at predicting missing node labels based on graph topology, semantics, and labeled nodes [5,11,25]. Earlier methods generate node embeddings with meta-paths [8,21] and apply a discriminative classifier like logistic regression on the node embeddings. Recently, GNNs have gained great popularity, which aim to address node classification in an end-to-end manner with more than 10% labeled nodes. HetGNN [34] can encode both structure and content heterogeneity. HAN [28] first applies node- and semantic-level attention mechanisms. Considering the limitations of the pre-defined meta-paths, recent methods start to generate meta-paths automatically. GTN [33] first performs graph convolution with the importance of all combinations of meta-paths within a length. HGT [13] proposes a mini-batch graph sampling method to learn personalized semantics. And ie-HGCN [31] performs object- and type-level aggregation and brings better interpretability. However, the above methods are designed for the semi-supervised node classification task, which cannot handle new classes with severely limited labels.

2.2   Few-Shot Learning

Few-shot learning [20] trains classifiers with only a few labeled samples. There are three popular types of approaches for few-shot learning: model-based, metric-based, and optimization-based approaches [29]. Meta network [19] consists of three components: meta knowledge acquisition, fast weight generation, and slow weight optimization. Prototypical network [23] uses the nearest neighbor classifier after learning the distance between the batch samples and the support samples. MAML [9] trains the parameters explicitly and yields good generalization performance with a small number of gradient steps. As for graph few-shot learning methods, Meta-GNN [38] applies meta-learning to GNNs for the first time and leverages MAML for the gradient updates. GPN [7] extracts meta-knowledge from attributed networks and further generates representative prototypes. MetaTNE [16] automatically learns prior knowledge with relations between graph topology and classes. Very recently, several few-shot methods start to consider meta-learning on heterogeneous graphs. MetaHIN [18] proposes a novel semantic-enhanced tasks constructor and a co-adaptation meta-learner. HG-Meta [36] aggregates heterogeneous information with meta-paths and leverages unlabelled information to alleviate the low-data problem. However, the above approaches can not improve both efficiency and robustness due to the lack of semantic knowledge. In this work, we are motivated to develop the metric-based method with heterogeneous GNNs to fully extract both semantic and task-related knowledge.

3   Preliminaries

In this section, we give definitions in this paper to describe our heterogeneous graph few-shot learning method.

Heterogeneous Graph. A heterogeneous graph $G = (\mathcal{V}, \mathcal{E}, A, X)$, contains a set of nodes $\mathcal{V}$, a set of edges $\mathcal{E}$, and an adjacency matrix set A. The following type mapping functions $\phi : \mathcal{V} \to \mathcal{A}$ and $\varphi : \mathcal{E} \to \mathcal{R}$ correspond to the type relations of nodes and edges respectively, in which $\mathcal{A}$ and $\mathcal{R}$ are predefined node and edge type sets, and $|\mathcal{A}| + |\mathcal{R}| > 2$. The features of nodes $\mathcal{V}$ are defined as X.

Heterogeneous Graph Few-shot Node Classification. Given a heterogeneous graph $G = (\mathcal{V}, \mathcal{E}, A, X)$ with a target type for node classification, the few-shot problem defines a training node set D_{train} which belongs to the class set C_1, and a testing node set D_{test} which belongs to the class set C_2, and $C_1 \cap C_2 = \emptyset$. In D_{test}, only N nodes have labels available in the support set, where N is a small number. Our objective is to classify the unlabeled nodes in the testing set.

Prototypical Networks. Prototypical Networks [23] compute a *prototype*, termed as $\mathbf{c}_k$, which is an M-dimensional embedding with $\mathbf{c}_k \in R^M$, for each task k by a small support set S_k of N labeled samples. This is achieved with the following function:

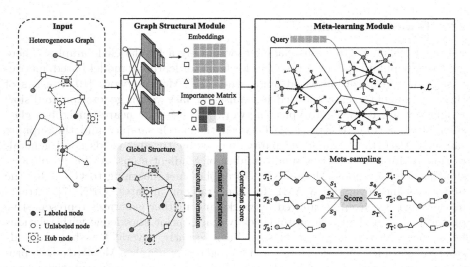

Fig. 1. The overall framework of HPN. HPN first encodes the heterogeneous graph to produce embeddings and a semantic importance matrix M. Then HPN samples meta-training classes and generate prototypes with induced subgraphs.

$$\mathbf{c}_k = \frac{1}{|S_k|} \sum_{(x_i, y_i) \in S_k} f_\phi(\mathbf{x}_i). \tag{1}$$

Prototypical Networks generate a distribution for a query set sample $\mathbf{x}$ by adopting a softmax function over distances to each label prototype:

$$p_\phi(y = k|\mathbf{x}) = \frac{\exp(-d(f_\phi(\mathbf{x}), \mathbf{c}_k))}{\sum_{k'=1}^{K} \exp(-d(f_\phi(\mathbf{x}), \mathbf{c}_{k'}))}, \tag{2}$$

where K denotes the number of training classes, and d is a distance metric. By minimizing the negative log-probability $J(\phi) = -\log p_\phi(y = k|\mathbf{x})$ of the true class k, the model is iteratively updated until stable.

4   Methodology

HPN aims to adapt both heterogeneous structure and semantic knowledge captured from several training sets to new tasks to fit few-shot learning settings. As illustrated in Fig. 1, HPN first extracts both structural and semantic correlations by GNN layers, then the meta-learning module samples the training classes with a semantic importance matrix M. HPN generates a prototype for each training class with heterogeneous substructures.

4.1   Graph Structural Module

We design a graph structural module to aggregate heterogeneous structures into the node embeddings and automatically extract multiple semantics to produce a semantic attention matrix that quantifies their importance. Inspired by ie-HGCN [31], we follow the graph projection architecture to project neighbors from heterogeneous domains into a new common semantic space before aggregating their information. For nodes $\mathcal{V}_q$ from node type $q \in \mathcal{A}$, the projection of $\mathcal{V}_q$ and its neighbors $\mathcal{V}_s$ from node type $s \in \mathcal{A}$ can be formulated as follows:

$$
\begin{aligned}
\mathbf{Y}^l_{q-q} &= \mathbf{H}^{l-1}_q \cdot \mathbf{W}_{q-q}, \\
\mathbf{Y}^l_{s \to q} &= \mathbf{H}^{l-1}_s \cdot \mathbf{W}_{s \to q}, s \in \mathcal{N}_q,
\end{aligned}
\tag{3}
$$

where $\mathbf{H}^l_q$ is the node representation of nodes $\mathcal{V}_q$ from type q at layer l, and the first layer can be defined as: $\mathbf{H}^0_q = X_q$ is the original input node features of $\mathcal{V}_q$. $\mathcal{N}_q$ is neighbors of nodes $\mathcal{V}_q$. Here $\mathbf{W}_{s \to q}$ is a projection matrix from node type s to type q, and $\mathbf{W}_{q-q}$ is designed as a self-projection matrix to project $\mathbf{H}^{l-1}_q$ from the original feature domain into a new joint space at layer l. $\mathbf{Y}^l_{q-q}$ and $\mathbf{Y}^l_{s \to q}$ are projected hidden representations of nodes $\mathcal{V}_q$ and neighbors $\mathcal{V}_s$, respectively.

After projecting all the neighbor node representations into the common semantic space, the neighbor information of $\mathcal{V}_q$ can be aggregated with the adjacency matrix set A. Specifically, the attention mechanism is employed to capture the importance of neighbors with different types of $\mathcal{V}_q$ automatically. Briefly, the higher-level representations of nodes $\mathcal{V}_q$ can be defined as follows:

$$
\mathbf{H}^l_q = \sigma \Big(\sum_{s \in \mathcal{N}_q} \alpha_{s \to q} \cdot \mathbf{Y}^l_{s \to q} \cdot \mathbf{A}_{s-q} + \alpha_{q-q} \cdot \mathbf{Y}^l_{q-q} \Big),
\tag{4}
$$

where σ is the nonlinearity, $\alpha_{s \to q}$ is normalized attention coefficients from node type s to the current node type q, and A_{s-q} is the corresponding adjacency matrix from set A between $\mathcal{V}_s$ and $\mathcal{V}_q$. After several layers, HPN can aggregate the structural and semantic dependencies in a heterogeneous graph. Here we define the final embeddings for nodes from all types as follows:

$$
\mathbf{H} = \{\mathbf{H}^L_1, \mathbf{H}^L_2, ..., \mathbf{H}^L_{|\mathcal{A}|}\}.
\tag{5}
$$

With the learned attention coefficients $\alpha_{s \to q}$ for each type pair from the node type set $\mathcal{A}$, HPN can calculate the mean attention distribution in each layer. Then we generate the semantic attention matrix $M \in R^{|\mathcal{A}| \times |\mathcal{A}|}$ containing the attention scores of each node type pair, based on which we can compute the importance scores of diverse meta-paths by matrix multiplication.

4.2   Meta-Learning Module

The meta-learning module aims to simulate the few-shot setting by sampling a few training nodes from D_{train}. In brief, HPN samples N_C node classes from training set C_1, where $N_C = |C_2|$. Then HPN samples N labeled nodes from each class to mimic the testing node classification settings. Each meta-training task $\mathcal{T}_t$ is defined as follows:

$$
\begin{aligned}
\mathcal{S}_t &= \{(v_1, y_1), (v_2, y_2), ..., (v_{N_C \times N}, y_{N_C \times N})\}, \\
\mathcal{Q}_t &= \{(v_1', y_1'), (v_2', y_2'), ..., (v_{N_C \times (D_k - N)}', y_{N_C \times (D_k - N)}')\}, \\
\mathcal{T}_t &= \mathcal{S}_t + \mathcal{Q}_t,
\end{aligned}
\tag{6}
$$

where $\mathcal{S}_t$ is the support set with N labeled nodes, while $\mathcal{Q}_t$ is the query set with $D_k - N$ unlabeled nodes, D_k is the number of nodes from c_k. We repeat the above sampling steps T times to generate T meta-training tasks.

Meta-Sampling. Previous approaches randomly sample classes from set C_1 for meta-training, which results in a high dependency of the model performance on the new test set. Worse, training on far-related node classes leads to a decrease in model classification efficiency. Therefore, we perform a global-level sampling to select meta-training classes from C_1 which are more related to the whole graph G per episode, making sure that the model can be explicitly trained for any testing task. Specifically, this is achieved by calculating the correlation score between nodes from the hub node set $\mathcal{V}_{hub}$ and nodes from each training class c_k. Hub nodes are structurally important nodes on the graph, which serve as the backbone of the graph and encode global graph information. Hub nodes can be measured by multiple node centrality scores such as node degree or PageRank [3]. We use PageRank in our experiments as an example.

With the hub node set $\mathcal{V}_{hub}$ and the adjacency matrix set A, we can count different types of neighbors of $\mathcal{V}_{hub}$ related by different meta-paths. Taking IMDB as an example, if movie m_1 and m_2 are directed by the same director d_1, then m_2 is one of m_1's "director"-type neighbors. Neighbors of different types are assigned different importance by the semantic importance matrix M.

With the number of neighbors from class c_k, the relevance score between c_k and the hub node set $\mathcal{V}_{hub}$ can be computed as follows:

$$
score(c_k, \mathcal{V}_{hub}) = \sum_{n_s^k \in \mathcal{N}_{\mathcal{V}_{hub}}} M_s \cdot |n_s^k|,
\tag{7}
$$

where n_s^k is the s-type neighbors of $\mathcal{V}_{hub}$, and n_s^k belongs to class c_k at the same time, M_s is the importance coefficient of type s from the semantic importance matrix M. Thus, class c_k is selected as the meta-training class from C_1 with the following probability:

$$
P_r(c_k) = \text{softmax}(score(c_k)) = \frac{\exp(score(c_k))}{\sum_{c_j \in C_1} \exp(score(c_j))}.
\tag{8}
$$

In this way, we can sample global-level meta-training classes with the probability $P_r(c_k)$. To guarantee task coverage and a high sampling rate, we design a fixed-length list $\mathcal{L}_{hist}$ with size $m = N_C \times |\mathcal{A}|$ to record up-to-m classes visited previously during the meta-sampling. Class c_k will be selected as the next meta-training class if c_k is not in $\mathcal{L}_{hist}$. Then we randomly select $\mathcal{S}$ and $\mathcal{Q}$ for each selected node class to generate meta-training tasks. The experimental results show that sampling global-level training nodes by $\mathcal{V}_{hub}$ can not only improve the model performance but also adjust the model for different testing tasks with different difficulties, which improves the model robustness.

Heterogeneous Prototype Generation. After generating the meta-training tasks, we propose to learn the class prototype based on the support set. Different from the prototypes in the homogeneous graphs that are computed as mean vectors of embeddings in the corresponding support set as GPN, the prototype in the heterogeneous graphs should aggregate multiple structures and semantic information. Therefore, HPN first extracts one-hop heterogeneous subgraphs G_k for nodes $\mathcal{V}_k$ in the support set $\mathcal{S}_k$ by extracting the first-order proximity in graph G. G_k contains multiple types of nodes and edges, which can be regarded as a local relational network of nodes $\mathcal{V}_k$. For each node type $q \in \mathcal{A}$ related to the current class c_k, the type-level prototype can be directly computed by:

$$\mathbf{c}_k^q = \frac{1}{|\mathcal{V}_q|} \sum_{v_q \in G_k} f_\phi(\mathbf{h}_q), \tag{9}$$

where $\mathbf{h}_q$ from set $\mathbf{H}$ is the embedding of node v_q from subgraph G_k, and $f_\phi(\cdot)$ is the prototype computation function, usually an average aggregator. Through the semantic importance matrix M, we can calculate the prototype of class c_k as a weighted average of each type of nodes from the subgraph G_k. Formally, the prototype of c_k can be defined as follows:

$$\mathbf{c}_k = M_q \cdot \sum_{q \in \mathcal{A}} \mathbf{c}_k^q, \tag{10}$$

where M_q is the importance score of node type q to the current classified type. Specifically, our heterogeneous prototype generation mechanism calculates class prototypes by both support set $\mathcal{S}_k$ and their local relational network G_k, which can aggregate multiple semantics and alleviate the low-data problem.

Model Optimization. Once we generate class prototypes in each meta-training task $\mathcal{T}_t$, these prototypes can be defined as predictors for nodes in the query set $\mathcal{Q}_t$, which assigns a probability $p(c_k|v_i')$ for every training class according to the vector distances between v_i' from $\mathcal{Q}_t$ and each class prototype vector $\mathbf{c}_k$:

$$p(c_k|v_i') = \mathrm{softmax}(-d(\mathbf{h}_i', \mathbf{c}_k)) = \frac{\exp(-d(\mathbf{h}_i', \mathbf{c}_k))}{\sum_{c_j} \exp(-d(\mathbf{h}_i', \mathbf{c}_j))}, \tag{11}$$

where $d(\cdot, \cdot)$ is a distance function, usually Euclidean distance.

By minimizing the classification loss $\mathcal{L}$, HPN trains each meta-training task $\mathcal{T}_t$ with the prediction of each node v_i' in $\mathcal{Q}_t$ and its ground-truth y_i'. Particularly, the final classification loss $\mathcal{L}$ is defined as the average negative log-likelihood probability of assigning correct labels, which can be calculated as follows:

$$\mathcal{L} = \frac{1}{N_C \times (D_k - N)} \sum_{v_i' \in \mathcal{Q}} -\mathrm{log}p(\mathbf{y}_i'|v_i'). \tag{12}$$

Compared with the previous approaches, our combination of heterogeneous information and node labels allows for an efficient and robust way to adapt to new classes.

4.3 Complexity Analysis

The proposed framework HPN consists of a graph structural module and a meta-learning module. The former module efficiently performs heterogeneous graph convolution in each layer without iterative multiplication of adjacency matrices. The computational complexity of graph convolution is $\mathcal{O}(|E| \times d \times d')$, where d and d' are the original node feature size and the node embedding size, respectively. Since the adjacency matrices A of real graphs are often sparse and the neighbors of nodes are typically very small compared to $|V|$ and $|E|$, the time complexity of the attention computation is $\mathcal{O}(|V| + |E|)$. In practice, the time complexity of our proposed HPN is linear to the number of nodes and edges of the input heterogeneous graph G, which indicates the efficiency of the model.

5   Experiments

In this section, we first introduce three real-world heterogeneous evaluation graphs and nine related competitor models. Then we compare our HPN with state-of-the-art models on several few-shot classification tasks and present the detailed experiment results. Finally, we provide further robustness analysis and ablation studies to prove the efficacy of HPN.

5.1   Datasets

We evaluate HPN on three real-world heterogeneous datasets from two domains. The detailed descriptions of three datasets are summarized in Table 1.

Aminer[1]. Academic heterogeneous graph Aminer consists of four types of nodes: 127,623 papers (P), 164,472 authors (A), 101 conferences (C), and 147,251 references (R). The target node set is classified into ten classes.

IMDB[2]. A subset of IMDB dataset in HetRec 2011 is extracted for evaluation. IMDB describes heterogeneous information of movies. IMDB contains 10,196

---

[1] https://www.aminer.cn/citation.
[2] https://grouplens.org/datasets/hetrec-2011/.

movies (M), 95,318 actors (A), and 4,059 directors (D). The target node set is classified into eighteen styles.

Douban[3]. Douban is another movie review network with four types of nodes: 9,754 movies (M), 6,013 actors (A), 2,310 directors (D), and 13,080 users (U). The target node set movie nodes contain ten classes.

Table 1. Statistics of three datasets. (src-dst) represents an edge type where src is the source node and dst is the destination node.

Dataset	(src-dst)	Num of src	Num of dst	Num of src-dst	Classes
Aminer	P − A	127623	164472	355072	10
	P − C	127623	101	127623	
	P − R	127623	101	392519	
IMDB	M − A	10196	95318	231736	18
	M − D	10196	4059	10154	
Douban	M − U	9754	13080	882298	10
	M − A	9754	6013	26346	
	M − D	9754	2310	8751	

For Aminer and Douban, we evaluate HPN on all combinations of two node classes as meta-testing classes. Similarly for the IMDB, except that $N_C = 3$ for meta-testing due to the relatively large number of unique node classes.

5.2 Baselines

We compare our model with nine traditional graph embedding models, along with state-of-art graph few-shot node classification approaches.

Semi-Supervised Methods. GCN [15] is a GNN-based model for homogeneous graph node classification. **metapath2vec** [8] and **HAN** [28] are heterogeneous node classification models based on meta-paths. **ie-HGCN** [31] develops GNN-based semi-supervised heterogeneous node classification models and automatically evaluates all combinations of meta-paths within a length limit.

Meta-learning Methods. Meta-GNN [38] first learns the graph few-shot node classification task with optimization-based MAML. **GPN** [7] derives representative class prototypes for meta-training with a network encoder and a node valuator. **AMM-GNN** [27] uses the attention mechanism to transform the support characteristics for meta-testing tasks. **G-Meta** [14] collects meta-gradients from local graphs surrounding the target nodes to update network parameters. **MetaHIN** [18] proposes a novel semantic-enhanced tasks constructor and a co-adaptation meta-learner for heterogeneous graphs.

---

[3] https://github.com/librahu/HIN-Datasets-for-Recommendation-and-Network-Embedding.

Implementation Details. For homogeneous graph classification methods, we apply them to the whole graph neglecting the graph heterogeneity and calculating all node and edge types equally. For metapath-based heterogeneous graph methods metapath2vec and HAN, we use the pre-defined meta-paths PAP, PCP and PRP on Aminer; MAM and MDM on IMDB; MAM, MUM and MDM on Douban. For semi-supervised models, we follow the way in paper [38] to adjust those methods to graph few-shot node classification settings. For Meta-GNN, GPN, AMM-GNN, G-Meta, and MetaHIN, we follow the settings from the original papers. For the proposed HPN, we rank all nodes by their PageRank scores [3] in descending order and select the top 5% nodes from the current type as hub nodes [17]. We select all possible meta-testing classes for each dataset and train the model over 200 episodes.

Table 2. Few-shot node classification results on three datasets.

Methods	Aminer						IMDB						Douban					
	5-shot		3-shot		1-shot		5-shot		3-shot		1-shot		5-shot		3-shot		1-shot	
	ACC	F1	ACC	F1	ACC	F1	ACC	F1	ACC	F1	ACC	F1	ACC	F1	ACC	F1	ACC	F1
GCN	0.504	0.530	0.500	0.518	0.515	0.497	0.248	0.207	0.346	0.294	0.300	0.209	0.639	0.660	0.454	0.440	0.480	0.454
metapath2vec	0.529	0.505	0.481	0.433	0.448	0.359	0.376	0.399	0.362	0.377	0.353	0.311	0.792	0.834	0.628	0.658	0.518	0.501
HAN	0.574	0.497	0.572	0.478	0.560	0.452	0.264	0.268	0.324	0.296	0.298	0.224	0.842	0.838	0.682	0.703	0.665	0.680
ie-HGCN	0.585	0.502	0.577	0.499	0.581	0.513	0.335	0.360	0.317	0.303	0.311	0.269	0.821	0.825	0.701	0.698	0.679	0.691
Meta-GNN	0.503	0.432	0.507	0.400	0.499	0.403	0.361	0.297	0.345	0.271	0.343	0.232	0.501	0.472	0.510	0.507	0.507	0.489
GPN	0.478	0.374	0.503	0.405	0.485	0.429	0.292	0.245	0.345	0.272	0.338	0.289	0.775	0.806	0.601	0.596	0.527	0.508
AMM-GNN	0.511	0.450	0.527	0.445	0.497	0.435	0.374	0.312	0.351	0.302	0.349	0.297	0.793	0.815	0.632	0.606	0.534	0.521
G-Meta	0.475	0.363	0.498	0.382	0.465	0.363	0.353	0.301	0.350	0.299	0.346	0.285	0.782	0.807	0.608	0.600	0.529	0.512
MetaHIN	0.579	0.548	0.570	0.503	0.569	0.516	0.395	0.387	0.360	0.342	0.361	0.320	0.839	0.806	0.645	0.624	0.613	0.577
HPN	**0.672**	**0.613**	**0.648**	**0.575**	**0.602**	**0.563**	**0.505**	**0.478**	**0.465**	**0.436**	**0.447**	**0.413**	**0.850**	**0.839**	**0.752**	**0.724**	**0.705**	**0.694**
Impr	0.087	0.065	0.071	0.057	0.021	0.047	0.110	0.079	0.103	0.059	0.086	0.093	0.008	0.001	0.051	0.021	0.026	0.003

5.3   Few-Shot Node Classification

To evaluate the performance of all the classification methods, we generate three few-shot tasks for each dataset: 5-shot, 3-shot, and 1-shot node classification, and report the Accuracy (ACC) and Micro-F1 (F1) of each method.

Results on all few-shot node classification tasks over three datasets are shown in Table 2. We can notice that the proposed HPN consistently and significantly outperforms all the baselines. Generally speaking, homogeneous graph few-shot approaches Meta-GNN, GPN, AMM-GNN, and G-Meta outperform the semi-supervised GCN, which could be readily overfitted with only a small number of labeled nodes. Obviously, adapting meta-learning into GNNs can alleviate the problem caused by few samples to a certain degree and achieve better performance. It is noteworthy that semi-supervised heterogeneous GNNs like HAN achieve higher ACC and F1-score than graph few-shot methods like AMM-GNN and GPN, even sometimes heterogeneous method MetaHIN, demonstrating the fact that the rich structural and semantic proximities in graphs are critical and beneficial to the node classification task. Compared with the best competitors, the 3-shot classification performance of HPN improves by 7.1%, 10.3%, and 5.1%

in ACC on the Aminer, IMDB, and Douban, respectively. HAN achieves a similar effect to HPN on the 5-shot classification task of the Douban dataset, mainly because of the rich interactions between users and movies. However, the performance of HAN suffers a significant decline on the sparse IMDB dataset due to the lack of meta-learning. While HPN can effectively extract graph structures and node labels as prior knowledge even if the graph is sparse. Significantly, HPN reaps more improvement on the challenging 1-shot task due to its capability of adapting to new classes, which verifies the effectiveness of our heterogeneous prototype generation.

5.4 Robustness Analysis

As declared in Sect. 1, the performance of the previous graph meta-learning methods is sensitive to the testing class selection. The meta-training task sampling of HPN aims to train a robust model that can handle different testing sets. Table 3 shows the standard deviation of three datasets. Obviously, the performances of the previous approaches differ largely across different testing sets. And in the experiments, we notice that the performance of most methods fluctuates greatly when the experiment is conducted on different support sets, which proves that semi-supervised GNNs is highly dependent on the node selection. While HPN achieves smaller std (< 0.05) on the three datasets since the graph meta-sampling of HPN is able to select effective training classes from C_1 by calculating the correlation score between $\mathcal{D}_{test}$ and each class in C_1.

Table 3. Robustness analysis. Standard deviation, denoted by std, is reported overall testing sets of the three datasets.

Methods	GCN	metapath2vec	HAN	ie-HGCN	Meta-GNN	GPN	AMM-GNN	G-Meta	MetaHIN	**HPN**
Aminer	0.036	0.091	0.063	0.132	0.048	0.089	0.074	0.088	0.056	**0.009**
IMDB	0.138	0.067	0.062	0.076	0.054	0.133	0.147	0.097	0.062	**0.009**
Douban	0.246	0.182	0.177	0.160	0.070	0.153	0.115	0.162	0.154	**0.032**

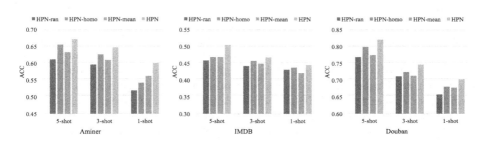

Fig. 2. Comparison of Variant Models.

5.5   Comparison of Variant Models

We conduct contrast experiments to verify the contribution of each component of our approach. We introduce three variant models based on HPN and report their performance on three datasets. The parameters of the three models are set to be the same as the proposed model HPN.

HPN$_{ran}$ is a variant of HPN that employs random selection in the meta-training task sampling part, without using the correlation score.

HPN$_{homo}$ is another simplified version of HPN that computes each class prototype as the mean vector of nodes from the homogeneous support set.

HPN$_{mean}$ replaces the semantic attention during the task sampling and prototype generation and only uses the adjacency matrix set A.

Figure 2 shows the ACC of the above models on three few-shot classification tasks. We can observe that the performance of HPN$_{ran}$ largely falls behind HPN and HPN$_{homo}$ in the Aminer and Douban datasets, which proves that sampling meta-training classes with global-level knowledge by the correlation score is able to adapt the model to different testing sets more rapidly. The performance of HPN$_{ran}$ is even worse in the 1-shot node classification task, which indicates that graph meta-sampling is especially essential in more challenging tasks. HPN$_{homo}$ outperforms the other variant models in most cases and is slightly worse than HPN. Without the heterogeneous subgraphs, the final class prototype contains only the implicit semantic information in the learned node embeddings, leading to a decrease in information utilization. Particularly, the effectiveness of HPN$_{homo}$ deteriorates rapidly in the 1-shot node classification on the Aminer dataset, indicating the contribution of the induced subgraphs to alleviate the shortage of training samples. The performance of HPN$_{mean}$ is obviously worse than HPN, which shows the superiority of the semantic importance matrix M learned by the graph structural module.

6   Conclusion

This paper formalizes the heterogeneous few-shot node classification problem and proposes a meta-learning scheme, termed Heterogeneous Graph Prototypical Networks (HPN) to solve it. Taking advantage of semantic and structural heterogeneities, we first devise a graph structural module for extracting representative node embeddings and a semantic importance matrix. Then a global-level meta-sampling mechanism is designed to generate meta-training classes with the semantic importance matrix and graph structure information as input. Moreover, we extend heterogeneous relational subgraphs for nodes in the support set to ensure the effectiveness of the learned class prototypes. Experimental results demonstrate that HPN significantly outperforms previous state-of-the-art methods on few-shot node classification tasks. Further ablation studies verify the effectiveness of graph meta-sampling, heterogeneous prototype generation, and semantic importance matrix, respectively. It is worth noting that our graph meta-sampling mechanism can avoid sampling redundant training classes through calculating the relevance between important hub nodes and each training class, which can effectively improve the performance and robustness of the trained model. For future work, we will focus on exploring representative class descriptions for nodes and extend our approach to addressing more challenging problems such as zero-shot node classification on heterogeneous graphs.

Acknowledgements. This work is funded by the National Key Research and Development Project (Grant No: 2022YFB2703100), the Starry Night Science Fund of Zhejiang University Shanghai Institute for Advanced Study (Grant No. SN-ZJU-SIAS-001), the Fundamental Research Funds for the Central Universities (2021FZZX001-23, 226-2023-00048), Shanghai Institute for Advanced Study of Zhejiang University, and ZJU-Bangsun Joint Research Center.

References

1. Allen, K., Shelhamer, E., Shin, H., Tenenbaum, J.: Infinite mixture prototypes for few-shot learning. In: International Conference on Machine Learning, pp. 232–241. PMLR (2019)
2. Bhagat, S., Cormode, G., Muthukrishnan, S.: Node classification in social networks. In: Aggarwal, C. (ed.) Social Network Data Analytics, Springer, Boston, MA (2011). https://doi.org/10.1007/978-1-4419-8462-3_5
3. Bianchini, M., Gori, M., Scarselli, F.: Inside pagerank. ACM Trans. Internet Technol. (TOIT) 5(1), 92–128 (2005)
4. Borgwardt, K.M., Ong, C.S., Schönauer, S., Vishwanathan, S., Smola, A.J., Kriegel, H.P.: Protein function prediction via graph kernels. Bioinformatics 21(suppl_1), i47–i56 (2005)
5. Cao, S., Lu, W., Xu, Q.: GraRep: learning graph representations with global structural information. In: Proceedings of the 24th ACM International on Conference on Information and Knowledge Management, pp. 891–900 (2015)
6. Chauhan, J., Nathani, D., Kaul, M.: Few-shot learning on graphs via super-classes based on graph spectral measures. arXiv preprint arXiv:2002.12815 (2020)
7. Ding, K., Wang, J., Li, J., Shu, K., Liu, C., Liu, H.: Graph prototypical networks for few-shot learning on attributed networks. In: Proceedings of the 29th ACM International Conference on Information & Knowledge Management, pp. 295–304 (2020)
8. Dong, Y., Chawla, N.V., Swami, A.: Metapath2vec: scalable representation learning for heterogeneous networks. In: Proceedings of the 23rd ACM SIGKDD International Conference on Knowledge Discovery and Data Mining, pp. 135–144 (2017)
9. Finn, C., Abbeel, P., Levine, S.: Model-agnostic meta-learning for fast adaptation of deep networks. In: International Conference on Machine Learning, pp. 1126–1135. PMLR (2017)
10. Fu, T.V., Lee, W.C., Lei, Z.: Hin2vec: explore meta-paths in heterogeneous information networks for representation learning. In: Proceedings of the 2017 ACM on Conference on Information and Knowledge Management, pp. 1797–1806 (2017)
11. Grover, A., Leskovec, J.: Node2vec: scalable feature learning for networks. In: Proceedings of the 22nd ACM SIGKDD International Conference on Knowledge Discovery and Data Mining, pp. 855–864 (2016)
12. Hamilton, W., Ying, Z., Leskovec, J.: Inductive representation learning on large graphs. In: Advances in Neural Information Processing Systems, pp. 1024–1034 (2017)

13. Hu, Z., Dong, Y., Wang, K., Sun, Y.: Heterogeneous graph transformer. In: Proceedings of The Web Conference 2020, pp. 2704–2710 (2020)
14. Huang, K., Zitnik, M.: Graph meta learning via local subgraphs. Adv. Neural. Inf. Process. Syst. **33**, 5862–5874 (2020)
15. Kipf, T.N., Welling, M.: Semi-supervised classification with graph convolutional networks. arXiv preprint arXiv:1609.02907 (2016)
16. Lan, L., Wang, P., Du, X., Song, K., Tao, J., Guan, X.: Node classification on graphs with few-shot novel labels via meta transformed network embedding. arXiv preprint arXiv:2007.02914 (2020)
17. Liu, Z., Fang, Y., Liu, C., Hoi, S.C.: Relative and absolute location embedding for few-shot node classification on graph
18. Lu, Y., Fang, Y., Shi, C.: Meta-learning on heterogeneous information networks for cold-start recommendation. In: Proceedings of the 26th ACM SIGKDD International Conference on Knowledge Discovery & Data Mining. pp. 1563–1573 (2020)
19. Munkhdalai, T., Yu, H.: Meta networks. In: International Conference on Machine Learning. pp. 2554–2563. PMLR (2017)
20. Ravi, S., Larochelle, H.: Optimization as a model for few-shot learning (2016)
21. Shi, C., Hu, B., Zhao, W.X., Philip, S.Y.: Heterogeneous information network embedding for recommendation. IEEE Trans. Knowl. Data Eng. **31**(2), 357–370 (2018)
22. Shi, C., Li, Y., Zhang, J., Sun, Y., Philip, S.Y.: A survey of heterogeneous information network analysis. IEEE Trans. Knowl. Data Eng. **29**(1), 17–37 (2016)
23. Snell, J., Swersky, K., Zemel, R.S.: Prototypical networks for few-shot learning. arXiv preprint arXiv:1703.05175 (2017)
24. Sung, F., Yang, Y., Zhang, L., Xiang, T., Torr, P.H., Hospedales, T.M.: Learning to compare: Relation network for few-shot learning. In: Proceedings of the IEEE conference on computer vision and pattern recognition. pp. 1199–1208 (2018)
25. Tang, J., Qu, M., Wang, M., Zhang, M., Yan, J., Mei, Q.: Line: Large-scale information network embedding. In: Proceedings of the 24th international conference on world wide web. pp. 1067–1077 (2015)
26. Veličković, P., Cucurull, G., Casanova, A., Romero, A., Lio, P., Bengio, Y.: Graph attention networks. arXiv preprint arXiv:1710.10903 (2017)
27. Wang, N., Luo, M., Ding, K., Zhang, L., Li, J., Zheng, Q.: Graph few-shot learning with attribute matching. In: Proceedings of the 29th ACM International Conference on Information & Knowledge Management. pp. 1545–1554 (2020)
28. Wang, X., Ji, H., Shi, C., Wang, B., Ye, Y., Cui, P., Yu, P.S.: Heterogeneous graph attention network. In: The World Wide Web Conference. pp. 2022–2032 (2019)
29. Wang, Y., Yao, Q., Kwok, J.T., Ni, L.M.: Generalizing from a few examples: A survey on few-shot learning. ACM Comput. Surv. 53(3) (Jun 2020)
30. Wu, Z., Pan, S., Chen, F., Long, G., Zhang, C., Philip, S.Y.: A comprehensive survey on graph neural networks. IEEE Transactions on Neural Networks and Learning Systems (2020)
31. Yang, Y., Guan, Z., Li, J., Huang, J., Zhao, W.: Interpretable and efficient heterogeneous graph convolutional network. arXiv preprint arXiv:2005.13183 (2020)
32. Yoon, S.W., Seo, J., Moon, J.: Tapnet: Neural network augmented with task-adaptive projection for few-shot learning. In: International Conference on Machine Learning. pp. 7115–7123. PMLR (2019)
33. Yun, S., Jeong, M., Kim, R., Kang, J., Kim, H.J.: Graph transformer networks. In: Advances in Neural Information Processing Systems. pp. 11983–11993 (2019)

34. Zhang, C., Song, D., Huang, C., Swami, A., Chawla, N.V.: Heterogeneous graph neural network. In: Proceedings of the 25th ACM SIGKDD International Conference on Knowledge Discovery & Data Mining. pp. 793–803 (2019)
35. Zhang, D., Yin, J., Zhu, X., Zhang, C.: Network representation learning: A survey. IEEE transactions on Big Data $6(1)$, 3–28 (2018)
36. Zhang, Q., Wu, X., Yang, Q., Zhang, C., Zhang, X.: HG-Meta: Graph Meta-learning over Heterogeneous Graphs, pp. 397–405
37. Zhang, S., Zhou, Z., Huang, Z., Wei, Z.: Few-shot classification on graphs with structural regularized gcns (2018)
38. Zhou, F., Cao, C., Zhang, K., Trajcevski, G., Zhong, T., Geng, J.: Meta-gnn: On few-shot node classification in graph meta-learning. In: Proceedings of the 28th ACM International Conference on Information and Knowledge Management. pp. 2357–2360 (2019)

Improving Deep Learning Powered Auction Design

Shuyuan You[1]([✉]) [iD], Zhiqiang Zhuang[1] [iD], Haiying Wu[1] [iD], Kewen Wang[2] [iD], and Zhe Wang[2] [iD]

[1] Tianjin University, Tianjin 300354, China
{tjuysy,zhuang,haiyingwu}@tju.edu.cn
[2] Griffith University, Brisbane, QLD 4111, Australia
{k.wang,zhe.wang}@griffith.edu.au

Abstract. Designing incentive-compatible and revenue-maximizing auctions is pivotal in mechanism design. Often referred to as optimal auction design, the area has seen little theoretical breakthrough since Myerson's 1981 seminal work. Not to mention general combinatorial auctions, we don't even know the optimal auction for selling as few as two distinct items to more than one bidder. In recent years, the stagnation of theoretical progress has promoted many in using deep learning models to find near-optimal auction mechanisms. In this paper, we provide two general methods to improve such deep learning models. Firstly, we propose a new data sampling method that achieves better coverage and utilisation of the possible data. Secondly, we propose a more fine-grained neural network architecture. Unlike existing models which output a single payment percentage for each bidder, the refined network outputs a separate payment percentage for each item. Such an item-wise approach captures the interaction among bidders at a granular level beyond previous models. We conducted comprehensive and in-depth experiments to test our methods and observed improvement in all tested models over their original design. Noticeably, we achieved state-of-the-art performance by applying our methods to an existing model.

Keywords: Auction · Mechanism Design · Deep Learning

1 Introduction

Designing revenue-maximizing and incentive-compatible auction mechanisms is an important topic in algorithmic game theory. The designing task is commonly termed as *optimal auction design*. Being able to identify the optimal auction in different settings is critical for many applications such as online advertising auction [14] and spectrum auction [5].

Myerson [15] solved the problem of optimal auction design for the single-item setting. Unfortunately, forty years after the publication of Myerson's seminal work, designing optimal auctions for the multi-item setting is still largely

B. Luo et al. (Eds.): ICONIP 2023, CCIS 1962, pp. 556–569, 2024.
https://doi.org/10.1007/978-981-99-8132-8_42

an open problem. Throughout the years, scholars have obtained some partial characterizations and approximations of the optimal auctions and in some very simple settings the optimal auction [1, 2, 4, 11, 16, 22, 23]. But until now we don't even know the optimal auction for selling two distinct items to more than one bidder.

Due to the apparent difficulty in manually designing optimal auctions, many scholars have attempted to formulate the task as a learning problem and use machine learning techniques to train a model that approximates optimal auctions. As pioneers in applying deep learning to optimal auction design, Dütting et al. [9] model an auction as a multi-layer neural network and name it as Regret-Net. Due to neural network's back-propagation and gradient descent process, achieving strict incentive compatibility is extremely hard if not impossible. For this reason, they relax the incentive compatibility constraint and quantified the violation of the constraint by the bidders' ex-post regret (i.e., the maximum utility increase by misreporting). RegretNet achieves near-optimal results in several known settings and obtain new mechanisms for settings where there is no known analytical solution. The design features of using regret as a relaxation of the incentive compatibility constraint, the separation of allocation and payment, and outputting a payment percentage rather than the actual payment are adopted by most of the follow-up works.

Despite the acceptable performance of RegretNet and the follow-up works, we can identify at least two aspects that can be improved in their general design. Firstly, existing models mostly sampled their data, that is the bidding profile, from continuous uniform distributions over $[0, 1]$. This means the precision of the bids is on the scale of 10^{-7} on the computer we used for testing. Such an unnecessarily high precision gives rise to an astronomical number of possible bid profiles. On the one hand, the bids are unrealistic and on the other hand, it confuses the neural network to distinguish bids that have differences in the scale as small as 10^{-7}. In any real-world auction mechanism, such bids are treated the same. When they are not, the tiny discrepancy may have a ripple effect that degenerates the trained neural network model. A more sophisticated approach is to specify more accurately the precision of bids that is appropriate for the auction environment. Therefore, we propose a discrete distribution sampling method to be able to specify the required precision. Empirical evidence suggests that this method enables better coverage of data which leads to better performance than the existing sampling method.

Secondly, to calculate the payment, a single payment percentage is generated for each bidder. At first sight, this is a natural choice. However, implicit in this choice is that all allocated items, despite their differences, are charged the same payment percentage. Therefore, this design ignores the different levels of competition between different items. As a result, the neural network cannot fully model the delicate interaction among the bidders competing for items of varying desirability. We propose to refine the payment network in an item-wise manner so that for each bidder, the network generates a separate payment percentage for each item. Empirical evidence confirms the superiority of such a more sophisticated design of payment network.

The aforementioned empirical evidence is gathered through comprehensive and in-depth experiments. We applied our methods to existing models on a wide range of auction settings. We found that our methods consistently achieve higher returns and lower regret than the baseline model. Actually, when combining our methods with RegretFormer, we achieve state-of-the-art performance.

It should be noted that our methods are not simply more advanced deep learning techniques. They are effective mostly because they better capture the essence of optimal auction design. Moreover, the methods are not restricted to the models we have tested, they are applicable to all auction models trained through neural networks. We believe the methods should be a standard design choice for such models. Finally, we want to point out that in real-world spectrum auctions, the bidding is over billions, a 1% increase in revenue means millions of more income for the seller. Therefore, a revenue increase of a mere 1% already make a huge impact.

2   Related Works

RegretNet [9] is the pioneering model using neural networks to find near-optimal auction mechanisms. It approximates well all existing optimal auctions and archives acceptable performance for settings in which the optimal auction is unknown. RegretNet encodes the allocation and payment rules of an auction mechanism as an *allocation network* and respectively a *payment network*. Both of them are fully connected feed forwards networks. After feeding the networks with a bidding profile, the allocation network outputs the probability of allocating each item to each bidder and the payment network outputs a payment percentage for each bidder from which the payment can be calculated. Most of the follow-up works adopted this design.

ALGNet [20] is a refinement of RegretNet that streamlines the training process. Following [3], the authors propose to evaluate auction mechanisms by the score $\sqrt{P} - \sqrt{R}$ where P is the expected revenue and R the expected regret. This leads to a loss function that is time-independent and hyperparameter-free which avoids the expensive hyperparameter search in RegretNet. Furthermore, the authors introduce a misreport network that computes the misreports that maximize bidders' utility. This design avoids the inner optimization loop for each valuation profile in RegretNet. With the refinements, ALGnet yield comparable or better results than RegretNet and circumvents the need for an expensive hyper-parameter search and a tedious optimization task. Due to the hassle-free and efficient training process of ALGNet, we conduct our experiments over ALGNet rather than RegretNet.

RegretFormer [12] is one of the latest neural network models. For the auction settings tested in [12], RegretNet either outperforms existing models or produces comparable performance. The authors propose a new network architecture that involves attention layers and a new loss function with the facility to pre-defined a regret budget. The input and output of RegretFomer are the same as RegretNet. They use an exchangeable layer to transform each bid into a vector containing information about other bids. And the main innovation is an item-wise and participant-wise self-attention layers for each feature vector corresponding to each bid. Our experiments confirm that combining our methods with RegretFormer achieves state-of-the-art performance.

After RegretNet, there are many other follow-up works that made valuable contributions. Feng et al. [10] extended RegretNet to deal with private budget constraints. Kuo et al. [13] introduced the notion of fairness. Peri et al. [17] introduced human preferences into auction design and use exemplars of desirable allocations to encode it. Rahme et al. [19] proposed a network architecture capable of recovering permutation-equivariant optimal mechanism. Curry et al. [7] revisited the problem of learning affine maximizer auctions and proposed an architecture that is perfectly strategyproof. Shenet et al. [21] and Dütting et al. [9] focus on architectures that are perfectly stretagyproof but restricted to one bidder. Curry et al. [6] made several modifications to the RegretNet architecture to obtain a more accurate regret. Qin et al. [18] articulated the advantages of permutation-equivariant neural networks. Duan et al. [8] incorporated public contextual information of bidders and items into the auction learning framework.

3 Auction Design Through Deep Learning

In this paper, we consider auctions with n bidders $N = \{1, 2, \ldots, n\}$ and m items $M = \{1, 2, \ldots, m\}$. Each bidder $i \in N$ has a valuation function $v_i : 2^M \rightarrow \mathbb{R}_{\geq 0}$. For simplicity, we write $v_i(\{j\})$ as $v_i(j)$. We focus on *additive* valuation functions, that is a bidder i's valuation for the set $S \subseteq M$ of items is the sum of the valuations for each item in S. Formally, $v_i(S) = \sum_{j \in S} v_i(j)$ for all $S \subseteq M$. Bidder i's valuation function is drawn from a distribution F_i over possible valuation functions V_i. We use the vector values of the function to represent it, and we denote the set of all possible valuation profiles as $V = \prod_{i=1}^{n} V_i$(Cartesian product) and $v = (v_1, v_2, \ldots, v_n) \in V$ a the valuation profile. We write $F = (F_1, F_2, \ldots, F_n)$ to denote all the possible distributions. The auctioneer knows about F but not the actual valuation functions

An auction mechanism (g, p) is a combination of allocation rules $g_i : V \rightarrow [0, 1]^M$ and payment rules $p_i : V \rightarrow \mathbb{R}_{\geq 0}$ for $1 \leq i \leq n$. Given a bid profile $b = (b_1, b_2, \ldots, b_n) \in V$ and a valuation v_i, bidder i's utility is such that $u_i(v_i, b) = v_i(g_i(b)) - p_i(b)$. Bidders may misreport their valuations in order to maximize utility. For a valuation profile v, we write v_{-i} as the valuation profile without v_i and similarly for b_{-i} and V_{-i}. An auction (g, p) is *dominant strategy incentive compatible* (DSIC) if a bidder's utility is maximized by reporting truthfully no matter what the others do, that is $u_i(v_i, (v_i, b_{-i})) \geq u_i(v_i, b)$ for all i, v_i, and b_i. An auction is *individually rational* (IR) if reporting truthfully always results in non-negative utility, that is $u_i(v_i, (v_i, b_{-i})) \geq 0$, for all v_i and b_{-i}. In a DSIC auction, since reporting truthfully maximises utility, the revenue of the auction over the valuation profile v is $\sum_{i=1}^{n} p_i(v)$. The purpose of the optimal auction design is to identity a DSIC and IR auction that maximizes the expected revenue.

Due to the difficulty of designing optimal auctions manually, many have attempted to find approximately optimal auctions through deep learning. Given a class of auction (g^ω, p^ω) parameterized by $\omega \in \mathbb{R}^d$ and a valuation distribution F, the goal is to identify an auction that minimizes the negated expected revenue $-\mathbb{E}_F[\sum_{i=1}^n p_i^\omega(v)]$ while satisfying DSIC and IR as much as possible. For the DSIC part, following [9], the common practice is to use the expected *ex-post regret* as a quantifiable relaxation. The regret measures to what extent an auction mechanism violates the DSIC property. Specifically, a bidder's regret is the maximum utility increase through misreporting while keeping the other bids fixed. Formally, the expected regret of bidder i under the parameterisation ω is

$$rgt_i(\omega) = \mathbb{E}[\max_{v_i' \in V_i} u_i^\omega(v_i, (v_i', v_{-i})) - u_i^\omega(v_i, (v_i, v_{-i}))]$$

The expected ex-post regret is approximated by the empirical one as follows:

$$\hat{rgt}_i(\omega) = \frac{1}{L} \sum_{\mathfrak{l}=1}^L \max_{v_i' \in V_i} (u_i^\omega(v_i^{(\mathfrak{l})}, (v_i', v_{-i}^{(\mathfrak{l})})) - u_i^\omega(v_i^{(\mathfrak{l})}, v^{(\mathfrak{l})}))$$

Then the learning problem for optimal auction design boils down to minimising the empirical negated revenue such that empirical ex-post regret is zero:

$$\min_{\omega \in \mathbb{R}^d} -\frac{1}{L} \sum_{\mathfrak{l}=1}^L \sum_{i=1}^n p_i^\omega(v^{(\mathfrak{l})})$$

$$s.t. \quad \hat{rgt}_i(\omega) = 0, \quad \forall i \in n$$

4   Methods

In this section, we introduce and elaborate on our methods to improve deep learning powered auction design. The first method is about a new data sampling process and the second one is about a new architecture of the payment network. We want to emphasize that these methods are not only applicable to Regret-Net, ALGNet and RegretFormer but to all auction mechanisms based on neural networks that compute the payment through a payment percentage.

4.1   Discrete Distribution Sampling

Our first method concerns how to best sample the bidders' valuations for training the neural network. We proved empirically that, instead of a continuous distribution, sampling over a discrete distribution (with an appropriate support) leads consistently to more revenue and less regret. We term the method discrete distribution sampling (DDS). Specifically, DDS is the process of discretizing a continuous distribution into a discrete one and then sampling data over the discrete distribution. For example, given a continuous uniform distribution over the interval $[0, 1]$, we can discretize it into a discrete one with a support

$\{0.0, 0.1, \ldots, 1.0\}$. So instead of an infinite number of possible values, sampling over the discrete distribution gives 11 possible values.

Apart from the empirical evidence, we can appreciate the superiority of DDS from another two perspectives. Firstly, it gives a more accurate modelling of a bidder's valuations in real-world auctions. While buying something, people care less about the lower value digits of the price than the higher ones. If a valuation is sampled from a continuous distribution, the precision of the sampled value is too high to be realistic. For example, if a bidder values an auctioned item as around a million dollars, the bidder couldn't care less if it is a million dollars and 20 cents or a million dollars and 40 cents. It is neither practical nor reasonable to differentiate bidders whose valuation difference is a few cents. They ought to be treated the same. When they are not, the minor differences may have a negative effect during the training process. DDS avoids such issues by specifying the appropriate precision of values through the support of a discrete distribution.

Secondly, DDS is more effective from the training perspective. In our experiment, the smallest number the computer can represent is roughly 10^{-7}. So, sampling through a continuous distribution over $[0, 1]$ leads to about 10^7 possible values for each bidder and each auctioned item. This is a huge number of values, but a large portion of them are redundant or irrelevant. In a real-world auction, the valuation difference in the scales of 10^{-7} makes no difference to the auction outcome and the bidders' payoffs. Alternatively, sampling through a discrete distribution over $\{0.0, 0.1, \ldots, 1.0\}$ leads to merely 11 possible values for each bidder and each item. Therefore, DDS on the one hand effectively reduces the space of possible values and on the other hand, preserving most of the essential ones, ones that matter to the bidders and sellers. Given an upper bound on the number of sampled data the computation resource allowed and considering the ratio of the number of sampled data over that of all possible data, DDS gives a much higher ratio thus providing much better coverage of the possible data.

Now we have explained and justified the inner mechanism of DDS at an intuitive level, we want to elaborate on how to make use of DDS in practice. More specifically, how to determine the appropriate support of the discrete distribution? Obviously, the values in the support have to be evenly spaced to eliminate bias towards any particular value. So the next question is how large the space between values should be. The larger the space the fewer possible values there are which means, given a fixed number of sampled data, a better ratio of data coverage. Intuitively, when the auction setting is more complex that is a larger number of bidders and items, the corresponding optimal mechanism is also more complex. Hence we need a more sophisticated neural network to approximate the optimal mechanism. The training of such a neural network will clearly require a better coverage of the data. Our experiments have verified this intuition that the size of the support should be inversely proportional to the complexity of the auction for the best performance. Essentially, this makes the size of the support another hyper-parameter for neural network training.

4.2   Item-Wise Payment Network

Following RegretNet, most approaches of deep learning powered combinatorial auction opt for a design that consists of an allocation network and a payment network handling the allocation and respectively the payment of an auction. Our second method concerns refining the payment network to make it more fine-grained.

To ensure IR, existing payment networks output a payment percentage $\tilde{p}_i$ for each bidder i from which i's payment p_i is determined as

$$p_i = \tilde{p}_i \sum_{j=1}^{m} z_{ij}\, b_{ij}$$

where z_{ij} is the probability of allocating item j to i and b_{ij} is i's valuation of item j. IR is guaranteed because a bidder's payment is always a fraction of its expected gain. This is a simple yet effective design, but we could do better.

Instead of a single payment percentage for each bidder, why not itemise the percentage for each auctioned item? We refine the payment network to output a payment percentage $\tilde{p}_{ij}$ for each bidder i and each item j such that bidder i's payment is

$$\sum_{j=1}^{m} \tilde{p}_{ij}\, z_{ij}\, b_{ij}$$

where $\sum_{j=1}^{m} \tilde{p}_{ij} \leq 1$ for all i. We refer to such a payment network as an item-wise payment network (IPN). It is easy to see that IPN guarantees IR as the payment never exceeds the expected gain. We have proved empirically that IPN leads consistently to better revenue and lower regret.

Actually, it is not hard to be convinced of the superiority of IPN even without the empirical evidence. To begin with, IPN has more representation power which means it is capable of representing the fine details of an auction mechanism the original design cannot. Thus, given sufficient training data and computational resources, we are expecting performance at least on par with the original design if not better. The increase in representation power puts IPN in a better position, but more importantly, IPN captures more accurately the interaction between the bidders competing for each item. In the original design, a bidder is charged a bundled price for all allocated items, so implicitly the payment percentage for each of these items are the same. Clearly, this design ignores the differences between the items and the different levels of competition for them. By specifying an item-wise payment for each allocated item, IPN takes such differences and competitions into account while determining the auction outcome and ultimately leading to a better approximation of the optimal mechanism.

5   Experiments

To verify the effectiveness of our methods, we apply them to ALGNet and Regret-Former. From here forward, we represent an auction setting as $n * m$ where n

is the number of bidders and m is the number of items. We observe improved or comparable performance across all tested auction settings for both ALGNet and RegretFormer.

5.1 ALGNet

We first test our methods on ALGNet which is referred to as the baseline model in this section. As mentioned, instead of RegretNet, we picked ALGNet for its more streamlined and efficient training process. Since existing models already achieved excellent results in some simple auction environments, we focus on the auctions settings of $3 * 10$ and $5 * 10$ with valuations sampled from the uniform distribution over $[0, 1]$. To ensure fairness, we add the coding for our method to those provided in [20] while keeping the original coding, hyperparameters and random seed unchanged.

We present the main experiment results in Table 1 which shows the changes in revenue and regret with or without our methods. Since ALGNet is trained by the evaluation metrics $\sqrt{P} - \sqrt{R}$, for fairness, we also report how the $\sqrt{P} - \sqrt{R}$ score changes.

Table 1. Results for applying DDS and IPN separately and simultaneously to ALGNet for the settings of $3 * 10$ and $5 * 10$.

Setting	Model	Revenue	Regret	$\sqrt{P} - \sqrt{R}$
3*10	Baseline	5.562 ($\pm$0.0308)	$1.93 * 10^{-3}$ ($\pm 0.33 * 10^{-3}$)	5.20888
	DDS	5.580 ($\pm$0.0229)	$1.60 * 10^{-3}$ ($\pm 0.21 * 10^{-3}$)	5.25749
	IPN	5.717 ($\pm$0.0418)	$1.44 * 10^{-3}$ ($\pm 0.08 * 10^{-3}$)	5.40701
	IPN & DDS	5.731 ($\pm$0.0097)	$1.43 * 10^{-3}$ ($\pm 0.08 * 10^{-3}$)	5.42169
5*10	Baseline	6.781 ($\pm$0.0504)	$3.85 * 10^{-3}$ ($\pm 0.43 * 10^{-3}$)	6.07766
	DDS	6.747 ($\pm$0.0276)	$2.85 * 10^{-3}$ ($\pm 0.33 * 10^{-3}$)	6.14111
	IPN	6.852 ($\pm$0.0369)	$3.95 * 10^{-3}$ ($\pm 1.01 * 10^{-3}$)	6.13601
	IPN & DDS	6.881 ($\pm$0.0480)	$3.59 * 10^{-3}$ ($\pm 0.45 * 10^{-3}$)	6.19558

We begin with the auction setting of 3*10. Applying only DDS to the baseline, we found a 17% decrease in regret, a similar revenue and a 0.93% increase in the $\sqrt{P} - \sqrt{R}$ score. Applying only IPN to the baseline, we found a 2.79% increase in revenue, a 25% decrease in regret, and a 3.80% increase in the $\sqrt{P} - \sqrt{R}$ score. Applying both DDS and IPN to the baseline, we found a 3.04% increase in revenue, a 26% decrease in regret, and a 4.08% increase in the $\sqrt{P} - \sqrt{R}$ score.

Moving on to the setting of $5 * 10$, Applying only DDS to the baseline, we found a 26% decrease in regret, a similar revenue, and a 1.04% increase in the $\sqrt{P} - \sqrt{R}$ score. Applying only IPN to the baseline, we found a 2.80% increase in revenue, a similar regret, and a 0.96% increase in the $\sqrt{P} - \sqrt{R}$ score. Applying both DDS and IPN to the baseline, we found a 1.47% increase in revenue, a 7% decrease in regret, and a 1.94% increase in the $\sqrt{P} - \sqrt{R}$ score.

It is clear from the results, that both DDS and IPN improve the performance of ALGNet in terms of more revenue and less regret and we achieve the best results when both methods are used. Looking at it separately, the results show that the main effect of DDS is to reduce regret, while the main effect of IPN is to improve revenue.

As we mentioned while introducing the method of DDS, the space between sampled values is actually a hyperparameter that has to be tuned for the best results. The results in Table 1 are obtained with the best hyperparameter setting for DDS. We want to dig into the tuning process to show how it works and more importantly to show how the space affects the end results. In fact, it could have a negative effect if it is not tuned properly.

In Fig. 1, we plotted the $\sqrt{P} - \sqrt{R}$ score against the size of the space used in DDS for the auction setting of $5 * 10$ and for when IPN is applied and when it is not. For example, the space size of 0.1 means the support for the discrete distribution is $\{0, 0.1, 0.2 \dots, 1\}$. We also indicated by a grey dotted line the $\sqrt{P} - \sqrt{R}$ score of ALGNet. The plotted graph shows that the $\sqrt{P} - \sqrt{R}$ score is at its peak when the space is 0.1 after which increasing the space size decreases the score. When the space is increased to 0.15 using DDS alone actually results in a worse score than that of the baseline.

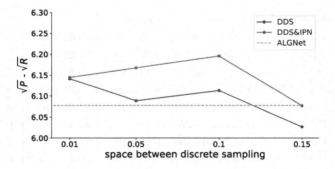

Fig. 1. The $\sqrt{P} - \sqrt{R}$ score against space between consecutive sampling values for the setting of $5 * 10$

Similarly, in Fig. 2, we plotted the graphs for the auctions setting of $3 * 10$. In this setting, the graphs show that applying DDS with or without IPN achieve better scores than the baseline. We also noticed that, in this simpler setting, the space used in DDS has a minor effect on the performance.

We have analysed the effects of the space size for DDS through the $\sqrt{P} - \sqrt{R}$ score because we need a single measure of the overall performance for both revenue and regret and the $\sqrt{P} - \sqrt{R}$ score has been verified to be such a good measure in ALGNet and in our experiments reported in Table 1.

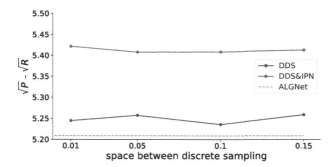

Fig. 2. The $\sqrt{P} - \sqrt{R}$ score against space between consecutive sampling values for the setting of $3 * 10$

So, for DDS we have to tune the space size for best performance. Do we have to do tuning as well for IPN? In IPN we propose to give a payment percentage for each item whereas the original design is to give a single percentage for all items. What about the intermediate setting? What will happen if say we give a percentage for every pair of items? For this, we conducted experiments for both the $3 * 10$ and $5 * 10$ settings with or without DDS. In the experiments, we output 5 payment percentages for each bidder, that is not every single item has a payment percentage but every two of them. We report the results in Table 2. For ease of comparison, we copied some results from Table 1. We find that as we increase the number of payment percentages the revenue keeps increasing and the regret and the $\sqrt{P} - \sqrt{R}$ score also gets better in most settings. The results indicate that in general increasing the granularity leads to better performance. In conclusion, IPN needs no tuning because it is already as granular as it gets.

5.2 RegretFormer

We also test our methods on RegretFormer which, to the best of our knowledge, bypasses other models in terms of revenue for many auction settings. Again, to ensure fairness, we didn't make adjustments to the hyperparameters and random seed of the original paper and made minimal changes to the original coding. Throughout our experiments, the regret budget is set to be $R_{max} = 10^{-3}$. In the original paper, RegretFormer is tested in settings in which the optimal mechanism is known and also in settings the optimal mechanism is unknown. Due to its extremely high computational cost in training and testing, RegretFormer is tested in relatively simpler auction settings where the most complex one is $3 * 10$. We repeat most of the tests after incorporating the original model with DDS and IPN.

Table 2. Results for applying IPN with varying numbers of payment percentages.

Setting	Model	Number	Revenue	Regret	$\sqrt{P} - \sqrt{R}$
3*10	Baseline	1	5.562 (±0.0308)	$1.93 * 10^{-3}$ (±0.33 * 10⁻³)	5.20888
		5	5.687 (±0.0353)	$2.12 * 10^{-3}$ (±0.26 * 10⁻³)	5.31299
		10	5.717 (±0.0418)	$1.44 * 10^{-3}$ (±0.08 * 10⁻³)	5.40701
	DDS	1	5.587 (±0.0465)	$1.65 * 10^{-3}$ (±0.59 * 10⁻³)	5.25935
		5	5.646 (±0.0412)	$1.74 * 10^{-3}$ (±0.11 * 10⁻³)	5.30787
		10	5.731 (±0.0097)	$1.43 * 10^{-3}$ (±0.08 * 10⁻³)	5.42169
5*10	Baseline	1	6.781 (±0.0504)	$3.85 * 10^{-3}$ (±0.43 * 10⁻³)	6.07766
		5	6.790 (±0.0355)	$3.26 * 10^{-3}$ (±0.36 * 10⁻³)	6.14093
		10	6.852 (±0.0369)	$3.95 * 10^{-3}$ (±1.01 * 10⁻³)	6.13601
	DDS	1	6.747 (±0.0276)	$2.85 * 10^{-3}$ (±0.33 * 10⁻³)	6.14111
		5	6.829 (±0.0990)	$3.45 * 10^{-3}$ (±0.86 * 10⁻³)	6.15909
		10	6.881 (±0.0480)	$3.59 * 10^{-3}$ (±0.45 * 10⁻³)	6.19558

We begin with the following settings of $1 * 2$ where there exist analytical solutions:

(A) A single bidder whose valuations of the two items v_1 and v_2 are drawn from a continuous uniform distribution over $[0, 1]$.

(B) A single bidder whose valuations of the two items v_1 and v_2 are drawn from a continuous uniform distribution over $[4, 16]$ and respectively from $[4, 7]$.

Table 3. Applying DDS and IPN to RegretFomer for the auction settings of (A) and (B).

Setting	Optimal	RegretFormer		DDS & IPN	
	Revenue	Revenue	Regret	Revenue	Regret
(A)	0.550	0.571	0.00075	0.572	0.00069
(B)	9.781	10.056	0.0099	10.071	0.0080

The results are reported in Table 3. We find that our model reduces the regret by 8% in setting (A) and by 19% in setting (B) and achieves similar revenue. Since the settings have optimal solutions, we also compare the models through allocation heatmaps as did in [12]. Plotted in Fig. 3 the heatmaps show that our allocation rules are slightly closer to the optimal allocation than RegretFormer in terms of allocation. We marked the area where our allocation rules are better in the black square. We conjecture that the improvement is because we effectively reduced the regret of the model.

Moving on to settings in which the optimal auction is unknown including $2 * 2$, $2 * 3$, $2 * 5$, and $3 * 10$, we report the results in Table 4.

We find that our methods can significantly reduce regret in most settings that is 11% in $2*2$, 29% in $2*3$ and 22% in $3*10$. We also improved the revenue for all settings, especially in the $3*10$ setting in which we have an 1% increase. In fact,

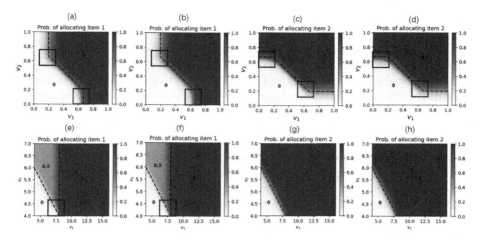

Fig. 3. The heatmaps represent the probability of item allocation in a $1 * 2$ setting: (a) and (c) correspond to applying our methods in setting (A) whereas (b) and (d) correspond to the original model for setting (A). (e) and (g) correspond to applying our methods in setting (A) whereas (f) and (h) correspond to the original model for setting (B). The solid regions describe the probability of the bidder obtaining item 1 under different valuation inputs. The optimal auctions are described by the regions separated by the dashed black lines, with the numbers in black the optimal probability of allocation in the region.

Table 4. Applying DDS and IPN to RegretFomer for the auction settings of $2 * 2$, $2 * 3$, $2 * 5$, and $3 * 10$.

Setting	Model	Revenue	Regret
2*2	Baseline	0.908 ($\pm$0.011)	0.00054
	DDS & IPN	0.912 ($\pm$0.004)	0.00048
2*3	Baseline	1.416 ($\pm$0.009)	0.00089
	DDS & IPN	1.422 ($\pm$0.0164)	0.00063
2*5	Baseline	2.453 ($\pm$0.0330)	0.00102
	DDS & IPN	2.474 ($\pm$0.0203)	0.00109
3*10	Baseline	6.121 ($\pm$0.0190)	0.00179
	DDS & IPN	6.184 ($\pm$0.0130)	0.00139

for these settings, the revenue and regret achieved by combining our methods with RegretFormer is state-of-the-art performance. For the improvement, we can attribute the reduction in regret to DDS and revenue improvement to IPN.

6 Conclusion

In this paper, we propose two refinements to the neural network models that aim to approximate an optimal auction. We first looked into the data sampling process for which the existing models unanimously drew their data from continuous distributions. By sampling through discrete distributions, the DDS method we proposed gives us control over the precision of the sampled data so that we can tune the level of data abstraction to the auction setting at hand for better performance. We then targeted the conventional design of existing models to produce a single payment percentage for calculating the payments. By giving an item-wise payment percentage, our IPN method is cable of better capturing the valuation difference between items and the different levels of competition for them. We conducted thorough experiments on ALGNet and RegretFormer to evaluate the effectiveness of our methods. The experiment results indicate stable and consistent improvement across all auction settings. The empirical evidence suggests that our methods of DDS and IPN should be in the toolbox of anyone aiming to design optimal auctions through deep learning.

So far we have tested our methods mostly for additive valuation functions. We believe our IPN method would be more effective when the valuation functions are more complex and intertwined. For future work, we plan to test the methods over such valuation functions.

Acknowledgements. This work was partially supported by National Natural Science Foundation of China(NSFC)(61976153).

References

1. Alaei, S., Fu, H., Haghpanah, N., Hartline, J.D., Malekian, A.: Bayesian optimal auctions via multi- to single-agent reduction. In: Proceedings of the 13th ACM Conference on Electronic Commerce, EC 2012, Valencia, Spain, 4–8 June 2012, p. 17 (2012)
2. Babaioff, M., Immorlica, N., Lucier, B., Weinberg, S.M.: A simple and approximately optimal mechanism for an additive buyer. J. ACM (JACM) **67**(4), 1–40 (2020)
3. Balcan, M.F., Blum, A., Hartline, J., Mansour, Y.: Mechanism design via machine learning. In: 46th Annual IEEE Symposium on Foundations of Computer Science (FOCS 2005), pp. 605–614 (2005)
4. Cai, Y., Zhao, M.: Simple mechanisms for subadditive buyers via duality. In: Proceedings of the 49th Annual ACM SIGACT Symposium on Theory of Computing, pp. 170–183 (2017)
5. Cramton, P.: Spectrum auction design. Rev. Ind. Organ. **42**, 161–190 (2013)
6. Curry, M., Chiang, P.Y., Goldstein, T., Dickerson, J.: Certifying strategyproof auction networks. Adv. Neural. Inf. Process. Syst. **33**, 4987–4998 (2020)
7. Curry, M., Sandholm, T., Dickerson, J.: Differentiable economics for randomized affine maximizer auctions. arXiv preprint arXiv:2202.02872 (2022)
8. Duan, Z., et al.: A context-integrated transformer-based neural network for auction design. In: International Conference on Machine Learning, pp. 5609–5626. PMLR (2022)

9. Dütting, P., Feng, Z., Narasimhan, H., Parkes, D., Ravindranath, S.S.: Optimal auctions through deep learning. In: International Conference on Machine Learning, pp. 1706–1715 (2019)

10. Feng, Z., Narasimhan, H., Parkes, D.C.: Deep learning for revenue-optimal auctions with budgets. In: Proceedings of the 17th International Conference on Autonomous Agents and Multiagent Systems, pp. 354–362 (2018)

11. Hart, S., Nisan, N.: Approximate revenue maximization with multiple items. J. Econ. Theor. **172**, 313–347 (2017)

12. Ivanov, D., Safiulin, I., Filippov, I., Balabaeva, K.: Optimal-ER auctions through attention. In: NeurIPS (2022)

13. Kuo, K., et al.: ProportionNet: balancing fairness and revenue for auction design with deep learning. arXiv preprint arXiv:2010.06398 (2020)

14. Liu, X., et al.: Neural auction: End-to-end learning of auction mechanisms for e-commerce advertising. In: Proceedings of the 27th ACM SIGKDD Conference on Knowledge Discovery & Data Mining, pp. 3354–3364 (2021)

15. Myerson, R.B.: Optimal auction design. Math. Oper. Res. **6**(1), 58–73 (1981)

16. Pavlov, G.: Optimal mechanism for selling two goods. BE J. Theor. Econ. **11**(1), 0000102202193517041664 (2011)

17. Peri, N., Curry, M., Dooley, S., Dickerson, J.: PreferenceNet: encoding human preferences in auction design with deep learning. Adv. Neural. Inf. Process. Syst. **34**, 17532–17542 (2021)

18. Qin, T., He, F., Shi, D., Huang, W., Tao, D.: Benefits of permutation-equivariance in auction mechanisms. In: Advances in Neural Information Processing Systems (2022)

19. Rahme, J., Jelassi, S., Bruna, J., Weinberg, S.M.: A permutation-equivariant neural network architecture for auction design. In: Proceedings of the AAAI Conference on Artificial Intelligence, vol. 35, pp. 5664–5672 (2021)

20. Rahme, J., Jelassi, S., Weinberg, S.M.: Auction learning as a two-player game. In: 9th International Conference on Learning Representations, ICLR 2021, Virtual Event, Austria, 3–7 May 2021 (2021)

21. Shen, W., Tang, P., Zuo, S.: Automated mechanism design via neural networks. In: Proceedings of the 18th International Conference on Autonomous Agents and MultiAgent Systems, pp. 215–223 (2019)

22. Wang, Z., Tang, P.: Optimal mechanisms with simple menus. In: Proceedings of the Fifteenth ACM conference on Economics and Computation, pp. 227–240 (2014)

23. Yao, A.C.C.: Dominant-strategy versus bayesian multi-item auctions: Maximum revenue determination and comparison. In: Proceedings of the 2017 ACM Conference on Economics and Computation, pp. 3–20 (2017)

Link Prediction with Simple Path-Aware Graph Neural Networks

Tuo Xu and Lei Zou$^{(\boxtimes)}$

Peking University, Beijing, China
{doujzc,zoulei}@pku.edu.cn

Abstract. Graph Neural Networks (GNNs) are expert in node classification and graph classification, but are relatively weak on link prediction due to their limited expressiveness. Recently, two popular GNN variants, namely higher-order GNNs and labeling trick are proposed to address the limitations of GNNs. Compared with plain GNNs, these variants provably capture inter-node patterns such as common neighbors which facilitates link prediction. However, we notice that these methods actually suffer from two critical problems. First, their algorithm complexities are impractical for large graphs. Second, we prove that although these methods can identify paths between target nodes, they cannot identify *simple paths*, which are very fundamental in the field of graph theory. To overcome these deficiencies, we systematically study the common advantages of previous link prediction GNNs and propose a novel GNN framework that summarizes these advantages while remaining simple and efficient. Various experiments show the effectiveness of our method.

Keywords: Link prediction · Graph neural networks · Simple paths

1 Introduction

Link prediction is a fundamental task in graph machine learning fields, and has been applied for solving various problems including knowledge graph completion (KGC) [25], recommender systems [15], drug discovery [11], influence detection [24], etc. Representative methods for link prediction includes traditional heuristic methods such as Common Neighbor, Adamic-Adar [2], Resource Allocation [43], knowledge graph embedding such as TransE [5], RotatE [30], and Graph Neural Networks (GNNs) [13,14]. Among them, GNNs are the fastest-growing deep learning models for learning graph representations.

However, plain GNNs are relatively weak on link prediction tasks. An example would be GAE [13]: given the target link (s, t), GAE first runs a GNN to collect node representations and then predict the link (s, t) by aggregating the representations of the nodes s, t individually. This approach has been proven to be insufficient for identifying simple link properties such as common neighbors, due to the automorphic node problem [42].

B. Luo et al. (Eds.): ICONIP 2023, CCIS 1962, pp. 570–582, 2024.
https://doi.org/10.1007/978-981-99-8132-8_43

Various GNN frameworks are proposed to overcome the problem, among which the most popular ones are the *labeling trick* [40] and *higher-order GNNs* [21]. Given a graph G and the target link (s, t), labeling trick methods first add additional labels on s, t to mark them differently from the rest of the nodes. Consider the graph (a) in Fig. 1. To predict the link (v_1, v_2), we first tag the nodes v_1 and v_2, resulting in the graph (b). Similarly, to predict the link (v_1, v_3), we tag the nodes v_1 and v_3 respectively, resulting in the graph (c). After that, we run a GNN on the induced graph and obtain the representation of (s, t) by aggregating the node representations of s and t respectively. Different from the labeling trick, higher-order GNNs mimic the k-Weisfeiler Lehman (k-WL) [21] hierarchy. They use node tuples as basic message passing units, which makes them significantly more expressive than ordinary GNNs, but also results in higher algorithm complexity.

Although previous works have extensively studied the expressive power of GNNs for link prediction, we find that there remains two critical problems: First, these GNN variants suffer from much higher algorithm complexity: compared with ordinary GCNs [14], these methods usually requires much more time for training and inference. Second, although these GNN variants are provably more expressive than GCNs, they still overlooked one simple and fundamental pattern in graph theory: they are unaware of simple paths. In this paper, we try to settle these problems within one framework. We first investigate previous GNN frameworks in the view of paths, and show that compared with ordinary GCNs (GAEs), labeling trick and higher-order GNNs are both aware of paths but unaware of simple paths. Then, we propose a novel GNN framework called Simple Path-aware GNN (SPGNN), which shares the common advantages of previous GNN variants but are much more efficient, and is able to distinguish simple paths.

To summarize, our contributions are:

- We conduct an comprehensive analysis of previous link prediction GNN architectures via the awareness of paths. We find that none of the GNN variants is able to distinguish simple paths.
- Based on the observation, we further propose a novel GNN framework that is able to distinguish both paths and simple paths, while being as efficient as GCNs [14] and perform well on link prediction tasks.

2   Related Work

Graph Neural Networks. GNNs are widely applied to graph-related machine learning tasks. Bruna et al. [6] first extend the concept of convolutional networks to graphs. Defferrard et al. [8] remove the expensive Laplacian eigendecomposition by using Chebyshev polynomials. After that, Kipf et al. [14] propose graph convolutional networks (GCNs), which further simplify graph convolutions with a redefined message-passing scheme. At each layer GCN propagates node representations through a normalized adjacency matrix S. Based on GCNs, Chen et al. [7] propose a more efficient GCN variant based on sampling. Wu et al. [34] remove all the non-linear parts of GCNs and obtain a extremely simple variant, namely SGC, which surprisingly show strong performance compared with non-linear GCNs. Our method is a GNN instance specially designed for link prediction tasks.

Higher-Order GNNs. Most GNNs focus on graph classification and node classification tasks, and they learn representations for each nodes (i.e., they are node level GNNs), thus their discriminative power are often restricted to 1-Weisfeiler-Lehman (1-WL) tests [35]. Noticing the similarity between GNN message-passing mechanism and WL tests, Morris et al. [21] propose a GNN model to parameterize the 2-WL and 3-WL tests. Similarly, Maron et al. [16] propose to parameterize the 2-FWL test with GNNs. These GNNs learn high-order representations, i.e. they learn representations for node tuples. This makes higher-order GNNs inherently suitable for link prediction tasks. However, previous high-order GNNs suffer from much higher time and space complexities. Our method, although not being an instance of higher-order GNNs, is partly inspired by the advantages of them.

Labeling Trick and Its Variants. To improve GNNs' performance on link prediction, Zhang et al. [40] propose SEAL, which uses the technique called labeling trick to predict links with plain GNNs. GraIL [31] extends SEAL to heterogeneous graphs such as knowledge graphs. Zhang et al. [42] further summarize these GNN variants into a unified framework and proven its expressive power. These methods tag both the source node and the target node to be different from the rest of the nodes, therefore enhance GNN's distinguish power. However, these methods also result in larger algorithm complexities and heavily rely on subgraph extraction to reduce the running time. Another series of works NBFNet [44], ID-GNN [38], etc. only tag the source node and learn representations for the target nodes. Compared with the original labeling trick, these methods are more efficient, and are able to learn full-graph representations. Our method, although not being an instance of labeling trick methods, is partly inspired by the advantages of them.

Other Link Prediction Methods. Other link prediction methods include rule-based methods and knowledge graph embedding. Representative rule-based includes earlier works MDIE [22], AMIE [10], etc. and recent neural-enhanced rule learning methods such as NeuralLP [37], DRUM [27] and RNNLogic [26]. These methods generally learn probabilistic logic rules to predict the existence of the target links. Knowledge graph embedding methods learn a distribution representation for each node and each type of edge in the graph by minimizing the distance between connected nodes. Representative methods include TransE [5], DistMult [36], RotatE [30] and many other works that propose new scoring functions [9,32]. Our method is less relevant with these methods.

3   Preliminaries

This paper considers the problem of link prediction. Given a (uncompleted) graph $G = (\mathcal{V}, \mathcal{E})$ with node set $\mathcal{V}$ and edge set $\mathcal{E}$, we are asked to predict whether there exists an edge between any node pair (u, v).

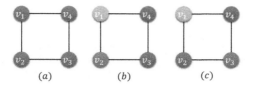

Fig. 1. An illustration of the labeling trick methods.

Message Passing Neural Networks. Message passing neural networks (MPNNs) are a dominant class of GNNs. Suppose $h_u^{(l)}$ is the node representation of node u at layer l, MPNNs compute the representations at layer $l+1$ as:

$$h_u^{(l+1)} = \phi\left(h_u^{(l)}, \psi\left(\left\{h_v^{(l)} \mid v \in \mathcal{N}(u)\right\}\right)\right),$$

where $\mathcal{N}(u)$ is the neighbors of u, ϕ and ψ are trainable functions. At the beginning layer, $h_u^{(0)}$ is initialized as the input node feature.

Graph Auto Encoders. Graph auto encoders (GAEs) [13] are perhaps the most simple method for link prediction with GNNs. To predict the link (u, v), GAEs first run a GNN to compute node representations. Then, the existence of (u, v) is computed as $p(u, v) = f(h_u, h_v)$, where f is a trainable function and h_u, h_v are node representations of u, v computed by the GNN respectively. GAEs are proven to be problematic in link prediction [42]: for example, it cannot distinguish the link (v_1, v_2) from (v_1, v_3) in Fig. 1(a), because each node shares the same node representation, which is called the automorphism problem.

Labeling Trick. Noticing the defects of GAEs, Zhang et al. [42] propose labeling trick. We briefly introduce the 0–1 node labeling for ordered node pair here for clarity. Given a graph $G = (\mathcal{V}_G, \mathcal{E}_G)$ and the target link (v_1, v_2), we add a new label l_u to each node u in G, where we have $l_{v_1} = 1, l_{v_2} = 2$ and $l_u = 0$ for the rest of the nodes u. It is proven that this approach avoids the automorphism problem of GAEs and performs better in link prediction. There are also variants of the labeling trick, among which the most popular one is used in NBFNet [44]. The only difference is that, NBFNet only tags the source node v_1 differently from the rest of the nodes.

Higher-Order GNNs. Higher-order GNNs mimic the k-WL paradigm [21]. We briefly introduce the Edge Transformer [4] here. Given a graph $G = (\mathcal{V}, \mathcal{E})$, it first sets up a representation $h_{uv}^{(0)}$ for each node pair (u, v) based on the input node feature and the edges. Clearly there are totally N^2 representations where N is the number of nodes in G. At each layer, it computes:

$$h_{uv}^{(l+1)} = \phi\left(h_{uv}^{(l)}, \psi\left(\left\{\left(h_{uw}^{(l)}, h_{wv}^{(l)}\right) \mid w \in \mathcal{V}\right\}\right)\right),$$

where ϕ, ψ are trainable functions. At the output layer L, $h_{uv}^{(L)}$ is used to compute the link property of (u, v). Higher-order GNNs can also avoid the automorphism problem of GAEs.

4   Methods

In this section we provide a thorough investigation of popular GNN variants in the link prediction literature including GAE [13], labeling trick methods (SEAL [40], GraIL [31]), variants of labeling trick methods (NBFNet [44], ID-GNN [38]), and higher-order GNNs [4,17,20,21]. Surprisingly, we find that all these works are not capable of detecting *simple paths* between the target nodes. We then propose an efficient GNN architecture which overcomes the above issues.

4.1   GAEs Are Not Good Enough: Awareness of Paths

Empirically, GAE methods perform extremely weak on link prediction tasks compared with other GNN variants (labeling trick, higher-order GNNs). We first try to dissect out the core functionality that enables these variants to surpass GAE.

It is known [42] that the automorphic node problem is the most several problem that hinders GAEs from obtaining feasible link representations. For example in Fig. 1, all nodes are isomorphic to each other, thus GAEs cannot distinguish the link (v_1, v_2) from (v_1, v_3). However, in most situations we can avoid this problem by simply considering the *paths* between the target nodes. In Fig. 1 the paths between (v_1, v_2) and (v_1, v_3) are clearly different, therefore we can avoid the automorphic node problem by considering the paths between the target nodes. This inspires us to investigate whether the GNN variants can capture path information and thus avoid the automorphic node problem. Our results are as follows.

GAEs are Unaware of Paths. Figure 1 provides a clear counterexample where GAEs fail to distinguish between (v_1, v_2) and (v_1, v_3).

Other GNN variants are Aware of Paths. We formally state this result in the following theorem.

Theorem 1. *Given any graphs $G = (\mathcal{V}_G, \mathcal{E}_G)$ and $H = (\mathcal{V}_H, \mathcal{E}_H)$, suppose $(v_1, v_2) \in \mathcal{V}_G^2$, $(w_1, w_2) \in \mathcal{V}_H^2$ are the target links. Then, all the other GNN variants with sufficient layers, including labeling trick, variants of the labeling trick, higher-order GNNs, would distinguish (v_1, v_2) from (w_1, w_2), if there exists some positive integer d such that the number of paths of length d from v_1 to v_2 is different from the number of paths of length d from w_1 to w_2.*

Proof. We first consider the labeling trick and its variants. Obviously, among these variants the partial labeling trick is the least expressive one [44], therefore we focus on this variant. Given $G = (\mathcal{V}_G, \mathcal{E}_G)$ and the target link (v_1, v_2), the partial labeling trick first assigns a unique label on v_1 and then apply a MPNN on G. We will refer to the additional label as IsSource where IsSource(v_1) = 1 and 0 otherwise. The procedure of counting paths can be described as $h_u^{(l+1)} = \sum_{w \in \mathcal{N}(u)} h_w^{(l)}$ and $h_u^{(0)} = $ IsSource(u), where $h_u^{(l)}$ corresponds to the number of paths of length l from v_1 to u. Obviously, this procedure is within the standard MPNN framework, thus can be computed by the partial labeling trick.

Next we consider the higher-order GNNs. We consider the variants that simulate the 3-WL [21] and 2-FWL [4]. Since the 3-WL and 2-FWL share the equivalent expressiveness, we focus on the 2-FWL variant here. Similarly as before, the procedure of counting paths can be described as $h_{uv}^{(l+1)} = \sum_{w \in V_g} h_{uw}^{(l)} h_{wv}^{(0)}$ and $h_{uv}^{(0)} = 1$ if $(u,v) \in \mathcal{E}_g$ and 0 otherwise. Obviously, this procedure counts the paths of length $l+1$ from u to v at layer l, and is within the 2-FWL framework, thus can be computed by the higher-order GNNs.

With Theorem 1 we can conclude that all these GNN variants except GAEs are aware of paths. Considering their performance on various link prediction benchmarks, we may assume that the awareness of paths is the key factor that helps these variants for learning better link representations.

4.2 What Existing GNNs Have Overlooked: Awareness of Simple Paths

With the discussions above, a natural question is that, are the GNN variants also aware of *simple paths*? Simple paths are a fundamental concept in graph theory, and play an important role in many graph-related problems. Therefore, one may simply assume that these GNN variants are also aware of simple paths. However, we find that this is not true: in fact, *none* of these variants can capture simple path patterns.

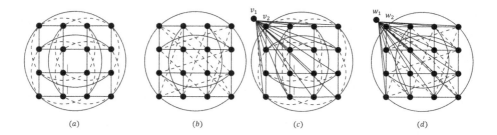

(a)   (b)   (c)   (d)

Fig. 2. Illustration of Rook's 4×4 and Shrikhande graph.

Constructing Counterexamples. A well-known non-isomorphic pair of graphs that is hard to distinguish is the Shrikhande Graph [28] and the Rook's 4×4 Graph [19]. The two graphs are shown in Fig. 2(a), (b). These graphs are strongly-regular and cannot be distinguished by the 3-WL test. Therefore, higher-order GNNs cannot distinguish between them. Moreover, we augment these two graphs with adding one node that's connected with the rest of the nodes, resulting in the graphs (c), (d) respectively. Obviously, (c) and (d) are non-isomorphic graphs. However, the labeling trick methods and their variants, and higher-order GNNs (which are within the 3-WL hierarchy) cannot distinguish (v_1, v_2) from (w_1, w_2).

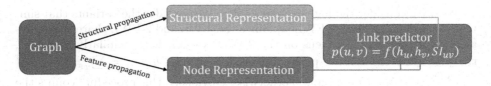

Fig. 3. Illustration the SPGNN framework. h_u, h_v are the node representations of nodes u, v computed in the node feature propagation procedure. SI_{uv} is the structural representation of (u, v) computed in the structural feature propagation procedure, which is aware of simple paths.

Generally, such strongly regular graphs are believed to be very hard to distinguish. However, we find that (c) and (d) can be simply distinguished by considering the *simple paths* between (v_1, v_2) and (w_1, w_2) respectively. That is, if we list the number of simple paths from v_1 to v_2 and from w_1 to w_2 respectively according to the path lengths, these numbers are clearly different (There are 13991683 paths from v_1 to v_2 while 14029327 from w_1 to w_2). This indicates that none of the previous GNN variants is able to capture simple paths.

4.3   Designing Efficient GNNs for Link Prediction

In this section we aim to design a more efficient GNN model for link prediction tasks. The above discussions have provided two intuitions: (1) the awareness of paths are important for link prediction and (2) existing GNN variants overlooked simple paths which might further help improve the link prediction performance. Based on these intuitions we propose a novel GNN framework, namely Simple Path-aware GNN (SPGNN) which both absorbs the information about paths and simple paths into link prediction as well as being extremely efficient.

The overall frameworks of SPGNN is illustrated in Fig. 3. Let $X \in \mathbb{R}^{N \times d}$ be the input node feature matrix where N is the number of nodes, d is the feature dimension. Let $S \in \mathbb{R}^{N \times N}$ be the adjacency matrix. Our simple path-aware GNNs contains two parts:

Node Feature Propagation. Denote $H^{(l)} \in \mathbb{R}^{N \times d}$ as the node representation matrix at layer l, with its i-th row being the representation of the i-th node $h_i^{(l)}$. We let $H^{(0)} = X$. At layer $l + 1$, we compute

$$H^{(l+1)} = \text{GNNLayer}\left(H^{(l)}\right),$$

Structural Propagation. Given the target link (u, v), consider the path counting algorithm. The number of paths of lengths l between any node pairs can be computed as A^l. However, this approach does not consider the node degree information. Inspired by GCNs, we generalize their convolutional operator to compute our structural representations. Given the node degree matrix D, the normalized adjacency matrix is computed as $S = D^{-\frac{1}{2}} A D^{-\frac{1}{2}}$. We then use S to compute the structural representations. We first compute powers of S: $(S, S^2, ..., S^l)$ to generalize path counting in the convolutional domain. After that, we also generalize simple path counting: we define

$$SP_{uv}^{(l)} = \sum_{(u, w_1, ..., w_{l-1}, v) \in \text{SimplePath}(u,v)} S_{uw_1} S_{w_1 w_2} ... S_{w_{l-1} v},$$

where l is the length of path, and S_{ij} is the (i, j)-th entrance of the normalized adjacency matrix S. Obviously, $SP_{uv}^{(l)}$ generalizes the counting of simple paths. We then compute the structural representation of (u, v) as

$$SI_{uv} = [S_{uv}, S_{uv}^2, ..., S_{uv}^l, SP_{uv}^{(1)}, ..., SP_{uv}^{(l)}].$$

Obviously, all the operations here are sparse and the structural representations can be efficiently precomputed before training.

Link Predictor. Both the above propagation steps are efficiently precomputed before predicting links, therefore SPGNN can efficiently be trained and inference. At the output layer L, we compute the prediction for the link (u, v) as

$$p(u, v) = f\left(h_u^{(L)}, h_v^{(L)}, SI_{uv}\right),$$

where f is a multi-layer perceptron. With the above method, we can efficiently compute link representations while capturing the simple path information between the target nodes.

Time Complexity. An advantage of our proposed method is that, compared with previous GNN-based link prediction methods, our method is more efficient. We compare the time complexites of our method with SEAL [40], a labeling trick method, NBFNet [44], a variant of the labeling trick method, and Edge Transformer [4], a higher-order GNN method. We assume we want to compute the representation of one link (u, v), given a graph with N nodes and E edges. The complexities are summarized in Table 1, which are only w.r.t. the size of the graph for clarity.

Table 1. Time complexities. We assume SGC [34] as the backend GNN model of our method, which is consistend with the experiments.

Complexity	SEAL	NBFNet	Edge Transformer	SPGNN
Preprocess	$O(1)$	$O(1)$	$O(1)$	$O(E)$
Inference	$O(E)$	$O(E + N)$	$O(N^3)$	$O(1)$

Higher-order GNNs inherently follow the k-WL hierarchy thus are less efficient. As for labeling trick methods, when predicting links with different sources/tail nodes, these methods need to relabel the graph first and then rerun the GNN on the relabeled graph, making them less efficient. Compared with these method, our method only needs to run the GNN once. The resulting node representations can then be efficiently utilized to predict any links of the graph, while considering the inter-node patterns between the target links.

Table 2. Dataset statistics.

Dataset	Cora	Citeseer	Pubmed	USAir	Celegans	Ecoli	PB
#Nodes	2708	3327	18717	332	297	1805	1222
#Edges	5278	4676	44327	2126	2148	14660	16714

Table 3. Main results. **Best results** are bold, and secondary results are underlined.

Method	Cora	Citeseer	Pubmed	USAir	Celegans	Ecoli	PB
Metric	Hit@100	Hit@100	Hit@100	AUC	AUC	AUC	AUC
CN	33.92 ± 0.46	29.79 ± 0.90	23.13 ± 0.15	92.80 ± 1.22	85.13 ± 1.61	93.71 ± 0.39	92.04 ± 0.35
AA	39.85 ± 1.34	35.19 ± 1.33	27.38 ± 0.11	95.06 ± 1.03	86.95 ± 1.40	95.36 ± 0.34	92.36 ± 0.34
RA	41.07 ± 0.48	33.56 ± 0.17	27.03 ± 0.35	95.77 ± 0.92	87.49 ± 1.41	95.95 ± 0.35	92.46 ± 0.37
GAE	66.79 ± 1.65	67.08 ± 2.94	53.02 ± 1.39	89.28 ± 1.99	81.80 ± 2.18	90.81 ± 0.63	90.70 ± 0.53
SEAL	$\underline{81.71} \pm 1.30$	83.89 ± 2.15	$\mathbf{75.54} \pm 1.32$	96.62 ± 1.22	$\underline{90.30} \pm 1.35$	$\mathbf{97.64} \pm 0.22$	$\underline{94.72} \pm 0.46$
NBFNet	71.65 ± 2.27	74.07 ± 1.75	58.73 ± 1.99	–	–	–	–
Neo-GNN	80.42 ± 1.31	$\underline{84.67} \pm 2.16$	73.93 ± 1.19	–	–	–	–
SPGNN	$\mathbf{88.63} \pm 1.23$	$\mathbf{93.20} \pm 1.25$	$\underline{74.14} \pm 0.72$	$\mathbf{97.28} \pm 1.17$	$\mathbf{92.51} \pm 1.23$	$\underline{96.14} \pm 0.29$	$\mathbf{94.84} \pm 0.37$

5   Experiments

Datasets. We consider two series of link prediction benchmarks. We report results for the most widely used Planetoid citation networks Cora [18], Citeseer [29] and Pubmed [23]. These datasets contain node features. We randomly remove 20% existing links as positive test data. Following a standard manner of learning-based link prediction, we randomly sample the same number of nonexistent links (unconnected node pairs) as negative testing data. We use the remaining 80% existing links as well as the same number of additionally sampled nonexistent links to construct the training data. Dataset statistics are summarized in Table 2.

We also conduct experiments on the widely used USAir [3], C.ele [33], Ecoli [41] and PB [1]. Different from the Planetoid datasets, these datasets contain no node features and thus we need to predict the existence of links based solely on graph structures. For these datasets we randomly remove 10% existing links from each dataset as positive testing data.

Baselines. We report results for three heuristics that are successful for link prediction, Common Neighbors (CN), Adamic-Adar (AA) [2], Resource Allocation (RA) [43]; a basic link prediction GNN model GAE [13]; a series of state-of-the-art link prediction GNNs considered in this paper SEAL [40], Neo-GNN [39] and NBFNet [44]. We select commonly used metrics in link prediction literature AUC and Hits@100 [42] as our evaluation metrics.

Implementation Details. We choose SGC [34] as our basic GNN layer. We select the number of layers to be 3 for all datasets. We select the hidden dimension to be 1024. The predictor f is implemented as a 2-layer MLP. If the dataset contains no node features we simply omit the node feature propagation step. We use the Adam [12] optimizer with learning rate 10^{-4}. The lengths of paths and simple paths are hyperparameters selected from $\{0, 1, 2, 3\}$ based on the validation set.

Results. Main results are in Table 3. SPGNN achieves the best results in 5 of the 7 datasets, and the secondary results in the remaining 2 datasets. We can see that these results align with our previous analysis: Cora, Citeseer and Pubmed datasets contain real-valued noisy node features, thus traditional heuristics CN, AA and RA perform poorly. GAE methods are weak in link prediction and cannot capture path information, therefore perform poorly in USAir, Celegans, Ecoli and PB datasets which contains no node features. SEAL is more expressive than GAEs and can identify paths between the target node pairs, which makes it the closest competitor. Compared with SEAL, our method is able to consider simple paths, therefore achieve better performance in most of the datasets. This implies the effectiveness of our structural propagation procedure which involves information about both paths and simple paths of the target links.

Runtimes. We compare SPGNN with SEAL, the closest competitor. We report wall times for both SEAL static (subgraphs are pre-generated) and SEAL dynamic (subgraphs are generated on the fly) on the PubMed dataset. The times are summarized in Table 4. SEAL dynamic needs to extract subgraphs during training, therefore is much slower than other variants. SEAL static first extract subgraphs before training, thus requires a preprocess procedure. Because SEAL needs to run the GNN on different subgraphs for predicting different links in training data, it still needs longer time for training. Compared with SEAL, GCN is much faster because we only need to run it once for predicting all target links in training data. Compared with GCN, SPGNN (with the utilization of SGC as the feature propagation model) precomputes both the structural representations and node representations before training, therefore requires an explicit preprocess procedure. This also makes SPGNN much faster compared with SEAL and GCN.

Ablation Studies. We aim to verify the two assumptions in this paper, i.e. (1) the awareness of paths is of vital importance in link prediction and (2) the awareness of simple paths can further improve the performance. We perform the ablation studies on Cora and Citeseer dataset and the results are in Table 5.

Table 4. Wall times for one epoch training.

time(s)	SEAL dyn	SEAL stat	GCN	SPGNN
preprocess	0	630	0	79
train	70	30	4	0.43

Table 5. Ablation studies. GAE: without paths. P: paths. SP: simple paths.

Dataset	GAE	P	P+SP
Cora	66.79	86.08	88.63
Citeseer	67.08	91.02	93.20

6   Conclusion

In this paper, we systematically investigate the expressive power of link prediction GNNs via the perspective of paths, and justify the superior performance of labeling trick and higher-order GNNs by showing their awareness of paths, of which GAEs lack. We then notice a fundamental property which all previous link prediction GNNs have overlooked, which is the awareness of simple paths.

The power and limits of existing GNNs inspire us to design a novel GNN framework that can provably capture both path and simple path patterns as well as being more efficient compared with existing labeling trick and higher-order GNNs. We test the performance and efficiency of our model via various common link prediction benchmarks.

References

1. Ackland, R., et al.: Mapping the us political blogosphere: are conservative bloggers more prominent? In: BlogTalk Downunder 2005 Conference, Sydney (2005)
2. Adamic, L.A., Adar, E.: Friends and neighbors on the web. Soc. Netw. **25**, 211–230 (2003)
3. Batagelj, V., Mrvar, A.: (2006). http://vlado.fmf.uni-lj.si/pub/networks/data/
4. Bergen, L., O'Donnell, T.J., Bahdanau, D.: Systematic generalization with edge transformers. In: NeurIPS (2021)
5. Bordes, A., Usunier, N., García-Durán, A., Weston, J., Yakhnenko, O.: Translating embeddings for modeling multi-relational data. In: NIPS (2013)
6. Bruna, J., Zaremba, W., Szlam, A.D., LeCun, Y.: Spectral networks and locally connected networks on graphs. CoRR abs/1312.6203 (2013)
7. Chen, J., Ma, T., Xiao, C.: FastGCN: fast learning with graph convolutional networks via importance sampling. ArXiv abs/1801.10247 (2018)
8. Defferrard, M., Bresson, X., Vandergheynst, P.: Convolutional neural networks on graphs with fast localized spectral filtering. In: NIPS (2016)
9. Dettmers, T., Minervini, P., Stenetorp, P., Riedel, S.: Convolutional 2d knowledge graph embeddings. In: AAAI (2018)
10. Galárraga, L., Teflioudi, C., Hose, K., Suchanek, F.M.: Fast rule mining in ontological knowledge bases with AMIE+. VLDB J. **24**, 707–730 (2015)
11. Ioannidis, V.N., Zheng, D., Karypis, G.: Few-shot link prediction via graph neural networks for COVID-19 drug-repurposing. ArXiv abs/2007.10261 (2020)
12. Kingma, D.P., Ba, J.: Adam: a method for stochastic optimization. arXiv preprint arXiv:1412.6980 (2014)

13. Kipf, T., Welling, M.: Variational graph auto-encoders. In: NeurIPS Workshop (2016)
14. Kipf, T., Welling, M.: Semi-supervised classification with graph convolutional networks. In: ICLR (2017)
15. Koren, Y., Bell, R.M., Volinsky, C.: Matrix factorization techniques for recommender systems. Computer **42**, 30–37 (2009)
16. Maron, H., Ben-Hamu, H., Serviansky, H., Lipman, Y.: Provably powerful graph networks. ArXiv abs/1905.11136 (2019)
17. Maron, H., Ben-Hamu, H., Serviansky, H., Lipman, Y.: Provably powerful graph networks. In: NeurIPS (2019)
18. McCallum, A., Nigam, K., Rennie, J.D.M., Seymore, K.: Automating the construction of internet portals with machine learning. Inf. Retrieval **3**, 127–163 (2000)
19. Moon, J.W.: On the line-graph of the complete bigraph. Ann. Math. Stat. **34**, 664–667 (1963)
20. Morris, C., Rattan, G., Mutzel, P.: Weisfeiler and leman go sparse: towards scalable higher-order graph embeddings. In: NeurIPS (2020)
21. Morris, C., et al.: Weisfeiler and leman go neural: higher-order graph neural networks. In: AAAI (2019)
22. Muggleton, S.: Inverse entailment and progol. N. Gener. Comput. **13**, 245–286 (2009)
23. Namata, G., London, B., Getoor, L., Huang, B., Edu, U.: Query-driven active surveying for collective classification. In: 10th International Workshop on Mining and Learning with Graphs, vol. 8, p. 1 (2012)
24. Nguyen, T., Phung, D.Q., Adams, B., Venkatesh, S.: Towards discovery of influence and personality traits through social link prediction. AAAI (2021)
25. Nickel, M., Murphy, K.P., Tresp, V., Gabrilovich, E.: A review of relational machine learning for knowledge graphs. Proc. IEEE **104**, 11–33 (2015)
26. Qu, M., Chen, J., Xhonneux, L.P., Bengio, Y., Tang, J.: RNNLogic: learning logic rules for reasoning on knowledge graphs. In: ICLR (2021)
27. Rocktäschel, T., Riedel, S.: End-to-end differentiable proving. In: NeurIPS, pp. 3791–3803 (2017)
28. Shrikhande, S.S.: The uniqueness of the l2 association scheme. Ann. Math. Stat., 781–798 (1959)
29. Sen, P., Namata, G., Bilgic, M., Getoor, L., Gallagher, B., Eliassi-Rad, T.: Collective classification in network data. AI Mag. **29**, 93 (2008)
30. Sun, Z., Deng, Z., Nie, J.Y., Tang, J.: Rotate: knowledge graph embedding by relational rotation in complex space. ArXiv abs/1902.10197 (2018)
31. Teru, K.K., Denis, E., Hamilton, W.L.: Inductive relation prediction by subgraph reasoning. In: ICML (2019)
32. Trouillon, T., Welbl, J., Riedel, S., Gaussier, É., Bouchard, G.: Complex embeddings for simple link prediction. In: ICML (2016)
33. Watts, D.J., Strogatz, S.H.: Collective dynamics of 'small-world' networks. Nature **393**(6684), 440–442 (1998)
34. Wu, F., Zhang, T., de Souza, A.H., Fifty, C., Yu, T., Weinberger, K.Q.: Simplifying graph convolutional networks. In: ICML (2019)
35. Xu, K., Hu, W., Leskovec, J., Jegelka, S.: How powerful are graph neural networks? ArXiv abs/1810.00826 (2018)
36. Yang, B., tau Yih, W., He, X., Gao, J., Deng, L.: Embedding entities and relations for learning and inference in knowledge bases. CoRR abs/1412.6575 (2015)
37. Yang, F., Yang, Z., Cohen, W.W.: Differentiable learning of logical rules for knowledge base completion. In: NeurIPS (2017)

38. You, J., Gomes-Selman, J.M., Ying, R., Leskovec, J.: Identity-aware graph neural networks. In: AAAI (2021)
39. Yun, S., Kim, S., Lee, J., Kang, J., Kim, H.J.: Neo-GNNs: neighborhood overlap-aware graph neural networks for link prediction. In: NeurIPS (2021)
40. Zhang, M., Chen, Y.: Link prediction based on graph neural networks. In: NeurIPS (2018)
41. Zhang, M., Cui, Z., Jiang, S., Chen, Y.: Beyond link prediction: predicting hyper-links in adjacency space. In: AAAI, vol. 32 (2018)
42. Zhang, M., Li, P., Xia, Y., Wang, K., Jin, L.: Labeling trick: a theory of using graph neural networks for multi-node representation learning. In: NeurIPS (2020)
43. Zhou, T., Lü, L., Zhang, Y.C.: Predicting missing links via local information. Eur. Phys. J. B **71**, 623–630 (2009)
44. Zhu, Z., Zhang, Z., Xhonneux, L.P., Tang, J.: Neural bellman-ford networks: a general graph neural network framework for link prediction. In: NeurIPS (2021)

Restore Translation Using Equivariant Neural Networks

Yihan Wang[1,3], Lijia Yu[2,3], and Xiao-Shan Gao[1,3(✉)]

[1] Academy of Mathematics and Systems Science, Chinese Academy of Sciences, Beijing, China
xgao@mmrc.iss.ac.cn
[2] SKLCS, Institute of Software, Chinese Academy of Sciences, Beijing, China
[3] University of Chinese Academy of Sciences, Beijing, China

Abstract. Invariance to spatial transformations such as translations is a desirable property and a basic design principle for classification neural networks. However, the commonly used convolutional neural networks (CNNs) are actually very sensitive to even small translations. There exist vast works to achieve exact or approximate transformation invariance by designing transformation-invariant models or assessing the transformations. These works usually make changes to the standard CNNs and harm the performance on standard datasets. In this paper, rather than modifying the classifier, we propose a pre-classifier restorer to recover translated inputs to the original ones which will be fed into any classifier for the same dataset. The restorer is based on a theoretical result which gives a sufficient and necessary condition for an affine operator to be translational equivariant on a tensor space.

Keywords: Equivariant network · Translation restorer

1 Introduction

Deep convolutional neural networks (CNNs) had outperformed humans in many computer vision tasks [9,12]. One of the key ideas in designing the CNNs is that the convolution layer is equivariant with respect to translations, which was emphasized both in the earlier work [5] and the modern CNN [12]. However, the commonly used components, such as pooling [7] and dropout [19,20], which help the network to extract features and generalize, actually make CNNs not equivariant to even small translations, as pointed out in [1,3]. As a comprehensive evaluation, Fig. 1 shows that two classification CNNs suffer the accuracy reductions of more than 11% and 59% respectively on CIFAR-10 and MNIST, when the inputs are horizontally and vertically translated at most 3 pixels.

Invariance to spatial transformations, including translations, rotations and scaling, is a desirable property for classification neural networks and the past few decades have witnessed thriving explorations on this topic. In general, there exist three ways to achieve exact or approximate invariance. The first is to design transformation-invariant neural network structures [2,6,8,10,15,16,18,21].

B. Luo et al. (Eds.): ICONIP 2023, CCIS 1962, pp. 583–603, 2024.
https://doi.org/10.1007/978-981-99-8132-8_44

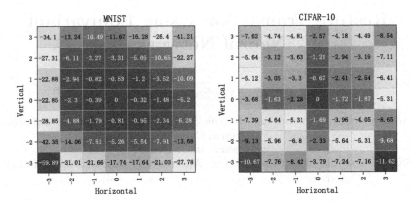

Fig. 1. The accuracy reduction after vertical and horizontal translations. The translation scope is $[-3, 3]$ pixels. Left: LeNet-5 on MNIST; Right: VGG-16 on CIFAR-10.

The second is to assess and approximate transformations via a learnable module [4,11] and then use the approximation to reduce the transformed inputs to "standard" ones. The third is data augmentation [1,3,17] by adding various transformations of the samples in the original dataset.

Those ad-hoc architectures to achieve invariance often bring extra parameters but harm the network performance on standard datasets. Moreover, the various designs with different purposes are not compatible with each other. Data augmentation is not a scalable method since the invariance that benefits from a certain augmentation protocol does not generalize to other transformations [1]. Including learnable modules such as the Spatial Transformer, all the three approaches require training the classifier from scratch and fail to endow existing trained networks with some invariance. It was indicated in [1] that "the problem of insuring invariance to small image transformations in neural networks while preserving high accuracy remains unsolved."

In this paper, rather than designing any in-classifier component to make the classifier invariant to some transformation, we propose a pre-classifier restorer to restore translated inputs to the original ones. The invariance is achieved by feeding the restored inputs into any following classifier. Our restorer depends only on the dataset instead of the classifier. Namely, the training processes of the restore and classifier are separate and a restore is universal to any classifier trained on the same dataset.

We split the whole restoration into two stages, transformation estimation and inverse transformation, see Fig. 2. In the first stage, we expect that standard inputs lead to standard outputs and the outputs of translated inputs reflect the translations. Naturally, what we need is a strictly translation-equivariant neural network. In Sect. 3, we investigate at the theoretical aspect the sufficient and necessary condition to construct a strictly equivariant affine operator on a tensor space. The condition results in *the circular filters*, see Definition 4, as the fundamental module to a strictly translation-equivariant neural network. We give the canonical architecture of translation-equivariant networks, see Eq. (2). In Sect. 4, details of the restorer are presented. We define a translation estimator, the core component of a restorer, as a strictly translation-equivariant neural network that guarantees the first component of every output on a dataset to be the largest component, see Definition 5. For a translated input, due to the strict equivariance, the largest component of the output reflects the translation. Thus we can translate it inversely in the second stage and obtain the original image. Though the restorer is independent of the following classifier, it indeed depends on the dataset. Given a dataset satisfying some reasonable conditions, i.e. *an aperiodic finite dataset*, see Definition 6, we prove the existence of a translation estimator, i.e. a restorer, with the canonical architecture for this dataset. Moreover, rotations can be viewed as translations by converting the Cartesian coordinates to polar coordinates and the rotation restorer arises in a similar way.

In Sect. 5, the experiments on MNIST, 3D-MNIST and CIFAR-10 show that our restorers not only visually restore the translated inputs but also largely eliminate the accuracy reduction phenomenon.

2   Related Works

As a generalization of convolutional neural networks, group-equivariant convolutional neural networks [2,6] exploited symmetries to endow networks with invariance to some group actions, such as the combination of translations and rotations by certain angles. The warped convolutions [10] converted some other spatial transformations into translations and thus obtain equivariance to these spatial transformations. Scale-invariance [8,15,21] was injected into networks by some ad-hoc components. Random transformations [16] of feature maps were introduced in order to prevent the dependencies of network outputs on specific poses of inputs. Similarly, probabilistic max pooling [18] of the hidden units over the set of transformations improved the invariance of networks in unsupervised learning. Moreover, local covariant feature detecting methods [14,22] were proposed to address the problem of extracting viewpoint invariant features from images.

Another approach to achieving invariance is "shiftable" down-sampling [13], in which any original pixel can be linearly interpolated from the pixels on the sampling grid. This "shiftable" down-sampling exists if and only if the sampling frequency is at least twice the highest frequency of the unsampled signal.

The Spatial Transformer [4,11], as a learnable module, produces a predictive transformation for each input image and then spatially transforms the input to a canonical pose to simplify the inference in the subsequent layers. Our restorers give input-specific transformations as well and adjust the input to alleviate the poor invariance of the following classifiers. Although the Spatial Transformers and our restorer are both learnable modules, the training of the former depends not only on data but also on the subsequent layers, while the latter are independent of the subsequent classifiers.

3   Equivariant Neural Networks

Though objects in nature have continuous properties, once captured and converted to digital signals, these properties are represented by real tensors. In this section, we study the equivariance of operators on a tensor space.

3.1   Equivariance in Tensor Space

Assume that a map $\tilde{x} : \mathbb{R}^d \to \mathbb{D}$ stands for a property of some d-dimensional object where $\mathbb{D} \subseteq \mathbb{R}$. Sampling $\tilde{x}$ over a $(n_1, n_2, \cdots, n_d)$-grid results in a tensor x in a tensor space

$$\mathcal{H} := \mathbb{D}^{n_1} \otimes \mathbb{D}^{n_2} \otimes \cdots \otimes \mathbb{D}^{n_d}. \tag{1}$$

We denote $[n] = [0, 1, \ldots, n-1]$ for $n \in \mathbb{Z}_+$ and assume $k \bmod n \in [n]$ for $k \in \mathbb{Z}$. For an index $I = (i_1, i_2, \cdots, i_d) \in \prod_{i=1}^d [n_i]$ and $x \in \mathcal{H}$, denote $x[I]$ to be the element of x with subscript $(i_1, i_2, \cdots, i_d)$. For convenience, we extend the index of $\mathcal{H}$ to $I = (i_1, i_2, \cdots, i_d) \in \mathbb{Z}^d$ by defining

$$x[I] = x[i_1 \bmod n_1, \cdots, i_d \bmod n_d].$$

Definition 1 (Translation). *A translation* $T^M : \mathcal{H} \to \mathcal{H}$ *with* $M \in \mathbb{Z}^d$ *is an invertible linear operator such that for all* $I \in \mathbb{Z}^d$ *and* $x \in \mathcal{H}$,

$$T^M(x)[I] = x[I - M].$$

The inverse of T^M *is clearly* T^{-M}.

Definition 2 (Equivariance). *A map* $w : \mathcal{H} \to \mathcal{H}$ *is called* equivariant with respect to translations *if for all* $x \in \mathcal{H}$ *and* $M \in \mathbb{Z}^d$,

$$T^M(w(x)) = w(T^M(x)).$$

Definition 3 (Vectorization). *A tensor* x *can be vectorized to* $X \in \vec{\mathcal{H}} = \mathbb{D}^N$ *with* $N = n_1 n_2 \cdots n_d$ *such that*

$$X(\delta(I)) := x[I],$$

where $\delta(I) := (i_1 \bmod n_1) n_2 n_3 \cdots n_d + (i_2 \bmod n_2) n_3 n_4 \cdots n_d + \cdots + (i_d \bmod n_d)$, *and we denote* $X = \vec{x}$. *Moreover, the translation* T^M *is vectorized as* $T^M(X) := \overrightarrow{T^M(x)}$.

3.2   Equivariant Operators

When $\mathbb{D} = \mathbb{R}$, the tensor space $\mathcal{H}$ is a Hilbert space by defining the inner product as $x \cdot z := \vec{x} \cdot \vec{z}$ which is the inner product in vector space $\vec{\mathcal{H}}$. In the rest of this section, we assume $\mathbb{D} = \mathbb{R}$.

According to Reize's representation theorem, there is a bijection between continuous linear operator space and tensor space. That is, a continuous linear operator $v : \mathcal{H} \to \mathbb{R}$ can be viewed as a tensor $v \in \mathcal{H}$ satisfying $v(x) = v \cdot x$. Now we can translate v by T^M and obtain $T^M(v) : \mathcal{H} \to \mathbb{R}$ such that $T^M(v)(x) = T^M(v) \cdot x$.

We consider a continuous linear operator $w : \mathcal{H} \to \mathcal{H}$. For $I \in \mathbb{Z}^d$ and $x \in \mathcal{H}$, denote $w_I(x) = w(x)[I]$. Then $w_I : \mathcal{H} \to \mathbb{R}$ is a continuous linear operator. An *affine operator* $\alpha : \mathcal{H} \to \mathcal{H}$ differs from a continuous linear operator w by a *bias tensor* c such that $\alpha(x) = w(x) + c$ for all $x \in \mathcal{H}$.

Theorem 1. *Let $\alpha(x) = w(x) + c : \mathcal{H} \to \mathcal{H}$ be an affine operator. Then, α is equivariant with respect to translations if and only if for all $M \in \mathbb{Z}^d$,*

$$w_M = T^M(w_0) \text{ and } c \propto \mathbf{1},$$

where $\mathbf{0}$ is the zero vector in $\mathbb{Z}^d$ and $c \propto \mathbf{1}$ means that c is a constant *tensor, that is, all of its entries are the same.*

Proof of Theorem 1 is given in Appendix A. Recall that $\vec{\mathcal{H}} = \mathbb{R}^N$ is the vectorization of $\mathcal{H}$ and T^M also translates vectors in $\vec{H}$. Each continuous linear operator on $\mathcal{H}$ corresponds to a matrix in $\mathbb{R}^{N \times N}$ and each bias operator corresponds to a vector in $\mathbb{R}^N$. Now we consider the translation equivariance in vector space.

Definition 4 (Circular filter). *Let $W = (W_0, W_1, \cdots, W_{N-1})^T$ be a matrix in $\mathbb{R}^{N \times N}$. W is called a* circular filter *if $W_{\delta(M)} = T^M(W_0)$ for all $M \in \mathbb{Z}^d$.*

As the vector version of Theorem 1, we have

Corollary 1. *Let $A : \mathbb{R}^N \to \mathbb{R}^N$ be an affine transformation such that*

$$A(X) = W \cdot X + C,$$

in which $W \in \mathbb{R}^{N \times N}$, $C \in \mathbb{R}^N$. Then, A is equivariant with respect to translations in the sense that for all $M \in \mathbb{Z}^d$

$$A(T^M(X)) = T^M(A(X))$$

if and only if W is a circular filter and $C \propto \mathbf{1}$.

This affine transformation is very similar to the commonly used convolutional layers [5,12] in terms of shared parameters and similar convolutional operations. But the strict equivariance calls for the same in-size and out-size, and circular convolutions, which are usually violated by CNNs.

3.3 Equivariant Neural Networks

To compose a strictly translation-equivariant network, the spatial sizes of the input and output in each layer must be the same and thus down-samplings are not allowed. Though Corollary 1 provides the fundamental component of a strictly translation-equivariant network, different compositions of this component lead to various equivariant networks. Here we give the *canonical architecture*. We construct the strictly translation-equivariant network F with L layers as

$$F(X) = F_L \circ F_{L-1} \circ \cdots \circ F_1(X). \tag{2}$$

The l-the layer F_l has n_l channels and for an input $X \in \mathbb{R}^{n_{l-1} \times N}$ we have

$$F_l(X) = \sigma(W[l] \cdot X + C[l]) \in \mathbb{R}^{n_l \times N}, \tag{3}$$

where

$$
\begin{aligned}
W[l] &= (W^1[l], \cdots, W^{n_l}[l]) \in \mathbb{R}^{n_l \times n_{l-1} \times N \times N}, \\
C[l] &= (C^1[l] \cdot \mathbf{1}, \cdots, C^{n_l}[l] \cdot \mathbf{1}), \\
W^k[l] &= (W^{k,1}[l], \cdots, W^{k,n_{l-1}}[l]) \in \mathbb{R}^{n_{l-1} \times N \times N}, \\
C^k[l] &= (C^{k,1}[l], \cdots, C^{k,n_{l-1}}[l]) \in \mathbb{R}^{n_{l-1}},
\end{aligned}
$$

σ is the activation, $W^{k,r}[l] \in \mathbb{R}^{N \times N}$ are circular filters, $C^{k,r}[l] \in \mathbb{R}$ are constant biases for $k = 1, \cdots, n_l$ and $r = 1, \cdots, n_{l-1}$, the $\cdot$ denotes the inner product and $\mathbf{1}$ is the vector whose all components are 1.

4   Translation Restorer

4.1   Method

In Sect. 3.3, we propose a strictly equivariant neural network architecture (2) such that any translation on the input will be reflected in the output. Generally speaking, once the outputs of an equivariant network on a dataset have some spatial structure, this structure shifts consistently as the input shifts. Thus, the translation parameter of a shifted input can be solved from its output. Finally, we can restore the input via the inverse translation. Figure 2 shows how a restorer works.

The whole restoration process splits into two stages, translation estimation and inverse translation. We first define the translation estimator which outputs a consistent and special structure on a dataset.

Definition 5. *Let $\mathcal{D} \subset \mathbb{D}^{P \times N}$ be a dataset with P channels. Then a translation-equivariant network*

$$F : \mathbb{R}^{P \times N} \to \mathbb{R}^N$$

is said to be a translation estimator for $\mathcal{D}$ if

$$F(X)[0] = \max_{i=0}^{N-1} F(X)[i],$$

where $F(X)[i]$ is the i-th component of $F(X)$.

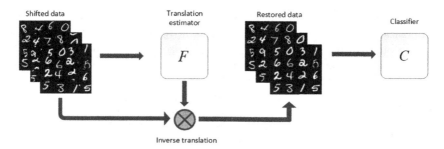

Fig. 2. The pre-classifier translation restorer. For a shifted data $T^M(x)$ as the input, the translation estimator obtains the translation M and restore the original data $T^{-M}(T^M(x)) = x$, which is feed into a pre-trained classifier.

Given such a translation estimator for dataset $\mathcal{D}$ and a shifted input $X' = T^M(X)$ for some $X \in \mathcal{D}$, we propagate X' through F and get the output $F(X') \in \mathbb{R}^N$. Since the first component of $F(X)$ is the largest, the location of the largest component of $F(X')$ is exactly the translation parameter:

$$\delta(M) = \mathrm{argmax}_{i=0}^{N-1} F_i(X').$$

Then we can restore $X = T^{-M}(X')$ by inverse translation. The restored inputs can be fed to any classifier trained on the dataset $\mathcal{D}$.

4.2   Existence of the Restorer

In this section, we show the existence of restorers, that is, the translation estimator. Note that our restorer is independent of the following classifier but dependent on the dataset. For translation, if a dataset contains both an image and a translated version of it, the estimator must be confused. We introduce aperiodic datasets to clarify such cases.

Definition 6 (Aperiodic dataset). *Let $\mathcal{D} \subset \mathbb{D}^{P \times N}$ be a finite dataset with P channels. We call $\mathcal{D}$ an aperiodic dataset if $\mathbf{0} \notin \mathcal{D}$ and*

$$T^M(X) = X' \iff M = \mathbf{0} \text{ and } X = X',$$

for $M \in \mathbb{Z}^{d+1}$ and $X, X' \in \mathcal{D}$. Here d is the spatial dimension and M decides the translation in the channel dimension in addition.

Let $\mathcal{D}$ be an aperiodic dataset. Given that $\mathbb{D} = [2^{Q+1}]$ which is the case in image classification, we prove the existence of the translation estimator for such an aperiodic dataset. The proof consists of two steps. The data are first mapped to their binary decompositions through a translation-equivariant network as Eq. (2) and then the existence of the translation-restorer in the form of Eq. (2) is proved for binary data.

Let $\mathbb{D} = [2^{Q+1}]$ and $\mathbb{B} = \{0,1\}$. We denote $\eta : \mathbb{D} \to \mathbb{B}^Q$ to be the binary decomposition, such as $\eta(2) = (0,1,0)$ and $\eta(3) = (1,0,1)$. We perform the binary decomposition on $X \in \mathbb{D}^{P \times N}$ element-wisely and obtain $\eta(X) \in \mathbb{B}^{G \times N}$, where $G = PQ$ is the number of channels in binary representation. A dataset $\mathcal{D} \subseteq \mathbb{D}^{P \times N}$ can be decomposed into $\mathcal{B} \subset \mathbb{B}^{G \times N}$. Note that the dataset $\mathcal{D}$ is aperiodic if and only if its binary decomposition $\mathcal{B}$ is aperiodic.

The following Lemma 1 demonstrates the existence of a translation-equivariant network which coincides with the binary decomposition η on $[2^{Q+1}]^{P \times N}$. Proof details are placed in Appendix B.

Lemma 1. *Let $\eta : [2^{Q+1}] \to \mathbb{B}$ be the binary decomposition. There exists a $(2Q + 2)$-layer network F in the form of Eq. (2) with ReLU activations and width at most $(Q + 1)N$ such that for $X \in [2^{Q+1}]^{P \times N}$*

$$F(X) = \eta(X).$$

The following Lemma 2 demonstrate the existence of a 2-layer translation restorer for an aperiodic binary dataset. Proof details are placed in Appendix C.

Lemma 2. *Let $\mathcal{B} = \{Z_s | s = 1, 2, \cdots, S\} \subset \mathbb{B}^{G \times N}$ be an aperiodic binary dataset. Then there exists a 2-layer network F in the form of Eq. (2) with ReLU activations and width at most SN such that for all $s = 1, 2, \cdots, S$,*

$$F(Z_s)[0] = \max_{i=0}^{N-1} F(Z_s)[i].$$

Given a $(2Q + 2)$-layer network F' obtained from Lemma 1 and a 2-layer network F'' obtained from Lemma 2, we stack them and have $F = F'' \circ F'$ which is exactly a translation restorer. We thus have proved the following theorem.

Theorem 2. *Let $\mathcal{D} = \{X_s | s = 1, 2, \cdots, S\} \subset [2^{Q+1}]^{P \times N}$ be an aperiodic dataset. Then there exists a network $F : \mathbb{R}^{P \times N} \to \mathbb{R}^N$ in the form of Eq. (2) with ReLU activations such that for $s = 1, 2, \cdots, S$,*

$$F(X_s)[0] = \max_{i=0}^{N-1} F(X_s)[i],$$

of which the depth is at most $2Q+4$ and the width is at most $\max(SN, (Q+1)N)$. Namely, this network is a translation restorer for $\mathcal{D}$.

5   Experiments

The core component of the restorer is the translation estimator which outputs the translation parameter of the shifted inputs.

We use the architecture described in Eq. (2) with $L = 6$, $n_l = 1$ for $l = 1, \cdots, L$ and ReLU activations. The training procedure aims at maximizing the first component of the outputs. Thus the max component of the output indicates the input shift. The experimental settings are given in Appendix D. We report four sets of experiments below.

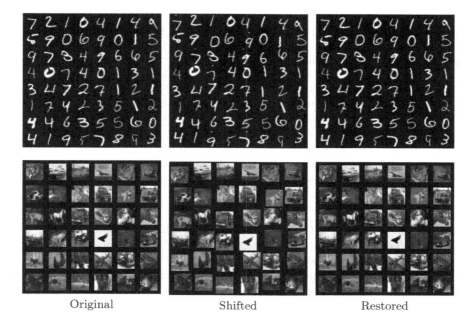

Original   Shifted   Restored

Fig. 3. The restorers for MNIST and CIFAR-10.

2D Images. We train translation restorers for MNIST and CIFAR-10. MNIST images are resized to 32×32 and CIFAR-10 images are padded 4 blank pixels at the edges.

In Fig. 3, the left column is the original images, the middle column is the randomly shifted images and the right column is the restored images. On both datasets, images are randomly shifted vertically and horizontally at most $\frac{1}{4}$ of its size. The shift is a circular shift where pixels shifted out of the figure appear on the other end. We can see that the shifted images are disorganized but the restored images are very alike the original images.

To evaluate the restoration performance of pretrained restorers, we train classifiers and test them on randomly shifted images and restored ones and the results are given in Table 1. When images are not shifted, the restorers lead to only 0.3% and 0.03% accuracy reduction on two datasets. Nevertheless, even if the translation scope is 1, restorers improve the accuracy. Moreover, no matter how the images are shifted, the restorer can repair them to the same status and result in the same classification accuracy, namely 98.59% and 88.18%, while the accuracies drop significantly without the restorer, and the larger the range of translation, the more obvious the restoration effect

Different Architectures. Our proposed restorer is an independent module that can be placed before any classifier. It is scalable to different architectures the subsequent classifier uses.

Table 1. Restoration performance on MNIST and CIFAR-10. Images are randomly shifted within the translation scope ranging from 0 to 8 in both vertical and horizontal directions. We use LeNet-5 on MNIST and ResNet-18 on CIFAR-10. "Res." and "Trans." stand for restorer and translation respectively.

	Res.\Trans.	0	1	2	3	4	5	6	7	8
MNIST	w/o	98.89	98.21	95.41	87.07	76.61	62.9	51.33	41.1	35.7
	w/	98.59	98.59	98.59	98.59	98.59	98.59	98.59	98.59	98.59
	Effect	-0.3	+0.38	+3.18	+11.52	+21.98	+35.69	+47.26	+57.49	+62.89
CIFAR-10	w/o	88.21	86.58	85.9	83.65	82.16	80.46	79.37	77.71	76.01
	w/	88.18	88.18	88.18	88.18	88.18	88.18	88.18	88.18	88.18
	Effect	−0.03	+1.6	+2.28	+4.53	+6.02	+7.72	+8.81	+10.47	+12.17

In Table 2, we evaluate the restoration performance on popular architectures including VGG-16, ResNet-18, DenseNet-121, and MobileNet v2. Translated mages (w/Trans.) are randomly shifted within scope 4 in both vertical and horizontal directions. The reduction of accuracy on original images is no more than 0.04% and the restorer improves the accuracy on shifted images by 1.66%–6.02%.

Table 2. Restoration performance on different architectures and CIFAR-10.

Res.\ Trans.	VGG-16		ResNet-18		DenseNet-121		MobileNet v2	
	w/o	w/	w/o	w/	w/o	w/	w/o	w/
w/o	89.27	83.40	88.21	82.16	92.14	90.46	88.10	83.36
w/	89.23	89.23	88.18	88.18	92.12	92.12	88.09	88.09
Effect	−0.04	+5.83	V0.03	+6.02	−0.02	+1.66	−0.01	+4.73

Translation Augmentation. Training with translation augmentation is another approach to improving the translational invariance of a model. However, translation augmentation is limited in a certain scope and thus cannot ensure the effectiveness for test images shifted out of the scope.

In Fig. 4, we compare the restoration performance on models not trained with translation augmentation (dash lines) and models trained with translation augmentation (solid lines). The augmentation scope is 10% of the image size, that is, 3 pixels for MNIST and 4 pixels for CIFAR-10. Translation augmentation indeed improves the translational invariance of the classifier on images shifted in the augmentation scope. However, when the shift is beyond the augmentation scope, the accuracy begins to degrade. In such a case, the pre-classifier restorer is still able to calibrate the shift and improve the accuracy of the classifier trained with augmentation.

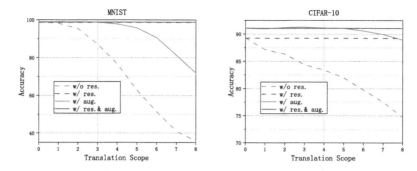

Fig. 4. Restoration performance on classifiers trained with or without translation augmentations. The models are LeNet-5 for MNIST and VGG-16 for CIFAR-10. "res. and "aug" stand for restorer and augmentation, respectively.

3D Voxelization Images. 3D-MNIST contains 3D point clouds generated from images of MNIST. The voxelization of the point clouds contains grayscale 3D tensors.

Figure 5 visualizes the restoration on 3D-MNIST. In the middle of each subfigure, the 3-dimensional digit is shifted in a fixed direction. This fixed direction is detected by the translation estimator and the restored digit is shown on the right.

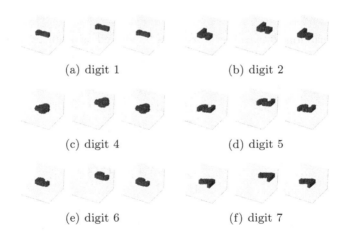

(a) digit 1 (b) digit 2

(c) digit 4 (d) digit 5

(e) digit 6 (f) digit 7

Fig. 5. The restorer on 3D-MNIST. In each sub-figure, the left is the original digit, the middle is the shifted digit, and the right is the restored digit.

6   Conclusion

This paper contributes to the equivalent neural networks in two aspects. Theoretically, we give the sufficient and necessary conditions for an affine operator $Wx+b$ to be translational equivariant, that is, $Wx+b$ is translational equivariant on a tensor space if and only if W has the high dimensional convolution structure and b is a constant tensor. It is well known that if W has the convolution structure, then Wx is equivariant to translations [5,9], and this is one of the basic principles behind the design of CNNs. Our work gives new insights into the convolutional structure used in CNNs in that, the convolution structure is also the necessary condition and hence the most general structure for translational equivariance. Practically, we propose the translational restorer to recover the original images from the translated ones. The restorer can be combined with any classifier to alleviate the performance reduction problem for translated images. As a limitation, training a restorer on a large dataset such as the ImageNet is still computationally difficult.

A   Proof of Theorem 1

We first prove a lemma.

Lemma 3. *Let $v : H \to \mathbb{R}$ be a continuous linear operator. We have*

$$v(T^M(x)) = T^{-M}(v)(x),$$

for all $x \in H$ and all $M \in \mathbb{Z}^d$.

Proof. A continuous linear operator v can be viewed as a tensor in H. We have

$$
\begin{aligned}
v(T^M(x)) &= v \cdot T^M(x) \\
&= \sum_{I \in \prod_{i=1}^d [n_i]} v[I] \cdot T^M(x)[I] \\
&= \sum_{I \in \prod_{i=1}^d [n_i]} v[I] \cdot x[I - M] \\
&= \sum_{I \in \prod_{i=1}^d [n_i]} v[I + M] \cdot x[I] \\
&= \sum_{I \in \prod_{i=1}^d [n_i]} T^{-M}(v)[I] \cdot x[I] \\
&= T^{-M}(v) \cdot x \\
&= T^{-M}(v)(x).
\end{aligned}
$$

Theorem 1. *Let $\alpha(x) = w(x) + c : \mathcal{H} \to \mathcal{H}$ be an affine operator. Then, α is equivariant with respect to translations if and only if for all $M \in \mathbb{Z}^d$,*

$$w_M = T^M(w_0) \text{ and } c \propto \mathbf{1},$$

where $\mathbf{0}$ is the zero vector in $\mathbb{Z}^d$ and $c \propto \mathbf{1}$ means that c is a constant tensor, that is, all of its entries are the same.

Proof. Here we denote the index by $I = (i_1, i_2, \cdots, i_d)$. On the one hand,

$$
\begin{aligned}
T^M(\alpha(x))[I] \\
= T^M(w(x))[I] + T^M(c)[I] \\
= w(x)[I - M] + c[I - M] \\
= w_{I-M}(x) + c[I - M].
\end{aligned}
$$

On the other hand,

$$
\begin{aligned}
\alpha(T^M(x))[I] \\
= w(T^M(x))[I] + c[I] \\
= w_I(T^M(x)) + c[I] \\
= T^{-M}(w_I)(x) + c[I],
\end{aligned}
$$

in which the last equation is from Lemma 3.

Sufficiency. Assume for all $M \in \mathbb{Z}^d$,

$$
w_M = T^M(w_{\mathbf{0}}) \text{ and } c \propto \mathbf{1}.
$$

We have

$$
\begin{aligned}
T^{-M}(w_I) = T^{-M}(T^I(w_{\mathbf{0}})) \\
= T^{I-M}(w_{\mathbf{0}}) \\
= w_{I-M}, \\
c[I - M] = c[\mathbf{0}] \\
= c[I], \\
T^M(\alpha(x))[I] = \alpha(T^M(x))[I].
\end{aligned}
$$

Thus,

$$
T^M(\alpha(x)) = \alpha(T^M(x)).
$$

Necessity. Assume α is equivariant with respect to translations in the sense that

$$
T^M(\alpha(x)) = \alpha(T^M(x)).
$$

We have

$$
w_{I-M}(x) - T^{-M}(w_I)(x) = c[I] - c[I - M].
$$

Fix indices $I = \mathbf{0}$ and obtain that for all $M \in \mathbb{Z}^d$,

$$
w_M(x) - T^M(w_{\mathbf{0}})(x) = c(\mathbf{0}) - c(M).
$$

Recall that a continuous linear operator can be viewed as a tensor where the operation is the inner product in tensor space. We have

$$c(\mathbf{0}) - c(M) = (w_M - T^M(w_\mathbf{0})) \cdot x = \overrightarrow{w_M - T^M(w_\mathbf{0})} \cdot \overrightarrow{x}.$$

For each fixed M, the left side is a constant scalar and the right side is a linear transformation on all vector $\overrightarrow{x} \in \overrightarrow{H}$. Thus, both sides are equal to zero tensor and we have

$$c(\mathbf{0}) = c(M),$$
$$\overrightarrow{w_M} = \overrightarrow{T^M(w_\mathbf{0})}.$$

That is, for all $M \in \mathbb{Z}^d$,

$$c \propto \mathbf{1},$$
$$w_M = T^M(w_\mathbf{0}).$$

B   Proof of Lemma 1

We first prove a lemma.

Lemma 4. *Let $\eta : [2^{Q+1}] \to \mathbb{B}$ be the binary decomposition. There exists a $(2Q + 2)$-layer network $f : \mathbb{R} \to \mathbb{R}^Q$ with ReLU activations and width at most $Q + 1$ such that for $x \in [2^{Q+1}]$*

$$f(x) = \eta(x).$$

Proof. We decompose $x \in [2^{Q+1}]$ as $x = x_0 + 2x_1 + \cdots + 2^Q x_Q$. Then for $q = 0, \cdots, Q$, we have

$$x_q = \sigma(1 - \sigma(2^q + 2^{q+1}x_{q+1} + \cdots 2^Q x_Q - x)).$$

Thus, for $q = 0, \cdots, Q - 1$, we construct

$$f_{2q+1}\left(\begin{pmatrix} x \\ x_Q \\ \vdots \\ x_{Q-q+1} \end{pmatrix}\right) = \sigma\left(\begin{pmatrix} 0 \\ \vdots \\ 0 \\ 2^{Q-q} \end{pmatrix} + \begin{pmatrix} 1 & \cdots & 0 & 0 \\ \vdots & \ddots & \vdots & \vdots \\ 0 & \cdots & 1 & 0 \\ -1 & 2^Q & \cdots & 2^{Q-q+1} \end{pmatrix} \begin{pmatrix} x \\ x_Q \\ \vdots \\ x_{Q-q+1} \end{pmatrix}\right) \in \mathbb{R}^{q+2},$$

$$f_{2q+2}\left(\begin{pmatrix} x \\ x_Q \\ \vdots \\ x_{Q-q} \end{pmatrix}\right) = \sigma\left(\begin{pmatrix} 0 \\ \vdots \\ 0 \\ 1 \end{pmatrix} + \begin{pmatrix} 1 & \cdots & 0 & 0 \\ \vdots & \ddots & \vdots & \vdots \\ 0 & \cdots & 1 & 0 \\ 0 & \cdots & 0 & -1 \end{pmatrix} \begin{pmatrix} x \\ x_Q \\ \vdots \\ x_{Q-q} \end{pmatrix}\right) \in \mathbb{R}^{q+2}.$$

The last 2 layers

$$f_{2Q+1}\left(\begin{pmatrix} x \\ x_Q \\ \vdots \\ x_1 \end{pmatrix}\right) = \sigma\left(\begin{pmatrix} 1 \\ 0 \\ \vdots \\ 0 \end{pmatrix} + \begin{pmatrix} -1 & 2^Q & \cdots & 2 \\ 0 & 1 & \cdots & 0 \\ \vdots & \vdots & \ddots & \vdots \\ 0 & 0 & \cdots & 1 \end{pmatrix} \begin{pmatrix} x \\ x_Q \\ \vdots \\ x_1 \end{pmatrix}\right) \mathbb{R}^{Q+1},$$

$$f_{2Q+2}\left(\begin{pmatrix} x_0 \\ x_Q \\ \vdots \\ x_1 \end{pmatrix}\right) = \sigma\left(\begin{pmatrix} 1 \\ 0 \\ \vdots \\ 0 \end{pmatrix} + \begin{pmatrix} -1 & 0 & \cdots & 0 \\ 0 & 1 & \cdots & 0 \\ \vdots & \vdots & \ddots & \vdots \\ 0 & 0 & \cdots & 1 \end{pmatrix} \begin{pmatrix} x_0 \\ x_Q \\ \vdots \\ x_1 \end{pmatrix}\right) \in \mathbb{R}^{Q+1}.$$

Let $f = f_{2Q+2} \circ \cdots \circ f_1$. For $x \in [2^{Q+1}]$ and $x = x_0 + 2x_1 + \cdots + 2^Q x_Q$ we have

$$f(x) = \begin{pmatrix} x_0 \\ x_Q \\ \vdots \\ x_1 \end{pmatrix}.$$

Lemma 1. *Let $\eta : [2^{Q+1}] \to \mathbb{B}$ be the binary decomposition. There exists a $(2Q + 2)$-layer network F in the form of Eq. (2) with ReLU activations and width at most $(Q + 1)N$ such that for $X \in [2^{Q+1}]^{P \times N}$*

$$F(X) = \eta(X).$$

Proof. From Lemma 4, there exists a network f such that for $x \in [2^{Q+1}]$, $f(x) = \eta(x)$. We denote the l-the layer of f by f_l for $l = 1, \cdots, 2Q+2$. Without loss of generality, we assume for $z \in \mathbb{R}^{K_{l-1}}$

$$f_l(z) = \sigma(w_l \cdot z + b_l),$$

where σ is ReLU activation and $w_l \in \mathbb{R}^{K_l \times K_{l-1}}, b_l \in \mathbb{R}^{K_l}$.

Now we construct a $(2Q + 2)$-layer network F in the form of Eq. (2). For $l = 1, \cdots, 2Q + 2$, let $n_l = K_l \times P$. We construct F_l in the form of Eq. (3) as

$$W^{k_l \times p, k_{l-1} \times r}[l] = \begin{cases} \mathrm{diag}(w_l[k_l, k_{l-1}]) & \text{if } p = r \\ 0 & \text{otherwise} \end{cases},$$

$$C^{k_l \times p, k_{l-1} \times r}[l] = \begin{cases} \dfrac{b_l[k_l]}{K_{l-1}} & \text{if } p = r \\ 0 & \text{otherwise} \end{cases},$$

for $k_l = 1, \cdots, K_l$, $k_{l-1} = 1, \cdots, K_{l-1}$ and $p, r = 1, \cdots, P$.

We can verify that for $X \in \mathbb{R}^{K_{l-1} \times P \times N}$, $k_l = 1, \cdots, K_l$, $p = 1, \cdots, P$ and $i = 0, \cdots, N-1$

$$
\begin{aligned}
F_l(X)[k_l, p, i] &= \sigma(W^{k_l \times p}[l] \cdot X + C^{k_l \times p}[l] \cdot \mathbf{1})[i] \\
&= \sigma(\sum_{k_{l-1}=1}^{K_{l-1}} W^{k_l \times p, k_{l-1} \times p} \cdot X[k_{l-1}, p, :] + C^{k_l \times p, k_{l-1} \times p})[i] \\
&= \sigma(\sum_{k_{l-1}=1}^{K_{l-1}} \operatorname{diag}(w_l[k_l, k_{l-1}]) \cdot X[k_{l-1}, p, :] + \frac{b_l[k_l]}{K_{l-1}})[i] \\
&= \sigma(b_l[k_l] + \sum_{k_{l-1}=1}^{K_{l-1}} w_l[k_l, k_{l-1}] \cdot X[k_{l-1}, p, i]) \\
&= \sigma(w_l[k_l, :] \cdot X[:, p, i] + b_l[k_l]) \\
&= f_l(X[:, p, i])[k_l].
\end{aligned}
$$

That is,

$$
F_l(X)[:, p, i] = f_l(X[:, p, i]).
$$

Thus, for $X \in \mathbb{R}^{P \times N}$ and $p = 1 \cdots P$, $i = 0, \cdots, N-1$

$$
\begin{aligned}
F(X)[:, p, i] &= f_{2Q+2}(F_{2Q+1} \circ \cdots \circ F_1(X)[:, p, i]) \\
&\vdots \\
&= f_{2Q+2} \circ \cdots \circ f_1(X[p, i]) \\
&= f(X[p, i]).
\end{aligned}
$$

For $Z \in [2^{Q+1}]^{P \times N}$,

$$
F(Z) = \eta(Z).
$$

C   Proof of Lemma 2

Assume a network $F = F_2 \circ F_1$ in the form of Eq. (2) with ReLU activations and $n_0 = G$, $n_1 = S$, $n_2 = 1$ satisfies that for $X \in G \times N$

$$
F(X) = \frac{1}{S} \sum_{s=1}^{S} \sigma(W^s[1] \cdot X + C^s[1] \cdot \mathbf{1}). \tag{4}
$$

Here, the weights and biases in F_2 degenerate as

$$
\begin{aligned}
W[2] &= W^1[2] = (\frac{1}{S}I, \cdots, \frac{1}{S}I), \\
C[2] &= 0.
\end{aligned}
$$

For convenience, in the rest of this section we simplify $W^s[1], C^s[1]$ to W^s, C^s. The following result is well known.

Lemma 5. *Let* $\mathcal{B} = \{Z_s | s = 1, 2, \cdots, S\} \subset \mathbb{B}^{G \times N}$ *be a G-channel aperiodic binary dataset. Let* $\|Z\|$ *be the* L_2*-norm of* Z.

1) $T^0(Z_s) \cdot Z_s = \|Z_s\|^2$.
2) $T^M(Z_s) \cdot Z_t \leq \|Z_s\|^2 \leq (GN)^2$ *for any* $M \in \mathbb{Z}^d$.
3) *For any* $M \in \mathbb{Z}^d$ *that* $M \bmod (n_1, n_2, \cdots, n_d) \neq \mathbf{0}$, $T^M(Z_s) \cdot Z_s \leq \|Z_s\|^2 - 1$.
4) *If* $\|Z_s\| = \|Z_t\|$, $T^M(Z_s) \cdot Z_t \leq \|Z_t\|^2 - 1$ *for any* $M \in \mathbb{Z}^d$.
5) *If* $\|Z_s\| > \|Z_t\|$, $\|Z_s\| \geq \sqrt{\|Z_t\|^2 + 1} \geq \|Z_t\| + \frac{1}{2GN}$.

The i-th component of $F(Z)$ in Eq. (4) is

$$F(Z)[i] = \frac{1}{S} \sum_{s=1}^{S} \sigma(W_i^k \cdot Z + C^k \cdot \mathbf{1}),$$

where $W_i^k = (W_i^{k,1}, \cdots, W_i^{k,G}) \in \mathbb{R}^{G \times N}$ and $W_i^{k,r}$ is the i-th row of $W^{k,r}$. Recall that each circular filter $W^{k,r} \in \mathbb{R}^{N \times N}$ in Eq. (4) is determined by its first row $W_0^{k,r} \in \mathbb{R}^N$ and $W_{\delta(M)}^{k,r} = T^M(W_0^{k,r})$. And the biases $C^{k,r}$ are actually scalars.

Lemma 6. *Let* $\mathcal{B} = \{Z_s | s = 1, 2, \cdots, S\} \subset \mathbb{B}^{G \times N}$ *be a G-channel aperiodic binary dataset. Endow* $\mathcal{B}$ *with an order that*

$$s \geq t \iff \|Z_s\| \geq \|Z_t\|.$$

Construct the filters and biases in Eq. (4) as

$$W_0^{s,r} = \frac{Z_s^r}{\|Z_s\|},$$

$$C^{s,r} = \frac{1}{G(2GN + 1)} - \frac{\|Z_{s-1}\|}{G},$$

for $s = 1, \cdots S, r = 1 \cdots, G$ *and set* $\|Z_0\| = \frac{1}{2GN+1}$.
    Then,

a) *if* $t < s$, *then* $\sigma(W_i^s \cdot Z_t + C^s \cdot \mathbf{1}) = 0, i = 0, 1, \cdots, GN - 1$;
b) *if* $t = s$, *then* $\sigma(W_0^s \cdot Z_s + C^s \cdot \mathbf{1}) - \sigma(W_i^s \cdot Z_s + C^s) > \frac{1}{2GN+1}, i = 1, 2, \cdots, GN - 1$;
c) *if* $t > s$, *then* $\sigma(W_i^s \cdot Z_t + C^s \cdot \mathbf{1}) < GN$.

Proof. This˙ proof uses Lemma 5.

a) Assuming $t < s$, we have

$$W_{\delta(M)}^s \cdot Z_t + C^s \cdot \mathbf{1} = T^M(Z_s) \cdot Z_t / \|Z_s\| + \frac{1}{2GN + 1} - \|Z_{s-1}\|,$$

$$\leq T^M(Z_s) \cdot Z_t / \|Z_s\| + \frac{1}{2GN + 1} - \|Z_t\|.$$

If $\|Z_s\| = \|Z_t\|$,

$$T^M(Z_s) \cdot Z_t / \|Z_s\| \le \|Z_t\| - \frac{1}{\|Z_t\|},$$

$$W^s_{\delta(M)} \cdot Z_t + C^s \cdot \mathbf{1} \le \frac{1}{2GN+1} - \frac{1}{\|Z_t\|} < 0.$$

If $\|Z_s\| > \|Z_t\|$,

$$T^M(Z_s) \cdot Z_t / \|Z_s\| \le \frac{\|Z_t\|^2}{\|Z_t\| + \frac{1}{2GN}},$$

$$W^s_{\delta(M)} \cdot Z_t + C^s \cdot \mathbf{1} \le \frac{\|Z_t\|^2}{\|Z_t\| + \frac{1}{2GN}} + \frac{1}{2GN+1} - \|Z_t\|$$

$$= \frac{1}{2GN+1} - \frac{1}{2GN + 1/\|Z_t\|}$$

$$< 0.$$

Thus, for all $M \in \mathbb{Z}^d$,

$$\sigma(W^s_{\delta(M)} \cdot Z_t + C^s \cdot \mathbf{1}) = 0.$$

b) We have

$$W^s_0 \cdot Z_s + C^s \cdot \mathbf{1} = \|Z_s\| - \|Z_{s-1}\| + \frac{1}{2GN+1},$$

$$W^s_{\delta(M)} \cdot Z_s + C^s \cdot \mathbf{1} = T^M(Z_s) \cdot Z_s / \|Z_s\| + \frac{1}{2GN+1} - \|Z_{s-1}\|$$

$$\le \frac{\|Z_s\|^2 - 1}{\|Z_s\|} + \frac{1}{2GN+1} - \|Z_{s-1}\|$$

$$= \|Z_s\| - \|Z_{s-1}\| + \frac{1}{2GN+1} - \frac{1}{\|Z_s\|}.$$

Since

$$\|Z_s\| - \|Z_{s-1}\| \ge 0 \text{ and } \frac{1}{2GN+1} - \frac{1}{\|Z_s\|} < 0,$$

we have

$$\sigma(\|Z_s\| - \|Z_{s-1}\| + \frac{1}{2GN+1}) - \sigma(\|Z_s\| - \|Z_{s-1}\| + \frac{1}{2GN+1} - \frac{1}{\|Z_s\|}) \ge \frac{1}{2GN+1}.$$

c) Assuming $t > s$, we have

$$W_{\delta(M)}^s \cdot Z_t + C^s \cdot \mathbf{1} = T^M(Z_s) \cdot Z_t / \|Z_s\| + \frac{1}{2GN+1} - \|Z_{s-1}\|$$

$$\leq \|Z_s\| - \frac{1}{\|Z_s\|} + \frac{1}{2GN+1} - \|Z_{s-1}\|$$

$$< \|Z_s\|$$

$$\leq GN,$$

$$\sigma(W_{\delta(M)}^s \cdot Z_t + C^s \cdot \mathbf{1}) < GN.$$

Lemma 2. Let $\mathcal{B} = \{Z_s | s = 1, 2, \cdots, S\} \subset \mathbb{B}^{G \times N}$ be an aperiodic binary dataset. Then there exists a 2-layer network F in the form of Eq. (2) with ReLU activations and width at most SN such that for all $s = 1, 2, \cdots, S$,

$$F(Z_s)[0] = \max_{i=0}^{N-1} F(Z_s)[i].$$

Proof. Without loss of generality, we assign an order to the dataset that

$$s \geq t \iff \|Z_s\| \geq \|Z_t\|.$$

We set $\alpha \geq 1 + GN + 2G^2N^2$ and construct F as Eq. (4) such that

$$W_0^{s,r} = \frac{\alpha^{s-1} Z_s^r}{\|Z_s\|},$$

$$C^{s,r} = \frac{\alpha^{s-1}}{G(2GN+1)} - \frac{\alpha^{s-1}\|Z_{s-1}\|}{G},$$

for $s = 1, \cdots S, r = 1 \cdots, G$ and set $\|Z_0\| = \frac{1}{2GN+1}$.

From Lemma 6, we have for $i = 1, 2, \cdots, GN - 1$,

$$S(F(Z_t)[0] - F(Z_t)[i])$$

$$= \sum_{s=1}^{S} \alpha^{s-1}[\sigma(W_0^s \cdot Z_t + C^s) - \sigma(W_i^s \cdot Z_t + C^s)]$$

$$= \sum_{s=1}^{t} \alpha^{s-1}[\sigma(W_0^s \cdot Z_t + C^s) - \sigma(W_i^s \cdot Z_t + C^s)]$$

$$\geq \frac{\alpha^{t-1}}{2GN+1} + \sum_{s=1}^{t-1} \alpha^{s-1}[\sigma(W_0^s \cdot Z_t + C^s) - \sigma(W_i^s \cdot Z_t + C^s)]$$

$$> \frac{\alpha^{t-1}}{2GN+1} - GN \sum_{s=1}^{t-1} \alpha^{s-1}$$

$$= \frac{\alpha^{t-1}}{2GN+1} - \frac{GN(1 - \alpha^{t-1})}{1 - \alpha}$$

$$= \frac{(\alpha - 2G^2N^2 - GN - 1)\alpha^{t-1} + 2G^2N^2 + GN}{(2GN+1)(\alpha - 1)}$$

$$> 0.$$

D   Experimental Settings

For CIFAR-10, we constantly pad 4 pixels with values 0 around images. For MNIST, we resize images to 32×32. For 3D-MNIST, we voxelize this dataset and constantly pad 8 pixels with value 0 around images.

We leverage restorers with 6 layers. In each layer, we use a sparse circular filter, for example, its kernel size is 9. Each layer outputs only one channel and has no bias parameter.

References

1. Azulay, A., Weiss, Y.: Why do deep convolutional networks generalize so poorly to small image transformations? J. Mach. Learn. Res. **20**, 1–25 (2019)
2. Cohen, T., Welling, M.: Group equivariant convolutional networks. In: International Conference on Machine Learning, pp. 2990–2999. PMLR (2016)
3. Engstrom, L., Tran, B., Tsipras, D., Schmidt, L., Madry, A.: A rotation and a translation suffice: fooling CNNs with simple transformations. arXiv preprint (2018)
4. Esteves, C., Allen-Blanchette, C., Zhou, X., Daniilidis, K.: Polar transformer networks. arXiv preprint arXiv:1709.01889 (2017)
5. Fukushima, K.: Neocognitron: a self-organizing neural network model for a mechanism of pattern recognition unaffected by shift in position. Biol. Cybern. **36**, 193–202 (1980)
6. Gens, R., Domingos, P.M.: Deep symmetry networks. Adv. Neural Inf. Process. Syst. **27** (2014)
7. Gholamalinezhad, H., Khosravi, H.: Pooling methods in deep neural networks, a review. arXiv preprint arXiv:2009.07485 (2020)
8. Ghosh, R., Gupta, A.K.: Scale steerable filters for locally scale-invariant convolutional neural networks. arXiv preprint arXiv:1906.03861 (2019)
9. He, K., Zhang, X., Ren, S., Sun, J.: Deep residual learning for image recognition. In: Proceedings of the IEEE Conference on Computer Vision and Pattern Recognition, pp. 770–778 (2016)
10. Henriques, J.F., Vedaldi, A.: Warped convolutions: efficient invariance to spatial transformations. In: International Conference on Machine Learning, pp. 1461–1469. PMLR (2017)
11. Jaderberg, M., Simonyan, K., Zisserman, A., et al.: Spatial transformer networks. Adv. Neural Inf. Process. Syst. **28** (2015)
12. LeCun, Y., Bottou, L., Bengio, Y., Haffner, P.: Gradient-based learning applied to document recognition. Proc. IEEE **86**(11), 2278–2324 (1998)
13. Lenc, K., Vedaldi, A.: Understanding image representations by measuring their equivariance and equivalence. In: Proceedings of the IEEE Conference on Computer Vision and Pattern Recognition, pp. 991–999 (2015)
14. Lenc, K., Vedaldi, A.: Learning covariant feature detectors. In: Hua, G., Jégou, H. (eds.) ECCV 2016. LNCS, vol. 9915, pp. 100–117. Springer, Cham (2016). https://doi.org/10.1007/978-3-319-49409-8_11
15. Marcos, D., Kellenberger, B., Lobry, S., Tuia, D.: Scale equivariance in CNNs with vector fields. arXiv preprint arXiv:1807.11783 (2018)
16. Shen, X., Tian, X., He, A., Sun, S., Tao, D.: Transform-invariant convolutional neural networks for image classification and search. In: Proceedings of the 24th ACM International Conference on Multimedia, pp. 1345–1354 (2016)

17. Shorten, C., Khoshgoftaar, T.M.: A survey on image data augmentation for deep learning. J. Big Data **6**(1), 1–48 (2019)
18. Sohn, K., Lee, H.: Learning invariant representations with local transformations. arXiv preprint arXiv:1206.6418 (2012)
19. Srivastava, N.: Improving neural networks with dropout. Univ. Toronto **182**(566), 7 (2013)
20. Srivastava, N., Hinton, G., Krizhevsky, A., Sutskever, I., Salakhutdinov, R.: Dropout: a simple way to prevent neural networks from overfitting. J. Mach. Learn. Res. **15**(1), 1929–1958 (2014)
21. Xu, Y., Xiao, T., Zhang, J., Yang, K., Zhang, Z.: Scale-invariant convolutional neural networks. arXiv preprint arXiv:1411.6369 (2014)
22. Zhang, X., Yu, F.X., Karaman, S., Chang, S.F.: Learning discriminative and transformation covariant local feature detectors. In: Proceedings of the IEEE Conference on Computer Vision and Pattern Recognition, pp. 6818–6826 (2017)

SDPSAT: Syntactic Dependency Parsing Structure-Guided Semi-Autoregressive Machine Translation

Xinran Chen, Yuran Zhao, Jianming Guo, Sufeng Duan, and Gongshen Liu[✉]

School of Electronic Information and Electrical Engineering,
Shanghai Jiao Tong University, Shanghai, China
{jasminechen123,zyr527,jsguojianming,1140339019dsf,lgshen}@sjtu.edu.cn

Abstract. The advent of non-autoregressive machine translation (NAT) accelerates the decoding superior to autoregressive machine translation (AT) significantly, while bringing about a performance decrease. Semi-autoregressive neural machine translation (SAT), as a compromise, enjoys the merits of both autoregressive and non-autoregressive decoding. However, current SAT methods face the challenges of information-limited initialization and rigorous termination. This paper develops a layer-and-length-based syntactic labeling method and introduces a syntactic dependency parsing structure-guided two-stage semi-autoregressive translation (SDPSAT) structure, which addresses the above challenges with a syntax-based initialization and termination. Additionally, we also present a Mixed Training strategy to shrink exposure bias. Experiments on six widely-used datasets reveal that our SDPSAT surpasses traditional SAT models with reduced word repetition and achieves competitive results with the AT baseline at a $2\times \sim 3\times$ speedup.

Keywords: Non-autoregressive · Machine translation · Syntactic dependency parsing

1 Introduction

While autoregressive neural machine translation (AT) maintains cutting-edge performance, its applications in large-scale and real-time scenarios are severely restricted by the slow inference speed [3,4]. In contrast, non-autoregressive (NAT) models, based on the independence hypothesis, significantly increase inference speed through parallel decoding but experience reduced performance [4,5,10]. As a compromise between AT and NAT, semi-autoregressive (SAT) models [15,21] utilize both autoregressive and non-autoregressive properties in their decoding, in which SAT models not only capture target-side dependencies more effectively than NAT models, but also enhance translation efficiency beyond AT model [15,21,22].

The primitive equal-length segmented SAT models in [15], which decode sentence segments non-autoregressively and generate words within those segments

B. Luo et al. (Eds.): ICONIP 2023, CCIS 1962, pp. 604–616, 2024.
https://doi.org/10.1007/978-981-99-8132-8_45

autoregressively, adopt the independence assumption and encounter the multimodality problem—the multi-modal distribution of target translations is difficult to capture [4]. To handle the multi-modality errors in primitive SAT of equal-length segmentation, [15] further proposes to expose the model to some unequal-length segmented samples in training and remove duplicate segments in inference. However, its mixed segmentation criterion is not clear enough, and its additional operation to remove repetitive segments is not straightforward. Section 5 shows detailed translation examples.

Different from [15], we attribute the multi-modality problem of SAT to two limitations: (i) **information-limited initialization:** during decoding, the SAT decoder which is initialized by a sequence of [BOS], lacks instructions on the subsequent prediction. (ii) **rigorous termination:** for SAT, dividing sentences into equal-length segments is simplest and most time-efficient [15], while it would lead to SAT learning the rigorous equal-length termination pattern. The multimodality problem leads to repeated or absent tokens. For example, in translation *" [BOS] There are lots of [BOS] of flowers outside. "*, the first segment decodes the final word *"of"* to keep the same length as the second segment, which starts decoding from *"of"*, and results in repetition. These same factors also contribute to the token missing, like *" [BOS] There are lots [BOS] flowers outside. "*, where *"of"* is omitted. Therefore, making initialization more informative and termination more reasonable in SAT is worth exploring.

In this work, we present a **S**yntactic **D**ependency **P**arsing structure-guided **S**emi-**A**utoregressive **T**ranslation model (**SDPSAT**) to overcome the two limitations above. In a syntactic dependency parsing tree, each branch corresponds to one sentence segment. As decoding branches from the root, corresponding sentence segments are decoded parallelly, which fits well with the global nonautoregressive and local autoregressive decoding scheme in SAT. Inspired by the decoding of syntactic dependency parsing tree and SAT, we design a layer-and-length-based tree transformation and traversal techniques to generate the syntactic structure labels. Specifically, the syntactic labels in the syntactic tree serve for two primary functions. Firstly, the syntactic structure labels are used as prediction guides, which offer content guidance for improved initialization during subsequent inferences. Secondly, syntactic structure labels also determine termination criteria in which sentences are divided into semantically consecutive segments throughout the training process. Additionally, we provide a Mixed Training technique to shrink the *exposure bias* between training and inference.

Our SDPSAT delivers the best translation quality within the SAT group and achieves a comparable translation quality with AT baselines on six benchmarks with a $2\times \sim 3\times$ speedup. Furthermore, SDPSAT achieves a low word repetition rate of about 0.30%–0.60%. The following is a summary of our contribution:

(i) We design a layer-and-length-based syntactic labeling method to integrate syntactic dependency parsing structure into SAT, which allows for a more flexible termination and a better initialization for the translation decoder.

(ii) We employ a Mixed Training strategy to mitigate the discrepancy between training and inference, ultimately enhancing translation performance.

(iii) According to experimental results on six widely-used datasets, SDPSAT not only expedites decoding but also improves translation quality with reduced repetition compared to NAT competitors.

2   Related Works

NAT accelerates machine translation inference by adopting the independence hypothesis. However, NAT faces a serious multi-modality problem [3,4], and numerous approaches have been presented to deal with it. Based on the decoding strategy, existing NAT works mainly fall into three categories: iterative NAT, fully NAT, and SAT [22]. Iterative NAT refines the translation in multiple steps at the cost of inference speed [3,5,6]. Fully NAT maintains a speed advantage by decoding in a single round and handles the multi-modality problem with the latent variables [1,10,16,27], new training objectives [8,23], and novel model architecture [26]. SAT [15,21] combines the properties of fully NAT and AT in decoding. [21] first proposes Semi-NAT with a globally AT but locally NAT decoding style, while [15] designs Semi-NAT with globally NAT but locally AT decoding methods and recovers from the repetition errors. The decoding pattern of our model aligns with that of [15].

As a fundamental technology for natural language processing (NLP), syntactic parsing analyzes the syntactic dependency between sentence components [2,7,25]. Previous researchers have introduced syntactic information to enhance NAT. [9] integrated syntactic tags at the word embedding level for NAT. [23] designed a novel training objective to eliminate the issue of syntactic multi-modality. [1] enhanced NAT by constituency parsing information. However, integrating syntactic dependency parsing into SAT is still uncharted territory. Different from the prior studies, we first present an innovative two-stage syntactic dependency parsing-based SAT framework. Our model is closely related to [1], but we differ in both aspects of modeling syntactic structure and decoding scheme: (i) they introduce constituency labels with chunk size, while we focus on dependency parsing and design a layer-and-length-based traversal to obtain syntactic labels with depth, which shrinks syntactic vocabulary and is more conducive to model learning; (ii) they adopt a single-round Mask-Predict decoding with length predetermined constituency, while we insist on Semi-NAT decoding to better comprehend the target-side dependency within segments and own a more flexible termination, which ultimately resolves the multi-modality problem.

3   Methodology

This section introduces SDPSAT in detail, including the layer-and-length-based method to get dependency labels from the syntactic dependency tree, the two-stage decoding of SDPSAT, and the Mixed Training strategy.

3.1   Layer-and-Length-Based Syntactic Labeling

We design the layer-and-length-based method to get the syntactically annotated sentence before training, and the syntactic labels are used to supervise the training of the parse decoder. Figure 1 gives an illustration of the whole process.

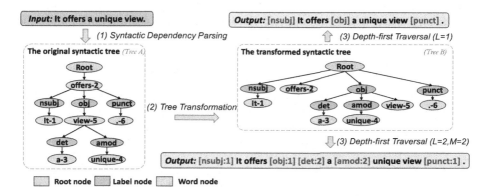

Fig. 1. Performing layer-and-length-based syntactic labeling for the input sentence. The layer and length denote the deepest dividing layer L and the maximum segment size M of the dependency parsing tree, respectively.

Dependency Parsing. Before tree transformation and traversal, we conduct syntactic dependency parsing. Each sentence is parsed into several tuples, which can be expressed formally as *"relation (governor, dependent)"*, where the **governor** represents the header, the **dependent** is the modifier of the header and the **relation** represents the syntactic dependency between the header and the dependent. Inspired by [7], we treat syntactic labels (relation) and words (governor, dependent) equally as tree nodes and represent each word node with token and positional index (*i.e.*, **It-1** in Fig. 1). The syntactic parsing tree is depicted in *Tree A* of Fig. 1.

We use *stanza* toolkits [13] to parse the target sentences, which provide leading-edge syntactic dependency parsing pre-trained models for 66 languages with highly accurate performance.

Tree Transformation and Traversal. The word and label nodes alternate in the original syntactic tree branches, and make the tree structure complicated. To match our proposed two-stage SAT decoding, we simplify the tree based on the deepest dividing layer L and the maximum segment size M. The label nodes deeper than L will be ignored in traversal for simplification. If the leaves of a $L - 1$ (note that $L > 1$) layer label node are equal to or more than M, it would be further divided according to its subsequent L layer label nodes. Restricted by the longest sentence segment, the SAT decoding could be accelerated with a larger L and an appropriate M.

Then we detail the tree transformation and traversal process. First, we move the word nodes to leaf nodes and keep the sentence sequential information for the subsequent traversal following [7]. In particular, if the positional index of the label' child is greater than its parent, the label node is adjusted to the right side of its parent, and vice versa. Second, based on L and M, we conduct a depth-first traversal on the transformed tree to get the annotated sequence. Specifically, when $L > 1$, we append the layer number of label nodes after the label to distinguish the layer (*e.g.*, [nsubj:1], [amod:2]). *Tree B* in Fig. 1 shows the process.

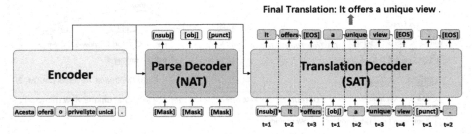

Fig. 2. An encoder and two decoders constitute SDPSAT. The translation decoder translates semi-autoregressively based on the syntactic structure given by the parse decoder. The parse decoder can be either AT or NAT. We display the NAT one here.

3.2   Two-Stage Decoding

As depicted in Fig. 2, the first SDPSAT decoding stage is to use the parse decoder to predict the syntactic structure of the target, and the second stage is to utilize the translation decoder to take the previous syntactic structure as the initialization and generate segments simultaneously.

Stage 1: Syntactic Labels Prediction. In this stage, we predict the syntactic labels of the transformed syntactic dependency parsing tree, which serve as syntactic structure hints for the next decoding stage. We design two feasible decoding schemes for this stage:

(i) **Fully NAT** We adopt Mask-Predict [3] to predict the label nodes of the transformed tree. Following the setting of [3], we employ a length predictor in the encoder to determine the length of the sentence before prediction. With the source sentence X, the probability distribution of the corresponding syntactic label Z is:

$$P(Z|X) = P(n|X) \prod_{t=1}^{n} P(z_t|X), \tag{1}$$

where n refers to the length of the syntactic label sequence. During training, we mask a specific percentage of tokens randomly and calculate the loss of the observed tokens, following [3]. As for inference, the parse decoder receives a sequence of [MASK] as input and generates the syntactic label sequence.

(ii) Fully AT A light fully-AT decoder, which only contains two layers, is also introduced to generate the syntactic label sequence, and its conditional probability could be expressed as:

$$P(Z|X) = \prod_{t=1}^{n} P(z_t|z_{<t}, X), \tag{2}$$

where $z_{<t}$ indicates the sequence generated prior to time-step t. We experimentally demonstrate that the AT decoder can better model the syntactic structure information, and its shallow decoder allows it to achieve decoding times close to those of NAT.

For simplicity, SDPSAT with an AT or a NAT parse decoder is abbreviated as SDPSAT (AT) or SDPSAT (NAT), respectively.

Stage 2: Words Prediction. With the guidance of syntactic structure labels generated by the parse decoder, the translation decoder predicts the target words semi-autoregressively, and the translation distribution could be written as:

$$P(Y|X) = \prod_{t=1}^{max\, n_i} P(S_t|S_{<t}, X, Z), \tag{3}$$

where S_t and $S_{<t}$ denote the predicted words in and before time-step t of all the segments, and Z presents the syntactic sequence. For a target sentence with segment lengths of $\{n_1, n_2, ..., n_k\}$, the total decoding step is $max\, n_i$.

3.3   Mixed Training Strategy

During training, the translation decoder is initialized with the ground truth syntactic labels, while in inference, it receives the generated syntactic labels of the parse decoder as initialization. This data distribution discrepancy, called *exposure bias* [17], may harm the translation performance.

Inspired by [24], we design a strategy called *Mixed Training* to shrink this discrepancy, which replaces the ground truth syntactic labels with predicted ones in the input of the translation decoder during training with an increasing probability. We define the probability p as follows:

$$p = \min \left\{ 0.5, \gamma \frac{t}{T} \right\}, \tag{4}$$

where t and T represent the current and maximum steps respectively, and γ indicates the rate controlling parameter. For better convergence, we conduct the strategy after training for 20 epochs.

For the parse decoder of AT, we adopt Force Decoding [24] to ensure the length of syntactic label prediction the same as the ground truth.

4   Experiments

4.1   Setup

Datasets Preprocess. The details of datasets preparation are as follows.

(i) **Datasets** We use the preprocessed datasets WMT14 En↔De (about 4.5M pairs), WMT16 En↔Ro (about 610k pairs) and WMT17 En↔Zh (about 20M pairs) from [3] for a fair comparison with prior studies. We use sequence-level knowledge distillation datasets for training. Specifically, the original target sentences are replaced with the translation of the standard AT Transformer to mitigate the multi-modality problem.

(ii) **Preprocess** The parallel sentences are preprocessed by Moses toolkits[1] and partitioned into subwords with Byte-Pair Encoding [18]. The target training set is processed by Stanza[2] to obtain the syntactic labels. We retrieve around 50 and 100 syntactic labels for each dataset under the conditions of $L = 1$ and $L = 2, M = 10$. The syntactic labels constitute the vocabulary of the parse decoder. The syntactic labels and words together make up the shared vocabulary of the encoder and the translation decoder.

Model Configurations. We use OpenNMT-py[3] as the framework. Our model is based on Transformer (base) [20] and we strictly follow its parameter settings. Specifically, we implement SDPSAT with a 6-layer encoder and a 6-layer translation decoder. As for the parse decoder, we adopt a 2-layer AT decoder and a 6-layer NAT decoder, respectively.

Key Baseline. AT Transformer and a series of NAT models are selected as baselines, including iterative NAT, fully NAT and Semi-NAT. Among Semi-NAT, there are two different decoding schemes, denoted as GALN and GNLA. "GNLA" is the abbreviation of globally non-autoregressive but locally autoregressive scheme, and "GALN" is the short for globally autoregressive but locally non-autoregressive scheme. GNLA-SAT divides the sentence into equal-length segments, and it is the common primitive model of RecoverSAT [15] and our SDPSAT. As the decoding paradigm of SDPSAT aligns with RecoverSAT and GNLA-SAT, we choose them as our key baselines and reimplement GNLA-SAT. GALN-SAT [21] decodes in the opposite way to ours.

4.2   Inference and Evaluation

We use the greedy search strategy for both first-stage (AT) and second-stage decoding. Following previous works, we evaluate the translation quality with BLEU [11] for all the datasets except WMT17 En→Zh, which is tested by Sacre-BLEU [12]. For the inference speed, we measure the averaged decoding latency with batch size set to 1 on the WMT14 En→De test set, using a NVIDIA GeForce RTX 3090 GPU.

---

[1] https://github.com/moses-smt/mosesdecoder.
[2] https://github.com/stanfordnlp/stanza.
[3] https://github.com/OpenNMT/OpenNMT-py.

4.3   Result

SDPSAT and a series of baselines are thoroughly compared in Table 1. The results lead to the following conclusions:

(i) SDPSAT attains state-of-the-art performance within Semi-NAT model group. SDPSAT outperforms the SAT baselines (GALN-SAT and GNLA-SAT) with better BLEU points and comparable speedup on all benchmark datasets. Also, SDPSAT achieves about 0.44/0.73 points improvement over its strong competitor RecoverSAT ($K = 2/5$) on average.

(ii) SDPSAT gains comparable performance with iterative NAT and fully NAT competitors. Compared with iterative NAT, SDPSAT (AT) achieves better results than CMLM and DisCo, while maintaining a greater speedup. As for fully NAT, except for REDER reaches a slightly higher BLEU on WMT16 Ro→En, SDPSAT achieves competitive performance with the remaining models.

Table 1. Performance comparison (BLEU scores and speedup rates) between SDPSAT and baselines. L denotes the deepest dividing layer, and M denotes the maximum segment size. K represents the segment number for SAT. G is the group size of GALN-SAT. NPD represents Noisy Parallel Decoding. n is the sample size of NPD. *iter* is an abbreviation for iteration. c is the maximum chunk size. * denotes the results under our implementation. The Mixed Training strategy is implemented for all of our SDPSAT models after training for 20 epochs for better convergence. "–" represents the data unreported.

Models		Speed	WMT14		WMT16	
			En→De	De→En	En→Ro	Ro→En
AT	Transformer	1.0×	27.30	–	–	–
	Transformer*	1.0×	27.48	31.88	33.69	33.98
Iterative NAT	CMLM (*iter* = 10) [3]	1.5×	27.03	30.53	33.08	33.31
	DisCo (*iter* = 10) [6]	1.5×	27.06	30.89	32.92	33.12
	LevT (*iter* = Adv.) [5]	4.0×	27.27	–	–	33.26
Fully NAT	NAT-FT+NPD (*n* = 100) [4]	2.4×	19.17	23.20	29.79	31.44
	SynSt (*c* = 6) [1]	4.86×	20.74	25.50	–	–
	FlowSeq+NPD (*n* = 30) [10]	1.1×	25.31	30.68	25.31	30.68
	CR-LaNMT [27]	11.0×	26.23	31.23	32.50	32.14
	ReorderNAT (AT) [16]	5.96×	26.49	31.13	31.70	31.99
	GLAT+NPD (*n* = 7) [14]	7.9×	26.55	31.02	32.87	33.51
	DCRF-NAT (rescoring 19) [19]	4.39×	26.80	30.04	–	–
	duplex REDER [26]	5.5×	27.36	31.10	33.60	**34.03**
Semi NAT	GALN-SAT (*G* = 2) [21]	2.07×	26.09	–	–	–
	GNLA-SAT (*K* = 5)*	4.49×	22.20	26.11	30.15	29.96
	GNLA-SAT (*K* = 2)*	1.97×	25.97	29.94	32.62	32.70
	RecoverSAT (*K* = 5) [15]	3.16×	26.91	31.22	32.81	32.80
	RecoverSAT (*K* = 2)	2.02×	27.11	31.67	32.92	33.19
	SDPSAT(NAT)($L = 2, M = 10$)	3.65×	25.30	30.51	32.84	32.64
	SDPSAT(NAT)($L = 1$)	2.94×	25.79	30.66	33.10	32.85
	SDPSAT(AT)($L = 2, M = 10$)	2.63×	26.88	31.12	33.39	33.32
	SDPSAT(AT)($L = 1$)	2.30×	**27.44**	**31.71**	**33.64**	33.85

In addition, SDPSAT enjoys a 2× ∼ 3× speed advantage over the AT baseline while maintains close translation quality.

(iii) The last group of Table 1 shows that SDPSAT (AT) gets a better translation quality than SDPSAT (NAT) with a slight speedup drop, indicating that SDPSAT (AT) better captures the syntactic information than SDPSAT (NAT). What' s more, the inference of models with a larger deepest dividing layer could be better expedited at the expense of slightly degraded performance.

We perform experiments on WMT17 En↔Zh datasets with a huge linguistic gap, to further verify the effectiveness of our SDP-SAT. Table 2 shows that SDPSAT achieves high BLEU scores in both directions, indicating that the syntactic decoder could provide syntactic structure hints for the target language on the basis of the source language despite the huge linguistic disparities.

Table 2. The performance comparison between SDPSAT and baselines on WMT17 En↔Zh.

Model	WMT17	
	Zh→En	En→Zh
AT Transformer*	24.30	35.44
CMLM($iter = 10$)	23.21	33.19
LevT($iter = $ Adv.)	23.30	33.90
DisCo($iter = 10$)	23.68	34.51
SDPSAT(AT)($L = 2, M = 10$)	23.78	33.60
SDPSAT(AT)($L = 1$)	24.00	34.53

5   Analysis and Discussion

The Effect of Mixed Training. To explore the effect of the rate controlling parameter γ in Mixed Training, we conducted experiments for SDPSAT(AT) ($L = 1$) on four validation sets. Table 3 shows that the parameter γ impacts the model's performance significantly. Mixed Training under appropriate parameter γ improves translation quality (*i.e.*, $\gamma = 10$), indicating that the strategy could alleviate *exposure bias*. However, too large γ (*i.e.*, $\gamma = 100$) may hurt the translation quality instead. With a larger γ, the model is more likely to be supervised incorrectly when it has not converged yet, which is detrimental to the training.

Table 3. The effect of γ on Mixed Training

γ	WMT14		WMT16	
	En→De	De→En	En→Ro	Ro→En
0	26.31	31.40	32.97	33.68
10	26.53	31.58	33.93	34.69
100	26.05	30.44	32.71	33.56

The Effect of Sentence Length.
We group the WMT14 En→De validation set according to source sentence lengths and calculate their
BLEU, respectively. Figure 3 shows
that SDPSAT (AT) outperforms
GNLA-SAT for most of the lengths.
Meanwhile, when the sentence length
is greater than 20 and less than 50,
the translation performance of SDP-
SAT (AT) is very close to the AT
model, which proves the effectiveness
of SDPSAT.

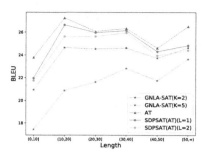

Fig. 3. Performance on various length groups.

The Effect of Repetition. Table 4
demonstrates that our GNLA-SAT
which adopts equal-length segmentation, suffers from highly severe
multi-modality problem, while SDP-
SAT reduces word repetition significantly. This finding indicates
that the syntactic structure could
make the meaning of each segment
clearer, leading to the improvement
of the translation quality. In addi-

Table 4. The performance (BLEU) and
repeated rates on two benchmark validation
sets.

Model	WMT14			
	En→De		De→En	
	BLEU	Reps	BLEU	Reps
SDPSAT (AT)	26.53	0.60%	31.58	0.30%
SDPSAT (NAT)	25.58	1.53%	30.54	1.43%
GNLA-SAT ($K = 2$)*	25.04	2.76%	29.93	2.97%
GNLA-SAT ($K = 5$)*	22.05	20.80%	26.53	19.07%
RecoverSAT ($K = 2$)*	26.17	1.43%	30.73	1.00%
RecoverSAT ($K = 5$)*	24.33	2.10%	28.65	1.77%

tion, SDPSAT (AT) reaches a lower repetition rate than the strong competitor
RecoverSAT.

Case Study. Table 5 shows a translation comparison between SDPSAT with the
SAT baselines. We find that (i) in GNLA-SAT, where the generated sentences are
consisted of equal-length segments, the multi-modality problem is severe and the
semantically consecutive words are often separated. In contrast, SDPSAT which
relies on the syntax structure as initialization and temination criterion, generates more semantically consistent segments and more fluent translation. (ii) in
RecoverSAT, due to implicit segmentation, the length of the generated segment
varies greatly, which harms the decoding efficiency. In RecoverSAT ($K = 2$), the
model generates an empty segment (*i.e.*, [BOS] [EOS]) and it even degrades to
AT model. What' s more, RecoverSAT requires additional operation of removing duplicate segments (*e.g.*, enough). In comparison, the decoding pattern of
SDPSAT is clearer and more straightforward.

Table 5. Translation comparison of SAT baselines and SDPSAT(AT)(L=1; $L = 2$, $M = 10$). [BOS] and syntactic label (e.g., [obl]) denotes the segment beginning. The [EOS] and [DEL] denotes the operation of keeping or deleting the segment for RecoverSAT. We use the double underscores (e.g., is) for repeated tokens, dashed underline (e.g., __) for missing words, and wave underline (e.g., targets) for semantic errors.

Source		Wer sich weniger als fünf Minuten ge@@ dul@@ det, wartet unter Umständen nicht lange genug, warnt Bec@@ ker und verweist auf einen Beschluss des Ober@@ land@@ es@@ geri@@ chts Ham@@ m
Reference		Those who tolerate less than five minutes may not wait long enough, warns Becker, referring to a decision of the Hamm Higher Regional Court
+Syntactic label		[nsubj:1] Those who tolerate less than five minutes [aux:1] may [advmod:1] not wait [advmod:1] long [advmod:1] enough [parataxis:1], warns Becker [conj:1] [cc:2] and refers [obl:2] to a decision of the Hamm Higher Regional Court [punct:1].
GNLA-SAT	$K = 2$	[BOS] Anyone who is tolerated less than five minutes may not wait long enough [BOS], warns Becker and refers to a decision by the Supreme __ Court __ Hamm
	$K = 5$	[BOS] Anyone who condonates less than five [BOS] five minutes may not wait for [BOS] wait for a long enough, [BOS] __ Becker and points to a Hamm [BOS] decision by the Supreme __ Cour
Recover-SAT	$K = 2$	[BOS] Anyone who tolerates less than five minutes may not wait long enough, warns Becker and refers to a decision of the Hamm Supreme __ Court. [EOS] [BOS] [EOS]
	$K = 5$	[BOS] Anyone who tolerates [EOS] [BOS] less than five minutes may not wait long [EOS] [BOS] enough, warns Becker and refers to a decision by the Supreme __ Court __ Hamm. [EOS] [BOS] enough [DEL] [BOS] [EOS]
Ours	$L = 1$	[nsubj] Anyone who tolerates less than five minutes [aux] may [advmod] not wait [advmod] long enough [conj], warns Becker [conj] and refers to a decision of the Higher Regional Court of Hamm [punct]
	$L = 2$	[nsubj:1] Those who tolerate less than five minutes [aux:1] may [advmod:1] not wait [advmod:1] long enough [parataxis:1], warns Becker [conj:1] [cc:2] and refer [obl:2] to a decision by the Hamm Supreme __ Court [punct:1]

6 Conclusion

We develop a layer-and-length-based syntactic labeling approach and present a novel two-stage SAT framework called SDPSAT, which enables a more flexible termination and a better initialization for SAT decoding. Besides, we present a Mixed Training strategy to diminish the exposure bias. Experimental results suggest that SDPSAT excels within the Semi-NAT group and achieves comparable translation performance with those strong NAT or AT competitors, while significantly alleviating the multi-modality problem.

Acknowledgements. This research work has been funded by Joint Funds of the National Natural Science Foundation of China (Grant No. U21B2020), and Shanghai Science and Technology Plan (Grant No. 22511104400).

References

1. Akoury, N., Krishna, K., Iyyer, M.: Syntactically supervised transformers for faster neural machine translation. In: Proceedings of the 57th Annual Meeting of the Association for Computational Linguistics, pp. 1269–1281 (2019)
2. Chen, K., Wang, R., Utiyama, M., Sumita, E., Zhao, T.: Syntax-directed attention for neural machine translation. In: Proceedings of the AAAI Conference on Artificial Intelligence, pp. 4792–4799 (2018)
3. Ghazvininejad, M., Levy, O., Liu, Y., Zettlemoyer, L.: Mask-predict: parallel decoding of conditional masked language models. In: Proceedings of the 2019 Conference on Empirical Methods in Natural Language Processing and the 9th International Joint Conference on Natural Language Processing (EMNLP-IJCNLP), pp. 6112–6121 (2019)
4. Gu, J., Bradbury, J., Xiong, C., Li, V.O.K., Socher, R.: Non-autoregressive neural machine translation. In: 6th International Conference on Learning Representations, ICLR 2018, Vancouver, BC, Canada, April 30 - May 3 2018, Conference Track Proceedings (2018)
5. Gu, J., Wang, C., Junbo, J.Z.: Levenshtein transformer. In: Proceedings of the 33rd International Conference on Neural Information Processing Systems, pp. 11181–11191 (2019)
6. Kasai, J., Cross, J., Ghazvininejad, M., Gu, J.: Non-autoregressive machine translation with disentangled context transformer. In: International Conference on Machine Learning, pp. 5144–5155 (2020)
7. Le, A.N., Martinez, A., Yoshimoto, A., Matsumoto, Y.: Improving sequence to sequence neural machine translation by utilizing syntactic dependency information. In: Proceedings of the Eighth International Joint Conference on Natural Language Processing (Volume 1: Long Papers), pp. 21–29 (2017)
8. Li, Y., Cui, L., Yin, Y., Zhang, Y.: Multi-granularity optimization for non-autoregressive translation. arXiv preprint arXiv:2210.11017 (2022)
9. Liu, Y., Wan, Y., Zhang, J., Zhao, W., Philip, S.Y.: Enriching non-autoregressive transformer with syntactic and semantic structures for neural machine translation. In: Proceedings of the 16th Conference of the European Chapter of the Association for Computational Linguistics: Main Volume, pp. 1235–1244 (2021)
10. Ma, X., Zhou, C., Li, X., Neubig, G., Hovy, E.: FlowSeq: non-autoregressive conditional sequence generation with generative flow. In: Proceedings of the 2019 Conference on Empirical Methods in Natural Language Processing and the 9th International Joint Conference on Natural Language Processing (EMNLP-IJCNLP), pp. 4282–4292 (2019)
11. Papineni, K., Roukos, S., Ward, T., Zhu, W.J.: Bleu: a method for automatic evaluation of machine translation. In: Proceedings of the 40th Annual Meeting of the Association for Computational Linguistics, pp. 311–318 (2002)
12. Post, M.: A call for clarity in reporting bleu scores. In: Proceedings of the Third Conference on Machine Translation: Research Papers, pp. 186–191 (2018)
13. Qi, P., Zhang, Y., Zhang, Y., Bolton, J., Manning, C.D.: Stanza: a python natural language processing toolkit for many human languages. In: Proceedings of the

58th Annual Meeting of the Association for Computational Linguistics: System Demonstrations, pp. 101–108 (2020)

14. Qian, L., et al.: Glancing transformer for non-autoregressive neural machine translation. In: Proceedings of the 59th Annual Meeting of the Association for Computational Linguistics and the 11th International Joint Conference on Natural Language Processing (Volume 1: Long Papers), pp. 1993–2003 (2021)

15. Ran, Q., Lin, Y., Li, P., Zhou, J.: Learning to recover from multi-modality errors for non-autoregressive neural machine translation. In: Proceedings of the 58th Annual Meeting of the Association for Computational Linguistics, pp. 3059–3069 (2020)

16. Ran, Q., Lin, Y., Li, P., Zhou, J.: Guiding non-autoregressive neural machine translation decoding with reordering information. In: Proceedings of the AAAI Conference on Artificial Intelligence, vol. 35, pp. 13727–13735 (2021)

17. Ranzato, M., Chopra, S., Auli, M., Zaremba, W.: Sequence level training with recurrent neural networks. In: 4th International Conference on Learning Representations, ICLR 2016, San Juan, Puerto Rico, 2–4 May 2016, Conference Track Proceedings (2016)

18. Sennrich, R., Haddow, B., Birch, A.: Neural machine translation of rare words with subword units. In: Proceedings of the 54th Annual Meeting of the Association for Computational Linguistics (Volume 1: Long Papers), pp. 1715–1725 (2016)

19. Sun, Z., Li, Z., Wang, H., He, D., Lin, Z., Deng, Z.: Fast structured decoding for sequence models. In: Advances in Neural Information Processing Systems 32: Annual Conference on Neural Information Processing Systems 2019, NeurIPS 2019, pp. 3011–3020 (2019)

20. Vaswani, A., et al.: Attention is all you need. In: Advances in Neural Information Processing Systems, pp. 5998–6008 (2017)

21. Wang, C., Zhang, J., Chen, H.: Semi-autoregressive neural machine translation. In: Proceedings of the 2018 Conference on Empirical Methods in Natural Language Processing, pp. 479–488 (2018)

22. Xiao, Y., et al.: A survey on non-autoregressive generation for neural machine translation and beyond. arXiv preprint arXiv:2204.09269 (2022)

23. Zhang, K., et al.: A study of syntactic multi-modality in non-autoregressive machine translation. In: Proceedings of the 2022 Conference of the North American Chapter of the Association for Computational Linguistics: Human Language Technologies, pp. 1747–1757 (2022)

24. Zhang, W., Feng, Y., Meng, F., You, D., Liu, Q.: Bridging the gap between training and inference for neural machine translation. In: Proceedings of the 57th Annual Meeting of the Association for Computational Linguistics, pp. 4334–4343 (2019)

25. Zhang, Z., Wu, Y., Zhou, J., Duan, S., Zhao, H., Wang, R.: SG-Net: syntax guided transformer for language representation. IEEE Trans. Pattern Anal. Mach. Intell. **44**(06), 3285–3299 (2022)

26. Zheng, Z., Zhou, H., Huang, S., Chen, J., Xu, J., Li, L.: Duplex sequence-to-sequence learning for reversible machine translation. Adv. Neural. Inf. Process. Syst. **34**, 21070–21084 (2021)

27. Zhu, M., Wang, J., Yan, C.: Non-autoregressive neural machine translation with consistency regularization optimized variational framework. In: Proceedings of the 2022 Conference of the North American Chapter of the Association for Computational Linguistics: Human Language Technologies, pp. 607–617 (2022)

Author Index

Printed in the United States
by Baker & Taylor Publisher Services